Challenges and Opportunities for the World's Forests in the 21st Century

FORESTRY SCIENCES

Volume 81

Series Editors:
Shri Mohan Jain
Hely Häggman
Alvin D. Yanchuk

For further volumes:
http://www.springer.com/series/5991

Trevor Fenning

Editor

Challenges and Opportunities for the World's Forests in the 21st Century

 Springer

Editor
Trevor Fenning
Center for Forest Resources and Management
UK Forestry Commission
Edinburgh, UK

ISSN 0924-5480   ISSN 1875-1334 (electronic)
ISBN 978-94-007-7075-1   ISBN 978-94-007-7076-8 (eBook)
DOI 10.1007/978-94-007-7076-8
Springer Dordrecht Heidelberg New York London

Library of Congress Control Number: 2013955070

Printed on acid-free paper

Springer is part of Springer Science+Business Media (www.springer.com)

*This book is dedicated to my wife Efi
and our two children, Anna and Gregory,
whose patience and forbearance over many
late nights and weekends enabled the
completion of this book, and also to the late
Duncan Greenwood CBE, FRS, who in life
was always an inspiration.*

Foreword

We operate in an increasingly interconnected and globalized society. Megaforces such as population growth, digital connectivity, accelerated consumption, and disparate prosperity combine with climate change, ecosystem decline, and growing resource scarcity to create far-reaching and often unpredictable impacts on the world in which we live and do business.

Forests play a central role in this complex interplay of global forces. They cover approximately 30 % of the world's land surface and provide essential resources for the survival of humans and other species – and also for the development of business and related operations. Forests supply the world with tangible, profitable resources like timber, and secondary products like fruits and pharmaceuticals. But perhaps more importantly, they provide employment and livelihoods for many communities, especially in emerging economies, offering a means to achieve economic independence and security.

At the same time, forests represent some of the most rich, diverse ecosystems on Earth. We are only just beginning to understand the extraordinary value and impact these global ecosystems have on the rest of the planet. Beyond tangible resources, healthy, intact forests perform critical ecosystem services such as stabilizing soil and soil nutrients, conserving and purifying water, and providing habitat for wildlife. There is also a growing awareness of the very significant role forests play in the global carbon cycle. The potential for mitigating climate change by increasing carbon uptake through reforestation, afforestation, and forest management, while decreasing carbon emissions caused by deforestation and forest degradation, is an exciting and promising area of forest science and policy.

In spite of the critical resources and services that forests provide, deforestation and forest degradation are occurring on a rapid scale globally, driven by a burgeoning human population and the corresponding economic needs and consumption patterns of both developed and developing nations. The area covered by primary forests has fallen by more than 40 million hectares since 2000, mainly because of logging and other human interventions including (but not limited to) commercial agriculture, mining, and biofuels production. If deforestation continues on its current trajectory, some scientists believe that the world's rainforests could completely disappear in 100 years.

To reverse this trend, we must focus on developing solutions to preserve, protect, manage, and sustainably use the planet's forest resources – keeping in mind the people and diverse species that directly depend on them for their livelihood and survival. Although forest management is generally well understood, the principles are poorly applied; most current approaches are fragmented and lack the necessary coordination to be truly effective. Successful solutions will come from better understanding the scientific, political, and socioeconomic drivers in various locations and integrating these perspectives to create workable forest management policies and tools. In compiling and communicating the current issues, research, and thinking on forest science and forestry, this book can help achieve a greater understanding of the critical role forests play in a complex, interconnected world.

Amsterdam/London   Yvo de Boer
                                                KPMG's Special Global Advisor on Climate
                                                          Change and Sustainability
                                           Former Executive Secretary to the United Nations
                                             Framework Convention on Climate Change

Acknowledgments

I would like to thank the many colleagues within *Forest Research* as well as within the rest of the UK *Forestry Commission* for their encouragement for this project, especially those who generously contributed to this book.

I would also like to thank the many other colleagues from the wider forest research community who have supported me in my work over many years, as well as for this book project. Finally, I would like to thank the *Max Planck Institute for Chemical Ecology* in Jena, Germany, as without the knowledge that I gained there, this book would have been impossible.

About the Book

- This book tackles the enormously important and complex issues facing the world's forests today, posed by climate change, conservation objectives, sustainable development needs and the growing demand for affordable energy, along with analyses of the difficulties and opportunities that these present.
- It contains contributions from numerous internationally recognised experts in these fields.
- These articles are explicitly intended to be accessible to a wide range of users from diverse disciplines and backgrounds, who may need to gain a rapid understanding of the key issues and opportunities facing the world's forests over the coming century, as well as for the rich contribution that forests themselves can make for dealing with these issues the world over.

This book addresses the urgent and complex threats and challenges to the world's forests posed by the four great problems of the age: climate change, conservation objectives, sustainable development needs and the growing demand for affordable energy. The intention is to outline the research and other efforts that are needed to understand how these issues will affect the world's forests along with the options and difficulties for dealing with them, as well as the opportunities that the world's forests and production forestry can offer for tackling these very issues.

This book includes sections on (1) sustainable forestry and conservation; (2) forest resources worldwide; (3) forests, forestry and climate change; (4) the economics of forestry; (5) tree breeding and commercial forestry; (6) biotechnological approaches; (7) genomic studies with forest trees; (8) bio-energy, lignin and wood; and (9) forest science including ecological studies. It therefore provides a wide-ranging and multidisciplinary view of the role and contribution of forests *and* forestry for addressing some of the most critical problems and issues facing humanity in the twenty-first century, as well as the interactions between them.

Contributions are provided by prominent organisations or individuals with an established record of achievement in these subjects, who have been asked to present their ideas on these topics in a manner that is readily accessible to professionals from disciplines other than their own. The aim being to provide a ready source of information and guidance on these topics for a wide range of scientists, policy makers and politicians for many years to come.

Contents

Part II   Forest Resources Worldwide

Part III   Forests, Forestry and Climate Change

Part IV   The Economics of Forestry

Part V   Tree Breeding and Commercial Forestry

Introduction

Trevor Fenning

The aim of this book is to contribute to a better understanding of the huge range of issues that affect forests, forestry and forest science worldwide, and to clarify the many benefits that they can offer for overcoming the environmental problems of the age. This book is intended to be of use both for scientists and forestry professionals who might not be familiar with all the work going on outside their immediate area of expertise, and also for policy makers, business and political leaders and who might be involved in making decisions about these issues.

The chapters of this book have been written by internationally renowned experts drawn from a range of disciplines relating to forests, forestry, forest science and climate change. The authors were given an open remit to discuss any aspect of their work that they considered most relevant to the underlying theme of the book, which is to identify and explain the benefits that forests can offer the world, in order to help ensure that these benefits are maximised now and for future generations. The authors were also asked to present their work in a way that is readily intelligible to professionals or interested parties from disciplines other than their own, rather than producing purely academic pieces that may be difficult for the non-specialist to understand.

This project was organised because the world is facing an unprecedented combination of challenges which not only pose a grave threat to the future of the world's forests, but also to ourselves. It is not yet sufficiently appreciated by non-specialists, however, that if properly managed, forests have the potential to reduce many of these problems to much more manageable levels and at a modest cost. Therefore the reasons why such policies have not been adopted need to be urgently addressed, especially the widespread ignorance of the huge influence that forests exert over the world's climate.

T. Fenning (✉)
Forest Research, Northern Research Station, Edinburgh, UK
e-mail: trevor.fenning@forestry.gsi.gov.uk

T. Fenning (ed.), *Challenges and Opportunities for the World's Forests in the 21st Century*, Forestry Sciences 81, DOI 10.1007/978-94-007-7076-8_1, © Crown Employees UK 2014

Examples of the problems facing the world at this time include the increasing demand for energy and the limited supply of fossil fuels (IEA 2012); the rapidly growing populations in many developing countries and the natural desire of the people living there to improve their standard of living; all of which is adding to the pressure on the natural world (UNFPA 2012). Because of this there is an urgent need to redress the damage that is being done to the natural environment and the eco-system services upon which we ultimately all depend, not least by the effects of climate change (Stern 2007; The 4th IPCC Report 2007; Read et al. 2009).

In a globalised world, environmental, economic and political problems such as these increasingly affect everyone everywhere, regardless of their precise cause or origin. From this it should be clear that we are all now dependant upon each others actions far more profoundly than at any other time in human history, and that these problems are too large and all encompassing for any one nation state to solve alone.

Large scale international cooperation is therefore essential if these problems are to be successfully tackled. Unfortunately it is also clear from the difficult debates about these issues at the international meetings, that there is as yet insufficient agreement about how best to respond to these problems, demonstrated by the failure of the Copenhagen and Doha Climate Change Conferences to reach any binding resolutions to reduce carbon emissions. These disagreements are presumably being driven in part by uncertainty as to what solutions there may be to these problems, and also how to share out the burdens of dealing with them.

Any one of these issues alone would present a major challenge to our future wellbeing, but taken together they are probably the most formidable combination of problems that humanity has ever faced. It is probably also no exaggeration to say that there is now more at stake with these issues than our mere wellbeing, so it is solutions that are needed rather than wrangling over who is to blame.

Unfortunately ignorance about how forests might be used to help address these problems is widespread, although they strongly influence all of these problems for better or worse. For instance, if properly used and protected, forests can be (and are) major carbon sinks, but if abused they can also be (and are) major emitters of carbon dioxide (CO_2). Forests already provide firewood for many of the world's poorest people, but have the potential to provide large quantities of new types of biofuels and other materials on a sustainable basis, that can substitute for fossil fuel intensive alternatives and in a manner that does not compromise the world's food supplies.

And this is without allowing for the fact that the world's forests are also major reservoirs of biodiversity and supply free of charge many of the eco-system services on which we all depend.

However, although there is a growing understanding of the role of forests in helping to offset these problems, the prospects for using them in this way is still not as widely appreciated as their potential merits. Ignorance of the benefits that forests can offer probably persists partly for historical reasons, and because it is difficult for non-specialists to grasp the sheer range of issues that affect forests, especially the amount of time required to work with them.

For instance, when the Kyoto Protocol was negotiated in 1997, it committed the participating countries to various reductions in fossil fuel emissions compared to the reference year of 1990, which was seen as a major step forward in the effort to

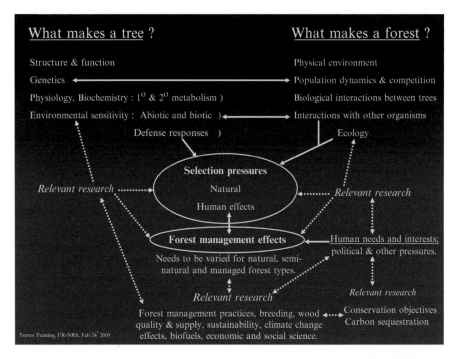

Fig. 1 The purpose of this diagram is to demonstrate the complexity of the biological, physical and human issues that affect natural and managed forests as well as the individual trees within them. The interactions between these issues being shown by solid arrows, while the underlying linkages of the different research themes to these fundamental biological issues, are indicated by dotted arrows.

For example, the structural, physiological, biochemical and genetic functions of individual trees or clones are worthy of study in their own right, but these factors will also be influenced by other organisms and environmental factors in the rest of the forest, including other trees. Such issues attract the interest of ecologists and population geneticists of course, but it should also be clear from this diagram that these matters cannot be understood properly in isolation, and that that it is necessary to understand these interactions in order to develop scientifically based forest management practices.

The interplay of selection pressures on trees, whether over evolutionary time scales, the lifetimes of individual trees, or as pursued by humans over our rather shorter lives, are likely to be quite subtle but still powerful nevertheless. Exceedingly little is known about what drives the interactions between forest trees, however, because they almost certainly take place over timescales that fall well outside normal human lifespans.

It is essential that experimental approaches and models are developed for studying these phenomena, however, because it is the interplay of various selection pressures on trees over assorted time scales, that drives the structure of natural forests that we observe today.

tackle climate change. Unfortunately, at the time there was considerable uncertainty about how to include forests in these plans, which led to numerous contradictions and so called "perverse incentives" being included in the treaty, some of which are still reflected in forest policies across the world today (Schulze et al. 2002, 2003).

Forests are far more complex than more heavily managed environments such as farmland, and there is still much that we do not understand about them (Fig. 1), because forests are essentially still wild, as are forest trees. Nevertheless, there is now a growing body of work showing that forests are key drivers of the world's climate (see the chapter in this book by Dave Schimel), with as much as a third of all the historic emissions of CO_2 being attributed to deforestation alone, as discussed by Canadell et al. (2007) and in the chapter of Grace et al. The role of forests was still overlooked by many within the wider scientific community up until the mid 2000s, however, such that it was still possible for the prestigious scientific journal *Nature* to run a special issue on climate change in 2005 that barely mentioned forests at all.

This began to change with the publication of the Stern Report in 2007, because although it was an economic analysis rather than a scientific one, it dedicated considerable space to the large role that forests play in greenhouse gas emissions, and it's suggestion that it would be worth paying the eight tropical countries with the worst rates of forest loss up to $11 billion per year to prevent it, attracted much attention.

This was followed shortly after by the 2007 Nobel Prize winning 4th Assessment Report of the Intergovernmental Panel on Climate Change (IPCC), which thoroughly discussed how forests influence climate. Its analysis concluded that if the current high levels of forest disturbance could be halted or even reversed by forest restoration projects, especially in the tropics, this alone could offset up to 30 % of global fossil fuel emissions over the next 50 years.

It also showed that much greater and longer term benefits could be obtained by using sustainably produced wood products to displace much of the current dependence on fossil fuels, through the production of bio-energy, fibre and bio-materials, as well as extending the traditional uses of wood. These issues are discussed in the chapters by John Grace et al.; Angela Karp; Ian Tubby, and Peter Hall and Michael Jack, which address the developing use of wood as a feedstock for the production of liquid biofuels.

Yet even these estimates are only based on what might be achieved at the current levels of productivity if the available methodologies were fully utilised, begging the question of how much more might forests and forestry be able to contribute, with modern technological developments and a little more coordination, as discussed in the chapter by Gardiner and Moore?

The adoption of modern technological methods to using forestry to meet carbon sequestration, mitigation and production goals is essential, as it will enable the development of novel goods and services that will offset the consumption of fossil fuels, and also help speed up existing approaches, which is needed because large amounts of time is a luxury that we do not have for solving today's environmental problems, as has been discussed previously by Fenning et al. (2008).

In order to fully appreciated the importance of forests globally, it needs to be realised that while forests cover about 8 % of the earth's total surface area, they account for as much as 30 % of the earth's atmospheric carbon cycle, which is incredibly strong leverage as compared to the oceans, which cover 70 % of the earth's surface, but 'only' account for 50 % of the atmospheric carbon cycle, as discussed in the chapters by Dave Schimel and Grace et al. Forests are very susceptible to human interference for better or worse, however, but fortunately the community of foresters and forest scientist know much more about how to manage forests for meeting multiple objectives, than ever before.

This encouraging prospect has not received as much attention as might be expected, however, and in most countries forests and forestry is often still consigned to the margins of the debate about how best to achieve global sustainability. Presumably this is because many people in a position to influence the debate about such issues simply do not understand the potential for using forests in this way.

This is not the case everywhere, however, as the excellent chapters by Nuyun Li and Jianjun Hu et al. make clear, where they describe the intense efforts that China is making to remedy many of it's environmental problems with the aid of it's forest sector, after the potential for using forests to help put China on a path towards sustainable development was recognised at the highest levels.

Nevertheless, although China's efforts have started from a low base and there is obviously still much to do, the high level of political support that these policies enjoy there is a pre-requisite for such ambitious plans to succeed. It seems that the failure to enact such plans elsewhere may be due at least in part to the failure to communicate to a much wider audience the hugely valuable contribution that forests and forestry can make for solving many of the world's most intractable problems. This indicates the urgent need for more effort in this area, although similar points have been made many times previously (e.g. Sutton 1999, 2000; Fenning and Gershenzon 2002).

However, despite any shortcomings in the delivery of forest services worldwide, at least compared to what is possible, there is now increasing evidence that using forest management and forest conservation strategies in this way is not only readily achievable, but is also a 'low risk' option from an economic point of view, as discussed by Valatin and Price in this book.

The urgent need to explain the huge value of forests and forestry to a much wider audience is therefore clear, and how new technology working in concert with more traditional approaches can be applied to solving long-standing problems in forestry and forest research.

This book can naturally only play a small part in addressing issues as large as these, but such an effort is still very much needed. Better coverage of the numerous issues that affect tropical and boreal forests would have been merited and how they might be affected by climate change, as well as the dangers posed by exotic diseases and pathogens to forests worldwide, which is becoming such a major issue that it would have been justified to dedicate even more space to it. Similarly, much more could be said on the subject of forest resources, economics and industrial trends, as well as the urgent need to conserve the germplasm of rare or endangered tree species, by *in situ* or *ex situ* gene banks, the complexity of which is often underestimated, but which affect many of the other issues addressed by this book.

Nevertheless, the subjects that are presented here do succeed in addressing a broad and comprehensive range of subjects about how forests and forestry affect the whole world, now and in the future. The articles in this book are written by leading forest professionals and scientists and will be an invaluable resource for anyone with an interest in forests, forestry or forest conservation.

The division of the sections in this book are somewhat arbitrary, in that many of the individual chapters contain elements that relate to more than one section. However, each chapter was assigned to its particular section on the basis of its principal subject matter and can be read as a stand alone piece. Where there is an obvious overlap with other sections or chapters, this is sometimes noted to encourage further reading where relevant to the interests of the reader. In addition, many chapters provide links to further reading on the subjects raised, to inform those readers who might want to find out more about certain issues.

Part I of the book addresses the topic of **Sustainable Forestry and Conservation Objectives**. **Paul McMahon** of the Princes Rainforests Project launches the book with a detailed analysis of the value of forests worldwide, but mainly addresses the human and environmental problems that affect tropical forests. In particular the failure of existing market or conservation mechanisms to generate sufficient protection for these forests is discussed, along with possible alternatives such as REDD and REDD+, as well as other initiatives supported by the Princes Rainforests Project.

Bill Mason and Jiaojun Zhu of Forest Research in the UK and the Chinese Academy of Sciences respectively, contrast and compare the histories of deforestation that affected both countries, together with recent efforts at re-afforestation. They reveal the similar experiences of the two countries, not least the initial tendencies to plant large, even-age blocks of a single species of trees, and the subsequent recognition that multiple objectives are often better met by using mixtures of species and ages of trees, in systems such as continuous cover forestry.

Such forests are undoubtedly more demanding to manage, but fortunately this indicates that important lessons in the development of sustainable forestry can be shared by these two countries and that other countries might also be able to benefit from this experience.

Mariella Marzano, Christopher P. Quine and Norman Dandy of Forest Research in the UK further develop the theme of multi-functional forests, based on the UK experience. In particular they make the case for developing a strong evidence base to inform decision-making regarding the balance and trade-offs that need to be made when choosing between different functions. They argue that the interactions between recreation and conservation objectives should be assessed at landscape scale, both in order to be sure that the benefits available from forests are felt as widely as possible, and because the support of the public is needed if these approaches are to be successfully maintained over the long term.

Simon Bell and Catharine Ward Thompson of Edinburgh University in the UK explore how forest planning decisions and policies are perceived by the public, and how people engage with them. They argue that with urban populations across the world now outnumbering rural populations for the first time in human history, such engagement is now a necessity.

Nuyun Li of the Department of Afforestation and Greening, state forestry administration in China, describes the various approaches to forestry that have been utilised in the People Republic of China since it was founded in 1949, from the focus on production orientated forestry in the early years, to today's efforts that are explicitly intended to develop fully sustainable forest and landscape management approaches, to offset its greenhouse gas emissions in accordance with international agreements.

This hugely ambitious programme is being developed within a framework of policies drawn up at the highest levels of the Chinese government, which were very publically promoted by the former President Hu Jintao. Nevertheless, as with much of the rest of the world, the pressure on the land is increasing, so it remains to be seen how far China can succeed with these efforts from its domestic forest base alone, but with the scale of the resources that the country is dedicating to this objective, and in combination with other renewable energy approaches and environmental management efforts, much more progress will surely be made than would be the case otherwise.

Debal Deb of the Centre for Interdisciplinary Studies in India meanwhile, gives a detailed but sometimes tragic account of how forest management schemes, whether for commercial or conservation objectives, have sometimes adversely affected indigenous forest peoples in India and elsewhere, past and present. Numerous examples are presented as to how even well meaning forest management efforts (let alone less well meaning ones) that failed to engage with the needs and different values of these peoples have sometimes been less effective than they might have been, or even resulted in serious damage to the affected forests, to the detriment of both their wild and human inhabitants.

Adrian Whiteman of the FAO in Rome begins **Part II (Forest Resources Worldwide)** with a thorough and comprehensive review of the current state of the world's forest resources, summarised from the FAO's published data (FAO 2011), along with a detailed analysis of the relevant trends and developments over the past 50 years. This data is, then extrapolated forward to determine how these same issues and concerns might develop in future, region by region.

Amongst the many interesting facts and trends that are described, one of the most notable is that while the world's tropical countries still have high levels of forest cover, they also have the highest rates of deforestation, probably because they are still at a relatively early stage of their economic development, and so it seems likely that their forests will continue to suffer from high rates of deforestation for some time to come.

However, Adrian Whiteman does note that developing effective evidence based forest policies to avert these high rates of deforestation can be successful, if rigorously enacted and supported at the governmental and regional levels, and there are signs that this is happening, as deforestation rates have fallen sharply across the tropical regions in recent years. This indicates both that using forest conservation and restoration projects to offset dangerous levels of climate change, as well as for addressing other serious environmental problems, may yet succeed and that concerted policy interventions can work if well directed. In short, it appears that the situation facing forests in the tropical regions is not as hopeless as the headlines sometimes suggest.

Wink Sutton of Plantation Focus Limited in New Zealand, who has long been known for his advocacy of plantation forestry (Sutton 1999, 2000), here describes some of the constraints to increasing the use of wood as a renewable resource.

He notes that although the current level of consumption of wood and wood products is much less than the potential supply available from managed forests, the ability to increase production by as much as might be desired in future will be severely constrained in the medium term due to falls in replanting rates across much of the world over the past 20 years. This illustrates the need for consistent policies in forestry, and how the consequences of poor decisions may continue for many decades after the short term political or financial considerations that caused them have been forgotten.

From this it should also be clear that although humanity is more dependant on the world's forests than perhaps some people realise, conventional market mechanisms and forestry do not always sit well together. For instance, the market for wood and biofuels completes with other energy sources and materials on price. So, in order to displace the use of concrete, steel, plastics or other fossil fuel dependent activities, wood needs to consistently beat them both on availability and cost. This is hard to do without reliable policies being in place to support the development of properly functioning markets in sustainably produced forest products.

There is no time to waste, however, because as the chapters within **Part III** indicate **(Forests, Forestry and Climate Change)**, climate change is rapidly becoming a real and active threat to many of the world's forests, which may ultimately limit our ability to use them to combat such problems.

David Schimel of the NASA Jet Propulsion lab in California gives an invaluable account of how the current state of knowledge developed, both about climate change itself and also of the huge leverage (or forcing) that forests exert over the earth's climate and atmospheric carbon cycle.

John Grace, James Morison and Michael Perks of the University of Edinburgh and Forest Research in the UK, build upon this excellent introduction to the subject with a comprehensive explanation of what is known about the effect of forests on climate and vice versa, and discuss the possibility of adapting forest management practices specifically for mitigating climate change.

They conclude that the prospects for using forests and forestry for mitigating many of the worst effects of climate change are generally good, provided the political will can be found to pursue this ambitious agenda and it can be reconciled with other demands for land, as also observed in the chapter by Adrian Whiteman.

Perhaps the greatest concern is that as Adrian Whiteman also indicated, this is likely to be hardest to achieve in the rapidly developing tropical belt, which is where John Grace and colleagues observe that the biggest gains from forest conservation and restoration need to be made, both for carbon sequestration by forests and mitigating fossil fuel emissions by providing sustainably produced alternatives.

The implication of this is that resolving this policy dilemma must be the top priority of international efforts if global climate change mitigation targets are to be met, as well as for meeting conservation objectives and sustainable forest production needs.

Patrick Tobin, Dylan Parry and Brian Aukema of the US Forest Service, the College of Environmental Science and Forestry, at the State University of New York, and the Dept. of Entomology at the University of Minnesota, respectively, discuss a possible mechanism by which climate change itself may exert devastating effects on forests the world over. The processes they describe are especially unpredictable, as they analyse how exotic organisms that have already been introduced into forests all over the world from elsewhere, may alter their behaviour as a regions climate begins to change. These organisms can then unexpectedly emerge as a major threat, where previously they might have been so harmless as to not even be noticed.

Most of the examples they cite are insects, but the same applies to other organisms as well, and so as ever more such cases become apparent, the value of the analysis presented here will become ever more pertinent.

Gregory Valatin of Forest Research in the UK and **Colin Price** who is a freelance academic living in North Wales in the UK, begin **Part IV (The Economics of Forestry)** by analysing the cost-effectiveness of using forestry for climate change mitigation, both in comparison to some of the alternatives under discussion at this time, and also in relation to the complex issues of carbon sequestration and pricing.

They note that forest restoration projects such as those being suggested for degraded land in the tropical belt, can only deliver their carbon benefit *once*, as the additional carbon sequestered by forests will decline as they mature. This is not a reason to object to such projects though, as the carbon benefits of any forest restoration project begun today would continue strongly for at least 50–100 years before tailing off, which is the most critical period for carbon mitigation efforts, while the benefits of using sustainably produced wood for offsetting carbon intensive activities would continue indefinitely.

The authors convincingly argue that the cost of using forests in this way the world over compares very favourably with even the best of the many alternatives that are being given attention currently.

Gregory Valatin continues with the theme of analysing the use of forests for climate change mitigation. In this chapter, however, the issue of "additionality" is explored, which is the accounting methodology by which the putative advantages of different carbon mitigation activities are compared to one another, as well as to the original activities which they may have been intended to replace.

The concept of additionality is an essential tool for the world's carbon markets, as it is needed for comparing the supposed benefits of different climate change mitigation activities, even if there is sometimes disagreement about how to apply the principles. Sources of uncertainty and trade-offs in practical application of the concept are highlighted, and the potential for perverse incentives also explored.

Much has indeed been written about the enormous influence and complexities of forest economics, but **David South** of Auburn University, Alabama, USA, addresses the often overlooked issue of how forest management objectives can be affected by the underlying attitude of the owners or managers of a forest to money, as well as to the forests themselves, and also the type of economic system employed by the land-owner or their government. Examples are presented that show how this can be a key driver of long term changes in forest management structures and outcomes, in ways that were not necessarily anticipated or intended when the management plans were first drawn up.

Part V (Tree Breeding and Commercial Forestry) begins with a chapter by **Gabriel Dehon Rezende, Dr. Marcos Deon Vilela de Resende and Teotônio Francisco de Assis** of the RIAZ-Forest and Paper Research Institute, Portugal, the Agricultural Research Corporation of the Federal University of Viçosa, Brazil, and Assistech Ltda., Brazil, who describe how their tree breeding programme for Eucalyptus ssp. has been organised from a scientific as well as a practical point of view. It is these programmes that have generated the tremendous gains in productivity in Brazil in recent years, with yields in excess of 80 $m^3.ha.yr^1$ on a 5–10 year rotation, which is helping to turn Brazil into the powerhouse of the world's forestry sector.

Forest conservation objectives will probably be much easier to achieve, if sustainably managed production orientated forests, can offer a good rate of return on the public or private capital invested. This will drive more investment into responsible forestry and away from depleting natural forests, than would be the case if the sector remains dependent on state subsidies, as the public funds available to support forestry will always be limited.

Yill-Sung Park of the Canadian Forest Service in Fredericton, Canada, describes the exciting gains that have been made there in recent years there, by integrating traditional tree breeding approaches with modern molecular and biotechnological tools, including advanced plant tissue culture methods such as somatic embryogenesis and the cryo-storage of rare or valuable cell lines.

Yill-Sung Park calls this combined approach *multi-varietal forestry* (or MVF), since it enables those trees with superior traits that have been produced by breeding or other improvement programmes to be identified, conserved and then propagated according to the needs of the forest industry at any one time. Considerable success has been achieved with this approach in Canada with the native white spruce species (*Picea glauca*), but it is also being extended to other conifer species and there is no reason in principle why it cannot be applied to almost any tree species. A version of this approach is indeed being used when breeding improved Eucalypts in Brazil, as discussed in the previous chapter.

Yousry El-Kassaby, Fikret Isik and Ross Whetten of the Department of Forest Sciences at the University of British Columbia, Canada and the Department of Forestry and Environmental Resources at North Carolina State University, USA, meanwhile introduce us to another aspect of modern tree breeding that has the potential to profoundly change the way that tree breeding is organised in future: Namely how the increasing availability of high throughput DNA sequencing will enable the genomes of many different organisms to be contrasted and compared quickly and cheaply, including numerous individuals within the same species. This will have tremendous effects on how research is organised in all areas of forest research and tree breeding, another example of which is given in the chapter by Andrew Groover and Stefan Jansson.

In combination with the ever more powerful computers used for analysing the vast amounts of data produced and the high density genetic maps that are being generated, many new approaches in tree breeding are becoming both possible and affordable. For instance, it is now becoming routine to reconstruct the pedigree of all the trees in an area of forest without any direct knowledge of either of the parents.

If the power of this technology seems intimidating, however, it should be noted that it will not replace the other approaches and methodologies discussed here, nor will it replace the knowledge or experience of foresters or other professionals, but instead will build on what they have to offer, as well as enabling new areas of cooperation to develop, such as between forest geneticists and remote sensing approaches.

Heidi Dungey, Alvin Yanchuk and Richard Burdon of Scion, New Zealand and the British Columbia Ministry of Forests, Canada, give a comprehensive review of how to organise a tree breeding programme from first principles, and how to manage all aspects of the process, including the changing expectations of foresters and funding agencies alike.

This chapter also reminds us that the funding bodies that oversee public research programmes also need to restrain themselves from imposing excessive burdens and costs on the actual business of tree breeding, such as by unnecessary or poorly thought through bureaucratic demands, which can end up impeding progress rather than improving efficiency. The authors strongly argue that a better balance needs to be struck between funding and administrative agencies, academic science and technical providers, if tree improvement programmes are going to continue to be successful in future.

Dag Lindgren who is an emeritus professor at the Swedish University of Agricultural Sciences in Umeå, Sweden, urges us not to forget the need to maintain the basic infrastructure of tree improvement and forest management programmes, noting that the huge gains in productivity and the excellent value for money they have given us, have been delivered by using seed orchards in support of tree breeding the world over.

Dag Lindgren notes that seed orchards will continue to make a large contribution to the future physical and ecological realities in our countries and also form the foundation of many more sophisticated forest based activities, as described elsewhere within this book. Dag Lindgren also asserts that forest activities such as these should therefore be seen as a moral activity, which might at first seem to be an unusual statement for a scientist to make about their profession, but such thoughts are probably a motivating factor for many people who work in the forest sector.

Part VI (Biotechnological Approaches) begins with a chapter by **Hely Häggman, Suvi Sutela, Christian Walter and Matthias Fladung** of the University of Oulu, Finland, Scion Research, New Zealand and the Institute of Forest Genetics, Grosshansdorf, Germany. They thoroughly review the scientific literature about the biosafety risks that transgenic trees might pose to the wider environment, if they were ever planted outside of closely monitored scientific trials for commercial or other purposes.

It is clear from this analysis that although GM trees will certainly not be suitable for every situation, to date no valid *scientific* objections to the responsible use of them outside of the controlled field trial have yet been identified or upheld. Political or cultural objections to their use may continue for some time, however, which cannot be ignored when drawing up forest management and conservation plans, but it can no longer be argued that GM trees present any special risk or danger to the environment that cannot be managed, merely on account of their transgenic status.

Armand Séguin, Denis Lachance, Annabelle Déjardin, Jean-Charles Leplé and Gilles Pilate of the Canadian Forest Service, Québec, Canada and the INRA Amélioration, Génétique et Physiologie Forestières institute at Ardon in France, continue with the theme of GM trees by giving a valuable account of how the technology of genetically engineering came about in the first place, before ultimately being applied to trees for the first time in 1987.

The authors make it clear that this is still far from being a routine procedure for trees, and that even now only relatively few clones of a limited number of species have proved amenable to genetic manipulation, and that the principle use of GM trees has so far been for research purposes only. However, they also discuss and advocate the adoption of this technology for environmentally responsible applications.

They also argue that scientifically validated field trials must be developed in tandem with work to improve the utility of GM technologies, so that the best scientific evidence for environmental risk management of the latest GM plants is immediately available to regulatory agencies as soon as it is needed, so that the numerous benefits that GM trees can offer are maximized, while retaining public confidence.

Kevan and Jill Gartland of Glasgow Caledonian University in Scotland, provide an overview of the present status and likely future trends in forest biotechnology in general, and not only as regards GM. They discusses the numerous objectives for which forest biotechnological tools are being developed and applied, as well as addressing the societal issues that affect our judgments about how to best use these emerging technologies, and whether we want to use them at all indeed.

They observe that there is a widespread support within the community of biotechnologists for utilising such methods in pursuit of environmental goals, although this article also highlights the different approaches being adopted by different countries to forest biotechnology, sometimes due to different ideas as to what is meant by 'sustainable development' or 'the precautionary principle', and how they are being implemented.

The need for biotechnological approaches for tackling some very specific but intractable forest conservation problems is also discussed, including efforts to save certain tree species that are threatened with extinction, either due to the introduction of exotic diseases or due to the negative effects of climate change. Examples include elms that are endangered by Dutch elm disease, and the American chestnut that has been devastated by chestnut blight.

Jianjun Hu, Lijuan Wang, Donghui Yan and Meng-Zhu Lu of the Biotechnology Lab, Research Institute of Forestry, Chinese Academy of Forestry, Beijing, China, meanwhile describes the hugely impressive progress and investment that is being made there, in utilising GM approaches for solving many of the most intractable problems that affect that country's forests. This chapter should be read in conjunction with the chapters by Bill Mason and Jiaojun Zhu and that of Nuyun Li, in order to understand the enormous effort that China is currently putting into its forest sector, of which the work described here is but a part.

The genetic engineering research programmes described by Hu and colleagues are mainly focused on conferring insect resistance traits to a limited range of clones of various *Populus* ssp. or hybrids, to help ensure their reliability for use as shelter belt trees or for restoring forest cover in environmentally degraded areas. However, the objectives of these programmes are being rapidly expanded to address other environmental problems, such as resistance to microbial pathogens and drought tolerance, and also the conservation of rare or endangered species, in anticipation of the adverse effects of climate change.

Part VII (Genomic Studies with Forest Trees) begins with an fascinating chapter by **Wellington Muchero, Jessy Labbé, Priya Ranjan, Stephen DiFazio & Gerald (Jerry) Tuskan** of the Department of Energy Oak Ridge National Laboratory, Tennessee, USA, the University of Tennessee at, Knoxville, USA and West Virginia University at Morgantown, USA, who describe some recent exciting discoveries emanating from the ongoing genomic studies with various *Populus* ssp.

The black cottonwood (*Populus trichocarpa*) became the first tree to have it's genome sequenced in 2006 and since then over 400 clones or hybrid genotypes have been resequenced as part of this ongoing programme. This has generated a genetic database of unparalleled depth for forest trees, enabling the construction of high density genetic maps for entire populations of this species.

This has led to the discovery that large-scale insertions and deletions (also called INDELs) are common in populations of these trees, which were not readily apparent from the individual sequencing results alone. The evolutionary relevance of these INDELs is discussed, including their likely role in pathogen recognition and signal cascades, which has implications for understanding the population dynamics of many tree species and not only *Populus* ssp.

Andrew Groover and Stefan Jansson of the USDA in Davis, USA and the Umeå Plant Science Centre, Sweden, address a different set of evolutionary insights that are emerging from the large scale sequencing efforts that are underway for many tree species. They remind us that trees are not a single phylogenetic group of plants, but that the tree habit has been gained and lost in different plant lineages many times in the earth's history, which has implications for our understanding of how certain traits that we usually associate with trees evolved, such as perennial growth and wood formation.

They argue that comparative genomic studies between distantly related tree species can be used to identify common mechanisms for regulating such traits more efficiently than could be achieved by working with a single species. The relevance of this approach for adding to our knowledge of wood formation is discussed, emphasising the need for more such studies in order to increase the pace of discoveries about the biology and ecology of trees.

Angela Karp of Rothamsted Research UK, begins **Part VIII (Bio-energy, Lignin and Wood)**, with an introduction to the numerous and diverse uses of willows (*Salix* ssp.), including the potential of using it as a feedstock for biofuels and industrial chemicals.

After a great deal of selection and breeding work, the outstanding potential of willows has at last been recognised, such that it is now providing commercial quantities of sustainably produced renewable biomass to the energy industry, including for co-firing with coal in electrical power stations and for dedicated biomass boilers of all sizes. These applications have already resulted in an increased acreage of willows being planted in the UK, and may soon result in very large scale plantings world-wide.

Meanwhile **Ian Tubby** of Forest Research, UK, foresees that although the developing energy markets for woodfuels such as willows have the potential to transform British forestry, many traditional broadleaved woodlands in the UK have fallen out of use in the decades since the Second World War, and that a considerable effort will be needed to bring them back into production.

This should not be seen as being bad for the environment, however, because this lack of management has caused many of these woodlands to become over-mature, which has contributed to a decline in biodiversity indicators. The emergence of viable markets for renewable woodfuels and the funds that will consequently flow into the forest sector, is therefore likely to stimulate a renewed interest in woodland management and so reverse the decline in the biodiversity of the UK's native woodlands.

However, the UK has a high population density and high per capita energy demand, and so it's existing forest resource can probably only supply a fraction of the country's total energy needs, albeit one still worth having. This will most likely lead to more woodlands being established, provided appropriate legal frameworks and policies in support of using renewable energy sources are put in place. Ian Tubby also predicts that the way wood is used to produce energy is likely to change in the future, as progress is made into converting wood into liquid fuels, and the use of biomass fired power generation is extended to help deliver carbon negative energy projects, which is the ultimate aim of all renewable energy plans today.

Peter Hall and Michael Jack of Scion, New Zealand give a detailed account of the opportunities and problems facing the forest sector there, which is similar in some ways to the situation facing the UK, and contrasting in others. The most notable difference between the two counties being that New Zealand has a modest population and a relatively large forested area.

Their analysis of New Zealand's bioenergy resources demonstrates that wood residues that are currently going to waste are the country's largest bioenergy resource, at least up until the 2040s. Projected falls in wood supply may constrain matters after this, however, due to the long term effects of low planting rates over the past 20 years (as discussed in the chapter by Sutton). But, this should be seen as a reason to support the forest sector, rather than being used as an excuse for further inaction.

Even without impacting the supply of wood to the rest of New Zealand's economy or postulating any increase in plantings over current trends, the authors note that this surplus volume of wood residues and pulp logs alone could produce sufficient liquid fuel to meet 5–6 % of the country's total demand, or >15 % of the requirement for diesel, and much more could be easily achieved with a moderate effort.

However, Peter Hall and Michael Jack also note that forestry currently suffers from a lack of organisation globally, which is having negative effects in forest planning, R&D in general and specifically for developing appropriate systems and processes that are needed to achieve the desired goal of global sustainability, and lastly the training of sufficient numbers of suitably qualified people for the forest sector to thrive. Providing these things requires the involvement of both private and public resources as well as long term planning, not unlike the original green revolution that succeeded in delivering food security in the aftermath of World War II.

Barry Gardiner and John Moore of INRA, France, and Scion, New Zealand, conclude this section with a thorough analysis of current wood consumption and supply patterns across the world, noting that the growing demand from the rapidly expanding economies of China, India and south-east Asia, is likely to drive an increase in the consumption of industrial roundwood globally of between 1.3–1.8 % per year up to 2030.

The authors continue by addressing what the needs of these emerging economies are likely to be and how they might best be met. They suggest that this growing demand can be best supplied by developing sophisticated tools such as remote sensing for fine tuning forest management and harvesting regimes, and tools for making detailed measurements of wood properties of individual trees in real time as they are being harvested. If systems can be put in place to track these results through the entire supply chain, they suggest that it should be possible to direct exactly the right type and quality of wood to each customer, and so drastically increase the efficiency of wood usage and reduce wastage.

Such technologies will not only result in large gains in the productivity and profitability for the world's wood industries, which in itself would be no mean achievement, but in so doing will improve the consistency of wood products and increase their competitiveness, further stimulating the demand for such products. Most impressive of all is the dramatic reduction of wastage this will make possible, with gains of up to 50 % in net value recovery being possible, showing that making better use of the world's existing wood production will go a long way towards overcoming any potential limitations in supply.

Part IX (Forest Science, Including Ecological Studies) begins with an excellent and insightful chapter by **Adam Wymore, Helen Bothwell, Zacchaeus Compson, Louis Lamit, Faith Walker, Scott Woolbright & Thomas Whitham**, that summarises many years of research findings in the field of community genetics at North Arizona University, USA. Community genetics being the study of the genetic variation of 'foundation species' such as trees, which drive key ecological processes and thus dominate the environment around them, including its biodiversity.

The main findings of this work are that different *genotypes* of an individual tree species may support *different* communities of other flora and fauna, and so individual genetic traits of a tree can influence many other ecosystem processes. These findings also correlate well with the recent discoveries of the poplar genome program, as described in the chapter by Wellington Muchero et al., which suggests a genetic mechanism by which these interactions are mediated. Primary productivity has a genetic component too and so also has effects on biodiversity as well as carbon sequestration, suggesting that the importance of studying community genetics will only increase with time.

Sanford Sillman of the University of Michigan, USA, discusses biogenic emissions into the atmosphere other than CO_2, such as the volatile organic compounds

(VOCs) that are often associated with forests, including the terpenoids that are primarily responsible for the characteristic smell of many trees and flowers. Unfortunately these volatiles can interact with anthropomorphic emissions such as nitrous oxides (NOx) produced by combustion processes, which contribute to the formation of ground level ozone and photochemical smog.

Smog and ground level ozone adversely affect both human health and the health of natural environments that are exposed to them, potentially including the very forests that may have produced the VOCs that have contributed to the problem. Unfortunately VOCs and NOx can also sometimes combine far downwind of the original sources of these compounds and so cause problems in apparently unpolluted areas, and so with urbanisation increasing rapidly across much of the world, these problems are likely to increase significantly and so cannot be ignored in forest planning.

With this in mind, in future it might even be regarded as irresponsible for forest planners to proceed with planting large numbers of trees that are known to be high emitters of VOCs near to urban centres known to be major sources of nitrous oxides. Such species include many coniferous and Eucalyptus species.

Hilke Schroeder and Riziero Tiberi of the Federal Institute for Forest Genetics, Germany and the University of Florence, Italy, discuss the complex ecological interactions of herbivorous forest insects with their hosts and predators, through the example of the lepidopteran oak leaf roller moth, *Tortrix viridana*, which is a major pest of various oak species throughout much of Europe. The ebb and flow of this species on its host species is an important example to study not only because it can be very destructive even in normal circumstances, but also because the frequency and severity of its periodic outbreaks and also its range are increasing, probably due to a warming climate.

The chapter by **Stephanos Diamandis** of the Forest Research Institute near Thessaloniki in Greece, notes that forests have survived severe changes in climate and other calamities many times in the past, and will undoubtedly also survive the present world crisis in some shape or form. However, they are also likely to be substantially changed by it and probably not for the better, or at least not in any time frame of interest to the current human population of this planet.

This chapter addresses the various mechanisms by which climate change may damage the world's forests over the coming decades, such as the increasing risk of major fires, but concentrates on the threat posed by insect pests and microbial pathogens. Numerous examples are given of alien invasive species that are causing increasing problems around the world at this time, resulting from the globalization of trade, which is resulting in huge costs to tax payers as governments try to control them. This chapter should be read in conjunction with that of Patrick Tobin et al. in Part III, which addresses how these problems might be anticipated, as the climate changes.

This chapter concludes by noting that the damage caused by these and other environmental problems can probably be contained, or at least limited, provided the political will exists to do so, but that they will continue to cause major problems, if the support for taking the necessary action is lacking.

Sarah Green, Bridget Laue, Reuben Nowell and Heather Steele of Forest Research and the University of Edinburgh, UK, provide a fascinating insight into how a single exotic pathogen that has appeared in Europe in the last decade, has devastated the European horse chestnut (*Aesculus hippocastanum*). Trees were observed to be suffering from a new form of bleeding canker on their stems which ultimately kills them, firstly in continental Europe and more recently across the UK as well.

The causal agent of the disease was identified as a new species of pathogenic bacterium; *Pseudomonas syringae* pv. *aesculi*. It is thought that this bacterium originated in India on the Indian horse chestnut, and was probably introduced to Europe via the plant trade (as highlighted by Brasier 2008). A vigorous programme of work has been initiated to study and tackle this disease, including the development of rapid PCR based diagnostic tests and the sequencing of the genome of this bacterium. These results have shown that there is only one strain of this organism across the whole of Europe, which suggests that the outbreak is probably the result of a single introduction event.

Whether this information can be used to develop an effective control strategy remains to be seen, but this work gives a fascinating insight into how a single act of carelessness can kill large numbers of trees across an entire continent.

Juan Suárez of Forest Research, UK, closes this book with a discussion of the use of airborne LiDAR and other remote sensing techniques which enable the accurate monitoring and measuring of forest parameters down to the level of the individual tree, as well as the modelling and mathematical approaches needed for analysing the data generated. The potential of such methods for assessing large areas of forest in real time and so helping set production or conservation priorities, or potentially even for identifying and monitoring disease outbreaks, should be self evident to anyone involved in forestry or forest science today.

Concluding Remarks

The forests of the world have much to offer for mitigating many of the most serious environmental and other concerns of the age, although much more needs to be done if this potential is ever to be fully realised. These benefits will also be greatly boosted by the application of modern approaches, which have tremendous implications for how forestry may be conducted in future.

Similar issues were encountered during the 1st green revolution, however, when the acute global food insecurity after World War II was overcome by the combined efforts of public and private research institutions, government action and subsidies in support of private investment, as well as concerted international cooperation. Although some of the lessons of that era may have been forgotten, it is at least conceivable that if a similarly vigorous effort were made in relation to managing our forests today, then at least some of the most serious concerns could be more readily dealt with than currently seems possible.

It has to be hoped that the ongoing efforts of foresters, forest scientists, bio-engineers, conservation groups, climatologists and the relevant regulatory authorities will be successful in dealing with these problems sooner rather than later. In particular, concerted efforts are needed to develop and apply the best forms of sustainable forest management worldwide, including appropriate regulatory oversight and good governance, because poor practice can not only damage the local environment where this occurs, but risks undermining public confidence in the whole enterprise, as may have happened in the past. And as discussed in the chapters of this book, some of these problems are both severe and urgent, however, and so we cannot afford to overlook the opportunity that forests can offer for helping to address these issues.

Although it would be incorrect to suggest that forests and forestry alone can be used solve all of the environmental problems that currently concern us, the influence of forests over many of these issues is so large that it is probably no exadgeration to claim that policies that put forests, forestry and forest research at the centre of their plans for tackling these problems will have a good chance of succeeding, while those that overlook them, will almost certainly fail.

Governments as well as scientific and funding insitutions all over the world need to grasp this hugely impressive potential, and respond positively to the challenges and opportunites that are being offered by forests and forestry in the twenty-first century.

References

Brasier CM (2008) The biosecurity threat to the UK and global environment from international trade in plants. Plant Pathol 57(5):792–808

Canadell JG, Le Quéré C, Raupach MR, Field CB, Buitenhuis ET, Ciais P, Conway TJ, Gillett NP, Houghton RA, Marland G (2007) Contributions to accelerating atmospheric CO_2 growth from economic activity, carbon intensity, and efficiency of natural sinks. Proc Natl Acad Sci 104(47):18866–18870

FAO (2011) State of the World's forests 2011. Food and Agriculture Organization of the United Nations, Rome

Fenning TM, Gershenzon J (2002) Where will the wood come from? Plantation forests and the role of biotechnology. Trends Biotechnol 20(7):291–296

Fenning TM, Walter C, Gartland KMA (2008) Forest biotech and climate change. Nat Biotechnol 26(6):615–617

IEA (International Energy Agency) (2012) http://www.iea.org/

Nature, special issue on climate change (2005) 438(7066):257–394

Read DJ, Freer-Smith PH, Morison JIL, Hanley N, West CC and Snowdon P (eds) (2009) Combating climate change – a role for UK forests: an assessment of the potential of the UK's trees and Woodlands to mitigate and adapt to climate change. The Stationery Office Limited, 26 Rutland Square, Edinburgh EH1 2BW: http://www.forestry.gov.uk/forestry/infd-7y4gn9

Schulze ED, Mollicone D, Achard F, Matteucci G, Federici S, Eva HD, Valentini R (2003) Making deforestation pay under the Kyoto protocol? Science 299(5613):1669

Schulze ED, Valentini R, Sanz MJ (2002) The long way from Kyoto to Marrakesh: implications of the Kyoto protocol negotiations for global ecology. Glob Chang Biol 8(6):505–518

Stern N (2007) The economics of climate change. Cambridge University Press, Cambridge, UK, http://webarchive.nationalarchives.gov.uk/+/http://www.hm-treasury.gov.uk/independent_ reviews/stern_review_economics_climate_change/stern_review_report.cfm

Sutton WRJ (1999) Does the world need planted forests? N Z J For 44(2):24–29

Sutton WRJ (2000) Wood in the third millennium. For Prod J 50(1):12–21

The 4th Assessment Report of the Intergovernmental Panel on Climate Change (IPCC) (2007) Working groups I, II, & III. Cambridge University Press, Cambridge, UK, http://www.ipcc.ch/

UNFPA web site (population projections checked October 2012) http://www.7billionactions.org/ data?gclid=COOUmKn0jLMCFSbItAodRngAyQ

Part I
Sustainable Forestry
and Conservation Objectives

A Burning Issue: Tropical Forests and the Health of Global Ecosystems

Paul McMahon

Abstract Tropical forests are in the front line of efforts to tackle climate change. This article provides an overview of the ecosystem services that tropical forests provide, the way in which rates of deforestation have changed over the past 20 years, and the economic drivers of tropical deforestation around the world. It describes recent intergovernmental efforts to reduce deforestation, both through a proposed UNFCCC REDD+ mechanism and through interim finance or 'Fast Start' partnerships that seek to achieve results between now and 2020. There has never been a better opportunity to forge international cooperation on this important environmental issue, but progress so far has been slow.

1 Introduction

Forests cover 3.7 billion hectares of the planet's surface, or 30 % of the global land area. Almost half of these forests are found in tropical areas (44 % of the total area), about one-third in boreal (34 %) and smaller amounts in temperate (13 %) and sub-tropical (9 %) domains. But whereas the amount of land under forest is growing in the boreal, temperate and sub-tropical zones, tropical forests are shrinking. Millions of hectares of forest in South America, Africa and Southeast Asia are cleared each year and converted to other uses (FAO 2011). These forests – in particular the humid tropical forests (or rainforests) which occupy approximately 1.2 billion hectares – constitute some of the most carbon-rich and biodiverse ecosystems in the world (Hansen et al. 2008). This puts tropical forests in the front line of the struggle against climate change.

P. McMahon (✉)
Princes Rainforests Project, Clarence House, London, UK
e-mail: findmcmahon@gmail.com

T. Fenning (ed.), *Challenges and Opportunities for the World's Forests*
in the 21st Century, Forestry Sciences 81, DOI 10.1007/978-94-007-7076-8_2,
© Crown Employees UK 2014

This article will describe the valuable ecosystem services that tropical forests provide to the world. It will look at the scale and distribution of tropical deforestation and seek to understand why the trees are being cut down. It will consider the types of solution that have been put forward to reduce tropical deforestation, both through the UN Framework Convention on Climate Change (UNFCCC) and through more flexible interim measures that are not dependent on a comprehensive climate deal. Progress has been slow but there is growing consensus on the types of international programmes that can work; there is a greater desire than ever among forest nations and the international community to put these solutions into practice; and, encouragingly, there is evidence that the rate of tropical deforestation is beginning to slow.

2 Why Rainforests Matter

Rainforests provide important ecosystem services to all of us. They store water, regulate rainfall and contain over half the planet's biodiversity. Most importantly, they play a crucial role in climate change, both as cause and as part of the potential solution. The continued destruction of these forests could have serious consequences for human well-being.

2.1 The Front Line of Climate Change

Forest ecosystems draw down atmospheric carbon dioxide through photosynthesis and store it in biomass and other carbon stocks. Rainforests are particularly carbon-rich. Huge amounts of carbon are stored in the trunks, branches and leaves of trees: there can be 100–300 tonnes of carbon in each hectare of above-ground biomass. The roots and soils below often contain even more carbon: for example, the rich, black peatlands in Indonesia can store almost 1,500 tonnes per hectare. In aggregate, there is more carbon stored in tropical forests than in the atmosphere. Deforestation and forest degradation – through the decomposition and burning of plant matter and the oxidation and burning of soils, especially peatlands – release this carbon into the atmosphere.

The Intergovernmental Panel of Climate Change estimates that the global forest sector accounts for 17 % of anthropogenic greenhouse gas emissions – approximately 7–8 Gigatonnes of CO_2 equivalent (CO_2e) each year (IPCC 2007). This would mean that forest emissions are greater than the entire transport sector, or larger than the annual emissions of the USA or China. More recent research indicates that forest emissions are even higher, accounting for more than a quarter of all emissions stemming from human activity between 1990 and 2007 (Pan et al. 2011). It is estimated that more than 95 % of these emissions are caused by tropical deforestation (Houghton 2003).

But that is not the full story. Healthy tropical forests keep on absorbing carbon dioxide from the atmosphere, drawing it down through photosynthesis and storing it in trees, plants and soils. One study estimates that tropical forests may soak up an extra 4.8 billion tonnes of CO_2e each year, close to 10 % of the emissions caused each year by human activities (Lewis et al. 2009). Therefore, as we pump more and more carbon into the air by burning fossil fuels, trees take some of it out of the atmosphere and store it away. If we destroy the forests we will lose this natural balancing mechanism, and our carbon emissions will run even further out of control. Governments around the world are now channeling billions of dollars into developing carbon capture and storage technologies for coal-fired power stations. Tropical forests do the same for free.

It will be extremely difficult to develop a sufficiently fast and adequate response to climate change that does not include an effective programme to reduce tropical deforestation. Research by McKinsey & Company indicates that in order to keep global warming below 2 °C by the end of the century – and therefore avoid the worst effects of climate change – the world will need to reduce its global CO_2e emissions, relative to business-as-usual, by 17 Gigatonnes per year by 2020. Action must be taken immediately, as each year of delay makes it more difficult to get on the right pathway. The forest sector offers one of the largest opportunities for carbon abatement. Reducing tropical deforestation could contribute over 5 Gigatonnes of CO_2e per year of avoided carbon emissions between now and 2020. It could also do so rapidly and at a low cost relative to other measures. Without addressing the issue of tropical deforestation, it is difficult to see how the world can achieve climate stability (McKinsey & Company 2009).

2.2 Biodiversity

Apart from regulating the carbon cycle, tropical forests provide many other vital ecosystem services. Rainforests are the most biologically rich ecosystems on our planet, the product of tens of millions of years of evolution. Although they cover only 5 % of the earth, they contain over half of the world's animal and plant species (The Prince's Rainforests Project 2009). This biodiversity has great medical and economic value. Rainforests have been the source of compounds vital to the discovery of modern medicines. According to the US National Cancer Institute, more than 70 % of plants with anti-cancer properties are found in the rainforests (National Geographic 2012). Agricultural scientists have also used wild rainforest plants to breed cultivated crops that have higher yields and more resistance to pests and diseases. Most of the species that exist in rainforests are still inadequately researched, their potential value to humanity and to the maintenance of environmental sustainability, as yet unknown. This biodiversity is being lost because of deforestation.

2.3   Rainforests and Water Regulation

Rainforests also help to regulate water cycles and rainfall patterns. During tropical storms, roots hold the soil together and absorb water, while during dry periods trees transpire vast amounts of water vapour from their leaves. They also release tiny particles, called volatile organic compounds, around which water droplets condense to form clouds and eventually rain. A rainforest acts like a huge sponge, absorbing water when it is plentiful and releasing water when it is scarce.

This action prevents catastrophic flooding and soil erosion during wet seasons and ensures a regular flow of clean water during dry seasons: this is why vast river systems, such as the Amazon and the Congo, never run completely dry. In contrast, deforestation can lead to flash floods and soil erosion, the drying of rivers and the silting of irrigation channels, with devastating consequences for those who live in these regions.

This water regulating effect can also be felt much further away. Moisture from the forest is carried by high-altitude winds, falling as rain on centres of population and farming thousands of miles from the forest. Some models suggest that the removal of rainforests could result in reductions in rainfall globally, including in the American Mid-West and parts of Central Asia (Avassar and Werth 2004, 2005). At a regional scale, water vapour from the Amazon contributes to rainfall patterns that are vital to the agricultural heartlands of southern Brazil and the La Plata Basin in Argentina, as well as to Brazil's hydro-electric power system (Morengo 2009).

2.4   Unique Human Cultures

An estimated 1.6 billion of the world's poorest people (those surviving on less than $2 per day) rely to some extent on forests for their welfare and livelihoods. About 300 million people depend on forests for their survival (World Bank 2008). These people include subsistence farmers, hunters, small-scale loggers, extractivists such as rubber-tappers, and harvesters of nuts, berries, fruits and medicinal plants. Wild products from the forest can be an important source of nutrition and income for local communities in developing countries, in particular during periods of food shortage.

The fate of indigenous people is especially closely linked to tropical forests. There are approximately 60 million indigenous people who rely on forests for their way of life (Secretariat of the Convention on Biological Diversity 2009). The destruction of tropical forests can have a catastrophic effect on indigenous people who live there. The encroachment of outsiders can lead to violence, land theft, the abuse of rights, and the destruction of the natural resources that provide sustenance. The introduction of 'new' diseases is sometimes the most devastating result. We have a duty to respect the rights of these people and to ensure that our demands do not lead to harm.

3   Rates and Causes of Deforestation

Whereas the area under forest is growing in temperate, boreal and sub-tropical regions, tropical forests are contracting. The latest figures from the UN Food and Agriculture Organization, based on analysis of satellite imagery, indicate that just over 8 million hectares of tropical forest were cleared each year between 1990 and 2000. The rate of deforestation rose to 10 million hectares per year between 2000 and 2005. (About 2 million hectares per year were also converted back to forest during this period, so net forest loss was slightly lower (FAO 2011).) Another study estimates that humid tropical forests, or rainforests, accounted for approximately 6 million hectares of deforestation per year between 2000 and 2005 (Hansen et al. 2008).

Tropical deforestation has slowed since then. One study estimates that the rate of deforestation has fallen by 42 % between 2006 and 2011. The biggest decline has taken place in Brazil, where the rate of deforestation has halved. This has a major impact on global figures, as Brazil accounted for three-quarters of tropical deforestation in 2005 (Wheeler et al. 2011). Indeed, Brazil's halving of deforestation represents the greatest single reduction in greenhouse gas emissions by any country over the past decade. The fall in deforestation is partly due to concerted government action at national and international level (progress that will be explored later in this article). But it is also a consequence of the slowdown in the global economy since 2008. This illustrates how tropical deforestation is increasingly driven by global commodity markets and global economic activity.

3.1   From Axes to Chainsaws

The nature of tropical deforestation has changed over the past three decades. Traditionally, deforestation was associated with the subsistence activities of local people. Poverty and land scarcity pushed farmers to clear native forest for agriculture, often using 'slash and burn' techniques. People chopped down trees to provide firewood, charcoal or timber for buildings. The products generated were either consumed by families or traded locally, but they did not reach foreign markets.

Increasingly, however, tropical deforestation is being driven by commercial operations linked to global markets. In Indonesia and Brazil, a growing proportion of deforestation is caused by export-led agricultural expansion. Palm oil, beef and soybeans are the key commodities. In other areas, cocoa, coffee and rubber production play a role, while mining and biofuels cause forest loss. The wood products industry is also a significant driver of destruction. Valuable trees are logged for hardwood timber and whole areas are clear-felled for pulp and paper factories. Much of the tropical timber – perhaps over half – is harvested illegally.

Rather than having a single cause, deforestation sometimes occurs because of the complex interplay between these activities. For instance, in South America land can be opened up with roads by logging companies, then slashed and burned by migrating subsistence farmers, cultivated for a few years, sold over to cattle ranchers and then bought by soybean farmers. Each stage can generate very different economic returns: the small-scale farmer may earn $2 or $3 per hectare, the cattle rancher $400 and the soybean farmer $3,000 (Grieg-Gran 2008). The potential to convert the land to more valuable uses motivates each individual, as well as the land speculators who act as intermediaries. Much of this activity may be illegal under national laws. Unclear ownership and user rights to forested land further complicates the picture.

The dynamics of deforestation are local, but the commodities go to feed global demand. Much of the beef, soya and palm oil produced in tropical countries is exported. It ends up on supermarket shelves, in restaurants, or – in the case of palm oil used to produce biodiesel – in the fuel-tanks of cars. Growing demand from fast-developing economies such as China is turbo-charging this consumptive process. The relentless growth in the world population – expected to increase from 7 to 9 billion by 2050 – will provide further impetus. For example, the UK Government's Gallagher Review estimates that growing demand for food, feed and biofuels is likely to require an additional 200–500 million hectares of agricultural land in the next decade (The Gallagher Review of the Indirect Effects of Biofuels Production 2008). This will place even more pressure on the tropical forests of the world.

3.2   Regional Differences

The rates and causes of deforestation differ from continent to continent. In recent years, most rainforest destruction has taken place in South and Central America, which has the largest area of rainforest in the world. In Brazil, cattle ranching and associated land speculation are widely recognized as being the main drivers of deforestation. The clearing of land for cattle by poor families bestows de facto ownership rights to land, albeit often illegal. Cattle ranchers' migration into the Amazon biome is also partly caused by the expansion of soybean cultivation in drier areas, which has pushed ranchers north into the forest frontier.

Southeast Asia has the highest rate of deforestation relative to the size of its forests. Logging for timber and pulp and paper, as well as subsistence and commercial agriculture, are the main drivers of deforestation. In Indonesia and Malaysia, logging, often followed by the establishment of palm oil or pulpwood plantations, is the main cause of the disappearance of forests.

The African continent had the lowest rate of tropical deforestation between 2000 and 2005, relative to the size of its forests. The relatively low rate of deforestation in this region can be explained by the lesser importance of commercial agriculture and logging as drivers of deforestation; instead, most forest is still cleared for subsistence agriculture or fuelwood. However, commercial logging activities are multiplying, facilitated by improved transport infrastructure. Large-scale agriculture is also increasing and is likely to account for more deforestation in the future, as land for agricultural expansion grows scarce on other continents (Data on Rates of Deforestation is Taken from Hansen M et al. 2008; Drivers of Deforestation is from Blaser J and Robledo C 2007).

3.3   A Market Failure

The causes of deforestation may differ from region to region, but they have one feature in common: the people clearing the rainforests are acting rationally, given the economic incentives they face. Deforestation allows rural populations to practice agriculture, landless people to acquire a patch of their own, companies to engage in profitable commodity production, and governments to generate tax revenue and foreign exchange. These people are responding to the market signals of an increasingly globalised world. It should be remembered that today's richest countries actively pursued deforestation and land conversion to agriculture in early phases of development for exactly these reasons.

Fundamentally, deforestation occurs because the world places more value on the commodities produced from deforested land than on the environmental services that tropical forests provide. The local returns from deforestation are specific and financial; the global benefits of preserving forests are diffuse and not valued in monetary terms. In the final calculation, the trees are worth more dead than alive. Unless a way is found to rebalance this equation, and value standing forests, the trees will continue to disappear.

Encouragingly, there does not have to be trade-off between forest conservation and economic development. Research by WWF shows that it will be possible to substantially increase food, fibre and biofuel production in tropical countries without touching the forests, mainly because there are opportunities to use non-forested land much more efficiently (WWF 2011). Similarly, The Prince's Rainforests Project has worked with the private sector in Brazil, Indonesia and West Africa to identify practical ways to intensify cattle, palm oil and cocoa production on degraded and non-forested land (The Prince's Rainforests Project 2010). This will require upfront investment and sustained effort over many years, which is why deforestation often remains an easier option. But appropriate finance from the international community could tip the balance in favour of these approaches. It could help forest nations make the investments that would be needed to pursue an alternative low carbon development trajectory. If used in this way, forest finance would not only achieve environmental goals but would also provide vital investment that could reduce poverty, enhance food security and accelerate 'green growth' in developing countries.

4   Initiatives to Address Deforestation

There has long been a consensus that the international community should work with the governments of forest nations to slow or halt the destruction of tropical forests. However, traditional donor programmes have been unable to compete with the economic drivers of deforestation outlined above. There is great hope that a new climate deal, agreed as part of the UNFCCC, will finally place an appropriate value on tropical forests, but such a deal is still many years off. This has created a gap that numerous international initiatives are now attempting to fill, with varying levels of success. It is still uncertain whether mechanisms of sufficient scale and ambition will emerge from this fragmented policy environment to have a significant impact on tropical deforestation in the next 10 years – but the opportunity exists.

4.1 Historical Approaches

Over the past three decades, a number of initiatives were established by the World Bank, UN agencies and donor countries to try to preserve forests in tropical countries. These included the Tropical Forestry Action Plan in the 1980s, National Environmental Action Plans from the 1990s, efforts to control international trade in illegal logs in the 2000s, and the integration of forestry into broader bi-lateral donor assistance programmes. In addition, dozens of international NGOs conducted project-level activities in tropical countries. While there have been some success stories, the overall results have been disappointing, as evidenced by the huge area of the world's tropical rainforests that has been cleared or heavily degraded during this period.

A number of reasons have been put forward to explain the failure of previous initiatives to reduce deforestation (The Prince's Rainforests Project 2009).

- **Narrow scope**: Initiatives focused only on the forestry sector rather than addressing the broader drivers of deforestation and failed to create alternative economic opportunities for local people.
- **Lack of political buy-in**: In many cases, neither governments nor local communities within forest nations shared the goals of international donors.
- **Uncommitted institutions**: The importance of forests was not always shared within development agencies, nor was there coordination between agencies.
- **Inadequate funding**: Historically, less than US$1 billion per year was available through Official Development Assistance for tropical forestry. This was never enough to compete with the drivers of deforestation.

In essence, political will was not been strong enough, nor sustained for long enough, to ensure the implementation of development approaches that could tackle the fundamental economic issues that caused deforestation in tropical countries. Clearing forests remained more lucrative than conserving them.

4.2 REDD+ and Climate Change

In recent years, the elevation of climate change to the top of the global policy-making agenda meant that, for the first time, there was a chance to harness enough political commitment and international funding to forge a long-lasting solution to tropical deforestation. As part of the UNFCCC, countries have agreed to set up a new mechanism for Reducing Emissions from Deforestation and Degradation (REDD) in developing countries. This concept has been expanded to REDD+, the 'plus' signifying that the mechanism should also support forest conservation, sustainable management of forests and enhancement of forest carbon stocks.

The principle behind REDD+ is that industrialized nations (Annex 1 countries in the UNFCCC protocol) should pay forest nations for verified reductions in greenhouse gas emissions that come about through reducing deforestation or preserving or enhancing forest stocks at a national level. This could be a government-to-government transaction, using public finance from Annex 1 countries, or it could involve private finance from carbon markets. The details are yet to be worked out, as are many other technical issues such as how to set appropriate reference levels against which to measure avoided deforestation, how to conduct monitoring and reporting, and how to ensure safeguards for vulnerable groups. But the goal is to generate sufficient flows of finance to forest countries to incentivize and to facilitate low-deforestation development paths.

The economic rationale for REDD+ is compelling. The Eliasch Review, a study on the role of forests in climate change commissioned by the UK Government, estimated that it would cost between US$17 billion and US$33 billion per year to halve deforestation. The net present value of this halving of deforestation, based on the global savings from reduced climate change minus the costs of forest finance, was calculated at a massive US$3.7 trillion (The Eliasch Review 2008). REDD+ is a good deal for Annex 1 countries looking to finance greenhouse gas reductions. It is potentially cheaper than most other mitigation options and could be achieved more rapidly. It should also be a good deal for forest nations, as it would provide much-needed finance for their development.

After much debate, at the UNFCCC conference in Cancun in December 2010 it was formally agreed that REDD+ would form part of the legally binding successor to the Kyoto Protocol. A technical working group was set up to work out the details of its operation. However, at the climate change conference in Durban in December 2011 it was agreed that any new protocol would be not be adopted before 2016 and would not come into effect before 2020. Therefore, the REDD+ mechanism will not come into operation for at least 7 years. Moreover, there is no guarantee that this timetable will be kept. So, a UNFCCC REDD+ mechanism still remains a solution of the future, not the present.

In the meantime, carbon markets have provided only a tiny measure of support to REDD+ projects in developing countries. The total value of the global forest carbon market is around US$149 million. About three-quarters of this funding has come from voluntary carbon markets such as the Chicago Climate Exchange. The rest is associated with forest projects approved under the Kyoto Protocol's Clean Development Mechanism. Combined, these projects cover an area of just 1.7 million hectares (Simula 2010). Private carbon finance is unlikely to be a significant factor before a REDD+ mechanism is agreed by governments as part of a global climate deal.

4.3   Interim REDD+ Finance

Delays in the implementation of a UNFCCC REDD+ mechanism were widely expected. As a result, steps have been taken to create programmes that could operate in the interim, outside of the formal UNFCCC process. In 2007 His Royal Highness The Prince of Wales established The Prince's Rainforests Project to help build consensus around near-term solutions to tropical deforestation. This project has worked with senior politicians, business leaders, non-governmental organizations and other interested stakeholders from around the world. On 1 April 2009 The Prince of Wales invited world leaders to a meeting in London at which it was agreed to establish an inter-governmental working group to develop proposals for a financing mechanism that could achieve rapid reductions in deforestation. The Informal Working Group on Interim Finance for REDD (IWG-IFR), representing 34 governments, produced its report in October 2009. Following a ministerial meeting in Paris in March 2010, a REDD+ Partnership was launched at the Oslo Climate and Forest Conference on 27 May 2010. The REDD+ Partnership is a voluntary, non-legally binding framework that brings together 58 countries committed to developing and implementing collaborative REDD+ efforts in the interim period before a UNFCCC agreement. It contains most tropical forest countries, as well as traditional donor countries. The latter made financing pledges exceeding US$4 billion for the 2010–2012 period (Norwegian Government 2010). These formed part of a broader pledge of 'Fast Start Finance' for climate mitigation and adaptation in developing countries, made a few months earlier by Annex 1 countries at the UNFCCC conference in Copenhagen.

What progress has been made with this interim finance? The programmes that have been started can be divided into two types: 'payment for performance' schemes and REDD+ preparatory schemes.

'Payment for Performance' Schemes

A small number of REDD+ partnerships have been formed under which funding countries agree to make payments to forest nations based on changes in actual deforestation rates from year to year. Norway has been the most active, agreeing partnerships with Brazil, Guyana and Indonesia that will provide US$2.25 billion in funding (although in the last case the 'payment for performance' component is not due to start until 2014). These schemes have some common features: the agreement of a price per tonne of CO_2 emissions abated (usually US$5 per tonne); the use of proxies to calculate emissions reductions (usually based on hectares of deforestation avoided); simple verification mechanisms that will build in sophistication over the years; and considerable freedom for forest nations to decide on how payments are used, within a framework of safeguards. There are also differences between the partnerships, especially in terms of the channels through which payments flow. The Norway-Guyana deal is also different in that it does not reward Guyana for reducing its deforestation rate compared to a historical baseline (very little deforestation has taken place in Guyana yet) but for making sure the deforestation rate does not rise in the future (Norad 2011).

Tropical forest nations have shown considerable initiative in driving these solutions. Brazil announced a goal to cut deforestation by 70 % by 2020, before signing up to a payment deal with Norway. The Indonesian government has committed to reduce its greenhouse gas emissions by 26 % through its own efforts and by 41 % if it receives international assistance. Guyana developed a comprehensive low carbon development plan before entering into its partnership with Norway. However, in each case the promise of international finance, in the form of payments for the ecosystem services provided by the forests, has been a catalyst for domestic action. It is too early to measure the full results of these partnerships but the recent falls in deforestation in Brazil, for example, indicate that positive steps are being taken on the ground.

Preparatory Schemes

The majority of interim finance has not gone to pay forest countries for actual reductions in deforestation but to fund programmes that are helping forest countries prepare themselves for a UNFCCC REDD+ mechanism or interim 'payment for performance' deals. These support activities such as strategy development, capacity building, institutional reform and establishment of forest monitoring systems. The argument is that forest countries need to be 'REDD+ ready' before they can engage with mechanisms that pay for performance. These activities are being funded by a wide range of donors through bi-lateral programmes, as well as by multi-lateral and regional programmes such as the UN-REDD Programme, the World Bank's Forest Carbon Partnership Facility, the Forest Investment Program or, in Africa, the Congo Basin Forest Partnership (Simula 2010).

4.4 Remaining Challenges

Through the REDD+ Partnership and Fast Start Finance, more public money has been committed to slowing tropical deforestation than ever before. Most forest nations are participating in at least one international REDD+ programme. However, a number of challenges remain. First, with the exception of the financial support pledged by Norway, most funding to date has focused on strategy development, capacity building or small-scale pilot projects. There are few 'payment for performance' schemes achieving real results. This is partly because of a 'chicken and egg' problem: Annex 1 countries are reluctant to commit until they see strong leadership and clear plans in forest nations; forest nations won't invest domestic political capital into developing these plans until they are certain there will be international funding available. Second, there have been many delays in REDD+ finance delivery and only a small proportion of the funds has actually been disbursed. One reason is that financing countries tend to view their support as Official Development Assistance, and have channeled their funds through traditional aid channels, whereas forest nations prefer to see this as a partnership, and prefer more flexible implementation mechanisms. Third, the Fast Start Finance pledges were barely adequate for the 2010–2012 period and certainly cannot cover the period until 2020, which is the earliest that a new UNFCCC protocol will come into effect. Further pledges will be required to fill the financing gap between now and then (The Prince's Charities' International Sustainability Unit 2011).

5   Conclusions

Tropical forests provide important ecosystem services to the world. They regulate rainfall, contain vast amounts of biodiversity and play a crucial role in the carbon cycle. It will be difficult to attain climate stability in this century without action to reduce deforestation. Tropical forests are cleared for many reasons but the fundamental cause is that there are strong economic incentives driving deforestation, often linked to global commodity demand.

Past attempts to conserve tropical forests have mostly failed because they have not been able to out-compete these drivers of deforestation. The prospect of a global climate deal opens up the possibility of a REDD+ 'grand bargain' under which industrialized countries would pay tropical nations for the ecosystem services that their forests provide, which would finally make the trees worth more alive than dead. The technical elements of this mechanism are becoming clearer but the global policy landscape is such that it will not be in place before 2020 at the earliest. As a result, a series of smaller bargains have emerged, as countries try out various interim approaches. Some of these partnerships are developing novel mechanisms to reward forest nations for actual reductions in deforestation between 2010 and 2020. However, most are more cautious, employing traditional aid approaches and focusing on building 'REDD+ readiness' rather than paying for results.

On a number of occasions since 2009 tropical forest nations have expressed their willingness to protect their forests, so long as they receive appropriate international support. There has never been a better opportunity for the international community to forge ambitious partnerships with forest nations to achieve substantial reductions in tropical deforestation. On the other hand, there is a risk that forest nations will be discouraged by the small amount of international finance available, the slowness of its disbursement, and the prevalence of uncoordinated, piecemeal approaches. Bolder steps will need to be taken to translate the promise of recent years into real results on the ground.

References

Avassar R, Werth D (2004) Global hydroclimatological teleconnections resulting from tropical deforestation. J Hydrometeorol 6:134–145

Avassar R, Werth D (2005) The local and global effects of African deforestation. Geophy Res Lett 32(L1270). http://onlinelibrary.wiley.com/doi/10.1029/2005GL022969/full

Data on rates of deforestation is taken from Hansen M et al. (2008) Humid tropical forest clearing from 2000 to 2005 quantified by using multitemporal and multiresolution remotely sensed data. PNAS 105(27):9439–9444

Drivers of Deforestation is from Blaser J and Robledo C (2007) Initial analysis on the mitigation Potential in the forestry Sector. Report prepared for the Secretariat of the UNFCCC

FAO (2011) Global forest land-use change from 1900 to 2005: initial results from a global remote sensing survey. FAO, Rome, http://foris.fao.org/static/data/fra2010/RSS_Summary_Report_lowres.pdf

Grieg-Gran M (2008) The cost of avoiding deforestation. International Institute for Environment and Development, London

Hansen M et al (2008) Humid tropical forest clearing from 2000 to 2005 quantified by using multitemporal and multiresolution remotely sensed data. PNAS 105(27):9439–9444

Houghton RA (2003) Revised estimates of the annual net flux of carbon to the atmosphere from changes in land use and land management 1850–2000. Tellus 55B:378–390

IPCC (2007) AR4 synthesis report. Intergovernmental Panel on Climate Change, Geneva

Lewis SL et al (2009) Increasing carbon storage in intact African tropical forests. Nature 457:1003–1006. doi:10.1038/nature07771

McKinsey & Company (2009) Global GHG abatement cost curve v2. ClimateWorks Foundation/McKinsey & Company, New York

Morengo J (2009) Climate change, extreme weather and climate events in Brazil. FBDS, Rio de Janeiro, http://www.lloyds.com/~/media/2ce4a37d36fd4fceab4c3fc876f412f7.ashx

National Geographic (2012) Rainforest: incubators of life http://environment.nationalgeographic.com/environment/habitats/rainforest-profile/

Norad (2011) Real-time evaluation of Norway's International Climate and Forest Initiative. Norad Evaluation Department, Oslo, http://www.regjeringen.no/upload/MD/2011/vedlegg/klima/klima_skogprosjektet/Evalueringsrapportene/Report_18_2010_Summary.pdf

Norwegian Government (2010) REDD+ partnership document. http://www.oslocfc2010.no/pop.cfm?FuseAction=Doc&pAction=View&pDocumentId=25019

Pan Y, Birdsey R, Fang J, Houghton R, Kauppi P, Kurz WA, Phillips OL, Shvidenko A, Lewis SL, Canadell JG, Ciais P, Jackson RB, Pacala S, McGuire AD, Piao S, Rautiainen A, Sitch S, Hayes D (2011) A large and persistent carbon sink in the world's forests. Science. doi:10.1126/science.1201609

Secretariat of the Convention on Biological Diversity (2009) Sustainable forest management, biodiversity and livelihoods: a good practice guide. Montreal

Simula M (2010) Analysis of REDD+ financing gaps and overlaps. Report for REDD+ Partnership. http://reddpluspartnership.org/25159-09eb378a8444ec149e8ab32e2f5671b11.pdf

The Eliasch Review (2008) Climate change: financing global forests. UK Government, London

The Gallagher Review of the Indirect Effects of Biofuels Production (2008) Renewable Fuels Agency, London

The Prince's Charities' International Sustainability Unit (2011) Emergency finance for tropical forests: two years on – is interim REDD+ finance being delivered as needed? http://www.pcfisu.org/wp-content/uploads/2011/11/Two-years-on_Is-interim-REDD+-Finance-being-delivered-as-needed.pdf

The Prince's Rainforests Project (2009) An emergency package for tropical forests. PRP, London

The Prince's Rainforests Project (2010) REDD+ and agriculture: proposed solutions from the private sector. PRP, London, http://www.rainforestsos.org/wp-content/uploads/pdfs/REDD-and-Agriculture-Proposed-Solutions-from-Private-Sector.pdf

Wheeler D, Kraft R, Hammer D (2011) Forest clearing in the pantropics: December 2005–August 2011. CGD Working Paper 283. Center for Global Development, Washington, DC. http://www.cgdev.org/content/publications/detail/1425835

World Bank (2008) Forest sourcebook. The World Bank, Washington, DC

WWF (2011) Living forests report: chapter 1. WWF, Gland

Silviculture of Planted Forests Managed for Multi-functional Objectives: Lessons from Chinese and British Experiences

W.L. Mason and J.J. Zhu

Abstract Planted forests are anticipated to increase in area and to supply an increasing proportion of the world's timber supplies in future decades. However, because of their increasing importance, management of these forests will need to pay greater attention to silvicultural practices which can help to sustain and improve the delivery of a range of other values besides timber production. Some indications about how this might be achieved can be gained from an examination of recent developments in the management of planted forests in Northeast China and the British Isles. Both regions share a common history of deforestation and unsustainable harvesting with forest cover being restored during the twentieth century by extensive reforestation programmes based on a few species that were robust to plantation silviculture. In response to changing societal pressures, these simple forests are now being managed to meet multifunctional objectives including biodiversity, recreation and landscape values. The most successful silvicultural methods for increasing the diversity of planted forests have involved the introduction of complementary species, either through planting or by natural regeneration, and the use of thinning to create a more open and varied stand structure. However, the relative merits of these approaches depend upon local conditions such as the light regime found within the planted forests and the occurrence of abiotic risks such as windthrow. An important lesson is the value of establishing long-term silvicultural trials to demonstrate the processes involved in the diversification of planted forests.

W.L. Mason (✉)
Forest Research, Northern Research Station, Roslin, Midlothian, Scotland EH25 9SY, UK
e-mail: bill.mason@forestry.gsi.gov.uk

J.J. Zhu
Forest Ecology and Management, Quingyuan Experimental Station of Forest Ecology, Institute of Applied Ecology, Chinese Academy of Sciences, No. 72, Wenhua Road, Shenyang 110016, China
e-mail: jiaojunzhu@iae.ac.cn

T. Fenning (ed.), *Challenges and Opportunities for the World's Forests in the 21st Century*, Forestry Sciences 81, DOI 10.1007/978-94-007-7076-8_3,
© Crown Employees UK 2014

1 Introduction

Planted forests are generally defined as being forests composed mainly of native or introduced tree species established either through planting or by deliberate seeding (Carle and Holmgren 2008). This is a broader concept than the more traditional term of 'plantation forests' which is normally applied to single species forests planted for timber production and frequently composed of exotic species (Kanninen 2010). Planted forests therefore also include mixed plantings of a range of species which are established for biodiversity, landscape or other non-commercial objectives. The increasing area of planted forests, and the growing importance of the plantation forest subset for future timber supplies (Paquette and Messier 2010), means that an improved understanding of the issues and methods pertaining to the management of planted forests will be of considerable importance in ensuring the sustainable management of the world's forest resources during the coming century. Parts of China and of the British Isles share a common history of deforestation or impoverishment of natural forest resources followed by their replacement by plantation forests of fast growing conifers. In both countries, the environmental and social effects of establishing single species plantations for the primary purpose of timber production have become controversial. With increasing recognition of the multi-functional objectives of sustainable forest management, forest policies and societal demands have gradually shifted from an emphasis on timber production to the provision of multi-functional ecosystem services from the planted forests (Zhu et al. 2010; Mason et al. 2011).

Planted forests in China cover 61.7 million hectares, representing 31.6 % of China's forest area or 23.4 % of the world's planted forest area (FAO 2010). China's planted forests are classified into productive and protective forests according to their functions. The productive forests for timber supply account for about 25 million hectares, or over 40 % of the area of planted forests. There has been a continuing increase in the planted forest area, benefiting from several key Chinese forestry programmes including: the Natural Forest Resources Protection (NFRP), Conversion of Cropland to Forest, Shelter Forest Construction in the Three-North and in the Upper and Middle Reaches of the Yangtze River, and the Development of Fast-growing and High-yielding Plantations (DFHP) (Xu et al. 2000; Li 2004). The major tree species planted include Chinese fir (*Cunninghamia lanceolata*) in South China and larches (*Larix* spp.) in North China. More than one-third of the national forests are located in Northeast China, and more than 60 % of these are secondary forests due to a century of excessive timber harvesting (Hu and Zhu 2008). These secondary forests are composed of broadleaved tree species such as *Fraxinus* spp., *Acer* spp., *Betula* spp., *Populus* spp., *Tilia* spp. and *Quercus mongolica*, with multi-purpose uses and high commercial values. However, in order to promote rapid economic development and to meet the objectives of the DFHP and NFRP programmes, since the 1950s a range of larch species (*Larix olgensis, Larix kaempferi* or *Larix gmelinii*) have been planted widely in Northeast China as commercial timber species to replace the secondary forests (Liu et al. 2005; Zhu et al. 2008).

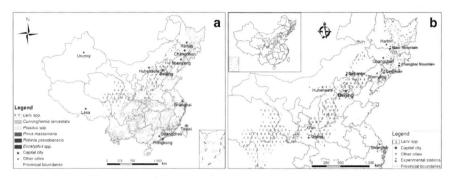

Fig. 1 Sketch map of productive plantation forests in China, about 25 million ha (**a**), and larch plantations in North China, about 3.78 million ha (**b**)

Currently, larch plantations amount to some 2 million hectares, around 55 % of the planted forests in Northeast China (Fig. 1). Many problems have been reported from these larch plantations in recent decades such as a decline in soil fertility, lower productivity in the next rotation, a lack of natural regeneration, and poor provision of ecosystem services (Liu et al. 1998; Yang et al. 2010a).

The countries of the British Isles consisting of the United Kingdom (England, Northern Ireland, Scotland and Wales) and the Republic of Ireland, fall within the Atlantic zone of north-western Europe with a native woodland cover composed mainly of temperate broadleaved species (e.g. *Quercus* spp., *Fraxinus excelsior*, *Betula* spp.), although there are outliers of the boreal forests in the Scots pine (*Pinus sylvestris*) dominated areas of northern Scotland. All countries share a similar history of centuries of deforestation and unsustainable forest management (e.g. Smout 2003) so that at the beginning of the twentieth century forest cover was very low, being less than 5 % in Great Britain (Forestry Commission 2011) and 1 % in Ireland (Horgan et al. 2003). The early years of the last century saw the start of concerted attempts to restore forest cover through afforestation and this has been continued until the present time with the result that the forest area of the British Isles is nearly three times as large (Table 1). The afforestation programme was primarily intended to provide timber for industrial purposes and was based on the use of fast growing and mostly non-native conifer species which could tolerate the exposed sites and impoverished soils where planting was concentrated (Mason 2007a). The result has been the transformation of the British forest resource from broadleaves to conifers in around 50 years (Mason 2007b) (Table 2). Thus the non-native conifer Sitka spruce (*Picea sitchensis*) is now the major tree species in Britain and in Ireland and accounts for a substantial proportion of the domestic timber supply.

During the latter half of the last century there were growing concerns in the British Isles about various impacts of the conifer afforestation programmes such as negative effects of planting and felling upon visual amenity (Foot 2003), conversion of native broadleaved woodlands to conifers, and the large-scale planting of upland habitats of high biodiversity value (Avery and Leslie 1992). These concerns resulted in substantial changes in policy during the last two decades of the last century.

Table 1 Forest area of different countries in the British Isles in 1905, 2010, and policy aspirations for the future

Country	Forest area (M ha) in 1905[a]	Forest area (M ha) in 2010[a]	Percent land area	Aspiration and target date	Sources for aspirations
England	0.7	1.3	10.0	Increase (2050)	DEFRA (2007)
Republic of Ireland	0.07	0.7	10.2	17 % (2030)	Forest Service (2007); Teagasc (2011)
Northern Ireland	0.02	0.1	6.5	13 % (2050)	NIFS (2006)
Scotland	0.4	1.4	17.8	25 % (2050 onwards)	FCS (2006)
Wales	0.09	0.3	14.7	Increase (2050 onwards)	FCW (2009)

Notes:

[a]Figures for 1905 and 2010 are derived from Forestry Commission (2011) and Forest Service (2007)

Table 2 Main tree species in Great Britain in 2010 by proportion of the forest area (after Forestry Commission 2011)

Species	Percent forest area
Pinus sylvestris	9.5
Pinus contorta	5.7
Picea sitchensis	29.1
Picea abies	3.2
Larix spp.	5.6
Other conifers[a]	5.8
Total conifers	**58.9**
Quercus spp.	9.3
Fagus sylvatica	3.5
Acer pseudoplatanus	2.8
Fraxinus excelsior	5.4
Betula spp.	6.7
Other broadleaves[b]	13.4
Total broadleaves	**41.1**

Notes:

[a]Other conifers includes *Pinus nigra* ssp. *laricio*, *Pseudotsuga menziesii*, *Abies grandis*, *Abies procera*, *Tsuga heterophylla*, and other introduced species

[b]Other broadleaves includes *Populus* spp., *Castanea sativa*, *Ulmus* spp., and areas of mixed broadleaves

A broadleaves policy was introduced in 1985 (Richards 2003), which essentially ensured that no further broadleaved woodland would be converted to conifers. In 1988 changes to the tax rules that had underpinned private forestry substantially reduced the amount of conifer afforestation (Foot 2003). The 1990s saw an increasing emphasis on the principle of sustainable forest management (SFM) for multiple benefits (often referred to as 'multi-purpose' forestry) exemplified in the widespread implementation of the certification process in British forestry (Mason 2007b). As a result of the changes of the last 30 years and the increasing emphasis upon SFM, there has been a move away from the industrial plantation management approach that dominated British forestry for over half a century. Both the Scottish (FCS 2006) and Welsh (FCW 2009) forestry strategies endorse the desirability of greater use of continuous cover forestry (CCF), whereby more irregular stand structures composed of mixed species are developed in all forests. However, a consequence of the history of deforestation followed by restoration through afforestation is that plantation forests dominate the forest resource of the British Isles to an extent that is in uncommon elsewhere in the world with nearly 90 % of forests in the Republic of Ireland and 68 % of forests in the United Kingdom being of plantation origin (Del Lungo et al. 2006). This means that these forests must not only provide wood for industry, but must also be managed to provide the range of other ecosystem services that is characteristic of forest management for multifunctional objectives (Mason 2007a; Mason and Meredieu 2011).

Current practices in China's larch forests and British Sitka spruce forests are outlined in Table 3. The typical structure that results from these regimes will be one of regular even-aged stands composed of very few species: thus 90 % of British forests had three species or less (MCPFE 2007). Until the 1990s, the main silvicultural approach employed in the conifer plantations of China and Britain was patch clearfelling (Matthews 1989) using coupe sizes of 5–50 ha with regeneration achieved mainly through planting. Rotation lengths have been primarily based upon predicted ages of maximum mean annual increment, although actual felling age in Britain was influenced by the anticipated risk of windthrow and determined by calculation of net discounted revenue (Johnston et al. 1967). A normal rotation would be between 35 years and 60 years, depending upon species and site productivity. Therefore, the larch and spruce forests of China and the British Isles are dominated by stands in the stand initiation and stem exclusion phases (after Oliver and Larson 1996) and changing them to the more complex and species rich structures envisaged by current policies represents at least as great a silvicultural challenge as the initial effort to restore forest cover on devastated and cutover lands (Mason and Quine 1995).

Adapting plantation forests to multifunctional management needs to consider the present structure of a forest and the services provided, the type of structure which might be most suitable for the provision of a range of ecosystem services to meet future needs, plus the site and climatic factors which influence silvicultural practice. At the stand level, there are five categories of silvicultural intervention which can be used to influence forest structure, namely: site preparation and the management of residues; species selection and the use of mixtures; stand density management especially the use of thinning; fostering of greater structural complexity; and adjustment of rotation length (after Bauhus and Schmerbeck 2010). In this chapter, we consider these different silvicultural options and their possible role in the development of more complex structures in the planted forests of China and the British Isles, with a focus on larches and Sitka spruce.

Table 3 A list of the main features of conventional management for Sitka spruce forests in Scotland and larch forests in north-east China (see note for data sources)

Feature	Sitka spruce	Larch
Use of planting stock from genetically improved material	10–30 % genetic gain available from first generation seed orchards through to cuttings of full sib families	15–40 % genetic gain can be obtained by using plants from hybrid larch seed orchards
Presence of other tree species	15–20 % of planting area includes other conifers or broadleaves	Broadleaved species from secondary forests surround larch plantations. A few areas are underplanted with other conifers when thinning is conducted
Site preparation	Mounding (scarification on some drier soils)	Hand screefing to remove shrubs and grass (often burnt) followed by pit planting. The pit dimensions are 50, 50 and 30 cm (length, width and depth)
Drainage	Yes	No
Planting density (trees ha^{-1})	2,700	2,200–2,500
Weeding	1–3 times in the first 3 years: chemical or hand	5 hand weedings within the first 3 years
Chemical protection	Against large pine weevil (*Hylobius abietis*)	Against shoot blight (dieback), needle cast and pine caterpillar (*Panthea* spp.)
Fertilizer inputs	Rare	Rare
Pruning	No	Yes. Carry out with thinning
Thinning	Only on 50 % of sites of lower wind risk; 2–4 times from years 20–25. Higher wind risk sites are unthinned	3–5 times for all sites from years 12–15 without considering the risk of wind or snow damage
Final stand density (trees ha^{-1})	300–1,000 (thinned – unthinned)	390–630 (according to the thinning intensity)
Rotation age (years)	35–50	40–60
Main silvicultural system and coupe size	Patch clearfelling; coupe size ranging from 5 to 40 ha depending upon aspects such as landscape impacts and wind risk	
Average productivity (m^3 ha^{-1} year^{-1})	14	6.5–8.5

Note: the main data sources used to compile this table were Mason and Meredieu (2011), Mason (2007a), Luo et al. (2005) and Zhang et al. (2009)

2 Silvicultural Options and Stand Structure

2.1 Site Management: Changes in Soil Fertility and Other Nutritional Problems in Plantations

The impacts of larch plantations on soil properties and nutrient cycling have been a particular concern in China. Although larch stands were planted on nutrient-rich soils that originally supported mixed broadleaved forests (Fig. 2), differences in soil properties were found during the first rotation of larch in Northeast China (Liu et al. 1998; Yang et al. 2010a, b). During a 3 year period of observation in a 25-year-old *L. gmelinii* plantation, there were consistent declines in soil available P (Gao 1983). Changing from natural secondary forests to larch plantations resulted in a significant decline in soil C and N concentrations, and a reduction in soil microbial biomass and nutrients (Yang et al. 2010b). The concentrations of microbial biomass P and labile organic P were significantly lower under larch than in natural secondary forest (Yang et al. 2010a). Evidence also showed that soil changes under pure larch plantations might continue in the second rotation, and this was accompanied by a decrease in productivity particularly where branches and other small material were removed for firewood (Liu et al. 1998). The litter accumulation in 20-year-old pure larch stands was higher ($23–28$ tonnes ha^{-1}) than in mixed broadleaved forests ($9–16$ tonnes ha^{-1}), and the thicker litter layer decomposed more slowly, inhibited thermal conduction and decreased soil temperature, so limiting soil microbial activities. By contrast, in Britain there has been little sign of detrimental changes in soil properties under repeated rotations of Sitka spruce (Mason and Quine 1995) and some evidence that growth has improved as a result of better silviculture, nitrogen deposition and climate warming (Cannell 2002). The only instances of decline in productivity over successive rotations have occurred following intensive residue and litter removal at time of clearfelling as in whole-tree harvesting (Mason et al. 2012).

In both China and Britain, fertilizers have been used to try to boost growth or to offset the effects of any changes in soil properties. An 8 year study in larch

Fig. 2 Images of the secondary forests (**a**) and larch plantations with broadleaved tree species (**b**)

(*L. kaempferi*) plantations indicated that fertilization (three levels of nitrogen and phosphorus applied in three consecutive years to a 15-year-old stand) did not significantly affect the yield increment (Dong 2000), because the effects were limited by site conditions, fertilization methods and the rotation length. The results of different types of fertilizer applications (mineral nitrogen, phosphorus, and animal manure) to 14-year-old and 34-year-old larch (*L. gmelinii*) plantations, showed that mineral fertilizers could increase soil fertility in the short term and improve the growth of the younger plantation, and also that organic fertilizer application improved long term soil fertility (Ji et al. 2004). Nitrogen is the most important fertilizer applied in larch plantations in China, but has been found to decrease fine root biomass, soil respiration and soil microbial biomass carbon and nitrogen in 16–40 years old stands (Son and Hwang 2003; Yu et al. 2007; Jia et al. 2010). The decreases in root biomass suggest that nitrogen addition can impact belowground biological processes, such as fine root dynamics, soil microbes, soil respiration etc., with potential consequences for soil carbon dynamics (Hu et al. 2010). In Britain, fertilizers were widely used in the establishment phase of the first rotation (Taylor 1991) to compensate for deficiencies due to previous agricultural practices. However, in recent decades their use has declined (Table 3) with increasing recognition that, on most sites, effective recycling of nutrients from the litter layer is key to sustained productivity (Miller 1995). Evans (2009) concluded that international experience suggested that care with harvesting, fire protection, conservation of organic matter, and careful management of weed growth were essential measures to avoid damage to soils under plantations.

2.2   Species Choice and Growing Tree Species in Mixture: Effects on Tree Growth and Soil Nutrients in Plantations

Larch has exhibited significantly greater growth in most mixed-species plantations when compared with pure stands. For example, in a 10-year-old mixed plantation of larch (*L. gmelinii*)-Manchurian ash (*F. mandshurica*) (6 larch: 4 ash), the mean diameter at breast height (DBH) and tree height (H) of larch was greater in mixture compared to pure stands (Zhang et al. 2001). A similar trend was found in a 13-year-old mixed plantation of larch (*L. gmelinii*)-Manchurian walnut (*Juglans mandshurica*) (6.2 larch: 3.8 walnut), where the volume of the pure larch was about 14 % less than that of the mixed plantation (Chen and Li 2004). Older (20–23 years) mixed plantations of larch (*L. gmelinii*)-Amur cork tree (*Phellodendron amurense*) (8 larch: 2 amur cork tree), larch (*L. kaempferi*)-alder (*Alnus tinctoria*) (7.5 larch: 2.5 alder) and larch (*L. kaempferi*)-birch (*Betula platyphylla*) (7.5 larch: 2.5 birch), had larch with significantly higher ($P<0.05$) mean DBH and H than those in pure larch plantations (Li et al. 2011).

Besides the better growth, the soil nutrient cycling in the mixed larch plantations was greatly improved. The comparison of leaf litter decomposition and nutrient return between pure stands of larch (*L. gmelinii*) or Manchurian walnut with mixed stands of

Fig. 3 Images of the mixed-species larch plantations. Larch + ash at Maoershan in Heilongjiang Province, NE China (**a**) and larch + broadleaved species at Qingyuan in Liaoning Province, NE China (**b**)

larch-Manchurian walnut ranked the decomposition rate of leaf litter as: larch < mixture of larch and Manchurian walnut < Manchurian walnut, while the returns of N, P and K via leaf litter were significantly higher in the mixed stand than in both pure stands (Chen and Li 2004). These results suggest that the introduction of broadleaved species into pure larch stands could increase the rates of litter decomposition and nutrient return, and so sustain forest productivity. The species used in mixture include Manchurian ash (Fig. 3), Manchurian walnut, Amur cork tree, Mongolian oak (*Q. mongolica*), maple (*Acer mono*), alder, birch, Korean pine (*Pinus koraiensis*), Yezo spruce (*Picea jezoensis*) and Korean spruce (*Picea koraiensis*). In the last two decades, many mixed larch plantations have been established. The ratio of larch to other species ranges between 3:2 and 4:1 while the initial stocking in the mixed-species plantations is similar to that in pure larch plantations (3,000–4,000 trees ha^{-1}).

By contrast, in the British Isles the use of mixtures in Sitka spruce stands has been limited, being largely confined to the planting of 'nursing' mixtures with pines (*Pinus contorta* or *P. sylvestris*) or larch (*L. kaempferi*) on soils of very poor nutrient status, where the spruce benefit from increased soil nutrient availability from mycorrhiza colonizing the roots of the nurse species (Taylor 1991). However, provided that the nursing effect is successful, these stands eventually self-thin to become pure stands of Sitka spruce. The greater leaf area of British spruce stands mean that, after canopy closure, the light levels in the understorey are insufficient for effective colonisation by other species until at least 40–50 years at the end of the 'stem exclusion' phase and the beginning of 'understorey reinitiation' (Hale 2001). This trend of Sitka spruce dominating other admixed species has been found in other experimental trials (Mason 2006) and has resulted in the recommendation that any long-term mixtures between Sitka spruce and other species require the latter to be planted in groups at least 0.05 ha in size. A consequence of this is that any attempt to increase species diversity in Sitka spruce forests, possibly as part of a strategy of adapting forests to projected climate change, would require other species to be planted in separate blocks of sufficient size to be self-sustaining.

2.3   The Use of Thinning to Increase Structural and Species Diversity in Conifer Plantations

Thinning is critical to the implementation of silvicultural approaches that foster greater structural diversity, such as CCF (Mason et al. 1999). This is because thinning will increase canopy openness and light availability, will modify soil temperature and moisture regimes, will affect the quantity and quality of litter, and by modifying the spatial distribution of trees can facilitate natural regeneration through gap formation (Malcolm et al. 2001; Puettmann et al. 2009; Mizunaga et al. 2010).

Trials of different thinning intensities were conducted in larch (*L. olgensis*) plantations in China to favour the regeneration of larch, but natural regeneration was poor even in the intensely thinned plantations (40 % thinned, 25.0 m^2 ha^{-1}) and the clearcut site (Zhu et al. 2008). The reason for this regeneration failure is that larch seedling emergence and survival were significantly reduced by the presence of litter and understory vegetation. For instance, all larch seedlings in both thinned stands and the clearcut disappeared within the first growing season when understory vegetation cover and litter cover were not removed; but they survived and grew in sites where ground cover was removed. Mason et al. (2011) also found that the presence of ground vegetation reduced the occurrence of European larch (*L. decidua*) regeneration in stands being transformed to CCF. Successful larch natural regeneration has occurred only after intense disturbances such as flood, windblow, and fire (Tsuyuzaki 1994). Thinning of larch forests can also be used to increase species diversity as shown by Chinese studies. For example, *L. olgensis* plantations between 30 and 40 years old were thinned to residual basal areas ranging from 25.9 to 38.8 m^2 ha^{-1} to examine the effects of thinning on natural regeneration (Zhu et al. 2010). Regardless of thinning intensity, there were sufficient naturally regenerated seedlings (height = 5–50 cm) and saplings (height = 50–500 cm) of tree species such as *Acer* spp., *Fraxinus* spp., *Cornus controversa*, *Quercus mongolica*, and *Pinus koraiensis* in the thinned stands. The mean density of regenerated seedlings ranged from 45,000 to 67,000 seedlings ha^{-1}, while the regenerated saplings reached 4,595 saplings ha^{-1} forming a regeneration layer with an average height of 2.0 m and basal diameter of 2.1 cm. If the Korean pine seedlings in the thinned stands could be conserved and promoted into the canopy layer, there would be a chance to transform the current larch-broadleaved stands into mixed broad-leaved-Korean pine forests which is one of the regional climax forests in Northeast China. Generally, the canopy openness of larch stands should be maintained between 25 % and 35 % after canopy closure for the overstorey development and microclimate amelioration (Liu et al. 1998; Zhu et al. 2010). Although selective thinning is normally used to promote regeneration, an alternative method is to open up strips within the larch stands through removing several rows of larch trees.

The role of thinning in Sitka spruce stands is more problematic because of considerable experience that opening up of the canopy may increase the incidence of windthrow, particularly if the operation is delayed or the stand is a zone of high wind risk (Quine et al. 1995). From the late 1970s, a non-thinning regime was adopted on British sites with a higher probability of wind damage which in some regions exceeded 50 % of the forest area (Quine et al. 1995). By contrast with the relatively light shade provided by larch plantations, the amount of light recorded in the understorey of Sitka spruce plantations managed under standard regimes is low and rarely favourable to colonization by woody species (Ni Dubhain 2010). Studies by Hale (2001, 2003) indicated that stocking densities in Sitka spruce stands needed to be reduced to 65 % or less of recommendations for plantation management to provide a favourable light environment for regeneration of a range of species (Mason et al. 2004). Investigation of regeneration in windblown gaps in an 80 years old Sitka spruce stand in northern England found that while spruce regeneration was present in all gaps, it only established in the larger ones (Quine and Malcolm 2007). In the same forest, Bertin et al. (2011) found poorer growth of natural regeneration in more densely stocked and higher basal area stands, where mean transmittance was about 17 % compared with better growth in more open stands when transmittance reached 23 %. Malcolm et al. (2001) analysed the light require- ments of different conifer species and proposed that gaps with a diameter:tree height ratio of at least 1.5–2.0 (equivalent to a gap area of 0.1–0.2 ha) were necessary to achieve satisfactory regeneration of Sitka spruce and other species of intermediate shade tolerance. Therefore, it appears that the key to the successful development of an irregular structure in Sitka spruce forests is a strategy of thinning, which opens the canopy to provide sufficient light to allow advance regeneration of under- planted trees to develop without compromising stand stability (Ni Dubhain 2010). The recent recognition that wind loading on individual trees is a function of tree size mediated by the density of the surrounding stand, i.e. local competition (Hale et al. 2012), offers the possibility of developing thinning strategies which favour the most stable trees within a stand. Similar approaches have been reported in stands of Norway spruce (*Picea abies*) in central Europe (Slodicak and Novak 2006), but they remain to be tested operationally under British and Irish conditions.

2.4 Adjustment of Rotation Length and Structural Diversity

There are three approaches used to increase species and structural diversity in planted forests. The first approach, and still the most widely used, is known as 'restructuring' (Hibberd 1985). In Britain, this was developed in the extensive for- ests on the Scottish-English border where the rolling topography and a high windthrow risk on the gley soils meant that traditional management often resulted in large unsightly felled areas with a loss of the forest microclimate. Restructuring involved zoning the forest into areas of greater or lesser visual sensitivity and adjust- ing coupe size accordingly with smaller coupes in more sensitive areas and in more sheltered zones. Felling schedules were then adjusted to ensure that adjacent coupes were not felled within 5 years or more of one another so that visual impact was further reduced. In addition, the range of species used in replanting was increased particularly in more sheltered zones with broadleaves introduced in riparian strips and other areas that proved difficult to harvest. This approach has been shown to reduce the visual and other impacts of clearfelling with limited effect on operational costs (Hibberd 1985; McIntosh 1995). In Northeast China, management practice in recently logged larch plantations is based on replanting using nursery plants of larches (*L. olgensis*, *L. kaempferi*, and *L. gmelinii*) or other local conifer tree species such as *Pinus koraiensis*, *Picea jezoensis*, *and Picea koraiensis* (Liu et al. 2005). While restructuring provides increased variation between adjacent coupes, the forest still tends to be composed of stands in the stand initiation and stem exclusion phases so that there is little change to the average rotation length over the whole forest. Structural features such as large veteran trees, gaps, irregular structures, and standing and falling deadwood, which are beneficial to biodiversity and attractive to visitors, tend to be rare in such forests (Peterken et al. 1992; Humphrey et al. 2003). Therefore, the certification standard for British forests (UKWAS 2012) requires a small proportion of a certified forest to be managed on longer rotations either as 'long-term retentions' or as 'natural reserves'.

The second approach is termed 'conversion' and describes the processes where management has decided to replace the dominant conifer species by a mixture of species. Conversion has been widely used where native broadleaved woodlands are being restored on sites that were planted with conifers during the last century (Thompson et al. 2004). In silvicultural terms the main differences from 'restructuring' are the wider range of species that are planted and the greater use of natural regeneration due to successional processes (Brockerhoff et al. 2008) than is usual in the management of plantation forests. In Northeast China, when light conditions are adequate to allow the establishment of broadleaved species, there is potential for *L. olgensis* plantations to be converted into mixed broadleaved-larch forests (Zhu et al. 2010). This is for three reasons: firstly there are enough seeds of broadleaved species coming from the secondary forests surrounding the larch plantations; secondly, the increased canopy openness in the thinned stands can provide the necessary light conditions for seedling survival and sapling growth; and thirdly, most of the regenerated broadleaved tree species have some degree of shade tolerance during early establishment. In general, conversion does not involve a substantial adjustment of rotation length, although individual trees may be retained beyond rotation age while the mixed stand desired for the future forest is formed.

The last approach is known as 'transformation' and differs from 'conversion' in that greater attention is paid to structural rather than species diversity. This approach is implemented through a range of silvicultural systems such as irregular shelterwood and group and single stem selection which are often collectively known as CCF or Close-to-Nature forestry (Mason 2003; Pommerening and Murphy 2004). A feature of CCF tends to be the prolonging of rotations beyond those normally practiced in plantation forests and the use of thinning to foster the development of advanced natural regeneration which is then promoted from the understorey to form part of the successor stand. This process results in forests which contain a higher proportion of stands in the later phases of stand development and thus provides a range of habitats for forest specialist species. In addition, CCF management does not cause the visual disruption associated with clearfelling and therefore this approach is increasingly favoured in areas of visual sensitivity or high recreational pressure (Mason et al. 2011). However, the implementation of CCF is dependant upon the feasibility of thinning stands and therefore is limited to sites at lower risk of windthrow. In addition, CCF does not necessarily result in the development of 'old-growth' type stand structures which are particularly valuable for biodiversity. Humphrey (2005) suggested that these conditions would not develop in British Sitka spruce stands until these were over 80 years of age.

3   Discussion

The two case studies described in this chapter provide an interesting contrast in the realities of manipulating planted forests to meet multifunctional purposes. When comparing Chinese and British experience of plantation forestry, it becomes clear that the canopy density of the favoured species, the resulting light regimes and local site conditions, have a major influence in determining which silvicultural practices are best suited to increasing species and structural diversity. Thus, in Northeast China, the comparatively light shade found under larch stands, plus the availability of seed sources of a range of native species, can facilitate the development of mixed species forests once the larch stands have received adequate thinning. Thus, the larch can be seen as a 'nurse' species enabling the restoration of a more natural community. Zerbe (2002) described a similar process whereby native broadleaved species colonise the understorey of Scots pine plantations in eastern Germany. By contrast, in the British Isles, the low light levels found under Sitka spruce, plus the risks of windthrow associated with thinning in more exposed areas, can make diversification of species or structure more difficult to achieve. An appropriate thinning strategy is therefore a critical silvicultural tool for promoting diversity in planted forests, particularly if a conversion or transformation approach is being adopted. Traditional thinning regimes that foster uniform spacing may no longer be the most appropriate, and regimes that develop more variable structures and spacing may need to be implemented (Puettmann et al. 2009).

A feature of experience with the larch forests of China has been the concern over changes in soil properties mainly due to the poor quality and slow decomposition of larch litter. Such effects can be exacerbated by other site management practices that impoverish the nutrient capital, as reported by Evans (2009) in plantations of Chinese fir in southern China. However, soil degradation can be overcome by introducing appropriate understory vegetation (Liu et al. 1998; Kim et al. 2010) while thinning can significantly increase the soil organic carbon concentrations by regulating soil chemical properties (Hwang et al. 2007). Furthermore, the available nutrients can be improved by thinning, thus helping to transform the pure larch plantations to uneven-aged forests with more complex stand structures through natural regeneration. These measures are also beneficial in alleviating the fragmentation of natural habitats and in maintaining the ecological functions of the forests. In Britain, such problems have occurred only on the most nutrient poor sites and only where experiments have sought to explore possible effects of intensive harvesting. The implications of these contrasting experiences are to highlight the importance of adjusting silvicultural practices to take account of site properties. The process of adjusting silvicultural practices to reflect site quality will become more important during the coming century in view of the need to adapt forests to projected climate change. Although neither larch nor Sitka spruce forests are considered to be immediately vulnerable to projected changes, there will be effects on productivity and some sites may become marginal for these species. For example, Sitka spruce may be at risk under the warmer and possibly drier climate recently suggested for eastern Scotland (Green et al. 2008).

One of the problems confronting foresters in both regions is a lack of reference forests to give an indication of desired stand structures for multifunctional management, their spatial pattern and species composition. Therefore an important requirement is the establishment of long-term silvicultural trials where the transformation of planted forests from simple to complex structures can be monitored over time. While several trials of this type exist in the British Isles and have been monitored over several decades (e.g. Cameron and Hands 2010; Kerr et al. 2010), there are no equivalent areas in China. Such trial areas can be of particular value both in training new generations of foresters and in educating other stakeholders how to adapt the management of planted forests to changing priorities, as shown by the AFI network of plots demonstrating irregular forest management in north western Europe (Susse et al. 2011). These trial areas can be used to demonstrate silvicultural options that will increase species and structural diversity and which will also enhance the resilience of forests against climate change, which will be essential to sustain the delivery of a range of ecosystem services from planted forests in the future.

Acknowledgements We are grateful for comments on an early draft of this paper from Drs. Bruce Nicoll and Trevor Fenning. Members of the 973 Programme (2012CB416900) provided helpful information on aspects of the management of larch plantations in China.

References

Avery M, Leslie R (1992) Birds and forestry. Poyser, London

Bauhus J, Schmerbeck J (2010) Silvicultural options to enhance and use forest plantation biodiversity. In: Bauhus J, van der Meer P, Kanninen M (eds) Ecosystem goods and services from plantation forests. Earthscan, London, pp 96–139

Bertin S, Palmroth S, Kim HS, Perks MP, Mencuccini M, Oren M (2011) Modelling understorey light for seedling regeneration in continuous cover forestry canopies. Forestry 84(4):397–409. doi:10.1093/forestry/cpr026

Brockerhoff EG, Jactel H, Parrotta JA, Quine CP, Sayer J (2008) Plantation forests and biodiversity: oxymoron or opportunity? Biodivers Conserv 17:925–951

Cameron AD, Hands MOR (2010) Developing a sustainable irregular structure: an evaluation of three inventories at 6-year intervals in an irregular mixed species stand in Scotland. Forestry 83(5):469–475. doi:10.1093/forestry/cpq029

Cannell MGR (2002) Impacts of climate change on forest growth. In: Broadmeadow M (ed) Climate change: impacts on UK forests, vol 125, Forestry Commission Bulletin. Forestry Commission, Edinburgh, pp 141–148

Carle J, Holmgren P (2008) Wood from planted forests: a global outlook 2005–2030. For Prod J 58:6–18

Chen YL, Li SL (2004) Study on the decomposition and nutrient return of the leaf litter under the pure and mixed plantations of *Juglans mandshrica* and *Larix gmelinni*. For Sci Technol 29(5):9–12 (in Chinese with English abstract)

DEFRA (2007) A strategy for England's trees, woods and forests. Department for Environment, Food and Rural Affairs, London, 43p

Del Lungo A, Ball J, Carle J (2006) Global planted forests thematic study: results and analysis. Working paper FP/38. FAO, Rome, 178p

Dong J (2000) Studies on fertilization effect of middle-aged plantations of *Larix kaempferi*. J Liaoning For Sci Technol 1:6–11 (in Chinese with English abstract)

Evans J (2009) Sustainable silviculture and management. In: Evans J (ed) Planted forests: uses, impacts and sustainability. FAO and CABI, Rome, pp 113–140

FAO (Food and Agriculture Organization of the United Nations) (2010) Global forest resources assessment 2010, country report China. Forestry Department, Rome, FRA2010/042

FCS (2006) The Scottish forestry strategy. Forestry Commission Scotland, Edinburgh, 86p

FCW (2009) Woodlands for Wales: the Welsh Assembly Government's strategy for woodlands and trees. Forestry Commission Wales, Aberystwyth, 62p

Foot D (2003) The twentieth century: forestry takes off. In: Smout TC (ed) People and woods in Scotland. Edinburgh University Press, Edinburgh, pp 158–194

Forest Service (2007) National forest inventory, Republic of Ireland. Forest Service, Wexford, 271p

Forestry Commission (2011) Forestry statistics 2011. http://www.forestry.gov.uk/pdf/ForestryStatistics2011.pdf/$FILE/ForestryStatistics2011.pdf. Accessed 22 Dec 2011

Gao YX (1983) Dynamic changes in soil water and nutrients in man-made larch forests. For Sci Technol 2:9–13 (in Chinese)

Green S, Hendry SJ, Redfern DB (2008) Drought damage to pole-stage Sitka spruce and other conifers in north-east Scotland. Scott For 62:10–18

Hale SE (2001) Light regime beneath Sitka spruce plantations in northern Britain: preliminary results. For Ecol Manag 151:61–66

Hale SE (2003) The effect of thinning intensity on the below-canopy light environment in a Sitka spruce plantation. For Ecol Manag 179:341–349

Hale SE, Gardiner BA, Wellpott A, Nicoll BC, Achim A (2012) Windloading of trees: influence of tree size and competition. Eur J For Res 131:203–217

Hibberd BG (1985) Restructuring of plantations in Kielder Forest District. Forestry 58:119–130

Horgan T, Keane M, McCarthy R, Lally M, Thompson D (2003) A guide to forest tree species selection and silviculture in Ireland. COFORD, Dublin, 256p

Hu LL, Zhu JJ (2008) Improving gap light index (GLI) to quickly calculate gap coordinates. Can J For Res 38:2337–2347

Hu YL, Zeng DH, Liu YX, Zhang YL, Chen ZH, Wang ZQ (2010) Responses of soil chemical and biological properties to nitrogen addition in a Dahurian larch plantation in Northeast China. Plant Soil 333:81–92

Humphrey JW (2005) Benefits to biodiversity from developing old-growth conditions in British upland spruce plantations: a review and recommendations. Forestry 78:33–53

Humphrey JR, Ferris R, Quine CP (eds) (2003) Biodiversity in Britain's planted forests. Forestry Commission, Edinburgh

Hwang JH, Son YH, Kim C, Yi MJ, Kim ZS, Lee WK, Hong SK (2007) Fine root dynamics in thinned and limed pitch pine and Japanese larch plantations. J Plant Nutr 30:1821–1839

Ji YZ, Feng WZ, Zhang XG (2004) Effects of fertilization on sustainability of soil fertility of larch plantations. Ecol Environ 13(3):376–378 (in Chinese with English abstract)

Jia SX, Wang ZQ, Li XP, Sun Y, Zhang XP, Liang AZ (2010) N fertilization affects on soil respiration, microbial biomass and root respiration in *Larix gmelinii* and *Fraxinus mandshurica* plantations in China. Plant Soil 333:325–336

Johnston DR, Grayson AJ, Bradley RT (1967) Forest planning. Faber and Faber Ltd., London

Kanninen M (2010) Plantation forests: global perspectives. In: Bauhus J, van der Meer P, Kanninen M (eds) Ecosystem goods and services from plantation forests. Earthscan, London, pp 1–15

Kerr G, Morgan G, Blyth J, Stokes V (2010) Transformation from even-aged plantations to an irregular forest: the world's longest running trial area at Glentress, Scotland. Forestry 83(3):329–344. doi:10.1093/forestry/cpq015

Kim C, Jeong J, Cho HS, Son Y (2010) Carbon and nitrogen status of litterfall, litter decomposition and soil in even-aged larch, red pine and rigitaeda pine plantations. J Plant Res 123:403–409

Li WH (2004) Degradation and restoration of forest ecosystems in China. For Ecol Manag 201:33–41

Li YZ, Luo GJ, Wang SP (2011) The preliminary study on the growth of the young forest mixed stand of Larch and *Phellodendron amurense*. J Agric Sci Yanbian Univ 33(1):64–67 (in Chinese with English abstract)

Liu SR, Li XM, Niu LM (1998) The degradation of soil fertility in pure larch plantations in the northeastern part of China. Ecol Eng 10:75–86

Liu ZG, Zhu JJ, Hu LL, Wang HX, Mao ZH, Li XF, Zhang LJ (2005) Effects of thinning on microsites and natural regeneration in a *Larix olgensis* plantation in mountainous regions of eastern Liaoning Province, China. J For Res 16:193–199

Luo X, Wang XQ, Zhang HG, Lan SB (2005) Growth performance and genetic gain of hybrid Larch. J Northeast For Univ 33(6):8–9 (in Chinese with English abstract)

Malcolm DC, Mason WL, Clarke GC (2001) The transformation of conifer forests in Britain – regeneration, gap size and silvicultural systems. For Ecol Manag 157:7–23

Mason WL (2003) Continuous Cover Forestry: developing close-to-nature forest management in conifer plantations in upland Britain. Scott For 57:141–149

Mason WL (2006) Managing mixed stands of conifers and broadleaves in upland forests in Britain, Forestry Commission Information Note 83. Forestry Commission, Edinburgh

Mason WL (2007a) Silviculture of Scottish forests at a time of change. J Sustain For 24:41–57

Mason WL (2007b) Changes in the management of British forests between 1945 and 2000 and possible future trends. Ibis 149(Suppl 2):41–52

Mason WL, Meredieu C (2011) Silvicultural strategies, sustainability, and adaptation to climate change in forests of the Atlantic region of Europe. J For Plan 16:1–11

Mason WL, Quine CP (1995) Silvicultural possibilities for increasing structural diversity in British spruce forests: the case of Kielder forest. For Ecol Manag 79:13–28

Mason WL, Kerr G, Simpson JMS (1999) What is continuous cover forestry? Forestry Commission Information Note 29. Forestry Commission, Edinburgh, 8pp

Mason WL, Edwards C, Hale SE (2004) Survival and early seedling growth of conifers with different shade tolerance in a Sitka spruce spacing trial and relationship to understorey light climate. Silva Fenn 38:357–370

Mason B, Edwards C, Hale S (2011) Continuous Cover Forestry in larch plantations: a case study in central Scotland. Scott For 65(3):14–22

Mason WL, McKay HM, Weatherall A, Connolly T, Harrison AJ (2012) The effects of whole-tree harvesting on three sites in upland Britain on the growth of Sitka spruce over ten years. Forestry 85(1):111–123. doi:10.1093/forestry/cpr064

Matthews JD (1989) Silvicultural systems. Oxford University Press, Oxford, 284p

McIntosh RM (1995) The history and multi-purpose management of Kielder forest. For Ecol Manag 79:1–11

MCPFE (2007) State of Europe's forests 2007. MCPFE Liaison unit, Warsaw, 247p

Miller HG (1995) The influence of stand development on nutrient demand, growth and allocation. Plant Soil 168–169:225–232

Mizunaga H, Nagaike T, Yoshida T, Valkonen S (2010) Feasibility of silviculture for complex stand structures: designing stand structures for sustainability and multiple objectives. J For Res 15:1–2

Ni Dubhain A (2010) An evaluation of continuous cover forestry in Ireland. COFORD, Dublin, 36pp

NIFS (2006) Northern Ireland Forestry: a strategy for sustainability and growth. Forest Service, Belfast, 20p

Oliver CD, Larson BC (1996) Forest stand dynamics. Wiley, New York, 520p

Paquette A, Messier C (2010) The role of plantations in managing the world's forests in the Anthropocene. Front Ecol Environ 8:27–34

Peterken GP, Ausherman D, Buchanan M, Forman RTT (1992) Old growth conservation within British upland conifer plantations. Forestry 65:127–144

Pommerening A, Murphy ST (2004) A review of the history, definitions and methods of continuous cover forestry with special attention to afforestation and restocking. Forestry 77:27–46

Puettmann KJ, Coates KD, Messier C (2009) A critique of silviculture: managing for complexity. Island Press, Washington, DC, 180pp

Quine CP, Malcolm DC (2007) Wind-driven gap development in Birkley Wood, a long term retention of planted Sitka spruce in upland Britain. Can J For Res 37:1787–1796

Quine CP, Coutts MP, Gardiner BA, Pyatt DG (1995) Forests and wind: management to minimise damage, Forestry Commission Bulletin 114. HMSO, London

Richards EG (2003) British forestry in the twentieth century, policy and achievements. Koninklijke Brill, Leiden

Slodicak M, Novak J (2006) Silvicultural measures to increase the mechanical stability of pure secondary Norway spruce stands before conversion. For Ecol Manag 224:252–257

Smout TC (2003) People and woods in British upland conifer plantations. Scotland: a history. Edinburgh University Press, Edinburgh, 244p

Son Y, Hwang JH (2003) Fine root biomass, production and turnover in a fertilized *Larix leptolepis* plantation in central Korea. Ecol Res 18:339–346

Susse R, Allegrini C, Bruciamacchie M, Burrus R (2011) Management of irregular forests. AFI, Besancon, 144p

Taylor CMA (1991) Forest fertilisation in Great Britain, Forestry Commission Bulletin 95. HMSO, London

Teagasc (2011) A brief overview of forestry in Ireland. http://www.teagasc.ie/forestry/technical_info/forestry_history.asp. Accessed 22 Dec 2011

Thompson R, Humphrey J, Harmer R, Ferris R (2004) Restoration of native woodlands on ancient woodland sites. Forestry Commission Practice Guide, Forestry Commission, Edinburgh

Tsuyuzaki S (1994) A thinned *Larix olgensis* forest on peatland in western part of Mount Changbai, China. Japan J Ecol 44:315–320 (in Japanese with English abstract)

UKWAS (2012) The UK Woodland Assurance Standard. Version 3.0. http://ukwas.org.uk/. Accessed 16 May 2012

Xu M, Qi Y, Gong P, Zhao G, Shao GF, Zhang P, Bai GX (2000) China's new forest policy. Science 289:2049–2050

Yang K, Zhu JJ, Yan QL, Sun OJ (2010a) Changes in soil P chemistry as affected by conversion of natural secondary forests to larch plantations. For Ecol Manag 260:422–428

Yang K, Zhu JJ, Zhang M, Yan QL, Sun OJ (2010b) Soil microbial biomass carbon and nitrogen in forest ecosystems of Northeast China: a comparison between natural secondary forest and larch plantation. J Plant Ecol 3:175–182

Yu LZ, Ding GQ, Zhu JJ, Shi JW, Yu SQ, Wang ZQ (2007) Effects of fertilization on biomass of fine roots in a *Larix kaempferi* plantation. Chin J Appl Ecol 18(4):713–720 (in Chinese with English abstract)

Zerbe S (2002) Restoration of natural broadleaved woodland in central Europe on sites with coniferous forest plantations. For Ecol Manag 167:27–42

Zhang YD, Shen YX, Wang ZQ, Bai SB (2001) Effects of the mixed on root growth and distribution of *Fraxinus mandshurica* and *Larix gmelinii*. Sci Silvae Sin 37(5):16–23 (in Chinese with English abstract)

Zhang YJ, Wu LY, Tan XR, Wang SK, Zhou Q, Ren FW (2009) Liaoning Province local standard: "Technical regulation for big-diameter timber cultivation of larch (*L. olgensis, L. kaempferi*) fast-growing forest plantations" (DB21/T 1724-2009). Quality and Technique Supervision Bureau, Liaoning Province (in Chinese)

Zhu JJ, Liu ZG, Wang HX, Yan QL, Fang HY, Hu LL, Yu LZ (2008) Effects of site preparation on emergence and early establishment of *Larix olgensis* in montane regions of northeastern China. New For 36:247–260

Zhu JJ, Yang K, Yan QL, Liu ZG, Wang HX (2010) The feasibility of implementing thinning in pure even-aged *Larix olgensis* plantations to establish uneven aged larch-broadleaved mixed forests. J For Res 15:70–81

Forests for All? Considering the Conservation Implications of Human-Species Interactions in the Context of Multifunctional Forestry

Mariella Marzano, Christopher P. Quine, and Norman Dandy

Abstract People and wildlife interact in many ways and recently there has been increased recognition that forests have multiple uses and management objectives providing opportunities for recreation, education, conservation and enhancement of biodiversity, carbon sequestration, and the production of timber and fuel. Whilst many benefits can perhaps be delivered simultaneously, there is potential for conflict between these objectives. In this chapter we focus on the possibly contradictory objectives of recreation and conservation. Forests are increasingly places where human-wildlife interactions occur through recreational activities such as walking, cycling, nature-watching, and general visits to relax, play and/or picnic. The pursuit of outdoor recreational activities can have substantial human benefits including improved mental and physical health; and there is evidence to suggest that positive recreational experiences are associated with natural [forest] settings where there is an opportunity to see or hear wildlife.

Public forest managers are charged with concurrently delivering broad ecological, social and economic benefits, which requires that they must, amongst other things, balance the impacts of public access for recreation with the requirement to conserve biological diversity. In the UK, the relatively recent focus on the 'multifunctionality' of the forest resource suggests that a strong evidence base is needed to inform decision-making regarding the balance and trade-offs between different functions. This chapter explores the key issues around human-species interactions in forests providing evidence on ecological impacts, but also highlighting major knowledge gaps on the social practices and attitudes tied to recreational activities and how these may be linked to greater awareness of wildlife, habitats and their needs. While

M. Marzano (✉) • C.P. Quine • N. Dandy
FR-NRS, Roslin & Alice Holt Lodge, Farnham, UK
e-mail: mariella.marzano@forestry.gsi.gov.uk; norman.dandy@forestry.gsi.gov.uk

T. Fenning (ed.), *Challenges and Opportunities for the World's Forests in the 21st Century*, Forestry Sciences 81, DOI 10.1007/978-94-007-7076-8_4,
© Crown Employees UK 2014

the majority of evidence is based at the site level, we argue that assessment of the interactions between recreation and conservation should also take place at landscape scale to facilitate the wider provision of public benefits.

1 Introduction

The Millennium Ecosystem Assessment highlighted the dependence of human well-being upon the services that ecosystems provide (MEA 2005). In the case of forests, more than 100 goods or services were identified as benefitting some societies. However, in many cultures the relationship between humans and forests has become distant, and the lack of 'woodland culture' has been noted (Independent Panel on Forestry 2012). Whilst trees are present in many urban and rural environments, the services provided may be less apparent or indirectly obtained; for example, street trees provide urban cooling benefits for many people but are commonly taken for granted, and timber and paper are sourced from suppliers with little reference to the distant forests from which they originate. Some suggest that, in parallel with societal changes, the practice of forestry has evolved through several phases (for example a focus on exploitative, custodial, sustained yield, production or ecological forestry (Seymour and Hunter 1999; Mather 2001)), but that the current trend is for a greater focus on the multiple values of forests – stimulated by global initiatives such as the Convention on Biological Diversity, interests in sustainability, and concern for the environment. As a result, in the past two decades there has been a shift in consideration of forests and their management to reflect (and seek to represent) the multiple values to societies. The concept of sustainable forest management reflects this.

A single forest can provide opportunities for recreation, education, conservation and enhancement of biodiversity, carbon sequestration and the production of timber and fuel. Consequently myriad individuals and social groups have 'stakes' in forests and their management. In this sense it might be said that forests are 'for all'. Management decisions influence the degree to which the forests are capable of delivering a number of these services and functions simultaneously, and appropriate management can often encourage this 'multifunctionality' (Otte et al. 2007). However, given that some of the forests' many functions can conflict, revised expectations and trade-offs are often needed. Owners vary in the extent to which their forests are managed in pursuit of multi-functionality, or whether those with a market price, such as timber, predominate and shape management decisions. Publicly owned forests (e.g. state forests) may be managed with a particular eye on multifunctionality and the delivery of public goods, but may nevertheless have to be financially self sustaining. Government agencies may seek to intervene through regulation or incentives to balance the objectives and encourage owners (private or public) to manage for those with public (and non market) as well as private (market) benefits. These interventions place strong demands upon evidence for the interactions between management, forest type and mix of services.

This chapter considers the balance and trade-offs between two major, potentially opposing, functions with high public benefit – the pursuit of outdoor recreation and the conservation of biodiversity. Using evidence primarily derived from a literature review on recreational disturbance in forests (Marzano and Dandy 2012b), we focus on the interaction between these functions, how this interaction is structured and influenced socially, and on the scale at which this interaction occurs and is conceptualised/studied. We argue that, whilst site level analyses of the interactions between recreation and conservation are common, understanding and assessing these at a landscape scale is also vital, and would aid the co-production of these public benefits. In addition, the need to understand human attitudes and behaviours broadly (i.e. beyond what individuals do at particular locations) is necessary to underpin policy and management decisions. Here we focus on forest use in the UK, but consider that the issues raised are relevant elsewhere.

2   Forests in the UK

Woodlands and forests currently make up 13 % of Britain's land cover representing nearly 3 million hectares (http://www.forestry.gov.uk/forestry/CMON-4UUM6R/). The extent of forests had reached an all time low of less than 5 % of the UK's land area in the early twentieth century (Forestry Commission 2011a) due to widespread felling and the demands of the World Wars (Rackham 2001). The UK Forestry Commission was established in 1919 to take forward re-afforestation that would meet the country's timber needs (http://www.forestry.gov.uk/forestry/CMON-4UUM6R/) through state planting and encouragement of private owners. The initial focus on restoration of a strategic reserve of timber has been adapted throughout the twentieth century to embrace a wider set of objectives. For example, the 1968 Countryside Act and the 1985 Wildlife and Countryside Act (Amdt) required public bodies to improve the amenity and conservation value of the countryside. For the Forestry Commission this meant a move towards multi-functional sustainable forestry balancing timber production with biodiversity conservation and improved public accessibility for recreation (Quine et al. 2004).

In recent decades forests across the UK and Europe have become popular venues for recreation (Martin 2007, 2008). This reflects and is a part of the broader transformation of rural landscapes from being primarily sites of agricultural and forestry production to also being 'leisure landscapes': places of amenity and recreational consumption (Aitchison et al. 2000; Mather 2001; Buijs et al. 2006). Numerous recreational activities take place in forests, but visitor surveys in the UK[1] show that four types of activity dominate – walking, cycling, nature watching and general visits (e.g. to relax or play). Amongst the many other activities are horse riding, camping, hunting, off-road vehicle driving, fishing, non-timber forest product collection and outdoor sports such as paintballing, rope trailing and orienteering (Edwards et al. 2008).

---

[1] www.forestry.gov.uk/website/forestry.nsf/byunique/infd-5pgazz

As with the provision of recreation, environmental protection and nature conservation is an important characteristic of sustainable forest management. Responsibility for biodiversity conservation in the UK is devolved to the constituent countries (England, Scotland, Wales and Northern Ireland), but the common goal is to tackle the loss of biodiversity across terrestrial and wetland areas (Defra 2007). The UK's approach to conservation is nested within international and European legislative frameworks and includes conservation agreements and directives such as the Wild Birds (1979) and Habitats Directives (1992) (Defra 2007; O'Connell and Yallop 2002). Around 10 % of land in the UK has been designated as Sites of Special Scientific Interest (SSSI) or Areas of Special Scientific Interest (ASSI): the primary protected area designation (DEFRA 2007). Within this, approximately 128,700 ha of woodland has SSSI designation, which is about 4 % of just more than 3 million hectares of woodland in the UK (Pryor and Peterken 2001; Forestry Commission 2011a). Beyond this type of designation, other woodlands in the UK (e.g. more than 550,000 ha of ancient woodland) are protected by additional forms of regulation and their conservation encouraged by a variety of incentives. For example, the England Woodland Grant Scheme EWGS has provided financial support to private owners for measures that enhance restoration of ancient woodland, and contribute to the conservation of particular species (such as red squirrel) or groups (e.g. woodland birds). Importantly, there is also a greater focus on landscape-scale action and management to better integrate forestry with other land uses and reduce fragmentation of ecosystems to improve the long-term viability of habitats and species (Defra 2007; Quine et al. 2010).

3   Why Do Recreationists Recreate in Forests?

Studies over the years have demonstrated that there are a variety of reasons for why people value forests for recreation. Forests are places where visitors can seek peace and quiet, relax, take pleasure in the natural and cultural heritage, enjoy the company of friends and family, walk their dog or engage in physical activity (Martin 2006; O'Brien and Morris 2013). Encounters with wildlife may or may not be a necessary element of the recreational experience (see also Carter et al. 2009, Newsome et al. 2002a). In the UK Public Opinion of Forestry survey (Forestry Commission 2011b), exercising (including walking, running and mountain biking) was the most popular activity to be undertaken in a forest followed by dog walking, relaxing and watching nature.

The benefits of spending time in forests and natural spaces have been widely documented (for examples, see Health Council of the Netherlands 2004; Pretty et al. 2005; O'Brien 2005; O'Brien and Morris 2009). Forests can contribute to improvements in mental, emotional and physical health through stress relief, stimulation of the senses and cognitive functioning, reduced blood pressure and help weight loss from exercising outdoors (O'Brien 2005). There is growing interest in the extent to which forests can contribute to health improvements with initiatives in the UK such as Branching Out (www.forestry.gov.uk/branchingout) and NHS Forest (nhsforest.org) receiving policy and institutional support.

4 Ecological Impacts of Recreation on Forest Habitats and Wildlife

Public interest in forests and their wildlife/biodiversity has been behind a number of the drivers to regulate for and protect special species and habitats. However, there are more direct (and negative) impacts and a substantial body of evidence has developed in recent decades regarding the impact of outdoor recreation on the conservation of wildlife. Impacts can be diverse, wide-ranging and in some cases strongly detrimental (e.g. Marzano and Dandy 2012a, b; Anderson and Radford 1992; Cole 2004). They can include direct impacts such as 'flight' and changing foraging and reproductive behaviour as well as indirect impacts such as habitat modification and potential introduction of pests, pathogens and weeds. In a recent review of the international literature on recreational disturbance Marzano and Dandy (2012b, p. 29) reported that:

> habitats can suffer from reduced plant and vegetation cover, plant damage and abrasion reducing growth and increasing premature leaf loss, reduced plant genetic and species diversity, modification of soil properties, soil removal and compaction, surface litter reduction, and damage to lichens and mosses. Wildlife can be crushed, hit and killed or disturbed through human or mechanical noise and/or close encounters. Recreational activities that interfere with feeding, breeding, travelling or resting behaviour can induce an alert or flight response affecting energy balances, social behaviour, increased vulnerability of the young or nest predation.

However, the review also highlights the somewhat patchy nature of both ecological and social evidence, particularly in UK forests; for example, whether the vertical vegetation structure in forests increases or reduces the disturbance of wildlife and whether the success or failure of species is related to the ecological conditions created by recreational use of forests. There have also been little or no systematic comparisons of forest types.

Much attention has focused on the impact of recreation on birdlife, vegetation and soils (e.g. Fernandez-Juricic et al. 2002; Banks and Bryant 2007; Kissling et al. 2009). A great deal of evidence available suggests that walkers can impact on habitats causing compacted soil, damaged or reduced vegetation cover, reduced plant species density, a decrease in leaf litter biomass and increased trail width and depth (e.g. Weaver and Dale 1978; Roovers et al. 2004; Wimpey and Marion 2010). Trampling of soils and vegetation and erosion of trails have also been documented in the literature on horse riding (Weaver and Dale 1978; Landsberg et al. 2001; Littlemore and Barlow 2005), mountain biking (Jacoby 1990; Geraghty 2000; White et al. 2006), camping (Leung and Marion 2004; Littlemore and Barlow 2005) and off road vehicle use (Buckley

2004). Off Road Vehicle use was found to negatively disturb vegetation (Buckley 2004) and wildlife in some instances (Naylor et al. 2009), although Bayfield (1986 in Newsome et al. 2002b) showed that in the Cairngorms, Scotland perceptions relating to the negative impact of Off Road Vehicles were greater than actual impacts measured.

Recreational activities can induce a 'flight' response in wildlife causing animals to flee from cover, impacting on their energy balances, feeding and reproductive behaviour as well as increasing the vulnerability of eggs or young. Walking (including with dogs) makes up a large proportion of the disturbance literature although much of this evidence is related to ground nesting birds or waterbirds in non-forest habitats (Langston et al. 2007; Showler et al. 2010; Rasmussen and Simpson 2010). The literature on camping, mountain biking and Off Road Vehicles provide evidence that these activities impact on wildlife behaviour. In addition to a 'flight' response, recreational use of forests can lead to food conditioning and habituation to people as well as avoidance or displacement into less favourable habitat (Liddle 1997; Blanc et al. 2006; Valentine and Birtles 2004; Marion et al. 2008).

There are potential harmful impacts resulting from the spread of non-native invasive weeds and pathogens through walker's boots, vehicle and bicycle tyres or horses' hooves, coats hair and dung (Jules et al. 2002; Cushman and Meentemeyer 2008; Landsberg et al. 2001). However, some suggest that in practice horses are not a significant vector as germination via dung, for example, is relatively low (Campbell and Gibson 2001), although soil disturbance can lead to creation of more suitable environments for invasive species (Newsome et al. 2002a). The available evidence linking biosecurity threats with recreational use is mixed. Some studies maintain that recreational use along hiking trails and forest paths by walkers and Off Road Vehicle users can have a detrimental impact if potentially infected mud and soil is not removed from tyres and hikers' boots (Turton 2005; Cushman and Meentemeyer 2008).

A number of studies have attempted to compare the disturbance impacts of different recreational uses. Much of this evidence is context specific, focussed on particular case studies, species or habitats and is therefore difficult to synthesise. Few studies have found any direct evidence suggesting that mountain biking can cause more disturbance to habitats than hiking or horse riding. However, it has been noted that mountain bikers are faster and can cover more ground thereby potentially disturbing more wildlife (Taylor and Knight 2003). Moreover, fast moving but quiet activities (i.e. no talking) are less predictable and can cause more stress to wildlife (Lathrop 2003; Sterl et al. 2008), while the popularity of mountain biking can significantly contribute to overuse of the countryside generally (Ruff and Mellors 1993). In their study of wildlife watching in Australia, Wolf and Croft (2010) found that wild Kangaroos were less disturbed by vehicles than tourists approaching on foot. However, Off Road Vehicle disturbance is very dependent on driving practices, the habitat in question and species type (Buckley 2004).

Many studies identified negative impact on forest wildlife and habitats from recreational use in the short-term (e.g. Banks and Bryant 2007; Smith-Castro and Rodewald 2010), but there are few longer term studies. Cole (2004) has noted that impacts are unavoidable following repeated recreational use and that recovery is generally slower than the pace at which impacts from disturbance can occur although this is dependent on physical (soil type, climate, structure and composition of vegetation, condition of trails) and social factors (type of use, frequency of use, season or time of day).

5 Management of Impacts

It is common for ecological studies of recreational disturbance to recommend some form of management. These include physical measures relating to site itself or management of people. Site management can involve visual vegetation screens, informational and/warning signs, trail maintenance and containment or dispersal of visitors to reduce impacts on wildlife and habitats at specific sites depending on the activity. Access restrictions can include buffer zones or minimum approach distances, which are calculated according to the area of influence; that is the area where human activity is most likely to disturb wildlife. Temporal restrictions can be seasonal or occur more regularly such as on a daily basis. Various codes of conduct such the UK Countryside Code as well as the universal 'leave no trace' policy (Littlemore and Barlow 2005) promote informed self-regulation. However, there are concerns that self-regulating systems will not work for vulnerable habitats and species and that restricting recreational use is more appropriate, especially where vulnerable species and habitats are concerned (Newsome et al. 2002b).

Attempts to influence the decisions and behaviour of visitors through education and interpretation are also a popular option – although the difficulties in being able to determine the effectiveness of such approaches has been noted (Higginbottom 2004). There is a concern that even low impact educational programmes are based on the assumption that individuals suffer from 'information deficit' such that negative behaviour is perceived to be the result of a lack of knowledge (Barr 2007). Marzano and Dandy (2012a) looked to the behaviour change literature, particularly cognitive theories and social practice theories in an attempt to understand key factors influencing recreationists' attitudes and behaviour. One central question is whether recreationists would change their behaviour in forests if they were aware that recreational activities could potentially negatively impact habitats and wildlife.

6 How Do Recreationists Perceive Their Impacts?

In many of the studies reviewed recreational disturbance is treated as an almost exclusively physical phenomenon such as the act of trampling or conversational noise. However, there is a significant gap concerning the social dimensions of human-wildlife interactions such as the social and cultural norms affecting recreationists' behaviour, how information is understood and acted upon, and attitudes towards impacts. Studies by Sterl et al. (2008), Taylor and Knight (2003) and Geraghty (2000) show that recreational users often have little awareness of the impacts their activities can have on habitats and wildlife. Sterl et al. (2008) surveyed 271 visitors consisting of dog walkers, on-trail walkers and off-trail walkers. Out of a list of 14 activities presented off-trail biking and dog walking were considered to have the highest impact on wildlife although 60 % of those surveyed did not believe that recreational activities created any disturbance. Similar findings were reported by Taylor and Knight (2003) where 50 % of the 640 trail users did not believe that recreation negatively impacted on wildlife. Factors influencing recreationists' belief that their activities do not disturb wildlife include engaging in unobtrusive behaviour, sticking to trails, staying quiet and obeying prescribed rules and regulations (Thompson et al. 1987 and Cooper et al. 1981 in Klein 1993). However, Taylor and Knight (2003) found that visitors were underestimating the impact of their activities, especially in terms of how close they felt they could approach wildlife without negative consequences.

There is some evidence that recreational users are aware that their activities can have some impact on habitats and wildlife, but generally they tend to hold others responsible for the majority of negative impacts. Geraghty's (2000) study in a UK country park with 73 mountain bikers, hikers and horse riders found that most participants believed other recreational groups were more responsible for damage caused to trails. Hikers, for example, felt that horse riders and mountain bikers equally had a negative impact on habitat with walking being the least damaging. Horse riders agreed with this view although it was felt that as mountain bikers are often concentrated in particular areas, the potential for significantly impacting on vegetation is greater. Mountain bikers believed that horse riding had the most negative impact. Similarly Heer et al. (2003) cite several studies where hikers believed that mountain bikers negatively impacted on habitats through the creation of new trails.

Overall, the range of evidence on the social dimensions of recreational use and its impact on wildlife and habitats is limited. We know very little about the cultural differences between recreationists related to their activities (King 2010), what social factors govern when and where impacts occur (e.g. decisions to recreate in particular forests at particular times) and how recreationists perceive their impacts and those of others. Research on the key social drivers influencing recreational users' values, attitudes and behaviour in natural settings is also sparse.

7   Pro-Environmental Behaviours: Would the 'Realisation' of Impacts Change Behaviour in Forests?

There is a body of work that has examined the link between pro-environmental attitudes and actual behaviour although none of these studies examine recreational disturbance. Findings are mixed (Bright and Porter 2001), with some authors finding non-existent or only weak associations between environmental values and pro-environmental behaviour (Lemelin and Wiersma 2007, see also Nord et al. 1998). In the sustainable consumption literature, Barr (2007) found that recycling behaviour was highly influenced by the fact that such behaviour is now a social norm while waste reduction and a re-use ethos were more influenced by environmental and citizenship values, concern about the impact of waste on society and personal welfare and knowledge of sustainability policy.

In early outdoor recreation literature, some studies explored the link between participation in outdoor recreation and environmental attitudes, values and behaviour. These attempted to classify different types of activity according to their expected impact on the environment such as 'consumptive' (activities that take something from the environment such as hunting and fishing), 'appreciative' (activities that involve appreciation of the environment without altering it such as walking, nature watching and camping), or 'abusive' (activities that can involve disturbance of habitat and wildlife such as mountain biking and Off Road Vehicle use) (Dunlap and Heffernan 1975; Geisler et al. 1977).

A more recent study argues against such broad classification, suggesting that each activity should be assessed individually (Teisl and O'Brien 2003), and there is now some evidence that participation in forest-based recreation is likely to increase pro-environmental behaviour – such as belonging to an environmental organisation, campaigning for environmental issues or participating in an environmentally friendly activity such as green consumerism (Nord et al. 1998; Teisl and O'Brien 2003; Hung Lee 2011). That is, there is a positive relationship between having the opportunity to experience, enjoy and learn about the natural world and pro-environmental behaviour (Larson et al. 2011). Moreover, enjoyment of forest-based recreational activities could potentially increase wider engagement with and support for conservation of forests. Hung Lee (2011) found that place attachment highly influenced a commitment to conservation and the likelihood of environmentally responsible behaviour amongst recreationists. However, such attitudes may not always prevail. In their survey of visitors to a nature reserve, Taylor and Knight (2003) asked participants to signal whether they would be supportive of certain management measures such as fewer trails, seasonal closure of trails, minimum approach distances, allowing only one type of recreation, requiring visitors to watch an educational video before embarking on their recreational activities and greater regulation. There was more support for the use of penalties against visitors who intentionally disturbed wildlife, but little backing for other measures.

Just as we have noted the distant relationship between humans and forests, Kareiva (2008) highlights concerns about a general decline in nature-based recreation in the US, but also other countries worldwide. As humans are becoming increasingly "disconnected from nature" there may be a negative impact on pro-environmental attitudes and behaviour. He argues that, "Just as we track trends in species loss and forest cover as key environmental indicators, we need to pay attention to trends in human behaviours and attitudes as the ultimate drivers of global change" (2008, p. 2757). The notion that experiencing and engaging with nature impacts upon outcome behaviour also lies behind popular well-being concepts such as 'nature-deficit disorder' (Louv 2005) and 'restorative environments' (Kaplan 1995). Nevertheless, while evidence suggests that exposure to nature does seem to have some impact on environmental values and behaviour, it is unclear whether recreationists' would change their behaviour even if they were aware of the impacts of their activities.

8   Conclusions

The shift towards 'multifunctionality' or sustainable forest management in recent decades reflects the many social, economic and environmental benefits that forests and other ecosystems provide. The conservation of biological diversity and outdoor recreation are the two important benefits explored in this chapter. There is evidence to suggest that people derive considerable benefits from observing or knowing that wildlife is present in forests. However, there are complex trade-offs involved with the potential that recreation can negatively impact upon the natural environment and its components – soil, vegetation, wildlife and water but recreation can also lead to pro-environmental behaviour that positively impacts on forest habitats and wildlife. With substantial legal conservation requirements and strong policy support for increasing outdoor recreation to improve the mental and physical health and well-being of society, managers are expected to makes informed choices regarding the mix of these functions (and associated benefits) in addition to others in their forests, not all of which have a market value or are quantifiable. The co-production of conservation and recreation benefits necessarily requires an understanding of the inter-relationship between humans and wildlife/habitats, but as we have highlighted, the availability of this information is currently sparse. While human-wildlife interactions take place across a range of habitats and spatial and temporal scales, much of the published evidence is short-term and small-scale, focussing on localised species/habitats and the physical impacts of recreational groups. Moreover, the interaction between humans and wildlife is a dynamic process that is rarely captured with wildlife additionally responding to the environmental changes (e.g. forest ageing, climate change) and human behaviour evolving in the light of socio-economic trends. We also currently have little evidence to suggest that recreationists' are aware of the impact that their activities may have on wildlife and habitats or that they are able or willing to change their behaviour.

The co-production of multiple benefits remains a challenge for forest managers with common management responses involving access restrictions such as buffer zones, time and site restrictions and visual screens. However, a debate around the potential for spatial segregation of different functions is under-developed. A common view perhaps is that co-production or multi-functionality is an aspiration for all forests while others highlight the benefits of separating these functions (Seymour and Hunter 1999). Greater scrutiny of the case for seeking multiple

Fig. 1 Dog walkers enjoying a stroll in the forest (Photos are reproduced courtesy of the Forestry Commission. © Crown Copyright 2012)

functions from every single hectare of forests would involve an exploration of whether spatial planning could produce more effective solutions at larger spatial scales. This may present opportunities for novel management approaches, such as sustainable intensification (Royal Society 2009), which would see the intensive or 'precise' (Lowe 2008) delivery of multiple forest functions, like recreation and conservation, at discrete points within a landscape. Some have suggested these decisions can be applied at the landscape scale; for example, Seymour and Hunter (1999) proposed a landscape triad – forests reserves, intense production areas, and broad-brush 'ecological forestry' – but how this can be achieved across multiple ownerships, multiple existing woodland types, and in the face of environmental change has yet to be established. It is also important to include the social dimension when considering the challenges of managing for multiple benefits for multiple stakeholders (e.g. Defra 2011). Our review has shown that there is a dearth of understanding/evidence on these dimensions, but management decisions involve more than just technical or economic considerations. 'Forests for all' is a worthy aspiration but there is some way to go before this has a sound evidence base (Figs. 1 and 2).

Fig. 2 Forests are important
places for mountain bikers
(Photos are reproduced
courtesy of the Forestry
Commission. © Crown
Copyright 2012)

References

Aitchison C, MacLeod NE, Shaw SJ (2000) Leisure and tourism landscapes: social and cultural geographies. Routledge, London

Anderson P, Radford E (1992) A review of the effects of recreation on woodland soils vegetation and flora. English Nature Research Reports, 27. Peterborough, English Nature

Banks PB, Bryant JV (2007) Four-legged friend or foe? Dog walking displaces native birds from natural areas. Biol Lett 3:611–613

Barr S (2007) Factors influencing environmental attitudes and behaviours. A U.K. caset study of household waste management. Environ Behav 39(4):435–473

Blanc R, Guillemain M, Mouronval JB, Desmonts D, Fritz H (2006) Effects of non-consumptive leisure disturbance to wildlife. Revue D Ecologie-La Terre Et La Vie 61(2):117–133

Bright AD, Porter R (2001) Wildlife-related recreation, meaning, and environmental concern. Hum Dimens Wildl 6(4):259–276

Buckley R (2004) Environmental impacts of ecotourism. CABI Publishing, New York

Buijs AE, Pedroli B, Luginbuhl Y (2006) From hiking through farmland to farming in a leisure landscape: changing social perceptions of the European landscape. Landsc Ecol 21:375–389

Campbell JE, Gibson DJ (2001) The effects of seeds of exotic species transported via horse dung on vegetation along trail corridors. Plant Ecol 157:23–35

Carter C, Lawrence A, Lovell R, O'Brien L (2009) The Forestry Commission public forest estate in England: social use, value and expectations. Final report. Forest Research

Cole DN (2004) Impacts of hiking and camping on soils and vegetation: a review. In: Buckley R (ed) Environmental impacts of ecotourism. CABI Publishing, New York, pp 41–60

Cushman JH, Meentemeyer RK (2008) Multi-scale patterns of human activity and the incidence of an exotic forest pathogen. J Ecol 96(4):766–776

Defra (2007) Conserving biodiversity – the UK approach. www.defra.gov.uk. Accessed 30 July 2012

Defra (2011) Biodiversity 2020: a strategy for England's wildlife and ecosystem services. Defra, London

Dunlap RE, Heffernan RB (1975) Outdoor recreation and environmental concern: an empirical examination. Rural Sociol 40(1):18–30

Edwards D, Morris J, O'Brien L, Sarajevs V, Valatin G (2008) The economic and social contribution of forestry for people in Scotland. Research Note FCRN102, Forestry Commission Scotland, Edinburgh

Fernandez-Juricic E, Jimenez MD, Lucas E (2002) Factors affecting intra- and inter-specific variations in the difference between alert distances and flight distances for birds in forested habitats. Can J Zool-Revue Canadienne De Zoologie 80(7):1212–1220

Forestry Commission (2011a) The UK forestry standard. Forestry Commission, Edinburgh, p 108

Forestry Commission (2011b) Forestry statistics 2011. Forestry Commission, Edinburgh

Forestry Commission (2011c) Public opinion of forestry 2011, UK and England. Results from the UK survey of public opinion of forestry. Forestry Commission, Edinburgh

Geisler CC, Martinson OB, Wilkening EA (1977) Outdoor recreation and environmental concern. A restudy. Rural Sociol 42(2):241–249

Geraghty T (2000) An examination of the physical and social aspects of mountain biking at Bestwood Park. Combined Studies in Science. Nottingham, Nottingham Trent. B.Sc. (Honours), p. 72

Health Council of the Netherlands and Dutch Advisory Council for Research in Spatial Planning, Nature and the Environment (2004) Nature and health. The influence of nature on social, psychological and physical well-being. The Hague: health Council of the Netherlands and RMNO publication no 2004/09E

Heer C, Rusterholz HP, Baur B (2003) Forest perception and knowledge of hikers and mountain bikers in two different areas in north-western Switzerland. Environ Manage 31(6):709–723

Higginbottom K (2004) Wildlife tourism: impacts, management and planning. Common Ground Pty Ltd, Altona

Hung Lee T (2011) How recreation involvement, place attachment and conservation commitment affect environmentally responsible behaviour. J Sustain Tour 19(7):895–915

Independent Panel on Forestry (2012) Final report. http://www.defra.gov.uk./forestrypanel

Jacoby J (1990) Mountain bikes: a new dilemma for wildland recreation managers? West Wildlands 16:360–368

Jules ES, Kauffman MJ, Ritts WD, Carroll AL (2002) Spread of an invasive pathogen over a variable landscape: a nonnative root rot on Port Orford cedar. Ecology 83(11):3167–3181

Kaplan S (1995) The restorative benefits of nature: towards an integrative framework. J Environ Psychol 15(3):169–182

Kareiva P (2008) Ominous trends in nature recreation. Natl Acad Sci USA 105(8):2757–2758

King K (2010) Lifestyle, identity and young people's experiences of mountain biking. Forestry Commission Research Note 007

Kissling M, Hegetschweiler K, Rusterhoz HP, Baur B (2009) Short-term and long-term effects of human trampling on above-ground vegetation, soil density, soil organic matter and soil microbial processes in suburban beech forests. Appl Soil Ecol 42(3):303–314

Klein ML (1993) Waterbird behavioural responses to human disturbances. Wildl Soc Bull 21(1):31–39

Landsberg J, Logan B, Shorthouse D (2001) Horse riding in urban conservation areas: reviewing scientific evidence to guide management. Ecol Manag Restor 2(1):36–46

Langston RHW, Liley D, Murison G, Woodfield E, Clarke RT (2007) What effects do walkers and dogs have on the distribution and productivity of breeding European Nightjar *Caprimulgus europaeus*? IBIS 149(s1):27–36

Larson LR, Whiting JW, Green GT (2011) Exploring the influence of outdoor recreation participation on pro environmental behaviour in a demographically diverse population. Local Environ 16(1):67–86

Lathrop J (2003) Ecological impacts of mountain biking: a critical literature review. University of Montana, Montana

Lemelin RH, Wiersma EC (2007) Perceptions of polar bear tourists: a qualitative analysis. Hum Dimens Wildl 12:45–52

Leung Y-F, Marion JL (2004) Managing impacts of camping. In: Buckley R (ed) Environmental impacts of ecotourism. CABI Publishing, New York, pp 245–258

Liddle M (1997) Recreation ecology. Chapman & Hall, London

Littlemore J, Barlow C (2005) Managing public access for wildlife in woodlands – ecological principles and guidelines for best practice. Q J For 99(4):271–285

Louv R (2005) Last child in the woods: saving our children from nature-deficit disorder. Algonquin, Chapel Hill

Lowe P (2008) Whose land is it anyway? ECSS lecture series, Birkbeck, 17th Oct

Marion JL, Dvorak RG, Manning RE (2008) Wildlife feeding in parks: methods for monitoring the effectiveness of educational interventions and wildlife food attraction behaviors. Hum Dimens Wildl 13(6):429–442

Martin S (2006) Leisure landscapes: exploring the role of forestry in tourism. Forest Research, Edinburgh

Martin S (2007) Leisure landscapes: exploring the role of forestry in tourism. Forestry Commission, Edinburgh

Martin S (2008) Developing woodlands for tourism: concepts, connections and challenges. J Sustain Tour 16:386–407

Marzano M, Dandy N (2012a) Recreationist behaviour in forests and the disturbance of wildlife. Biodivers Conserv 21(11):2967–2986

Marzano M, Dandy N (2012b) Recreational use of forests and disturbance of wildlife – a literature review. Forestry Commission Research Report. Forestry Commission, Edinburgh

Mather AS (2001) Forests of consumption: postproductivism, postmaterialism and the postindustrial forest. Environ Plan C, Gov Pol 19:567–585

Millennium Ecosystem Assessment (2005) Ecosystems and human well-being: synthesis. Island Press, Washington, DC

Naylor LM, Wisdom MJ, Anthony RG (2009) Behavioral responses of North American elk to recreational activity. J Wildl Manag 73(3):328–338

Newsome D, Milewski A et al (2002a) Effects of horse riding on national parks and other natural ecosystems in Australia: implications for management. J Ecotourism 1(1):52–74

Newsome D, Moore SA, Dowling RK (2002b) Natural area tourism: ecology, impacts and management. Channel View Publications, Sydney

Nord M, Luloff AE, Bridger J (1998) The association of forest recreation with environmentalism. Environ Behav 30(2):235–246

O'Brien L (2005) Trees and woodlands: nature's health service. Forest Research, Farnham

O'Brien L, Morris J (2009) Active England. The woodland projects. Forest Research, Farnham

O'Brien L, Morris J (2013) Well-being for all? The social distribution of benefits gained from woodlands and forests in Britain. Local Environ. doi:10.1080/13549839.2013.790354

O'Connell M, Yallop M (2002) Research needs in relation to the conservation of biodiversity in the UK. Biol Conserv 103:115–123

Otte A, Simmering D, Wolters V (2007) Biodiversity at the landscape level: recent concepts and perspectives for multifunctional land use. Landsc Ecol 22:639–642

Pretty J, Griffin M, Peacock J, Hine R, Sellens M, South N (2005) A countryside for health and well-being: the physical and mental health benefits of green exercise. A report for the Countryside Recreation Network, UK

Pryor SN, Peterken GF (2001) Protected forest areas in the UK: report for the WWF and forestry commission. Oxford Forestry Institute, Oxford

Quine CP, Humphrey JW (2010) Plantations of exotic tree species in Britain: irrelevant for biodiversity or novel habitat for native species? Biodivers Conserv 19:1503–1512

Quine CP, Humphrey JW, Watts K (2004) Biodiversity in the UK's forests – recent policy developments and future research challenges. In: Paivanen R, Franc A (eds) Towards the sustainable use of Europe's forests – forest ecosystem and landscape research: scientific challenges and opportunities. European Forest Institute, Joensuu, pp 237–248

Rackham O (2001) Trees and woodlands in the British landscape: the complete history of Britain's trees, woods and hedgerows. (Revised Edition). Phoenix Press, London

Rasmussen H, Simpson S (2010) Disturbance of waterfowl by boaters on pool 4 of the upper Mississippi river national wildlife and fish refuge. Soc Nat Resour 23:322–331

Roovers P, Verheyen K, Hermy M, Gulinck H (2004) Experimental trampling and vegetation recovery in some forest and healthland communities. Appl Veg Sci 7(1):111–118

Royal society (2009) Reaping the benefits: science and the sustainable intensification of global agriculture. Royal Society, London

Ruff A, Mellors O (1993) The mountain bike – the dream machine? Landsc Res 18:104–109

Seymour RS, Hunter ML Jr (1999) Principles of ecological forestry. In: Hunter ML Jr (ed) Maintaining biodiversity in forested ecosystems. Cambridge University Press, Cambridge, pp 22–61

Showler DA, Stewart GB, Sutherland WJ, Pullin AS (2010) What is the impact of public access on the breeding success of ground-nesting and cliff-nesting birds? Systematic review CEE 05–010, Collaboration for Environmental Evidence

Smith-Castro J, Rodewald A (2010) Behavioral responses of nesting birds to human disturbance along recreational trails. J Field Ornithol 82(2):130–138

Sterl P, Brandenburg C, Arnberger A (2008) Visitors' awareness and assessment of recreational disturbance of wildlife in the Donau-Auen National Park. J Nat Conserv 16(3):135–145

Taylor AR, Knight RL (2003) Wildlife response to recreational and associated visitor perceptions. Ecol Appl 13:951–963

Teisl MF, O'Brien K (2003) Who cares and who acts? Different types of outdoor recreationists exhibit different levels of environmental concern and behaviour. Environ Behav 35:506–522

Turton SM (2005) Managing environmental impacts of recreation and tourism in rainforests of the wet tropics of Queensland World Heritage Area. Geogr Res 43(2):140–151

Valentine P, Birtles A (2004) Wildlife watching. In: Higginbottom K (ed) Wildlife tourism: impacts, management and planning. Common Ground Publishing, Altona, pp 15–34

Weaver T, Dale D (1978) Trampling effects of hikers, motorcycles and horses in meadows and forests. J Appl Ecol 15:451–457

White DD, Waskey MT, Brodehl GP, Foti PE (2006) A comparative study of impacts to mountain bike trails in five common ecological regions of the Southwestern U.S. J Park Recreat Adm 24(2):21–41

Wimpey JF, Marion JL (2010) The influence of use, environmental and managerial factors on the width of recreational trails. J Environ Manage 91(10):2028–2037

Wolf ID, Croft DB (2010) Minimizing disturbance to wildlife by tourists approaching on foot or in a car: a study of kangaroos in the Australian rangelands. Appl Anim Behav Sci 126(1–2):75–84

Human Engagement with Forest Environments: Implications for Physical and Mental Health and Wellbeing

Simon Bell and Catharine Ward Thompson

Abstract Humans have always had close relationships with forest but since 2005 most people live in cities or urbanised areas. This has profoundly shifted the kind of relationship people have with forests. Instead of being sources of utilitarian products, forests are increasingly valued for their benefits for physical and mental health and wellbeing. Different forest types, such as wilderness areas, managed production forests or urban forests offer different possibilities. Effects on the physical environment of urban areas, such as pollution reduction or temperature mitigation help to provide healthier places to live while the ability to see or to visit green areas close to home helps to reduce stress, to provide settings for relaxation, socialising and physical exercise and provides aesthetic pleasure. Much recent research has strengthened the evidence base for these benefits and public health policy in many Western countries is beginning to take the results seriously. However the research is focussed at present in Western countries, while the mega-cities with poorest environments are in developing countries. It is here that the gaps in research and the challenges for the future are to be found.

1 Introduction

1.1 The Social Context: An Urbanised and Aging Population

Sometime around 2008 the proportion of the world's population living in urban areas reached and passed 50 % for the first time in human history and this proportion continues to grow (UN Population Fund 2007). While most developed countries became urbanised some time ago, starting in the late nineteenth century,

S. Bell (✉) • C. Ward Thompson
Edinburgh College of Art, Edinburgh University, Edinburgh EH3 9DF, UK
e-mail: s.bell@ed.ac.uk; c.ward-thompson@eca.ac.uk

T. Fenning (ed.), *Challenges and Opportunities for the World's Forests in the 21st Century*, Forestry Sciences 81, DOI 10.1007/978-94-007-7076-8_5, © Springer Science+Business Media Dordrecht 2014

through processes such as industrialisation, and accelerating after the Second World War, for developing countries this trend is relatively recent. The major cities in developing countries also tend to be much larger in scale (mega-cities such as Sao Paolo, Dacca, Mumbai), less-well planned (with many people living in slums) and lacking in many amenities such as clean water, satisfactory drainage and green areas. Even when people do not live in strictly urban areas, many nevertheless have urban values and lifestyles. Thus, the twenty-first century marks the beginning of an era when we can speak of "urbanised societies" across the globe.

Life in a city has many advantages – more employment opportunities, better access to educational, cultural and leisure resources, better transport and (at least potentially) a life with more leisure time to enjoy these amenities. It can also have many negative aspects ranging from overcrowding, the easy spread of communicable diseases, pollution from gases and particulate matter, large amounts of refuse to dispose of, higher levels of crime and a greater feeling of a lack of security as well as a lower sense of community. The city also provides a constant stream of often unwelcome stimulation – visual, auditory and olfactory – which can be stressful.

The move from a rural area, where life probably consisted of often strenuous manual work and a diet of simple, unprocessed foods, to an urban area, often also leads to a much more sedentary urban lifestyle and a diet of plentiful and cheap food but which may also be high in fat, sugar and salt and low in fibre. The resulting lack of exercise relative to calorie intake has caused a so-called "epidemic of obesity" together with other associated lifestyle diseases or health conditions, such as Type 2 diabetes (Stein and Colditz 2004). These are factors which place major burdens on health services, more so in the most developed urban societies in Europe and North America, for example, than elsewhere, but this is now increasingly common throughout the world.

The lack of green spaces in early industrial cities in the UK, Europe and the USA was recognised as far back as the 1830s, when public health concerns and a recognition that fresh air and exercise was good for everyone, led to the development of public parks (Ward Thompson 1998, 2011). This concern has been with us ever since but urban expansion, uncontrolled sprawl and recent policy shifts towards denser and more compact cities means that green areas are not easily accessible to everyone living in an urban area. In deprived urban areas, even in affluent countries, factors such as poor housing, low-paid work, low levels of educational achievement, high rates of unemployment and poor availability of green areas, frequently go together, leading to a poor quality of life (Department for Communities and Local Government 2007).

We can make a distinction between purely urban societies – that is the people who live in urban areas and for whom access to green space within the urban fabric might vary according to social and economic as well as spatial factors – and urbanised people in general who may live in more rural or small-town areas, yet at the same time lead an urban lifestyle, whether by choice or necessity (mill towns and mining communities, often now in a post-industrial decline in the western world, are typical of the latter) (Ward Thompson et al. 2004). For both groups, whether they live in so-called 'developed' countries or, as is increasingly the case, in urbanising, developing countries such as China, Brazil or India, a common feature of the urban lifestyle is a loss or diminution of connection with "nature" – that sense of familiarity and frequent engagement (physically and emotionally) with non-urban rural or forest landscapes. Green areas in the urban environment increasingly mean managed parks, gardens and tree-lined streets and perhaps less-managed transport corridors, derelict land or other "unofficial" green spaces. Wildlife may live in these places but they may also be heavily used by people for all sorts of recreation and be dominated by urban sights and sounds, making it difficult to achieve any sense of escape from the urban environment.

The phenomenon of urbanisation also involves large-scale movement of people from the countryside to the cities within countries and from rural areas in one country or continent to urban areas in another (Bell et al. 2010). This is an age of mass-migration, with frequent dislocation of people from their home culture and community into a new and stressful environment. People newly arrived in cities in a strange country may feel they have progressed socially and economically by moving away from rural poverty and they may associate rural areas and forests with the environment they are grateful to have left behind. Others may find that the childhood memories of the countryside or forest lead them to seek out similar places in their new neighbourhood, in order to feel a connection with the home and countries they have left (Silveirinha de Oliveira 2012). Yet second and third generations of people from rural backgrounds, whether immigrant into a new country or not, may be fully assimilated into their new city and have become thoroughly urbanised, feeling no such association with nature.

One of the great social and medical advances of the last century has been the progressively rising life expectancy of people in developed countries, which has been accompanied by reduced birth rates. This has led to the phenomenon of the "ageing society" where increasing proportions of the population are over 65 years of age (UK Department for Work and Pensions 2012). This longevity, as people progress into their 80s or 90s, is accompanied by increasing physical and mental health incapacities which need to be addressed by health services and social support systems. Remaining physically and mentally active into old age is one way to defer such problems and there is increasing interest in environments which offer opportunities for a wide range of activities in which older people can participate.

1.2   The Forest and the City

The forest, in one sense, represents the antithesis of the city. Writing in the eighteenth century, in his *Scienza Nuova* (*New Science*), Giambattista Vico claimed "This is the order of human institutions: first, the forests, after that the huts, then the villages, next the cities, and finally the academies" (Vico 1725, p. 239). This provides a metaphor for the human relationship with forests which were, and continue to be, cleared to make way for agriculture which has enabled civilisations to arise and so to develop into the urban-dominated environment of today. However, it also suggests that we have a historic relationship with forests which, apart from supplying timber, fuel and forage, also provide us with other values – spiritual, aesthetic or recreational, for example. Patrick Geddes, the Scot widely credited as the 'father of town planning', developed the valley section in 1909 as a similar, abstracted expression of what a city in its region represented, with hills or mountains and forested landscape in the uplands, giving way to pastoral and cultivated landscapes before the development of the city, close to the sea. This diagrammatic section represented changes in time as well as in space, but linked the city to the 'natural' or basic occupations of forester, hunter or shepherd, for example, as an illustration of what the ideal city still might be (Ward Thompson 2006). This link with the natural world has been recognised as important for urban dwellers for as long as cities have existed, it would seem, reflected in Martial's concept of *rus in urbe* in ancient Roman times and in mediaeval discussions of the virtues of access to green and wooded landscapes for good health (Ward Thompson 2011).

The development of the urban parks movement in the nineteenth century was in large part a response to the cramped and polluted living conditions of factory workers crowding from the countryside into rapidly industrialising cities. Access to parks was not only seen as contributing to physical health and prevention of disease, but also to the psychological and spiritual renewal of the urban working classes (Ward Thompson 1998). In the mid-nineteenth century, the term "lungs of the city" was cited repeatedly in the service of arguments to develop public parks, whether in Berlin, Paris or New York City. The original use of the term "forest" (in English at least) was as a hunting ground and not as the generally extensively wooded areas as we understand the term today. It is no surprise, then, that for many urban dwellers the forests or wooded areas which were most easily accessible to them for recreation were former royal hunting forests such as the Tiergarten in Berlin, the Vienna Woods or the Bois de Boulogne outside Paris (Bell et al. 2005; Ward Thompson 2011) (Fig. 1).

Today we recognise that urban dwellers have certain spiritual, recreational and health and wellbeing needs that can at least in part be fulfilled by natural areas in general, including forests. An increasing amount of research has been undertaken to try to understand how these benefits arise, how important they are and how best to ensure that as many people as possible can obtain them (Ward Thompson 2011). What stands out is that most research has been carried out in developed countries and that there are great gaps in evidence and understanding, especially in the context of the swiftly urbanising mega-city regions, where little is known of local or regional preferences, needs, demands or supply of social benefits from forests.

2   Different Forest Types for Human Engagement and Social Use

In recent decades a wide range of types of forest have been recognised, which present a varying set of opportunities and constraints to maximising the social and health benefits they are capable of supplying. Equally, there are different cultural

Fig. 1 People of Turkish background having a picnic in the Tiergarten in Berlin, Germany, a good example of a former hunting forest now accessible to everyone. People of ethnic minorities frequently use forests differently from the original population (Source: Simon Bell)

regions where the bonds with forests vary considerably. Adding to the complexity of these inter-relationships, increasing migration means that people may live in countries or regions where their links with the kind of nature they grew up with cannot be maintained. Nonetheless, a useful starting point is to recognise the nature of the forest resource that is potentially available for social use.

Firstly, there are protected "natural" forests, variously categorised as reserves, wilderness areas or parks. These are landscapes where it is possible for those who have the time and resources, as well as interest, to make the effort to travel there to immerse themselves in as wild or natural an environment as it is possible to find. People using such forests for recreation can also get away from crowds and (whether by choice or necessity) communication technology, find a solitary experience if they wish and practice self-reliance. These kinds of places are epitomised by the designated "Wilderness Areas" in the USA, which are generally managed by the US Forest Service (United States Forest Service 2012). Since it usually requires a considerable effort to get there, relatively small numbers of people generally benefit but those that do gain considerably from the experience.

Secondly, there are vast areas of managed natural forests, which may be old primary growth or mainly second growth, and which may be managed under industrial conditions primarily for timber or under multiple-objective regimes. These are usually some distance away from urban areas, although some may be within a relatively short drive. Their natural or aesthetic qualities may have been compromised by decades of management, including logging, but if they are managed as certified sustainable forests then they are often accessible (perhaps through their legal status as public land or through a general recognition of the right to public access even to private land for recreation and exercise – "every-man's rights"). They may be equipped with extensive networks of trails for hiking, skiing or cycling and with parking areas, picnic sites and so on. These kinds of forests are epitomised by those of Finland or Sweden, where outdoor recreation is highly popular and where forests may be within reach of urban centres by public transport, for example the Central Park in Helsinki, Finland, which is a wedge of forest penetrating into the very centre of the city. These countries are also characterised by large amounts of forests relative to small populations. The culture is also strongly associated with forests in all sorts of ways, e.g. through traditions of berry and mushroom picking at different times of year (Vistad et al. 2010).

Thirdly, there are countries and regions where historical clearances of forests for agriculture and urban expansion have led to more recent reforestation programmes, initially for timber production but increasingly to provide multi-purpose forests for a wide range of social and ecological purposes. Some of these forests may be domi-nated by exotic species of trees while recent trends are to use native ones. Countries with small amounts of forest but large urban populations, such as the UK, the Netherlands or Denmark, are typical of this type. The cultural associations with forests are typically weak in these countries as a result of the long period without substantive woodland cover since clearances took place, as well as the long history of urbanisation (Elands et al. 2010) (Fig. 2).

Fourthly, there are forests close to or within urban areas. These may be tracts of forest which were protected in some way, such as former hunting parks or for-ests that have become incorporated into the urban fabric, or they may be areas specially planted to benefit the urban population, such as the Black Country Urban Forest in the former industrial heartland of the English Midlands. The concept of the "urban forest" includes not only these kind of wooded areas but also park, street and garden trees – in fact any trees found within the urban envelope (Konijnendijk et al. 2005). This approach has been a significant focus of develop-ment in research and urban green space management practice over the last 20 or so years, in both America and Europe, and has also found a role in Asia and else-where. With the advance of urbanisation it seems clear that the concept of urban forestry has much to offer and it is unsurprising that such forests are a focus of recent research efforts. Clearly, these are the areas that are likely to be under most pressure from the forces of urbanisation and also from over-use by urban popula-tions. They also offer possibilities for providing both direct and indirect benefits of many kinds, ranging from benign effects on the urban environment's atmo-sphere, hydrology and ecology to aesthetic, recreational and health benefits, as will be demonstrated further below.

Fig. 2 The Vestskogen – or West Forest – in Copenhagen, Denmark, is an example of a relatively new urban forest, planted to provide recreational access as part of the famous Copenhagen "Fingerplan" for the development of the city (Source: Simon Bell)

3 The Evidence on Health and Wellbeing Benefits from Human Engagement with Forests

Human engagement with forests clearly goes back many millennia, as Vico succinctly pointed out (Vico, ibid), but in the context of urbanised societies we are less concerned with the contribution that forests may make to basic requirements of survival and more with what Maslow and others have termed 'higher level' needs (Maslow 1954) associated with wellbeing, fulfilment and pleasure. Engagement with forests for health and wellbeing can be direct or indirect, active or passive, external (viewing the forest as part of the landscape) or internal (being within it) or a combination of these. There have been several recent, wide-ranging initiatives to try to grasp the multiple facets of these different modes of engagement. The initiatives include three European Union funded COST Actions[1] covering "Urban Forests and Trees", "Forest Recreation and Nature Tourism" and "Forests,

---

[1] COST stands for Cooperation in Science and Technology and the system facilitates networking primarily among researchers but also includes practitioners and policy makers.

Trees and Human Health and Wellbeing". Each of these produced a range of publications and reports which brought together research and practice, often from widely diverse fields, thereby emphasising the multi-disciplinary character of the subjects (Konijnendijk et al. 2005; Bell et al. 2007b, 2009; Pröbstl et al. 2010; Nilsson et al. 2011). IUFRO[2] has also focused on this area recently, establishing a Task Force on 'Forests and Health' and also the new thematic research agenda of 'Forests for People' which held its first conference in Austria in May 2012[3]. A Nordic-Baltic network called CARE-FOR-US (Centre for Advanced Research in Forestry Serving Urbanised Societies)[4] is also into its second period of funding and is focussing on various aspects of environmental, social and health benefits in the Nordic-Baltic region.

There is thus a considerable, and growing, literature on themes related to wellbeing and human engagement with forest environments, and an ongoing programme of research to develop better understandings on this theme. This chapter can only touch briefly on each of the different aspects of engagement and wellbeing benefit. Expertise on specific fields is associated with particular research networks linking universities and forest institutes and is often clustered in specific locations or regions around the world.

3.1   Effects on the Human Physical Environment

It is well-known that urban areas develop their own micro-environment. This results in phenomena such as: the so-called "heat island" effect, where temperatures in an urban area can be several degrees higher than in the surrounding rural landscape; increased run-off and localised flooding due to an increase in sealed surfaces; increased air turbulence and windiness, due to the effect of air flows around tall buildings being more extreme; and gaseous and particulate air pollution from vehicle exhaust gases, industrial emissions and domestic heating and cooking sources (particularly in developing countries) (Tyrväinen et al. 2005). These effects can be detrimental to human health and safety and all can be mitigated to a certain extent by the presence of greater proportions of green areas in general and trees in particular, although with some limitations (de Vries 2010). Shade and evapotranspiration help to cool the urban micro-climate; trees intercept rain and green areas help sub-surface infiltration of water; trees create shelter from or reduce the variability and turbulence of wind gusts, depending on the species and location; and, finally, trees and vegetation trap dust and reduce many aspects of air pollution. However, these positive effects are not universal

---

[2] IUFRO is the International Union of Forest Research Organisations with its secretariat in Vienna. Division 6 is concerned with social aspects of forestry.

[3] http://ffp2012.boku.ac.at/

[4] http://www.nordicforestresearch.org/care-for-us2/

Fig. 3 Street trees in Bordeaux, France, are part of the "urban forest" and help to provide much needed shade and to trap dust, among other benefits to the urban microclimate (Source: Simon Bell)

and in some circumstances there can be negative effects, such as when poorly sited trees trap pollutants at street level and prevent them from dissipating (Sieghardt et al. 2005) or where trees produce aromatic hydrocarbons or pollen or harbour insects which may provoke allergic reactions in some people. Trees planted in order to provide benefit in one way may cause a problem in another due to poor positioning or poor choice of species, so there remains a need for research to improve the efficiency and minimise unwanted side-effects of the urban forest in this regard (Fig. 3).

Despite certain limitations of the sort just described, the many benefits of trees in cities are clear, and may seem self-evident to forestry and tree professionals. Yet their overriding benefits may not be so evident to engineers, planners or architects who may yet need to be convinced of the case. With the current, widely-applied policy focus on urban densification (Rogers et al. 1999), green areas may be squeezed out of cities. Yet if urban areas become more compact, with a higher density of residents, then the need for a better urban micro-environment becomes greater, not less, and the case for a strong green infrastructure which is not just mown grass, i.e. that includes trees and their many benefits, needs to be made more strongly. Further work on establishing the economic justification of urban green and urban forest areas in this area of benefits is also needed.

3.2   Aesthetic Aspects of Forests and Trees

Forests and trees are major elements of the landscape. In this context, the term 'landscape' is taken from that of the European Landscape Convention – "An area of land as perceived by people which has arisen as the result of natural and/or human processes…" (Council of Europe 2000). This emphasis on perception is an important aspect and should be understood as differentiating the term landscape as used here from that commonly used in landscape ecology. Aesthetic preferences for both external views – the forest as scenery – and internal views – as experienced from paths or tracks within the forest – have been researched for many years and a significant body of practice has also been built up that draws on understandings of preference, especially in those countries where forest management activities which cause major changes to the landscape have been, or continue to be, common (Bell and Apostol 2008). This principally means practice that pays attention to the landscape implications of logging and associated activities such as road construction in managed natural forests and planting, logging and road construction in plantation forests.

The main foci of research into, and implementation of, aesthetic or visual design principles have been the USA, the UK, Canada, Australia and Scandinavia. The main concerns have been about the perceived reduction in "naturalness" by the introduction of plantation elements with geometric shapes into the landscape as well as the impact in closer views of logging activities (such as debris). Aesthetics as a subject lies at the interface of philosophy, environmental psychology and landscape design and involves many factors. Primarily it is concerned with the pleasure obtained from looking at or experiencing the landscape (Bell 2012). Forests, because of the way that trees enclose the observer, screen or focus views and contribute to a multi-sensory "engagement" with the environment, create specific aesthetic conditions. Much empirical research has involved carefully controlled studies of preferences for different scenes or management interventions, mainly using photographs or visualisations of different landscape types or forest management activities (Daniel and Boster 1976; Karjalainen and Tyrvainen 2002). Most studies reveal similar results across a range of cultures, although there are differences (Kohsaka and Flitner 2004; Han 2007). However, the focus of work in Western countries leaves many gaps in understanding the role of forest aesthetics in other cultures and also how such concerns emerge as countries become more developed, as tourism values associated with scenery take on more importance and as middle classes emerge who have time and inclination to be concerned about aesthetic qualities. Thus there is considerable research needed to extend the understanding of these issues in different cultural contexts across the globe (Ward Thompson 2012) (Fig. 4).

The urban forest can play a different aesthetic role, above and beyond those considered above. Here the presence of trees in streets, parks and gardens can serve to introduce natural and green elements into an environment otherwise dominated by built forms and artificial elements (Bell et al. 2005). The wealthier residential

Fig. 4 The aesthetic enjoyment of being in a forest – in this case a managed one in Estonia – enables us to relax and provides a valuable setting for other forms of engagement, both physical and psychological (source: Simon Bell)

areas in almost all cities are also leafy or have views from houses and apartments which include substantial amounts of green vegetation, with trees having a dominant role due to their height and longevity. Studies have shown that houses with views of trees and green spaces tend to sell for higher prices so, in this respect, trees as aesthetic objects can also have a market value which can be calculated (Tyrväinen and Miettinen 2000). However, these studies are limited in number and geographic coverage and more research is needed in order to be able to obtain a better idea of the degree to which these relationships can be generalised across countries and cultures.

3.3 Restorative Environments and the Psychological Benefits of Nature Experiences

A comparatively recent field of research focuses on the "restorative" effects of the natural environment (Hartig 2007; Hartig et al. 2011). The attention restoration theory of Rachel and Stephen Kaplan (Kaplan and Kaplan 1989; Kaplan 1995) is one approach to explaining why certain types of environments, particularly natural ones, appear to be effective in stress reduction and restoration from fatigue. The theory suggests that directed attention used in coping with complex patterns of daily life, including work, is a highly limited resource, easily exhausted if there are not opportunities for recovery. People recover best in environments where this system can rest and where they can use another type of attention – involuntary attention or 'soft fascination' – which the natural environment is particularly well-suited to supporting. There is evidence to show that, for people suffering from mental fatigue, looking out of a window at a scene which includes natural elements such as trees, or perhaps sitting for a while in a garden surrounded by plants and trees, speeds up the restoration compared with a break where only built elements are visible. This demonstrates the potential for urban trees to contribute to improved wellbeing, to say nothing of employees' productivity, when more people work in jobs which require sustained periods of concentration. As the number of people working in such jobs – often involving IT – increases globally, the potential benefits are significant yet the research has only begun to scratch the surface.

It is not only work which can be stressful; merely living in an urban environment can include exposure to many psychological stressors, as noted in the introduction. "The struggle to pay attention in cluttered and confusing environments (such as crowded urban ones) turns out to be central to what is experienced as mental fatigue" (Kaplan and Kaplan 1989, p. 182). Research is beginning to show how viewing a forest scene compared with an urban scene can have positive effects on physiological indicators of stress such as blood pressure and hormone levels.

Such evidence builds on theories such as the biophilia hypothesis (Kellert and Wilson 1993) and theories and models suggesting that the human response to the environment is strongly rooted in our evolutionary origins (Orians and Heerwagen 1992; Ulrich 1999). A psycho-evolutionary basis for the benefits of engagement with natural environments is supported by the strong evidence for physiological and psychological responses to perceiving nature which are thought to take place via psychoneuroendochrine mechanisms, independent of conscious activity choice. Being in or viewing wooded or natural environments has been shown to reduce physiological measures of stress including blood pressure, heart rate, skin conductance and muscle tension (Ulrich et al. 1991; Hartig et al. 2003; Ottosson and Grahn 2005). In Japan, a study exploring the effect of a walk in the forest ('Shinrin-yoku' – taking in the forest atmosphere) has shown that such environments can promote lower concentrations of cortisol, lower pulse rate, lower blood pressure, greater parasympathetic nerve activity and lower sympathetic nerve activity when compared to city environments (Park et al. 2007, 2010; Lee et al. 2011).

Recent, innovative work in the UK has shown how higher levels of green and natural space in deprived urban communities are linked with lower stress, measured both subjectively and objectively, using diurnal patterns of cortisol secretion (Ward Thompson et al. 2012). Diurnal cortisol patterns indicate everyday circadian rhythms of health and are sensitive to the longer term effects of stressors in the social and physical environment; the patterns will reflect any hormone dysregulation that is associated with conditions such as clinical depression or chronic stress. Thus, an association between access to natural environments and healthy cortisol cycles is particularly compelling in suggesting that spending time in

Fig. 5 Pre-school children in Austria attending a forest kindergarten. They are able to become familiar with the forest and nature and studies have shown that there are benefits from such engagement from an early age (Source: Simon Bell)

natural environments such as forests may help to reduce mental and psychological ill-health. An epidemiological study of a large, general population sample from across Scotland has shown that physical activity in natural environments is associated with a reduction in the risk of poor mental health to a greater extent than physical activity in other environments, and that regular users of woodlands or forests for physical activity were at about half the risk of poor mental health of non-users (Mitchell 2012).

Forest settings have also been shown to benefit children and adolescents, especially those with behavioural problems and mental disorders who can be highly disruptive in conventional school settings. A recent, comparative study with children aged from 10 to 13 years showed that the forest setting was advantageous to mood in all behaviour groups but particularly in those children suffering from 'mental disorder' (Roe and Aspinall 2011a). Detailed analysis of a small group of children with severe behavioural problems illustrated how the forest setting offered opportunities for curiosity, creativity, exploration and challenge, opportunities largely missing in these young people's lives to date, and how the setting allowed therapeutic processes to occur naturally, without professional intervention (Roe and Aspinall 2011b). These are important findings for today's urbanized society, where a childhood without access to woodlands and natural places may contribute to the difficulties young people and adults experience later in life (Ward Thompson et al. 2008) (Fig. 5).

3.4   Forests, Physical Activity and Health

The sedentary lifestyle of modern Western societies is contributing to the epidemic of obesity and poor cardio-vascular health, such that a major focus of public health organisations is to increase physical activity (Pate et al. 1995; Department of Health 2004; U.S. Department of Health and Human Services 2008; Department of Health, Physical Activity, Health Improvement and Protection 2011). Motivating people to take more exercise is proving difficult in many circumstances and it seems that part of this has to do with the environment in which the exercise takes place as well as barriers such as cost or the accessibility of places in which to exercise. Findings from a cross-European study showed that the quality of the landscape appears to influence physical activity (Ellaway et al. 2005). In urban areas, people may go to parks or other open spaces for exercise and these may be very busy in some cities. The attractiveness of a forest is that it provides a much more pleasant ambience for physical activity and, if it is large enough, it can accommodate a range of different forms of recreation which promote physical health, mental restoration and aesthetic pleasure all at once. Research on local woodland use by communities in Scotland showed that the predominant use is for walking and cycling, suggesting that forests can play a key role in maintaining healthy and active lifestyles for all ages and both sexes (Ward Thompson et al. 2004, 2005). Compared with streets or small open spaces, forests provide a much more pleasant environment with less pollution, cleaner air to breathe while exercising and potentially the beneficial effects of aromatic hydrocarbons.

In countries with a strong cultural association with forests, such as Finland, forests may be the location of choice for outdoor recreation and exercise and children are brought up able to ski cross-country and be comfortable with the forest environment. This familiarity means that they frequently develop the habit of visiting forests and gaining multiple benefits from this. In other countries where there is less forest and less of a cultural connection, forests may be less attractive, so that people may miss out on the benefits. Fear of crime, of being attacked or getting lost may be a major barrier, especially to women in some urban environments, while children may be discouraged or prevented from playing in woods owing to safety concerns (Ward Thompson et al. 2004). However, a North American study of women's experience of exercise such as running or cycling showed that it was more enjoyable and meaningful in a large park with many woodland trees, compared to in the street, because of the beautiful scenery and the therapeutic or spiritual experience associated with the park's aesthetic qualities. The park afforded a traffic-free environment where women felt freer to dress comfortably and generally less susceptible to unwelcome remarks (Krenichyn 2006).

4   Broader Aspects of Forest Recreation

Forests, because of public ownership or everyman's right of access, often provide extremely valuable opportunities for general recreation across the spectrum of potential activities, from relaxing, sitting or walking the dog to jogging, cycling or skiing, etc., which may include obtaining aesthetic pleasure, restoration, reduction of stress and obtaining physical exercise. Forests also provide a good setting for family or friends' groups and social activities such as children's play, picnicking or having a barbecue. Indeed, the woodland environment seems to offer particularly rich opportunities for inter-generational activities, such as grandparents and their grandchildren sharing knowledge about the natural environment, and for children's play that has different, multisensory qualities that distinguish it from play in conventional urban environments or playgrounds (Ward Thompson et al. 2004, 2008). As mentioned earlier, forests can provide opportunities for activities and freedom of movement for teenagers and young adults that may be important for their development and difficult to provide for in the conventional urban environment (Bell et al. 2003).

The kind of activity taking place in forests varies according to social and cultural contexts. In countries with large numbers of people of immigrant origin from different cultural backgrounds, there may be rather different demands for recreation than those commonly undertaken by the majority population. In some cases, as noted in the introduction, some ethnic groups may not use forests or other outdoor space at all. This may be for a variety of reasons and attempts to understand this phenomenon and to increase the use of green spaces including forests and woods among minority ethnic groups have met with mixed success (CABE 2010). In some cultures, such as Muslim communities, young girls and women generally stay at home or indoors and, if they go to parks or green areas, they need to be there in the company either of male relatives or only of other women. This clearly limits the possibilities of these groups participating in outdoor recreation, unless special areas are set aside for them or the environment offers opportunities for separate use of different spaces, a demand highlighted in some cities in the UK (CABE 2010). Forests and wooded parks have the benefit of providing visual screening that can offer such spatial differentiation and may therefore be particularly valuable in providing opportunities for different cultural groups to enjoy outdoor recreation within comparatively private family groupings.

The ageing demographic of the global population means there is increasing interest in environments that support healthy activity into old age. Maintaining physical activity is an important way for older people to remain fit and healthy and for preventing or reducing the impact of a range of problems such as osteoporosis in women (Simonsick et al. 2005; Sugiyama and Ward Thompson 2008). Studies have shown that the presence of attractive trees and vegetation in a local park may be a significant attraction for walking in an older population, as for younger age groups. However, the physical conditions of accessibility to natural areas such as forests must be considered. The surfaces and gradients of paths, the provision of benches at regular intervals, hand rails along paths with steeper gradients and provision of sufficient information explaining the lengths and conditions of paths have been found to be important aspects to be considered when planning to make areas accessible for older people (Aspinall et al. 2010) (Fig. 6).

There are also people with a wide range of disabilities whose needs must be considered. This includes people with mobility impairment, making walking difficult or impossible, people who are blind or visually impaired, those who are deaf or

Fig. 6 Older people benefit a lot from physical exercise – here older Dutch people are taking it seriously in a state forest (Source: Jan Blok)

hard of hearing, and people with a range of cognitive impairments or mental disabilities. People with some form of disability are often poorly provided for so that they are either unable to visit or to fully enjoy the outdoors. Enough is now known about how to provide support for many aspects of impairment so that people with a disability can benefit from the rich experience of a forest visit; the main problems are probably lack of awareness among planners and managers and lack of funds to implement special facilities (Fig. 7).

5   Social and Community Engagement with Forests

In many countries so-called "community forests" are an important element of the forest estate. In developing countries these may be substantial areas managed for the benefit of local communities and be important sources of food, fodder, grazing and fuel and also providers of non-timber forest products with an economic value. In developed, Western countries, such areas tend to be smaller and to fulfil different objectives. A community forest generally means one managed by, as well as for, the benefit of the local community and the focus may be on recreation, nature conservation and social engagement. When woodlands in and around towns are planned and managed by local people they offer many benefits over and above the recreation and health ones already discussed above. These may include the possibilities to work together in a socially valuable endeavour, to make a personal or

Fig. 7 Disabled people may need special equipment or better surfaces to access the forest but they gain a lot from being able to visit (Source: Simon Bell)

communal investment into something with long-lasting and growing benefits for future generations, for communities from different backgrounds to mix and work together, to introduce children to nature and generally to obtain altruistic feelings of caring for nature and the environment. Such projects include the Forestry Commission Scotland's programme for 'Woods In and Around Towns' (WIAT), focused on the more deprived urban communities in Scotland, where social capital and cohesion may arise from investment in physical and social infrastructure and community engagement with forest maintenance and woodcraft skills, above and beyond any benefits from mental or physical health (Roe and Ward Thompson 2010; Ward Thompson et al. under review) (Fig. 8).

6 Challenges for the Future

The ranges of benefits to health and wellbeing discussed in this short overview reveal a developing field of knowledge which has not been ignored by policy makers or practitioners in some countries but which still needs much more research and promotion. The challenges can be considered under two aspects: breadth and depth.

Fig. 8 People engaging with the forest – Lochend Woods, in Dunbar in Scotland – as part of a community group, helping to manage a forest and to take part in activities celebrating it (Source: Simon Bell)

The breadth of the understanding of and research into human engagement with forests, as has been noted above, has tended to be focussed in major developed countries such as Europe, North America, Australasia and Japan. The influence of organisations such as IUFRO has made sure that forest research agencies and institutions in a wide range of countries can network and focus on all fields of forest research but, understandably, given the different priorities in developing countries, the area of human engagement as discussed in this chapter has been lower on the list of priorities. However, with rapid rates of development in countries such as India and China, the rapid urbanisation as noted in the introduction and the increasing numbers of middle-class people employed in jobs which produce the same sedentary conditions and same mental stresses as those experienced in developed post-industrial regions, it seems clear that the research and policy focus described in this chapter should be expanded more widely. The research findings from the regions where this has been carried out may be generally transferable but there may be important environmental and cultural differences which mean that repeating studies in different conditions would add to the evidence base in valuable and distinctive ways.

The issue of depth of research arises from the increasing need to establish evidence of a quality that is recognised as valid in the world of public health and medical research, where standards and expectations of research methods, sample sizes, experimental design and statistical validity are frequently much more demanding than those of forestry, for the simple reason that human lives may be at risk from poor research conduct. This issue has already been recognised and some innovative, trans-disciplinary research is being undertaken, but it is still an important issue for the future (Ward Thompson 2011; Ward Thompson et al. 2010). Identifying research priorities for green and public space, including urban forests, has been carried out already, e.g. by Bell et al. (2007a), and recently the European Forestry Institute conducted a research foresight exercise about urban forestry, the results of which will be published at some point[5].

A further challenge is that of translating research into aspects such as the health benefits of forest environments into practice, and in using practice to inform research. The obstacles arise as a result of the different traditions, expectations, languages and limited experience of these disciplines working together (Van Herzele et al. 2011). There are some examples from Europe but these are sporadic and not necessarily sustainable (Bell et al. 2011). More work, especially using action research approaches would help to strengthen the evidence base.

7 Conclusions

The subject field of human engagement with forest environments for physical and mental health and wellbeing is, in some senses, surprisingly well-developed and has seen an upsurge in research and application in the last decade or so. The issue is also rising up the research and policy agenda because the world is becoming much more urbanised and the results of urban lifestyles are causing concern in terms of health, wellbeing and quality of life. Forests have been demonstrated to play a key role, along with other forms of green space, in many aspects of environmental support for healthier lifestyles. However, the focus of research in developed countries and the still weak connections with the health-related professions are issues that need to be addressed and which cause complexities and difficulties for achieving the best possible outcomes. Awareness that the natural environment, and forests in particular, might play a role in enhancing health, and perhaps prevent illness at a fraction of the cost of post hoc medical intervention suggests that research into the salutogenic aspects of forest environments for human society is an urgent need in an urban world.

References

Aspinall PA, Thompson CW, Alves S, Sugiyama T, Vickers A, Brice R (2010) Preference and relative importance for environmental attributes of neighbourhood open space in older people. Environ Plan B: Plan Design 37(6):1022–1039

---

[5] http://www.nbforest.info/news/efinord-and-care-us-networks-join-forces

Bell S (2012) Landscape: pattern, perception and process, 2nd edn. Routledge, London

Bell S, Apostol D (2008) Designing sustainable forest landscapes. Taylor and Francis, London

Bell S, Ward Thompson C, Travlou P (2003) Contested views of freedom and control: children, teenagers and urban fringe woodlands in Central Scotland. Urban For Urban Gree 2:87–100

Bell S, Blom D, Rautamaki M, Castel-Branco C, Simson S, Olsen EA (2005) Design of urban forests. In: Konijnendijk C, Nilsson K, Randrup TB, Schipperijn J (eds) Urban forests and trees. Springer, Berlin

Bell S, Montarzino A, Travlou P (2007a) Mapping research priorities for green and public urban space in the UK. Urban For Urban Gree 6:103–115

Bell S, Tyrvainen L, Sievanen T, Pröbstl U, Simpson M (2007b) Outdoor recreation and nature tourism: a European perspective. Living Rev Landsc Res 1:46

Bell S, Simpson M, Tyrvainen L, Sievanen T, Pröbstl U (eds) (2009) European forest recreation and tourism: a handbook. Taylor and Francis, London

Bell S, Alves S, Silveirinha de Oliveira E, Zuin A (2010) Migration and land use change in Europe: a review. Living Rev Landsc Res 4:2

Bell S, van Zon R, Van Herzele A, Hartig T (2011) Health benefits of nature experience: implications of practice for research. In: Nilsson K, Sangster M, Gallis C, Hartig T, de Vries S, Seeland K, Schipperijn J (eds) Forests, trees and human health. Springer, Berlin

CABE (2010) Community green: using local spaces to tackle inequality and improve health. London: CABE. http://www.cabe.org.uk/publications/community-green. Accessed 4 Dec 2010

Council of Europe (2000) European landscape convention. http://conventions.coe.int/Treaty/en/Treaties/Html/176.htm. Accessed 19 Aug 2012

Daniel T, Boster RS (1976) Measuring landscape esthetics: the scenic beauty estimation method, Research Paper RM-167. U.S. Department of Agriculture, Forest Service, Rocky Mountain Forest and Range Experiment Station. Fort Collins, CO, London

De Vries S (2010) Nearby nature and human health: looking at the mechanisms and their implications. In: Thompson CW, Aspinall P, Bell S (eds) Innovative approaches to researching landscape and health: open space: people space 2. Routledge, Abingdon, pp 75–94

Department for Communities and Local Government (2007) Indices of multiple deprivation. http://www.communities.gov.uk/publications/communities/indiciesdeprivation07. Accessed 26 July 2012

Department for Work and Pensions (2012) The ageing society. http://www.dwp.gov.uk/policy/ageing-society/. Accessed 26 July 2012

Department of health (2004)At least five a week: Evidence on the impact of physical activity and its relationship to health. A report from the chief medical officer. Department of Health, London

Department of Health, Physical Activity, Health Improvement and Protection (2011) Start active, stay active: a report on physical activity from the four home countries' chief medical officers. Department of Health, London, www.dh.gov.uk. Accessed 26 Oct 2011

Elands B, Bell S, Blok J, Colson V, Curl S, Kaae BC, Van Langenhove G, McCornmack A, Murphy W, Petersson JG, Praestholm S, Roovers P, Worthington R (2010) Atlantic region. In: Pröbstl U, Wirth V, Elands B, Bell S (eds) Management of recreation and nature-based tourism in European forests. Springer, Heidelberg

Ellaway A, Macintyre S, Bonnefoy X (2005) Graffiti, greenery, and obesity in adults: secondary analysis of European cross sectional survey. Brit Med J 331:611–612

Han KT (2007) Responses to six major terrestrial biomes in terms of scenic beauty, preference, and restorativeness. Environ Behav 39(4):529–556

Hartig T (2007) Three steps to understanding restorative environments as health resources. In: Ward Thompson C, Travlou P (eds) Open space: people space. Taylor and Francis, Abingdon, pp 163–179

Hartig T, Evans GW, Jamner LD, Davies DS, Gärling T (2003) Tracking restoration in natural and urban field settings. J Environ Psychol 23:109–123

Hartig T, van den Berg A, Hagerhall C, Tomalak M, Bauer A, Hansmann R, Ojala A, Syngollitou E, Carrus G, van Herzele A, Bell S, Camilleri Podesta MT, Waaseth G (2011) Health benefits of nature experience: psychological, social and cultural processes. In: Nilsson K, Sangster M,

Gallis C, Hartig T, de Vries S, Seeland K, Schipperijn J (eds) Forests, trees and human health. Springer, Dordrecht

Kaplan S (1995) The restorative benefits of nature: toward an integrative framework. J Environ Psychol 15:169–182

Kaplan R, Kaplan S (1989) The experience of nature: a psychological perspective. Cambridge University Press, Cambridge

Karjalainen E, Tyrvainen L (2002) Visualization in forest landscape preference research: a Finnish perspective. Landsc Urban Plan 59(1):13–28

Kellert SR, Wilson EO (eds) (1993) The biophilia hypothesis. Island Press, Washington, DC

Kohsaka R, Flitner M (2004) Exploring forest aesthetics using forestry photo contests: case studies examining Japanese and German public preferences. For Policy Econ 6(3–4):289–299

Konijnendijk C, Nilsson K, Randrup TB, Schipperijn J (eds) (2005) Urban forests and trees. Springer, Berlin

Krenichyn K (2006) "The only place to go and be in the city": women talk about exercise, being outdoors and the meanings of a large urban park. Health & Place 12:631–643

Lee J, Park BJ, Tsunetsugu Y, Ohira T, Kagawa T, Miyazaki Y (2011) Effect of forest bathing on physiological and psychological responses in young Japanese male subjects. Public Health 125(2):93–100

Maslow A (1954) Motivation and personality. Harper and Row, New York

Mitchell R (2012) Short report: is physical activity in natural environments better for mental health than physical activity in other environments? Soc Sci Med 66:1238. doi:10.1016/j.socscimed.2012.04.012

Nilsson K, Sangster M, Gallis C, Hartig T, de Vries S, Seeland K, Schipperijn J (eds) (2011) Forests, trees and human health. Springer, Dordrecht

Orians GH, Heerwagen JH (1992) Evolved responses to landscapes. In: Barkow JH, Cosmides L, Tooby J (eds) The adapted mind. Oxford University Press, Oxford, pp 555–579

Ottosson J, Grahn P (2005) A comparison of leisure time spent in a garden with leisure time spent indoors: on measures of restoration in residents in geriatric care. Landsc Res 30(1):23–55

Park BJ, Tsunetsugu Y, Kasetani T, Hirano H, Kagawa T, Sato M, Miyazaki Y (2007) Physiological effects of Shinrin-yoku (taking in the atmosphere of the forest) – using salivary cortisol and cerebral activity as indicators. J Physiol Anthropol 26(2):123–128

Park BJ, Tsunetsugu Y, Kasetani T, Kagawa T, Miyazaki Y (2010) The physiological effects of Shinrin-yoku (taking in the forest atmosphere or forest bathing): evidence from field experiments in 24 forests across Japan. Environ Health Prev Med 15:18–26

Pate RR, Pratt M, Blair SN, Haskell WL, Macera CA, Bouchard C, Buchner D, Ettinger W, Heath GW, King AC et al (1995) Physical activity and public health: a recommendation from the centers for disease control and prevention and the American college of sports medicine. J Am Med Assoc 273:402–407

UN Population Fund (2007) State of the world population 2007: unleashing the potential of urban growth. UN, New York

Pröbstl U, Wirth V, Elands B, Bell S (2010) Management of recreation and nature-based tourism in European forests. Springer, Heidelberg

Roe J, Aspinall P (2011a) The restorative outcomes of forest versus indoor settings in young people with varying behaviour states. Urban For Urban Green 10:205–212

Roe J, Aspinall P (2011b) The emotional affordances of forest settings: an investigation in boys with extreme behavioural problems. Landsc Res 36:535–552

Roe J, Ward Thompson C (2010) Contextual background to the WIAT (Woods In and Around Towns) longitudinal survey, 2006–2009. OPENspace Research Centre report for Forestry Commission, August 2010. http://www.forestry.gov.uk/website/forestry.nsf/byunique/infd-8a2dmd. Accessed 04 Mar 2012

Rogers R, Urban Task Force et al (1999) Towards an urban renaissance: final report of the urban task force chaired by Lord Rogers of Riverside. Department of the Environment, Transport and the Regions, London

Sieghardt M, Mursch-Radlgruber E, Paoletti E, Couenberg E, Dmitrakopolous A, Rego F, Hatzisthasis A, Randrup TB (2005) The abiotic environment: impact on growing conditions on urban vegetation. In: Konijnendijk C, Nilsson K, Randrup TB, Schipperijn J (eds) Urban forests and trees. Springer, Berlin

Silveirinha de Oliveira E (2012) Immigrants and Public Open Spaces: attitudes, preferences and uses. PhD Thesis, Edinburgh: University of Edinburgh

Simonsick EM, Guralnik JM, Volpato S, Balfour J, Fried LP (2005) Just get out the door! Importance of walking outside the home for maintaining mobility: findings from the women's health and aging study. J Am Geriatr Soc 53:198–203

Stein CA, Colditz GA (2004) The epidemic of obesity. J Clin Endocrinol Metab 89(6):2522–2525

Sugiyama T, Thompson CW (2008) Associations between characteristics of neighbourhood open space and older people's walking. Urban For Urban Green 7:41–51

Tyrväinen L, Miettinen A (2000) Property prices and urban forest amenities. J Environ Econ Manag 39(2):205–223

Tyrväinen L, Pauleit S, Seeland K, de Vries S (2005) Benefits and uses of urban trees and forests. In: Konijnendijk C, Nilsson K, Randrup TB, Schipperijn J (eds) Urban forests and trees. Springer, Berlin

U.S. Department of Health and Human Services (2008) *2008* Physical activity guidelines for Americans. www.health.gov/paguidelines. Accessed 2 Nov 2010

Ulrich RS (1999) Effects of gardens on health outcomes: theory and research. In: Cooper MC, Barnes M (eds) Healing gardens. Therapeutic benefits and design recommendations. Wiley, New York

Ulrich RS, Simons RF, Losito BD, Fiorito E, Miles MA, Zelson M (1991) Stress recovery during exposure to natural and urban environments. J Environ Psychol 11:201–230

United States Forest Service (2012) Wilderness areas. http://www.fs.fed.us/recreation/programs/cda/wilderness.shtml. Accessed 26 July 2012

Van Herzele A, Bell S, Hartig T, Camilleri Podesta MT, van Zon R (2011) Health benefits of nature experience: the challenge of linking practice and research. In: Nilsson K, Sangster M, Gallis C, Hartig T, de Vries S, Seeland K, Schipperijn J (eds) Forests, trees and human health. Springer, Dordrecht

Vico G (1725) Scienza Nuova (the new Science). Stamperia Museana, Napoli

Vistad OI, Erkkonen J, Rydberg D (2010) Nordic region. In: Pröbstl U, Wirth V, Elands B, Bell S (eds) Management of recreation and nature-based tourism in European forests. Springer, Heidelberg

Ward Thompson C (1998) Historic American parks and contemporary needs. Landsc J 17(1):1–25

Ward Thompson C (2006) Patrick Geddes and the Edinburgh zoological garden: expressing universal processes through local place. Landsc J 25(1):80–93

Ward Thompson C (2011) Linking landscape and health: the recurring theme. Landsc Urban Plan 99(3):187–195

Ward Thompson C (2012) Landscape perception and environmental psychology. In: Howard P, Thompson I, Waterton E (eds) Companion to landscape studies. Routledge, Abingdon

Ward Thompson C, Aspinall P, Bell S, Findlay C, Wherrett J, Travlou P (2004) Open space and social inclusion: local woodland use in Central Scotland. Forestry Commission, Edinburgh

Ward Thompson C, Aspinall P, Bell S, Findlay C (2005) "It gets you away from everyday life": local woodlands and community use – what makes a difference? Landsc Res 30(1):109–146

Ward Thompson C, Aspinall P, Montarzino A (2008) The childhood factor: adult visits to green places and the significance of childhood experience. Environ Behav 40(1):111–143

Ward Thompson C, Aspinall P, Bell S (eds) (2010) Innovative approaches to researching landscape and health: open space: people space II. Routledge, Abingdon

Ward Thompson C, Roe J, Aspinall P, Mitchell R, Clow A, Miller D (2012) More green space is linked to less stress in deprived communities: evidence from salivary cortisol patterns. Landsc Urban Plan 105:221–229

Ward Thompson C, Roe J, Aspinall P (under review) Woodland improvements in deprived urban communities: what impact do they have on people's activities and quality of life? Landsc Urban Plan

Sustainable Forest Management in China: Achievements in the Past and Challenges Ahead

Nuyun Li

Abstract China has made great strides in increasing its forested area and standing volume over the past two decades, largely attributed to sustainable forest management. Forest management in China has gone through two stages, i.e. timber-oriented forest management that once caused several problems such as depletion of forest resources, soil erosion and desertification before 1990s, and sustainable forest management after the 1990s. The practice of sustainable forest management in China mainly includes continuous efforts through government initiatives to restore degraded landscapes and encourage production orientated plantations through its six key forestry programs, providing policies, encouraging community-based forest management, and individualizing forest management in collective forest regions. In spite of the achievements made in sustainable forest management, China is still faced with challenges, including limited suitable areas for afforestation/reforestation, balancing the demands between economic requirements and ecological needs from forests, improving the quality of forests to improve its protective functions and ecosystem services. Therefore, innovations need to be explored for the country's sustainable forest management in the future.

On November 17, 2009, State Forestry Administration of China (SFA) officially released the 7th (2004–2008) National Forest Inventory (NFI) data to the public in Beijing (SFA 2009a). The outcomes of NFI show that China's forest area and forest stock volume (or the standing inventory of trees) has been increasing steadily since the forest area and stock volume increase of the 4th NFI (1989–1993). China's sustainable forest management system has therefore made great contributions to the "dual increase" of both the forest area and forest stock volume.

N. Li (✉)
Department of Afforestation and Greening Management, State Forestry Administration (SFA), P.R. China, 18 Hepingli Dongjie, Dongcheng District, Beijing 100714, China
e-mail: linuyun516@vip.sina.com

T. Fenning (ed.), *Challenges and Opportunities for the World's Forests in the 21st Century*, Forestry Sciences 81, DOI 10.1007/978-94-007-7076-8_6,
© Springer Science+Business Media Dordrecht 2014

1 Introduction

China has paid great importance to forest sustainable management in its national development strategies. Nowadays, sustainable development is a general concern of the international community to which the Chinese government has always attached great importance. As a large developing country, China is fully aware of its obligations and potential roles in protection of global ecosystems and environment, and the Chinese government gives top priority to the global issue of the twin goals of environmental protection and sustainable development. China is striving to use natural resources more efficiently, is speeding up ecological construction and environmental protection in its national strategies, according to the principles of Agenda 21 of the Rio Declaration on Environment and Development, Statement of Forest Principles, of the United Nations Conference on Environment and Development (UNCED), Rio Janeiro, Brazil in 1992. Since the beginning of the twenty-first century, China has established a scientific outlook for the implementing its economic and social development, with its guiding principle and basic requirement being to adhere to the people-oriented, comprehensive coordinated and sustainable development. Forestry has naturally become one of the most important national sustainable development strategies. The Chinese government, based on the deep understanding of the role of forests to ecology, economics and society, has realized that forests can contribute to four major objectives in the national development and overall strategy planning, i.e. they have a considerable role in supporting the sustainable development strategy, play a primary role in ecological system building, are fundamental to Western Expansion and are critical to tackling climate change (Wen 2009).

1.1 China's Basic Information and Its Forest Distribution

Basic Information of China

China is situated in eastern Asia and bounded by the Pacific Ocean to the east. It is the third largest country in the world, it has an area of 9.6 million square kilometers, which is one-fifteenth of the world's landmass. The Chinese border stretches over 22,000 km on land and its coastline extends well over 18,000 km, washed by the waters of the Bohai Sea, the Huanghai, the East China and the South China Seas. The Bohai Sea is China's only inland sea. There are 6,536 islands larger than 500 m^2, the largest being Taiwan, with a total area of about 36,000 km^2, and the second, Hainan. The South China Sea Islands are the southernmost island group of China (more details see map of China). At the end of 2011, China's total population in the mainland amounted to 1347.35 million with a population density per sq km of 141 people (Statistics China 2012).

Map of China (http://www.chinapage.com/map/china_pol01.jpg)

Forest Resources Distribution in China

China's forest resources distribute into five major regions due to its diversity of climate styles and landscapes across the whole nation.

Northeast Inner Mongolia Forest Region

This forest region is located in the provinces (autonomous region) of Heilongjiang, Jilin and Inner Mongonia, covers the mountains of Daxing'anling, Xiaoxing'anling, Wandashan, Zhangguangcailing and Changbaishan. With humid climate and mild mountainous topography, the forest region is an ecological barrier for the

Songnen Plain, Sanjiang Plain and Hulun Buir Grassland in the northeast China and an important timber production base as well. The forest coverage is high at 67.10 % and the forest area is 35.90 million ha and forest stock volume is 3,213 million cubic meters in the forest region (SFA 2009a).

Southwest Mountainous Forest Region

This forest region covers some parts of the provinces of Yunnan and Sichuan and the Tibet autonomous region. Characterized by low latitudes, high elevation, topographic variation, favorable conditions of water and heat, and high levels of biodiversity. This region is known as "the kingdom of animals and plants" and China's treasure house for it's riches of unique wild fauna and flora. Forest coverage is 23.00 % and the forest area is 43.48 million ha and forest stock volume is 5,090 million cubic meters in the forest region (SFA 2009a).

Southeast Low Mountain and Hilly Forest Region

This forest region covers all or some parts of the provinces or autonomous regions of Jiangxi, Fujian, Zhejiang, Anhui, Hubei, Hunan, Guangdong, Guangxi, Guizhou, Sichuan, Chongqing and Shanxi. It lies in the subtropical zone with a mild climate and plentiful rainfall, which provide favorable natural conditions suitable for growing trees including of Chinese Fir, Masson Pine, and Gum Tree. This region is known as an important region for growing timber and the economic value of its forests. The forest coverage is again high at 51.97 %, the forest area is 57.81 million hectares and forest stock volume is 2,565 million cubic meters in the forest region (SFA 2009a).

Northwest Mountainous Forest Region

This forest region consists of forests growing in the mountainous areas such as Tianshan and Altai in Xinjiang Uygur Autonomous Region, Qilianshan, Bailongjiang and Ziwuling in Gansu Province, Qinling and Bashan in Shangxi Province. Most forests of the region are natural and distributed in high mountains with good water conditions. This region plays a decisive role in ecological environmental protection and socio-economic development in northwest China. In this forest region, the forest coverage is 39.14 %, the forest area is 5.09 million ha and forest stock volume is 5,310 million cubic meters (SFA 2009a).

Tropical Forest Region

The forest region, featuring parts of Yunnan, Guangdong and Hainan Provinces and the Guangxi and Tibet Autonomous Regions, is blessed with warm climate, plentiful rainfall and long growing period for diverse types of tropic forest and monsoon rain forest with an evergreen-crown and large-diameter trees in a complex structure.

Table 1 Forest resource distribution of five major forest areas. Unit: 10,000 ha, 10,000 m³, %

Forest regions	Forest coverage	Forest area	Percentage of the forest area of in the whole country	Forest stock volume	Percentage of the forest stock volume of in the whole country
Northeast Inner Mongolia	67.10	3,590.09	16.68	321,269.48	24.04
Southwest Mountainous	23.00	4,348.02	20.20	509,024.38	38.09
Southeast Low Mountain and Hilly	51.97	5,780.65	26.86	256,458.50	19.19
Northwest Mountainous	39.14	508.86	2.36	53,093.23	3.97
Tropical	44.57	1,180.37	5.48	86,260.91	6.46

Source: SFA 2009a

In this forest region, the forest coverage is 44.57 %, the forest area is 11.80 million hectares and forest stock volume is 863 million cubic meters (SFA 2009a).

The forest categories feature a layered transition from north to south: frigid-temperate zone coniferous forest, coniferous and broad-leaved forests, deciduous broad-leaved forest and coniferous forest in warm temperate zone, subtropical ever-green broad-leaved forest and coniferous forest, tropical monsoon forest and rain forest (Table 1) (SFA 2009a).

1.2   Definition of Sustainable Forest Management

There is currently no agreed definition as to what constitutes sustainable forest man-agement (SFM) in the global community. The FAO states that several international meetings have suggested that the extent of forest resources, biological diversity, forest health and vitality, productive functions of forest resources, protective func-tions of forest resources, socio-economic functions, legal, policy and institutional framework thematic elements are key components of SFM.

According to the statement of FAO above, we know that SFM aims to ensure that the goods and services derived from the forest meet present-day needs, while at the same time securing their continued availability and contribution to long-term development. In its broadest sense, forest management encompasses the administrative, legal, technical, economic, social and environmental aspects of the conservation and use of forests. It implies various degrees of deliberate human intervention, ranging from actions aimed at safeguarding and maintaining the for-est ecosystem and its functions, to favoring specific socially or economically valuable species or groups of species for the improved production of goods and services (FAO 2012)

1.3   Goals of China's Sustainable Forest Management

With the biggest population of the world, both ecological benefits and commercial functions from forestry are essential to Chinese society. A well developed forestry sector can also benefit the global community.

Based on the multiple ecological functions of forests for water and soil conservation, carbon fixation and oxygen release, as wind breaks and for stopping drifting sand, purify air and protecting biodiversity, as well as its important economical functions such as providing constructions materials, forest products, non-forest products and leisure facilities etc. China has always tried its best to keep the balance between ecological welfare and the economic benefits it derives from its forestry sector. The ideal goals of SFM for China are to facilitate sustainable, rapid and healthy development of forestry, for the purpose of setting up a relatively complete forestry ecosystem and a relatively well developed forestry-related industry system with the protection and improvement of the eco-environment as the priority by classifying management forests, plantation and rehabilitation as the main measures (Zhou 2000).

2   China's Practice of Sustainable Forest Management with Its Achievements

The People Republic of China was founded in 1949. In order to give readers a clear picture of China's Sustainable Forest Management practice, we here divided the practice into two periods of from 1949 to 1997 and 1998 to the present.

2.1   The First Stage of SFM Practice

At the foundation of the new China, the national leaders already understood the importance of forestry, since forests provide ecological shelters that stimulate high and stable yields in agriculture and animal husbandry, and secure the conservation of water resources and guarantee the long-term functions of natural water sheds. Xi Liang, a forestry expert, was appointed as the first minister of Ministry of Forestry, which was unusual at that special point as most ministers were generals back from the battle fields.

A Brief Review of China's Forest Management from 1948 to 1997

The Period from the People's Public of China Was Founded to 1978

During this special period, China's forest management practice was characterized more by taking than giving and ignoring the ecological functions of forest. When the People's Public of China was established in 1949, China was one of the countries most deficient in forest resources in the world. In November 1950, ministry of forestry of China announced its policy of forest management to regulate China's forest practice was "regarding the administration of forests as basis" and the idea of "afforestation is the key task" and "rational harvesting and comprehensively use forest resource". However, from 1950s to late 1970s, the national construction mainly demanded timber and forest products. As the fundamental industry of national economy, the main task of forestry was to produce timber. Providing timber was the core task of forest management practice for a long time during this period. Timber yield had reached 28,000,000 m^3 in 1957, up from 5,500,000 m^3 in 1950 (Cao 2000). Before the reform and the policy of opening-up was adopted in 1979 in China, forest was experiencing the real principle of "taking more than giving" instead the theoretical one mentioned above, the forestry construction was in effect on the road of sacrificing the forest's ecological interests to supporting national economy construction by centering on wood production, with neither the Chinese governments nor its public considering the ecological functions of forests.

After China's Reform and Opening-Up Policy in 1979–1998

Between 1979 and 1998, China's forest sector made good changes in its ecological management, as well as its industrial development.

Forestry Ecological System Framework Initially Established and Forests Diversified Management Was Introduced

The launch of the "Three North Shelterbelt Development Program" in November 1978 marks that start of key ecological programs in China, including such programs as the "Shelterbelt Development Program" the "Upper and Middle Reaches of the Yangtze River", the "Coastal Shelterbelt Development Program", the "National Desertification Prevention Program", the "Taihang Mountain Greening Program", the "Plain Greening Program" and the "Shelterbelt Development Program for the Integrated Management In the Basin of Liaohe River". These ten programs cover the major environmentally fragile regions of soil erosion, windy and sandy, and sanitized areas in China, with a total project area of 7.056 million square kilometers, taking up 73 % of the total land territory. The planned total establishment area of the above programs, except the Plain Greening Program, reaches 63.048 million hectares. A total of 35.934 ha of plantation had been established under these ten programs by the end of 1998 (SFA 2000).

The Ministry of Forestry delivered the Notification on the Experiments of Diversified Forestry Management Reform in 1996, to start the trials on diversified forestry management. Forests are classified into forests of commercial purpose and forests of public benefits according to different functions and uses, and managed according to their own specialties and rules. As a public cause, forests of public benefits, which have environmental development as their main purpose, will rely on government input and will be financed through various levels of the government as appropriate, while ecosystem services value of a forest will be actively explored in order to pursue the principle of "the one who benefits should foot the bill". Commercial forests, which focus on economic use, will be supported according to market economy rules of the best economic return through intensified management. Diversified management reform is a long-term reform of importance to all forestry development, the basis for fundamental change in economic regime and economic growth of forestry, as well as the objective of developing forest management practices in accordance with the associated industrial systems (SFA 2000).

Expanding Forestry Industrial System

Ever since the reform and opening-up, constant readjustment in forestry industrial structure has taken place along with the change in market demand and the condition of the forest resources. The forest management policy of "forest focused integrated utilization, diversified management, stereo-typed development and advancement in full scale" was put forward in 1986. After these principles were put forward, focal points and targets for forestry industrial restructuring were put forward in the measures by the Ministry of Forestry, in order to carry out the Decision on Key Issues regarding Current Industrial Policy of the State Council. The structural readjustment of traditional forestry industries such as timber production and processing, forestry chemistry industry and forestry machinery manufacturing was further accelerated and the support to developing new industries such as trees as cash crops, bamboo forest, flower, medicinal herb, forest food and forest tourism was strengthened. Efforts have been made to develop forest reserves, and to develop forestry processing industry and establish new industries while keeping a steady increase in timber production. A distinct improvement has been achieved in forestry industrial structure, and the industrial timber supply capacity is expanding. Progress has also been made in expanding plantation resources, forestry industries, diversifying the utilization of forest resources in forested regions, as well as tertiary industries, such as forest tourism. The gross product value of the forestry sector by 1998 had increased by 20.7 times over that of 1978 (SFA 2000).

Forest Resourced Protection and Management System Is Gradually Perfected

Improved systems of forest fire control, forest pest and disease control and forestry policy administration have been established. Upon the initiation of the National Headquarters for Forest Fire Control in 1987, a relatively complete forest fire control system was set up consisting of fire control centers in various local authorities, a professional firefighter team, a fire prevention network integrating communication, fire-watching, monitoring and fire breaks, and set of firefighting machines. Since this time, the level of fire damage in China has been maintained at less than 0.5 % except in one specific year, and is well below the world average of 1 %. Following the principle of prevention first and integrated management, an integrated forest pest and disease control system based on the idea of ecological control incorporated with biological control, bionic control and chemical control has been set up.

The success rate in forest pest and disease control rose from 48.9 % in 1978 to 71.2 % in 1998 (SFA 2000).

A wildlife conservation framework has also been initiated. Starting from ten in the early 1970s, the number of nature reserves had reached 926 in 1998. A series of programs were started by the State since the 1980s, including the Giant Panda Rescue Program, the Crested Ibis Rescue Program, the Chinese Alligator Protection and Development Program, the Breeding Program for Eld's Deer, the David's Deer Breeding and Naturalization Program, the Wild horse Breeding and Naturalization Program and Saiga Breeding Program. As part of these, there are now over 60 successful programs for breeding endangered wildlife species (SFA 2000).

A forest resource quota system was also introduced. The state began the implementation of the forest resources quota system in 1987, and started to transfer the quota system for commercial timber to a complete control on forest resources consumption in 1991.

Forest resources supervision agencies in the major forested provinces or autonomous regions and key forestry industrial enterprises were introduced. A forest resource supervision office was established and forest resources supervision commissioners were sent to Inner Mongolia, Jilin, Heilongjiang, Fujian, Yunnan province (autonomous regions) and Daxinganling Forestry Corporation (Group) since 1989. By the end of 1998, 110 commissioners and their subordinates were dispatched and the total number of staff involved hit 3,000 (SFA 2000).

Legislation and Enforcement to Strength Forestry Support
and Safeguard System

Legislative capacity building and law enforcement have made remarkable progress. In total four pieces of law and legislative document concerning forestry, 11 forestry administrative regulations, over 60 directives and some 200 local legislations and regulations have been issued during the period from the promulgation of the Forest Law of China (Trial Edition) in 1979 to the end of 1998. So far the forestry legal system was initiated with the Forest Law and the Wildlife Protection Law as the core, and incorporated a series of relevant laws, administrative regulations, sector rules and local laws and regulations. Legislative systems for forest resource protection and development, regarding forest land and forest tenure arrangements, afforestation and silviculture, regeneration of logged sites, logging quotas, timber transportation and inspection, wildlife import and export management and forest plant quarantine, etc., were established to give a powerful legal basis for forestry development. In law enforcement aspects also apply; forestry authorities in 30 provinces have set up forest public security agencies by the end of 1998, with a total 50,000 forest police staff. According to statistics, forest police have investigated 1,398,000 cases related to forest and wildlife, and penalized 2.275 million abusers, with 0.18 billion dollars direct economic loss retrieved (SFA 2000).

Extended service networks at the grass root level were also strengthened. A three-degree grass root service network of forestry technology station and seed and seedling nursery at the county level and forestry administration in towns have been established (SFA 2000).

Transformation of Forest Management Policy

Timber production has long been regarded as the guiding principle due to the need of timber supply for national economic development and the limitations in management policy since the founding of the People's Republic of China. Over-quota cutting has not only put forestry regions into the survival and development difficulties, but also caused a series of problems such as ecological deterioration. A growing focus on these issues at the end of 1970s finally led to a series of transformation trials in forestry development objectives and structure. The start of the Three North Shelterbelt System Program in 1978 marked the trial of a new approach of large-scale forestry environmental programs. Over the past 20 years, a series of forestry environmental programs have been rolled out nationwide with ten major forestry ecological programs as the core. The National Tree Planting Compulsory Campaign was initiated since 1981, and land greening activity was incorporated into legislation in order to encourage the participation of the general public. In 1993, the policy of "forestry development with Chinese character to increase resources, vitality and benefits so as to green, vitalize and enrich the nation" was put forward. In 1995, a diversified management strategy was raised, that set the development industrial system as the defined long-term objective. In 1997, under the call of "holding high the banner of ecological environment construction", the focal point of forestry's course was shifted to the protection and improvement of eco-environment. The 15th Committee of the CPC called for "tree-planting, soil and water conservation, combating desertification and eco-environment improvement" as the main content of cross-century development strategy. The 3rd Plenary Session of the 15th Committee of the CPC stressed again that "eco-environment improvement is a long-term plan associated with the subsistence and development of Chinese nation, and a basic approach in combating natural calamity such as draught and flood." After the severe flood occurred in the basin areas of the Yangtze, Songhua and Nenjiang Rivers, the Central Committee of the CPC and the State Council listed "mountain closure for forest regeneration and conversion of cropland for forest and grassland" as the major solution for reconstruction and flood control thereafter. Modern forestry development thinking targeting at the transformation of forest management concept, readjustment of forest management objectives and the transformation of forestry growth approach was gradually formed in this process (SFA 2000).

Advancing Forestry Institutional Reform

Decentralization marked the initiation of property rights reform. In March, 1981, the Central Committee of the CPC and the State Council issued Decisions of Several Points regarding Forest Conservation and Forestry Development, which government urged to "stabilize the present ownership structures of hills and forests, allocate hillsides to farmers for their private use, define the forestry production responsibility system", in order to grant the forest to farmers as the main body for management through awarding forest ownership certificate to the forest farmers. Ever since then, new approaches such as contract tree-planting by specialized households or joint ventures and cooperative tree-planting have been exploited for forestry development. The appearance of shareholding cooperative forest farms has further promoted the diversity of forest management entities. After the socialist market economy was set as the target for economic reform, new vigor was stimulated in the forestry market economy by auction of the "four barrens" and contract management. In state forest region forestry enterprises started the contract and responsibility system in 1982, and the director responsibility system in full scale and experiments on forest pricing were launched in 1985–1987. Based on the Provisional Ordinance on Contract and Responsibility System for Enterprises Owned by the Whole People, forestry enterprises put forward various contracting measures on the principle of "six Contracts and three link-ups" at large. By reconfirming the main body of state-run forestry enterprises in forestry management, this has successfully transformed the management structure through carrying out such systems as whole-process contract, series contract, etc. Following the request to "concentrate on the big ones and neglect the small ones" by the State so as to revitalize large and medium-sized state-run enterprises, state-run forestry enterprises in north-east China and Inner Mongolia were consolidated to four large forestry industrial groups, and thus became four of 56 pilot groups of the State for reorganization according to modern enterprise system. With regard to forestry market system development, the Central Committee of the CPC and the State Council issued in January 1985 the Ten Policy Measures Regarding Further Activating Rural Economy, which were also called the No. 1 document. Timber price controls were gradually relaxed in collective forest regions and timber market was liberated with certain limitations. Planned timber distribution was shrinking and self-trade was expanding on a yearly basis in state-owned forestry region. By 1998, China has 995 timber markets with a trading area of 16.217 million square meters and annual turnover of US$2.5 billion. Meanwhile, markets for forestry technology, information, forest insurance and mid and young-age stand transfer are forming in certain areas. Government plays an important role in regulating forestry development. The State has allocated substantial financial inputs to forestry ecological programs with public benefits, and has promulgated relevant policies regarding forestry industrial development. The forestry public service system and legal framework have also been strengthened (SFA 2000).

Main Problems of Forest Sector in Late 1990s

Since the reform and opening-out policy was adopted, China has made many effective efforts to improve its natural environment by providing extra investment, reforming forest management and so on in its forest sector. However, there were still some large problems existing in the late 1990s.

Clear cutting has caused the qualities of forests deteriorate. Since 1949, demand for timber has resulted in extensive cutting of forests, and timber harvests increased from 20 million cubic meter/year in the 1950s to 63 million cubic meter/year in the 1990s (MOF 1997). Government policy did not require that native tree species be replanted after logging, but promoted planting of fast-growing tree species, such as larch (*Larix* sp.), poplar (*Populus* sp.), and Chinese fir (*Cunninghamia lanceolata*). Although a large-scale increase of plantation-style forests in non-forested areas increased total forest coverage in China from 5.2 % in 1950 to 16.55 % by 1998 (Richardson 1990), natural forests declined to 30 % of the total forest area in China and unit-area stocking of natural forests decreased by 32 % (Zhang et al. 1999).

Land erosion was very serious. China's human population has increased about 2.5 times over the past 50 years, yet the human population in forested areas has increased fivefold (Zhang et al. 1999). Scientists foresaw the potential conflict between human population growth and forest resource use and began advocating changes in China's forest policy as early as the 1960s (Liu 1963). They had little success. In the 1970s, government policy limited clear-cut areas to <10 ha in northeastern China (Shao and Zhao 1998), but departures from this standard were routine in field conditions. During the 1990s, eroded lands continued to increase by >10,000 km^2 annually, with the result that 38 % of China's total land area is now considered badly eroded (Zhang 1999).

Environmental risks were at high level. The sharp decline in the quantity and quality of natural forests resulted in loss and fragmentation of natural habitats. At least 200 plant species have become extinct in China since the 1950s, and >61 % of wildlife species have suffered severe habitat losses (Li 1993). Valuable and rare species, such as ginseng (*Panax ginseng*), are threatened with extinction. Changes in forest composition have also caused severe ecological and environmental disasters. Insect infestations have damaged >9.3 million hectares of forests per year, causing >10 million meters of timber losses (Zhang et al. 1999). Flash flooding, in part the result of the loss of natural vegetative cover, caused a total loss of US$20 billion in the summer of 1998 alone (Qu 1999).

The quality of afforestation was not good enough. According to statistics of Ministry of Forestry, China afforested a cumulative total of 160 million hectare from 1949 to 1998. The forest coverage rate was supposed to be 30 % by 1998, but was just 16.55 % according to the data of the China's Fifth National Forest Resource Inventory (1994–1998) (Hu 2000).

2.2   The SFM Practice in China from 1998 Up to Now

In the late 1990s, the sustainable development strategy had been defined as one of the basic strategies of China, the drastic flood disasters which happened in 1998 and the frequent sand storms that happened every year in some areas of China, attracted the attention of the whole country. In 1997, the President of China called on all Chinese people to rebuild a harmonious motherland with beautiful mountains and purified rivers. A nation level strategy called the Great West Development Strategy, of which ecological protection was one of its major goals, was implemented in the next year. This was a milestone for China's forest conservation, and the whole society paid more attention to its environment. Since that time, China's forests have played a more and more important role in protecting and improving the natural environment and are facilitating the sustainable economic and social development of the country (SFA 2000). Since 1998, China has made tremendous investments in its forest sector for environmental issues. Over the same time period, a series of innovative policies have been brought forward to the forestry sector, to speed up the process of sustainable forest management and have had tremendous achievement.

Large Scale Plantation and Rehabilitation Projects

Starting in 1998, China has gradually integrated its existing ecological programs into six new large scale plantation and rehabilitation projects, in addition to the National Compulsory Tree Planting Campaign etc. Those new programs are known as six key forest programs, and include Natural Forest Protection Program (NFPP), Conversion of Cropland to Forest Program (CCFP), Sand Control Programs for areas in the vicinity of Beijing and Tianjin (SCP), Forest Industrial Base Development Program (FIBDP), Three North Shelterbelt Development Program and Shelterbelt Development Program along the Yangtze River Basin (3Ns&YRB) and Wildlife Conservation and Nature Reserves Development Program (WCNRDP).

The Natural Forest Protection Program-Strategic Move to Restore Natural Forest

The initial purpose of NFPP was to protect natural forests by implementing a logging ban in ecologically sensitive forests of the middle and upper branches of the Yangtze and Yellow Rivers, and by reducing the timber harvest in State-owned forest regions in northeast China and Inner Mongolia. These measures have reduced the timber supply from domestic forests. The program also requires State-owned forest enterprises and farms to cultivate commercial forests intensively, establish industrial forests, and cultivate precious species and large-diameter timber forests in order to maintain the timber supply.

The project involves 734 counties and 163 industry bureaus in 17 provinces, autonomous regions or municipalities directly under the Central Government in the upper reaches of the Yangtze River, upper and middle reaches of the Yellow River, Northeast China, Inner Mongolia and other key state-owned forest zones. It is designed to prohibit and alleviate commercial cutting to help the forests with restoration and sustained growth.

A total of 14.11 million hectares of natural forest had been established by the end of 2011, including 2.97 million hectares of artificial plantation and 3.35 million hectares of aerial seeding plantings, and 7.79 million hectares of newly protected mountain areas for natural regeneration. The forest area under management and protection has now reached 115.96 million hectares (DPFM SFA 2012).

Natural forest in Northeastern China (Credit: Jianwei Chen)

Conversion of Cropland to Forest Program (CCFP)-Fundamental Solution to
Water Loss and Soil Erosion

CCFP was being carried out in 25 provinces, including 1897 counties. Most of these
areas were in western and central China; the aim being to reduce soil erosion in
critical areas and then to improve the ecological environment and enhance local
people's timber supply in the long term. Natural forest accounts for 80 % of the
converted lands. The plan is to return 14.7 million hectares of farmland to forests
and afforest 17.3 million hectares of barren hills and other suitable waste land
between 1999 and 2010.

A total of 26.87 million hectares of forest had been established by the end of
2011, 9.06 million hectares of which was planted on converted cropland and 15.34
million hectares were planted on barren hills and waste land. An additional cumulative 2.47 million hectares of mountain (sandy land) has been protected for natural
regeneration. Eighty percent of the plantation established on converted cropland
was natural forest (DPFM SFA 2012).

A project site of Conversion of Cropland to Forest (Credit: Jianwei Chen)

Sand Control Program for Areas in the Vicinity of Beijing and Tianjin (SCP)-
Green Shelters of Beijing and Tianjin

The SCP program covers 75 counties in five provinces (autonomous regions and municipal cities) with an area of 460,000 km². Its aim is to reduce the hazard of sandstorms in areas surrounding Beijing. It is planned to return 2.63 million hectares of farmland to forests, afforest 4.94 million hectares of land, develop 10.63 million hectares of grassland, build 113,800 supporting water conservation facilities, regulate 23,000 km of drainage areas and resettle 180,000 people for environmental improvement purposes between 2001 and 2010. This program requires inputs equivalent to about US$6.8 billion.

From the beginning of 2000 to 2010, a total of 4.02 million hectares plantation was established by artificial plantation and by aerial seeding. An additional 2.29 million hectares of mountain (sandy land) was protected for natural regeneration. The treatment area totaled 9.67 million hectares, including 2.35 million hectares of grassland and 1 million hectares of watershed areas (DPFM SFA 2012).

Three North Shelterbelt Development Program and Shelterbelt Development Program Along the Yangtze River Basin (3Ns&YRB) (Fourth Stage of the Project): Rehabilitate the Degraded and Desertified Land

The aim of this program is to rehabilitate the degraded and desertified land. It is planned to afforest 9.46 million hectares of land and rehabilitate 1.3 million hectares of desertified land between 2001 and 2010. By program completion, the forest cover in the program area will be increased by 1.84 %, and 12.66 million hectares of desertified, salinized, and degraded grasslands will have been protected and rehabilitated. In the lower middle reaches of the Yangtze River, it is planned to afforest 18 million hectares of land, to improve 7.33 million hectares of low-efficiency shelterbelts, and to regulate and protect 37.33 million hectares of existing forests during the period in 2001–2010.

From 2001 to 2011, 12.40 million hectares were afforested including 7.62 million hectares of artificial plantation, 0.29 million hectares of plantation completed by aerial seeding and 4.49 million hectares of mountains (sandy land) was enclosed for natural regeneration. The targets of 3Ns&YRB have been completed successfully (DPFM SFA 2012).

Forest shelterbelts of roads, farmlands and city (Credit: Jianwei Chen)

Wildlife Conservation and Nature Reserves Development Program (WCNRDP)-Rescue Action for Wild Species

The main goal is to increase the conservation of critical species. Priorities are being given to three projects between 2001 and 2010. The first involves completing 15 wild fauna and flora protection projects, including those for the giant panda, golden monkey, Tibetan antelope, and plants in the orchid family. The second involves completing 200 nature reserve projects in forests; land affected by desertification, and wetland ecosystems, 32 wetland conservation and comprehensive utilization demonstration projects, and 50,000 nature reserve districts. The third project involves completing the germplasm pools for wild fauna and flora conservation, the wild fauna and flora national research system and relevant monitoring networks. By 2010, there should be 1,800 nature reserves, including 220 state-level ones, covering 16 % of China's total land area [i.e., double the figure suggested by (S1)]. The protection of these reserves is enforced, but some economic activities are permitted (such as the harvesting of bamboo).

By the end of 2011, the nature reserves in forestry sector amounted to 2126 with a protected area of 123 million hectares, accounting for 12.78 % of the national terrestrial area. Included in which, there are 263 national nature reserves with an area of 76.29 million hectares, with 90 % of the terrestrial ecosystem types, 85 % of wild animal populations and 65 % of the higher plant community currently under effective protection (DPFM SFA 2012).

Aerial view to nature reserve in Xishuangbanna, Yunnan Province, China (Credit: Jianwei Chen)

Fast-Growing and High-Yielding Forest Project (FHFP)-Major Approach to Timber Supply

In order to ease the shortage of timber supply and reduce the pressure of timber demands on forest resources, the FHFP program has been implemented since 2001. The project covers 886 counties (cities or districts) in 18 provinces (autonomous regions) such as Hebei, Guangxi, Hainan and 114 forestry bureaus (farms). It is planned to establish 13 million hectares of fast-growing, high-yield timber plantations in three phases between 2001 and 2015.

By the end of 2011, at total of 1.872 million ha of FHFP forests have been planted and the timber industrial belt for national industrial raw materials has taken shape, which has played an important role in increasing the supply of domestic timber.

In addition to those six key forest projects descript above, during the same period of time, China also introduced some new projects, such as wetland protection project, comprehensive control project of stony desertification, and wild fauna and flora protection and nature reserve projects etc., as well as continuing to pursue previous projects, and afforestation projects in plains areas as well. Those projects with much higher investment standards have speeded up the process of forest area increase and vegetation rehabilitation (DPFM SFA 2011).

Strict Regulations on Forest Resources Management

The SFA has also implemented many regulations to manage forest resources in compliance with the requirements of the Forest Law of China and other relevant Chinese laws.

Forest Land Protection and Management

To strengthen the forest land management, the Forest Law of China stipulates that forest land shall not be occupied or a reduced amount of forest land shall be occupied when reconnaissance, mining or development projects are carried out. If the occupation or requisition of forest land is needed, it must be reviewed and approved by the forestry department in charge under the people's government above county level, in accordance with the law and regulations relevant to land management. The land user must pay a forest vegetation restoration fee, a compensation fee for forest land and trees and settlement allowance. In 2010, the Chinese government promulgated the "Guidelines on Planning of Protection and Use of Forest Land in China". In compliance with the principles of "control of total amount, quota management, economizing on the use of land and balance between occupation and compensation", it was decided that the total area for occupation and requisition of forest land in the country from 2011 to 2020 shall be limited to less than 1.055 million hectares. The reserved forest land will increase from 303.78 million hectares to 312 million hectares, up to 32.5 % of the total land area of the country, of which the total forest area will increase from 195 million hectares to 223 million hectares. The forest coverage shall reach above 23 % and the total forest stocking volume shall increase to 15.8 billion cubic meters (SFA 2010). It is beholden upon these main measures to prevent forest lands being converted into other uses.

Forest Utilization Management

Enhancing forest resource protection has been used as one of the main ways to achieve sustainable forestry development in China. The Chinese government sets quotas for forest logging operations so as to bring resource consumption under strict control. The Government has formulated and promulgated relevant laws, regulations and policies which strictly enforce logging quotas as the central element of forest management and also set quotas for forest resource consumption. China's forest utilization management system mainly comprises cutting quotas, license-based harvesting, license-based transport and license-based management (processing) of timber. The cutting quota scheme is the core of China's forest utilization management. Following the principle of consumption of timber forest being less than growth, the government establishes a national cutting quota every 5 years. License-based harvesting is referred to as the cutting license and has to be applied for prior to the commencement of harvesting operations, which is the legal certificate for carrying out cutting activities. License-based transport means that with the license the timber can be transported out of the forest area. The transport license is reviewed and issued by the forestry department in charge. There are currently more than 4,000 timber check points across the country, responsible for checking timber transport. License-based management/processing of timber means that such a license must be applied for if the management/processing of timber are conducted in the forest area. The implementation of the above-mentioned measures has played an important role in deterring the over-consumption of forest resources, continuously increasing forest area and stocking volume and ensuring ample supply of forest products (Wang 2011).

Forest Resources Supervision Management

In order to effectively control forest consumption and protect the existing resources according to the principles of forest laws and regulations relevant to forest resource protection and management, China introduced a forest resources supervision management system in 1989. The national headquarter of forest resources supervision was established in 2003 at SFA. Since 1989, the SFA has set up 15 forest resources supervision sub-agencies across the country to conduct supervision covering the whole territory except Hong Kong, Macao and Taiwan. Those agencies have exercised supervision over the forest resources management in their individual regions. At the same time, they have set up such supervision systems as case handling, harvest quota checking, occupation and requisition of forest land checking and checking of forest silviculture, which has greatly strengthened the enforcement of the laws and regulations relevant to forest resource protection and management (Wang 2011).

National Forest Inventory System (NFIs)

In order to get continuous real time information about forest resource changes nation-wide, the National Forest Inventory System was initiated in China in the early 1950s. With several decades' of effort and development, the NFIs have been generally established. NFI is organized by the central government (referred to as the 1st class inventory). The NFI is conducted at five yearly intervals with the province (autonomous region, municipality) as a population. For the NFI, permanent sample plots and remote sensing sample plots are systematically laid out. Through field measurement of those sample plots, data on current state, consumption and growth of forest resources have been collected. The 8th NFI is now under way. Starting from the 7th NFI, macro-level assessment of forest functions has been added. Such an assessment has been performed for the first time for the natural functions and benefits of forests across the country. Biomass models are built for 100 species of trees, which has enabled the monitoring of forest carbon sequestration in the country (Wang 2011).

Establish Motivations Mechanism to Protect Forest Resources

In addition to the strict laws and regulations on forest resources protection, China has introduced a series of incentive mechanisms to encourage people to protect the forest resources.

Grant Forest Property Right to Farmers

Ownership of forest land in China has two forms: state ownership and collective ownership, while the actual forest has three forms: state ownership, collective ownership and individual ownership. In order to make the collective ownership of forest and forest land closer to the farmers, the central committee of the Chinese Communist Party and the State Council decided to carry out a comprehensive reform in the system of collective ownership of forest and forest land in 2008. The core of this reform is the premise of keeping the ownership of collective forest land unchanged and in compliance with laws, but to contract the management of the forest land and the ownership of the forest to the farmers of the same collective organization by means of the individual farmers who have requested the managerial rights over the forest land. By the end of 2011, the area of the collective-owned forest that had clarified their property and contracted its management to individual households had reached 173 million hectares, accounting for 95 % of the total area of 183 million hectares of the collective-owned forest, over 83.79 million farmers had obtained the certificate of forest and forest tenure rights and more than 300 million had gained direct benefits from this reform. The reform has inspired the enthusiasm of the farmers in afforestation and the protection of forest and has facilitated the process of democratizing the management of the rural communities (Jia 2011).

Established Ecological Benefit Compensation

According to the different intended purpose of the forests, China has divided the forest into two categories: one is public welfare forest land focusing on achieving ecological benefits and the other is commercial forest land oriented to providing wood products. Each type of forest adopts different policy measures and different modes of operation and management. The purpose of the public welfare forest is to achieve ecological benefits and commercial logging is not allowed. In order to compensate the managers of the public welfare forest, the central government started to explore the building of public welfare forest compensation mechanism in 2001, and in 2004 officially set up the forest ecological benefit compensation fund supported by the central budget to grant economic compensation to the 83.93 million hectares of State-level public welfare forests from the central budget. By 2011 the State-owned welfare forests were being granted RMB 75 Yuan/hectare/year and the collectively and individually owned ones were granted RMB 150 Yuan/hectare/year. In recent years, according to their own economic conditions, the local governments at all levels have set up local compensation funds, developed and made innovations to this mechanism. Beijing municipality, for example, has set up development funds since 2010 to compensate the ecological benefits of the mountainous public welfare forests, with the standard of RMB 600 Yuan/hectare/year, ranking as the highest in the whole nation. Besides, it is stated that the standard should be increased once every 5 years.

Meanwhile, a financial subsidy system for the forestry sector has also been established and upgraded. The central financial subsidy system for afforestation and forest tending has been upgraded, and more effort has been taken to gradually expand the pilot area and enhance subsidy support for key wetland protection. Improved variety subsidy system has been set up to subsidize key tree breeding farms and the production of improved seedlings. The financial policy has been instituted to deepen collective forest tenure reform and promote understory economic development. More supports have been given to the development of woody oil industry like oil-tea camellia and walnut and forestry bio-energy as well as to the monitoring and control of harmful organisms. Financial and taxation policy support to forestry has been improved as well. These measures have sped up China's forest based sustainable development (Zhang 2011).

Achievements of China's Sustainable Forest Management

China has made tremendous achievements in its forest sector in term of sustainable forest management, especially over the last 10 years. According to "Global Forest Resources Assessment 2005" published by UN FAO, China's forest area accounts for 4.95 % of the world's total, ranking the fifth behind Russia, Brazil, Canada and the United States of America. China ranks the sixth in the world in terms of forest stock volume, following Brazil, Russia, Canada, the United States of America and Democratic Republic of Congo (FAO 2005). The total area of plantations ranks

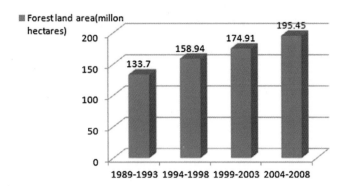

Fig. 1 Changes of forest land area in China

China the first in the world. "State of the World's Forests 2011" published by UN-FAO reiterated the achievements of China's forest sector: In the Asia and the Pacific region as a whole, forests were lost at a rate of 0.7 million hectares per year in the 1990s, but grew by 1.4 million hectares per year over the period of 2000–2010. This was primarily due to large-scale afforestation efforts in China, where the forest area increased by 2 million hectares per year in the 1990s and by an average of 3 million hectares per year since 2000. Planted forests (i.e. forests established through planting and/or deliberate seeding of native or introduced tree species) made up 16 % of the forest area in the region. Planted forests experienced a substantial increase within the last 10 years in the Asia and the Pacific region. Most of the region's planted forests were established through afforestation programs. China contributed the bulk of this growth through several large programs that aimed to expand its forest resources and protect watersheds, control soil erosion and desertification, and maintain biodiversity (FAO 2011). Compared to the results of the 6th NFI (1999–2003), the results of the 7th NFI indicate the forest characters as main achievements of the SFM China had in the last 10 years were the following six aspects (SFA 2009a).

Achieved Continuous Growth of Both Forest Land Area and Forest Stocking Volume

Since the 1990s, China has achieved growth in both forest land area and forest stocking volume. The 7th NFI (2004–2008) results show that the forest land area in the country has reached 195 million hectares, forest coverage 20.36 %, forest stocking volume 13.721 billion cubic meters, reserved man-made forest area 62 million hectares with a stocking volume of 1.961 billion cubic meters. The results of 7th NFI indicate that China's forest resource has entered a period of rapid development.

The net increase of the forest area is 20.5430 million hectares, and the forest cover has gone up from 18.21 % to 20.36 %, an increase of 2.15 %. The net increase of the forest standing volume is 1.128 billion cubic meters and that of the forest stock volume is 1.123 billion cubic meters (Figs. 1 and 2).

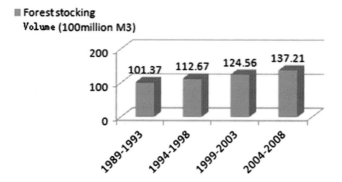

Fig. 2 Changes of forest stocking volume (Source: SFA 2009a)

The Area and Stock Volume of the Natural Forests Increase Significantly, Especially in the Project Areas of the Natural Forest Protection Program

The net increase of the natural forest area is 3.9305 million hectares and the natural forest stock volume is 676 million cubic meters. In the project areas of the NFPP, the net increase of the forest area in the 7th NFI is 26.37 % more than that in the 6th NFI, and the net increase of the stock volume in the 7th NFI is 2.23 times above that in the 6th NFI.

The Plantation Area and Stock Volume Grow Quickly and the Reserved Forest Resources Show a Trend of Increase

The net increase of plantation area is 8.4311 ha with the stock volume of 447 million cubic meters. The area of young and immature plantation is 10.4618 million hectares, of which, the area of commercial tree species is 6.3701 million hectares, an increase of 30.17 % compared with that in the 6th NFI.

The Growth Rate of the Forest Stock Volume Increment Is High and the Logging Is Being Gradually Transferred from Natural Forests to Plantations

The annual net increment of the forest stock volume is now 572 million cubic meters, the annual logging for consumption is 379 million cubic meters, and the increment is now also continuously larger than the consumption, and the surplus between increment and consumption is continuing to expand. The logging of natural forests is decreasing and that of plantation is increasing, now delivering 39.44 % of the total, a rise of 12.27 %.

The Forest Quality Has Been Improved Slightly and the Forest Ecological
Function Increased Continuously

The stock volume of the commercial forests has increased by 1.15 m³/ha with the
average increment of 0.30 m³/ha and the proportion of the mixed forests up by
9.17 %. Among the forested land, the ratio of the public benefit forests is 52.41 %
of the total, an increase of 15.64 %. The forest ecological value has been further
enhanced due to the increase of forest resources and the improvement of forest
structure and quality.

The Proportion of the Forest Area Managed by the Individuals Is Rising, and
the Achievement Obtained by the Reform of the Collective Forest Tenure
System Is Significant

Among the existing forest land, the proportion of the forest area managed by individuals is 32.08 % of the total, a rise of 11.39 %. The area of plantations managed
by the individuals is 59.21 % of the total, and the area of young and immature plantations managed by individuals is 68.51 %. The farmers, who are the main body of
forest managers, have become the backbone of forestry development in China.

In addition to these achievements by China, its forestry industry has also developed steadily during the same period. There are two significant developments in
China's forestry sector in the last decade. Firstly, it has transformed from a mainly
primary industry into a compound sector combining the primary, secondary and tertiary industries. On the basis of a steady output of timber, wood-based panels, rosin,
furniture, economic forest and other traditional products, non-timber-based industries such as bamboo, rattan and flower products, forest tourism, forest food and
medicine have seen rapid growth. New industrial sectors such as wildlife breeding
and development, biomass energy and bio-materials are growing with unabated
momentum. Secondly, the sector has grown steadily larger. In 2011, the forestry sector had created an annual value of over 486 billion US dollars (at current values) and
forestry-related trade reached more than 90 billion US dollars, which is more than
double and quadruple their respective values in 2000 after adjustment for inflation.
Rosin, wood-based panels and furniture output of China is now second to none in the
world, while forest tourism and woody cereal- and oil-yielding plants are also growing rapidly, playing a major role in creating jobs and added value, promoting regional
economic prosperity and ensuring forestry products supply(DPFM SFA 2012).

The progress described above show that China's forest resources have entered
a period of rapid development. The key forestry development programs have
been progressing steadily, the total amount of forest resources has been increasing continuously and the multiple functions and benefits have manifested themselves gradually. The supply capacity of timber, forest products, ecological
products and ecological cultural products has also been strengthened. This has
laid a solid foundation for the sustainable development of China's forestry sector
far into the future.

3 Challenges and Strategies for China's SFM

Currently, China's forestry organizations shoulder important functions of developing and protecting forest ecosystems, managing and rehabilitating wetlands ecosystem, improving and treating desert ecosystem and protecting and developing biodiversity, as well as ensuring the supply of forest products. Thus forestry has become an important sector to China's economic, social and environmental development. The main challenges China's forestry sector face must be dealt with comprehensive strategies in the future.

3.1 Main Challenges to China's Forestry

China's current forest resources face the following main issues.

The Forest Resources are Limited

The forest coverage rate in China is just one third of the world's average; its per-capita forest area of 0.145 ha is only one fourth of the world's average and the per-capita forest stocking volume of 10.151 m³ is less than one seventh of the world's average (SFA 2009a).

The Area of Forest Is More Difficult to Increase

The land area suitable for planting trees with good conditions is only 13.21 %, while poor conditions account for 52 %. Among those suitable lands for plantation, 60 % of them located in Inner-Mongolia and the Northwestern China with the annual perception less than 400 mm, which makes it difficult for trees to survive. More efforts must be put into the same tasks of plantation than ever before (SFA 2009a).

Forest Quality Is Comparatively Low

China's forest per unit stock volume is 85.88 m³/ha, is approximately 78 % of the world average of 114 m³/ha, with the plantations forests being 49.01 m³ which is still very low by world average level The average annual stock volume growth (or increment) of China's forests is 3.85 m³/ha, average of canopy closure is 0.56 and the average forest stand diameter is only 13.3 cm (SFA 2009a).

The Structure of Forests Is Not Rational

The proportion of pure plantation forest is 63 %, of which the proportion of pure forest planted with commercial species is as high as 87 %. In terms of age class structure, the young and semi-mature forests account around 60 % of the existing forest stand, and majority of them belong to the plantations and secondary forest with low biomass (SFA 2009a).

The Pressure on Forest Land Protection Is Getting Tight

During the 7th NFI period, 8.3173 million hectares land assigned for forest use was found to have been converted into non-forest land; including 3.77 million hectares of actual forest. The conversion of forests into other land uses without approval still occurs in certain areas, and the occupation of forest land has increased in some areas by unauthorized reclamation. The demand for land in China will continue to increase, since China's urbanization and industrialization is still underway (SFA 2009a).

The Gap Between Forest Products Demand and Supply Remains

China's rapid economic growth, increased capital investment, and growing middle-class consumption have driven up the demand for wood products. China not only needs wood to meet domestic demand, it also has a growing and very successful export industry. In 2009, the import volume of wooden forest products in China was equivalent to 184.3662 million cubic meters of timber. There is also an immediate shortfall in supply of about 2 million cubic meters of timber and this will not be solved in a short term. Meanwhile, there are not that many investors interested in the FHFP. Ten years passed which is about two third of the project period of time, but less than one sixth of the plantation of the FHFP target has been met so far (Wang 2011).

3.2   Targets and Strategies of China's Sustainable Forest Management

The Chinese government has long attached great importance to forest development and protection. Since the global community has reached the consensus that forests can provide a major mitigation and adaptation function to combat climate change, as the international negotiation process of United Nations Framework Convention on Climate Change (UNFCCC) moves on. The Chinese government has paid more efforts to develop forestry and improve ecological environment through forest measures. In September 2009 at the United Nation's Climate Change Summit, President Hu Jintao announced to the world community that "China will greatly increase its forest carbon sink by increasing 40 million hectares of forest area and 1.3 billion cubic meters of forest stock volume by 2020 from that of 2005 level" (hereafter referred to as Double Increases). In November 2009, the "Double Increases" became one of the three voluntary commitments made by the Chinese Government in controlling the greenhouse gas emission. According to the "Double Increases" and other planning related to China's forestry development, the following targets have been laid out.

Objectives of the 12th Five-Year (2011–2015) Forestry Plan

In order to reach the "Double Increases" targets by the end of the year of 2015, the main targets of China's forestry development are as follows:

From 2011 to 2015, the area of new plantations will be 30 million hectares, and 35 million hectares of other intensively managed forests. At the same period of time, 12 billion trees will be planted through National Compulsory Tree Planting Campaign. At the end of the 12th Five-Year Plan, the forest coverage will be 21.66 % with a forest stock volume above of 14.3 billion cubic meters. The carbon storage of forest vegetation is expected to be 8.4 billion tons. The sustainable management of key regional ecological resources can obtain remarkable results, not to mention that the forest industries output value will reach 3.5 trillion RMB Yuan (550bn US$) and the forestry industrial structure and productivity layout will be more reasonable (SFA 2011).

Objectives of the 13th Five-Year (2016–2020) Forestry Plan

The area of annual afforestation (including protected areas such as mountains set aside for natural regeneration) will be over 5 million hectares, with the forest coverage of 23 % and the forest stock volume of 14 billion cubic meters. The area of desertified lands coming under active management will be over 50 % of those suitable for control. About 110 million hectares of national key public welfare forests will be effectively protected. More than 60 % of natural wetlands will be taken under sound protection. The "Double Increases" will be completed successfully. At that time, the overall functions of forest ecosystem to store carbon will be further strengthened, and the capability of forests to sequestrate carbon will be further improved (SFA 2009b).

3.3 Strategies of China's SFM

As mentioned previously, however, China's forest sector still faces many challenges. In order to reach its "Double Increases" on time and close the gaps between ecosystem service functions and the demand for commercial forestry products generated by industry and the whole society, China's forestry sector will have to adopt some comprehensive strategies to further develop its sustainable forest management.

The first of these will have to be to increase the intensity of the tree planting and afforestation to achieve an efficient afforestation area expansion.

Implementing the existing national key forest ecological programs continue in order to construct China's north green ecological shelter belt and the green ecological shelter belt in coastal areas. With the dissemination of "Increase Green Around You" action, compulsory tree planting for all citizens will be deepened and the gross forest areas will thus be rapidly expanded. At the same time the forest silviculture will be strengthened to ensure "plant one seedling, one survivor and one future tree crop". By the year 2020, China's forest coverage is expected to have increased from present 20.36 % to 23 % as the twin targets of the "Double Increases" are accomplished.

The second is to do a good job in silviculture to improve the forest quality of standing crop.

The improvement of standing crop quality can be realized by the way of adjusting the structure of forest stand, intensifying silviculture to increase per unit standing volume and the increment volume, improving forest stand quality, maximizing forest land productivity.

The third is to intensify the protection of forest resources and prevent forest degradation.

Such measures are needed to prohibit various kinds of deforestation activities such as unauthorized use and expropriation of forest land and to control the conversion of forest land to non-forest land. Efforts to conserve the vegetation and the soil, timber logging operational procedures will be updated to reduce the damage done to the ground cover and forest soils from logging operations. Early warning systems for forest fire and forest disease and pest prevention shall be established to effectively limit the occurrences and areas affected by forest fires as well as pests and diseases. Integrated measures such as these will insure China's forest resources against degradation and guarantee the efficiency of forest carbon sequestration.

The fourth will be to expand the timber utilization and develop the 'green economy'.

In order to expand timber utilization, reform is needed of the timber logging license approval procedure and accelerate the establishment of industrial plantation, promote the incentive mechanism of timber processing and formulate subsidy policy for using timber to replace energy-intensive materials, broaden the forest management and timber use market, promote the "use of timber instead of plastic, iron or steel", by which green economy will be developed, and green employment promoted, while forest carbon stocks are increased.

The fifth strategy needs to actuate all approaches to mobilize the whole society's involvement in forestry development.

That the main investor is currently always the government need to change. Various investment and financing mechanisms that match the collective forest tenure reforms are needed, along with changes in insurance instruments, and the subsidy and allowances that mobilize farmers' initiative in afforestation investment also need to be upgraded. China's eco-system services market will be fostered, especially for forests and the carbon trade, and flourish by scientifically operating the forest carbon mechanism, to let the broad base of forest farmers become the beneficiaries of collective forest tenure reform. The reform of state-owned forestry enterprises also need to be put on the government's agenda to improve the living standard of people living in forest communities and protect the precious natural forest resources. Such innovative policies should attract more private investors to the forest industry and so close the gap between forest products supply and the demand from the society.

Some things are easier said than done, however, and. China's forest issues can't be solved by forestry sector alone. Forestry sector definitely needs cooperation from other sectors even the whole society's support to reach its comprehensive goals. We also hope to share other countries' experience of SFM.

References

Cao Y (2000) Fifty years of China forestry—history. Forestry China 8:4–5 (in Chinese)

Division of Development Planning and Funds Management (DPFM) of SFA. Statistics analysis report on National Forestry in 2010. Forestry Economics No. 5 May 2011(No. 226), Beijing, p. 46

Division of Development Planning and Funds Management (DPFM) of SFA. Statistics analysis report on National Forestry in 2011 (in Chinese). State Forestry Administration. May 15, 2012. http://www.forestry.gov.cn/uploadfile/main/2012-5/file/2012-5-15-374ecfc1461842f685d6db 6b9d47fea1.pdf

FAO (2005) Forestry Paper 147: Global forest resources assessment 2005.Progress towards sustainable forest management. P13-14 Chapter 2. Food and Agriculture Organization of the United Nations, Rome, 2005 http://www.fao.org/docrep/008/a0400e/a0400e00.htm

FAO (2011) State of the world's forests 2011. Food and Agriculture Organization of the United Nations, Rome, 2011. http://www.fao.org/docrep/013/i2000e/i2000e00.htm

FAO (2012) Towards sustainable forest management. Springer, Berlin, http://www.fao.org/forestry/sfm/en/

Hu W (2000) Fifty years of China forestry—theory. Forestry China 8:6–9 (in Chinese)

Li W (1993) China's biodiversity. China's Scientific Publishing House, Beijing p. 168 (in Chinese)

Liu S (1963) Selected-cutting and natural regeneration of Korean Pine forests. Forestry of China 10:8–10 (in Chinese)

Ministry of Forestry (MOF) (1997) China's forestry yearbook. China Forestry Publishing House, Beijing, 1997 (in Chinese)

Qu G (1999) Environmental protection knowledge. China's Red Flag Publishing House, Beijing, p 354 (in Chinese)

Richardson SD (1990) Forest and forestry in China. Island Press, Washington, DC

SFA (2000) 2000 China forestry development report. State forestry administration. China Forestry Publishing Housing, Beijing, 2000.12

SFA (2009a) Forest resources in China-The 7th national forest inventory. State Forestry Administration, Beijing, November 17, 2009 http://www.forestry.gov.cn/ZhuantiAction.do?dispatch=index&name=slzyqc7

SFA (2009b) The forestry action plan to address climate Chang (in Chinese and English). State Forestry Administration, P.R. China. China Forestry Publishing House, Beijing, November 6, 2009

SFA (2010) Guidelines on planning of protection and use of forest land in China (2010–2020). State Forestry Administration. Aug.25, 2010. http://www.forestry.gov.cn/uploadfile/main/2010-8/file/2010-8-25-782d45dbdeea41398ff31b1023814c13.pdf

SFA (2011) 12th Five-year (2011–2015) forestry plan (Abstract) (in Chinese). State Forestry Administration, P.R. China. November 1, 2011 http://www.forestry.gov.cn/

Shao G, Zhao G (1998) Protecting versus harvesting of old-growth forests on the Changbai Mountain (China and North Korea): a remote sensing application. Nat Areas J 18:334–341

Statistics China (2012) Statistical communiqué of the People's Republic of China on the 2011 national economic and social development. National bureau of statistics of China, Beijing, February 22, 2012. http://www.stats.gov.cn/english/newsandcomingevents/t20120222_402786587.htm

Wang X (2011) Forest resources management in China. In: Conference report on the Second Asia-Pacific Forestry Week, Beijing, 2011

Wen J (2009) Speech on central forestry conference on June 22, 2009. People Newsnet. http://cpc.people.com.cn/GB/64093/64094/9530784.html

Zhang P (1999) Discussion on the national programme for natural forests conservation. Sci Silvae Sin 35:124 (in Chinese)

Zhang Y (2011) Forestry development and international trade in forest product in China. The second Asia-Pacific Forestry Week, Beijing. Nov 9, 2011

Zhang P, Zhou X, Wang F (1999) Introduction to natural forest conservation program. China's Forestry Publishing House, Beijing, p 388 (in Chinese)

Zhibang J (2011) To enhance the multi-functions of forest to promote green growth and continuously push forward the construction of modern forestry (in Chinese). Speech on the Conference of Directors of the National Forestry Department, Beijing. December 29, 2011

Zhou S (2000) Preface. 2000 China forestry development report. State forestry administration. China Forestry Publishing Housing, Beijing, 2000.12

The Value of Forest: An Ecological Economic Examination of Forest People's Perspective

Debal Deb

Abstract While a comprehensive economic valuation of all use and non-use values of the forest is impossible, indigenous societies seem to have a clear, albeit inchoate idea of the value of the forest on which they depend for their material and cultural existence. Forests were valued in all ancient civilizations, and often carefully preserved for subsistence as well as esoteric uses. Following the rise of capitalism, governments in Europe and her colonies considered forests first as wastelands, and then a valuable resource for economic development, and abrogated the customary rights of indigenous forest villagers. All governments of ex-colonies have passed laws to conserve forests as national assets, but often consider them as an obstacle to economic prosperity, whenever profits from industrial land use appear to exceed the instrumental value of the forest. Throughout this cycle of the loss and gain of economic importance of the forest, indigenous people and their perspectives are pushed into oblivion.

Indigenous forest people consider the forest's existence value to be as important as its use value, and as the bedrock of their cultural and political identity. Bereft of ownership and management rights to the state-owned forest, indigenous villagers have created their own forests on their private and community lands – both as "non-forest" vegetations for biomass removal, and as sacred groves, which uphold the non-use value of the land. Several tree species are planted and maintained along roadside, at home gardens and in sacred groves, regardless of their use values. Many rare and endangered trees that have disappeared from the state forest now exist only in these folk forests. These "worthless" trees and forests highlight the indigenous ecological economic perspective, in which the cultural significance of the forest transcends its instrumental value. This perspective of the value of the forest underlies the cultural-political motive for forest conservation, in opposition to the profit motive of industry and the development agenda of the state.

D. Deb (✉)
Centre for Interdisciplinary Studies, Barrackpore, Kolkata, West Bengal 700123, India
e-mail: info@cintdis.org

T. Fenning (ed.), *Challenges and Opportunities for the World's Forests in the 21st Century*, Forestry Sciences 81, DOI 10.1007/978-94-007-7076-8_7,
© Springer Science+Business Media Dordrecht 2014

1   Introduction

The role of biodiversity in economic performance and the impact of economic activities on biodiversity – both used to be external to the economist's realm of interest. Eager to draw attention of economists to the importance of biodiversity, many biologists have highlighted the economic value of biodiversity. Their assumption was that if people realize the value of biodiversity, they would likely preserve it for long term direct benefits. E. O. Wilson, among the most influential biologists, pointed out that biodiversity is a huge stock of materials that we can mine to obtain food, medicines, raw materials for industry, and make money. Why should the destruction of biodiversity concern us? In addition to the loss of much of its current utility, "vast potential biological wealth will be destroyed. Still undeveloped medicines, crops, pharmaceuticals, timber, fibers, pulp, soil-restoring vegetation, petroleum substitutes, and other products and amenities will never come to light" (Wilson 1992: 331). All this translates into the economic value of biodiversity from its potential future uses.

The appeal for biodiversity conservation based on its lucrative, and mostly untapped, commercial value led to the launching of several projects of biodiversity prospecting – the search for new commercial products among naturally occurring organisms – as both a mechanism and an argument for conserving biodiversity in tropical rainforests (Wilson 1992; Reid et al. 1993). Such bio-prospecting projects have generated controversy over, and interest in, intellectual property rights of indigenous peoples and nation states over biodiversity (ETC. Group 1994; Medaglia 2007). However, the value of biodiversity *per se* – beyond the obvious instrumental value of commercialized plants and animals – scarcely made a topic of mainstream economic discourse, because nature and non-instrumental value of life lay beyond the scope of neoclassical economics. "If Judeo-Christian monotheism took nature out of religion, Anglo-American economists (after about 1880) took nature out of economics" (McNeill 2000: 336). It is only after the emergence of 'ecological economics' as a discipline that nature seems to get accommodated in mainstream economic literature, albeit with considerable econometric inconveniences.

Over the past few decades, economic valuation of forest ecosystems and their 'services' has become one of the major exercises of economists in order to convert biodiversity into a new capital, the 'natural capital'. The forest is already a major source of revenue for the state and profit for industries, drawing on forest goods (e.g. timber and various non-timber forest products [NTFP]) and services (e.g. hunting, tourism). In the current environmental economic approach, the object of forest valuation is chiefly to internalize the cost of forest loss in the price structure which, economists expect, will likely entail conservation.

In contrast, the cultural-ecological perspective recognizes the non-market, non-consumptive, and non-use values of forest, transcending commercial returns (Hanley and Milne 1996). From this perspective, sustainable forest use conducing conservation is based on people's perception of the value of the forest, but the depth and extent of that perception is determined by the cultural-political as well as economic importance of the forest to the local user community.

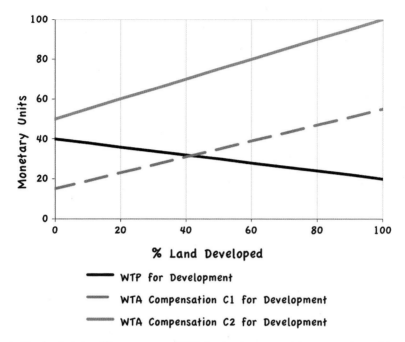

Fig. 1 The developer's willingness to pay (WTP) for development and the community's willingness to accept (WTA) different amounts of compensation C1 and C2 (see text in Box 1)

While indigenous forest villagers understandably attach a great value to the forest due to their economic dependence on it for their subsistence (Malhotra et al. 1992; Conroy et al. 2002), research in community forest management shows that cultural-political linkages of the user community to the forest, rather than the market value of forest goods and services, are the key factors that ensure continuation of the forest resource use over generations (Bernard and Young 1997; Springate-Baginski and Blaikie 2007; Cronkleton et al. 2011). Specifically, the forest users' sense of proprietorship and rights to use the forest are crucial to their civic participation in protecting the forest from profligate use by their own community members, and from expropriation by the state and/or industry. All indigenous land rights movements presuppose civic action by the user community, based on a shared idea of justice and rights, rather than individual gains over others. The existence of individuals with rights-based beliefs invalidate the principal foundation of cost-benefit analysis as a means of making judgments over whether a resource re-allocation improves or reduces social welfare. This foundation of economic benefit calculations, as Hanley and Milne (1996) point out, is the Kaldor-Hicks principle, "which states that a resource re-allocation is welfare-improving if the gainers could compensate the losers and still be better off. But if the losers suffer environmental costs for which, due to their holding of rights-based beliefs, they would reject any amount of money compensation, then the possibility for compensation is violated and the principle can no longer be applied." (Hanley and Milne 1996: 258). Figure 1 illustrates this point with the Menominee forest resource as a key instance. Rights-based beliefs, which also espouse the appreciation of the bequest value of forests, are not negotiable benefits, and therefore financial compensation for the forest loss, to these cultures, is inconceivable and impracticable.[1]

The existence of customary restraints over the use and management of the commons in contemporary pre-industrial societies thus lies beyond the explanatory reach of the Kaldor-Hicks principle – and related assumptions of neoclassical economics. The *homo oeconomicus* model of people pursuing short-term selfish interests also fails to explain the persisting drive of humans for social justice and "interests in common welfare, as in the maintenance of a democratic constitution or sustainable development" (Faber et al. 2002: 325). In contrast, real-life people, the *homo politicus*, seeks "agreement on justice and the common good with his surrounding community and, hence, tries to act and behave in a way such that he receives approval" (Faber et al. 2002: 329). The value of the forest as perceived by the *homo politicus* therefore transcends the aggregate of all tangible economic values calculated by the *homo oeconomicus*.

This ecological-cultural perspective of forest value subsumes a political ecological understanding of the forest, and relies on a broader etymological connotation of 'value' – significance – to the user community. In this connotation, the value of the forest is understood in its differential ontological importance to different stakeholders/user communities. To hunter-gatherers and shifting cultivators, the forest connotes a habitat as well as a material resource for their survival, and also forms part of the landscape shaping their cultural identity. To a timber merchant, the value of the forest consists in its commercial utility, depending on the market price of the extracted timber. To the children of forest-fringe villagers, the value of the forest would consist of the physical space for exploration and entertainment as well as the fruits and mushrooms to eat. To the urban picnicker, the forest is a place of relaxation and enjoyment; to them, a forest with large canopy trees and charismatic animals would be more valuable than a monoculture plantation of *Eucalyptus* or teak.

In this paper I will examine forest people's perception of the value of the forest, and from this perspective, show the limitation of the economist's focus on the value confined to monetary terms. I would argue that any attempt at economic valuation of the forest (or any other natural resource) is fated to be incomplete and parochial, simply because there are no markets for many of the goods and services, and also because the prevailing prices of goods are often determined and distorted by extra-market factors. I contend here that the ecosystem people's view of the forest captures the ecological and cultural interfaces more comprehensively than does the standard economic view of the forest as a stock of commodities. Yet, this perspective remains absent in the forestry discourse and the forest economic valuation literature. I intend to show the general conflict of authoritarian standard ('mainstream')

---

[1] This explains why most ecosystem peoples fail to claim compensation for the loss of their forest habitats due to development projects – they cannot imagine the possibility of money compensation for the loss.

perspective with the ecosystem people's perspective of the forest, and the contrasting ecological and economic consequences of the two different worldviews.

2   The Forest's Worth

Over millennia, hunter-gatherers and shifting cultivators have largely maintained their forests as their habitat, but were displaced whenever the elite classes usurped this resource base. Archaeological evidence and mythologies depict episodes of forest destruction due to elite extravagance, political conflicts and invasion, alongside a legacy of indigenous conservation ethic (e.g. Hughes 2001; Diamond 2005; Turner and Berkes 2006). Many rulers had realized the adverse economic consequences of deforestation, and maintained royal forest reserves. Emperor Ashoka (269–231 BCE) expressed the value of the forest on his famous stone edicts, prohibiting gratuitous destruction of forest and wildlife. Despite this long history of people's appreciation of the value of the forest, monetary valuation of the totality of forest goods and services has never been a veritable project in economic history until the 1990s. It is entirely a new idea of late capitalism, emerging from what Escobar (1995: 203) calls the "semiotic conquest of nature" – it enables capital to engulf nature and the cultural meanings of nature (Deb 2009).

With the rise of ecological economics as a distinctive discipline, a considerable volume of literature relating to the classification of value and valuation methods began to be published in the 1990s. The values of forest and other ecosystems are broadly grouped into direct, indirect, and non-use value, with some overlaps between these categories. A comprehensive valuation of a natural entity not only involves identification of the market and non-market value of the entity, but also of the non-consumptive use value, ethical value, bequest value, and existence value (Randall and Stoll 1983; Manoka 1997), such that the total economic value (TEV) of the forest is:

$$TEV = UV\left(use\,value\right) + NUV\left(nonuse\,value\right) \tag{1}$$

where UV = consumptive use value + non-consumptive (e.g. aesthetic) use value + user's option (for future use) value + nonuser's option value + ethico-religious (e.g. sacred) value; and NUV = existence value + cognitive value + bequest value[2]. Table 1 shows a summary of the categories of value that comprise the TEV of a tropical forest.

Broadly speaking, direct use value spans all known uses of the forest – such as food, fodder, fuel, and raw materials for construction and implements, articles of furniture, and medicine. Direct use of forest also includes the use of forest litter for mulch in agricultural fields, and the use of the forest space and forest products for religious, ritual and aesthetic purposes. Clearly, all these uses usually have a definite market price. In spite of the mismatch between market price and economic value,

---

[2] Manoka (1997) describes bequest value as a form of use value (to posterity).

Table 1 Components of total economic value of a tropical forest

Use values			Non-use values
Direct value	Indirect value	Option value	Non-use values
Timber and pulpwood	Water storage and water table maintenance	Future (both direct and indirect) uses of currently-used items	Existence value
Non timber forest products	Nutrient cycling	Future uses of yet-unused items	Cultural heritage and Bequest value
Recreation and tourism	Air pollution reduction		
Human habitat	Climatic functions (e.g. albedo reduction, "bringing rain")		
Plant and animal genetics	Crop resistance to pathogens		
Education and research	Predation on crop pests		
Ritual & Religious uses	Crop pollination		
Vicarious enjoyment	Storm barrier		

Modified after Bann (1997)

the direct use value of a forest is commonly estimated from the aggregate market price (or in some cases, shadow price) of the different forest products.

Indirect use value incorporates all uses that people make only incidentally, or with no explicit purpose of using. The shade of forest canopy, or the reduction of albedo effect is a benefit from the forest, which is an incidental, yet important service of the forest, and all humans obtain this service without any effort or expenditure. Indirect uses of the forest are related to the natural functions of the forest biodiversity (primary production, inter-specific interactions, etc.) that provide various services to the economy (pollination, pest control, water table maintenance, etc.). People in the vicinity of the forest breath the oxygen produced from the forest, without any deliberate effort to use that oxygen.[3] Like the forest products (goods), the indirect uses (services) have a consumptive value. The value of such indirect uses is difficult to estimate, yet may be approximated by employing some imputed 'shadow prices' or option values of the service in the absence of the forest.

It is necessary here to distinguish between non-consumptive use value and non-use value. The former refers to certain forms of cultural uses that entails no quantifiable consumption, yet has some use. Examples include the aesthetic use of a landscape, which people visit for enjoyment, or which adds to the value of

---

[3] Oxygen will continue to be available for people to breathe also in the absence of any particular forest, so ecological economists talk about oxygen balance maintenance instead of oxygen production as a forest ecosystem service.

adjacent properties, or the use of flowers in ritual or ceremonial decoration. All these uses can be, however, valued in monetary terms. The value of aesthetic use of a place may be estimated using travel cost method, for instance, while the ritual use may be valued using the market price of the item or its nearest substitute, and therefore should be included as a "direct use value".

In contrast, the non-use value (NUV) is impossible to estimate with precision. Examples cited in literature include "existence/intrinsic value" and "bequest value". We may add "sanctity" value, alluding to the cultural-religious concepts of sacredness attached to an organism or a place. People may never use any part of a sacred species, or never visit a holy place in the forest for pilgrimage, yet hold it in reverence. Because a tree or a forest patch held sacred in a local culture cannot be substituted except with a similar sacred tree or sacred grove, its monetary value is therefore practically inestimable. Thus, a flower used in a religious ceremony construes its religious use, which is a type of direct use value, while a sacred tree, no part of which is used for any purpose other than its being held sacred, is an example of the NUV.

Non-use value is the most formidable conceptual difficulty for economic valuation, because in classical economics, a thing of no utility has no value. Consequently, there is no reliable economic tool to estimate the value of non-use of anything. Non-users' willingness to pay (WTP) for the forest's existence is the only gross approach to approximating the NUV. The WTP for existence value, however, is sure to vary widely with the social and economic status, location of the residence, cultural background, and awareness level of individual respondents, and even the type of questions asked to estimate the respondent's WTP.

Many indigenous societies in South Asia and Africa conserve small patches of vegetations, from which no living matter is extracted (Malhotra et al. 2007; Sheridan and Nyamweru 2008). These sacred groves are conserved for generations, regardless of the direct or indirect use value of the component biota (Deb and Malhotra 2001; Deb 2008a). Somewhat overlapping this sanctity value, an important NUV of the forest is its existence value, which is independent of the utility of the object being valued to the valuer, or its use to someone other than the valuer. Thus, a forest is of value to the non-users if they are happy to know that the forest exists there, and will be unhappy if it is destroyed. Many forest people also feel obliged to leave at least parts of their forest intact and bequeath the asset to posterity for possible future use, so its bequest value accrues to future generations.

The biggest difficulty in assessing TEV of the forest (or any species, ecosystem, or landscape) is that there are interactive influences between different components of TEV (Tisdell 2005). The biodiversity of the forest has its existence value, but also has a great use value, because all the forest products and services are a function of biodiversity. Thus, the total value of biodiversity subsumes the values of timber, fruits, mushrooms, medicinals, game animals, as well as the value of tree shades (in albedo reduction) and photosynthesis (in carbon fixation). Similarly, if villagers consider a part of the forest as a sacred space, it would have a cultural-religious value *and* a bequest value of the cultural heritage. Because a range of interactive influences between components of the TEV exists, "the additivity assumption" in Eq. (1) is invalidated (Tisdell 2005: 112). Therefore, the worth of the forest cannot be surmised from the sum of the values of all its goods, services, and opportunities, "any more than the value of a mountain can be estimated from the sum of the market value of its constituent blocks of stones" (Deb 2009: 68).

Non-instrumental/non-market values are often incompatible with the view of economic rationality and dissonant with the conventional economic valuation rationale. In particular, conventional econometric tools can hardly assess the NUV of a forest in absolute terms, because the NUV of the object varies with the extent of information available to, and awareness level of, the non-user. A conservation activist may value the existence of the Amazonian rainforest a lot higher than someone who never heard of it. The existence value of the forest in this case would tend to increase with the individual's education levels and awareness of the current direct and indirect use values of the forest. The non-user's knowledge of corporate investments in bioprospecting of the rainforest to tap its potential use values would enhance the forest's bequest value to the individual.

The NUV serves to signify a type of value that is not adequately captured in neoclassical economics, and can be used as a handle to grasp the deontological worldview of many indigenous societies. Recognition of the NUV of biodiversity is embedded in cultural, ethical and normative orientations of different indigenous cultures (Deb and Malhotra 2001; Deb 2009), and this constitutes the cultural-political dimension of value. Thus, the recognition of NUV construes a deontological view of nature, which implies that acts (such as the planting of trees) are more important than the consequences of acts (utilities obtained from the plantation). "This is the opposite viewpoint to the teleological perspective which lies behind cost-benefit analysis" (Hanley and Milne 1996: 257). This deontological worldview is ensconced in the indigenous land ethic and cultural identity.

The forest's existence is also important to numerous people who believe it brings rain, purifies the air, or is an abode of spirits. Regardless of empirical validity, such beliefs espouse a value that transcends the instrumental value of the forest, and also posits it as an essential component of the landscape that cannot be traded. This perception underlies, and legitimizes, the indigenous conception of the forest (and land in general) as a gift of nature to support life. This concept is incompatible with the modern concept of forest as a resource or commodity, and contrary to price-tagging of the land. As a leader of the indigenous civil society movement in eastern India said: "The land is given to us by Sing Bonga (God). There can be no tax on land nor can it be bought and sold!" (Tilka Manjhi, cited in Siemenpuu Foundation 2008: 75). Translated into mainstream economic jargon, Manjhi's statement implies that the value of the rent from the land cannot be discounted. With a zero rate of discount, the price of the forest becomes infinite[4] – so no one can sell or buy the forest.

---

[4]The standard economic view usually assumes that a good can be either free or have a finite price. From the ecological perspective, however, most environmental services have an *infinite price* — no sum of money, however large it may be, should be permitted to purchase the right to destroy these services (Deb 2009). In other words, if the rent from the forest land as renewable resource is R and discount rate is d, then the price (P) of the forest land, as a renewable resource, will be

$$P = R / d$$

At a discount rate $d=0$, the price of the land becomes infinite, so no one will be able to buy the land.

Box 1 The Menominee Forest: Too Valuable to Develop

For the indigenous forest people, the forest is not just a resource to subsist on, but a place of inheritance, history, and ethno-cultural identity. The Menominee tribe, occupying a reservation in central Wisconsin, USA, have been logging their 220,000 acre of old-growth forest for the past 157 years, yet have never exceeded their logging quota for the market, never logged for economic growth. The timber mill of the Menominee Tribal Enterprise (MTE) is run to generate employment and subsistence-level income, not for making profit. The MTE never changed the age-old harvesting regulations in response to fluctuations in market demands. As a Menominee leader put it, "the forest drives the sawmill, the sawmill doesn't drive the forest" (Pecore 2005: 177). The stock of their forest, in terms of both biodiversity and timber quality, has significantly improved over a century as a result of their "sustained yield" forestry, of which the MTE is the best model of ecoforestry (Bernard and Young 1997). The Menominee forest has been described as "an island of timber in an ocean of cleared land", and is a complex mixture of diverse species and stand ages. Each forest stand is structurally complex, "containing seedlings, saplings, and saw timber trees as well as dead material in various stages of decomposition" (Pecore 2005: 176).

Proceeds from saw timber sale are equally divided among all members of the Menominee Nation, and everyone earns a modest living, but "no one can imagine to cut another tree in order to become a little bit wealthier" (personal notes from my interview with the Board members, Menominee Tribal Enterprise, 22 November 2009). The Menominee possess an asset worth more than a billion dollars, yet do not aim to attain "economic salvation" at the cost of this asset because, as a member proclaims: "The forest is the Menominee people… Our history lives here. We have always tried to hang on because this is where we belong." (cited in Bernard and Young 1997: 98).

The Menominee traditionally own most of the forestland in common, and few individuals hold title to land. This arrangement was perturbed in the 1960s, following the passage of the Menominee Termination Bill in June 17, 1954, which terminated the federal protection of the Menominee Indian reservation. Upon Termination, every enrolled Menominee man and woman received a payment of $1,500 and the payments for children and incompetents were held in trust by the government. Soon afterwards, unaccustomed to making profit, the Menominee ran their timber business in continual loss, and failed to pay taxes. In a government bid to secure a tax base, a portion of their pristine lakeside land went in 1965 to a development company. This alarmed and united all members of the Menominee tribe: "Educated Menominee who had left the reservation and were living in Milwaukee and Chicago; young people emboldened by the events of the 60s; elder tribal members who had stayed close to the reservation and more traditional ways – all united to fight the conversion and loss of their forest" (Bernard and Young 1997: 106).

(continued)

Box 1 (continued)

After a series of legal battles and advocacy campaign, the Menominee won back their independence and sovereignty over their land. The US Congress officially terminated the policy of Termination on December 22, 1973, to reinstate the Menominee tribe (Bernard and Young 1997).

This history of a brief period of termination and subsequent reinstatement of the tribe is a brilliant testimony to the immense value of the forest to the forest people. To the Menominee, the value of their forest far exceeded any monetary value that economists strive to fathom with people's "willingness to accept compensation" for the loss of their land. Figure 1 illustrates the difference in marginal evaluation of land between the perspective of forest people (in this case, the Menominee) and the developing agency (in this case, the US government). The black line shows the government's willingness to pay (WTP) some amount of monetary compensation for developing the Menominee forestland during the period of Termination. If the community had agreed to accept a compensation C1 (red line) for the loss of their land, development of a certain proportion of land would be optimal. The optimal proportion is at the intersection of the government's WTP and the community's willingness to accept (WTA) lines. However, during the termination period, the Menominee were unwilling to accept the level of compensation and/or economic benefits to be accrued from development of their land. Even decades after the revocation of the Termination, the Menominee continue to insist: "no money could buy our forest, which is our existence, our past, our future" (Marshall Pecore, personal interview dtd. 22 Nov 2009). This situation may be described as the Menominee's willingness to accept, hypothetically, an arbitrarily high level of compensation C2 (green line), which is far above the WTP line. As no intersection of this green WTA line and WTP is possible, no development is optimal. Thus, development of the Menominee land could not take place, and the Menominee's dissent led to the reinstatement of the tribe and their commons.

In India, all hunter-gatherers and shifting cultivators – from the Onge and Jarawa of the Andaman Islands (Venkateswar 2004) to the Warli of Maharashtra (Prabhu 2003) and the Dongaria Kondh of Odisha (Padel and Das 2010) – who have not yet been extirpated from their traditional economies and cultures – value their forests incomparably higher than any amount of money could possibly compensate for the loss of their forest. To all these forest people, the NUV is as important as the direct use value of the forest. The perception of the NUV of the forest becomes prominent in indigenous people' land rights movements, which are shaped by their cultural and political institutions, such that the "valuation languages" of the ecosystem peoples and the people of industrial societies inexorably diverge (Martinez-Alier et al. 2010).

3 The Transformation of Value: A History of Forest Management

The history of forest management in modern India is the history of enclosure of the forest commons, and of the expropriation and decimation of the resource base of the ecosystem people. In modern India, the commercial silvicultural agenda of forest administration has, over the past 150 years, destroyed and fragmented ecosystems, transformed traditional land use systems, and economically and culturally marginalized all 'ecosystem peoples' (Gadgil and Guha 1995). This profligate land use system is an heir to the European land use policy of the past centuries.

In the seventeenth century, Europe began to witness "a slow, tentative and contested movement away from feudal entitlements, where land was held 'of' others, to a more recognizably modern conception of land as a basis for secure entitlements that could be rented, used, sold and willed" (Blomley 2007: 2). This movement left no room for the commons and their customary management by rural communities, which began to disintegrate in response to the concretion of the concept of private property and supportive legislations. In the nineteenth century, most of the commons (also termed "wastes" or "wastelands") in England were enclosed for either the state or private ownership, which terminated all ancient regimes of community ownership (Neeson 1993; Blomley 2007), as a result of which peasants were "first forcibly expropriated from the soil, driven from their homes, turned into vagabonds, and then whipped, branded, tortured by laws grotesquely terrible, into the discipline necessary for the wage system" (Marx 1887: 688). The emergence of a new land use system in nineteenth century England essentially redefined human relations with forests and pastures, in order to foster industrial development at the expense of subsistence needs of the rural agro-pastoral communities. The emergence of European forestry thus led to "the creation of a technological, and often, technocratic elite with a monopoly over decision making on forest use" (Rajan 1998: 326).

This part of European history is necessary for an understanding of the orientation of forest governance in modern India, because the European empires exported their land use policies to all their colonies in order to replace native customary land use systems with statutory institutions, and in the process destroyed the commons (Bryant 1996; Grove et al. 1998; Spence 1999; Rocheleau et al. 2000; Chiuri 2005). All colonial governments legislated for the enclosure of forests and other commons[5], and proscribed traditional indigenous forest economies – hunting-gathering, and shifting cultivation. In India, the enclosure of the forest not only decimated the natural wealth of biodiversity, but also brought about unprecedented displacement, poverty and criminalization of numerous indigenous societies (Gadgil and Guha 1995; Grove et al. 1998; Sivaramakrishnan 1999; Davis 2001; Equations 2007), and established a legacy of mistrust and conflict

---

[5]The same model of forestry is by and large continuing in ex-colonies after their independence (Colchester 2006; Bose et al. 2012).

between the state and ecosystem peoples across generations (Kapoor 2007; Springate-Baginski and Blaikie 2007; Deb 2008b).

In the early decades of British rule in India, forests were considered to be unproductive wastelands, a "bar to the prosperity of the Empire" (Ribbentrop 1900: 60). With an aim to converting the forest into revenue-generating farmland, a *Baze Zamin Daftar* (Department of Wastelands) was instituted (Sivaramakrishnan 1999). The government encouraged cultivators to clear forests, and endorsed sacrifice of forests "without hesitation" in cases where demand for arable land could only be met through such sacrifice (Flint 1998: 432). Throughout the nineteenth century and early twentieth century, European finance capital dictated clearing of forests, in all colonies, in order to raise extensive commercial plantations of cash crop – indigo, opium and tea in India, rubber in Malaysia, cacao in Africa, coffee in Brazil, bananas in Central America, and cotton in the southern States of USA (Davis 2001; Goldewijk et al. 2001).

The empire, however, later deemed the forest to be a valuable resource when railway expansion and shipbuilding put an increasing demand of timber. 'Scientific forestry' was employed from the 1850s on with the objective of ensuring the supply of timber in the service of imperial economy,[6] and the imperial Forest Department was established in 1864. Throughout India, forests that had already been depleted by the railways, were "overworked to provide timber and fodder for military use between 1914 and 1918, and exploited even more excessively to meet wartime wood needs between 1940 and 1944" (Flint 1998: 441).

Prior to the state's enclosure, all forests of the Indian subcontinent, with a few exceptions of royal hunting reserves, were customarily managed by societies of hunter-gatherers and shifting cultivators. In various indigenous hunter-gatherer and shifting cultivator societies, still surviving in different parts of the subcontinent, a matrix of taboos, customary quotas on harvest, and closed seasons for fishing, hunting and gathering reflect tacit cultural restraints on exhaustive resource use, and shape the indigenous ecological ethic, which has conserved the indigenous resource base over millennia (Deb and Malhotra 2001; Maffi 2004; Turner and Berkes 2006). The traditional conservation ethics of the indigenous societies, wherever they survive, continue to prohibit exhaustive use of the resource base over generations – until these societies become 'modernized' through education and commerce.

The ancient customary natural resource management regime in India began to collapse in the 1870s, when forest legislations appropriated all forests of the country and abolished the customary rights of forest villagers over 'protected' and 'reserved' forests. This act of enclosure of forests constituted a major watershed in India's ecological and political history, marked by episodes of dispossession

---

[6] The science of forestry was born in the nineteenth century with the explicit objective of contributing to the continuation of economic progress, not wildlife or nature conservation. As Gifford Pinchot put it succinctly: "The object of our forest policy is not to preserve the forests because they are beautiful ... or because they are refuges for the wild creatures of the wilderness ... but the making of prosperous homes.... Every other consideration comes as secondary" (cited in Robinson 1975: 55).

and displacement of ecosystem people, which in turn elicited over 70 major adivasi (indigenous peoples) uprisings across the country,[7] between the Malpahariya revolt in Bihar in 1772 and Lakshman Naik's insurgency in Orissa in 1942 (Equations 2007). Most of these rebellions were brutally quelled, although in a few instances the British acceded to the immediate adivasi demands, as in the case of the Bhil revolt of 1809 in central India and the Naik revolt of 1838 in Gujarat. These uprisings reinforced the image of the rebellious forest tribes as savages, whom the government sought to bring into its domain of civilization. The British government enlisted Lombrosovian criminology to pass the Criminal Tribes Act in 1871, which stigmatized about 150 recalcitrant adivasi groups of nomadic hunter-gatherers and shifting cultivators, who chose to remain outside the pale of agrarian economy (Equations 2007; Deb 2009). The state sought to tame both the forest and the "savage" forest people, by law and legitimized coercion. "Coercion by the paternalistic state, like force-feeding a detesting sick child, was intended for the healthy development of the savage. From the progressivist perspective, the fact that binding the savage down to the civilized life of settled agriculture generated revenue and a labour force for the state appears incidental, yet proves the benefits of development." (Deb 2009: 32–33).

After India's independence, the Forest Policy of 1952 reiterated the state's absolute ownership of forests and the promotion of state capitalism in the forestry sector. The Clause 7 of the 1952 Forest Policy stated: "The accident of a village being situated close to a forest does not prejudice the right of the country as a whole to receive the benefits of the national asset." (GoI 1952). To protect this national asset, the policy urged for a "scientific conservation of the forest" that "inevitably involves the regulation of rights and the restriction of privileges of users ... however irksome such restraints may be to the neighbouring areas" (*ibid.*). Forestry was geared to supply raw material to industry at the expense of the objective of biodiversity conservation, and the livelihood needs of forest villagers (Gadgil and Guha 1995; Sivaramakrishnan 1999; Joshi 2010). The forest department used to sell bamboos at 1 rupee per ton to paper industry for five decades after independence, while the local artisans, such as basket makers, had to pay Rs. 3,000 per ton (Gadgil and Guha 1995). From the late 1970s onwards, the Forest Department (FD) raised large-scale plantations of exotic quick-growing species (QGS) on forest lands, and in so doing displaced numerous native flora and fauna, thus violating the professed policy objective of conservation. In the sal (*Shorea robusta*) forests of southwestern districts of Bengal, even the root stocks of native trees were uprooted for raising eucalypt plantations (Deb 2008b). The harvest of QGS in short rotations continues throughout the country to ensure the supply of pulpwood and poles for paper and mining industry, respectively. The indigenous forest villagers find it both logically and ethically unacceptable to witness the FD organizing large-scale commercial

---

[7] Some of the milestones were the insurrections and revolts of the Tamar (1795), Bhumij (1798–1799), Munda (1819–1820), Paik (1817), Bhil (1818–1831), Santal (1955–1956), Kharia (1860–1880) Kairwar (1871), Munda (1895), Sardari (1859–1895), and the Oraon (1915–1920).

harvest of timber and bamboo, which are forbidden to them for use in local craft economies. As a Kondh woman from Odisha expressed,

"[T]he government says the land on which you have *patta* (title) is yours and the rest is the government's! This is not our way .... The water for instance, does not belong to anyone like the government thinks—it is given by God for the forests, the animals and humans alike. But the government would not understand this. This soil does not belong to the government or the government's parents (*sarkar kimba aur tanko bapar ko mati nahi tho*). They have been given to us by God through the ages. Who is this government (*e sarkar kee?*) that lets the paper mills take the longest bamboo and best wood and then asks us for royalty and taxes for small cuts for poles?"

(*cited in* Kapoor 2007: 23)

After independence, India's land use policy has served the political and economic interests of the urban industrial- bureaucratic-political elite (Gadgil and Guha 1995; Bose et al. 2012). Forest villagers and ecosystem people were continually divested of their resources and their right to managing the resources, while industry and commerce were given priority. The neoliberal land use policy, designed to foster industrialization, has not only decimated India's forests, but also systematically violated indigenous land rights. In spite of the Constitutional recognition in the 5th and 6th Schedules of tribal ownership rights over land and forests in Scheduled (protected) Areas, "contradictory legal provisions and failure to implement or translate Constitutional Provisions into reality" continue to undermine the indigenous people's land rights (ACHR 2005: 4–5). For example, forest laws that confer usufruct rights to use non-timber forest products without right to ownership contradict the ownership rights provisions of the 5th Schedule. Forest villagers are divested of their traditional land rights through various bureaucratic strategies of the forest directorate, which include: forest reservations, leasing of land to industrialists, and the eviction of forest villagers as encroachers into national parks and sanctuaries. The indigenous land right does not exist under the Forest Conservation Act (1980), the Wild Life (Protection) Act (1972), and the Land Acquisition Act (1894): "The government has the sovereign right to evict people for undefined public interest or 'larger interest', but the affected people do not have the right to question the decision of the government on forced evictions" (ACHR 2005: 9).

Development projects in India have cleared vast tracts of forest, evicted millions of forest villagers, and have even de-notified several national parks and sanctuaries (Kothari et al. 1995; Shrivastava and Kothari 2012). The industrial policy in various States has sought to attract the inflow of investment and investment promises with numerous industrial projects. For recent instances, a plethora of projects such as Sterlite-Vedanta's bauxite plant at Niyamgiri Hills, and steel plants of Posco at Paradeep, Tata at Kalinga Nagar and Arcelor-Mittal at Keonjhar – all in the State of Odisha – have over the past decade destroyed the traditional resource base of the indigenous residents on the project sites. Under the neoliberal land policies, "traditional occupiers of land under customary law confront the prospect and reality of becoming illegal encroachers on land they have cultivated and sustained for generations" (Pimple and Sethi 2005: 239).

In India, as in most ex-colonies, forest policies and laws have facilitated exploitation of forest villagers both by the FD and industry (Pimple and Sethi 2005; Colchester 2006). Low wages, arbitrary high-handed treatments by forest officials, extortion of "gifts" from the villagers, and departmental corruption in collusion with the rural elite are common throughout the country (Deb 2007b; Joshi 2010). The national Working Group of Welfare of Backward Classes for the Fourth Plan acknowledged in 1967 that the manner in which the existing forest policy is understood and implemented had placed the forest villagers at a complete disadvantage (cited in Joshi 2010).

This protracted process of systematic violation of rights of forest villagers and their estrangement from forest management has over decades severed their cultural linkages to the forest and nurtured a sense of what Marx (1862) termed alienation (*Entfremdung*) of people from their life and nature at large. The Koya people of Malkangiri district, who had always cared for the wild bamboo, are now employed by the Odisha Forest Development Corporation to destroy their own bamboo groves (Sainath 1996: 100). To these alienated forest villagers, the value of the forest is largely confined to the forest products they can extract for subsistence. Like the urban industrial-bureaucratic elite, the disentitled villagers have little recognition of the value of the forest beyond its direct use value. Bereft of the traditional value-linkages, the cultural meaning of the forest is presently confined to its instrumental value to the villagers, who in the liberal market arrangements would pursue their individual ends "relatively unfettered and oblivious to the social consequences" (Shutkin 2000: 41). Furthermore, bereft of community ownership and accountability, the forest is relegated from its common property status to open access status, with few restraints to observe. When access to this open resource is denied, the disgruntled ecosystem people "turn their anger against the very forest that sustained them for centuries, which everyone else exploits, and which forest officials 'protect'" (Padel and Das 2010: 430). Enclosure has finally turned the wealth into a waste.

The state's 'protection' of the forest resource for the benefit of industry and commerce over the past two centuries have taught people to value forest only as a commercial resource. The FD's de-notification of protected forest areas in the past decades has already proven the priority of the industrial imperative over conservation needs (Kothari et al. 1995). While the new Forest Policy of 2008 upholds the conservation objective and undermines the commercial agenda, a series of big industrial projects (Vedanta, Birla Cement, BALCO, Tata Steel) have received government approval to clear forest lands for mining and factory sites. Between the passage of the Forest (Conservation) Act in 1980 and 2007, approximately 1,140,177 ha of forest land was cleared for industry, urban sprawl, roadways, agricultural expansion, and encroachment by development refugees. The era of Globalization has enormously accelerated this process of deforestation: more than a quarter – approximately 3,11,220 ha – of the total of 26 years of clearance was lost between 2003 and 2007 (Wani and Kothari 2008). This neoliberal turn of India's development only reiterates the early colonial view that forests are "unproductive wastelands", and "a bar to the prosperity of the empire" – this time, the empire of industry. The state's perception of the value of forest vis-a-vis development has thus come full circle.

The empire of industry continues to destroy all ancient social arrangements that used to prioritize the interests of the community over private interests, and reinforces what Harvey (1996) calls the "standard view" of development. This standard view, deeply entrenched in India's ruling ideology, creates new consumer desires, and shapes all citizens into consumers, who are inculcated to value nature only as a resource to exploit and exhaust for quick dividends (Deb 2009). This developmentality is in fact an outcome of European rule in colonies, occupied for creating and expanding markets, for which creation of "wants" was strategically essential. Thus, a senior civil servant of the English East India Company, sent from Madras to subjugate the Konds of Odisha under British rule in the nineteenth century, advocated creating "new wants", which would eventually make the self-sufficient tribal societies "dependent upon us for what will, in time, become necessities of [their] life" (cited in Padel and Das 2010: 179).

The state's authority over the forest, as well as the industrial rationale of forest depletion, is challenged when local people are entitled to managing their resources as commons. Reinstatement of the commons invariably reverses the process of exhaustive use and destruction of the forest, and individual profit motives are subjugated to the community interest. Wherever local village communities acquire the authority to govern their forest, the condition of the forest demonstrably improves – in terms of species diversity, stand architecture, and use value. In northeastern States, 75–90 % of the forests are owned and managed by autonomous tribal councils, and the forests in these States are the richest in biodiversity, housing numerous rare and endemic biota. The most efficiently managed sustained-yield forest in the country is the "people's forest" at Lapanga in Sambalpur district of Odisha, which the villagers created in the 1930s by sacrificing their own farmlands (Mishra 2008). Villagers have been autonomously managing this forest, covering 418 acres, with their own community regulations for the past 80 years.[8] In the districts of Nayagarh, Deogarh and Mayurbhanj of Odisha, many villagers have formed autonomous protection committees to protect and regulate harvest of trees in large tracts of forest, with no technical inputs from statutory institutions. These self-initiated community forest management efforts have resulted in spectacular regeneration of degraded forests (Conroy et al. 2002).

In spite of the state's priority given to industrial needs, and the supremacy of the State Forest Department's authority, villagers in many districts continue to protect remnants of ancient vegetations as sacred groves, which are an important part of the village landscape in many parts of India (Malhotra et al. 2007). In the northeastern State of Mizoram, the institution of sacred groves is reincarnated as "safety forests", which are maintained without the need of any official agency to enforce conservation (Deb 2009). In these instances, traditional regulations and customary edicts are more successful in conserving the commons than similar regulations imposed by external authorities and statutory institutions. In all societies, various

---

[8] This idyllic forest land is now under threat of destruction from Hindalco's aluminium mining. The Lapanga villagers won in a legal battle against the company, yet the State government connives at the company's mischief, in defiance of the judicial order.

cultural and moral drives nurture self-restraint in resource use, which are invoked in norms of the commons, to override individual interests and resist the temptation of immediate economic benefits from altered land use (Burke 2001; Oström 2005, 2009; Baron 2010).

The commons can also be reinstated by enlightened state policies and laws that may empower people to democratically exercise their customary rights. The new Forest People's Rights Act (FRA) of 2006 is a brilliant instance of enlightened legislation[9]. For the first time in history of India's forest governance, the FRA recognizes, at least on paper, indigenous forest people's rights, which include *nistari* (community forests), minor forest produce, fish and other aquatic products, grazing land, traditional seasonal resource access for nomadic or pastoralist communities, community intellectual property rights, and traditional knowledge pertaining to biodiversity and cultural diversity. After the passage of the FRA, many villagers in north Bengal re-invoked their right to manage their own forests, and started exercising governance control over forests. In Coochbehar and Kurseong Forest Divisions, gram sabhas (village councils) stopped the Forest Department's felling activities. In May and June 2008 the villagers at Chilapata forest area of Coochbehar Forest Division, and the villages of Kurseong Forest Division blockaded the FD's timber depots at various places, demanding proper implementation of the FRA. The villagers also stopped all forestry activities by the forest department without permission from the gram sabha (Banerjee et al. 2010).

4   Value Perceptions of Forest Users and Non-users

No matter how economists value the forest goods and services, people have always valued the forest as a resource to meet the various needs of livelihoods and luxuries. To everyone, including forest villagers, the direct use value (like timber and NTFP) is all too obvious, and the value perceptions of the academic valuer and of the woman who gathers NTFP for sale on market are no different. Their value perceptions begin

---

[9] In July 2004, India's Ministry of Environment and Forests, in an affidavit filed at the Supreme Court, admitted:

"That, for most areas in India, especially the tribal areas, record of rights did not exist due to which rights of the tribals could not be settled during the process of consolidation of forests in the country. Therefore, the rural people, especially tribals who have been living in the forests since time immemorial, were deprived of their traditional rights and livelihood and, consequently, these tribals have become encroachers in the eyes of law. That these guidelines, dated 5 February 2004, are based on the recognition that the historical injustice done to the tribal forest dwellers through non-recognition of their traditional rights must be finally rectified.… Further, that because of the absence of legal recognition of their traditional rights, the adjoining forests have become 'open access' resource as such for the dispossessed tribals, leading to forest degradation in a classic manifestation of the tragedy of commons" (cited in Blaikie and Springate-Baginski, 2007: 77–78). With this background recognition of forest people's rights, the new FRA was passed in 2006.

Table 2 Diversity of forest biota and their use by forest-fringe villagers of west Bengal

Type of Use	Plants			Animals	
	Angiosperms	Ferns	Fungi	Invertebrates	Vertebrates
Food	25	1	13	2	14
Fodder	5				
Fuelwood	32				
Household articles	6				
Construction	8				
Medicinals	46	1		2	1
Decorative/Ornamental	3	1			
Religious/Ritual	5				
Commercial	27				
Total No. of Spp. in Use	98	2	13	4	14

Data adopted from Malhotra et al. (1992)

to diverge for the forest goods and services that are not sold on market. Nevertheless, it is not difficult to find forest villagers who can articulate their cognizance of the indirect use values. Although no forest flower is sold on the local market, most villagers in our surveys can estimate, when asked, some "opportunity costs" of the wild flowers and leaves, which they use in religious ceremonies and cultural festivals.

Villagers who have witnessed disappearance of streams after deforestation are partly, or inchoately aware of the hydrological service of the forest, although they may not tell the money value of that service. Owing to the spread of environmental awareness, villagers also recount the value of the forest as source of oxygen and "purifier of air". Many also value the forest as a place of beauty, an important place of cultural performances, social gatherings, picnicking and pastime. Clearly, there are overlaps in consumptive and non-consumptive use value perceptions. We attempt here to discuss the expressions of these value perceptions under separate rubrics.

4.1   The Consumptive User's Perception

To all forest users, the value of the forest is primarily determined by its use value, or what Millennium Ecosystem Assessment (2005) labels as 'provisioning services', which in essence are the goods from the forest (Table 2). The forest goods and their uses listed in Table 2 are neither exhaustive nor typical of the goods and services from the forest. Forest biota has different spectra of uses in different ethnic communities and cultures, because the region-specific biodiversity shapes the material culture of a society. For example, *Combretum decandrum* vines are harvested from the forest in southwest Bengal for making termite-resistant baskets, which are not used in other parts of the same State. The tassar (*Bombyx mori*) pupae and leaf stitching ants (*Oecophylla smaragdyna*) from the forest are distinctive items of food in the local tribal food culture (Malhotra et al. 1992), but are not consumed in other districts.

Similarly, people in this area harvest numerous plants for different medicinal uses, which may be unknown in other parts of the country.

In most cases, the gathering of most items of NTFP from the forest is unrestricted both by law and custom. Nevertheless, individual gatherers tend to observe certain cultural restraints on profligate harvest during their regular gathering expeditions. Thus, women from forest villages in southwest Bengal collect just enough mushrooms to fulfill their daily requirement (Deb and Malhotra 2001). These customary 'harvesting quotas' generally serve to keep the bioresources available over long time horizons. Harvesting quotas do not exceed the quantity needed to fulfill the average household needs, and so are maintained by peer pressure, self-control ethics and community norms. While the individual's actual behavior can and does deviate from the ideal and the ethical, the community norms inhibit entirely selfish behaviour in indigenous societies (Burke 2001; Gurven 2004).

Harvesting quotas however disappear in commercial gathering expeditions, where profit maximization in the shortest time is the objective, not long term fulfillment of subsistence needs. When resources are transformed from common property to a market commodity, their non-commercial characteristics are ignored in the accounting of costs and benefits to the users. "As long as their uses are entirely commercial and designed for the accumulation of capital, there is no mechanism for conservation" (Marchak 1989: 20). Unregulated depletion of the resource is inevitable when there is a drive for quick extraction of profit, because that is economically rational. The low level of extraction required to ensure sustainable harvesting of wild populations of American ginseng (*Panax quinquefolium*) and wild leeks (*Allium triccocum*) in Canada (Nantel et al. 1996) and amla (*Phyllanthus emblica*) fruit in India (Shankar et al. 1996) suggests that "at current prices, sustainable harvest levels for these species are not an economic proposition for commercial gatherers" (SCBD 2001: 15).

4.2　The Non-consumptive User's Perception

The forest has a plethora of goods and services to humans, but many people make no use of the forest. Fuelwood from the forest is the most important and common necessity for all forest-fringe village households, but families with livestock and a sizeable home garden can meet their daily need of fuel from their farm residues, brushwood from within the village, and cow dung. While it would be impossible to find a person in any forest-fringe village who has never made at least an indirect use of the forest, an urban elite may make no use of the forest (barring the global ecosystem services of the forest, such as carbon sequestration). To the villager who lives too far from a forest to make a trip, the forest appears to be an esoteric entity that exists only in her imagination. She may, nevertheless, consider the forest as a nice place for some future use, such as a prospective pleasure trip. Table 3 gives a working list of reasons why individuals living far from the forest area long to visit the forest.

Table 3 Why people from a non-forest habitat want to visit a forest

Category of use	Reason for visit to a forest
Aesthetic/Inspirational	Feel Nature/Relate with Mother Nature
Aesthetic	Pleasure trip/picnic
Aesthetic	Enjoy nature/see old trees/flowers
Intellectual	Gather knowledge about forest
Intellectual	View rare/interesting animals/plants
Intellectual	Match imagined forest with a real forest
Ethical	See a metaphor of life/world
Ethical/educational	See a place of equality/beauty/tranquility
Intellectual/educational	Understand ecological principles/science of life
Spiritual	Find solace/peace of mind
Spiritual	Communicate with a deity/personal God
Religious	Fulfill a vow to a divinity
Intellectual	Share experience of a forest with others
Political	Win others' admiration/appreciation of the experience of wilderness
Health	Breathe pure air/Detoxify oneself
Religious	(In case of a sacred site) Pilgrimage

4.3 The Non-user's Perception

Sacred groves are a paramount example of a forested area maintained and valued for its non-use value as well as non-consumptive use value. The conferment of a religious value to a species or an ecosystem, regardless of its consumptive end-uses, seems to be a symbolic recognition by local cultures of its "existence value". In most indigenous cultures, norms against callous or cruel conduct toward animals and excessive and gratuitous exploitation of plant resources are often motivated by "sentiments of affinity", and are unrelated to a calculated empiricism (Kellert 1996: 151). The sacred karam tree (*Adina cordifolia*), and the shrub manasa (*Euphorbia neriifolia*) that have no direct use values, were nevertheless deified in various local cultures in eastern India. Similarly, a pond at Chhandar village in Bankura district, West Bengal, is not used by villagers for bathing, washing, fishing or any purposes, and yet is held sacred for over 600 years (Deb and Malhotra 2001; Deb 2008a). The concept of sacred in local cultures thus implies a recognition of the existence value of living objects, over and above their use values, and a moral attitude towards nature in general. This attitude is what Fromm (1973) calls *biophilia* – an innate love and respect for life and creatures. Biophilia tends to be reflected in the entire belief system of indigenous societies. Furthermore, "Biophilous ethics have their own principles of good and evil. Good is all that serves life; evil is all that serves death. Good is reverence for life, all that enhances life, growth, unfolding. Evil is all that stifles life, narrows it down, cuts it into pieces." (Fromm 1973: 365–6). Good and evil omens may thus assume special semiotic significance with respect to biophilia. For example, the Santal consider as good omens the sighting of footprints of

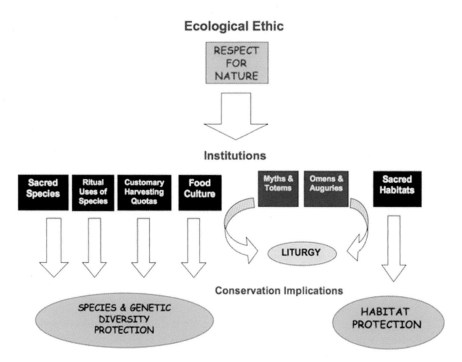

Fig. 2 The conservation implications of the indigenous bioethics (after Deb 2009)

cattle, tiger and of leopard during a marriage ceremony. Likewise, the sighting of cattle, fox, and mango are auspicious signs to the Munda. Ill omens, for the Santal, include the sight of a headload of fuelwood, for the Munda, that of felling of a tree, and in Hindu folk culture, include the sighting of a hunted turtle, and of cut fuelwood (Deb and Malhotra 2001) – signs that are carefully noticed during the rites of passage. These omens, and related auguries are an expression of the underlying belief that the existence of a variety of life forms, regardless of their utility, around people is a sign for "good living", whereas the acts of destruction of life are considered as bad for human life. A consequence of biophilia is *in situ* conservation of diverse species and their habitats (Fig. 2).

Biophilia is an ethical dimension of the recognition of the existence value of life forms in indigenous societies. Many indigenous cultures tend to ensure the very existence of trees and animals, although they may have no practical use. The recognition of the existence value is an outcome of a combination of ethical and philosophical obligation of a group of people to objects and places, not an expectation of benefits, and therefore cannot be assigned an instrumental value in the manner of a typical item of consumption. The idea of existence value, as Sagoff (2004: 46) argues, "depends on deeply held moral convictions and religious beliefs that underlie social policy". And because "beliefs are not benefits" (Sagoff 2004: 47), the benefit to society of behaving according to a belief system or moral judgment cannot be assigned a price tag. Economists have recently employed new sophisticated concepts like abstract benefits from carbon sequestration services and other global environmental services in forest valuation exercises, but these concepts are unknown to the indigenous societies, whose biophilous ethic protects their sacred trees, sacred groves and ponds, even when they make use of none of them.

Moreover, people have a tacit understanding of the bequest value of the forest upon which they depend for livelihood. Thus, the head of a forest protection committee of Khandkelgaon village of Odisha believes that even if the villagers do not need to gather anything from a given patch of forest (assuming that all their material needs are fulfilled from other parts of the forest), he would still want to protect it, so that "posterity would benefit from its services" (personal interview with Prem Bhoi, at Khandkelgaon, Saintala, on 6 May 2012). Such statements indicate the awareness of the ecosystem people about both the option value and bequest value of the forest.

4.4   The Victims' Perception

The existence of the forest itself may appear undesirable, if it threatens a person's or a household's security of life and/or livelihood. Crop losses, personal injuries and loss of life from attacks of wild animals are a common experience of ecosystem people living in forest-fringe villages. Furthermore, living in or visiting a forest may expose humans to the vectors of many serious diseases (e.g. Lyme disease in USA, malaria and Kysanur forest disease in India, sleeping sickness in sub-Saharan Africa, Legionnaires' disease in Europe). These maladies and risks constitute the "negative services" of the forest, or simply "disservices" which are hardly mentioned in valuation literature (Lelé et al. 2013). Death tolls from attacks of man-eating tigers in the mangrove islands of Sunderban, or that of rogue elephants in West Medinipur forest villages are often followed by bureaucratic harassments to families of the "wildlife victims". Many relatives of the victims of snake bites, wild bee attacks, and diseases contracted from forest visits often portray the forest as the abode of evils, or Evil itself, and want the forest to be cleared, or at least purged of all dangerous creatures.

We interviewed several survivors from wildlife attacks, to inquire into their experiences in retrospect, and asked if they were hostile to the forest. The interviews were formally structured but informally conducted, spanned 5 years, in five different forest areas of West Bengal, and were originally meant for a participatory study of human-wildlife conflicts in forest-fringe villages. Table 4 summarizes the responses to the interviews. Our limited sample size notwithstanding, only one person expressed his abhorrence to the forest and regretted having to spend his life in proximity to the forest. All others were rather stoic and recounted the attacks as "misfortunes" that should be anticipated for anyone entering the forest. The general opinion of most of the survivors was: "The forest is the habitat of wild animals, and it's the people who are intruders, so people must be responsible for their actions."

Table 4 Perceptions and attitudes of selected forest-fringe villagers who survived wildlife attacks inside the forest

Culprit animal	District/region (in West Bengal)	No. of respondents			Perceived cause of attack				Attitude toward the culprit animal	
					Misfortune		Self			
		M	F	Total	Accident	Divine design	Was naïve/tactless	Was intruder	Hostile	Tolerant
Tiger	Sunderban	5	0	5	2	1	1	1	3	2
Elephant	Bankura and West Medinipur	13	8	21	16	3	1	1	6	15
Snakes	Bankura and West Medinipur	6	9	15	10	2	0	3	15	0
Bees & Wasps	Bankura and Puruliya	11	7	18	14	1	1	2	16	2
Crocodile	Sunderban	1	2	3	2	1	0	0	3	0
Scorpion	Puruliya & West Medinipur	2	5	7	5	0	0	2	7	0
Total		38	31	69	49	8	3	9	50	19
Overall %		55	45	100	71	12	4	13	72	28

A Sunderban *mouli* (honey hunter) who survived a tiger attack inside the tiger reserve, recounted after 3 weeks of intensive hospital care: "One of our team members must have defiled the habitat in some way, so the tiger came to punish us." The general Indian religious attitude of viewing most misfortunes as divinely ordained is reflected in the responses of a significant proportion (12 %) of wildlife victims. About 13 % of the survivors saw their own fault in their horrid experiences – either they had not known the tactics to avoid the encounter with the animal, or intruded into the territory or the path of movement of the animal. Two out of seven scorpion victims and one-fifth of the snake victims thought their intrusion into the animals' path of movement evoked the defensive attack from the animal. Amongst the five tiger victims, one considered the incident was provoked by the men's entry into the predator's territory.

Most of the wildlife victims felt the existence of 'culprit' animals were undesirable. In particular, crocodile victims in the Sunderban villages believed that the forest department was unnecessarily maintaining the crocodiles, threatening the fishers' lives and livelihood. The victims' total abhorrence of venomous snakes seems to reflect the general attitude of people towards snakes, and linked to the widespread herpetophobia. However, about a quarter of the victims interviewed in our study considered the 'culprit' animals should exist (last column in Table 4), notwithstanding mortal risks to human life. This remarkable tolerant attitude seems to be shaped by certain cultural factors. For instance, a majority (ca. 40 %) of the elephant victims considered the elephants' should not disappear, because people in south-western districts of West Bengal identify the elephant as an embodiment of a benevolent deity. Furthermore, many people show fondness of the animal, owing to its mythical charm and intelligence. Two out of three tiger victims also considered tigers should exist, because the tiger is the associate of the forest goddess *Ban-bibi* of the Sunderban.

Knowledge of certain animals' ecological functions is also important in shaping people's tolerant attitude toward the animals, despite the risks to human life. A few victims of bee and wasp attacks were aware of the pollination service of the insects, and in spite of their near-death experience, wanted them to exist.[10] In contrast, ecological services and functions of crocodiles and scorpions were unknown to the population under study, and therefore all the victim survivors wanted the 'culprit' species to disappear from the area.

Regardless of people's fear or abhorrence of dangerous animals from the forest, none of the victims wanted to do away with the forest. All the wildlife victims reported their horrid encounters with the animals inside the forest, and all of them knew the forest was the habitat of all the 'culprit' species. Nevertheless, none of the victims had any qualms about the forest's existence. A majority (71 %) of the victims took the painful experiences in their stride as "accidents", which did not distort their overall perception of the forest's value, although they recognized the forest

---

[10] My own grief for the death of a thousand bees who stung me to anaphylaxis on January 1, 2011 was shaped by my knowledge of the bees' ecological services and functions, in addition to my knowledge of their fascinating eusociality and semiobiology.

as the habitat of dangerous animals. The forest was not only the basis of their economic existence, but also a pivot of their cultural life (see previous sections), and their permanent address. This point was most poignantly stated by Bishnupada Mandal of Satjelia, to explain why he had not considered relocating from the Sunderban even after his father's violent death from a tiger's attack: "One doesn't abandon his house after his father dies in the house" (personal interview on 22 August 1994 at WWF-India Eastern Region office, Kolkata). Like the survivors, their relatives also felt no particular animosity to the forest. Although they learned to take more caution to encounter forest animals, the forest *per se* was not perceived as a horrific object to them.

5   The Value of 'Non-forest' Vegetation

All over India, trees outside forests make a major contribution to vegetation cover. An estimated 2.68 billion trees outside forests contribute an equivalent additional area of 9.99 Mha, contributing to about 13 % of the country's total area under tree cover (FSI 2003). Stands of vegetation constitute important elements of the rural landscape, which seems to portray the traditional Indian context for living. Especially in the remote regions that are not yet modernized through industrial and urban development, the typical village landscape (see Fig. 4) includes houses surrounded by diverse trees, interspersed by ponds surrounded by palm trees, and at least a small sacred grove, containing old-growth trees. Home gardens are another considerable hotspot of tree diversity, which contribute substantially to the household economy. Tropical home gardens are by definition polycultures (cultivation of multiple crops and fruit trees on the same farm plot), with high species diversity valued for its consistent productive utility and convenience (Mazarolli 2011).

Indigenous people always tend to plant trees that have direct economic uses – for food, fodder, materials for implements and household articles, and for medicinal and ornamental purposes – on the homestead land and village commons. Fruit trees and trees for timber are most common in home gardens, but sacred trees, of no direct economic use, are also planted in conformity with local customs. A typical village with 200 households may contain 200–1,000 mature woody trees per square kilometers, planted on roadsides, homestead lands and pond margins.

Our study conducted in three forest-fringe villages (Bhagabandh, Arjunpur and Baghmara), and two villages (Garia and Ahmedpur) with no adjacent forest, in West Bengal, reveals that neither the species richness (S) nor density (d) of woody trees (GBH > 10 cm) in the village landscape is significantly related to the proximity to forest. The S, d, and Hill's index of diversity (e^H) show wide variations in the region (Table 5), and the difference in both S and d, between the villages with presence and in absence of adjacent forest, are not significant ($p > 0.20$). The vegetation profiles show that planted diversity of woody trees in home gardens and pond margins is considerably rich, while that of the roadside plantations is comparatively less (Fig. 3).

Table 5 Species count (*S*), density (No. km^{-2}) and hill's species diversity (e^H) of woody trees in five villages of West Bengal, compared to adjacent forest tracts

District	Village	Home-stead S	Pond margins S	Road-side S	Sacred Grove S	Village total S	No./km^2	Exp(H')	No. of spp. not found in Forest
Bankura	Bhagabandh	29	27	24	15	43	552	26.9	11
	Arjunpur	36	33	30	8	50	883	16.6	12
Puruliya	Baghmara	29	27	27	12	47	626	24.7	8
Birbhum	Garia	38	7	26	9	44	1,289	16.7	–
Hugli	Ahmedpur	21	9	10	2	28	202	17.6	–

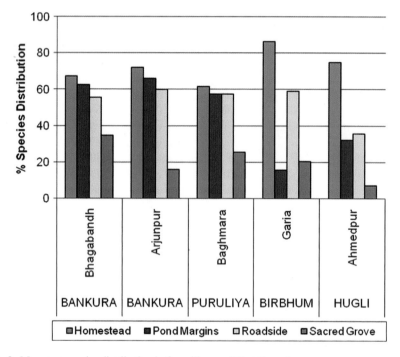

Fig. 3 Mean tree species distribution in five villages of West Bengal

The explicit purpose of the plantations is to meet household needs for fruits (mango, jackfruit, *Aegle marmelos*, *Moringa oleifera*), cattle fodder (leaves of jackfruit, *Feronia limonia*, *Syzygium cumini*), fuel (leaves of *Acacia auriculiformes*, twigs of *A. nilotica*), medicinals (*Azadirachta indica*, *Aegle marmelos*), wood for construction (*Syzygium cumini*, *Holoptelea integrifolia*, *Alangium lamarckii*, *Gmelina arborea*) and household articles such as mats and baskets (leaves of *Borassus flabolifer*, *Phoenix acaulis*). However, a small fraction of this husbanded biodiversity has no economic use whatsoever. In addition to the ornamental (*Delonyx regia*, *Nyctanthes arbor-tristis*) and sacred species (*Ficus religiosa*, *Adina cordifolia*), for which these indirect services can be obtained, there are a few species that are planted purely for their existence value. While economists can surmise some indirect use value for such trees, including carbon sequestration, albedo reduction, biodiversity enrichment, etc., villagers are not aware of these global ecological services while planting them. Rather, the existence value and bequest value of the trees with no uses are prominent. In our survey of four districts (Table 5), at least three species (*Crateva nurvala, Oroxylon indica* and *Streblus asper*) have no economic use, including ornamental or religious use, yet are frequently maintained on roadsides, pond margins, and even on homestead lands. This testifies to the fact that people in these villages tend to plant or nurture many trees regardless of their direct or indirect uses, because people like to see the *existence* of diverse trees in their surroundings.

Much of the diversity of tree species in the village landscape owes its existence to plantings by villagers, but a largely natural floral diversity is customarily maintained in village sacred groves. In remote rural areas of most Indian states, sacred groves are an important feature of the rural landscape, and exist as remnants of ancient forest patches, consecrated to a local deity or ancestral spirit. Extraction of any living matter from the sacred groves (SG) premises is generally prohibited, although there are some locale-specific exceptions to this rule. The entire village community usually observes the rules of protection of the SG, whose physical boundaries are customarily demarcated. In southwestern districts of West Bengal and southern Jharkhand, every village has at least one SG, each maintained by an ethnic community (Figs. 4c and 5).

The most remarkable characteristic of the SGs is that the constituent flora and fauna are protected, regardless of their economic use value. SGs may contain some sacred trees (e.g. *Ficus religiosa, Adina cordifolia, Aegle marmelos*) or sacred animals (e.g. Hanuman langur, bats or storks), which in many indigenous societies are protected both inside and outside SG premises. However, in some SGs, all the constituent plants and animals, regardless of their "sacred" status, are protected by generations of the local ethnic community. Thus, the institution of SG seems to testify that indigenous societies tend to sanctify and protect the entire habitat for its ecosystem value, rather than for the economic value of its constituent species.

A consequence of the conservation of the SG habitat is that a considerable number of rare plants and animals have been protected from extinction. Many rare and endangered biota have disappeared from the entire region, including the forest under state management, but persist in a few SGs (Malhotra et al. 2007). Such point endemism is demonstrated in a number of SGs in India; for instance, a SG in Kerala is the last habitat of the endemic liana *Kunstleria keralensis* (Mohanan and Nair 1981); the endemic frogs *Philautus sanctisilvaticus* and *Leptolalax khasiorum* were discovered from single SGs in Madhya Pradesh (Das and Chanda 1997) and Meghalaya (Das et al. 2010), respectively.

A comparative floristic survey in SGs and protected forests in West Bengal reveals that the SG relics are the last bastions of at least ten rare tree species, none of which are found outside the SGs (Table 6). These ten species have no direct economic use value, nevertheless are protected within the SGs. The indigenous

Fig. 4 (**a**) Trees and shrubs planted around a house beside the owner's farm, in Rayagada district, Odisha. (**b**) View of a tribal hut in Balangir district, Odisha. A live tree is used as a component of the fence. (**c**) A typical sacred grove in Bankura district, West Bengal with an image of deity and votives. (**d**) Vegetation in a village in Koraput district, Odisha

conservation ethic is illustrated in two fragmented sacred groves in Bankura district, which contain presumably the last adult specimens of *Vitex glabrata* R. Br. in West Bengal (Deb 2008a). Extensive search for a third specimen of this species in four adjoining States of eastern India and in Bangladesh has so far proved futile, indicating that the species metapopulation has collapsed in its prior natural habitat range due to agricultural extension, urbanization and industrialization. Similarly, there are no more than four adult specimens of the sitapatra tree (*Premna* sp.) within the 500 km^2 area surveyed, two of which occur in two derelict SGs of West Medinipur district. Based on the IUCN (2001) criteria of extremely narrow extent of occurrence, highly fragmented habitats, and the precariously small number of mature individuals, these species may be declared as Critically Endangered species in eastern India.

Rare and endangered trees recorded in this study include *Tamilnadia uliginosa* (Retz.) Tirveng. & Sastre, *Suregada multiflora* (Juss) Baill. and *Cordia myxa* Linn. These species, like *Vitex glabrata* R. Br., do not occur in the forest tracts of the districts surveyed for this study, and do not occur in any greater density than 1/100 km^2 in the study area. Clearly, these "inferior species" were exterminated by the State Forest Department's silvicultural practices. Standard literature of Indian flora (Prain 1903; Parkinson 1923) and herbarial records, preserved at J C Bose Botanic Garden, Shibpur, India and at the Royal Botanic Garden, Kew, UK indicate

Fig. 5 A pristine sacred pond, adjacent to an ancient sacred grove in Chilapata, West Bengal

Table 6 Distribution of threatened plants in West Bengal

Species	Occurrence[a] in SG	Forest	LOC	IND	AOO	Inferred status[b]
Ternanthera monopetala	Yes	Yes	< 300	< 1,000	< 20 km²	R
Crateva nurvala	Yes	Yes	< 200	< 800	< 10 km²	EN
Morinda angustifolia	Yes	Yes	< 200	< 800	< 10 km²	EN
Tamilnadia uliginosa	Yes	No	< 200	< 200	< 10 km²	EN
Ventilago madras-patana	Yes	Yes	< 150	< 200	< 10 km²	EN
Cordia myxa	Yes	No	< 100	< 100	< 5 km²	EN
Suregada multiflora	Yes	No	18	23	< 1 km²	EN
Sitapatra (*Premna* sp.)	Yes	Yes	4	4	< 25 m²	CR
Ficus krishnii	Yes	No	4	4	< 25 m²	CR
Vitex glabrata	Yes	No	2	2	< 10 m²	CR

Keys: [a] Within the extent of 500 km² surveyed. AOO Area of occupancy, LOC No. of sites of occurrence, IND No. of mature individuals. Categories inferred following IUCN (2001)
[b] For the critically endangered species, we followed the IUCN Criteria (Version 3.1):
B1a. Severely fragmented; and
B2. Area of occupancy < 10 km²; and
B2a. Severely fragmented populations; and
B2b. Continuing decline, observed, inferred or projected, in (i) extent of occurrence, (ii) area of occupancy, (iii) number of locations or subpopulations, (iv) number of mature individuals; and
C. <250 mature individuals

that these species were once widely distributed in eastern India, Bangladesh, and South East Asia about 100 years ago.[11]

Taken together, the composite vegetations grown and maintained in these villages, outside the formally recognized forest are apt to be called "folk forest" – comprising a rich biodiversity. Several tree species that have become rare, and at least locally endangered, exist in the folk forests, including the sacred groves. Conservation of rare taxa is an incidental consequence of the customary plantation and maintenance of vegetation by the villagers, regardless of the use values of the trees.

6 Concluding Discussion

All economic valuation of forest ecosystem goods and services are fated to be incomplete, even inappropriate, because the entire discourse of resource economics centers on the utilitarian, instrumental view derived from the ubiquity of market forces and utility-maximizing motives of homo oeconomicus. Such economic assumptions underlying the economic valuation methodologies "lack scientific validation of human behavior and subsequent decision-making processes of ecosystem management" (Kumar and Kumar 2008: 818). The economist's concern with the instrumental value of positional goods ignores the world of human cognitive and cultural complexity behind appreciation of the relational value of natural entities and their cultural meanings that transcend market and utilitarian considerations (Sagoff 2004; Tisdell 2005; Kumar and Kumar 2008). For this reason, the value perceptions and valuation languages of the ecosystem people differ from those of the economist.

Ecosystem people of course value the forest as a resource for its economic use value – the many kinds of forest goods and services derived from the forest, but they also value the forest for its non-use value – not as a resource but as an asset of nature, and for its various cultural meanings. However, the perception of this value is distorted under the modern economic arrangement of monetary values – a price tag, which captures only a fraction of its uses while concealing a legion of others. It is truism that the value of a forest is more than the sum total of the worth of all its component trees, but it is not always obvious that the market price of a 'valuable' tree does not capture the value of all its goods and services. The price of a tree log captures only its value on timber market, but omits the value of its fruits (as food to humans and/or other animals) and that of leaves (as organs to fix carbon), which were sacrificed when the tree was logged. This is because firstly, goods and services that are not directly consumed are difficult to value, and secondly, when a thing is placed on the market, its exchange value becomes non-additive – it only estimates

---

[11] Once widely distributed in India, Bangladesh and South East Asia, *V. glabrata* has now become rare in eastern India, and is nearly extinct in Thailand due to its over-extraction from the wild for its use in traditional Thai medicine (Chamnipa et al. 2012).

the value of that particular utility, to the exclusion of all other use values. Non-consumptive use value and non-use value of the forest therefore have appeared in the mainstream economic literature only recently, although attempts at assessment of these values are fraught with great uncertainty.

In contrast, non-market values of the forest as well as individual trees are acknowledged in pre-capitalist economies of contemporary hunter-gatherer and agrarian societies. These societies have, or used to have, various cultural mechanisms to appreciate non-use values, which are reflected along metaphorical corridors of the indigenous *Weltanscauung* – namely, sanctification of certain trees and groves, taboos prohibiting profligate resource use, and omens and auguries highlighting the significance of the existence value of biota (Fig. 2).

The notion that a thing is not valuable if it is not useful, took shape with the emergence in Europe of capitalism, which equated development of civilization with the advent of market that treated all natural objects as commodities. This had far-reaching implications for the perception of the value of forest in India and other colonies. Insofar as the forest was not "productive" of revenue or marketable goods, the political economy of development used to define it as "wastelands", and sought to "develop" it by agriculture and other "civilized" forms of economy. However, the economy of development before long discovered value in the forest as a resource, as soon as the forest provided materials for railways and shipbuilding. This discovery of value entailed enclosure of the valued asset. In the process, native ecosystem people and their perception of the value of forest were extirpated and pushed into oblivion.

However, as the value of forest was largely determined by the prime mover of development – private profit, whenever alternative means of generating more profit are discovered, the forest – or any other 'natural capital' – is disposed of. Thus, assured profits from aluminium mines, oil wells, and cement factories as a rule necessitate forest clearing, and entail de-regulation of the forest to circumvent the state's conservation objectives and mandates. The motive of maximizing profits still propels the economies of both ex-empires and ex-colonies, which continue with the land use policy of fostering industrial development. In modern India, large tracts of forest have been sacrificed, and national parks have been 'denotified' – to make way for factories and mines, and forest products are supplied with massive subsidies to industry – to safeguard the 'national interest' of industrial growth. Concomitantly, customary user rights are abrogated to give land titles to entrepreneurs and corporations who want to work and develop 'unproductive' forests. This legalizes a system of disenfranchising the traditional user community, who consequently lose all interests to conserve them, not least because their inter-generational right to use the resource is abrogated. The local user community that is divested of right to use the enclosed forest is apt to perceive the commons as an open access resource at best (Ostrőm 2005), or now the enemy's property at worst (Deb 2008b). The result has been decimation of the resource, either by rapid resource extraction for individual profit, or gratuitous plunder, as a token of protest by the traditional user community.

It transpires that the forest, divorced from the traditional community of users, and managed by a statutory authority, whether for industry and commerce or even for well intended pure conservation aims, tends to lose all non-consumptive use and non-use values, but instead acquires an overriding instrumental value based on the crude market value of its goods (or provisioning services, *sensu* MEA [2005]). Thus, forest villagers who have no rights over the extraction of valuable forest goods tend to be profligate with the extraction of the resource. Combined with the market inroads, the insidious intrusion of non-tribal cultural mores is steadily replacing the local dialects, altering food and dressing habits, changing belief systems, modes of resource governance and livelihood, and "has led to a serious decline of the very same mores and norms that symbolized the adivasi spirit of intrinsic regard for harmony and celebration of life processes" (Narendra 2009: 5). Today, the adivasi societies demand the return of their forests, not necessarily for their cultural value, as in the past, but in order to harvest them for commerce, often in a profligate manner. They tend to extract forest goods for sale on market for quick money, and remove the trees that have no commercial value (personal obsv.). The industry-bureaucracy clique often abets this vicious scrambling for the forest resource, and takes advantage of the popular movements to plunder the forest. The individual's pursuit of profit, at the expense of the public good and quality of life, is encouraged by "liberalism's minimalist state and the ability of individuals to amass great quantities of property or capital" (Shutkin 2000: 42).

In contrast, when a community of local users manages the forest as a true commons, members of the community tend to observe certain tacit rules that ultimately serve to prohibit exhaustion and gratuitous destruction of the resource (Oström 2005, 2009). A plethora of studies show that the users of the commons make considerable investments in "designing and implementing costly governance systems to increase the likelihood of sustaining them" (Oström 2009: 210). Such governance systems foster self-regulation among individual users, which is motivated by altruistic moral preferences that are reciprocal, and tend to be stronger with socioeconomic proximity between citizens (Baron 2010). The socioeconomic distance between members of the community is reduced when they participate in managing the commons with a "shared vision of good life" (Bernard and Young 1997: 205). All stories of successful management of the forest commons, where individual restraint is invoked for common good, are stories of a shared vision of the community.

This shared vision is an outcome of shared history and cultural identity, which characterizes indigenous ecosystem people, who consider the forest as a *sine qua non* of their material wellbeing and cultural life. Forest tribes such as the Warlis believe that once the forests go, their cultures will go as well. To the Warlis, the forest is the means in the present through which they can transmit the collective wisdom of their ancestors and their culture to the future, the new generation born and unborn (Prabhu 2003: 77). Like the Menominee of Wisconsin (see Box 1), the Kondh of Odisha proclaim, "Our forests are our history and our culture (*amor jangalo, amaro itihas ote avom a thi amoro sanskriti*)" (cited in Kapoor 2007: 23).

This organic, cultural link of the indigenous user community to the forest is destroyed when the ownership and customary right of governance of the forest are wrested from the community's hands. There is ample empirical evidence to indicate that both private or state ownership of resources invariably leads to the collapse of the commons, and that resources are best conserved so long as their management is in the common ownership of the user community (Bernard and Young 1997; Shutkin 2000; Oström 2005; Deb 2007a, 2009). Indeed, revival of traditional community forest management systems has in several instances rescued the forest from destruction. To cite a few, the Kollaba in the Indian State of Karnataka revived their tradition of community forest management (CFM) in the 1920s, in response to the forest destruction under management of the state forest department; the Mizo tribe in Mizoram reinstated in 1990 their tradition of community-managed "safety forest" to recuperate their forests (Deb 2009); more recently, self-initiated CFM in Odisha have regenerated and restocked degraded forests (Conroy et al. 2002). In all these cases, the forest people's perception of the value of forest transcended the instrumental value of the forest goods, and incorporated the intrinsic value of, and restored cultural links to, the forest. It is not too late to reorient national forest policies and state management systems, in tune with the robust lesson from history, that the forest must be liberated from statutory enclosures, and brought back to its 'commons' status, in order to ensure its conservation by the people, for the people.

Acknowledgements I am grateful to Debdulal Bhattacharya, Bhairab Saini, Arun Ram, Debashis Mukherjee, Subrata Das, Nirmal Mandal, Bishnu Mandal, Shanti Roy and the late Rabi Mahato for diligent assistance in gathering data about non-forest vegetaion and wildlife victims in 5 districts of West Bengal.

References

ACHR (Asian Centre for Human Rights) (2005) Promising picture or broken future? Commentary and recommendations on the 'Draft National Policy on Tribals' of the Government of India. www.achrweb.org

Banerjee, Ghosh AS, Springate-Baginski O (2010) Obstructed access to forest justice in West Bengal: state violations in the mis-implementation of the Forest Rights Act 2006. IPPG discussion paper #49. Improving Institutions for Pro-Poor Growth Research Programme, Manchester, UK. http://www.ippg.org.uk/papers/dp49.pdf

Bann C (1997) The economic valuation of tropical forest land use options: a manual for researchers. EEPFSA/IDRC, Singapore. http://www.idrc.org.sg/eepsea

Baron DP (2010) Morally motivated self-regulation. Am Econ Rev 100:1299–1329

Bernard T, Young J (1997) The ecology of hope: communities collaborate for sustainability. New Society Publishers, East Haven

Blaikie P, Springate-Baginski O (2007) Understanding the policy process. In: Springate-Baginski O, Blaikie P (eds) Forest, people and power: the political ecology of reform in South Asia. Earthscan, London, pp 61–91

Blomley N (2007) Making private property: enclosure, common right and the work of hedges. Rural Hist 18(1):1–21

Bose P, Arts B, van Dijk H (2012) 'Forest governmentality': a genealogy of subject-making of forest-dependent 'scheduled tribes' in India. Land Use Policy 29:664–673

Bryant R (1996) The political ecology of forestry in Burma 1824–1994. University of Hawaii Press, Honolulu

Burke BE (2001) Hardin revisited: a critical look at perception and the logic of the commons. Hum Ecol 29:449–476

Chamnipa N, Thanonkeo S, Thanonkeo P (2012) Enhanced production of 20-hydroxyecdysone in cell suspension cultures of *Vitex glabrata* R.Br. by elicitor feeding. J Med Plant Res 6:3317–3323

Chiuri W (2005) Planning sustainable development in Sub-Saharan Africa: is there room for indigenous land tenure and knowledge systems? A case study of the Agikuyu people in Kenya, pp 39–60. In: Sustainable Development Institute, sharing indigenous wisdom: an international dialogue on sustainable development. SDI, College of Menominee Nation, Green Bay

Colchester M (2006) Justice in the forests: rural livelihoods and forest law enforcement. CIFOR, Bogor

Conroy C, Mishra A, Rai A (2002) Learning from self-initiated community forest management in Orissa, India. Forest Policy Econ 4:227–237

Cronkleton P, Bray DB, Medina G (2011) Community forest management and the emergence of multi-scale governance institutions: lessons for REDD + development from Mexico, Brazil and Bolivia. Forests 2:451–473

Das I, Chanda SK (1997) Phialutus sanctisilvaticus (Anura: Rhachophoridae), a new frog from the sacred grove of Amarkantak, central India. Hamadryad 22:21–27

Das I, Lyngdoh Tron RK, Duwaki R, Hooroo RNK (2010) A new species of Leptolalax (Anura: Megophryidae) from the sacred groves of Mawphlang, Meghalaya, north-eastern India. Zootaxa 2339:44–56

Davis M (2001) Late Victorian holocausts: El Niño, famines and the making of the third world. Verso, London

Deb D (2007a) Sacred groves of West Bengal: a model of community forest management. Working Paper No 8. Understanding livelihood impacts of participatory forest management implementation in India and Nepal. (Ser. Ed. Oliver Sprigate-Baginski). University of East Anglia, Norwich. www.uea.ac.uk/dev/People/staffresearch/ospringate-baginskiresearch/PFM-Nepal-India/8-sacred-groves-west-bengal

Deb D (2007b) Personal experience of researching sacred groves adjacent to joint forest management forests. In: Sprigate-Baginski O, Blaikie P (eds) Forest, people and power. Earthscan, London, pp 254–255

Deb D (2008a) Sacred ecosystems of West Bengal. In: Ghosh AK (ed) Status of environment in West Bengal: a citizens' report. ENDEV Society for Environment and Development, Kolkata, pp 117–126

Deb D (2008b) Joint forest management. In: Ghosh AK (ed) Status of environment in West Bengal: a citizens' report. ENDEV Society for Environment and Development, Kolkata, pp 96–104

Deb D (2009) Beyond developmentality: constructing inclusive freedom and sustainability. Earthscan, London

Deb D, Malhotra KC (2001) Conservation ethos in local traditions: the West Bengal heritage. Soc Nat Resour 14:711–724

Diamond JM (2005) Collapse. Penguin, London

Equations (2007) This is our homeland: A collection of essays on the betrayal of adivasi rights in India. Equations, Bangalore. Available at www.equitabletourism.org

Escobar A (1995) Encountering development: the making and unmaking of the Third World. Princeton University Press, Princeton

ETC. Group (1994) Bioprospecting/biopiracy and indigenous peoples. http://www.etcgroup.org/en/node/482# Last accessed on 26 Mar 2012

Faber M, Petersen T, Schiller J (2002) 'Homo oeconomicus and homo politicus in ecological economics. Ecol Econ 40:323–333

Flint EP (1998) Deforestation and land use in northern India with a focus on sal (Shorea robusta) forests 1880–1980. In: Grove R, Vinita D, Sangwan S (eds) Nature and the orient. Oxford University Press, New Delhi, pp 421–483

Fromm E (1973) The anatomy of human destructiveness. Holt, Reinehart and Winston, London

FSI (2003) State of the forest report 2003. Forest Survey of India, Ministry of Environment and Forest, Government of India, Dehradun

Gadgil M, Guha R (1995) Ecology and equity. Psnguin India, New Delhi

Goldewijk KK, Navin R (2001) Land use, land cover and soil sciences, vol I, Land use changes during the past 300 years. UNESCO, Paris

GoI (Government of India) (1952) National forest policy. Government of India, New Delhi

Grove R, Damodaran V, Sangwan S (eds) (1998) Nature and the orient. Oxford University Press, New Delhi

Gurven M (2004) To give and to give not: the behavioral ecology of human food transfers. Behav Brain Sci 27:543–583

Hanley N, Milne J (1996) Ethical beliefs and behaviour in contingent valuation surveys. J Environ Plan Manag 39:255–272

Harvey D (1996) Justice, nature and the geography of difference. Blackwell, London

Hughes JD (2001) An environmental history of the world. Routledge, London

IUCN (2001) Red list – categories and criteria (version 3.1). IUCN/SSC Red List Programme, Cambridge. http://www.iucnredlist.org/apps/redlist/static/categories_criteria_3_1#critical

Joshi G (2010) Forest policy and tribal development. Cultural survival (March 2010). http://www.culturalsurvival.org/ourpublications/csq/article/forest-policy-and-tribal-development

Kapoor D (2007) Subaltern social movement learning and the decolonization of space in India. Int Educ 37(1):10–41

Kellert SR (1996) The value of life. Island Press, Washington, DC

Kothari A, Suri S, Singh N (1995) People and protected areas: rethinking conservation in India. Ecologist 25(5):288–294

Kumar M, Kumar P (2008) Valuation of the ecosystem services: a psycho-cultural perspective. Ecol Econ 64:808–819

Lelé S, Springate-Baginski O, Springate-Baginski O, Lakerveld R, Deb D, Dash P (2013) Ecosystem services: origins, contributions, pitfalls, and alternatives. Conserv Soc (in press)

McNeill JR (2000) Something new under the sun: an environmental history of the twentieth century world. W.W. Norton, New York

Maffi L (2004) Conservation and the "two cultures": bridging the gap. Policy Matters Issue 13 History Cult Conserv :256–266

Malhotra KC, Deb D, Dutta M, Adhikari M, Yadav G (1992) The role of non-timber forest products in village economies of southwest Bengal. Rural Development Network Paper 15d. (Summer 1993). Overseas Development Institute, London. http://www.mekonginfo.org/mrc/rdf-odi/english/papers/rdfn/15d-i.pdf

Malhotra KC, Gokhale Y, Chatterjee S, Srivastava S (2007) Sacred groves in India: an overview. Indira Gandhi National Museum of Mankind/Aryan Books International, New Delhi

Manoka B (1997) Existence value: a re-appraisal and cross-cultural comparison. IDRC, Ottawa. http://203.116.43.77/publications/research1/ACF26B.html

Marchak P (1989) What happens when common property becomes uncommon? BC Stud 80 (Winter 1988–1989):1–21.

Martinez-Alier J, Kallis G, Veuthey S, Walter M, Temper L (2010) Social metabolism, ecological distribution conflicts and valuation languages. Ecol Econ 72:153–158

Marx, Karl (1862) [1989]. Theories of Surplus Value. Part 1. Progress Publishers, Moscow

Marx K (1887) [1954] Capital, vol 1. Progress, Moscow

Mazarolli DM (2011) The Benefits of tropical homegardens. The Overstory 239. Permanent Agriculture Resources, Holualoa. http://www.overstory.org

Medaglia JC (2007) Bioprocessing partnerships in practice: a decade of experiences at INBio in Costa Rica. In: Phillips PWB, Onwuekwe CB (eds) Accessing and sharing the benefits of the genomics revolution. Springer, New York, pp 183–195

Millennium Ecosystem Assessment (2005) Ecosystems and human well-being. Island Press, Washington, DC

Mishra PK (2008) Globalisation and deforestation: a case study of Lapanga (Sambalpur, Orissa). In: Global forest coalition 2008. Report of the national workshop on underlying causes of deforestation and forest degradation in India, Bhubaneswar, Orissa: 26th to 28th January 2008, pp 101–118. http://vh-gfc.dpi.nl/img/userpics/File/UnderlyingCauses/India-Report-Underlying-Causes-Workshop.pdf

Mohanan CN, Nair NC (1981) *Kunstleria* Prain – a new genus record for India and a new species in the genus. Proc Indian Natl Sci Acad B 90:207–210

Nantel P, Gagnon D, Nault A (1996) Population viability analysis of American ginseng and wild leek harvested in stochastic environments. Conserv Biol 10:608–621

Narendra (2009) Adivasi culture and civil society. North South Perspect 2(1) (August 2009):4–5. http://www.siemenpuu.org/download/5529

Neeson JW (1993) Commoners: common right, enclosure, and social change in England, 1700–1820. Cambridge University Press, Cambridge

Oström E (2005) Understanding institutional diversity. Princeton University Press, Princeton

Oström E (2009) A general framework for analyzing sustainability of social-ecological systems. Science 325:419–422

Padel F, Das S (2010) Out of this earth: East India Adivasis and the aluminium cartel. Orient Black Swan, New Delhi

Parkinson CE (1923) A forest flora of the Andaman Islands. Bishen Singh Mahendra Pal Singh, Dehra Dun

Pecore M (2005) Menominee sustainable forestry. In: Sustainable Development Institute, sharing indigenous wisdom: an international dialogue on sustainable development. SDI, College of Menominee Nation, Green Bay, pp 175–179

Pimple M, Sethi M (2005) Occupation of land in India: experiences and challenges. In: Moyo S, Yeras P (eds) Reclaiming land: the resurgence of rural movements in Africa, Asia and Latin America. Zed, London, pp 235–256

Prabhu P (2003) Nature, culture and diversity: the indigenous way of life. In: Kothari S, Ahmad I, Reifeld H (eds) The value of nature: ecological politics in India. Konrad Adanauer Stiftung/Rainbow, New Delhi, pp 39–82

Prain D (1903) Bengal plants,vols I & II. W. Newman, London

Rajan R (1998) Imperial environmentalism or environmental imperialism? European forestry, colonial foresters and the agendas for forest management in British India 1800–1900. In: Grove R, Damodaran V, Sangwan S (eds) Nature and the orient. Oxford University Press, New Delhi, pp 324–371

Randall A, Stoll JR (1983) Existence value in a total valuation framework. In: Rowe RD, Chestnut LG (eds) Managing air quality and scenic resources at national parks and wilderness areas. Westview Press, Boulder

Reid WV, Laird SA, Meyer CA, Gamez R, Sittenfeld A, Janzen DH, Gollin MA, Juma C (eds) (1993) Biodiversity prospecting: using genetic resources for sustainable development. World Resources Institute, Washington, DC

von Ribbentrop B (1900) Forestry in British India. Office of the Superintendent of the Government Printing, Calcutta

Robinson G (1975) The forest service. Johns Hopkins University Press, Baltimore

Rocheleau D, Ross L, Morrobel J, Hernandez R (2000) Community, ecology, and landscape change in Zambrana-Chacuey. In: Harris JM (ed) Rethinking sustainability. University of Michigan Press, Ann Arbor, pp 249–286

Sagoff M (2004) Price, principle, and the environment. Cambridge University Press, Cambridge

Sainath P (1996) Everyone likes a good drought: stories from India's poorest districts. Penguin, London/Delhi

SCBD (2001). Sustainable management of non-timber forest resources. CBD Technical Series no. 6. Secretariat of the Convention on Biological Diversity, Montreal. http://www.biodiv.org

Siemenpuu Foundation (2008) Wild forests: making sense with people. Siemenpuu Foundation/CEDA Trust/FoEI, Helsinki

Shankar U, Murali KS, Shaanker RU, Ganeshaiah KN, Bawa KS (1996) Extraction of non-timber forest products in the forests of Biligiri Rangan hills, India. 3. Productivity, extraction and prospects of sustainable harvest of Amla (Phyllanthus emblica), Euphorbiaceae. Econ Bot 50:270–279

Sheridan MJ, Nyamweru C (eds) (2008) African sacred groves: ecological dynamics and social change. James Currey, Oxford

Shutkin WA (2000) The land that could be: environmentalism and democracy in the twenty-first century. MIT Press, Manchester

Shrivastava A, Kothari A (2012) Churning the earth: the making of global India. Penguin/Viking, New Delhi

Sivaramakrishnan K (1999) Modern forests: statemaking and environmental change in colonial eastern India. Oxford University Press, New Delhi

Spence MD (1999) Dispossessing the wilderness: Indian removal and the making of the National Parks. Oxford University Press, New York

Springate-Baginski O, Blaikie P (eds) (2007) Forest, people and power: the political ecology of reform in South Asia. Earthscan, London

Tisdell C (2005) Economics of environmental conservation, 2nd edn. Edward Elgar, Cheltenham, UK

Turner N, Berkes F (2006) Coming to understanding: developing conservation through incremental learning in the Pacific Northwest. Hum Ecol 34:495–513

Venkateswar S (2004) Development and ethnocide: colonial practices in the Andaman Islands. IWGIA Document No. 111. International Work Group on Indigenous Affairs, Copenhagen

Wani, Milind and Ashish Kothari (2008). Globalisation vs India's forests. Economic and Political Weekly Sept 13: 19–22

Wilson EO (1992) The diversity of life. Penguin, London

Part II
Forest Resources Worldwide

Global Trends and Outlook for Forest Resources

Adrian Whiteman

Abstract With a growing population, the demands placed on the World's forest resources are likely to continue to expand. This presents forest owners and managers with many challenges although, as this chapter suggests, some trends in the future may benefit forests and forestry.

This chapter starts by describing some recent trends in forest resources, such as trends in forest area and characteristics, forest management and use. It shows that forests are now being managed more sustainably in many respects, but that progress has been mixed in some regions and with respect to some indicators of sustainable forest management. It also shows that forests are currently managed and used for many diverse benefits and that the importance of these different benefits has changed over time.

The chapter then explains how some of the main driving forces (population trends, economic growth, government policies, etc.) have led to differences in forest area, forest management and use in different parts of the World. In particular, this focuses on how these forces have affected land-use change (i.e. forest conversion) and demand for forest products and services, as well as how government policies have attempted to change these demands.

The chapter concludes by describing what might happen to the driving forces in the future and how this would affect forest resources and forest management. One key message is that demographic trends (towards smaller and older rural populations) will significantly alter the processes leading to land-use change. The intensity of forest management will also have to increase in the future to meet the expanding demands for forest goods and services, but it is unclear how this can be done for anything other than wood production. Finally, it describes how forestry policies and

A. Whiteman (✉)
FAO, Rome, Italy
e-mail: adrian.whiteman@fao.org

T. Fenning (ed.), *Challenges and Opportunities for the World's Forests in the 21st Century*, Forestry Sciences 81, DOI 10.1007/978-94-007-7076-8_8,

institutions might adapt to these changes if they are to meet the rising expectations and continue to support sustainability in the sector.

1 Introduction

The management and use of forest resources varies greatly across the World, depending on factors such as the amount and type of forests present in a country, local social and economic circumstances, history, traditions and government policies both within and outside the sector. Furthermore, forest management and use continue to evolve over time in response to changes in these external factors as well as changes in the characteristics of the resource.

Forest management is also complex because forests can produce such a wide variety of goods and services. Many of these outputs can be produced simultaneously, but often there are also trade-offs between them, especially between the commercial and non-market outputs from forests. While the demands for these outputs varies between countries and over time, it is probably true to say that they are mostly increasing, leading to ever more complicated and difficult decisions for forest managers and policymakers wishing to satisfy these competing demands.

This chapter will focus on some of the social and economic dimensions of forest management and use and describe how they have affected forest resources in the past and may affect them in the future. This will include a discussion of how policies and extra-sectoral developments affect forests and forest management and, in particular, a discussion of how forest management may adapt to changing circumstances in the future.

The chapter is divided into three sections examining the trends in forest management and use and the driving forces affecting forests, then concluding by presenting a tentative outlook for the future. It will examine developments at the global level, so the analysis will be broad and not necessarily apply to all countries. It will, however, examine differences between regions and groups of countries where that is useful.

2 Trends in Forest Management and Use

Since the mid-1980s, the Food and Agriculture Organization of the United Nations (FAO) has been systematically collecting information about the trends and current status of forest resources in every country in the World. Gradually, the amount of information collected in each global Forest Resource Assessment (FRA) has expanded to cover many different and diverse measures of the extent of forest resources, their characteristics and information about how they are managed and used. The results of these assessments (plus information from other sources) are presented below to highlight some of the most significant changes in forests that have occurred since the 1980s.

Table 1 Recent trends in forest area

Region	Forest area (m ha)			Change in forest area (%/yr)		
	1990	2000	2010	1980–1990	1990–2000	2000–2010
Europe	989	998	1,005	0.0	0.1	0.1
North America	677	677	679	0.0	0.0	0.0
Africa	749	709	674	−0.7	−0.6	−0.5
Asia-Pacific	775	769	784	−0.3	−0.1	0.2
Latin America and Caribbean	978	933	891	−0.7	−0.5	−0.5
World	4,168	4,085	4,033	−0.4	−0.2	−0.1

Source: FAO (1995, 2010a)
Note: The 1980–1990 change in forest area is calculated using slightly different definitions to the later figures. Mexico is included as part of North America

2.1 Forest Area and Characteristics

Table 1 shows recent trends in the area of forests at the global level and in each of the five main regions of the World. At present, forests cover slightly more than 4 billion hectares (equal to 31 % of the global land area). An additional 1.1 billion hectares (9 % of the land area) is classified as "other wooded land" (for details of the definitions used in the FRA, see FAO 2007).

About 25 % of the global forest area is located in Europe (including the Russian Federation) and another 22 % is in Latin America and the Caribbean. These two regions also have the highest levels of forest cover (44 % and 48 % respectively). The remaining area is divided roughly equally between North America (Canada, Mexico and the United States of America), Africa and the Asia-Pacific region, but these three regions have lower and quite different levels of forest cover (at 31 %, 22 % and 19 % respectively).

At the global level, forest area has declined over each of the last three decades, but this decline appears to have slowed over the period. This is partly because levels of deforestation have reduced in a number of countries, but it also reflects an expansion of forest area (mostly due to afforestation) in other countries.

The changes in forest area in each of the five main regions are very different, but all show that the trends are gradually improving. Forest area has declined in Africa and Latin America and the Caribbean, but the loss of forest each decade has become smaller. Conversely, forest area has increased slightly in Europe and North America over the last three decades. The Asia-Pacific region is unique in that the change in forest area has reversed from a loss in earlier decades to an increase in the period 2000–2010.

The FRA also asks countries about the area of forests in three different categories: primary forest; other naturally regenerated forest; and planted forest. The latter two categories also each have a sub-category for the area where introduced species predominate. These forest characteristics represent, in a simple way, the degree to which forests have been altered by humans (e.g. no major indications of human

Table 2 Forest characteristics in 2010

Region	Primary forest (m ha)	(%)	Other naturally regenerated forest (m ha)	(%)	Planted forest (m ha)	(%)	Not specified (m ha)	Total
Europe	262	26	669	67	69	7	4	1,005
North America	275	41	366	54	38	6	0	679
Africa	48	7	437	65	15	2	174	674
Asia-Pacific	145	19	510	65	127	16	2	784
Latin America and Caribbean	629	71	199	22	15	2	48	891
World	1,359	34	2,182	54	264	7	229	4,033

Source: FAO (2010a). See FAO (2007) for definitions of the different forest characteristics

activities in natural forest and, at the other end of the scale, trees established by planting or deliberate seeding in planted forests).

Globally, just over half of the World's forests fall into the middle category and about a third of the area is classified as primary forest (see Table 2). Planted forests amount to only about 7 % of the total and a similar amount is unclassified. At the regional level, forest characteristics in Europe, North America and the Asia-Pacific region are quite similar to the global average, although the Asia-Pacific region has a relatively large area of planted forest and a small proportion of primary forest. In Africa, most forest is in the middle category, indicating that very little primary forest remains and that there has also not been much investment in forest plantations. By far the largest area of primary forest is located in Latin America and the Caribbean (i.e. the Amazon Basin).

Historical figures for the area of primary and planted forests have also been provided by most countries and, from these, the area of other naturally regenerated forest can be deduced. Changes in the area of each of these are shown in Table 3. As this table confirms, the change in total forest area is a combination of changes (mostly losses) in primary and other naturally regenerated forest and increases in the area of planted forest. In particular, outside Europe and North America, the other three regions show losses of primary forest that are higher than the global average. The area of planted forest is increasing in all regions, with quite a high growth rate. However, this only has a significant impact on the overall change in forest area in the Asia-Pacific region, where planted forests are relatively important.

In addition to forest area, the quality or condition of forests is another important dimension that should be examined when looking at trends in forest resources. However, changes in this dimension (most commonly referred to as "forest degradation" – i.e. for a negative change in condition) are very difficult to define and measure. For example, Lanly (2003) noted a decade ago that many studies of forest degradation have used imprecise and often subjective interpretations of the term that make it difficult to analyse trends and examine the causes and effects of such changes.

Table 3 Change in forest area (percent per year) by forest characteristics and region

Region	Primary forest		Other forest		Planted forest		All forest	
	90–00	00–10	90–00	00–10	90–00	00–10	90–00	00–10
Europe	0.7	−0.1	−0.2	0.1	1.0	0.6	0.1	0.1
North America	0.1	0.1	−0.2	−0.1	3.9	1.9	0.1	0.1
Africa	−1.2	−1.0	−0.6	−0.6	1.1	1.8	−0.6	−0.6
Asia-Pacific	−0.2	−0.5	−0.6	−0.1	2.0	2.8	−0.1	0.3
Latin America and Caribbean	−0.5	−0.5	−0.5	−0.6	2.8	4.3	−0.5	−0.4
World	−0.2	−0.3	−0.4	−0.2	1.9	2.1	−0.2	−0.1

Source: FAO (2010a)

Note: this data only covers countries reporting area of primary and planted forests over the whole period

There have been a few global assessments of land degradation that have followed rigorous scientific methodologies (e.g. Bai et al 2008) and these show where the amount of vegetation has changed over time. However, they do not separate changes in forest cover from changes in the condition of forests and they do not examine other important aspects of forest degradation, such as losses of biodiversity, productivity and ecosystem functions.

The FRA does not collect data on forest degradation, but FAO has reviewed the many existing definitions, measures and indicators (FAO 2011a), with a view to collecting comparable data in the future. One partial indicator of the condition of forest resources is the growing stock volume per hectare and some countries do report their growing stock volume to the FRA as well as forest area (so this can be calculated). However, in some cases, it appears that countries simply calculate growing stock volume in each year as the forest area multiplied by a single figure (i.e. their estimate of volume per hectare), rendering these figures useless for assessing trends in forest stocking.

The countries that do seem to provide independent estimates of forest area and growing stock are mostly temperate countries and in many of them growing stock per hectare appears to be increasing. A similar trend since 1950 was also noted for most European countries in first European Forest Sector Outlook Study (EFSOS) published in 2005 (Gold 2003; UN 2005). Similarly reliable and comprehensive information is not available for most tropical countries, but it is commonly thought that forest degradation or a decline in volume per hectare is the most likely trend in many of these countries.

2.2 Forest Management

Since the United Nations Conference on Environment and Development (UNCED) in Rio de Janeiro in 1992, sustainable forest management (SFM) has been the aim of forestry policymakers all over the World. SFM follows the broad principles of sustainable development, in that forests should be managed in a way that meets the needs of the present without compromising the ability of future generations to meet their needs. SFM is an extension of the much earlier concept of sustained yield familiar to most foresters and first described 300 years ago in Germany (Grober 2007). However, it is concerned with sustaining the social, environmental and economic values or outputs from forests rather than just timber yields (for one commonly accepted definition of SFM, see: MCPFE 1993).

To assess progress towards SFM, a number of different organisations have developed methodologies (criteria and indicators) for measuring the multiple values or outputs from forests. There are currently nine different international processes working on criteria and indicators, with some of them working at the national level while others are applied at the level of forest management units (e.g. for forest certification).

At the national level, most of the World's forests are now covered by one of the three main sets of criteria and indicators (see Box 1). However, there are still many challenges with respect to measuring sustainability in the sector, including the following:

- **Data quality**: the quality and availability of data required to measure some of the indicators is limited, particularly in terms of how recently data has been collected (e.g. a lot of information about forest resources is collected in forest inventories that may not be updated very frequently). Furthermore, some of the indicators that appear to be most valid are also the ones where it seems to be most difficult to obtain recent and reliable data.
- **Measurement at the global and regional level**: the three main national criteria and indicator systems currently cover 90 countries, but this is still only about half of the countries in the World. In particular, these systems cover very few small island states and arid zone countries (e.g. in North Africa, the Near East, Central Asia and South Asia) and a few significant countries, such as India and South Africa, are not included under the three main systems. In addition, due to the different indicators used in each system, many of the measures are not directly comparable and can not be aggregated to give a regional or global assessment.
- **Measurement validity**: in a number of areas, criteria and indicators measure inputs rather than outputs (e.g. the proportion of a country's forest area covered by management plans, the proportion of area managed for different purposes or designated as protected areas). These indicators may be a good proxy for achievement in some cases, but some of them are quite subjective and may not reflect real achievements in progress towards SFM on the ground. A related problem is that it is not always clear whether an increase in some of the measures reflects an improvement in performance or not. This is particularly the case with some of the social indicators (where changes may occur due to external factors) and it is somewhat ambiguous whether, say, an increase in collection of non-wood forest products represents an improvement in social welfare. Similarly, the economic indicators that measure the value of production may reflect changes in market conditions much more than changes in the way that forests are managed.

Box 1 Criteria and Indicators for Sustainable Forest Management

The three main systems for measuring progress towards SFM at the national level are as follows:

- **Criteria and indicators for the sustainable management of tropical forests.** These were developed by the International Tropical Timber Organization (ITTO) and issued in 1992, with revisions in 1998 and 2005 (ITTO 2005). They cover all ITTO producer countries (most major producers of forest products in the tropical zone). The indicators are used, amongst other things, to produce periodic assessments of how much of the permanent forest estate in the tropics is sustainably managed (ITTO 2011).

- **Pan-European indicators for sustainable forest management.** These were developed by the Ministerial Conference on the Protection of Forests in Europe (MCPFE – now renamed "Forest Europe") and issued in 1994, with a revision in 2002 (MCPFE 2002). They are used in 46 European countries including Turkey and the Russian Federation (which is also included in the Montréal Process). Reports are produced periodically showing progress under each criterion and indicator for individual countries and European sub-regions (Forest Europe 2011).

- **Montréal Process criteria and indicators for the conservation and sustainable management of temperate and boreal forests.** These were first issued in 1995, with revisions in 2007 and 2009 (Montréal Process 2009). They are used in 12 countries accounting for the majority of temperate and boreal forests outside Europe.

Each system contains a number of criteria that describe different aspects of sustainability in the economic, social or environmental dimension and, for each criteria, one or more indicators. These indicators attempt to measure performance of the forestry sector over time against each of the criteria.

- **Aggregation and comparison across indicators**: one final issue concerning the utility of criteria and indicators is that they can not easily be aggregated to give an overall measure of progress in a country.[1] Reflecting the fact that forests produce multiple outputs, they have been specifically designed to try to capture progress in the different dimensions of SFM, but this makes it very difficult to communicate the results to non-specialists. They also do little to help assess trade-offs or provide an overall picture of whether forests are being managed more sustainably or not. For example, forest area in a country may be declining

---

[1] The one exception to this is the ITTO indicators, where the ITTO reports the area of tropical forests that is sustainably managed (although it is somewhat unclear how the indicators are used to arrive at the final figures presented in their reports).

at the same time as many other indicators appear to be improving, but there would probably be little agreement about whether such a country was making progress towards SFM or not.

The FRA also presents information about progress towards SFM in seven thematic areas: extent of forest resources; biodiversity; forest health; productive, protective and socio-economic functions (three separate elements); and legal, policy and institutional frameworks. While many of the issues described above also apply to the information presented in the FRA, there is agreement that these thematic elements broadly cover the main aspects of SFM measured in the various criteria and indicator processes (FAO 2004; UN 2008) and the data collected in the FRA attempts to cover every country in the World.

Table 4 presents the global results from the FRA for the last two decades, with an assumption that an average annual change of more than 0.5 % represents a major change in an indicator (shown as either ++ or -- in the table). As the table shows, there have been major increases in the areas of forests where biodiversity conservation and soil and water protection are the main management objectives, areas affected by fires and insects have declined significantly, protected areas in forests and the area of privately owned forests has increased, as has the area of forests covered by management plans.

Two of the variables reflecting the extent of forest resources show a negative trend and the area of primary forests has also declined, although these changes have been less than 0.5 % per year over the period. Employment and staffing of public forest administrations has also declined significantly, but the number of students graduating in forestry has increased a lot.

As noted previously, it is difficult to say from this evidence whether there has been significant progress towards SFM at the global level, other than to say that many aspects of forest management appear to have improved significantly while a few have not improved.

2.3   Forest Uses

One important subset of information collected in the FRA is the extent to which forests are used for different purposes. In the FRA, this is referred to as the "*primary designated function*", which is defined as follows:

> The primary function or management objective assigned to a management unit either by legal prescription, documented decision of the landowner or manager, or evidence provided by documented studies of forest management practices and customary use (FAO 2007).

Although this definition is partly based on the management objective for an area of forest (rather than its actual use), it can be assumed that the main use of a forest and its management objectives are closely aligned. Therefore, it should give a rough indication of the relative importance of different forest uses over time and, possibly, between countries and regions. However, it should also be noted again that this is an

Table 4 Progress towards sustainable forest management at the global level, 1990–2010

Thematic element	FRA 2010 variables and data availability		Annual change rate (%)				Annual change		Unit
			1990–2000		2000–2010		1990–2000	2000–2010	
Extent of forest resources	Area of forest	H	+/–	–0.20	+/–	–0.13	–8,323	–5,211	1,000 ha
	Growing stock of forests	H	+/–	0.13	+/–	0.14	n.s.	n.s.	m^3/ha
	Forest carbon stock in living biomass	H	+/–	–0.18	+/–	–0.17	–538	–502	million t
Forest biological diversity	Primary forest area	M	+/–	–0.40	+/–	–0.37	–4,666	–4,188	1,000 ha
	Area for biodiversity conservation	H	++	1.14	++	1.92	3,250	6,334	1,000 ha
	Forest area within protected areas	H	++	1.09	++	1.97	3,040	6,384	1,000 ha
Forest health	Area of forest area affected by fire	M	--	–1.89	--	–2.15	–345	–338	1,000 ha
	Forest area affected by insects	L	--	–1.88	--	–0.70	–699	–231	1,000 ha
Productive functions	Area for production	H	+/–	–0.18	+/–	–0.25	–2,125	–2,911	1,000 ha
	Planted forest area	H	++	1.90	++	2.09	3,688	4,925	1,000 ha
	Wood removals	H	+/–	–0.50	++	1.08	–15,616	33,701	1,000 m^3
Protective and socio-economic functions	Area for soil and water protection	H	++	1.23	++	0.97	3,127	2,768	1,000 ha
	Privately owned forest area	H	++	0.75	++	2.56	3,958	14,718	1,000 ha
	Value of wood removals	M	+/–	–0.32	++	5.77	–241	4,713	million $
	Forestry employment (FTE)	M	--	–1.20	+/–	–0.11	–126	–10	1,000
Policy and institutional framework	Forest area with management plan	M	++	0.51	++	1.07	6,964	15,716	1,000 ha
	Staff in public forest institutions	L	--	–1.94	+/–	0.07	–23,568	830	number
	Students graduating in forestry	L	++	15.67	++	8.83	4,384	4,081	number

Source: FAO (2010a)

Notes: H/M/L (high/medium/low) indicates that countries providing data for a variable represent 75–100, 50–74, or 25–49 % of the global forest area and the symbols (++, +/–, --) indicate whether each variable has increased or fallen by more than 0.5 % (++ or --) or within the range of +/– 0.5 %

indicator of inputs rather than outputs, so it does not capture the intensity of use. It will also not reflect the importance of some forest uses that can be significant even if they are not the main focus of management (e.g. forest recreation or soil and water protection in forests managed for production).

Information is collected in the FRA about the area of forests managed in a country for each of the following primary designated functions:

- **Production**: Forest areas designated primarily for production of wood, fibre, bioenergy and/or non-wood forest products (including subsistence collection of these products).
- **Biodiversity conservation**: Forest areas designated primarily for conservation of biological diversity including (but not limited to) forests in legally protected areas.
- **Soil and water protection**: Forest areas designated primarily for protection of soil and water, including: areas where there are specific restrictions aimed at maintaining tree cover or not damaging vegetation that protects the soil; buffer zones to protect watercourses; areas with steep slopes where forest harvesting is restricted; and areas managed for combating desertification.
- **Social services**: Forest areas designated primarily for social services, including activities such as: recreation; tourism; education; research; and conservation of cultural or spiritual sites.

In addition to the above, countries may also report areas managed for other purposes or for multiple-uses (where no one management objective predominates), or they may report areas with no specific management objectives at all.

Figure 1 presents the trends in forest uses over the last two decades by region and at the global level. As the figure shows, production is the most significant use of forests in most regions and at the global level (accounting for 28 % of the forest area in 2010). Production also accounts for over half of the total area designated for the four main functions in all regions except Latin America and the Caribbean. The proportions of the forest area used for soil and water protection and biodiversity conservation are similar at the global level (7 % and 9 % respectively in 2010) and these are the next two most important functions. Apart from Latin America, very few countries report significant areas of forest designated for social services, with many countries stating that most of their forests are used for these purposes at the same time as they are used for one of the other primary functions, or that areas designated as multiple-use forests are used for this purpose.

The figure shows that the decline in the global forest area used for production (reported in Table 4) has also led to a slight decline in the proportion of forest used for production. However, it also shows that there are large differences between the regions in terms of the areas used for different functions and the trends in these variables.

The proportion of the forest area used for production in Europe, Asia-Pacific and Africa has fallen, but is still relatively high compared to the other two regions. Conversely, in North America and Latin America it has increased quite a lot, but from a low base. These increases are probably due to the rapid expansion of

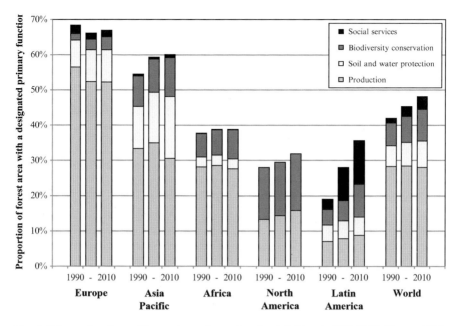

Fig. 1 Primary functions of forests by region and at the global level, 1990–2010 (Source: FAO (2010a))

planted forests in some countries in these two regions (e.g. Brazil and United States of America). In addition, the relatively small areas designated for any primary function are because almost all forest in Canada is designated as multiple-use forest (87 % in 2010) and only half of Brazil's forest (51 %) has a clearly identified primary function.

The proportion of forests used for biodiversity conservation has increased over the last two decades at the global level and in all regions. The highest proportion is reported for North America (16 % in 2010) and the region where this has increased the most is Latin America and the Caribbean (from 4 % in 1990 to 9 % in 2010). The proportion managed for soil and water conservation has also increased slightly at the global level (from 6 % to 7 %), although increases at the regional level have only occurred in Europe and the Asia-Pacific region. A number of countries (e.g. United States of America) reported that no forests are designated specifically for protection of soil and water, because this function is embedded in legislation governing the management of all forests. Thus, the importance of soil and water protection as a forest use is no doubt underestimated by these figures.

The FRA also collects information about the proportion of forests in legally designated protected areas and data was provided by 135 countries (covering 91 % of the global forest area). Countries were asked to include only forest areas in protected areas falling under IUCN categories I–IV (Dudley 2008), where the emphasis of management is protection of non-productive functions rather than sustainable use. These figures suggest that more than 10 % of the total forest area is now in protected areas in most countries and regions (FAO 2010a). This is less than the proportion of forest area managed for the three non-productive functions identified above (20 %), because not all of these areas will be legally designated as protected areas. However, it suggests that non-productive uses of forests are sufficiently important to be protected by specific legislation and formal designation in many cases.

One final interesting observation about the areas reported for each designated function is the seemingly high proportion of the World's forests that are not clearly associated with one of the four specific uses or functions mentioned above. Approximately one-quarter of the global forest area is designated as multiple-use forest and another one-quarter falls into the other or unknown categories. In many countries, it appeared that some forest areas could be put clearly into the first four use categories and the remaining forest area was simply put into the three non-specific categories because of a lack of information. However, in others, specific justifications and information were provided about why some forest areas should be classified as multiple-use, other or unknown (see Box 2).

Based on the information about forest functions presented in the FRA, a number of more general observations about how forests are managed and used can be made:

- **Competing forest uses**: many countries were able to identify and report areas of forest used for production and biodiversity conservation with little ambiguity. This does not mean that areas used for one of these functions are not used for the other as well, but it does suggest that these two uses are to some extent mutually exclusive. For example, planted forests in almost all countries were designated for production (and where they weren't, they are mostly used for soil and water protection). Conversely, about half of the forest areas used for biodiversity conservation were in legally protected areas that generally place very strict limitations on any harvesting activities.
- **Compatible forest uses**: in some countries, part of the forest estate is specifically designated for soil and water protection, implying that this is so important that protection and production can be viewed as mutually exclusive uses. This was particularly true for arid and mountainous countries. However, many countries reported that soil and water protection was a basic requirement of forest management and reported no areas designated for this function. Even more countries made the same statement about forests used for social services. Thus, it appears that these uses are compatible with production and biodiversity conservation (and may even be complementary to the latter).
- **Management for multiple-uses or specialisation**: the FRA results also give an insight into different forest management philosophies around the World. For example, Canada reported very small forest areas used for production and biodiversity conservation and a huge proportion of the forest designated for multiple uses. A few other countries (Mexico, Norway, United States of America) also reported significant areas of multiple-use forest. At the other extreme, in New Zealand, almost all natural forest was designated for biodiversity conservation

Box 2 Different Approaches to the Estimation of Areas with Primary Designated Functions

The FRA country reports show how countries classified their forests (in terms of designated functions), based on information about legal status, tenure, management objectives and ownership, as well as national forest inventory results and other literature. A summary of some of the approaches is as follows:

- **Australia**: Classification was based on tenure statistics and national forest inventory results. The forest plantation area was classified as production and the area of Nature Conservation Reserves was put in the biodiversity conservation category. Leasehold areas were classified as other and the relatively small areas of unresolved tenure were classified as unknown. All remaining forest was classified as multiple-use forest. Together, the non-specific designations (multiple-use, other and unknown) accounted for 84 % of the total in 2010.

- **Brazil**: Based on the National System of Conservation Units and the Brazilian Forestry Code, 15 national designations were identified. Forest plantations and National Forests were classified as production, Permanent Preservation Areas were put in the soil and water protection category, Environmental Protection Areas were designated as multiple-use and other designated areas were put into the biodiversity conservation or social services category. Forest areas without any national designation were classified as unknown (49 % of the total forest area).

- **Canada**: In Canada, forest area is divided into 25 classes of forest ownership and legal status. Most of the private industrial forest was classified as production and federal, provincial and territorial reserved forests were placed in the biodiversity conservation category. Most of the remaining forest was designated as multiple-use, with a relatively small amount placed in the unknown category. These last two classes accounted for 94 % of the total in 2010.

- **China**: National forest inventories in China recognise 11 forest management functions and areas meeting each of these were reclassified into the FRA categories. In the data reported for 2010, 76 % of the forest area was placed into one of the four specific categories used in the FRA and the remaining 24 % was all classified as multiple-use forest.

- **New Zealand**: In New Zealand, classification into designated functions was based on forest ownership and forest type (planted or indigenous). All planted forest was classified as production and all state-owned indigenous forest was put into the biodiversity conservation category (for the year 2010). Most of the privately owned indigenous forest was also put into the biodiversity conservation category, with a small amount classified as production.

(continued)

Box 2 (continued)

- **Russian Federation**: The Russian Forest Code (2006) divides the forest estate into three types: operational forests; protective forests; and reserve forests. For the FRA, operational forests were classified as production and the area of protective forests was divided into the other three specific designated functions and multiple-use forest. Reserve forests are forests in remote areas that are unlikely to be developed for at least another 20 years and these were all classified as other.

- **South Africa**: Forests in South Africa were divided into legally declared wilderness areas, forest plantations and other forest areas (calculated as the residual forest). The forest plantation area was classified as production, wilderness areas were placed in the biodiversity conservation category and the remaining area (71 % of the total) was classified as multiple-use forest.

- **United States of America**: In the United States of America, none of the national systems used to classify forest land could be easily translated into the FRA categories, so a combination of data and assumptions was used to provide this information. All planted forests and a proportion of some natural forest types was classified as production. Several other types of forest were placed in the biodiversity conservation category and the remaining area (46 % of the total) was classified as multiple-use forest.

Source: FAO (2010b)

and almost all of the plantation area was designated as production forest (with a tiny amount for soil and water protection). This apparently strict and clear separation of function by forest type in New Zealand has developed over many years and after much public and political debate (Reid 2001; Perley 2003).

Countries such as China, Vietnam and Russian Federation also appeared to have clearly defined forest areas for different uses, as did many West European countries. The latter result is, perhaps, surprising considering that Europe has a strong tradition of managing forests for multiple-uses (see, for example: CEC 2006). For the European Union as a whole, only 16 % of the forest area was designated as multiple-use and only Germany and Spain reported significant areas (and proportions) of forest with a multiple-use designation.

- **Planning for the future**: the significant area of forests where there are currently no clearly identified forest uses suggests that planning for their use will be an important task in the future. This is particularly true in countries where there are many pressures to convert forests to other land-uses. However, this lack of any designation also reflects a lack of information and weaknesses in the capacity of forest administrations in many cases. It seems likely that these countries should be a high priority for capacity building and technical assistance if they are to make further progress towards SFM.

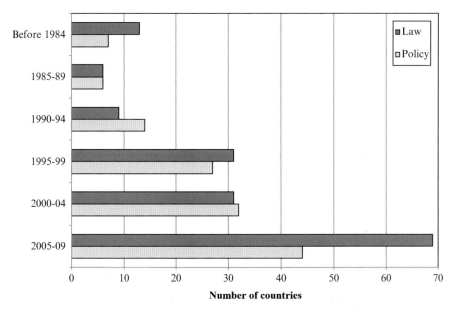

Fig. 2 Primary functions of forests by region and at the global level, 1990–2010 (Source: FAO (2010a))

2.4   Legal, Policy and Institutional Trends

The FRA also presents information about trends in legal and policy frameworks and forestry institutions in countries. Although changes in these aspects are often more qualitative than quantitative, this information is useful for showing some other general trends in the way that forests and forest management have developed.

With respect to policies and laws, almost all countries have a national forestry policy (or regional policies in federated states) or a national forest programme. In fact, many countries have both. The main exceptions are the Russian Federation (which does not have a forest policy but does have a national forest programme) and Canada and United States of America (which have policies, but not national forest programmes – probably because they are federated states). In addition, almost all countries have some sort of legislation concerning forests (either a specific forest law or other laws that include forestry).

The FRA also shows that many countries have updated their forestry policies and laws quite recently. For example, about two-thirds of countries have enacted, issued or revised their forest policies and legislation in the last decade (see Fig. 2) and three-quarters have developed national forest programmes during the same period. In many cases, these policies, laws and programmes have been developed or revised to take into account international commitments (such as the main environmental conventions on biodiversity, climate change and desertification) and to

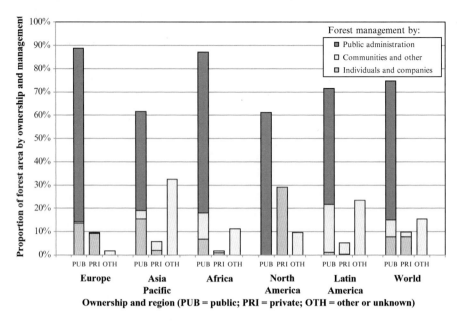

Fig. 3 Forest ownership and management by region and at the global level in 2010 (Source: FAO (2010a))

broaden the scope of policies and laws so that they more adequately reflect the aims of sustainable forest management. However, it is also worth noting that about 10 % of countries still have forestry policies and legislation that date from the 1980s or even earlier.

Another aspect of the legal and institutional framework for forest management is forest ownership and the related issue of who is actually managing forests. In the FRA enquiry, three ownership categories were used (public private and other) and countries were asked about the type of owners of private forest (individuals; businesses or institutions; and communities). For public forests, the FRA also asked about who holds the rights to manage and use forest resources (public administrations; individuals; businesses; communities; and other), because public forests are sometimes managed by non-state actors. This information can be combined to show roughly the proportion of forests that are managed by the state, the private sector and communities and this is shown in Fig. 3.

The first major observation from these results is that a significant proportion of the World's forests (75 %) are owned by the state. Public forest ownership is highest in Europe and Africa, although the results for Europe are heavily influenced by the Russian Federation (where all forests are owned by the state). Excluding the Russian Federation, public forest ownership in Europe is only about 45 % and private ownership is slightly higher at about 50 %. Information about ownership has only been collected in the FRA for 2005 and 2010, but it shows that public ownership has declined at the global level (it was about 80 % in 2005). This is due to privatisation of forests in some regions (e.g. Europe) and establishment of privately-owned planted forests. The one region where there has been little change is Africa, where the state continues to own a significant proportion of the forest estate.

The next largest share of forest resources (about 15 %) falls into the "other ownership" category. This includes areas where ownership is disputed or unclear or where the ownership arrangements do not fit the definitions of public or private. In reality, it is likely that many of these areas are under the control of local communities (although often with unclear tenure rights). This category is significant in Latin America and the Caribbean and the Asia-Pacific region (and Mexico, which accounts for all of the area designated as other ownership in North America). The remaining 10 % of the World's forest resources are privately owned, with the majority of this area located in North America and Europe.

Combining the ownership and management information, it appears that about 60 % of the World's forests are owned and managed by the state although, as noted previously, this proportion seems to be declining. If it is assumed that communities manage most of the forest designated as "other ownership", then communities manage just under 25 % of the World's forests. They both own and manage a small proportion of this (about 2 % of the global area) and have legally recognised management rights in public forests (another 7 % of the global area), but the majority of this area falls into the other ownership category. Community ownership and management also appears to be increasing, especially in Latin America and the Caribbean and the Asia-Pacific region. For example, in Latin America and the Caribbean, the proportion of forests managed by communities across all ownership categories is close to the proportion owned and managed by the state (both about 45 %).

The proportion of forests managed by individuals and businesses (the formal private sector) is slightly more than 15 %, with this area divided equally between areas owned and managed by the private sector and areas of public forests where the private sector have management rights (i.e. forest concessions). This has also been increasing due to privatisation, expansion of forest concessions in a few places (e.g. Russian Federation) and expansion of planted forests.

Another point worth noting is that there appears to be a very clear distinction in ownership and management in Europe and North America (excluding Mexico), where almost all forest ownership and management is designated as either public or private. This no doubt reflects the well-developed legal frameworks for tenure and ownership in these regions compared to much of the developing World.

3   Driving Forces Affecting Forests

In very broad terms, the main driving forces affecting forest resources are socio-economic trends such as economic growth and demography. Environmental conditions are also important but, until recently (with the prospect of climate change), they largely determined differences between regions in terms of factors such as the productivity of land and forests. The impacts of climate change on forests are largely

Table 5 Income, population density and population structure in 2010

Region	GDP per capita (USD)	Population density (persons/km²)		Rural Population (percent)	Economically active population (percent)
		Land	Forest		
Europe	25,838	33	73	27	50
North America	37,199	22	67	19	50
Africa	1,699	34	152	60	39
Asia-Pacific	5,140	107	536	58	49
Latin America and Caribbean	8,687	26	54	20	49
World	9,220	53	171	49	48

Source: FAO (2010a, 2012a)

outside the scope of this chapter,[2] but they may also have major consequences for forests and people over the coming decades that will require the sector to adapt.

Trends in these underlying socio-economic variables result in changes that are more directly observable, such as increased demand for land, increased demand for forest products, changes in the availability and cost of capital and labour and changes in production, consumption and trade flows. In response to these trends, the forestry sector (like the rest of the economy) adapts by introducing new technology and altering management practices. Governments also play a part in this process, when government policies – both within forestry and in other sectors – seek to promote some activities or discourage others in an attempt to increase efficiency or equity for the benefit of societies as a whole.

3.1 Socio-Economic Trends

Despite the fact that forests provide many environmental and social benefits, the management of forests and associated land is still largely driven by economic considerations. Forests are one land-use out of many alternatives and land is one of the three basic factors of production in the economy (land, labour and capital). Thus, like any productive activity, the relative abundance of these three basic factors of production explains quite a lot of the differences in forest management between countries and over time (including why forests are sometimes converted to other land-uses).

For example, Table 5 presents information about income and population in the five main global regions in 2010. Comparing the two more developed regions (Europe and North America), both regions have good access to capital (as shown by the high gross domestic product or GDP per capita), but Europe has a relatively higher population density and higher proportion of the population living in rural

---

[2] See Sect. 3 of this book, especially the chapter by Grace et al.

areas (and both of these figures would be much higher if the Russian Federation was considered separately).

Thus, in Europe, capital is abundant and land is relatively scarce, so land is managed intensively on farms and in forests that are quite small in size and with a comparatively large number of people working in forestry and agriculture. Conversely, in North America, land and forests are managed more extensively and at a larger scale, with generally larger landholdings, fewer people working the land and relatively more capital used in agriculture and forestry. Similar differences between these two regions also appear in the management of forest resources with, for example, a higher proportion of planted forests and higher roundwood production per hectare in Europe (excluding the Russian Federation) compared to North America.

Comparing the other three regions, the table also shows that the abundance of the three basic factors of production are very different, suggesting that land and forest resources in each of these regions may be on very different development paths.

Latin America has relatively good access to capital and a low population density, suggesting that forestry and agriculture in many of these countries could develop along a similar path to what has already occurred in North America. Conversely, the Asia-Pacific region may follow more of a European path, with more intensive management of smaller landholdings and a gradual substitution of capital for labour when people leave rural areas to migrate to cities. Already in forestry there is some evidence of these changes taking place with, for example, the growing importance of planted forests in Asia.

In Africa, capital is scarce and land is abundant (as shown by the low population density although, of course, a large part of North Africa is uninhabitable). Thus, agriculture and (to some extent) forest management in Africa is typically small-scale and labour intensive, but these activities do not utilise all of the land and forest resources available on the continent. This relative abundance of land in Africa and the currently low levels of investment in land management there partly explain why there is so much interest at the moment in leasing African land for agricultural development (Deininger et al. 2010).

Of course, this regional description does not capture the many different socio-economic situations in countries within each region. However, the broad differences described above are also likely to affect forestry development at the country level. For example, the current and future development path for forests in sparsely populated Laos is likely to be very different to neighbouring Thailand and Vietnam. Similarly, the development of agriculture and forestry in the southern part of South America has been very different to, say, the countries of Central America.

3.2 Competing Land-Uses

The brief discussion above focused on forest and land management, but the same factors also affect land-use change and there is a growing body of literature describing how economic and demographic trends result in land-use changes over time.

One of the first attempts to examine this was the paper by Shafik (1994), which presented the hypothesis that the relationship between environmental quality and income follows the path of an inverted U-curve or "Kuznets Curve". This postulates that environmental degradation tends to get worse as economies develop until average income reaches a certain point, then changes in behaviour and mitigation measures result in a gradually improving environment. The paper examined trends in a number of environmental indicators (including deforestation), but presented mixed results. Some indicators followed the inverted U-curve as expected, but others continued to get worse as incomes rose or continuously improved.

In Shafik's paper, the relationship between deforestation and income was one that followed an inverted U-curve, but the results were not significant due to specification and measurement problems. Others have examined the evidence for a deforestation Kuznets Curve in more detail and have arrived at similarly inconclusive results (Mather and Needle 1999; Ehrardt-Martinez et al 2002). These studies have concluded that the hypothesis of a Kuznets Curve for deforestation is broadly correct (i.e. that deforestation tends to occur in poor countries rather than rich countries), but that the socio-economic drivers of deforestation are far more complex than a simple relationship between income and forest conversion.

More recent studies have included labour and commodity costs in the analytical framework and have proposed that forests go through four transitional stages as economies develop (see Box 3). Similar to the inverted Kuznets Curve hypothesis, these studies also suggest that deforestation becomes gradually worse as incomes rise (from a low level), but then starts to improve and eventually reverse. However, they provide a better explanation of why these changes might occur.

In particular, as described by Rudel et al. (2005), there are two hypotheses about why deforestation might reverse as economies develop and incomes rise:

- **Rising rural labour costs**: With increasing incomes and urbanisation, rural labour costs will rise and farmers will abandon their more remote, less productive fields and pastures to concentrate labour on their more productive and profitable landholdings.
- **Increasing forest scarcity**: As forest area declines in a country, the potential supply of forest products (from domestic resources) also falls, leading to increased prices of forest products. This improves the financial returns from afforestation and forest management, leading to an expansion of forests and reversals in forest degradation.

The paper by Rudel et al. describes how a number of countries appear to have moved from the late-transition to post-transition stages since 1990 (for each of the two reasons described above). However, it also describes why some other countries apparently made much less progress towards the post-transition stage for a variety of reasons.[3]

---

[3] See also the chapter by South in Sect. 4 of this book, which addresses the different types of forest ownership that exist and describes how this might affect how forests are used, now and in future.

Box 3 The Four Stages of Forest Transition

Based on analyses of forest cover and changes in forest area (i.e. deforestation and afforestation) in different countries, a number of authors have proposed that countries typically move through four stages of forest transition as they develop economically (for a fuller description of the development of the forest transition hypothesis, see: Rudel et al 2005). The four stages are described as follows:

- **Pre-transition**: Pre-transition countries have high forest cover and low deforestation rates. Population densities are generally low and there is little pressure to convert forests to other land uses.

- **Early-transition**: Early-transition countries have relatively high forest cover and high deforestation rates. Forests are converted to agriculture to feed growing populations, often after access to forests has been increased due to investments in roads for forest harvesting. Forests may also be cleared for other purposes such as mining, infrastructure or urban development.

- **Late-transition**: Late-transition countries have relatively low forest cover, with low (and slowing) deforestation rates.

- **Post-transition**: In post-transition countries, forest cover starts to increase either due to afforestation or natural regeneration on abandoned agricultural land.

The level of forest cover at which deforestation starts to slow and, eventually, reverse will depend on various factors such as population density, the structure of an economy and environmental factors.

A more elaborate model developed by Hyde (2012) follows a similar logic, to explain why forests in a country might be cleared, degraded or sustainably managed, due to a number of economic forces (including the two mentioned above). This model takes the idea of forest transition a stage further to explain how both land-use change and forest degradation develop over time and how sustainable forest management eventually arises when the value of forest products reaches a level sufficient to justify protection and management rather than simply exploitation of the resource.

Another factor affecting the forest transition is the availability of capital and improved technology. For example, if capital is readily available, then farmers could mechanise if labour costs rise and this would not lead to a reversal in deforestation. On the other hand, the same conditions would tend to support afforestation in the face of forest scarcity. This is why many countries that have successfully reversed deforestation trends have supported reforestation and afforestation with tax incentives, grants and other subsidy schemes to overcome the initial capital costs of tree planting.

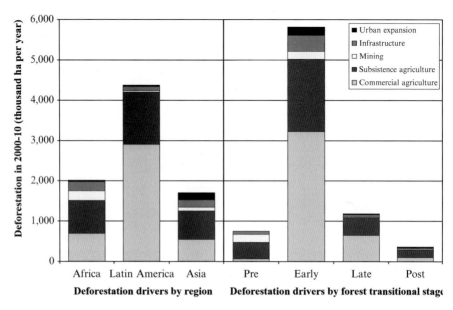

Fig. 4 Estimated drivers of deforestation in 46 tropical and sub-tropical countries, 2000–2010 (Source: Hosonuma et al. (2012))

The impacts of improved technology at the forest-agriculture interface has been studied by a number of authors (for a comprehensive review, see: Angelsen and Kaimowitz 2001). Similar to many other economic developments, improvements in technology have both an income and substitution effect that may or may not lead to an expansion of agriculture into forest areas, depending on which effect is stronger. On the one hand, improved technology enables farmers to produce more output at lower cost, but this tends to depress product prices and result in less expansion of agriculture as farmers focus efforts on their best land (a substitution effect, similar to the rising labour cost effect described above). However, by lowering production costs, these developments also raise the profitability per hectare of agriculture (an income effect), which would tend to encourage expansion of agriculture into other areas. With expanding markets for agricultural products (another important impact of rising incomes), the latter effect is likely to be dominant in most places, leading to conversion of forests into agricultural land.

Based on a comprehensive review of documents prepared for REDD[4] projects, Hosonuma et al. (2012) recently estimated the amount of deforestation in tropical countries that can be attributed to different types of land conversion. The results of this analysis are shown in Fig. 4 for the three tropical regions and four stages of forest

---

[4] Reduced Emissions from Deforestation and Degradation – an approach to climate change mitigation in the forestry sector.

transition. The results suggest that commercial agriculture accounted for 40 % of deforestation in the last decade, followed by local or subsistence agriculture (33 %). Infrastructure and urban expansion account for a further 10 % each and mining the remaining 7 %.

These results are consistent with earlier studies which show that agricultural expansion is by far the predominant driver of deforestation. However, in contrast to many previous studies, this study shows that commercial agriculture rather than subsistence agriculture is now the most important driver. The study shows that conversion of forests for cattle ranching, soybeans and oil palm has grown in importance in recent years, especially in Latin America and Asia. The study confirms the simple analysis presented earlier, showing that commercial agriculture is expanding mostly in Latin America (where land and capital are relatively abundant), while subsistence agriculture remains the most important driver of deforestation in Africa. It also shows how the availability of capital (e.g. for commercial agriculture) is increasing in importance as a driving force behind forest conversion.

The presentation of results by transitional stage is, perhaps, the most worrying indication for the outlook for forest resources. This estimates that about three-quarters of deforestation is currently occurring in countries in the early transitional stage (high forest cover and high deforestation rate), suggesting that many countries will continue to convert forests to agriculture for some years to come before deforestation rates fall or reverse.

With the expectation that forest conversion will continue for some time to come, there is a growing debate about how to reconcile the need for more agricultural land with the need to protect important biodiversity and ecosystems. In the same way that forests can be managed for multiple uses, some have suggested that "land sharing" – where environmentally-friendly agriculture occurs in multi-functional landscapes – can meet the need to produce more food while protecting biodiversity. The alternative approach is specialisation and intensification to produce more food from existing land and reduce the need to expand agriculture into other areas such as forests ("land sparing").

As with many other studies of land-use change, there is evidence to suggest that both approaches have strengths and weaknesses. For example, a study by Phalan et al. (2011) examined the density of tree and bird species in Ghana and India with different levels of forest cover and agricultural intensification. Their results suggested that land sparing would be a better strategy for minimising the negative impacts of increased food production. However, another recent study of the Peruvian Amazon (Gutíerrez-Velez et al 2011) showed that 75 % of the land used for expansion of high yield oil palm plantations was obtained by clearing primary forests compared to only 30 % for the expansion of small-scale, low-intensity production. The authors suggested that land tenure and fiscal incentives partly explained these differences, showing how government policies are also important driving forces in the conversion of forests to other land-uses.

3.3   Increasing Demand for Forest Products and Services

In addition to increased demand for agricultural land, socio-economic trends also lead to increased demand for forest products and services. However, at the same time, these trends also lead to improvements in technology, resource management and utilisation that can expand supply and reduce the impacts of increased demand on the forest resource.

Trends in the production of forest products over the last two decades suggest that many countries have been successful in reducing the amount of wood and fibre from forests used to manufacture forest products. At the global level and in most regions and product categories, production of processed forest products has increased since 1990, but there has been no upward trend in the consumption of roundwood by the wood processing sector.

One reason for this has been the continual increase in the recovery and re-use of forest products that has occurred since the 1970s, starting first with recovered paper (or wastepaper) and, more recently, the growth in recovery and recycling of solid wood products in regions such as Europe. Greater use of wood residues from solid wood manufacturing (i.e. sawnwood and plywood production) has also contributed to this effect. For example, the European Forest Sector Outlook Study (UN 2005) noted that less than half of the wood and fibre now used to manufacture forest products in Western Europe comes from forests and trees, with the majority coming from the use of wood residues and recycled wood and fibre.

Another contributing factor has been the substitution of reconstituted wood panels (i.e. particleboard and fibreboard) for sawnwood and plywood in many end-uses. These products are more efficient in the use of wood raw materials (i.e. less wood is required to manufacture one cubic metre of product compared to sawnwood and plywood) and they can be manufactured with wood residues and recovered wood, as well as smaller-sized trees and species that would not be considered commercially viable for manufacturing sawnwood and plywood.

In the markets for paper, rising incomes and technological change are starting to reduce demand for some types of paper, as people in developed countries switch to electronic media, book readers and communicate more by email. For example, newsprint consumption in North America and some European countries has now fallen to levels last seen in the 1960s. However, with improvements in income and literacy and the development of a service sector, demand for newsprint and printing and writing paper in most developing countries is still growing very fast and it is unlikely that this trend will slow down for many years to come.

The regional trends in paper consumption mentioned above reflect a more general pattern of forest product consumption that has appeared over the last two decades. In most developed countries now, per capita consumption of forest products is relatively high, but growing only slowly and, in some cases (e.g. newsprint), declining. Indeed, it appears as though per capita consumption of some forest products is reaching a natural limit with little potential for further growth. In addition, as noted above, much of this consumption is recycled into new products, reducing the demands placed on the forest resource for wood and fibre.

In contrast, consumption in many developing countries is increasing rapidly (although from a much lower base) and recycling is less well developed in most places. The increases in demand expected in developing countries in the future will result in greater demand for wood and fibre. However, with increasing forest scarcity and the large potential to increase the efficiency of resource use, it seems likely that improvements in technology, recovery and recycling will occur (as they have already in developed countries). Thus, it can be expected that such improvements will make a significant contribution to mitigating the impacts of higher product consumption on the demand for wood raw materials.

Another shift that has occurred in the last few decades is the rapid and significant increase in the area of planted forests and the growing importance of these resources for wood supply. For many decades, five countries with significant forest resources (Canada, United States of America, Sweden, Finland and Russia) accounted for over half of the global production of forest products. However, in the last two decades, the development of fast-growing forest plantations has reduced the dominance of these countries in global forest product markets. In particular, countries in the Southern Hemisphere – such as: Brazil, Chile, South Africa, New Zealand and Australia – have grown to become important suppliers and exporters of forest products based, to a large extent, on their planted forest resources. These countries have recently been joined by China, Indonesia and others that are developing significant planted forests estates.

As noted by Brown (2000), the comparative advantage in production of forest products appears to have shifted somewhat, away from countries with lots of trees to places where trees grow fast. This shift can also been seen within countries and regions with, for example, the increasing importance of forest plantations in the south-eastern United States of America as a source of wood supply and increased wood production in Europe from forest plantations in places such as Ireland, Spain and Portugal. These shifts have not only been supported by research and development in tree breeding, plantation establishment and management, but also by the changes in processing technology and end-uses (mentioned previously) that allow a wider range of wood raw materials to be used in manufacturing.

One of the important impacts of the development of planted forests is that it enables the growing demand for forest products to be satisfied by relatively small areas of forest. The contribution of planted forests to wood supply is not known, because roundwood production statistics are not disaggregated by forest type. However, the potential production from planted forests can be estimated from the planted forest area statistics collected in the FRA, estimates of yield and other information about management regimes.[5]

One of the first attempts to estimate potential wood supply from planted forests was produced by Brown (2000), based on the results of the 1995 FRA. Using a very

---

[5] For example, whether the planted forests in a country are likely to be grown and managed on short rotations for pulpwood or longer rotations for sawlogs or a mixture of pulpwood and sawlogs (based on species, market conditions and other factors).

narrow definition of planted forests,[6] the study estimated that the 3.5 % of the global forest area classified as industrial forest plantations (in 1995) had the potential to produce 22 % of the industrial roundwood used in that year. An update to this estimate was produced by Carle and Holmgren (2008), based on the results of the 2005 FRA. Using a much broader definition of planted forests, they estimated that forest plantations (covering 7 % of the global forest area) had the potential to supply two-thirds of global wood production in 2005. Most recently, Indufor (2012) presented an analysis suggesting that there are 54 million hectares of fast-growing forest plantations managed for commercial purposes (1.5 % of the global forest area), with the potential to produce about one-third of current industrial wood supplies.

The large difference between these estimates of plantation wood supply occurs because some planted forest areas may not be used for wood production. Furthermore, actual production may not reach the level of potential production because of a number of operational factors (e.g. lack of local demand, high harvesting costs, tree mortality). However, all of these studies and a number of other similar reports have highlighted how these relatively small areas of planted forest could meet much of the World's demand for wood and fibre at present and in the future.

Socio-economic trends have one other major impact on the demand for forest products and services and this is that people start to demand a broader range of benefits from forests and forest products when incomes rise. Similar to Maslow's hierarchy of human needs (Maslow 1943), at low income levels demand focuses first on meeting basic human needs, then it expands (as incomes rise) to include demands for products and services that meet higher needs (e.g. information and education) and, eventually, products and services that help people to achieve esteem and self-actualisation. In the markets for forest products, these differences in the structure of demand can be seen in the high proportion of wood used as fuelwood and sawnwood (to meet the basic needs of energy and shelter) in poorer countries, with more demand for higher-value wood products (e.g. furniture) and paper at higher levels of income and demand for legal, sustainable and "green" products at the highest levels of income.

As noted in FAO's State of the World's Forests 2011 (FAO 2011b), many developed countries have already reached a point where concerns about the sustainability of forest products and interest in issues such as corporate social responsibility and the so-called green economy are becoming important factors affecting the future development of the forest industry. The same trend is also leading to increased demand for forest services such as recreation (e.g. in Western Europe – see: UN 2005) and is reflected in a number of recent developments such as forest certification and legislation to strengthen forest law enforcement.

---

[6] The early FRA definition of forest plantations was: "Forest stands established by planting or/and seeding in the process of afforestation or reforestation, with either introduced species (all planted stands) or intensively managed stands of indigenous species, which meet all the following criteria: one or two species at planting; even age class; and regular spacing (FAO 1998). Later versions of the FRA introduced the broader concept of "planted forests", which included the above plus other forests of planted trees that did not meet the criteria set-out above.

3.4 Government Policies

In addition to socio-economic trends, the other major driving force affecting forest resources is the overall policy environment in a country. This includes sectoral policies (both within forestry and in other sectors) as well as broader macroeconomic and social policies. For many years, forestry policy discussions have emphasised how forests and forestry are often influenced more by policies in other sectors than by policies within the forestry sector itself. However, a particularly important feature of the last two decades has been the increased liberalisation of markets, the growth of free trade and the globalisation of the World's economy.

The main impact of market liberalisation and globalisation has been the increase in economic growth that has occurred in the last two decades. In particular, the economies of many developing countries and countries in transition have grown dramatically, due to closer integration into global markets and reductions in central planning and state ownership and control of economic activities. Globalisation has also supported the regional shifts in roundwood supply (noted earlier) and, more recently, shifts in forest industry investment towards developing regions, where labour costs are lower and demand growth is high.

These shifts in production, demand and investment have altered the economics of forest management all over the World, making it more difficult to cover the costs of SFM in some regions. Increased globalisation also means that changes in forest management in one location are more likely to lead to unexpected impacts elsewhere. For example, concerns have been expressed that measures to alter forest management or reduce wood production (in countries where production levels are unsustainable) may simply lead to undesirable changes in other places. These spillover effects are becoming increasingly important for policymakers attempting to improve sustainability through measures such as logging bans and changes in forest law enforcement or by developing new projects for bioenergy production or climate change mitigation (e.g. REDD projects).

The other broad policy area that has a significant impact on forest resources is the arrangements for land-use, land tenure and land ownership in countries. For example, Contreras-Hermosilla (1999) described many reasons why insecure land tenure may result in forest conversion or forest degradation and also why strengthening tenure may not always lead to reductions in these problems unless such measures are complemented by other policy reforms. The results of the FRA show that forest tenure and ownership is still uncertain in a significant proportion of the World's forests (particularly in tropical countries), but that the situation is gradually improving (FAO 2011c). In addition, measures such as the recently produced "Voluntary Guidelines on Tenure Governance" (FAO 2012b) should help to strengthen and clarify tenure if implemented by countries.

With respect to cross-sectoral linkages, a number of papers have highlighted how policies in a few specific sectors appear to have the most significant impacts on forests and forestry (Contreras-Hermosilla 1999; Broadhead 2001; Dubé and Schmithüsen 2007). Policies promoting agriculture are the most obvious example of a cross-sectoral impact that tends to have a negative impact on forestry, followed by transport policies (which tend to increase access to forest areas, thus indirectly facilitating forest conversion). Policies in the energy sector are another major source of cross-sectoral impacts, although the impacts on forests may be positive or negative depending on the aim of the energy policy. Conversely, it is little surprise that policies supporting nature conservation and environmental improvement or protection tend to have major beneficial impacts on forests.[7]

The growing concern about climate change is the most recent development that has significant potential to affect forests and forestry. Forests can contribute to climate change mitigation in a number of ways. First, as carbon sinks, they can sequester and store carbon from the atmosphere through afforestation, reforestation or changes in forest management. Alternatively, carbon emissions from forests can be reduced if deforestation and forest degradation is reduced (REDD). Secondly, wood products can be used as substitutes for other products manufactured using more energy (i.e. wood products generally contain less embodied energy than other products), with the additional benefit that the wood products may also store carbon for long periods of time. Woodfuel may also be used in place of fossil fuels, with benefits in terms of lower net carbon emissions if the carbon emitted from combustion is replaced by forest regrowth or if the wood would have decomposed anyway.

It might be expected that policies supporting climate change mitigation should benefit forests and forestry overall, but achievements have been quite limited so far. For example, afforestation is one option to obtain credits for emission reductions, but very few projects have met the rigorous standards required to obtain such credits. Emission reductions through REDD is also being piloted in numerous countries at the moment, but countries have yet to agree if or how credits for REDD activities can be incorporated into global carbon financing mechanisms; and even if there is agreement on an international mechanism for REDD financing, the project preparation, transaction and monitoring costs may result in limited uptake. Beyond forests, the carbon stored in wood products can now be included in a country's carbon accounts, but very few countries have any policies or incentives to promote wood products as low-carbon alternatives to other materials.

The promotion of bioenergy is the one relevant area where many countries have implemented targets, policies and laws in recent years. For example, Cushion et al. (2010) reported that 57 countries now have targets for bioenergy use (as a share of total energy supply or fuel use), including all developed countries and most of the large developing countries. However, by focusing heavily on energy consumption, the policies for achieving these targets have contradictory implications for the outlook for forest resources.

---

[7] It is also interesting to note that these papers all show that broad policies supporting development or strengthening institutional frameworks appear to have as much (if not more) of an impact on forestry than policies in specific sectors. This suggests that planning and trying to mitigate the impacts (on forests) of broad policy changes may be a more suitable approach than the currently high level of interest in how to address cross-sectoral impacts.

On the one hand, by stimulating demand for wood, they improve the economics of forest management, providing more income for forest owners and supporting more intensive and active management of forests. However, there is concern that the potentially huge amounts of wood required to meet these targets may be difficult to supply sustainably. Furthermore, because non-wood crops are often more economically attractive sources of bioenergy or biofuel (e.g. sugar crops or oilseeds for liquid biofuel production), these policies have actually increased the pressure to convert forests to other uses in order to meet these new demands. As noted above, these pressures may also occur in places far away from the countries where bioenergy consumption is increasing.

Within the forestry sector, many countries have forestry policies that have been recently updated to reflect international commitments and current best practices. However, while forestry policies may be well designed and strongly supportive of SFM, there remains a lack of capacity in many countries to implement these ever more complex and complicated policies.

The FRA shows that the number of staff in forestry administrations has declined over the last two decades (see Table 4), but it also shows that there is a wide gap between developed and developing countries in the amount of public funding available to support forestry policy implementation. For example, the latest FRA estimated that public expenditure on forestry in 2005 amounted to at least USD 19 billion (or USD 7.30 per hectare),[8] but also that Europe, North America and China accounted for over 85 % of this total. Expenditure per hectare in all three developing regions (excluding China) was below the global average and far below the average expenditure in developed countries (FAO 2010a).

This lack of implementation capacity may partly explain why policies outside the sector seem to have much more influence on what happens to forests than forestry policies. It has also led to a long-running international debate about how to finance SFM, although this – like REDD-financing discussions – has yet to result in significant financial flows to support SFM. At the same time, forest administrations in many developing countries continue to find it difficult to raise income from the management of public forests (e.g. through higher forest charges) and to compete with other parts of the public sector for funding to support SFM implementation.

The challenges mentioned above could imply that forestry policy is not such an important driving force for the outlook for forest resources, but the lack of capacity in some developing countries may actually result in many unintended (and detrimental) consequences for forests. For example, where weak capacity collides with complex forestry regulations, the tendency for corruption may be higher. This increases uncertainty and reduces the economic viability of SFM, making it more likely that forests will be degraded or converted to other uses in the long-run. Alternatively, if countries try to address this problem by implementing simple policy measures (such as bans on hunting, logging or pitsawing, which are quite common), this may reduce the potential for corruption, but it can

---

[8] Only 103 countries (covering 64 % of the global forest area) reported this information to the FRA, so total global expenditure will be somewhat higher than this figure.

also remove any incentive for local people (as well as forest officers) to protect and conserve forests. To put it simply, where institutions are weak, more forestry policies, laws and regulations may increase the pressure to convert forests to other uses that have clearer and more certain policy and institutional frameworks. This issue of capacity will be vitally important for the future of forests, as well as for recent policy measures that try to strengthen forest governance (e.g. the US Lacey Act and EU Timber Regulation).

4   Outlook for Forest Resources

The preceding analysis has shown that the global forest area is declining (although at a slowing rate) and that forest degradation is probably increasing in many, but not all, places. However, at the same time, many indicators of SFM appear to be improving and the demands placed upon forests are expanding (both in terms of quantity and the range of goods and services desired).

The links between the main driving forces (economic growth, population change, increased demands), forest characteristics (area, forest type, use, etc.) and forest management are difficult to assess in detail but, at least in the case of forest area, competing demands for land appear to be a major factor affecting the outlook for forest resources. With respect to forest management, intensification (e.g. expansion of forest plantations) and expansion of protected areas appear to be two of the strategies most commonly employed by countries to try to meet the increasing demands placed on the sector.

The FRA has shown that the state is still the dominant force in forestry in most countries, due to its regulation of the sector and, in many places, through its ownership and management of forest resources. However, there are also indications that state forest ownership and management is declining. In addition, the capacity of forestry administrations to implement policies and legislation is quite weak in many places.

In light of the above, this final section will describe the outlook for some of the driving forces affecting forests then examine three main issues: how much forest will there be in the future; what will it be used for; and how will it be managed? This will include some brief thoughts about how stakeholders in the sector might respond to these changes.

4.1   Outlook for the Socio-Economic Driving Forces

At present, the short-term outlook for the global economy looks uncertain, so detailed projections of economic growth will not be presented here. However, it is possible to present some general statements about likely economic developments over the next few decades.

First, over the last century or so, the historical economic growth rate in most developed countries has been about 2.5 % per year on average. This rate has been driven by a mixture of population growth, technology and innovations and increased efficiencies in economies. Once the current economic problems have been overcome, it seems likely that growth in these economies will return to a more stable and predictable path. However, with ageing populations and little or no future population growth, a more subdued growth rate (e.g. 1.5–2.0 % per year) may become the norm in many of these countries.

Outside developed regions, long-term economic growth trends have been much less predictable in the past with, for example, some developing countries suddenly expanding very rapidly and others experiencing prolonged periods of little or no growth or even shrinking economies. However, over the last 10–15 years, stable positive growth has gradually appeared in many of these countries. Some of the so-called developing countries already have a relatively long history of solid economic growth that has propelled them into the ranks of the high-income economies (e.g. Republic of Korea, Singapore), with others following closely behind (e.g. Chile, Mexico, Malaysia). In addition, four of the larger countries (the so-called "BRIC" economies of Brazil, Russian Federation, India and China) have expanded to become major forces in the global economy and others (e.g. Mexico, Indonesia, Thailand) may soon join them.

Continued relatively high economic growth of around 2.0–4.0 % per year seems likely to continue (in most of these countries) for the foreseeable future. The countries with even higher rates of growth (e.g. China) may be able to continue growing at these levels for some time, but growth will slow down eventually, due to rising labour costs and demographic changes (see below). The one exception might be Africa, where economic growth rates in many countries have reached unprecedented levels in recent years. Average incomes in Africa are far behind the other global regions, but much of the continent has started to benefit from closer integration into the global economy, high demand for commodities and natural resources, favourable demographic trends and some improvements in governance, peace and security. If these improvements can be sustained, then above average growth could continue for some time to come.

The two main impacts of the economic outlook are as follows. First, growth in the global economy will continue to shift away from the developed regions of Europe and North America to the rest of the World, particularly Asia and Latin America and the Caribbean. Average incomes in these latter two regions will continue to approach the income levels in Europe and North America, although differences between countries within these regions may remain quite large. Average incomes in Africa are likely to continue to remain some way behind the rest of the World, due to the extremely low levels of income there at present.

The second impact will be a significant expansion in the middle-income population (or "middle class") in many developing countries. As already noted, this will lead to some major shifts in consumer demand, as people move up their needs hierarchy towards higher value products, more consumption of services and greater expenditure on luxury goods and leisure activities. It will also raise labour costs,

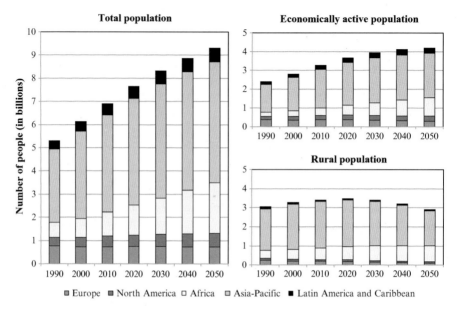

Fig. 5 Trends and projections for global population (medium fertility variant), 1990–2050 (Source: UN (2011), with economically active population from ILO (2011) to 2020 and estimated thereafter)

resulting in more substitution of capital for labour, increased labour productivity and a large reduction in people relying on the informal economy to make a living. These latter changes, combined with demographic trends, could have some profound impacts on the future of forest resources.

Demographic projections are available to the year 2100 (UN 2011) and are generally more reliable than long-term projections for economic growth. So, current projections for total population in each of the main global regions, along with estimates of the rural population and economically active population are presented in Fig. 5.

The global population is projected to increase by over two billion people over the next four decades to reach about 9.3 billion in 2050. Almost all of this increase will occur outside Europe and North America, although population growth is also expected to slow dramatically by the end of the period in Latin America and the Caribbean and the Asia-Pacific region. In contrast, Africa is the one region where population will increase most dramatically (from 1.0 billion in 2010 to 2.2 billion in 2050) and with very little decrease in the population growth rate in the future.

The population living in rural areas is currently (2010) about 3.4 billion people or roughly half the total population. This is expected to remain about the same over the next few decades then gradually reduce to less than three billion in 2050. The global trend mostly reflects an expected decrease of 0.6 billion in the rural population in the Asia-Pacific region in the coming decades. The rural population in Europe and the Americas is currently relatively small and is also expected to decline slightly in the future. In Africa, it is expected to increase slightly from 0.6 billion to 0.8 billion (2010 to 2050), but start to fall at the end of the period.

The economically active population (ILO 2011) is currently about 3.3 billion people and this is expected to increase to 4.2 billion people in 2050. However, the age structure of the population is shifting towards higher proportions of older people (including people beyond retirement ages) in most regions. Thus, the economically active population is expected to decline in Europe and remain roughly the same in the Americas. It will increase in the Asia-Pacific region in the next two decades – from 2.0 billion in 2010 to 2.4 billion in 2040 – but will not increase any further. Again, Africa is the one exception to these regional trends, where the economically active population is projected to increase continuously, from 0.4 billion in 2010 to about 1.0 billion in 2050. This is due to the current age structure (many young people) and high fertility that is expected to continue for some time to come.

In light of current thinking about the drivers behind forest transitions, these population projections are likely to have major implications for the future of forest resources in different parts of the World. In the Asia-Pacific region, the expected decline in the rural population and population ageing will result in far fewer people living in forest areas and engaging in activities that have traditionally been viewed as drivers of deforestation and degradation (e.g. shifting cultivation, grazing, fuelwood and non-wood forest product collection). So, to some extent, demography may "save" the forest. However, as also noted above, more capital will be employed in land and natural resource management to compensate for these shrinking and more expensive labour supplies. This will continue to create pressure to replace forests with large-scale commercial agriculture, but it will also encourage more forest plantation establishment (in areas where commercial agriculture is not viable) to meet the growing urban demand for wood products. In other words, the transition will really be one from a small-scale production or subsistence landscape to one that is more actively managed at a larger-scale and for more commercial production. This will not necessarily be detrimental to forest resources in all places, but could have both positive and negative implications for forests and forestry.

Parts of Latin America (where population density is relatively high) could follow a similar path to the Asia-Pacific region, because most of these demographic shifts are also expected to occur there. However, in the parts of this region where population densities are low, significant areas of land and natural resources are already very actively managed (although extensively and at a larger scale) with a strong focus on commercial production of forest and agricultural outputs. In these places, the transition will not be from a subsistence to managed landscape, but more likely an expansion of managed landscapes into areas that are currently lightly used.

These same demographic trends are also expected in Europe and North America, as well as Australia, Japan and New Zealand. In these countries, most land is already actively managed, land-use change is often regulated (e.g. through local planning systems) and land-uses tend to be more stable. An ageing and shrinking rural population may continue to encourage forest expansion in some areas for a variety of reasons. For example, farmer retirement and out-migration can lead to land abandonment (and natural regeneration) or deliberate tree planting to continue to use the land but with much lower input requirements. Alternatively, land ownership can change, replacing ownership by working farmers with new owners that may have more of an interest in tree planting (e.g. for long-term investment or non-commercial purposes). Management intensity (of both agricultural land and forests) might also increase to meet increasing product demand with lower labour availability, but much will depend on the policies and incentives present in a country.

The one clear exception to the future scenarios described above is Africa. In Africa, all of the demographic trends are the opposite, with an increasing total population, rural population and economically active population expected in the future. Here, the pressure to convert forests to other land-uses will increase. However, two factors could mitigate the impact of demographic changes on forest resources.

The first is that population density is quite low (especially in most of the countries with significant forest areas). Thus, deforestation may be quite limited in the countries with large populations (because they already typically have few remaining forest resources), while the demographic pressures may be reflected more in forest degradation (rather than deforestation) in the countries with high forest cover.

Secondly, there is still much potential to increase the productivity of agriculture in Africa. As noted by Langyintuo (2011), African cereal yields are less than a quarter of the global average, the use of fertilisers is minimal (often leading to soil degradation) and average yields of individual crops are between a half and one-fifth of what is technically feasible. Similarly, Africa has a lot of potential for the development of forest plantations (South Africa being a notable example of this), but the continent has not so far been able to take advantage of this.

Numerous studies have highlighted three main constraints to the development of agriculture, forestry and fisheries in Africa, namely: lack of access to investment capital; poorly developed markets; and unclear land tenure arrangements (FAO 2003, 2012c). The first constraint may ease in the future with rising incomes and foreign investment in the sector, but the other two issues require broad policy reforms that will need to be addressed if the continent is to achieve its potential.

4.2 Outlook for Future Demand for Products and Services from Forestry and Agriculture

The more direct forces affecting the outlook for forest resources are the growing demands for products, land and other environmental services expected in the future. This includes not only forest products but also the products from alternative land-uses and forest services or what the Organisation for Economic Co-operation and Development (OECD) has referred to as the "Five Fs" – food, feed, fibre, fuel and forest conservation (see: OECD 2009). Recent projections for some of these outputs are available and are presented below.

FAO regularly produces projections of future supply and demand for food and the results of the latest projections to the year 2050 are shown in Table 6. The figures

Table 6 Trends and projections for global production of agricultural products, 1990–2050

Region	1990	2000	2010	2020	2030	2040	2050
All food and non-food commodities							
Developed countries	93	96	100	107	115	118	122
Near East and North Africa	62	80	100	117	137	155	174
Sub-Saharan Africa	55	75	100	128	164	202	248
South Asia	62	80	100	121	146	166	189
East Asia	49	73	100	114	129	136	143
Latin America and Caribbean	53	75	100	118	140	152	164
World	66	83	100	114	129	140	152
Meat							
Developed countries	88	94	100	107	115	117	119
Near East and North Africa	49	71	100	127	161	193	233
Sub-Saharan Africa	56	75	100	133	177	236	314
South Asia	67	79	100	154	237	329	457
East Asia	45	69	100	121	146	154	164
Latin America and Caribbean	45	70	100	118	140	149	159
World	64	81	100	116	135	148	163

Source: derived from Alexandratos and Bruinsma (2012)
Note: the units above are index numbers, based on 2010 = 100

in the top of the table represent the total amount of production of all food and non-food commodities converted to an index number (with 2010 set to 100). Thus, this includes products such as cereals, oilseeds, fruit, vegetables, meat, milk and eggs, roots and tubers, as well as non-food products such as tobacco.

The information is presented using regions that are slightly different to those shown earlier, but the link between population trends and food production is quite clear. Production of food in developed countries, where population growth is minimal, is only expected to increase by about 22 % by 2050. In contrast, an increase of almost 150 % is expected in Sub-Saharan Africa and increases of about 50 % or more are expected in all other regions. In addition to the regional trends, these increases in production are likely to decline over the next 40 years, with a lower growth rate (in percent) from 2030 to 2050 compared to the period 2010–2030.

The lower part of the table shows the projected production of one commodity – meat – over the same time period. An important impact of rising incomes is that people will increasingly be able to afford more expensive food products (such as meat) in the future, so diets will change. This will have beneficial effects on nutrition, but it will also increase the demand for natural resources such as land and water, because production of meat is resource intensive compared to other types of food (i.e. for a given level of calories, production of meat generally requires more land, water and other inputs). As the bottom of the table shows, meat production is expected to increase by even more than total food production in all regions except developed countries and Latin America, with three or fourfold increases expected in some places by 2050.

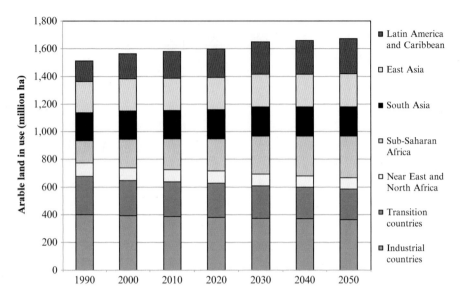

Fig. 6 Trends and projections for the global area of arable land in use, 1990–2050 (Source: derived from Bruinsma (2011))

Increases in agricultural production can be achieved in three main ways, namely: increases in yields (production per hectare during one crop cycle); increases in intensity of production (reducing fallow or unused areas and increasing the number of crop cycles achieved in a year); and expanding the area of arable land. Historically, improvements in yields have contributed to most of the increases in agricultural production (about 77 % of the total increase in global production from 1961 to 2007), with increased cropping intensity and expansion of agricultural land accounting almost equally for the remaining share. The two main exceptions have been Sub-Saharan Africa and Latin America and the Caribbean, where expansion of arable land has been more important (for further details, see: Alexandratos and Bruinsma 2012).

The projections for food production have been combined with projections for technological improvements (i.e. cropping intensities and yields) to produce projections of how much arable land might be used for agriculture in the future and these are shown in Fig. 6. Overall, the global area of arable land is expected to increase by almost 100 million hectares in the next 40 years, from 1,580 million hectares in 2010 to 1,675 million hectares in 2050.[9]

---

[9]It should be noted that the text above only refers to arable land. Other land used in agriculture includes permanent crops, grazing land and land used for shifting cultivation. Including these, the total area of land used for agriculture is 3–4 times greater than the area of arable land. However, projections for future requirements for these other types of land are not currently available.

These projections are consistent with the explanations of the driving forces and land-use development paths described earlier. Thus, for example, increases in food production in Asia are expected to be achieved by increases in yields and cropping intensity (due to more investment in inputs, capital and better management), with almost no expansion of arable land. In industrialised and transition countries, these same improvements will actually expand production per hectare faster than the projected increase in total food production, leading to a fall in the area of arable land required for food production.

The two regions where arable land is expected to increase are Sub-Saharan Africa (with an increase from 220 million hectares to 300 million hectares from 2010 to 2050) and Latin America and the Caribbean (with an increase from 195 million hectares to 255 million hectares). In the case of Latin America, land expansion is relatively more important because the potential to increased yields is limited (i.e. yields are already quite high, because industrial-scale agriculture is already well developed in parts of this region). In Africa, yields will increase, but they will be limited by the availability of finance for investment in agriculture and they will not be able to keep up with the huge increases in production expected in the future.

The projections above do include an expansion in production of biofuel feed-stocks (oilseeds, sugar, cassava and cereal crops), but they only assume an increase in production to the year 2020, then production held at that level for the rest of the period. They also do not present any projections of the use of wood for bioenergy. A more comprehensive study on bioenergy produced by the Word Bank (Cushion et al 2010) shows that bioenergy production will continue to increase to 2030 (if all current bioenergy targets are achieved) and scenarios by the IEA suggest that production could continue to grow significantly after this if these policies are continued (IEA 2012).

Figure 7 presents a projection for total primary energy supply (TPES) from solid and liquid biofuels, from the analysis of Cushion et al. (2010) and an extrapolation to 2050. The amounts are shown in million tonnes oil equivalent (MTOE) and one MTOE is equivalent to the energy content of about 3.8 million tonnes of woodfuel (on average).

The figure presents a number of important features in the trends for current and future bioenergy use:

- **Declining traditional uses of wood energy**: At present, most wood energy is consumed in developing countries in the form of fuelwood and charcoal used for cooking and, sometimes, for heating.[10] This is very different to what is often called "modern" uses of woodfuel, such as burning wood in power stations or modern heating boilers. So-called modern uses of wood energy also include the use of wood in sawmills and pulpmills for heat and energy generation (although this is not a modern phenomenon and has been done for many years).

---

[10] Until recently, FAO statistics presented data about production of fuelwood and charcoal. However, the data now is presented for woodfuel and charcoal. Woodfuel is the wood burned directly for energy (what used to be called fuelwood) plus the wood used for making charcoal.

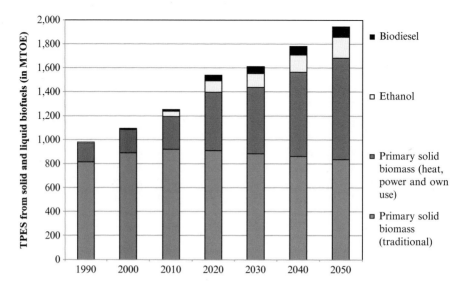

Fig. 7 Global trends and projections for solid and liquid biofuel production, 1990–2050 (Source: trends and projections to 2030 derived from Cushion et al. (2010), with an extrapolation to 2050 based on IEA (2012). Note that these figures are for all biofuel and not just biofuels made from wood. MTOE = million tonnes oil equivalent)

With rising incomes and urbanisation in many developing countries, growth in traditional consumption of woodfuel is slowing, because people now have greater access to other fuels (e.g. kerosene, bottled gas) in towns and cities and the income to purchase these more expensive fuels. Current analyses suggest that traditional use of woodfuel will not increase any more at the global level and may even decline slightly in coming decades.

The one region that is an exception to this is Africa, where urbanisation will not be accompanied by sufficient increases in income to encourage switching to non-wood fuels. In Africa, people will switch from fuelwood to a more convenient fuel, but this will be charcoal. Producing one metric tonne of charcoal uses about 6 m³ of wood (much more than if the wood was used directly for energy), so woodfuel consumption is expected to rise dramatically in this region in the future.

- **Increase in modern uses of wood energy**: Until recently, most industrial users of wood energy were in the forest products industry and growth in wood energy use was largely a by-product of growth in the production of these products. However, with the implementation of renewable energy policies and targets, wood is increasingly being used in power stations either for co-firing with other fuels or for use in facilities designed specifically for wood power generation. In addition, installation of new, modern wood heating boilers has increased significantly in the last few years.

If these policies drive growth in consumption as expected, wood used in this part of the bioenergy sector could double by 2020 (the target date for achievement of many of these policies) and may continue to increase by about the same amount again by 2050. Taking into account that almost all of this bioenergy is produced from wood, this use of woodfuel will become much more important than traditional uses in the future.[11]

- **Use of wood for liquid biofuels**: Agricultural crops are currently used for the production of almost all liquid biofuels. However, a number of countries have policies and programmes specifically to support and encourage the development of liquid biofuels made from other types of biomass (usually wood, but also including grasses and, sometimes, algae).

  Liquid biofuel production is likely to double by 2020 then increase again (by the same amount) over the period 2020–2050. The proportion of this that will be produced from non-food crops is highly uncertain, but most governments seem to be aiming for about one-third as a long-term ambition.

Out of all these development, liquid biofuels are likely to have most of an impact on land-use change, because they are mostly made from agricultural crops. However, not all of these changes will necessarily involve the clearance of forests for biofuels. For example, Whiteman and Cushion (2009) showed that less than half of the additional land needed for biofuels might come from forests (about 45 million hectares) and that 25 million hectares of this could be development of energy crops (i.e. non-food biomass crops) that might include wood grown on short-rotations for this purpose (see Table 7). Furthermore, their estimate did not include any assumptions about increases in yield which, as shown by Alexandratos and Bruinsma (2012), would also significantly reduce this new demand for land.

Assuming that yield and other productivity increases might meet half of the growing demand for biofuels, it appears that maybe about 10 million hectares of forest land might be converted to produce crops for biofuels by 2030 and another 10 million hectares for the period 2030–2050. Most of this conversion would occur in tropical regions as a direct or indirect consequence of expansion of sugar and oil-seed production.

Land required for the first period is already included in the projections presented earlier (showing arable land increasing by about 100 million hectares from 2010 to 2050 – see: Fig. 6), so the second amount would be in addition to this and would be required for the later years of this projection. In addition to this, 25 million hectares of forest might be converted to energy crops by 2030 and the same amount in the

---

[11] Comparing FAO woodfuel statistics and IEA solid biomass energy statistics, it appears that wood currently accounts for only about half of the traditional production of energy from solid biomass (woodfuel production in 2010 was 1,880 million cubic metres, equal to about 500 MTOE). Other materials used in this sub-sector include dung, crop residues and other forms of biomass. For heat, power and own use, wood accounts for almost all production and the only other significant biofuel used is bagasse. It is suspected that very little of this is captured in FAO's current woodfuel statistics because, until recently, most of this came from the unreported use of black liquor, wood residues and waste and very little came from forest harvesting.

Table 7 Additional land required for biofuel production by 2030 (at current yields)

Region	Potential land-use change and impact on forests (in million ha) by crop type and type of land likely to be used						
	Existing agricultural land		Degraded land	Potential conversion from forest to bioenergy use			
	Sugar beet and cereals	Temperate oilseeds	Jatropha, cassava, sorghum	Biomass energy crops	Sugar cane	Tropical oilseeds	Total for all crops
Net biofuel importers							
North America	11.5	6.3		10.0			27.9
European Union	8.9	12.2		15.0			36.2
East Asia and Pacific	1.0	5.2	5.9		1.4	3.5	17.0
South Asia			6.8		0.4		7.3
Net biofuel exporters							
Europe and Central Asia		3.0					3.0
Latin America					4.3	8.0	12.3
Africa			1.4		1.3	2.8	5.5
World	21.5	26.8	14.2	25.0	7.4	14.2	109.1

Source: Whiteman and Cushion (2009)
Note: the above figures assume no increases in yield

second period (2030–2050), although much of this might remain as intensively managed forest plantations if that is the preferred option.

With respect to the use of woodfuel, additional annual production of up to 250 million cubic metres would be needed by 2030 to meet the expected demand for liquid biofuels made from energy crops (depending on how much wood or other types of cellulose are used for this). A second increase of a similar amount would be required during the following two decades. However, a far greater expansion in wood use is expected for heat and power generation. The projected increases shown in Fig. 7, when converted from million tonnes oil equivalent (MTOE) to wood use, amount to a new additional demand for about one billion cubic metres of wood in each of the two 20-year periods. Some of this will be met from non-wood sources and increased use of black liquor, but the majority is likely to come from increased use of wood and fibre.

In addition to increased demand for woodfuel, production of other forest products will rise in the future as populations and economies expand, creating additional demand for industrial roundwood. The last global projections of production and consumption of forest products were published in FAO's State of the World s Forests 2009 (FAO 2009). These have since been negated by recent economic events, but a simple revision of the outlook for industrial roundwood production (also extrapolated to 2050) is shown in Fig. 8.

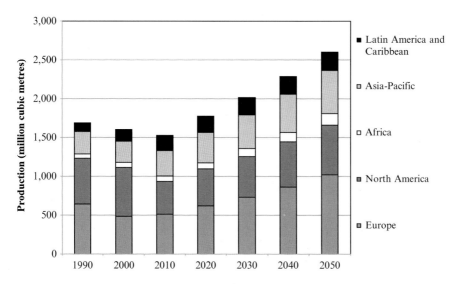

Fig. 8 Trends and projections for industrial roundwood production, 1990–2050 (Source: trends and projections to 2030 derived from FAO (2009), with an extrapolation to 2050)

For both of the next two 20-year periods, production of industrial roundwood might increase by slightly more than 500 million cubic metres, leading to a projected level of production in 2050 of 2.6 billion cubic metres, compared to the current (2010) level of production of about 1.5 billion cubic metres. Europe is expected to account for about half of this increase (mostly due to increased production in the Russian Federation), with the Asia-Pacific region and North America accounting for most growth in the rest of the World.

Coming to the last of the "Five Fs", it is difficult to predict exactly how the increasing demand for more forest conservation will affect forest resources. Schmitt et al. (2009) have recently produced an independent analysis of forest protected areas (based on satellite data), which shows that only 7.7 % of the World's forests are within protected areas in IUCN categories I–IV. For a number of reasons (data sources, definition of forest, date of the study), this figure is lower than the result from the FRA, but it does suggest that more areas of forest might be put into legally protected areas in the future. However, the FRA also showed that a much larger area of forest is already managed for conservation (with or without legal protection), so it seems likely that protected areas will mostly expand into these areas rather than other areas that are currently used for production or may be converted to other uses.

A more important question for the future is whether increased demand for biodiversity conservation will lead to broader measures that affect forests used for production (either now or in the future). Protected species legislation can result in large forest areas being taken out of production, as can stricter limitations on areas that can be harvested (e.g. on slopes, near watercourses or in particular forest types). More general conservation measures (e.g. requirements for environmental impact assessments and other planning measures) can also increase the costs of wood production, making some areas less economically viable for production. In some cases, countries may also simply choose to ban all production in certain types of forest (e.g. logging bans in natural forests, which are quite common in some regions). What does seem likely is that current forestry policy and management trends are likely to continue, discouraging harvesting in natural forest and pushing countries towards relying more heavily on planted forests for future wood production.

4.3   Future Scenarios for the Global Forest Area

Constructing projections for the future area of forests in the World is difficult because of the complex forces driving land-use change and the limited availability and quality of spatially referenced data on land-use, soils, climate, population and other relevant variables. Furthermore, compared to producing projections for production and consumption, it is more difficult to model land-use change in a deterministic way. Thus, most attempts to project land-use change have tried to combine various spatial models (for population, economics, trade, climate, land use, etc.) and use a range of possible changes in underlying variables to produce different scenarios for the future (see, for example: PBL 2012).

The three most recent global outlooks for land-use change have been produced for the Millennium Ecosystem Assessment (MEA 2005), the fourth Global Environmental Outlook GEO-4 (UNEP 2007) and the OECD Environmental Outlook (OECD 2012). Although all three of these studies used different definitions of forests and different land-use data sets, they all produced projections of forest area to 2050 under a variety of different scenarios. The projections from these three studies are shown in Fig. 9, with the forest area figures from each of them converted to an index number (2010 = 100) for comparability.

The nine scenarios presented in Fig. 9 show some significant differences in terms of the forest area expected in the future. The OECD projection shows a forest area in 2050 about 6 % higher than at present (equal to an additional 230 million hectares, based on the FRA estimate of total forest area in 2010). The MEA scenarios range from an increase of 4 % to a decrease of 3 % (+170 million hectares to −100 million hectares), with three of the four scenarios suggesting that forest area will increase. The GEO-4 scenarios are all much more pessimistic about the future, showing a decrease in forest area of 2–16 % (or a loss of about 90–660 million hectares).

Despite these differences, all of the scenarios also show some similarities. For example, they all suggest that the forest area in Europe, North America and the Asia-Pacific region is likely to stay roughly constant or increase in the future and that most deforestation will occur in Africa and Latin America. The MEA and GEO-4 scenarios also show how changes in assumptions about policies can have significant impacts on land-use change.

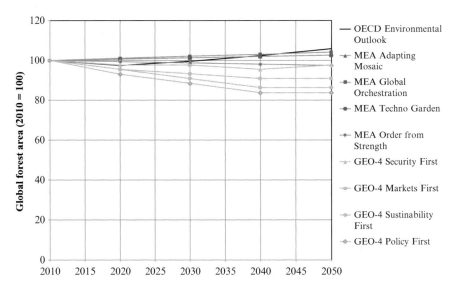

Fig. 9 Scenarios for future forest area, 2010–2050 (Source: derived from OECD (2012), UNEP (2007) and MEA (2005))

One of the main differences in the scenarios is the extent to which afforestation and forest restoration appear in the future. For example, the OECD projection suggests that deforestation will slow in most tropical countries over the next 10–20 years and that the expansion of forest plantations will increase in the future in temperate countries and places such as India, Brazil and China. This leads to the high projected increase in global forest area over the next four decades. A similar development occurs in the higher projections from the MEA. In contrast, expansion of forest plantations is not a major factor in the GEO-4 projections, suggesting that their projection model does not include a strong link between economic growth and forest plantation expansion.

From the scenarios presented above, it is difficult to get a clear idea of exactly how much forest there might be in the future, although it looks like there will probably be less forest in Africa and Latin America and more forest in the rest of the World (or at least, about the same amount as now). This general impression is also consistent with the historical trends reported in the FRA and the projections for arable land shown earlier.

Another important message from these scenarios is that continued deforestation in tropical countries is no longer as certain as it once seemed, with four out of the nine scenarios suggesting that forest area might actually increase in the future. The uncertainty here will be whether the socio-economic drivers (rising wood demand and falling rural populations) will be sufficient to overcome some of the current obstacles to afforestation and reforestation. Improvements in governance, land tenure and other factors enabling investment would seem to be crucial in this respect.

4.4   Some Final Thoughts About Forest Management and Use in the Future?

The analysis presented here confirms the statements in the introduction to this chapter that the demands placed upon forests are expanding in terms of both the quantity and variety of benefits that people expect to obtain from forests. Many of the implications of this for forest management and use have been described in the previous sections on driving forces and outlook. This final section presents three broader implications for consideration by stakeholders in the sector.

Intensity of forest management: The projections for forest product demand suggest that woodfuel consumption might increase by about half in the next four decades and industrial roundwood consumption may increase by about two-thirds. Increases in demand for other forest benefits can not be estimated precisely, but the growth in global population alone suggests that they might also increase significantly. At the same time, most projections of forest area show a decline or, at best, a very modest increase in forest area in the future. Thus, in order to satisfy these demands, the intensity of forest management will have to increase.

The challenge here will be to adapt forestry policies and management techniques to intensify management in ways that are sustainable and economically viable. For wood production, some fairly clear strategies are already available and well-tested (e.g. establish more forest plantations and increase recovery and recycling of wood and fibre). Furthermore, to a large extent, developments such as these will occur naturally in the future, being driven forward by market forces. However, when it comes to increasing the intensity of forest management for other social and environmental benefits, the techniques and strategies that can be employed are not so clear and their effectiveness is far less certain.

For example, in the case of biodiversity conservation, the most common approach currently used by many countries to meet this demand is simply to increase the area of forests in legally protected areas. Indeed, there is a specific, globally agreed target to do this. Taking a similar approach, conservation measures used in other forest areas are also often focused on restricting activities to protect biodiversity (although, in some cases, countries find it difficult to enforce these rules). While these measures no doubt, in principle, go some way towards helping to conserve biodiversity, they are not the same as actively and intensively managing forests to increase biodiversity. Some developed countries have the knowledge and resources to take a more active approach to biodiversity conservation in forests (and are doing so), but many other countries lack this capacity and, unfortunately, these are often exactly the places where demands for more biodiversity conservation are highest.

With other more easily measured environmental factors (such as carbon and water), scientific knowledge about how to manage forests for these outputs is more advanced. However, there is still probably quite a long way to go before this knowledge can or will be applied in the field in many countries. In particular, the benefits of storing carbon in forests and wood products are relatively easy to understand and measure, but only a limited number of options are widely discussed at present, clear management prescriptions are not generally available and incentive structures are not yet in place to make a significant difference to current practices.

In the social dimension, the situation is a little better, because many countries have a lot of experiences of different approaches and techniques they have used to try to increase the social benefits provided by forests. Of course, many of these benefits are highly context-specific and their production probably depends as much on local socio-economic situations as it does on how forests are managed. However, beyond a few general prescriptions (e.g. increase access to forests, strengthen tenure and local governance), there has not really been a systematic assessment of the linkages between forest management and policies and the delivery of these benefits.

If the forestry sector is to meet the growing expectations of society for more forest benefits, then some of these challenges will have to be overcome, particularly if public funding is required to invest in production of some of these benefits. This will require not only better management of forests for these benefits, but also better communication of the results. If not, the sector will continue to remain vulnerable to developments in other sectors and will fail to achieve its potential to contribute to a more sustainable future for the World.

Managing the forest transition: The text earlier described forest transitions in terms of forest cover and deforestation rates, but a broader transition is also taking place (at least in many developing countries) and this is the increasing importance of economic or commercial factors in decisions about land-use and land management. As shown in Fig. 4, commercial agriculture is becoming relatively more important as a driver of deforestation, so the sector will have to adapt to this. In particular, forestry administrations in many places may have to start placing less emphasis on defending forests from shifting cultivators and focus more on developing forest land-use plans that justify why areas should not be converted to other uses.

With the expected developments in agriculture and food production, some forest clearance in the future will be inevitable and desirable. More intensive forest management for wood production will also result in changes in forest characteristics (i.e. more forest plantations). Thus, it will become increasingly important to identify the management objectives for different forest areas, so that planned and rational land-use change can take place. As recent experiences in Brazil have shown (The Economist 2012), attempts to try to formalise and rationalise land-use change are likely to be publicly and politically sensitive, but many more countries are likely to have to face these challenges in the future.

A related issue is that increasing the commercial benefits from forest management is likely to become even more important in the future if forests are to compete with other land-uses. This means that forestry stakeholders will have to explain more effectively how using wood creates value for forests and helps to protect them (rather than destroy them), as well as supporting rural livelihoods. It also means that forests and forest management will have to deliver more local economic benefits to the people living in and around forests, not only as a social benefit but as an incentive to keep land under forest cover.

This transition to more commercial landscapes will also present a number of opportunities for the forestry sector. For example, many different types of land will be unsuitable for commercial agriculture (e.g. steep slopes, stream banks, rocky areas) and it is likely that many commercial landowners may be interested in planting trees as an additional business to use their spare land. For generations small-scale tree planting on farms has been common in Europe and North America (for a variety of reasons) and, more recently, tree planting for wood production has accelerated dramatically in places such as India and China. In a similar way, with smaller rural populations and less shifting cultivation, there may be much more potential in some places to promote forest restoration. If these opportunities can be identified, forestry administrations could play a very positive and active role in supporting such developments in the future.

The outlook for forest resources in Africa: One final observation concerns the outlook for forest resources in Africa. While much of the analysis above has shown some remarkable similarities between regions, Africa is notable as the one global region that has a very different socio-economic outlook. Here, the population is expected to continue to grow quite rapidly in the future. It will also remain quite a young population, but with relatively low income and a significant number of people remaining in rural areas. In addition, despite having rather abundant forest and land resources, the projections for production of food and forest products suggest that Africa is still some way from achieving its full potential.

Here, the outlook for forest resources is much less predictable, but it seems likely that the need to demonstrate local economic benefits from forest management will be more important than elsewhere, if forests are to compete with other land uses. Furthermore, given the huge demands that will be placed upon governments to provide public goods and services to these fast-growing populations, the dominance of the public sector in the control and management of forest resources looks likely to become increasingly untenable in the future.

If forests and forest management in Africa are to progress towards meeting their full potential, it may be necessary for forest policies and institutions there to create more space for small-scale, private-sector development within the sector. Already, some countries have made significant progress with community forestry and small-scale enterprises for production of non-wood forest products. A number of countries have also shown how partnerships between large companies and small-scale producers can be successful (in agriculture, forestry and plantation crops). Building upon these successes and replicating them in other countries may be the best way to generate local economic benefits and support forest management in the future.

References

Alexandratos N, Bruinsma J (2012) World agriculture towards 2030/2050: The 2012 revision, ESA Working Paper No. 12–03, Food and Agriculture Organization of the United Nations, Rome, Italy

Angelsen A, Kaimowitz D (2001) Agricultural technologies and tropical deforestation. CABI Publishing, London

Bai ZG, Dent DL, Olsson L, Schaepman ME (2008) Global assessment of land degradation and improvement – 1. Identification by remote sensing, Report 2008/01, ISRIC – World Soil Information, Wageningen, Netherlands

Broadhead J (2001) Cross-sectoral policy impacts in forestry – examples from within and outside FAO. Food and Agriculture Organization of the United Nations, Rome

Brown C (2000) The global outlook for future wood supply from forest plantations, Global Forest Products Outlook Study Working Paper GFPOS/WP/03, Food and Agriculture Organization of the United Nations, Rome, Italy

Bruinsma J (2011) The resources outlook: by how much do land, water and crop yields need to increase by 2050? In: Conforti P (ed) Looking ahead in world food and agriculture: perspectives to 2050. Food and Agriculture Organization of the United Nations, Rome

Carle J, Holmgren P (2008) Wood from planted forests: a global outlook 2005–2030. For Prod J 58(12):6–18

CEC (2006) Communication from the Commission to the Council and the European Parliament on an EU forest action plan, COM(2006) 302 final. Commission of the European Communities, Brussels

Contreras-Hermosilla A (1999) Towards sustainable forest management: an examination of the technical, economic and institutional feasibility of improving management of the global forest estate, FAO Paper for the World Bank Forest Policy Implementation Review and Strategy FAO/FPIRS/01, Food and Agriculture Organization of the United Nations, Rome, Italy

Cushion E, Whiteman A, Dieterle G (2010) Bioenergy development: issues and impacts for poverty and natural resource management. The World Bank, Washington, DC

Deininger K, Byerlee D, Lindsay J (2010) Rising global interest in farmland: can it yield sustainable and equitable benefits? The World Bank, Washington, DC

Dubé YC, Schmithüsen F (eds) (2007) Cross-sectoral policy developments in forestry. CABI, Wallingford

Dudley N (ed) (2008) Guidelines for applying protected area management categories. IUCN – International Union for Conservation of Nature, Gland

Ehrardt-Martinez K, Crenshaw E, Jenkins J (2002) Deforestation and the environmental Kuznets curve: cross-national evaluation of intervening mechanisms. Soc Sci Q 83:226–243

FAO (1995) Forest resources assessment 1990: global synthesis, FAO Forestry Paper 124, Food and Agriculture Organization of the United Nations, Rome, Italy

FAO (1998) FRA 2000 terms and definitions, Forest Resources Assessment Programme Working Paper No 1, Food and Agriculture Organization of the United Nations, Rome, Italy

FAO (2003) Forestry outlook study for Africa – regional report – opportunities and challenges towards 2020, FAO Forestry Paper 141, Food and Agriculture Organization of the United Nations, Rome, Italy

FAO (2004) FAO/ITTO Expert consultation on criteria and indicators for sustainable forest management, 2–4 Mar 2004, Cebu City, Philippines

FAO (2007) Specification of national reporting tables for FRA 2010, Forest Resources Assessment Programme Working Paper 135, Food and Agriculture Organization of the United Nations, Rome, Italy

FAO (2009) State of the world's forests 2009. Food and Agriculture Organization of the United Nations, Rome

FAO (2010a) Global forest resources assessment 2010: Main report, FAO Forestry Paper 163, Food and Agriculture Organization of the United Nations, Rome, Italy

FAO (2010b) FRA 2010 country reports (available at: http://www.fao.org/forestry/fra/67090/en), Food and Agriculture Organization of the United Nations, Rome, Italy

FAO (2011a) Assessing forest degradation: Towards the development of globally applicable guidelines, Forest Resources Assessment Working Paper 177, Food and Agriculture Organization of the United Nations, Rome, Italy

FAO (2011b) State of the world's forests 2011. Food and Agriculture Organization of the United Nations, Rome

FAO (2011c) Reforming forest tenure: issues, principles and process, FAO Forestry Paper 165, Food and Agriculture Organization of the United Nations, Rome, Italy

FAO (2012a) FAOSTAT – FAO's online statistical database for food and agriculture, (available at: http://faostat.fao.org), Food and Agriculture Organization of the United Nations, Rome, Italy

FAO (2012b) Voluntary guidelines on the responsible governance of tenure of land, fisheries and forests in the context of national food security. Food and Agriculture Organization of the United Nations, Rome

FAO (2012c) The state of food and agriculture 2012: investing in agriculture for a better future. Food and Agriculture Organization of the United Nations, Rome

Forest Europe (2011) State of Europe's forests 2011: status and trends in sustainable forest management in Europe. Forest Europe Liaison Unit, Oslo

Gold S (2003) The development of European forest resources, 1950 to 2000: a better information base, Geneva Timber and Forest Discussion Paper – ECE/TIM/DP/31, United Nations, Geneva, Switzerland

Grober U (2007) Deep roots – A conceptual history of "sustainable development". (Nachhaltigkeit), Wissenschaftszentrum Berlin für Sozialforschung (WZB), Berlin

Gutierrez-Velez VH, de Fries RS, Pinedo-Vasquez M, Uriarte M, Padoch C, Baethgen W, Fernandes K, Lim Y (2011) High-yield oil palm expansion spares land at the expense of forests in the Peruvian Amazon. Environ Res Lett 6(044029):5

Hosonuma N, Herold M, de Sy V, de Fries RS, Brockhaus M, Verchot L, Angelsen A, Romijn E (2012) An assessment of deforestation and forest degradation drivers in developing countries. Environ Res Lett 7(044009):12

Hyde WF (2012) The global economics of forestry. RFF Press, New York

IEA (2012) Energy technology perspectives 2012: pathways to a clean energy system. International Energy Agency, Paris

ILO (2011) Economically active population, estimates and projections, 6th edn. International Labour Office, Geneva, October 2011

Indufor (2012) Strategic review on the future of forest plantations. Indufor, Helsinki

ITTO (2005) Revised ITTO criteria and indicators for the sustainable management of tropical forests including reporting format, ITTO Policy Development Series No 15, International Tropical Timber Organization, Yokohama, Japan

ITTO (2011) Status of tropical forest management 2011, ITTO Technical Series No 38, International Tropical Timber Organization, Yokohama, Japan

Langyintuo A (2011) African agriculture and productivity, paper presented to the Sharing Knowledge across the Mediterranean (6) Conference, Villa Bighi, Malta, 5–8 May 2011

Lanly J (2003) Deforestation and forest degradation factors, paper presented to the XII World Forestry Congress, 21–28 Sept 2003, Quebec City, Canada

Maslow AH (1943) A theory of human motivation. Psychol Rev 50(4):370–396

Mather A, Needle C (1999) Environmental Kuznets curves and forest trends. Geography 84:55–65

MCPFE (1993) Resolution H1: general guidelines for the sustainable management of forests in Europe. Second Ministerial Conference on the Protection of Forests in Europe, Helsinki, 16–17 June 1993

MCPFE (2002) Improved Pan-European indicators for sustainable forest management. MCPFE Expert Level Meeting, Vienna, 7–8 Oct 2002

MEA (2005) Millennium ecosystem assessment: ecosystems and human well-being (volume 2). Findings of the Scenarios Working Group, Millennium Ecosystem Assessment, Island Press, Washington, DC

Montréal Process (2009) Montréal Process criteria and indicators for the conservation and sustainable management of temperate and boreal forests – technical notes on implementation of the Montréal process criteria and indicators, Criteria 1-7, 3rd edn, June 2009, 20th Montréal Process Working Group Meeting, 8–12 June 2009, Jeju Island, Korea

OECD (2009) Symposium on what future for the agriculture and food sector in an increasingly globalised world? 30–31 Mar 2009, Paris, France

OECD (2012) OECD environmental outlook to 2050: the consequences of inaction. Organisation for Economic Co-operation and Development, France

PBL (2012) IMAGE model site: Integrated Model to Assess the Global Environment, (available at: http://themasites.pbl.nl/tridion/en/themasites/image/index.html), Netherlands Environmental Assessment Agency, The Hague, Netherlands

Perley CJK (2003) Resourcism to preservationism in New Zealand forest management: implications for the future, paper prepared for Australia and New Zealand Institute of Forestry Conference, 28 April–1 May 2003, Queenstown, New Zealand

Phalan B, Onial M, Balmford A, Green RE (2011) Reconciling food production and biodiversity conservation: land sharing and land sparing compared. Science 333:1289–1291

Reid A (2001) Impacts and effectiveness of logging bans in natural forests: New Zealand, in "Forests out of bounds: Impacts and effectiveness of logging bans in natural forests in Asia-Pacific", RAP Publication 2001/08, Food and Agriculture Organization of the United Nations, Bangkok, Thailand

Rudel TK, Coomes OT, Moran E, Achard F, Angelsen A, Xu J, Lambin E (2005) Forest transitions: towards a global understanding of land use change. Glob Environ Chang 15:23–31

Schmitt CB, Belokurov A, Besançon C, Boisrobert L, Burgess ND, Campbell A, Coad L, Fish L, Gliddon D, Humphries K, Kapos V, Loucks C, Lysenko I, Miles L, Mills C, Minnemeyer S, Pistorius T, Ravilious C, Steininger M, Winkel G (2009) Global ecological forest classification and forest protected area gap analysis: analyses and recommendations in view of the 10% target for forest protection under the Convention on Biological Diversity (CBD), 2nd Revised Edition. Freiburg University Press, Freiburg

Shafik N (1994) Economic development and environmental quality: an econometric analysis, Oxford Economic Papers, New Series, vol 46, Special Issue on Environmental Economics (Oct. 1994), pp. 757–773

The Economist (2012) Environmental law in Brazil – compromise or deadlock? The Economist Print Edition, 2 June 2012

UN (2005) European forest sector outlook study 1960–2000–2020 – main report, Geneva Timber and Forest Special Paper – ECE/TIM/SP/20, United Nations, Geneva, Switzerland

UN (2008) Resolution adopted by the General Assembly 62/98 – non-legally binding instrument on all types of forests, Resolution A/RES/62/98, United Nations, New York, United States of America

UN (2011) World population prospects: the 2010 revision, CD-ROM edition, United Nations, New York, United States of America

UNEP (2007) Global environment outlook 4 (GEO-4). United Nations Environment Programme, Nairobi

Whiteman A, Cushion E (2009) The outlook for bioenergy and implications for the forestry sector, paper prepared for 13th World Forestry Congress, 18–23 October 2009, Buenos Aires, Argentina

Save the Forests: Use More Wood

William Ronald James "Wink" Sutton

Abstract Wood, either as a fuel or as an industrial product, is and always has been one of the world's most important raw materials. However, wood is almost always ignored in any evaluation of global resources. More wood is used globally than wheat, maize or rice for instance.

In 2010 the world harvested just over twice as much wood as it did in 1920. However, over those 90 years the global population has increased almost fourfold so the per capita wood consumption has almost halved This is despite wood being both renewable and sustainable as well as requiring little energy for its conversion into products.

In spite an over a century of predictions of a global wood famine, the world's supply of wood has increased and could even be slightly increased further.

Tree growing requires at least 20–30 years before they are large enough to be harvested. We can only return to the per capita consumption levels of 1920s if there is a huge investment (many tens or even hundreds) of billions of dollars, even without allowing for compound interest) in plantations of fast growing tree species. Only pension funds and large companies have the scale as well as both the vision and the financial resources to invest in such a long term, capital intensive industry as forestry.

The well meaning global environmental movement is often opposed to forest harvesting, even where the forests are responsibly managed. This is a totally misguided belief as trees are living organisms. Locking up forests and preventing forest management is a sure way of ensuring forests will eventually collapse (often with tragic consequences – e. g. a major fire). All forests are very capable of recovering from a catastrophic disaster (forest clearance, mega-fire, volcanic eruption, etc.). If forest preservation is the objective, the best means of achieving

W.R.J. Sutton (✉)
Plantation Focus Limited, Rotorua, New Zealand
e-mail: winkbev@xtra.co.nz

T. Fenning (ed.), *Challenges and Opportunities for the World's Forests in the 21st Century*, Forestry Sciences 81, DOI 10.1007/978-94-007-7076-8_9,
© Springer Science+Business Media Dordrecht 2014

this is with responsible management and tree harvesting. Many environmentalists also are opposed to plantations, especially those of introduced tree species, but these can be very productive and unlike natural forests are likely to attract funding and be self-financing thereafter.

Forest harvesting is often portrayed as deforestation, but we have the apparent paradox that those countries with the greatest wood harvest also have the least deforestation (Europe and North America).

Is there any product that is more renewable, more sustainable, and more environmentally friendly, than wood?

1 Introduction

Over the last 100 years the global population has grown from 1.8 billion in 1910 to almost seven billion in 2010. The global population is predicted to exceed nine billion within the next 40 years. Understandably, everyone wants to enjoy a lifestyle similar to those in developed countries.

The world is almost certainly approaching the supply peaks of not just fossil fuels (except coal) but almost all resources, if they have not already been passed – i.e. peak everything (Heinburg 2010). It is impossible for a population of seven, let alone nine, billion to go on consuming non-renewable/non-sustainable resources at the present rate. Meadows et al. in their 1972 book *Limits to Growth* were among the first to raise global awareness that the 1970s level of consumption was unsustainable. Since then there have been increasing concerns about the coming energy and material shortages, as well as questions about the sustainability of a system based on increasing economic growth. Most impressive is the detailed research by Jeremy Grantham of GMO (a global investment management firm). Grantham "concludes that the world has undergone a permanent "paradigm shift" in which the number of people on planet Earth has finally and permanently outstripped the planet's ability to support us….[T]he phenomenon of ever-more humans using a finite supply of natural resources cannot continue forever – and the prices of metals, hydrocarbons (oil), and food are now beginning to reflect that". In other words, as Grantham says, "…it *is* different this time". Later he wrote a paper "We're Headed For A Disaster Of Biblical Proportions". Other commentators are not so pessimistic e.g. Gilding in The Great Disruption (2011).

2 Wood Is a Major Commodity

Wood and forestry is usually excluded from any study of global resources. This is hard to credit as more wood is still used, and always has been, than any of the major food items or industrial products (steel, plastics, aluminium). A greater weight of cement is now used than wood, but nearly 60 % of the world's cement production is currently produced in China (as is 45 % of the world's steel production). Annual

Table 1 Annual global production of wood, food and industrial products for 2000 and 2010

Wood (millions of cubic metres or tonnes)				Food (millions of tonnes)			Industrial products (in million of tonnes)		
	2000		2010						
Product	m³	Tonnes	m³	Tonnes	Product	2000	2010	Product	2000	2010
Industrial wood	1,603	721	1,537	692	Maize	592	844	Cement	1,560	3,400
					Wheat	586	651	Crude steel	850	1,428
Fuelwood	1,807	813	1,868	841	Rice	599	672	Plastics	180	269
					Potatoes	327	324	Aluminium	21	27
All wood	3,410	1,534	3,405	1,532						

Sources: FOA and/or International Organisations
Wood cubic metres converted to tonnes assuming an average dry specific gravity of 0.45

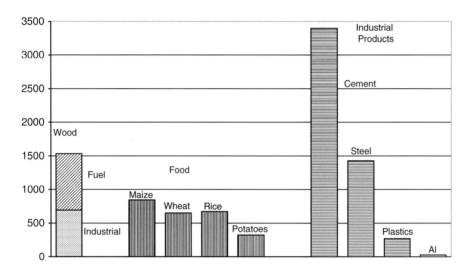

Fig. 1 2010 Annual global production (in 000 of tonnes)

production for the years 2000 and 2010 of the major food items and the major industrial products (including wood) are given in Table 1 and, for 2010 only, Fig. 1.

We are fortunate to have long trend data of global wood production going back to 1920. For that year the USDA Forest Service (Zon and Sparhawk 1923) assembled a detailed and comprehensive global study of wood production of both fuelwood and sawlogs (in 1920 sawlogs were by far the dominant end-use of industrial wood). Since the late 1940s FAO (UN Food and Agriculture Organisation) has complied annual wood production statistics. Global wood production data from the 1920 to 2010 are given in Table 2 together with the global population and per capita consumption trends. The trends in harvest levels are presented in Fig. 2. The trends in per capita consumption is presented in Fig. 3.

Table 2 1920–2010 Global population (in billions), Wood harvest (in billion m³ per year) and per capita consumption (in m³ per person per year)

Year	Global population (In billions)	Industrial wood Harvest	Per capita	Fuelwood Harvest	Per capita	Total wood consumption Harvest	Per capita
1920	1.86	0.74	0.40	0.85	0.46	1.59	0.85
1950	2.52	0.72	0.29	0.70	0.28	1.42	0.56
1960	3.00	1.03	0.34	0.76	0.25	1.79	0.60
1970	3.65	1.28	0.35	1.54	0.42	2.82	0.77
1980	4.45	1.45	0.33	1.68	0.38	3.13	0.70
1990	5.29	1.70	0.32	1.83	0.35	3.53	0.67
2000	6.04	1.62	0.27	1.81	0.30	3.43	0.57
2010	6.89	1.54	0.22	1.87	0.27	3.41	0.49

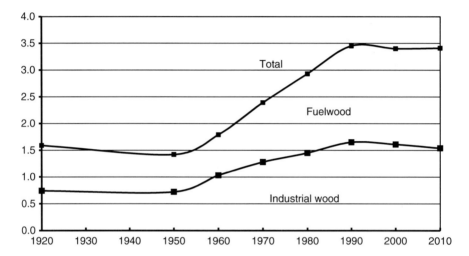

Fig. 2 1920–2010 Annual global wood harvest (in billion of m³)

Our use of wood (per capita consumption) was high in the 1920s (when the global population was less than two billion), reached another peak through the 1960s to 1990 and has since declined, especially in the first decade of the twenty-first century. In 2010 the world consumed, on a per capita basis, for both fuel-wood and industrial wood, almost half as much wood as we did in 1920.

There is little doubt that the world is heading for major material supply problems. Energy is a very major problem – if there is an endless supply of very cheap, renewable and sustainable energy we can do almost anything – grow tomatoes at the poles, extract essential minerals even at very low concentrations, desalinate saltwater, etc. Other than geothermal or atomic energy, our only renewable energy source is the sun. How then is the best way to utilise the sun's energy? Wind, hydro, wave, photovoltaic methods are often advocated. Almost always overlooked are forests, or

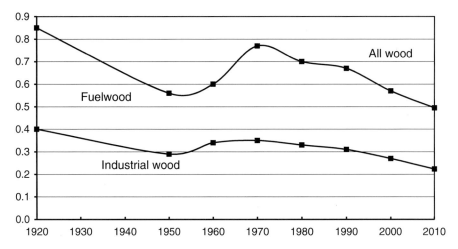

Fig. 3 Global per capita wood use 1920–2010 (m³ per person per year)

more specifically, wood. A 100 year old tree contains 100 years of stored solar energy. Bio-energy has been advocated, but wood has much more to offer the world than just energy – it is far more logical to use energy efficient and environmentally friendly solid wood products to displace the use of other fossil-fuel intensive alternatives (Sutton 2010), and to then use the arising wood residues for pulp, reconstituted wood products or energy production. Because solid wood offers so much, it should be used to replace the use of metals, concrete and plastics wherever possible (other than those made from wood residues).

3   The Ability of Trees to Absorb Carbon Dioxide Is Only Temporary

Because wood is little more than water and carbon dioxide, the potential of the forest to sequester of atmospheric carbon is often seen as an important contribution (helping to mitigate climate change). However, any carbon sequestration by forests is only temporary – forests cannot be net sequesters of carbon for more than a few decades, centuries at best. The exception is where wood is converted to biochar and incorporated into the soil. The carbon is therefore sequested for at least 1,000 years. All trees are living organisms and as such they must eventually die. When this happens, most of a tree's sequestered carbon will eventually be released back into the atmosphere. Wood recycles atmospheric carbon dioxide, whereas the use of fossil fuels permanently adds carbon dioxide to the atmosphere. It is far more logical for the world to substitute energy intensive materials such as metals, concrete and plastics with energy efficient and renewable wood products. If wood is incorporated into structures or buildings that last decades or even centuries, the carbon they contain will not be is not released back into the atmosphere until they are finally burnt or decay. Thereby buying the world valuable time for further weaning itself off its current addiction to fossil fuels.

4   The Future Wood Supply Is Predictable

It is not generally appreciated that, since they take at least 25 year to mature, unless grown in plantations on very short rotations crops for chipping for pulp, bio-energy, reconstituted wood products, etc. as with Eucalypts for instance. Other than any thinnings any tree now being harvested originates from previously unharvested forests or from a forest planted or otherwise regenerated at least 25 years earlier. It follows that almost all the trees that could be harvested within the next 20–30 years are already growing. Since tree growth rates, especially for plantations, are well known, we are able with a fair degree of confidence, to predict possible harvest levels for the next 20 years, which is not something that can be done for most agricultural crops. Several predictions of the possible future wood harvests have been published – not surprisingly they are very similar. One of the more recent is the 2006 study by New Zealand's Scion (Turner et al. 2006). Given that any new plantation establishment will not increase global solid wood supplies for at least the next 20–30 years, it follows that the wood supply cannot be significantly increased at least for the same period. The global wood harvest in the next 20–30 years could only increase if there is more exploitation of previously unharvested forests. The writer is not aware of any proposals to do this, but there is major environmental pressure to **reduce** harvesting in these forests (e.g., the Australian greens are proposing to stop all harvesting their indigenous eucalypt forests). This means that globally there may be less wood available for harvesting than predicted.

The Scion projections of future wood harvest (which presumably includes wood from previously unharvested forests, regenerated and managed native forests and plantations, are summarized in Table 3.

Although the maximum global wood supply can be predicted we cannot predict future wood demand with the same degree of confidence. Consequently, few predictions have even been attempted, especially in the last decade. Because wood is a major (and possibly the only) raw material that is not only energy efficient, but also renewable and sustainable it is very likely that it might well be desirable to increase

Table 3 Scion (2006) projected global wood harvest (in billions of m^3)

End-use	For the years		
	2010	2020	2030
Fuelwood	1.91	2.06	2.18
Industrial	1.88	2.33	2.93
Total	3.79	4.39	5.11

Table 4 Estimates (in billions of m³) of global wood use (assuming per capita wood use of 1980 and 1920)

Base year	End use	Per capita (m³)	For a global population (in billions) of		
			7	8	9
2010	Fuelwood	0.27	1.89	2.16	2.43
	Industrial wood	0.22	1.54	1.76	1.98
	Total	0.49	3.43	3.92	4.41
1980	Fuelwood	0.38	2.66	3.04	3.42
	Industrial wood	0.33	2.31	2.64	2.97
	Total	0.66	4.97	5.68	6.39
1920	Fuelwood	0.46	3.22	3.68	4.14
	Industrial wood	0.40	2.80	3.20	3.60
	Total	0.86	6.02	6.88	7.74

the use of wood in future, and by implication the future demand for wood also. Recently, there have been major advances in wood construction technology – houses, apartments, buildings up to at least 30 floors, small bridges, large warehouses, airport hangers, impressive public structures, etc. Wooden buildings are generally much "safer" than those built with concrete and steel, as they are far less likely to collapse in major earthquakes. Because a wooden building has only about a fifth of the weight of is its equivalent in concrete and steel there is also less need of extensive foundations. Confirmation of the environmental friendliness of wood comes from a just published USA review (Ritter et al. 2011).

While we cannot predict the future wood demand with any certainty. We can make estimates based on past wood consumption levels. In Table 4 consumption estimates for a global population of 7, 8, and 9 billion assuming that the population uses as much wood per capita as was used in 2010,1980, and 1920.

Further discussion in this chapter is limited to only industrial wood, because although fuelwood may become very important as an energy source, it will be covered by other chapters and requires very different forest management techniques. High per capita fuelwood use tend to be in countries that are not major industrial wood producers. Bio-energy, wood pellets, etc. are included in the totals for Industrial wood, however.

In Fig. 4 estimates of future consumption (for industrial wood only) for per capita consumption levels of 2010, 1980, and 1920 and assuming a global population of eight billion, are compared with Scion's supply predictions for 2020. The conclusion is that if the world continues to consume wood at the level it did in 2010 there will be no foreseeable wood shortage. However, if there is major increase in wood demand, which is likely or at least possible, there could be serious supply problems that would take many years to rectify.

The 2006 Scion projections of wood yield also include projections for 2010. That prediction was that 1.88 billion cubic meters of industrial wood would be available for harvest, while the actual global industrial wood harvest was only 1.54 billion cubic meters. In other words nearly 0.35 billion cubic meters of wood available for

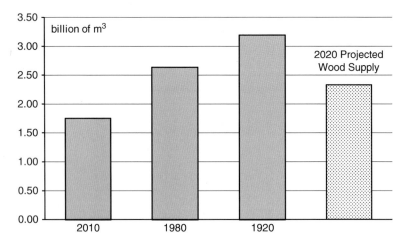

Fig. 4 Projected global industrial wood demand assuming a population of eight billion and per capita consumptions of 2010, 1920, and 1920 with the Scion supply projection for 2020

harvest but was never actually harvested. This implies an oversupplied market and possibly low wood prices. Certainly low wood prices have been experienced in many markets. Low wood prices have very serious implications for future wood yields, as forests are already often seen as a poor investment option, further weakening the incentive to invest private or public capital in new plantations. But a sudden increase in wood prices is unlikely to be sustained, as an over supplied market implies that there is a wood surplus. Fortunately forest owners have some flexibility when it comes to tree harvesting, (an option that is not usually open to agricultural producers) for when prices are depressed, they can often delay harvesting without serious financial consequences for several years in the hope of more favourable times. If wood prices begin to increase, there is likely to be a flood of wood into the market, so depressing prices again unless and until there is a significant and permanent increase in the overall volume of wood consumed. That will have the effect of enabling a sustained increase in wood prices.

5   Forests Are Not Only Very Expensive but They Are Also Very Capital Intensive

Very few outside (and even many of those within) the forestry industry are aware that production forestry including plantations, is much more capital intensive as measured by investment per employee, than industries that are typically regarded as being capital intensive, such as aircraft manufacture, railways, pulp and paper mills, etc. (Fenton 1970). Even though their actual management costs may be low, their generally slower growth rates, lower yields and longer rotations means that managed indigenous forests are even more capital intensive than intensively managed fast growing plantations. The reason is that most of the capital cost is interest. For fast growing plantations, McKenzie (1991, personal communication) commented that "the capital value of a production forest will exceed that of all the utilisation capital put in place to use the log out turn."

Much depends of course on what interest rate is used. By way of illustration – let us assume a forest in Siberia has just been harvested and although the forest is be naturally regenerated we have just one initial inspection at a cost of $10 a hectare. There are to be no further costs, no additional management costs, no fire protection, no administration costs, but that initial $10 cost is to remain unpaid until the regenerated forest is eventually harvested. This could be in 150 years time. At just 1 % interest that $10 compounds up to a charge of $44 – at 2 % and 5 % it becomes $195 and $15,000 respectively! The interest rate charged is usually higher than 1 %. Even if that $10 had been invested in establishment of fast growing plantations on a 25 or 35 year rotation then at 5 % that initial $10 investment at maturity would have compounded to a charge of $34 and $55 respectively. In practice early development costs are often ignored or written off (or subsidised) but those costs are still real and should be treated as such. The value of the forest at harvest time is a critical factor. If a forest (including a tree plantation) is perceived to offer low returns at final harvest it is doubtful if there will be much interest in investing initially in such a venture. Investors in forestry do so because they perceive that their investment will reward them when the tree crop is finally harvested. If there is a perception that the crop will have a low value at harvest time there will be few investors. Current wood prices are at or are near an historic low and currently there is little new investment. In some countries little or no value is placed on managed indigenous forests. Any costs are simply absorbed into the economy.

The future of wood might be very bright but the current low wood price is a major global disincentive to forest or plantation investments. Governments might have a perception that there is an urgent need for a greater wood volume in the future but as most Governments are heavily in debt they are most unlikely, unwilling or unable to put money into such a long-term capital intensive investment such as forestry.

How much investment money are we talking about? Almost certainly many tens, possibly hundreds, of billions of US$. The Republic of Congo recently announced proposals to seek "donor and international investment of US$2,600 million across 1 million hectares". A total investment of US$2,600 /ha over the life of a plantation is comparable with direct costs (i. e. without any compound interest charges) for fast growing plantations in countries such as New Zealand. The Congo proposal (which, if it happens, will only increase the global supply of industrial wood by 1–2 %) has been criticised by environmentalists because they claim these plantations will be low yielding, sequester significantly less carbon as well as supporting less biodiversity. Claims that are not supported by experience in other countries.

Plantations and managed indigenous forests require such huge long-term investments that only large companies or pension funds have both the necessary financial resources as well as the long term vision required. However, that investment will only be forthcoming if it is perceived that the investment will be profitable.

Unless there is a great increase in the per capita use of wood there will not be a wood supply problem in the short term (say the next 20 years). But beyond that there could very serious supply problems that would then take many years to rectify. To ensure future wood supplies, the world needs to greatly increase its investments in forestry and forestry related enterprises. But the world faces a dilemma: An increased investment in forestry may be required but the benefits of this won't be felt for 20–30 years, that may well cause the price of wood to increase over the short term.

6   But Shouldn't We Now Be Experiencing a Global Wood Famine?

Over a Century ago in 1908 President Theodore Roosevelt declared to a White House Conference of Governors: "We are on the verge of a timber famine in this country…" (Clepper 1967). Admittedly, Roosevelt confined his claim to the USA, but parallel global claims have often been made both before and since. But, as shown in Table 2, the world is currently using at least twice as much wood as it did in the 1920 (we have no estimate of how much wood was actually consumed in the 1908 but it would probably have been less than that used in 1920). Current projections are that will be no wood famine for at least the next 20 years, although this assumes that there will be no major increase in wood use over this time period.

The reasons advanced for there being not being a global wood shortage are:

Improved economic factors mean that we can now harvest trees that were once considered unmerchantable.
Responsible and improved forest management.
An increased wood supply from plantations.

We can now process trees that were once considered either too small or were of an undesirable species. Access around the world has greatly improved and we can now profitably harvest trees on sites that once were considered so remote that profitable harvesting was not possible. We can therefore profitably utilise a greater volume of the standing inventory of wood and this acts as a buffer against short term demand – sometimes at the cost of the indigenous forest.

Forests are probably better managed than 100 years ago. Most forests, especially those in countries with advanced economies, North America, Europe, etc., are responsibly managed and most forests are managed on a sustained yield basis.

Plantations whether of indigenous or introduced tree species are increasing important. FAO (Carle and Holmgren 2008) estimates that plantations currently supply about 35 % of the industrial wood supply, but my own estimate is that the plantation supply, especially for saw and peeler logs, is less than the FAO prediction. The reason there is so much interest in plantations is that they are generally much more productive than indigenous forests. The native forests of the Amazon have a greater average standing volume of wood than most other tropical forest (120 cubic metres per hectare of trees greater than 30 cms DBH). However, less than 10 % (i. e. less than 12 cubic metres per hectare) consists of "high value" tree species (FAO 1997). Plantations, however, can be expected to grow at least twice this volume every year. Not only are plantations much more productive globally but there is very little evidence of any decline in productivity with subsequent rotations (Evans 1999).

However, although such developments have been important, they almost certainly cannot account for the subsequent major increase in tree harvest volumes that have been achieved over the past 90 years. For all of the last 50–60 years the global tree harvest has been at least double that when President Roosevelt made his prediction in 1908. Assuming there is no great increase in per capita wood consumption, then it is unlikely that we would face a critical wood famine for at least the next 20–25 years.

As these factors are unable to explain why the increase in the wood harvest has not resulted in a global wood femine there must be another contributing factor(s).

The present environmental movement have portrayed, and continues to portray, the harvested forest as the equivalent of "deforestation" and therefore incapable of recovery. The early conservationists such as the those foresters that influenced President Roosevelt also considered that poor harvesting not followed by careful forest management results in forest degradation.

7   Forests Have a Remarkable Ability to Recover from Catastrophe

Before giving examples of forest recovery following large catastrophes, it is relevant to recall the work of E.W. Jones in the mid 1940s.

The concept of what is a "virgin" forest appears to vary. Some are of the opinion that past human interference should always have been totally absent. Others claim that as long there has not been any attempt to modify the forest (even though there may have been some tree utilisation) it still can be regarded as "virgin".

E. W. Jones (a lecturer in silviculture at the Oxford School of Forestry) was a passionate supporter of Selection forestry (a mixture of species and age classes as practiced in Switzerland and France – the equivalent of the current "continuous cover forestry"). In the 1940s Jones evaluated the few European forests that had not been "devastated or changed by man". Because there were so few examples of untouched European forests, Jones extended his study to include North America. In the northern, central and southern USA there were few "virgin" forests left but there were still large areas left in the west. Jones, much to his surprise, observed that "… all aged forests with irregular canopy answering to the forester's picture of 'selection forest' type appear be rare". Jones records that most of the "virgin" forests were even-aged monocultures or near monocultures. Jones reasoned that the prime cause of the resultant forest structure was fire – caused by lightening strikes. My interpretation of these observations by Jones is that "virgin" forests ultimately end in catastrophe, such as by fire, insect outbreaks, hurricanes, volcanic eruptions, etc. Jones concluded that the concept of climax vegetation is "... a concept only, never existing in practice..." (Jones 1945).

Over millions of years forests have experienced countless catastrophes, and so it is therefore not surprising that natural forests are exceptional in their ability to recover. To illustrate the ability of forest to recover from a catastrophe I will give just three examples:

- In the nineteenth century there was massive forest clearing in the USA – in the North-East for farming and in the South for cotton growing. From early in the twentieth century many of these cleared and farmed areas began to be abandoned. Even without any deliberate tree planting efforts, almost all of that abandoned land has now reverted to forest. For example: in the first 50 years of the nineteenth century the State of Massachusetts was reduced to 25 % forest cover: today forest covers 85 % of the state. In the Southeast USA abandoned farms were invaded by Southern pines to create a huge forest resource that now accounts for more than half of the USA's wood supply. The Eastern USA has also been subjected to periodic hurricanes – not just in the southeast but also in the north-east. For example, the great New England Hurricane of Sept 1938 killed 600 people and destroyed 25,000 homes.

- Over 2 days -August 20 and 21 in 1910 – 3 million acres (one and a quarter million hectares) – an area the size of Connecticut of forest in North-eastern Washington, northern Idaho and western Montana was engulfed in the mega-fire, which also killed 87 people plus countless forest wildlife Pictures taken after the fire show a desolate landscape completely devoid of anything living above ground (Anon 1987). The recovered forest is now abounds with wildlife – black bears, moose, elk, deer, mountain goats, wolverines, cougars and wolves. So complete has been the recovery that there has been increasing pressure since 1980 for about 275,000 acres (12,000 ha) to be declared the Great Burn Wilderness area, and yet 100 years ago the original forest cover was totally destroyed.

- On the 27th August 1883 the volcanic Island of Krakatau (between the Indonesian Islands of Java and Sumatra erupted and expelled 21 km^3 of ash (an explosion 7–10 times greater than the volume expelled by Mt Helens in May 1980). The explosion was heard nearly 5,000 km away and it generated a 40 m tsunami that was recorded in tidal gauges 7,000 km away. Raup (1991, p. 152) records that the eruption put so much (debris, ash and aerosols) into the atmosphere "... that the effects were noticeable for several years. The global temperature dropped by a few degrees ..." The resulting tsunami killed over 30,000 people and wiped out 165 coastal villages. No living organism on the island survived. Yet today the Island is covered with a dense canopy of trees. Inventory assessments show that so far only about 10 % of the original tree species have returned (80 species compared with an expected 800–1,000 species in "non-disturbed tropical forest"). But Krakatau is an island and little more than a century is probably too short a time for all original tree species to become re-established. Also, much of the neighbouring mainland area was also devastated.

Where there has been total forest destruction (and even forest clearance with agricultural use for a century or longer) forest cover will soon be restored once the area is left alone.

Far less destructive to the forest is tree harvesting. The environmental movement is very keen to publish photographs of recently logged forest. To the inexperienced eye the logged site can look like a disaster area. But give the site a few years and the site is green with healthy regeneration. Planting the site can hasten the recovery process. I once heard an agriculture speaker comparing livestock farming with forest harvesting. He was grateful that the farmer need only show cows grazing in green fields followed by steaks frying. This negated the need to display in public the animal slaughter and the cutting up of the carcass. If the public were constantly reminded of the slaughter process there might be less meat eaters. Forest owners are not so fortunate as forest clearance is very visible and may initially give the impression that an environmental disaster has been perpetrated that will never recover.

The environmental movement should be more honest and not limit their publicity to very recently logged areas. As well showing photos of recently harvested forests they should also show how exceptional forests are at recovery after logging. In the mid 1990s the Forest Alliance of BC published a book (Burch et al. 1995) showing many photos of extensive clearcuts taken very soon after logging and matched these with photos from the same photographic location several decades later. The recovery of the forest was most impressive.

There have been many claims that forest harvesting results in the extinction of species. Dr Patrick Moore, a Greenpeace founder, at least 15 years ago issued a challenge to name just one species that had become extinct through responsible forest utilisation. So far not one species has been named. Dr E.O. Wilson of Harvard has repeatedly made the claim, which has been endless repeated by others, that we are currently losing species at the rate of 50,000 per year. I emailed Dr Wilson and asked him to name ten species that had become extinct in the last decade (i.e. one species per year) as a result of tree harvesting. I had not realised that it is very difficult to claim that a species has become extinct (as well as what was the cause). It appears that Dr Wilson's claim is theoretical.

It is widely assumed that industrial wood harvesting results in deforestation. Dr Peter Ince of the US Forest Service Forest Products Laboratory conducted a comprehensive global study to test this hypothesis. Ince found that "In general, the data show that global regions with the highest levels of industrial timber harvest and forest product output are also regions with the lowest rates of deforestation" [e. g., North America and Europe]. Ince reasoned that "…industrial roundwood demands provide revenue and policy incentives to support sustainable forest management (Ince 2010).

Ince is credited with the claim "Save a Forest: Print your Emails". His argument is that only if forest owners perceive realisable returns from a forest will they have the incentive to responsibly and sustainably manage their forests.

8 A Serious Challenge to the Global Environmental Movement

Worldwide the environmental movement is generally opposed to forest harvesting in indigenous forests. Harvesting is portrayed as being very destructive. But forests are exceptional in their ability to recover from what appears at first to be a catastrophe. Forests are not "destroyed" by harvesting. If left alone (even if the logging has been carried out irresponsibly) the forest will eventually recover, but recovery is faster if harvesting is carried out responsibly and/or if the forest is replanted. Forests are only "destroyed" if the area is permanently cleared for agriculture or some other land-use. Many environmentalists are also opposed to plantations, especially where the tree species used is not indigenous. Surprisingly, the same logic is not applied to agriculture.

Wood is (probably the only) renewable and sustainable raw material the world has. Currently we use a great volume of wood, but to maintain our life-style we should be using more wood. Without responsible harvesting of both our native forests and our plantations now there will probably be less wood available in future, when we really need it. Before advocating any moves towards using even less wood than we do now, environmental lobby groups need to give serious consideration as to whether there are any realistic alternatives to such an adaptable material that can so readily and sustainably be produced on the scale needed, and even if there is that it won't cause more problems that it solves. It can be readily demonstrated that an increasing the use of wood will go a long way towards solving many of the world's environmental problems. The question arrises, "why not use it?"

9 Concluding Remarks

The world uses a great deal of wood but it's importance is almost always ignored.

Wood is the world's only large scale renewable, sustainable and environmentally friendly raw material and more systematic use of its potential needs to be made at the global level, if the aim of achieving true sustainability for the world is to be met.

Although there are definitely problems in some regions, overall the world is not over-exploiting its forests. Unless there is a large increase in the global per capita consumption of wood, no global shortage is foreseen for at least the next 20 years. However problems are likely to encountered thereafter.

Forest management is very capital intensive and worldwide requires billions of investment dollars. Investments in forests, especially productive plantations, will only be forthcoming if that investment is perceived as being profitable. The current low price for wood is a major disincentive.

Forests are not being "destroyed" by responsible forms of wood harvesting. Forests have an exceptional ability to recover from catastrophe, including total and long term clearance for farming, as well as after major volcanic eruptions, fires or storms.

Finally a plea to environmentalists to stop opposing responsible forest management and tree harvesting. Pressure to lock-up of all old growth forests is not necessarily desirable, as most will inevitably end in catastrophic collapse. There is frankly no serious alternative to the large scale utilisation of wood for meeting the world's long term needs for bio-energy, fibre and building materials in an environmentally responsible and sustainable manner, and the sooner everyone realises it, the better.

Acknowledgements The writer is indebted to Dr Trevor Fenning and Dr David South for their comments on the initial draft.

References

Anon (1987) The great burn – up from the ashes of 1910. Montana Magazine October 1987, pp 72–76

Burch G, Walker A, Robson PA (1995) The working forest of British Columbia – the working forest project. Harbour Publishing, Columbia

Carle J, Holmgren P (2008) Wood from planted forest – a global outlook 2005–2030. For Prod J 58(12):5–18

Clepper H (1967) Tree farming in America. Unasylva 85:9–16

Evans J (1999) Sustainability of forest plantations – the evidence. Report commissioned by the Department for International Development. London

FAO (1997) Global forest resources 2000…. Chapter 43 Tropical South America Page 3 of the internet download

Fenton R (1970) Criteria for production forestry. NZ J For 15(2):150–157

Gilding P (2011) The great disruption: why the climate crisis will bring on the end of shopping and the birth of a new world. Bloomsbury Press, New York

Heinberg H (2010) Peak everything: waking up to the century of decline. New Society Publishers. P. O. Box189, Gabriola Island, BC V0R 1X0, Canada, p 240

Ince P (2010) Global sustainable timber supply and demand – chapter 2, pp 29–41 in Sustainable Development in the Forest Products Industry (sourced via the internet)

Jones EW (1945) The structure and reproduction of the virgin forest of the North Temperate Zone. New Phytol 44(2):130–148

Meadows DH, Meadows DL, Randers J, Behram WW III (1972) The limits to growth. A report for the club of Rome's project on the predicament of mankind. Universe Books, New York, p 205

Raup DM (1991) Extinction – bad genes or bad luck? W. W Norton & Company Inc., Pennsylvania, p 210

Ritter MA, Skeg K, Bergman R (2011) Science supporting the economic and environmental benefits of using wood in green building construction. USDA Forest Service Forest Products Lab. General Technical Report FPL – GTR – 206, p 18

Sutton WRJ (2010) Wood – the world's most sustainable raw material. NZ J For 55(1):22–26

Turner JA, Boungiorno J, Maplesden F, Zhu S, Bates S, Li R (2006) World wood industries outlook 2005–2030. Forest Res Bull 230, Scion, Rotorua, New Zealand

Zon R, Sparhawk WN (1923) Forest resources of the world. McGraw Hill, New Jersey

References and Further Reading

A large number of references were used in the preparation of this chapter. Where the material came from published sources and these are usually quoted below. However some material, especially that "published" since 1995, was sourced via the internet. For example, internet sources supplied almost all the information used in the three examples of forest recovery following a catastrophic event. None of those information sources are referenced. Websites are often changed or simply disappear. Far better to "google" key words. Where possible, data came from authoritative sources (often by the internet) – FAO, Trade organizations, Governments, etc

Part III
Forests, Forestry and Climate Change

Forests in the Global Carbon Cycle

David Schimel

Abstract Forests play a major role in the global carbon cycle. Deforestation is a major source of carbon to the atmosphere, and forest regrowth is a major sink for carbon from the atmosphere. This understanding comes from several decades of intense research focusing on explaining patterns of variation in atmospheric CO_2 over time and space. The observed patterns reflect the influences of industrial humanity, ocean processes and ecosystem dynamics and understanding these patterns required pooling the techniques of many scientific disciplines. Understanding the role of forests in the global carbon cycle required the expertise of atmospheric scientists, oceanographers, remote sensing pioneers, as well as the full range of forest science disciplines, from biometrics to history. Atmospheric measurements, satellite images and automated sensors have become part of the standard tools of forest scientists, leading to the inclusion of forest dynamics in models of the global system. These Earth System models suggest that the future of our planet is inextricably connected to the fate of its forests.

1 Introduction

Today, it is widely recognized that forests play a major role in the global carbon cycle. This has not always been the case, and, just decades ago, the oceans and atmosphere were assumed to dominate the carbon cycle. This is, at least in part, because forest science lacked the tools to describe forest dynamics as a global phenomenon, even though the tools and techniques to describe individual stands and

D. Schimel (✉)
NASA Jet Propulsion Lab, Pasadena, CA, USA
e-mail: dschimel@jpl.nasa.gov

T. Fenning (ed.), *Challenges and Opportunities for the World's Forests in the 21st Century*, Forestry Sciences 81, DOI 10.1007/978-94-007-7076-8_10, © Springer Science+Business Media Dordrecht 2014

tracts have long been highly developed. Over the past four decades, atmospheric, instrumental and remote sensing techniques have revolutionized our knowledge of global forests, and provided a view of the role of forest carbon in the climate system, past, present and future.

How did the scientific community come to understand and quantify the role of forests in the global cycles? Charles David (Dave) Keeling began measuring CO_2 at the summit of Mauna Loa in Hawai'I in 1958 as part of the International Geophysical Year (Keeling 1998). When these measurements began, no-one knew what would happen to fossil fuel CO_2 but most assumed it would either accumulate in the atmosphere or be taken up by the oceans. There was no expectation that the data would reveal a trend over a short record and a main goal was simply to obtain a highly accurate baseline at a site remote from potential local influences.

Keeling's measurements, however, immediately showed both a year-to-year trend and a strong seasonal cycle. However, from Keeling's measurements of the trend, and knowledge of global energy production, it was apparent early that the measured atmospheric accumulation was lower than the estimates of emission. Significant amounts of carbon were being absorbed by the Earth System. While initially most discussion of sinks for fossil carbon focused on oceanic processes, oceanographers quickly converged on an estimate of ocean uptake of roughly half of the missing carbon, requiring an additional sink.

Work on the global carbon cycle was initiated in the atmospheric and oceanographic communities, as an outcome of the IGY, but quickly ecologists became involved. In the 1970s, oceanographers were gaining confidence from models and tracer oceanography that the ocean uptake could not explain the difference between atmospheric accumulation and fossil fuel use. The geophysical community argued from this basis that the "missing sink" had to be on land and developed a theory that increasing CO_2 was fertilizing land plants and causing the uptake required by the budget. George Woodwell, who had been working of forest metabolism for the US Department of Energy became interested in the carbon cycle. Woodwell and his colleague Richard Houghton had developed a prescient micrometeorological technique for measuring forest metabolism, using CO_2 changes in the nocturnal boundary layer, and were also keenly aware of the accelerating pace of development of tropical forest regions.

Instead, Woodwell and colleagues argued that deforestation was likely causing the land biosphere to be an additional source of carbon to the atmosphere (Woodwell et al. 1977). Many ecologists also argued that land plants were not generally limited by CO_2, but in an argument based on Liebig's law of the minimum, argued they were typically limited by water or nutrients. Woodwell and Houghton went on to develop (and continue to maintain) a global analysis of deforestation rates that showed forests, initially in the Northern Hemisphere but after the mid-twentieth century, mainly in the tropics to be a significant source of carbon to the atmosphere (Houghton et al. 1983). Their values for this flux showed it to be significant, though contrary to some early back-of-the-envelope calculations, rather smaller than the fossil fuel flux to the atmosphere.

2   Land Use, Carbon Fertilization and the Atmospheric Constraint

By the early 1980s, much had been learned about the global carbon cycle, but enormous unknowns loomed. The atmospheric increase was well-defined, based on decades of observation. Oceangraphers were increasingly confident in bounding ocean uptake at a level too low to explain the imbalance in the carbon cycle. Ecologists confronted two very different scientific challenges. First, what was the magnitude of current and historical deforestation and how large a carbon flux could result from land use? Second, could CO_2 fertilization create a sink sufficiently large to account for the imbalance in the carbon budget and compensate for additional fluxes from deforestation? For many years, a great deal of ecological research focused on these two topics, seeking on the one hand, improved quantification of land use change impacts on carbon budgets, and on the other, seeking to understand the magnitude of and controls over CO_2 fertilization.

Much of the effort to better quantify deforestation focused on tracking area cleared and its fate using remote sensing techniques. While clearing triggers carbon losses, the subsequent management of the land has major consequences. What happens to the original biomass? Is it immediately burned and returned to the atmosphere or turned into long-lived products, creating storage? What happens to the land? Is it converted to permanent, low-carbon storage farmland or allowed to regrow and begin to recover carbon stocks? What were the original carbon stocks? Many of the early ecological data on tropical carbon storage were obtained from research sites, with pristine and often optimal conditions, which may bias carbon storage high, relative to an entire biome. In other cases, forest products data are used, and these data have uncertainties and biases all of their own. While the original Houghton et al. (1983) global estimate was based largely on reported statistical data, subsequent efforts focused more on using remote sensing. The first satellite-based large-area estimate was Skole and Tucker (1993), who used newly available Landsat time series to estimates rates of deforestation, and its spatial structure. Since then, improved satellite data and data analysis techniques have allowed the continual refinement of space-based estimates of deforestation and regrowth (Asner et al. 2009).

Today, estimates of deforestation rates and carbon emissions draw on remote sensing, national surveys of biomass and spaceborne measures of biomass, as well as areas of cleared and intact forest (Saatchi et al. 2011). About half of the world's forests have been deforested over the past 150 years, resulting in about 20 % of human-caused emissions to the atmosphere. In the 1990s, deforestation emissions peaked at about 1.5 Gt C per year (±0.7). Subsequently, the rates appear to drop to closer to 0.9 Gt C (±0.7), compared to fossil emissions over 8 Gt C per year.

The global consequences of CO_2 fertilization have proved very difficult to quantify. Experiments increasing atmospheric CO_2 have been done for decades, initially in lab settings or using small enclosures in the field. Lab experiments almost universally show a significant response of photosynthesis and net primary productivity to increasing CO_2, but field studies have shown more equivocal results (Long et al. 2006). Some systems respond initially, but responses decline over time. Other systems appear to respond more strongly. In an important synthesis, Mooney et al. (1991), presented an integrating hypothesis that guides CO_2 fertilization research to this day. Expanding on the Law of the Minimum argument, Mooney suggested that the relative impact of CO_2 fertilization on carbon uptake would increase as other limitations (especially water and nutrients) decreased. In other words, carbon storage in wet, nutrient-rich ecosystems should be most responsive to CO_2, and dry, nutrient poor systems should be least responsive. The Mooney Hypothesis suggested that the effects might be largest in tropical rainforests, and should generally speaking be proportional to NPP (which also increases as water and nutrients increase: Schimel et al. 1997). This hypothesis emerged just in time to be tested in a most unexpected way.

As the Keeling record expanded from Mauna Loa to a global network of measurements, Keeling and others noticed another regularity. Concentrations of CO_2 were elevated in the Northern Hemisphere relative to the Southern Hemisphere. Since the bulk of fossil emissions are in the Northern Hemisphere this result was not unexpected, given that the mixing time between the hemispheres is about a year. By this time, maps of oceanic CO_2 were available and could be used to estimate where the ocean absorbed and released carbon (Takahshi 1986). Combining ocean carbon exchange maps with the atmospheric gradient of CO_2 Inez Fung, Taro Takahashi and Pieter Tans used a model of atmospheric transport to estimate global fluxes of carbon dioxide (Tans et al. 1990). Their analysis predicted an excess of CO_2 in the Northern Hemisphere atmosphere relative to observations, meaning that in the real world, some process was taking up carbon in the Northern hemisphere, and constrained by the ocean observations, the model predicted it had to be on land.

3   Land Use as a Sink

Tans, Fung and Takahashi estimated a significant Northern Hemisphere mid-latitude uptake of carbon by solving a mass balance equation for the atmosphere spatial distribution of the inputs and removals was constrained by the observations of atmospheric concentrations latitudinally, and by Takahashi's map of ocean CO_2 uptake. This prediction contradicted the Mooney et al. hypothesis: if, based on ecophysiological theory and limited experimental evidence, CO_2 fertilization was a significant force, it should result in a tropical forest sink, or at least a sink distributed proportionally to NPP. The atmospheric evidence appeared to contradict that prediction. Following the disconnect between the bottom-up experimental approach of the forest ecology and ecophyiological communities and the top-down geophysical approach, ecologists sought explanations for the apparent global distribution of terrestrial carbon storage, the *mid-latitude sink*.

In Houghton et al. (1983), they had tried to take into account regrowth of mid-latitude forests. Between 1800 and the mid-twentieth century, large portions of the

Northern Hemisphere's forests had been harvested or cleared for agriculture. With the intensification of agriculture, the abandonment of wood as a fuel and the loss of fertility in many soils, by the mid-twentieth century, much of this area had been allowed to regrow (Pacala et al. 2001). In addition, management practices such as fire suppression caused increased in carbon density in many regions. Other factors may have also contributed to increased carbon storage in mid-latitudes forests, such as relief of nitrogen limitation by air pollution-derived nitrogen, the development of protected areas and even trends in climate (Dai and Fung 1993). While Houghton et al. (1983) had tried to take into account these factors, with increased impetus provided by the mid-latitude sink hypothesis, as more data accumulated, the possibility that a land use-driven sink existed in the mid-latitudes gained more credence (Schimel 1995; Schimel et al. 2001; Pacala et al. 2001). In the 1990s, it was likely that the legacy of historical forest use caused terrestrial uptake of a gigaton or more of carbon each year. Interestingly, much of this recovery of forest carbon may have been occurring in hilly and mountainous landscapes, with agriculture and urbanization dominating lower ground (Schimel et al. 2002).

4   The Future of Forest Carbon Stocks

While forests have played a major role in terrestrial carbon-climate coupling in the past, they may also play a major role in causing feedbacks as the climate begins to change in response to continuing fossil fuel emissions. Climatologists and ecologists realized that there were important physical feedbacks between the land surface and the atmosphere and have began to build models focused on these interactions (Land Surface Models). Initially, these models focused on evapotranspiration, albedo and other purely physical (albeit with some biological control) feedbacks, but now include many components of the terrestrial carbon cycle. These models, operating as components of climate or Earth System models are referred to as *coupled* models. In the 2000s, the land biosphere was found to be taking up between 1 and 2 Gt of carbon per year, and the oceans take up a similar but less variable amount. In the future, these fluxes could decrease, leading to an increase in the rate of atmospheric concentration increase, or even change sign, releasing carbon to the atmosphere (Woodwell et al. 1995). When we evaluate the sensitivity of the carbon cycle to climate (and other perturbations), we compare it to these baseline present-day fluxes. Models are calibrated to match observed present-day terrestrial uptake, and models then diverge from the present values as a function of climate and other environmental and assumed land-use changes in the future.

In most coupled models, increasing atmospheric CO_2 tends to cause increases in ecosystem carbon storage (Friedlingstein et al. 2006). Warmer temperatures increase carbon uptake in moist forests, but decrease carbon uptake in drier forests. Drier conditions tend to cause carbon loss, while increases in moisture cause enhanced carbon uptake. Temperature effects dominate in wet forests, where reductions in soil moisture are insignificant or compensated for by warmer conditions and longer growing seasons with opposite effects in dry regions. In regions with intermediate levels of rainfall, warmer conditions may increase evaporative demand, and, in effect, cause forests in these areas to become drier and less productive. Geographically, this tends to result in increased carbon uptake in the cooler forests of the Northern Hemisphere, but accelerating carbon loss from tropical forests as they become drier (Fung et al. 2005). However, the relative strength of these two effects differs between models, with some having stronger or weaker effects overall or in one or the other hemisphere. These differences are due to the physical climate the model simulates, the sensitivity of the simulated forests and even to the interaction between ecosystem change and climate processes.

Essentially all terrestrial carbon models show *positive* feedback. As temperatures warm, ecosystems take up less carbon (or lose more) and so contribute to increasing atmospheric concentrations. The extent to which this occurs depends on the model (some are more sensitive to positive CO_2 effects, others are affected more by water or temperature stress). The impact of climate also depends on the climate simulated in the physical model. Some climate models project more extreme responses of temperature or larger changes to precipitation than others. It can be difficult to separate out these effects (of the climate model versus the carbon component) in a coupled model.

As climate affects Earth System processes, *complex carbon-climate feedbacks can ensue* (Woodwell et al. 1995). Some of these feedbacks are captured by today's models, but others reflect complicated and multidimensional local conditions. For example, in 1997, the growth rate of CO_2 was twice the normal rate, implying that terrestrial uptake dropped to zero for that year. Terrestrial uptake may have been normal that year, but the tropical droughts that frequently accompany El Nino (1997 was an El Nino year) were severe that year (Wang and Schimel 2003). In Indonesia, tropical peatlandforest are being harvested, and to harvest them, the loggers cut canals and lower the water table. The El Nino droughts in 1997 caused wildfires in the Kalimantan region, resulting in vast amounts of stored carbon being lost to the atmosphere.

5   Testing Model Predictions

While forest science today has a fairly robust understanding of the processes of photosynthesis and respiration, and their control by climate, nutrients and water, the interactions of ecosystem physiology with disturbance, forest demographic processes and complex land management practices is in its infancy. The atmosphere allowed robust testing of some macrohypotheses about the operation of the terrestrial biosphere, but many models with very different formulations of the land biosphere approximate observed trends in atmospheric CO_2 roughly equally well. That is, the global atmosphere does not allow discrimination between correct and fallacious models of emerging key processes, even though it served that function for an earlier generation of hypotheses (e.g. land use versus CO_2 fertilization). In order to evaluate today's models and ensure that processes critical to prediction of future carbon-climate feedbacks, new observing and experimental strategies are required.

Unfortunately, the emerging questions about forest carbon storage and how it will change and interact as the climate changes require understanding of finer scale processes than those so powerfully constrained by global atmospheric analysis. On the other hand, vast amounts of data must be amassed if local studies and even national forest inventories are to be effectively extrapolated to the globe! Fortunately, a new class of intermediate scale methods are emerging, using the atmosphere as an integrator, but over smaller areas better corresponding to forest types and regions, and over time scales better corresponding to forest processes. Desai et al. (2011) showed that diurnal sampling of the atmosphere allowed regional estimation of gross photosynthesis and its seasonal variation. By using the daily drawdown in boundary layer CO_2, they were able to estimate photosynthesis over thousands of square kilometers, in a region dominated by just a few forest types, and evaluate whether the regional climate sensitivity of carbon uptake in those forests was similar to, or different from that estimated at a single, intensively studies site. By doing so, they were able to evaluate model performance at scales much larger than traditional plot-and-tower based techniques, but small enough that variations could be linked to more specific processes. At this intermediate scale, similar to the globe, the atmosphere cannot reveal mechanism but it does allow some hypotheses to be tested! The future of forest carbon science will involve more and more of these multi-scaled observing strategies and increased collaboration between forest scientists and scientists in other disciplines.

6   Conclusions

The world's forests play a major role in the global carbon cycle and the climate system. They store vast amounts of carbon, sufficient to significantly impact climate, and have played a significant role in the centennial trends in atmospheric CO_2, in the interannual variability of atmospheric CO_2 and will play a major role in the future of atmospheric CO_2 and climate. Some of the most important discoveries about forests in the Earth System have come from the confrontation between the fine-grained local knowledge of forest scientists and signals in the global atmosphere. One the one hand, this shows the great value of geophysical constraints on predictions about global-scale forest behavior. On the other hand, the global-scale signals of forest carbon exchange, spatially, seasonally and interannually testify to the quantitative significance of the world's forests in the Earth System. This assessment of the importance of forests on climate is just focused on their carbon balance, and neglects their impacts on albedo and the surface energy balance, evapotranspiration and the global hydrological cycle and the biogeochemistry of the other trace gases (Schimel 2013). Forests remain a vital source of ecosystem services and a wide range of important commodities. Foresters remain the stewards of these increasingly diverse resources, but while the goals and priorities of the profession remain consistent with its centuries-long legacy, the issues, scientific tools and challenges are large in scale and technological in nature, bringing new challenges and opportunities to the field!

References

Asner GP, Knapp DE, Balaji A, Paez-Acosta G (2009) Automated mapping of tropical deforestation and forest degradation: CLASlite. J Appl Remote Sens 3:033543

Dai A, Fung I (1993) Can climate variability contribute to the "missing" CO_2 sink? Global Biogeochem Cycles 7:599–609

Desai AR, Moore DJP, Ahue W, Wilkes PTV, De Wekker S, Brooks BG, Campos T, Stephens BB, Monson RK, Burns S, Quaife T, Aulenbach S, Schimel DS (2011) Seasonal patterns of regional carbon balance in the central Rocky Mountains. J Geophys Res Biogeosciences 116, G04009. doi:10.1029/2011JG001655

Friedlingstein P, Cox P, Betts R, Bopp L, von Bloh W, Brovkin V, Cadule P, Doney S, Eby M, Fung I, Govindasamy B, John J, Jones C, Joos F, Kato T, Kawamiya M, Knorr W, Lindsay K, Matthews HD, Raddatz T, Rayner P, Reick C, Roeckner E, Schnitzler K-G, Schnur R, Strassmann K, Weaver AJ, Yoshikawa C, Zeng N (2006) Climate – carbon cycle feedback analysis, results from the C4MIP model intercomparison. J Clim 19:3337–3353

Fung IY, Doney SC, Lindsay K, John J (2005) Evolution of carbon sinks in a changing climate. Proc Natl Acad Sci USA 102.32:11201–11206

Houghton RA, Hobbie JE, Melillo JM, Moore B, Peterson BJ, Shaver GR, Woodwell GM (1983) Changes in the carbon content of terrestrial biota and soils between 1860 and 1980: a net release of CO"2 to the atmosphere. Ecol Monogr 53:235–262

Keeling CD (1998) Rewards and penalties of monitoring the earth. Annu Rev Energy Environ 23:25–82

Long SP, Ainsworth EA, Leakey ADB, Nosberger J, Ort DR (2006) Food for thought: lower-than-expected crop yield stimulation with rising CO_2 concentrations. Science 312:1918–1921

Mooney HA, Drake BG, Luxmoore RJ, Oechel WC, Pitelka LF (1991) How will terrestrial ecosystems inter-act with the changing CO_2 concentration of the atmosphere and anticipated climate change? Bioscience 41:96–104

Pacala SW, Hurtt GC, Houghton RA, Birdsey RA, Heath L, Sundquist ET, Stallard RF, Baker D, Peylin P, Moorcroft P, Caspersen J, Shevliakova E, Harmon ME, Fan S-M, Sarmiento JL, Goodale C, Field CB, Gloor M, Schimel D (2001) Consistent land – and atmosphere-based U.S. carbon sink estimates. Science 292:2316–2320

Saatchi SS, Harris NL, Brown S, Lefsky M, Mitchard ETA, Salas W, Zutta BR, Buermann W, Lewis SL, Hagen S, Petrova S, White L, Silman M, Morel A (2011) Benchmarkmap of forest carbon stocks in tropical regions across three continents. Proc Natl Acad Sci U S A 108(24):98999904, June 14

Schimel DS (1995) Terrestrial ecosystems and the carbon cycle. Glob Chang Biol 1:77–91

Schimel DS, Braswell BH, Parton WJ (1997) Equilibration of the terrestrial water, nitrogen, and carbon cycles. Proc Natl Acad Sci 94(16):8280–8283

Schimel D (2013) Climate and ecosystems. Primers in climate. Princeton University Press, Princeton NJ

Schimel DS, House JI, Hibbard KA, Bousquet P, Ciais P, Peylin P, Braswell BH, Apps MJ, Baker D, Bondeau A, Canadell J, Churkina G, Cramer W, Denning AS, Field CB, Friedlingstein P, Goodale C, Heimann M, Houghton RA, Melillo JM, Moore B III, Murdiyarso D, Noble I, Pacala SW, Prentice IC, Raupach MR, Rayner PJ, Scholes RJ, Steffen WL, Wirth C (2001) Recent patterns and mechanisms of carbon exchange by terrestrial ecosystems. Nature 414:169–172

Schimel D, Kittel TGF, Running S, Monson R, Turnispeed A, Anderson D (2002) Carbon sequestration studied in western U.S. mountains, Eos Trans AGU 83(40). doi:10.1029/2002EO000314

Skole DL, Tucker CJ (1993) Tropical deforestation and habitat fragmentation in the Amazon: satellite data from 1978 to 1988. Science 260:1905–1910

Takahashi T, Goddard J, Sutherland S, Chipman DW, Breeze C (1986) Seasonal and geographical variability of carbon dioxide sink/source in the oceanic areas: observations in the North and Equatorial Pacific Ocean, 1984–1986 and global summary. Final Technical Report for Contract, MRETTA 19X-89675C (10964). U. S. Department of Energy, Lamont -Doherty Geological Obs, Palisades, p 66

Tans PP, Fung IY, Takahashi T (1990) Observational constraints on the global atmospheric CO_2 budget. Science 247:1431–1439

Wang G, Schimel D (2003) Climate change, climate modes and climate impacts. Annu Rev Environ Resour 28:1–28

Woodwell GM, Houghton RA (1977) Biotic influences on the world carbon budget. In: Global chemical cycles and their alterations by man, pp 61–72

Woodwell GM, Mackenzie FT, Houghton RA, Apps MJ, Gorham E, Davidson EA (1995) Will the warming feed the warming? In: Woodwell GM, Mackenzie FT (eds) In biotic feedbacks in the global climatic system. Oxford University Press, New York, pp 393–411

References and Further Reading

Houghton RA (2003) Revised estimates of the annual net flux of carbon to the atmosphere from changes in land use and land management 1850–2000. Tellus 55B:378–390

Forests, Forestry and Climate Change

J. Grace, J.I.L. Morison, and M.P. Perks

Abstract Temperate and boreal forests represent substantial stocks of carbon as biomass and organic matter. Depending on the extent of climate change and the nature of future management practices, these forests may lose, retain or accumulate carbon in future decades. We review data on the impact of management on forest carbon, and we lay out some management options including longer rotations, lower disturbance, nitrogen fertilization and the afforestation of new land. We ask whether the carbon gains made possible by such practices will be at the expense of timber production and also we examine the impact of these practices on a range of environmental services (nature conservation, watershed protection, public amenity value, bio-fuel production). We look forward to a period of warming, which favours more rapid growth of trees in northern regions, but predictions from ecosystem models must be tempered by the likelihood that there may be more extremes, including droughts, storms and outbreaks of new pests and diseases.

1 Introduction

The natural cover of much of today's land surface is still forest. Forests or woodlands may occur wherever the precipitation exceeds about 500 mm per year and the summer temperatures are greater than 10 °C. This includes around 50 % of all the land area on the planet. However, today's forest cover only about 30 % of the land, as vast areas have been cleared by humans. The clearing of forests has taken place over many hundreds of years, mostly to make way for people and their agriculture.

J. Grace • J.I.L. Morison • M.P. Perks (✉)
University of Edinburgh & Forest Research, Edinburgh, Scotland
e-mail: mike.perks@forestry.gsi.gov.uk

T. Fenning (ed.), *Challenges and Opportunities for the World's Forests in the 21st Century*, Forestry Sciences 81, DOI 10.1007/978-94-007-7076-8_11,
© Crown Employees UK 2014

These two forms of land cover, forest and agriculture, are in marked contrast to each other, not just in the way they are used and valued by humans but also in the way they exchange energy, mass (particularly carbon and water) and momentum with the atmosphere. There are two-way links between vegetation and the atmosphere, so that any widespread change in land use may bring about a change in the climate system. In this chapter we explore the relationship between forests, land use, climate and energy use.

We discuss three main aspects: (1) the effect of forests on climate (through carbon and greenhouse gas exchanges and key biophysical interactions) (2) the impact of climate change on forests, and (3) the possible impact of management on the 'climate function' of forests including afforestation and deforestation.

2 Carbon Cycle, Greenhouse Gas Balance and Land Use Change

2.1 Global Scale Carbon Balance

One of the most important linkages between forests and the climate occurs because atmospheric carbon dioxide (CO_2) is both the source of most of the biomass contained in green plants and the most important greenhouse gas. Moreover, the terrestrial biosphere contains about three times as much carbon as that contained in the atmosphere as CO_2 (Fig. 1), most of it being stored in forests, woodlands and peat-lands. Thus, one way to curtail the rise of atmospheric CO_2 is to conserve forests and peat-lands, and where possible plant more trees. This view is embodied in the Kyoto Protocol, which (in some circumstances) allows carbon captured in forests to be used to offset emissions from fossil-fuel burning. Moreover, some countries, including the UK, have a policy to increase the rate of tree planting to fight global warming (e.g. Read et al. 2009).

Measurement of the CO_2 in the atmosphere shows an inexorable rise in concentration over the last 100 years, caused by the burning of fossil fuels and also, to a lesser extent, by deforestation. It seems obvious that planting trees and avoiding deforestation should be an important step towards prevention of dangerous climate change, along with decarbonising the energy supply by developing every possible kind of renewable energy source.

So far, there are very few signs that this is being achieved. Fossil fuels are still being burned at an increasing rate whilst further deposits of oil and gas are being discovered. Only a few of the industrialised countries within the Kyoto Protocol have honoured their commitments to reduce emissions whilst rapidly developing countries such as India and China, which are not part of the Kyoto process, have hugely increased their emissions. The global CO_2 total emissions coming from fossil fuel burning has increased from 6 billion tonnes of carbon in 1990 to around 9 billion tonnes today.

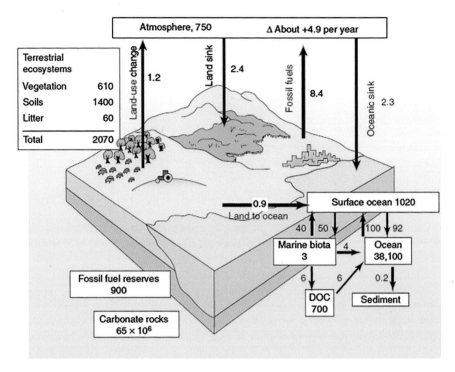

Fig. 1 Global carbon cycle, updated for 2010 (http://www.globalcarbonproject.org/). Units for carbon stocks are billions of tonnes, units for fluxes are billions of tonnes per year. One billion tonnes $= 10^9$ tonnes $= 1$ Gigatonne $= 1$ Petagramme. Data for stocks on land and in the ocean are highly uncertain. The term for fossil fuels includes emissions from cement manufacture. Data for emissions have the following uncertainty: fossil fuel 8.4 ± 0.5 and land use change 1.2 ± 0.7

However, not all the released carbon results in increased CO_2 concentrations in the atmosphere. Some of it dissolves in the oceans and another fraction is absorbed by vegetation, becoming biomass and soil organic matter. This net uptake in CO_2 by vegetation ('the terrestrial carbon sink') reflects an imbalance between the capture of carbon by photosynthesis (believed to be in excess of 120 billion tonnes per year) and the losses that occur during plant and microbial respiration (each is probably around 60 billion tonnes per year). It is generally considered that with an un-changing climate the photosynthetic capture would be balanced annually by respira-tory losses i.e. a hypothetical 'steady state'. However, we know that the conditions for plant growth are generally improving, as both CO_2 and nitrogen compounds in the atmosphere (industrially-produced active nitrogen – ammonium and nitrate) act inadvertently as fertilizers, stimulating photosynthesis. They reach the land surface and its vegetation by wet and dry deposition, and enter into terrestrial ecosystems where their effects are far-reaching. Thus, photosynthesis is thought to be increasing rather rapidly whilst the respiratory processes are increasing more slowly. This theory of the imbalance between global photosynthesis and respiration is helpful as an explanation of the terrestrial carbon sink. We know there is indeed a terrestrial sink from analysis of the mass of CO_2 measured in the atmosphere, and the comparison between this and the known rates of emissions from fossil fuel burning and deforestation: the most up-to-date results from such analysis are expressed in Fig. 1. Some 8.4 billion tonnes of carbon are emitted from fossil fuel burning (2010 figures) and a further 1–2 billion tonnes from tropical deforestation, yet only about 5 billion tonnes stays in the atmosphere. The terrestrial carbon sink is currently about 2.3 billion tonnes of carbon per year, similar in magnitude to the ocean sink but with greater inter-annual variability caused by irregularities in the climate system

From a comparison of the geographical pattern of these sinks, detected from global patterns of CO_2 concentration, it is clear that strong uptake occur wherever there is forest cover: North America, Eurasia and the remaining intact forests of the tropics. Were it not for these sinks, atmospheric concentrations of CO_2 would rise at an even more alarming rate than at the present, and cause even more global warming. However, carbon sinks may fail, if for example forests are degraded or if warming increases respiratory processes sufficiently.

Globally, forestry therefore has an important role in managing the C-cycle (Table 1). It has been estimated that a concerted attempt to maximise tree planting could achieve a new sink of up to 1–3 billion tonnes of carbon per year (Cannell 2003; Nabuurs et al. 2007), but land is in increasingly short supply because of the requirements to grow food, fibre and biofuels. Increasing the planting rate would be only part of a forestry strategy. Greater benefits would come from reducing the deforestation rate, given that tropical forests are both carbon stocks and carbon sinks. Additional reductions in net emissions can come from the effective use of forest products, as it is important to realise that felling of trees is not necessarily bad for the longer-term carbon balance, if wood products are used to substitute for fossil-fuel intensive energy generation and materials like concrete and steel. This does depend on trees being replanted and good management being employed so that forest ecosystem function is not damaged (see Sect. 5.3). In contrast, agricultural systems are generally not effective in storing carbon: even though they usually have high rates of photosynthesis during their growth period, the biomass produced is retained for only weeks and months because a much smaller fraction is in the form of lignin, the decay-resistant carbon polymer that all trees manufacture. We now move to consider the carbon capture process in forest ecosystems, at a local scale.

2.2 Local Scale Measurements of Forest C Balance

It is clear that forests can be established to create carbon sinks and build medium to long term C stocks. Forests hold substantial stocks of carbon (Table 2). Everyone knows, in general terms, that a young forest plantation accumulates carbon as biomass in the stems and in the organic matter in the soil. But how much, and for how long? Forest mensuration has provided us with data on the growth rates of plantations, but this is only a part of the story. The complete analysis of carbon flows in forests has been facilitated in the last two decades by the use of new techniques,

Table 1 Carbon stocks and fluxes, summarised from the data base of the FAO (line 2 of the table) and Luyssaert et al. (2007) for lines 3, 4, and 5. Carbon fluxes are given as Net Ecosystem Production, NEP, i.e. what remains of the Gross Primary Productivity after subtracting plant and microbial respiration; it is measured directly by eddy covariance. The numerical values are means with estimated standard seviations. On the assumption that the sample is representative we estimate the total carbon stock in the biomass of forests to be 406 billion of tonnes of carbon

	Boreal	Temperate evergreen	Temperate deciduous	Tropical humid
Latitude (degrees, with range)	58 ± 7	44 ± 8	44 ± 9	14
Estimated area (millions km^2)	13.7	1.0	9.4	13.4[a]
Above ground biomass (tC ha^{-1})	57 ± 37	149 ± 135	108 ± 56	113 ± 58
Below ground biomass (tC ha^{-1})	14 ± 8	4.6 ± 4.6	2.5 ± 2.6	29 ± 22
Total biomass (tC ha^{-1})	72	154	110	142
Total biomass (Gt C)	98.6	15.4	103.4	190.3
NEP (tC ha^{-1} year^{-1})	1.34 ± 0.14	4.02 ± 0.53	3.10 ± 0.29	4.10 ± 0.07

[a]http://www.fao.org/docrep/014/i2247e/i2247e00.pdf

Table 2 Typical parameters for models of forest and grassland

	Grassland or cropland	Forest	Influence on climate of transition to forest
Height (m)	0.1–2.0	5–60	
Roughness length (m)	0.01–0.2	0.5–6.0	Cooling
Albedo (dimensionless)	0.15–0.25	0.08–0.20	Warming
Canopy resistance, dry (s m^{-1})	20	50	Warming
Aerodynamic resistance (s m^{-1})	50	5	Cooling
Carbon stock (tC ha^{-1})	1–10	60–150	
Carbon accumulation rates (tC ha^{-1} y^{-1})	0	−7 to 0	Strong cooling

of which eddy covariance is the most important. It enables a sensor, supported on a mast above any land surface cover, to measure the gains and losses of carbon by an ecosystem rather directly. This is more revealing than simply measuring the growth of the plant material, as the sensor measures *all* the C-fluxes including the very large respiratory efflux from the heterotrophic organisms including the microbes in the soil. Moreover, the fluxes are observed to vary from day to day, depending on the weather and time-of-year, and so it is possible to derive models that may be used in a diagnostic and predictive way.

There are now several hundred local sites in the world where investigations of CO_2 exchange are on-going, and a number of summary papers have been published, in which the difference between tropical, temperate and boreal forests are very clear (e.g. Luyssaert et al. 2007).

One of the surprises from a meta-analysis of the literature which considered many carefully measured sites around the world, is that even old forests continue to be carbon sinks, albeit smaller sinks than young plantations (e.g. Luyssaert et al. 2008). This general observation is very important in evaluating the impact of deforestation. The undisturbed forests of the tropics, for example, are important not only because they contain large stocks of carbon but also because they continue to act as carbon sinks.

2.3 Other Important Gases

Forests and agriculture differ in their exchange of the other key greenhouse gases: methane and nitrous oxide. Both are more effective than CO_2 in radiative forcing when compared on an equal mass basis, although in absolute terms they are less important because they are found at lower concentrations in the atmosphere. To calculate the warming effect of their emissions, their masses must be multiplied by a factor called the Global Warming Potential (GWP) which expresses their warming effect (per kg) relative to that of CO_2. Expressed over a 20 years' timeframe, the GWP of methane is 72 and that of nitrous oxide is 289.

Methane is produced when organic matter is decomposed in the absence of oxygen. Globally, anthropogenic sources are estimated as 307–428 million tonnes a year, exceeding natural sources which are 145–260. Agricultural systems (ruminant animals and rice growing) account for more than one-third of the anthropogenic emissions. Forests produce rather little methane, except when they are being burned or when they are flooded, and when dry they may even be net sinks for methane. Substantial methane concentrations are seen by satellite over the major regions of tropical rain forests in the wet season.

Most nitrous oxide is produced naturally in soils through the microbial processes of denitrification and nitrification. Most natural woodlands and forests produce rather low emissions of nitrous oxide. Even under management, the use of artificial fertilizer in forestry is much lower than that used in intensive forms of agriculture, which can be well over 100 kg-N ha^{-1} year^{-1}. The GHG emissions from modern agriculture are large: in a recent study it was demonstrated that whilst European ecosystems absorb about 10 % of the Europe's fossil fuel CO_2, their benefit is cancelled out entirely when agricultural emissions of CH_4 and N_2O are taken into account Schulze et al. (2009).

Recent discussions on the use of biofuel from oil palm (e.g. Committee on Climate Change 2011) have highlighted that growing biofuel in the humid tropics usually leads to substantial greenhouse gas emissions because the plantations are heavily fertilized with nitrogen and so they emit substantial amounts of N_2O (Crutzen et al. 2008). Hence, the fuel is not as 'green' as most people believe, especially where rain forest has been destroyed to make way for oil palm, and the very large C stocks in the forest released.

2.4 CO₂ Equivalent Emissions

Data on national GHG emissions are available from official statistics. All the greenhouse gas data (CO_2, CH_4, N_2O and some industrial gases) are summarised as CO_2 equivalent (CO_2eq) to satisfy international reporting rules. Taking Europe as an example, the 27 member States together are estimated to have emitted 4,409 Teragrams (Tg = 1 million tonne) of CO_2eq in 2010,[1] of which agriculture was a source of 462 Tg

---

[1] www.eea.europa.eu/publications/european-union-greenhouse-gas-inventory-2012

and land-use change, some of which is represented by afforestation, was a sink of 312 Tg. The role of natural changes in the greenhouse gas budget may not be accurately represented by such estimates.

2.5   Volatile Organic Compounds

In assessing the impact of vegetation on the climate, volatile compounds of carbon need to be considered. Plants emit a wide range of organic molecules, many of which participate in the chemistry of the atmosphere. Of these, isoprene (C_5H_8) is one of the most important. It is produced by trees and shrubs in relatively large quantities and it reacts in the atmosphere with free radicals (e.g. OH), producing CO_2. The two gases, C_5H_8 and CH_4, in effect, compete for the available OH. Some of the isoprene is converted to larger organic molecules which participate in the formation of aerosols and form water droplets leading to haze and clouds. This is one of the less-understood ways in which forests may affect the climate system, and is not included at the moment in climate models.

3   Feedbacks Involving Forest, Land Use Change and the Climate System

3.1   Land Surface Characteristics

Clearly forests are involved directly in the global GHG balance. In a very general way we can expect that increasing forest cover will take up more CO_2, and thus help reduce climate change. However, there are more complex interactions between forests, other land uses and the climate. They arise because of key differences in their biophysical characteristics:

- *Energy*. The quantity *albedo* defines the fraction of solar energy that is reflected back into space. It is also called the shortwave reflectance. Surfaces such as fresh snow have an albedo as high as 0.95 whilst some bare soils have an albedo of 0.05. There have been numerous comparisons of the albedo of different land covers; forests and many shrublands have an albedo in the range 0.08–0.20 whilst crops and grasslands are higher, in the range 0.15–0.25. Once energy is absorbed, it is dissipated to the air above as sensible heat (increase in air temperature, or 'warmth') and evaporation ('moisture'). Here again, there are contrasts: forests tend to dissipate less energy as sensible heat and more as evaporation.
- *Momentum*. Roughness expresses the tendency of a surface to absorb the momentum of the fluid (air, water) that flows over it. The roughness of the land surface affects the development of the atmospheric boundary layer, the lowest zone in the atmospheric where the exchanges with the surface take place, and the genera-

tion of turbulence. Thus, forests present a rough surface that slows down the wind speed (the horizontal momentum of the wind is converted to motion of the trees). At a local scale it means that forest plantations can be used to provide shelter for humans, livestock and crops; at a large (regional) scale it means that wind speed will tend to be reduced over forested continents.

• *Mass.* The enhancement of turbulence by rough (forest) surfaces has implications for the exchange of materials including water evaporating from the leaf surface and pollutants being deposited from the air: the rough surfaces of forest generates large scale turbulence and mixes the surface air with the air aloft. This tendency is parameterised by the wind-speed dependent aerodynamic resistance (Table 2). The increase in transfer is greatest for the evaporation of liquid water deposited on the leaves; for transpiration and CO_2 the effect is less because the diffusing molecules encounter the additional resistance of the stomatal pores.

• *Parameters.* In Table 1 we summarise typical parameters that are used in model formulations, contrasting forests and grass or croplands. Where possible, we indicate whether the difference between forest and croplands will imply warming or cooling. It is clear that most parameters will lead to the conclusion that forests cool the planet compared to agriculture, but that the albedo effect alone suggests that forests warm the planet.

3.2   Vegetation-Atmosphere Feedbacks

Interactions between these biophysical processes and other ecosystem functioning results in various complex feedbacks; some that are positive (likely to destabilise the climate system) and some that are negative feedbacks (tending to stabilize the climate system). Some of these have been widely discussed, but they are not yet fully understood. The list is not exhaustive.

Positive Feedbacks

1. Warming will increase microbial decomposition, releasing greenhouse gases from the carbon stock which is presently locked away as soil organic matter, and thus increasing the warming rate.
2. Tropical deforestation will decrease tropical evaporation, leading to drier and warmer conditions in the tropics (see Sect. 3.3) with additional smaller effects in other parts of the world.
3. Under warming, boreal forest will spread northwards, decreasing solar reflectance, thus leading to warming.
4. Warmer, drier conditions in Mediterranean and seasonal tropics will result in loss of savanna-type woodlands, more wildfires, and desertification.
5. Increased fires contribute to GHG release and soot formation, leading to warming.

Negative Feedbacks

1. Forest expansion to higher latitudes and altitudes will create a stronger carbon sink than existing vegetation.
2. Replacement of boreal (conifer) forest by warmth-loving deciduous forest will increase solar reflectance, leading to cooling.
3. Warming will accelerate breakdown of organic matter in the soil, thus releasing locked-away nitrogen and phosphorus, acting as a fertilizer to stimulate productivity, and a stronger carbon 'sink'.
4. Increased leaf area and transpiration in a warmer world will lead to more clouds, thus cooling the planet.

To understand the net effect of such feedbacks, Global Circulation Models have been developed to incorporate land cover (some representation of the carbon cycle). Such models are often called Earth System Models (ECMs).

3.3   Forests and Overall Energy Balance

In 1992, at the time of inception of the United Nations Framework Convention on Climate Change, our knowledge of processes and interactions was appreciably less than it is today. Thus, the 1997 Kyoto Protocol, which is intended to fight climate change, considers forests only inasmuch as they constitute stocks and sinks of carbon. A general realisation that radiative and evaporative differences between forest and cropland are important came somewhat later (Bonan 2008; Betts et al. 2000; Anderson et al. 2011) even though comparative measurements of albedo over different land covers had been in progress for half a century (for example, they were tabulated in an early text book, Geiger 1966).

When a forest is planted to replace grass or cropland, it does indeed begin to capture carbon and store it as biomass and organic matter in the soil. Large areas of planting should thus remove CO_2 from the atmosphere and slow down the warming rate. However, the albedo of the land cover will fall because most forest has a lower albedo than most types of cropland; hence, a warming tendency must be considered alongside the cooling effect. Where the forest is in the extreme northern regions, the albedo effect is likely to be especially large because conifer forests lose their snow cover more rapidly than other land uses (e.g. croplands, grasslands) through sublimation, evaporation and mechanical shedding. For individual stands or forests it is possible to calculate the relative contribution of CO_2 uptake and albedo change from afforestation by comparing the 'radiative forcing'; for example in New Zealand the net radiative forcing benefit of afforestation of grassland with a radiata pine stand was reduced by some 17–24 % over the whole length of the rotation because of the lower albedo of forest (13 %, compared to 20 %, Kirschbaum et al. 2011). Such calculations depend on the albedo differences, the rate of tree growth and hence CO_2 uptake, the time period and assumptions about longevity of the carbon sink.

However, to estimate the full climatic effect of planting new forests (afforestation) or replacing forests that have been lost (re-forestation), any Earth System Model should capture all relevant processes. There are evident difficulties in making it so; for example, not all known processes are well understood, and, moreover, it is likely that there are many processes and interactions which still have not been discovered. Here we focus on one such study (Arora and Montenegro 2011), where an Earth System Model, incorporating a Global Circulation Model and a representation of the terrestrial carbon cycle was used. The experiment tested the effect of replacing 50 % or 100 % of existing croplands with forest, over a 50 year time period. It also tested the effect of making the 50 % replacement of croplands with forest in (a) the boreal region (b) in the north temperate zone or (c) in the tropics. Afforestation resulted in a larger land carbon uptake and lower atmospheric CO_2 concentration, but the resulting cooling effect was partly or totally offset by the warming effect of the new low albedo surface. The authors find that afforestation of 50 % of all croplands over 50 years would cool the planet only by 0.45 °C compared to the 'business as usual' control (Fig. 2). They conclude "the direct temperature effects provided by afforestation are marginal" though they recognise the other benefits of afforestation such as timber supply, habitat and fossil fuel substitution. However, they did find that tropical afforestation is more effective in cooling the planet than boreal or temperate afforestation, a result also found by Betts (2000). This general observation gives additional weight to the current emphasis on avoiding further deforestation of the tropics.

3.4   The Case of Tropical Deforestation

The tropical rain forest has received most attention from modellers because of its rapid rate of loss over the last 40 years, although tropical woodlands are probably being removed or modified at an even higher percentage rate. Of all the rain forests, it is the Amazon that has been investigated most thoroughly (Table 3). Using Global Climate Models which contain a version of the carbon cycle, experimental modelling of deforestation of the entire Amazon suggest that precipitation will decrease and temperatures will rise (Table 3). The extent of the effect is not the same in all the models. This variation occurs because the models are not parameterised in exactly the same way, and they differ somewhat in their structure.

Several studies indicate that the impact of deforestation on such a scale as the Amazon will also influence the regional climate elsewhere. Avissar and Werth (2005) propose that tropical deforestation of the Amazon disrupts the rainfall in Argentina, with severe implications for agriculture. These long-range effects are termed *teleconnections*.

Perturbations in the climate system have also occurred before the Industrial Revolution. Pongratz and Caldeira (2012) report that 20–40 % of LUC emissions have been estimated to have occurred pre- 1850 due to rapid agricultural expansion resulting in albedo and energy balance changes, as discussed above.

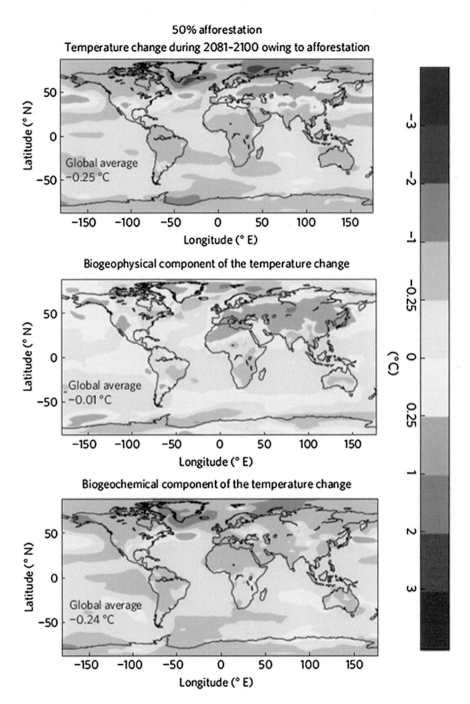

Fig. 2 The effect of converting 50 % of all croplands to forest over a 50 year period (Arora and Montenegro 2011). Red denotes warming, blue shows cooling (see vertical bar for the key). The top panel shows the overall effect of the 50 % afforestation, the middle and lower panels show respectively the effect of the physical changes (mostly albedo) and effect of transfer of carbon from the atmosphere to the ecosystem

Table 3 Amazonian deforestation experiments with large-scale climate models. The results in columns 3 and 4 refer to the change in the Amazonian climate caused by deforestation

Authors	Model resolution (lat.×long.)	Change in rainfall (% of control)	Change in temperature (°C)
Shukla et al.[a]	1.8°×2.8°	−21	+2.5
Lean and Warringlow[b]	2.5°×3.75°	−7	+3
Costa and Foley[c]	4.5°×4.75°	−12	+1.4
Voldoire and Royer[d]	2.8°×2.8°	−8	−0.01
Medvigy et al.[e]	25 km	−2.4	0 to +4

[a]Shukla et al. (1990)
[b]Lean and Warringlow (1989)
[c]Costa and Foley (2000)
[d]Voldoire and Royer (2004)
[e]Medvigy et al. (2011)

4   Effects of Climate Change on Forests: Regional Variations

According to the *Fourth Assessment Report* of the Intergovernmental Panel on Climate Change (IPCC 2007) global warming is now unequivocal, being evident in rising air and ocean temperatures, rising sea level and melting of ice and snow. Especially relevant to forests, *extreme events* are likely to occur more often (IPCC 2012; Coumou and Rahmstorf 2012). However, there are clear regional variations in the current trends, the distribution of extremes, and in the outlook for the future. The effects of climate change are likely to include changes in forest productivity and changes to the geographic range of certain tree species, as well as other components of the forest flora and fauna. Sustainable forest management practices are already taking into account species suitability to predicted future climate and bioclimatic modelling is being used to assist in ensuring species are well adapted to predicted changes. Changes in forest health are predicted because of changes in rainfall patterns and temperatures affecting trees and forest ecosystems directly and because their pests and pathogens are also likely to change in their prevalence, persistence and impact Kurz et al. (2010). Thus, climate change has rather different implications for tropical, temperate and boreal forests (Fig. 3).

4.1   Boreal Forests

Warming is predicted to be most pronounced in the boreal region, largely as a result of the positive feedback involving the reduction in the snow and sea-ice cover, and the consequent greater absorption of solar energy. Precipitation will tend to increase, especially in the winter, and in some places this might increase the formation of peat ('paludification'). These changes will have important implications for both agriculture and forestry, as spring will be earlier and the growing season will be extended; land which is now treeless (at high latitude or elevation) may support tree growth. Overall an increase in the biological productivity of the boreal zone is expected,

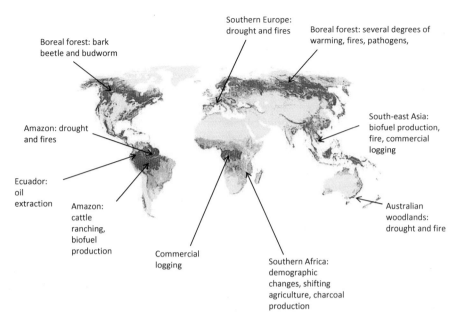

Southern Europe:
drought and fires

Boreal forest: several degrees of
warming, fires, pathogens,

Boreal forest: bark
beetle and budworm

South-east Asia:
biofuel production,
fire, commercial
logging

Amazon: drought
and fires

Ecuador:
oil
extraction

Amazon:
cattle
ranching,
biofuel
production

Commercial
logging

Southern Africa:
demographic
changes, shifting
agriculture, charcoal
production

Australian
woodlands:
drought and fire

Fig. 3 Threats to forests and woodlands caused directly or indirectly by climate change and increasing demand for land and timber products

although there may be threats posed from increased activity of pests and diseases Bentz et al. (2010). Warming may significantly enhance the breakdown of organic matter, releasing locked-up nitrogen and phosphorus to support plant growth but also transferring more CO_2 from the predominately organic and organo-mineral soils to the atmosphere and speeding up global warming. On the other hand, it is expected that the incidence of insect and fungal pests will increase, as has happened already in parts of North America (see below).

4.2 Temperate Forests

Warming is expected to be less intense than in the boreal region, but still significant and similar remarks apply. Bud-break already occurs earlier in the spring and this trend is likely to continue; a metanalysis of phenological records from across Europe shows a 2.5 day advance in bud burst for every decade (Menzel et al. 2006). There is abundant evidence already of greater overall productivity of forests, but this may be the result (at least in part) of increased deposition of active nitrogen and increased CO_2 concentration. There have been several attempts to untangle these influences, but they have not been conclusive. Outbreaks of pests and diseases have already increased, exacerbated by the trade of young trees and timber materials between countries.

Summer drought is likely to be increasingly frequent because of increased temperatures causing higher evaporation and, in some regions, decline in summer rainfall. The effect on trees of the 2003 drought in Europe was thoroughly analysed: in that year and the subsequent year there were widespread reductions in growth and CO_2 uptake (Ciais et al. 2005). If there are successive years of drought, forests may be permanently damaged. Tree quality may be affected, through morphology and stem drought cracks; mortality may occur of particular cohorts or species, leading to a change in forest structure or species composition. Drought also increases the susceptibility to biotic attack with, for example, increased incidence of root-borne fungal disease (e.g. *Armillaria*, *Heterobasidium*), insect–vectored fungal disease (e.g. blue stain fungi) and wind-vectored disease (e.g. *Dothistroma*, *Phytophthora*).

4.3   Mediterranean

For the Mediterranean a pronounced warming is expected, particularly in the summer months, associated with a reduction in precipitation. This will increase the probability of wildfire occurrence, extend the length of the fires season and the area at risk, and increase the probability of large fires so it is hard to escape the conclusion that the Mediterranean will suffer more fire-related damage to its ecosystems. Indeed, almost all studies point to the Mediterranean becoming an especially vulnerable region. These conclusions apply to Mediterranean climates elsewhere in the world: parts of the west coast of North America, Chile, southern Africa and southern Australia. In these parts of the world, fire management is a key concern.

4.4   Tropical Forests

The rain forests of South America are expected to become warmer and drier; but the changes predicted in the other two regions, SE Asia and Africa are less dramatic. Until now, there has been good observational evidence from repeated measurements of sample plots that the biomass of rain forests throughout the tropics is increasing (Phillips et al. 1998; Lewis et al. 2009) supporting the hypothesis that the humid tropics are a carbon sink strong enough to offset the carbon losses from regions where rain forests are being rapidly cleared. This increased growth is generally attributed to the effect of rising CO_2, although the evidence for this is not strong. However, model simulations and some field observations suggest that rising temperature and increasing incidence of drought over future decades will reverse this trend (Phillips et al. 2009), so that in the Amazon region, rain forest may be replaced by more open savanna-like vegetation. On the other hand, two experimental field studies on the impact of drought on rain forests (achieved by excluding 50 % of the rainfall for several years) have demonstrated that such forests are rather resilient. Similar experiments have not yet been done on the savanna woodlands. This issue of the impact of drought remains an important research area – it is too early to draw firm conclusions.

5   Managing Forests for Carbon and GHG Balance Benefits

There is general agreement that forest management can contribute to carbon emissions mitigation. For example, the Intergovernmental panel on Climate Change stated in 2007 that "*in the long term, a sustainable forest management strategy aimed at maintaining or increasing forest carbon stocks, while producing an annual sustained yield of timber, fibre or energy from the forest, will generate the largest sustained mitigation benefit*" (IPCC 2007). However, within that apparently simple statement are several complexities. Essentially, there are a many different ways to manage forests and their products, and a key question is 'what is the best use of wood?'. To a large extent the answer will depend on the local situation (Lippke et al. 2011), but we discuss some key aspects below.

5.1   Different Forest Management Types

Forests can be managed for multiple different objectives, and increasingly the emphasis in forestry policy is on sustainable management and multi-functional forestry. For example, the member countries of the Ministerial Conference on the Protection of Forests in Europe (MCPFE) agreed in 2011: "*To shape a future where all European forests are vital, productive and multifunctional. Where forests contribute effectively to sustainable development, through ensuring human well-being, a healthy environment and economic development in Europe and across the globe. Where the forests' unique potential to support a green economy, livelihoods, climate change mitigation, biodiversity conservation, enhancing water quality and combating desertification is realised to the benefit of society*".

A classification system which captures the different broad management objectives in forestry has been developed using a framework of 'forest management approaches' (FMA's) (Duncker et al. 2012, Table 4), and has been applied across Europe. The five FMA's represent the continuum of management practices from non-intervention to intensive production forest management, namely: (1) forest nature reserve, (2) close-to-nature forestry, (3) combined objective forestry, (4) intensive even-aged forestry, and (5) wood biomass production. This continuum also reflects (declining) degree of naturalness (Edwards et al. 2011).

After establishment, carbon sequestration by a woodland increases substantially as growth rates increase as the canopy develops, before slowing down as the trees reach maturity, but old growth stands, which have been wooded for >100 years, continue to show net carbon sequestration with a build up of litter, dead wood and soil carbon (Luyssaert et al. 2008). Old growth forests are FMA type 1; in Europe they are rare and tend to occur in isolated regions where access for harvesting is difficult. Close-to-nature forestry (FMA type 2) occurs throughout Europe, with substantial areas of high timber value native broadleaf forest in some locations, which are managed commercially with extraction on single-tree or small group selection systems the most commonly adopted approaches. The close-to-nature forestry

Table 4 Definition of forest management alternatives and management approach (after Duncker et al. 2012)

Management alternatives	Description and management objective	Tree species rules	Site management and cultivation rules	Harvest and stand management rules
[1] Forest nature reserve	No management, natural disturbances and succession is the driver of development	Natural	Not applicable	Not applicable
[2] Close to nature forestry	Only minor management. Could also be called "low intervention"	Natural or adapted	Mostly natural regeneration without soil tillage. Minimal chemical or physical site manipulations	Thinnings are extensive; final harvesting often according to target diameter
[3] Multifunctional forestry	Alternative characterised by inclusion of several considerations and goals, e.g. social, environmental and economic	Often natural or adapted	Cultivation might be artificial after site and soil preparations. Chemical manipulation like fertilisation and use of pesticides and other physical manipulation is little	Thinnings and stand regulation are often performed. Rotation length often lengthened because of environmental and social considerations
[4] Intensive even-aged forestry	Focus mostly on saw-logs. Other products or externalities are of second priority	Optimal according to production or purpose	No restrictions within legislation	No restrictions within legislation
[5] Wood-biomass production	Focus is only on production of lignin typically for energy or pulp. Often short rotation	No restrictions. Optimal according to purpose	No restrictions within legislation	No restrictions. Often short rotation and none or few thinnings. Roots and stumps may be harvested as well

Increasing degree of manipulation

management approach in the UK is exemplified by 'continuous cover forestry' (CCF) or 'low-impact silviculture' (LISS). The approach favours natural regeneration and aims to promote age and species diversity and is growing in popularity, partly driven by promotion of its benefits for climate change adaptation and improving the resilience of woodland ecosystems. Both FMA2 and FMA3 promote a diversity of species, though not necessarily in intimate mixtures, and reflect moves in forest science, management and policy towards improving climate resilience, adaptive capacity and enhanced structural (through mixed age distributions) and biological diversity (Linder et al. 2010), as well as addressing issues of aesthetics and land-scape value. To provide GHG emissions mitigation benefits over short (i.e. decadal) timescales a recent silvicultural strategy, commonly termed 'short rotation forestry' (SRF), has been proposed where a shortened rotation (15–25 years) focused on woody biomass provision for woodfuel is the primary management goal. Short rotation forestry is a FMA type 5 system.

5.2 Implications of Forest Management Types for GHG Balances

Increasing interest in the potential for net GHG emissions mitigation through forestry has led to assessments of FMA's for forest carbon storage, carbon seques-tration and fossil fuel substitution benefits. These assessments have included more complete life cycle analysis (LCA) of forestry and forest products to quantify fossil fuel use, and consequent emissions. In general, forestry operations from tree nursery stages, through establishment to harvest, and the infrastructure of forest roads, only result in small emissions from direct fossil fuel use, compared to the GHG benefits of forest products, and timber transport is usually the largest component, although still small (for an analysis in the UK, see Morison et al. 2012). While the carbon storage and sequestration benefits of forestry are evident, both 'in-forest' in biomass and soil organic carbon, and 'outside the forest' in product stocks in use or in land-fill, the substitution benefits are less clear. In principle, GHG emissions may be reduced by using wood as a fuel (direct substitution) or in place of other more fossil-fuel intensive materials (indirect substitution). However, these potential substitution benefits need to be examined closely. Woodfuel substitution will only produce a net reduction in GHG emissions if forests regrow and re-fix the CO_2 lost during wood-fuel production, combustion and energy generation, and the calculation of benefit needs to take into account the higher emissions per unit energy output of woodfuel than many fossil fuels. Material substitution will only produce a net reduction in GHG emissions if, for example, material life times are taken into account in the assessment; material substitution by wood only takes place at the end of life; and if the fossil-fuel intensive material being substituted does not find an alternative usage (referred to as 'leakage').

In managed woodlands, the substitution benefit derived from timber utilisation typically increases as larger timber is produced, because this can be used for structural components with longer residency times. The soil processes which determine organic carbon turnover slow down after initial establishment and forests continue to sequester carbon for long periods. Therefore, the choice of management options and species has a significant impact on the potential of woodland to store carbon or provide other mitigation benefits (cf. Broadmeadow and Matthews 2003). If focussing on carbon sequestration for climate mitigation, minimal intervention (FMA1 & 2) favours the accrual of forest carbon stocks and there are fewer emissions from forestry operations. FMA1 evokes management of a woodland as a 'carbon reserve', whereas FMA2 typifies 'selective intervention' (Broadmeadow and Matthews 2003). However, if the objective is to rapidly sequester carbon, then choosing highly productive, fast growing species on fertile land is likely to be the best option. When the fate of harvested wood products from the forest is accounted for, productive forestry (FMA4) and woody biomass production (FMA5) provide both considerable sequestration and substitution benefits, although the in-forest C stock of FMA5 is usually lower than FMA4.

A minimum intervention approach is more likely to be chosen for slow growing stands, heritage forest areas and remote locations where there is little demand for wood products or where conservation objectives take precedence. Commercial woodland management can result in GHG emissions through fuel use, woody waste decomposition and soil disturbance, but delivers mitigation benefits through use of the wood products. In this case, the combined wood product provision and sequestration benefits will vary according to species and site factors. In the UK around 69 Mt CO_2eq are found outside forests in wood products where the substitution of more fossil fuel intensive products by wood-based products in effect means very long term storage, with replacement by sustainably-produced wood saving 7 Mt CO_2 y^{-1} (Read et al. 2009).

Current policy in the UK and EU is likely to promote diversification away from FMA4 (intensive even-aged forestry) with increasing emphasis on FMAs 2–3 in areas which were, in previous rotations, productive single species conifer plantations. However, with productive species well matched to site conditions CCF has the capacity to provide similar or slightly improved mitigation benefits than conventional forestry, according to recent model based investigations of the two silvicultural systems (Seidl et al. 2008).

5.3   Other Management Drivers that Affect GHG Balances

The major biophysical constraint to optimal management of forests in the UK, and large areas of central to northern Europe is wind damage, and considerable effort to understand and manage risk of wind damage has been made often using models and software tools (e.g. ForestGales) across Europe. There has been increasing interest in recent years due to some notable storms with substantial impacts on forest productivity. For example, one modelled estimate for forest losses in the 1999 storm 'Lothar' in Central Europe was 3 million tones carbon (11 million tCO_2) equivalent to 30 % of net biome production for Europe (Lindroth et al. 2009). The primary influences of forest stand age and site factors (soil type, rooting depth etc.,) in wind risk management for commercial plantations are well understood, including the requirement for early thinning to promote stand stability. Management must be cognizant of wind risk where moves to more irregular stand structures, for other benefits, are envisaged. Similarly, extending forestry rotation lengths is recognised to increase carbon stocks sequestered in the biomass (Nabuurs et al. 2008), and is also likely to increase the quality and longevity of timber products obtained from the forest, although it has implications for windthrow risk. Conversely, regular thinning interventions to improve stand stability and extend rotation length will result in a reduction in the sequestration potential of the stand for a period, until the leaf area is regained. In a productive conifer stand (age ~20 years) the impact of a commercial thinning was to reduce forest sequestration rate by 20 % for 12 months (Clement et al. 2003). These intermittent losses will be more than compensated by extending the rotation and improved timber utilization.

Although a move away from FMA4 type management is envisaged due, to structural and landscape diversification and biodiversity drivers, productive conifers are still likely to provide significant benefits due to good growth, timber quality and utilisation (Mason and Perks 2011). In the last decade there has been an expansion in engineered wood products, which use lamination, impregnation etc., to provide a product that can replace steel or concrete, particularly in building construction. Furthermore, there is also potential benefit from targeted tree breeding approaches which aim to improve the quality of timber, which will further increase the quality and volume of structural timber attained from extension of the rotation length.

5.4 Forest Residues and GHG Balances

In many managed forests, residue harvesting is either practiced routinely or under active consideration. In principle, forest residues (e.g. 'lop and top', brash, stumps) resulting from thinning or harvesting could be used as either fuel or materials, and thus could contribute to emissions mitigation. However, there are several environmental issues that need to be considered, particularly around nutrient losses, soil disturbance and subsequent forest productivity. These concerns have led in some countries to guidance and controls to promote 'best practice' for sustainable management. For example, as most nutrients are in leaves and needles, not in woody tissues, retaining those on site can prevent damaging nutrient loss. While it might seem attractive to derive energy from woody material that would otherwise decompose, the net effects on medium and long-term forest sustainability, C stocks and GHG balance needs to be considered carefully and for each particular situation (Morison et al. 2012).

The growing interest in forest residue utilisation over recent years has seen considerable research into the potential benefits of 'biochar'. Biochar is made by heating organic material, including wood, under controlled oxygen conditions producing gases ('syngas') and liquids ('bio-oil'), and yields a solid product, which, if intended for use as a soil amendment, is named 'biochar' (Shackley and Sohi 2010). The pyrolysis production systems can therefore derive some of the energy requirement from syngas, bio-oil can be combusted or used as a bio-chemical feedstock and the solid soil amendment can improve soil organic carbon stabilisation, though the net benefit is currently uncertain (Hammond et al. 2011). In addition biochar can provide additional soil benefits through improved water retention and nutritional inputs to aid growth. Although the development of biochar-supported systems is in its infancy it is of specific interest to forestry. Potential utilisation is likely to focus particularly upon soil remediation for woodland creation on brownfield sites and the production of seedling trees with biochar used as a growing media amendment. Furthermore, using wood waste and forest wood residue feedstocks for biochar production provide the highest GHG emissions abatement (Shackley and Sohi 2010).

5.5   Afforestation and Woodland Creation Benefits

As discussed above, afforestation and reforestation can provide substantial climate change mitigation benefits. At a local level, plantations of exotic species can achieve very high rates of carbon sequestration (Clement et al. 2012). At a global level, the potential of the forestry sector to reduced net GHG emissions has been evaluated by the IPCC as 6.7 GtCO$_2$ y^{-1} in 2030 (Nabuurs et al. 2007). Although the carbon sequestration and eventual substitution benefits have to be carefully assessed against the 'biophysical' effects (Sect. 3.3), attempts to expand forest cover, particularly in the temperate zone, are central to much forestry policy, in response to climate change and biodiversity losses. Several studies have concluded that afforestation measures are one of the more cost effective emissions reduction mechanisms (e.g. Read et al. 2009).

However, there are several important points that need to be noted and several uncertainties. Firstly, afforestation only provides a 'one-time opportunity' to increase C stock on that land area (Lippke et al. 2011), because once mature forest is established, rates of sequestration are substantially reduced. However, sustained management of new forests can subsequently continue to provide carbon mitigation benefits. Secondly, there remains considerable uncertainty over the soil carbon impacts of afforestation. Although most forest soils contain substantially more organic carbon than agricultural soils, and deforestation usually results in substantial soil C loss, it is surprisingly difficult to establish unequivocally what the rates of soil C accumulation are likely to be after afforestation. Several recent meta-analyses (e.g. Laganière et al. 2010; Li et al. 2012) have shown that key determinants of soil C accumulation are (1) the previous land use, with larger accumulation after afforestation of croplands rather than natural grasslands or pastures (2) soil type, with clay-rich soils accumulating more, (3) soil disturbance during planting, with the least disturbance resulting in the most accumulation, (4) tree species (or broad type e.g. conifers vs broadleaves, evergreens v. deciduous) and (5) climate. A key control on soil C accumulation is the soil nitrogen balance, as C-N interactions are major determinants of both vegetation growth and decomposition rates. Furthermore, soil C accumulation is not linear over time, and, particularly in higher organic content soils may show a decline over the first decade or so. Even in mineral soils, organic C components that accumulated during agricultural cropping may decline after tree planting, before a net accumulation is subsequently evident (Li et al. 2012).

Land Availability and Land Use Competition

It is evident that for afforestation to make a significant contribution to net GHG emissions reduction, it requires a land use change, usually from land already in some form of agricultural management, with the exception of urban greenspace. While there are many additional benefits of urban woodland creation, net areas are small and unlikely to have a major effect on GHG balances.

Clearly, there is competition with both arable and pastoral agriculture, given continuing demand for crops and improved meat and dairy production with a rapidly increasing global population, and thus increasing commodity prices. Land that is agriculturally most productive is unlikely to be available for forestry such that assessing the potential land resource available for woodland requires 'constraint mapping' to identify suitable areas. Such an assessment might proceed by identifying three broad categories of land: land that is predominantly not available for woodland expansion due to physical and biological factors; land that is affected by national designations and policies which impose varying degrees of constraint on woodland expansion; and the remainder of the land which has potential for woodland expansion (e.g. WEAG 2012). Physical and biological suitability can be assessed using GIS. Further constraints include urban areas, industry and infrastructure, prime agricultural land and existing woodland. National and international conservation designations that might restrict woodland creation are national parks, nature reserves, and landscapes with particular aesthetic, cultural or heritage characteristics (e.g. World Heritage Sites). Finally, policy and socio-economic measures (e.g. the EU Common Agricultural Policy) are also critical in that, for example, government support mechanisms will certainly affect land use profitability and thus the economic availability of land for afforestation.

A primary constraint for woodland creation presently across much of the temperate zone is economic. Setting land aside for tree growth usually reduces income compared to agriculture, and the preparation, planting and establishment costs can usually only be supported if there is financial assistance through grants. The economics could, however, be changed dramatically by the establishment of a market for the carbon sequestered, or by other income, perhaps through 'payments for ecosystem services' that could follow from the widespread adoption by government and agencies of this framework for valuing nature. Such a development would emphasise the co-benefits of afforestation and woodland creation, for example for biodiversity, nutrient cycling, soil restoration, in addition to the socio-economic benefits of air and noise pollution control, recreation opportunities and health benefits. It could also help reconcile potentially conflicting objectives – for example, conservation of particular species requiring old-growth forest, versus higher CO_2 sequestration rates from younger forests.

To preserve tropical forests, woodlands and mangroves for the protection of the climate system is a gargantuan task. In 2005 the Coalition of Rainfall Nations introduced the concept of REDD (Reduced Emissions from Deforestation and Degradation). The idea is that the carbon storage services provided by these high carbon ecosystems are of benefit to the world as a whole, and therefore those who have damaged the climate system by emitting greenhouse gases should pay for the task of protection. This might be done by subscriptions made from rich countries to a common fund administered by the World Bank, from which payments to the governments of tropical countries would be made. However, it has proved difficult to agree on details; moreover, there is a general concern expressed by some environmental groups that such an arrangement effectively passes on the responsibility of absorbing pollution to the poor countries, which are not large emitters, would be deeply unethical and impinge on national sovereignty.

5.6 A Multipronged- Approach

It is clear that forests can be used in many ways to mitigate climate change, and it is possible to indentify a range of actions which are now required as part of a strategy to reduce the rate at which greenhouse gases accumulate in the atmosphere (Fig. 4). But the lesson to be learned from a decade of REDD negotiations, and two decades of the Kyoto Protocol, is that international agreements are not easy to achieve despite the demonstration that forestry is one of the cheapest and easiest technologies to reduce GHG's (Read et al. 2009).

6 Conclusions

(i) The changing climate (warming, decreases and increases in rainfall, increased overall variability) is likely to impact upon forests and woodlands in different ways in various regions of the world. The effects are not simple, as they involve many processes acting upon the physiology of the plants themselves, their pests and diseases, and the whole ecosystem. Moreover, increasing CO_2 and enhanced deposition of nitrogen are likely to affect the responses. In some cases, species migration to new latitudes and higher altitudes are likely.

(ii) Known biophysical and biogeochemical feedbacks through exchanges of energy, water and CO_2, mean that land use can affect the climate at large scales. Model analysis suggests that this effect also varies between geographical regions. For example the conversion of croplands to forests, or the afforestation of degraded lands has a larger cooling effect in the tropics than in northern regions.

(iii) Forests are large stores of carbon, and when undisturbed they are also effective sinks for carbon. Most forests are managed for timber production, and timber and woodfuel production have GHG emission mitigation benefits, so it is

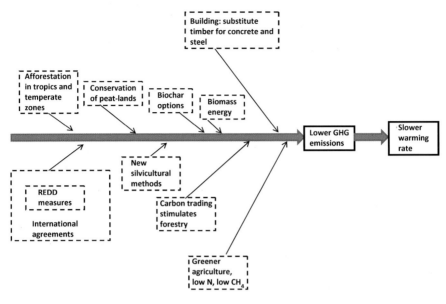

Fig. 4 Mechanisms by which forestry can help reduce the rate of climate change

essential to chose the best options for management of forests and their products. We also need to develop silvicultural methods involving less disturbance, and better use of harvested wood products to maximise forest mitigation of climate change. Several studies have showed that forestry is one of the most cost effective methods to capture carbon from the atmosphere.

References

Anderson RG, Canadell JG, Randerson JT, Jackson RB, Hungate BA, Baldocchi DD, Ban-Weiss GA, Bonan GB, Caldeira K, Cao L, Diffenbaugh NS, Gurney KR, Kueppers LM, Law BE, Luyssaert S, O'Halloran TL (2011) Biophysical considerations in forestry for climate protection. Front Ecol Environ 9:174–182, http://dx.doi.org/10.1890/090179

Arora VK, Montenegro A (2011) Small temperature benefits provided by realistic afforestation efforts. Nat Geosci 4:514–518

Avissar R, Werth D (2005) Global hydroclimatological teleconnections resulting from tropical deforestation. J Hydrometeorol 6:134–145

Bentz BJ, Jacques R, Fettig CJ, Hansen EM, Hayes JL, Hicke JA, Kelsey RG, Negrón JF, Seybold SJ (2010) Climate change and bark beetles of the Western United States and Canada: direct and indirect effects. Bioscience 60(8):602–613

Betts RA (2000) Offset of the potential carbon sink from boreal forestation by decreases in surface albedo. Nature 408:187–190

Bonan GB (2008) Forests and climate change: forcings, feedbacks, and the climate benefits of forests. Science 320:1444–1449

Broadmeadow M, Matthews R (2003) Forests, carbon and climate change: the UK contribution. Forestry Commission, Edinburgh, p 12, FCIN48

Cannell MGR (2003) Carbon sequestration and biomass energy offset: theoretical, potential and achievable capacities globally, in Europe and the UK. Biomass Bioenergy 24:97–116

Ciais P, Reichstein M, Viovy N, Granier A, Ogee J, Allard V, Aubinet M, Buchmann N, Bernhofer C, Carrara A, Chevallier F, De Noblet N, Friend AD, Friedlingstein P, Grunwald T, Heinesch B, Keronen P, Knohl A, Krinner G, Loustau D, Manca G, Matteucci G, Miglietta F, Ourcival JM, Papale D, Pilegaard K, Rambal S, Seufert G, Soussana JF, Sanz MJ, Schulze ED, Vesala T, Valentini R (2005) Europe-wide reduction in primary productivity caused by the heat and drought in 2003. Nature 437(7058):529–533

Clement R, Moncrieff JB, Jarvis PG (2003) Net carbon productivity of Sitka Spruce forest in Scotland. Scott For 57:5–10

Clement RJ, Jarvis PG, Moncrieff JM (2012) Carbon dioxide exchange of a Sitka spruce plantation in Scotland over five years. Agric For Meteorol 153:106–123

Committee on Climate Change (2011) Bioenergy review, London, p 90

Costa MH, Foley JA (2000) J Clim 13:18–34

Coumou D, Rahmstorf S (2012) A decade of weather extremes. Nat Clim Chang 2:491–496

Crutzen PJ, Mosier AR, Smith KA, Winiwarter W (2008) N_2O release from agro-biofuel production negates global warming reduction by replacing fossil fuels. Atmos Chem Phys 8:389–395

Duncker PH, Barreiro S, Hengeveld G, Lind T, Mason W, Ambrozy S, Spiecker H (2012) Classification of Forest Management Approaches: a new methodological framework and its applicability to European forestry. Ecol Soc 17(4):51 http://dx.doi.org/10.5751/ES-05262-170451

Edwards D, Jensen F, Marzano M, Mason WL, Pizzirani S, Schelhaas M-J (2011) A theoretical framework to assess the impacts of forest management on the recreational value of European forests. Ecol Indic 11:81–89

Geiger R (1966) Climate near the ground. Harvard University Press. (translation of the 4th edition of Das Klima der bodennahen Luftschicht published 1961)

Hammond J, Shackley S, Sohi SP, Brownsort PA (2011) Prospective life cycle carbon abatement for pyrolysis biochar systems in the UK. Energy Policy 39:2646–2655

IPCC (2007) Contribution of working group III to the fourth assessment report of the Intergovernmental Panel on climate change. Cambridge University Press, Cambridge

IPCC (2012) Special report: managing the risks of extreme events and disasters to advance climate change adaptation. Cambridge University Press, Cambridge

Kirschbaum MUF, Whitehead D, Dean SM, Beets PN, Shepherd JD, Ausseil A-GE (2011) Implications of albedo changes following afforestation on the benefits of forests as carbon sinks. Biogeosciences 8:3687–3696

Kurz WA, Dymond CC, Stinson G, Rampley GJ, Neilson ET, Carroll AL, Ebata T, Safranyik L (2008) Mountain pine beetle and forest carbon: feedback to climate change. Nature 454:987–990

Laganière J, Angers DA, Paré D (2010) Carbon accumulation in agricultural soils after afforestation: a meta-analysis. Glob Chang Biol 16:439–453

Lean J, Warringlow DA (1989) Nature 342:311–413

Lewis SL, Lopez-Gonzalez G, Sonké B, Affum-Baffo K, Baker TR, Ojo LO, Phillips OL, Reitsma J, White L, Comiskey JA, Ewango C, Feldpausch TR, Hamilton AC, Gloor E, Hart T, Hladik A, Kamdem M-ND, Lloyd J, Lovett JC, Makana J-R, Malhi Y, Mbago FM, Ndangalasi HJ, Peacock J, Peh KS-H, Sheil D, Sunderland T, Swaine MD, Taplin J, Taylor D, Thomas SC, Votere R, Wöll H (2009) Increasing carbon storage in intact African tropical forests. Nature 477:1003–1006

Li D, Niu S, Luo Y (2012) Global patterns of the dynamics of soil carbon adn nitrogen stocks following afforestation: a meta-analysis. New Phytol 195:172–181

Linder M, Maroschek M, Netherer S, Kremer A, Barbati A, Garcia-Gonzalo J, Seidl R, Delzon S, Corona P, Kolström M, Lexer MJ, Marchetti M (2010) Climate change impacts adaptive capacity, and vulnerability of European forest ecosystems. For Ecol Manag 259:698–709

Lindroth A, Lagergren F, Grelle A, Klemedtsson L, Langvall O, Weslien P, Tuulik J (2009) Storms can cause Europe-wide reduction in forest carbon sink. Glob Chang Biol 15:346–355

Lippke B, Oneil E, Harrison R, Skog K, Gustavsson L, Sathre R (2011) Life cycle impacts of forest management and wood utilization on carbon mitigation: knowns and unknowns. Carbon Manag 2(3):303–333

Luyssaert S, Inglima I, Jung M, Richardson AD, Reichstein M, Papale D, Piao SL, Schulze ED, Wingate L, Matteucci G, Aragao LEOC, Aubinet M, Beers C, Bernhofer C, Black GK, Bonal D, Bonnefond JM, Chambers J, Ciais P, Cook B, Davis KS, Dolman AJ, Gielen B, Goulden M, Grace J, Granier A, Grelle A, Griffis T, Grünwald T, Guidolotti G, Hanson PJ, Harding R, Hollinger DY, Hutyra LR, Kolari P, Kruijt B, Kutsch WL, Lagergren F, Laurila T, Law B, Le Maire G, Lindroth A, Loustau D, Malhi Y, Mateu J, Migliavacca M, Misson L, Montagnani L, Moncrieff J, Moors EJ, Munger JW, Nikinmaa E, Ollinger SV, Pita G, Rebmann C, Roupsard O (2007) CO_2 balance of boreal, temperate, and tropical forests derived from a global database. Glob Chang Biol 13:2509–2537

Luyssaert S, Schulze E-D, Börner A, Knohl A, Hessenmöller D, Law BE, Ciais P, Grace J (2008) Old-growth forests as global carbon sinks. Nature 455:213–215

Mason WL, Perks M (2011) Sitka spruce (Picea sitchensis) forests in Atlantic Europe: forest management approach and projected climate change. Scand J Forest Res 26:72–81

Medvigy D, Walko RL, Avissar R (2011) J Clim 24:2147–2163

Menzel A, Sparks TH, Estrella N, Koch E, Aasa A, Ahas R, Alm-Kübler K, Bissolli P, Braslavská O, Briede A, Chmielewski FM, Crepinsek Z, Curnel Y, Dahl Å, Defila C, Donnelly A, Filella Y, Jatczak K, Måge F, Mestre A, Nordli Ø, Peñuelas J, Pirinen P, Remišová V, Scheifinger H, Striz M, Susnik A, Van Vliet AJH, Wielgolaski F-E, Zach S, Zust A (2006) European phenological response to climate change matches the warming pattern. Glob Chang Biol 12:1969–1976. doi:10.1111/j.1365-2486.2006.01193.x

Morison J, Matthews R, Miller G, Perks M, Randle T, Vanguelova E, White M, Yamulki S (2012) Understanding the carbon and greenhouse gas balance of forests in Britain, Forestry Commission Research Report. Forestry Commission, Edinburgh, p 149

Nabuurs G-J, Masera O, Andrasko K, Benitez–Ponce P et al (2007) Forestry, in contribution of working group III to the fourth assessment report of the Intergovernmental Panel on climate change. Cambridge University Press, Cambridge

Nabuurs GJ, Thurig E, Heidema N, Armolaitis K, Biber P, Ciencala E, Kaufmann E, Makipaa R, Nilsen P, Petritsch R, Pristova T, Rock J, Schelhaas MJ, Sievanen R, Somogyi Z, Vallet P (2008) Hotspots of the European forests carbon cycle. For Ecol Manag 25:194–200

Phillips OL, Malhi Y, Higuchi N, Laurance WF, Nuñez VP, Vásquez MR, Laurance SG, Ferreira LV, Stern M, Brown S, Grace J (1998) Changes in the carbon balance of tropical forest: evidence from long-term plots. Science 282:439–442

Phillips OL, Aragao LEOC, Lewis SL, Fisher JB, Lloyd J, Lopez-Gonzalez G, Malhi Y, Monteagudo A, Peacock J, Quesada CA, van der Heijden G, Almeida S, Amaral I, Arroyo L, Aymard G, Baker TR, Banki O, Blanc L, Bonal D, Brando P, Chave J, de Oliveira ACA, Cardozo ND, Czimczik CI, Feldpausch TR, Freitas MA, Gloor E, Higuchi N, Jimenez E, Lloyd G, Meir P, Mendoza C, Morel A, Neill DA, Nepstad D, Patino S, Penuela MC, Prieto A, Ramirez F, Schwarz M, Silva J, Silveira M, Thomas AS, ter Steege H, Stropp J, Vasquez R, Zelazowski P, Davila EA, Andelman S, Andrade A, Chao K-J, Erwin T, DiFiore A, Honorio CE, Keeling H, Killeen TJ, Laurance WF, Cruz AP, Pitman NCA, Vargas PN, Ramirez-Angulo H, Rudas A, Salamao R, Silva N, Terborgh J, Torres-Lezama A (2009) Drought sensitivity of the Amazon rainforest. Science 323:1344–1347

Pongratz J, Caldeira K (2012) Attribution of atmospheric CO2 and temperature increases to regions: importance of preindustrial land use change. Environ Res Lett 7, 034001:8. doi:10.1088/1748-9326/7/3/034001

Read DJ, Freer-Smith PH, Morison JIL, West CC HN, Snowdon P (eds) (2009) Combating climate change- a role for UK forests. An assessment of the potential of the UK's trees and woodlands to mitigate and adapt to climate change. The synthesis report. The Stationery Office, Edinburgh

Schulze ED, Luyssaert S, Ciais P et al (2009) Importance of methane and nitrous oxide for Europe's terrestrial greenhouse-gas balance. Nat Geosci 2:842–850. doi:10.1038/ngeo686

Seidl R, Rammer W, Lasch P, Badeck FW, Lexer MJ (2008) Does conversion of even-aged, secondary coniferous forests affect carbon sequestration? A simulation study under changing environmental conditions. Silva Fennica 42:369–386

Shackley S, Sohi SP (2010) An assessment of the benefits and issues associated with the application of biochar to soil. Defra, London, p 132

Shukla J, Nobre CA, Sellers P (1990) Science 247:1322–1325

Voldoire A, Royer JF (2004) Clim Dyn 22:857–874

WEAG (2012) Report of the woodland expansion advisory group to Scottish Government, www.forestry.gov.uk/weag. p 92

Useful Internet Sources

http://www.decc.gov.uk/assets/decc/11/stats/climate-change/4817-2011-uk-greenhouse-gas-emissions-provisional-figur.pdf(UK greenhouse gas inventory)

http://www.globalcarbonproject.org/(annual report on the global carbon cycle)

http://www.esrl.noaa.gov/gmd/ccgg/carbontracker/(summary of carbon fluxes derived from atmospheric observations)

http://www.carbontracker.org/(aligning the capital markets with climate change objectives)

http://www.fao.org/forestry/sofo/en/(biennual review of the areas of forests by region, data on deforestation rates)

http://www.official-documents.gov.uk/document/other/9780108507632/9780108507632.pdf (The Eliasch Review: *Climate Change*: *Financing Global Forests*)

http://www.defra.gov.uk/forestrypanel/reports/(DEFRA's Independent Panel on Forestry, to inform and advise the UK government)

The Influence of Climate Change on Insect Invasions in Temperate Forest Ecosystems

Patrick C. Tobin, Dylan Parry, and Brian H. Aukema

Abstract Climate change could potentially become one of the most important influences on forest ecosystem function and diversity due to its profound effect on many biotic processes. Additionally, climate change could interact with other anthropogenically driven agents of forest alteration, such as non-native invasive species. Although their arrival is primarily facilitated by global trade and travel, climate and changes to climate have affected and will likely continue to affect rates of invasive species establishment, range expansion, and impact to native ecosystems. In this chapter, we attempt to synthesize broadly the interaction between climate change and non-native insect invasions in temperate forest ecosystems. We highlight four primary effects: changes in distributional ranges, outbreak frequency and intensity, seasonality and voltinism, and trophic interactions. A paucity of data for some processes necessitated the use of exemplar native species in native ranges, and their extrapolation to non-native species. Future studies should give greater attention to the complexity associated with these interacting forces of change in forest ecosystems.

P.C. Tobin (✉)
Forest Service, United States Department of Agriculture,
Northern Research Station, Morgantown, WV 26505, USA
e-mail: ptobin@fs.fed.us

D. Parry
College of Environmental Science and Forestry,
State University of New York, Syracuse, NY 13210, USA

B.H. Aukema
Department of Entomology, University of Minnesota,
St. Paul, MN 55108, USA

T. Fenning (ed.), *Challenges and Opportunities for the World's Forests in the 21st Century*, Forestry Sciences 81, DOI 10.1007/978-94-007-7076-8_12,
© US Government 2014

1   Introduction

Climate change and invasions by non-native species each constitute major threats to forest ecosystems worldwide (Dale et al. 2001), yet attempts to synthesize broadly their interacting impacts are limited (Engel et al. 2011; Hellmann et al. 2008; Walther et al. 2009). The inherent complexity of each and the potential confounding and compounding effects when considered jointly have hampered analysis of actual and potential effects on forest ecosystems. Another consideration is that the paramount stage of biological invasions, the arrival stage, is predominately a result not of climate change but of increasing global trade through which new species 'hitchhike' on products, nursery stock, packing materials, and in ship hulls and ballast water (Hulme et al. 2008; Lockwood et al. 2007). Consumer demand for many foreign manufactured goods is also not generally dependent upon or influenced by climate, at least not in the short-term. In fact, one could argue that due to changing climates, there could be a greater emphasis on reducing carbon footprints associated with foreign imports with a consequent increase in the desire to purchase locally available products that in turn could conceivably reduce the volume of imports and species that are introduced with these imports.

In the absence of an environmental revolution and major shifts in economic patterns and in governmental policy, however, the arrival of new species will likely continue to increase as long as global trade increases. Once a new species arrives to a novel habitat, the other stages of the invasion process, establishment, spread, and impact, could be directly and indirectly influenced positively or negatively by climate and changes in climate. In this chapter, we first briefly describe the causes and dynamics of biological invasions and climate change independently. We then focus on how these forces interact to affect forest ecosystems by mediating range shifts, altering population outbreak frequency and intensity, changing seasonality and voltinism, and decoupling interactions between and among trophic levels. In some cases, we rely on examples from native species in native ranges, and then extrapolate these observations to non-native species in non-native habitats. We lastly consider the economic ramifications of the interaction between biological invasions and climate change.

2   Dynamics of Biological Invasions

Biological invasions consist of four distinct stages: arrival, establishment, spread, and impacts (Lockwood et al. 2007). The arrival stage is defined as the movement of a species from an area in which it is established to a novel habitat. Although the interchange of biota among biogeographic regions is integral to the history of life (Crosby 1986; di Castri 1989), arrival rates of new species have increased dramatically in recent decades due to increases in global trade and travel (Aukema et al. 2010). Establishment after arrival is a critical transition, and it is believed that most arriving species fail to establish (Simberloff and Gibbons 2004). One of the most important factors influencing the establishment of newly arrived species is propagule pressure: the number of individuals in the arriving founder population, the number of independent introductions, or a combination of both (Lockwood et al. 2007). For example, low-density populations tend to be subject to environmental and demographic stochasticity, and Allee effects (positive density-dependence, Stephens et al. 1999). Allee effects can arise when individuals in sparse populations encounter difficulties finding suitable mates, satiating natural enemies, and overcoming host defense mechanisms, all of which can exacerbate the challenges that small founder populations already face during the establishment phase (Liebhold and Tobin 2008; Taylor and Hastings 2005; Tobin et al. 2011). Other factors influencing establishment success include the level of genetic diversity in founder populations, availability of resources, degree of habitat disturbance, and presence or absence of competitors, mutualists, and natural enemies (Lockwood et al. 2007). Another factor that is important to consider in the context of this chapter is climate suitability (Beaumont et al. 2009; Hayes and Barry 2008; Thuiller et al. 2005), and thus shifting climatic patterns could have critical consequences for the establishment success of invaders.

Following successful establishment, a species often begins to expand its geographic range, often through a process known as stratified dispersal in which local growth and diffusive spread is coupled with long-distance population 'jumps' (Hengeveld 1989). These population jumps are often facilitated through anthropogenic (e.g., Gilbert et al. 2004), hydrological (e.g., Davidson et al. 2005), and atmospheric transport mechanisms (e.g., Isard et al. 2005), and can greatly accelerate the rates of spread of invading species (Shigesada and Kawasaki 1997). Long-range dispersal can also serially initialize new invasions, which are effectively subject to the same constraints that affect the establishment success of any newly-arrived species in a non-native environment (Liebhold and Tobin 2008; Taylor et al. 2004).

The fourth stage of the invasion process is the ecological and economic impacts of non-native species (Lockwood et al. 2007). Impacts due to invasive species can vary dramatically among species, and within a species depending upon the region being invaded. A recent analysis on non-native insects in forests within the continental United States has suggested that only a minority of introduced and established insects (≈14 %) since 1860 have caused major damage (Aukema et al. 2010), yet this minority of species can still account for several billion USD in costs (Aukema et al. 2011; Holmes et al. 2009; Pimentel et al. 2005). The costs to the United States of the non-native emerald ash borer, *Agrilus planipennis* Fairmaire, which are primarily associated with the treatment, removal, and replacement of ash, *Fraxinus* spp., is predicted to be 10.7 billion USD alone (Kovacs et al. 2010). It is also likely that the impacts and consequent costs due to non-native species will be affected by climate and changes in those regimes.

3   Climate Change

Recent increases in the concentrations of atmospheric greenhouse gases, most notably carbon dioxide, methane, and nitrous oxide, have led to a corresponding change in local and global climates. Global mean surface temperatures have increased by

0.3-0.6 °C over the last century (Mann et al. 1998), and global temperatures are projected to continue to increase by the end of the next century (Intergovernmental Panel on Climate Change 2007). The projected increase in global surface temperatures is thought to range from 1 °C under a low (B1) greenhouse gas emission scenario, which assumes substantial mitigation and reductions in greenhouse gas emissions, to 6 °C under a high (A1fi) greenhouse gas emission scenario, which assumes temperatures under the current conditions and without any mitigating strategies (Hayhoe et al. 2007; Kunkel et al. 2008).

The increase in global surface temperatures can also result in a number of cascading effects. For example, as temperatures warm, there could be changes in hydrological cycles leading to increases or decreases in precipitation patterns, such that some places could be more prone to drought while others more prone to flooding. Changes in climatic variability, including increases in storm event frequency and intensity, could also be the result of recent warming trends (Rosenzweig et al. 2001). More subtle changes, such as the diminishment of winter snow pack, could also have important ecological ramifications for both insects and trees. Consequently, climate change has the potential to destabilize many ecosystem functions, and cause major changes to the dynamics of individual species and to those communities in which they reside.

Insect species are particularly sensitive to climate change because many of their physiological processes are temperature-dependent. Insects could respond to changing climates by going extinct, moving into areas with a more tolerable weather regime, or adapting in situ. Climate change can affect insect species differentially depending upon the latitude at which they live; for example, temperate insects that evolved under strong seasonality could inherently have greater phenotypic plasticity for withstanding the pressures of climate change relative to tropical species already near their thermal tolerances, and thus could be more capable of adapting to warming trends (Deutsch et al. 2008). Such changes could include accelerated developmental rates with a consequent change in the timing of year-to-year phenological events, and/or in the number of generations per year. Species abundance can be directly affected by temperature, resulting in population density extremes, from extinction to outbreaks. The distributional ranges of insects can also be affected by the removal of biogeographic boundaries formed by climatic factors.

4 Ecological Interactions Between Forest Insect Invasions and Climate Change

We now turn our attention to the effects, both documented and those projected to occur, of climate change on insect invasions in forest ecosystems. Although non-native species across a diversity of taxa have been introduced to novel areas (Pimentel 2002; Simberloff and Rejmánek 2011), forest insects, such as xylophagous species, represent a particularly important group of invaders because of their propensity for importation in solid wood packaging material, wood dunnage, and wood pallets (Bartell and Nair 2003; Brockerhoff et al. 2006a; Liebhold et al. 1995; Wingfield et al. 2010), all of which are important components in global trade. Forest insects can also be introduced through imports of infested timber and forest products, as well as on infested live plants (Brasier 2008; Liebhold et al. 2012; Reichard and White 2001). In recent decades, a number of economically damaging, non-native forest insect species have been or were likely introduced through global trade routes, such as *A. planipennis* (Poland and McCullough 2006) and *Anoplophora glabripennis* (Motschulsky) (Smith et al. 2009) in North America, *Dendroctonus valens* LeConte in China (Yan et al. 2005), and *Sirex noctilio* F. in Australasia, South Africa, South America and most recently, North America (Slippers et al. 2012a).

4.1 Changes in Geographic Ranges

There exists a rich literature around bioclimatic envelope modeling documenting how the distribution and abundance of forest insects are restricted to (or released from) suitable habitat(s) by various climatic factors in time and space (e.g., Hlasny and Turcani 2009; Jonsson et al. 2011; Logan et al. 2003; Robinet et al. 2007; Williams and Liebhold 1995a). Changes in climate compared to historic norms have relaxed these boundaries, allowing for rapid range expansion in some species, and range retraction in others. In this section, we will discuss three types of changes in the geographic ranges of forest insects: expansions in latitudinal range, expansions in geographic range, and range retractions. We restrict our treatment to natural dispersal, even though human-assisted transport may be a confounding factor in many range expansions (Brockerhoff et al. 2006b) that could be detected post hoc within the genetic structure of subpopulations (Kerdelhué et al. 2009).

In montane ecosystems, range expansions frequently become first apparent in altitudinal changes rather than latitudinal changes. Simulation studies have shown that the mountain pine beetle, *Dendroctonus ponderosae* Hopkins, which is native to North America, responds to increases in temperature first by increasing its altitudinal range before becoming more apparent at more northern locations in western Canada (Sambaraju et al. 2012). In the Greater Yellowstone Ecosystem of the western United States, *D. ponderosae* has had a long association with lodgepole pine forests with infrequent outbreaks at climatically-inhospitable high elevations. Recently, however, it has exhibited behavior outside of its observed range of natural variability, and threatens to decimate high-elevation five needle pines (Logan et al. 2010; Logan and Powell 2001), a host species that does not appear to have a co-evolved relationship with this insect. These pronounced shifts in elevation are not restricted to bark beetles. In recent years, for example, the winter moth, *Operophtera brumata* (L.), has defoliated stands of mountain birch at elevations up to the tree line in coastal areas of northern Norway (Hagen et al. 2007). Historically, *O. brumata* outbreaks have occurred at lower elevations due to bioclimatic and/or competitive effects with a sympatric species, the autumn moth *Epirrita autumnata*

(Borkhausen) (Ammunet et al. 2010). The gypsy moth, *Lymantria dispar* (L.), also a defoliator, is expected to continue a progression into regions of higher elevations in Europe as it doubles its range compared to 50 years ago (Hlasny and Turcani 2009).

Latitudinal range shifts are perhaps more dramatic than altitudinal shifts, as humans may notice defoliated or killed trees more quickly in regions where such activity was historically unapparent. There exist a number of examples of lepidopteran defoliators in Europe where this phenomenon has occurred over the past decade. While *O. brumata* has expanded into higher elevations and moved northeast across Norway, the more cold-hardy *E. autumnata* has exhibited a concomitant increase into colder, continental areas (Jepsen et al. 2008). Recent phenological changes in budburst have facilitated a rapid northward range expansion of another geometrid, the scarce umber moth, *Agriopis aurantiaria* Hübner, in the same region (Jepsen et al. 2011). In France and Portugal, higher elevation and northward latitudinal incursions have been documented for the pine processionary moth, *Thaumetopoea pityocampa* (Denis and Schiffermüller), a frequent defoliator of pine and cedar (Arnaldo et al. 2011; Battisti et al. 2005). In North America, a warming climate could facilitate the expansion of *L. dispar* into areas where current overwintering temperatures are too cold to permit survival (Régnière et al. 2009). However, a constraint in the southern geographic range of *L. dispar* is the lack of sufficient cooling periods to terminate diapause (Gray 2004); thus, although warming temperatures could allow for northern expansion of *L. dispar*, there could a concomitant restriction in its potential and realized southern range.

Latitudinal range expansion has not been limited to Coleoptera and Lepidoptera. For example, damage from the poplar woolly adelgid, *Phloemyzus passerinii* (Signoret), was recently recorded for the first time in northern France (Rouault et al. 2006). The European pine sawfly, *Neodiprion sertifer* (Geoffroy), is currently limited in its range by the degree of freeze tolerance of its eggs, is also expected to move northwards over the next decades (Veteli et al. 2005). Lethal mortality to overwintering life stages is frequently cited as a key delimiter for range margins (Bale and Hayward 2010), which, once ameliorated, could permit rapid expansion in a number of species.

A notable example of climatic release permitting range expansion has been the recent outbreak of *D. ponderosae* in western Canada (Aukema et al. 2006). Mountain pine beetle has been historically limited in range by a −40 °C thermal isocline, restricting its Canadian range to a line primarily west of the Rocky Mountains (Safranyik et al. 1975). Recently, this insect breached the historic geoclimatic divide and is now reproducing in lodgepole pine forests of northwestern Alberta (de la Giroday et al. 2011, 2012; Robertson et al. 2009). This is of grave concern because range expansions of this magnitude could permit access to new hosts and new habitat corridors. For example, *D. ponderosae* is now reproducing in an area where lodgepole pine hybridizes with jack pine (Cerezke 1995; Cullingham et al. 2011), which could permit further range expansion eastward through the boreal forest of North America. This insect seems well adapted to jack pine, and exhibits elevated reproductive rates in evolutionarily naïve hosts (Cudmore et al. 2010).

Of course, climatic suitability across the extent of a new range is never assured, and different models could have contradicting outcomes. Projections of habitat suitability for *D. ponderosae* across the boreal forest of Canada through global simulation models show a wide range of projections, from high to low climatic suitability, depending upon the scenario applied (Safranyik et al. 2010). Projections of climatic suitability for the same insect in the Rocky Mountain region of the western United States, however, are more uniform, predicting decreases in suitable habitat by up to 50 % by the year 2050. In contrast, projections for the same regions for a different insect, the western pine beetle, *Dendroctonus brevicomis* LeConte, indicate increases or decreases in suitable habitat, depending on the scenario (Evangelista et al. 2011). Potential range retractions are not restricted to bark beetles. *Lymantria dispar* and the nun moth, *Lymantria monacha* (L.), are projected to lose up to 900 km from their southern European ranges as temperatures warm (Vanhanen et al. 2007). Many projections of climate warming demonstrate either rapidly altered ranges of the host trees (Refehldt et al. 2006) or maladaptive seasonal phenology that disrupts insect development and host procurement (Williams and Liebhold 1995a).

Climatically-mediated range shifts could also lead to changes in physiological and morphological traits of insects. Indeed, latitudinal clines in physiological, reproductive, and morphological traits of insects are well known (Blanckenhorn and Demont 2004). For example, wing size tends to be larger at higher latitudes in *Drosophila subobscura* Collin (Gilchrist et al. 2001) and genetic change in this species appears to be tracking climate change (Balanya et al. 2006). Furthermore, invasive populations of this species formed a latitudinal gradient in wing size very similar to that in native populations in only 20 years, which suggests that some invasive species could evolve rapidly in response to elevated temperatures. A number of widespread native forest Lepidoptera exhibit similar adult size across their latitudinal range but express reductions in fecundity and (or) larger offspring with increasing latitude (Ayres and Scriber 1994; Harvey 1983; Parry et al. 2001). A latitudinal shift in a number of fitness parameters is evident in introduced populations of the fall webworm, *Hyphantrea cunea* (Drury), in Japan (Gomi 2007; see Sect. 4.3). Because such latitudinal clines appear to be common in many species, selection could be expected to form clines in invasive species as their distributional range expands.

4.2 Changes in Outbreak Frequency and Intensity

The study of forest insect outbreaks has a long and storied history, and many of the species that have been the focal point of these studies have broadly served as model systems for studying the conceptual and mechanistic processes of insect population dynamics (e.g., Barbosa and Schultz 1987). Several forest insect species undergo regular cycles in population density, from innocuous to outbreak and back. Apart from being inherently fascinating, forest insect outbreaks often cause considerable economic and ecological damage by affecting nutrient cycles, animal and plant populations (Frost and Hunter 2004; Payette et al. 2000; Work and McCullough 2000), and human uses of forests for timber and recreation (Coyle et al. 2005). Another aspect of many forest insect outbreaks is that they can often be spatially synchronized (Haynes et al. 2009; Johnson et al. 2005; Peltonen et al. 2002), in which the congruence in the temporal variation of abundance among geographically distinct populations results in outbreaks over a large spatial scale (Bjørnstad et al. 1999). Geographically widespread outbreaks can be particularly important for a number of reasons. First, spatially synchronous outbreaks could dilute regulating effects of natural enemies that could otherwise provide local control (Royama 1984). Second, they can reduce the ecological landscape's ability for buffering because most areas within an ecosystem experience simultaneous disturbance (Lovett et al. 2002). Lastly, they can overwhelm the budgetary and logistical efforts available for protection of economic assets or ecological functions through suppression programs (Tobin et al. 2012).

Many non-native species are innocuous in their native ranges where they evolved in concert with host tree defensives and the regulatory effects of natural enemies, but can be problematic when introduced into naïve environment and/or to naïve host species (Liebhold et al. 1995; Hu et al. 2009). Such a contrast is apparent when comparing the insect borers *A. planipennis* and *Agrilus anxius* Gory. The latter species is native to North America where it is relatively innocuous to North American birch species unless coupled with severe stress such as drought. In contrast, *A. anxius* causes significant mortality in non-native birch species planted in North America, whether trees were stressed or not (Nielsen 1989). Similarly, *A. planipennis* rarely causes mortality in ash in its native range in Asia, while all North American ash species appear to be extremely susceptible (Poland and McCullough 2006). Non-native species often lack natural enemies in a new area, and consequently, the 'enemy release' hypothesis has been suggested to play a role in the ability of species to invade novel habitats and reach outbreaking densities (Keane and Crawley 2002; Torchin et al. 2003).

The effect of climate change could result in changes to both the outbreak intensity and the periodicity of forest insect outbreaks (Logan et al. 2003; Volney and Fleming 2000). Trends in climate warming are thought to have had direct effect on the development, intensity, and geographic extent of outbreaks of *D. ponderosae* Hopkins, in North America (Kurz et al. 2008). Although this species is native to western North America, it is now achieving outbreak densities in northern British Columbian forests where it had never previously been found, at least not over the last few centuries of direct observation (see Sect. 4.1). Similarly, outbreaks of other eruptive bark beetle species such as the southern pine beetle, *Dendroctonus frontalis* Zimmermann, the Mexican pine beetle, *Dendroctonus mexicanus* (Hopkins), and the European spruce bark beetle, *Ips typographus* (L.), are expected to exhibit outbreaks of increased magnitude as temperature and precipitation regimes change (Kausrud et al. 2012; Waring et al. 2009).

In Fennoscandia, climate warming is thought to have shifted the geographic distribution of outbreaks of *O. brumata* (Hagen et al. 2007). Although this species is native to Fennoscandia, it, like *D. ponderosae* in North America, is believed to have crossed altitudinal barriers that were previously impassable due to climate that was historically unfavorable to its winter survival. In addition to the expanse of areas experiencing outbreaks, the duration of outbreaks is thought to have increased due to climate warming (Jepsen et al. 2008). Examples of the interplay between climate change and outbreaks by non-native species are less documented, likely due to the fact that invasions by non-native forest insects are a more recent phenomenon (Aukema et al. 2010), relative to the time scale that native species have existed in their native ranges. Indeed, many important invasive species are still in the active range expansion phase of their colonization, making it impossible to partition climatic influences separately from other processes driving spread. However, the change in outbreak dynamics in native species is likely not unique and will likely result in related changes in the outbreak dynamics of non-native species.

Not all forest insects necessarily benefit from recent trends in climate change, adding complexity to our efforts to understand expected patterns. For example, regular outbreaks have been recorded for the larch budmoth, *Zeiraphera diniana* Guénée, in the European Alps (Bjørnstad et al. 2002). A recent dendrological reconstruction of these outbreaks has suggested that this cyclical behavior had been occurring for at least 1,173 years and during previous climatic events, such as periods of warming during the Middle Ages and cooling during the Little Ice Age (Esper et al. 2007). However, since 1981, *Z. diniana* outbreaks have been conspicuously absent with the supposition that recent trends in climatic warming have upset the balance of a system that previously had exhibited remarkable stability (Esper et al. 2007). However, because the absence of *Z. diniana* outbreaks is thought to be due to climate-mediated disruption of the stability of this system, non-native species, which are agents of disturbance in themselves, may or may not be less prone to collapse in forest ecosystems that are also experiencing disturbance due to changing climates.

4.3   Changes in Seasonality and Voltinism

Increasing temperatures will have direct consequences for insects including alterations to life cycle duration (developmental rate) and changes in voltinism (the number of generations per year). While most insects are capable of increased growth rates at elevated temperatures, a key factor is during which part of a particular insect's life history that the temperature change occurs. Thus, generalizations concerning the response of insect growth rate and development to global climate change must be tempered with knowledge that a species may behave idiosyncratically with respect to temperature.

At the level of individual insect species, a major determinant of the response to climatic shifts is the type of life-cycle and the developmental strategy employed. Danks (2006) suggested that insect development could be viewed as either an active default, where it proceeds until some reliable environmental cue signals it to stop, or a passive default, where development stops at a preset point irrespective of current environmental conditions and does not resume again until an appropriate cue is received. A good example of this dichotomy would be a multivoltine species that produces additional generations as long as diapause-inducing cues are absent, while an univoltine species would develop faster in its single generation, but would still be constrained by an obligate diapause to one generation annually. Insects using an active developmental default will likely receive greater benefits from warming temperatures than those with passive default development systems. Because many multivoltine insect species use photoperiodic cues to initiate diapause, which do not change in response to changing climates (Tauber and Tauber 1976), sufficient increases in temperature prior to the onset of diapause-inducing photoperiods could be a key determinant to the number of generations possible per year under future climate scenarios (Chen et al. 2011; Tobin et al. 2008).

Many geographically widespread insects exhibit latitudinal gradients in voltinism (Wolda 1988); thus, shifting temperatures should slide the boundaries between voltinism states in predictable directions. Voltinism may be relatively plastic, and in some species governed by photoperiod, temperature, and host plants, whereas in others it is fixed (Tauber and Tauber 1976). For those species with flexible voltinism, warming may be advantageous, permitting faster growth and additional generations annually (Bale et al. 2002; Tobin et al. 2008). In Europe, extensive data sets encompassing hundreds or even thousands of species of Lepidoptera have allowed comparisons among different time periods. For butterflies and moths with the capacity for multivoltinism, there have been significant increases in the frequency of species exhibiting bi- or multivoltine life cycles, with much of this increase occurring in the last two decades (Altermatt 2010; Pöyry et al. 2011).

Increased voltinism could promote faster population growth because more offspring are being produced per seasonal time period, thus increasing the likelihood of outbreaks of pest species or elevating non-pests or minor pests to a more economically important stature (Steinbauer et al. 2004; van Asch and Visser 2007). In forested ecosystems, changes in the voltinism of Lepidoptera and of Coleoptera (particularly scolytid bark beetles), are of concern as these groups contain some of the most economically damaging forest pests. A number of native bark beetle species in both Europe and North America have shifted lifecycles by adding annual generations, in an apparent response to moderating temperatures at higher latitudes or altitudes (Berg et al. 2006; Jonsson et al. 2009; Werner et al. 2006). With respect to the spruce beetle, *Dendroctonus rufipennis* (Kirby), this change was associated with devastating outbreaks in Alaska's Kenai Peninsula (Sherriff et al. 2011).

Despite the apparent high frequency of shifts in voltinism in native insects in many temperate zones globally, documentation of such shifts in invasive forest insects has thus far been rare. It is not known if the apparent rarity in forest insects is an artifact or represents real patterns. There are several possibilities for this absence. First, some invasive species could have a fixed voltinism, such as *L. dispar*, which is exclusively univoltine. Another possibility is that simply too little is known about the biology of many invasive species. Additionally, invasive species could lack sufficient genetic diversity or plasticity to respond to climate shifts, owing to small founder population size and the relatively short interval of observation. Successful species could have also been introduced with, and indeed may owe their success to, the expression of an appropriate voltinism for a given region.

One well-known case of shifting voltinism concerns *H. cunea*, a relatively benign defoliating lepidopteran accidentally introduced from North America to Europe and Asia where it has become a major pest. In Japan, the founding population was bivoltine, but within 50 years of introduction and coupled with a southward spread, trivoltine populations became the norm in warmer areas (Gomi and Takeda 1996). The shift to a trivoltine life-cycle was associated with a subtle but biologically significant change in sensitivity to photoperiod (Gomi 2007).

Multivoltinism appears to be rare in eruptive folivores, as most appear to be constrained to the nutritionally superior, but inherently risky (see Sect. 4.4) early season foliage of woody plants (Hunter 1991, 1995). For invasive folivorous species with obligate diapause, such as many univoltine spring-feeders, increasing temperatures could provide respite from natural enemies because development will accelerate through the vulnerable larval period, a function of escape from the trap of slow-growth and high mortality (Benrey and Denno 1997; Zalucki et al. 2002). This, however, makes the assumption that natural enemies will not respond similarly to elevated temperatures, or that they will not quickly adapt to a seasonal shift in prey abundance. Some, but not all, studies have suggested that *L. dispar* outbreaks are correlated with warmer spring temperatures in the year of, and the year prior to, defoliation (Elkinton and Liebhold 1990), although the mechanism underlying the pattern is not known. Communities of forest Lepidoptera irrespective of taxonomic affinity, especially those that are spring-feeders, exhibit concordant population dynamics, suggesting commonality of either positive or negative responses to a significant environmental driver like meteorological conditions (Raimondo et al. 2004; Stange et al. 2011).

A major concern from an invasive species perspective could be the response of wood borers. Unlike folivorous insects, many wood and cambium feeders have considerable plasticity with respect to voltinism and are constrained mainly by the combination of wood as a nutritionally poor resource and the relatively low temperatures in temperate and boreal forests. Increased voltinism observed in variety of native bark beetles (e.g., Faccoli 2009) could be a harbinger of what to expect in this particular guild. Voltinism in *A. glabripennis* is a function of latitude in China with southern populations requiring only a single year to complete development (Hu et al. 2009), while populations introduced into North American are variable, with both semivoltine or univoltine emergence recorded. Another invasive cerambycid, the brown spruce long-horned beetle, *Tetropium fuscum* (Fabricius), currently has a univoltine lifecycle but has been recorded as bivoltine in parts of its native range. In China, *D. valens* has already devastated vast tracts of Chinese red pine. In its native range in the southern United States, this species has up to three generations per year, but only one and perhaps a partial second generation has been recorded in China (Sun et al. 2004), suggesting that this insect could become an even greater threat in its introduced range under warming temperatures. These examples highlight the flexible voltinism apparent in many wood feeding insects and thus a high propensity to benefit from climatic warming.

4.4   Decoupling Species Interactions

One of the first and probably best-documented effects of anthropogenic driven climate change has been a phenological shift in the seasonal occurrence of a diverse array of organisms. Phenology, the seasonally influenced timing of developmental processes (e.g., Visser et al. 2010), is strongly correlated with temperature regime for many organisms including plants, insects, and vertebrates (Parmesan 2006; Root et al. 2003). In temperate regions, a large number of species have shifted seasonal biological activities such as onset of bud break, flowering time, emergence, or migrating earlier or maintaining activity later in the season as a response to recent changes to the onset of spring and the increasing length of the growing seasons, respectively. For example, the spring phenology of European Lepidoptera has advanced significantly over the past four decades (Altermatt 2010; Roy and Sparks 2000; Stefanescu et al. 2003) as it has or will for other insects (Hassall et al. 2007; Logan et al. 2003; Masters et al. 1998); these changes are apparently correlated with an increase in degree-day availability early in the season (Parmesan 2006).

Although changes in the phenology of individual species are well-described (Menzel et al. 2006; Robinet and Roques 2010), less attention has been paid to climatically driven mismatches to the trophic relationships of interacting species, despite predictions about the important negative consequences of asynchrony and its resultant decoupling (Donnelly et al. 2011; Singer and Parmesan 2010). Climatically driven decoupling is expected when synchrony between species is disrupted in time or space (Stenseth and Mysterud 2002). Decoupling can be viewed from either a temporal or spatial perspective. Spatially, rapid range expansion by a species could decouple relationships between predator and prey (Menendez et al. 2008; see Sect. 4.1), whereas temporally, differential response to shifting temperatures could lead to a phenological decoupling of a species relationship, be it plant-herbivore, predator-prey, or tritrophic interactions.

Whether or not a system will become phenologically decoupled depends on the response of the participant species to climatic drivers. For example, no net change could occur if the interacting species respond similarly to the same environmental cues or to different environmental cues in a way that is highly correlated. However, decoupling might be expected where species are responding to specific cues that become less correlated as temperatures and/or seasonality change. For example, a photoperiodic response by one species could lead to a divergent phenology if an interacting species responds primarily to degree-day accumulations. In tri-trophic relationships, elucidating the effects of climatic shift will be difficult and the relative changes in responses by an herbivorous insect, its host plant, and its natural enemies could be neutral, negative, or positive depending on the degree of decoupling and the nature of the decoupled mechanism(s).

The potential for climatically driven phenological decoupling of herbivorous insects and their host plants has long been recognized (Buse and Good 1996; Dewar and Watt 1992; Harrington et al. 1999), but has been investigated extensively in only a few systems. The importance of phenological synchrony of insect herbivores with host plants varies between and among species, functional feeding guild, and the seasonal activity period of a species with effects likely to be neutral or negative, as a positive effect seems implausible. The same mechanisms that drive asynchrony and decoupling, however, could allow insects to utilize hosts that were previously outside of their phenological range as climatic change differentially alters seasonal timing of tree and herbivore (e.g., Jepsen et al. 2011).

Sensitivity to phenological change is likely to be greatest for spring feeding species (Forkner et al. 2008), but could also affect other seasonal guilds depending on the nature and magnitude of change. Increased voltinism (see Sect. 4.3) may push some phenologically insensitive species into more vulnerable early or late season envelopes. Species whose activity (i.e., egg hatch, larval emergence) is timed to bud burst of host trees may be susceptible to even relatively small alterations to synchrony. These species often have a narrow window of opportunity to maximize growth because they are constrained by starvation if they emerge too early, and by declining nutritional value and increasing secondary phytochemical concentration of maturing leaves should development be delayed until later in spring (Ayres and MacLean 1987; Feeny 1970; Hunter 1993; Jones and Despland 2006; Martel and Kause 2002; Parry et al. 1998). While the effects of phenological asynchrony are best known for Lepidoptera, negative consequences have been shown in many other insect herbivores including Homoptera, Diptera, Coleoptera, and Hymenoptera (Dixon 1976; Fox et al. 1997; Martel et al. 2001; Yukawa and Akimoto 2006).

We know of no study that has specifically addressed phenological decoupling of an insect-plant interaction in the context of biological invasions, but it seems unlikely that introduced species would differ substantially from native species. Although direct research is lacking, extrapolation is possible from a few well-studied native insect-plant interactions. One exemplar insect, *O. brumata*, could be particularly instructive in elucidating consequences of climate change on phenological decoupling as it is well-studied in its native range, and is currently invading North America.

The winter moth is a univoltine spring feeder native to Great Britain and Europe, but was accidentally introduced to Nova Scotia (1940s), British Columbia (1960s), and more recently Massachusetts (2000s) and other New England states (Elkinton et al. 2010; Roland and Embree 1995). Flightless females ascend host trees such as oaks in the fall, and oviposit on branches and twigs in the canopy. Eggs hatch in the following spring in close proximity to bud break. The fitness consequences of synchrony with bud break are significant (van Asch and Visser 2007) as the newly hatched larvae have a limited ability to survive starvation if emergence is early but suffer from the declining value of maturing foliage if emergence is late.

Winter moth has been extensively studied, especially in Great Britain (e.g., Varley et al. 1973) and the existence of several long term data sets allows insight to the effects of climatic change on *O. brumata* with pendunculate oak. In the Netherlands, the onset of winter moth egg hatch and bud break of this oak species have advanced considerably over a quarter century (Visser and Holleman 2001). Egg hatch, however, has advanced more than bud break, decreasing synchrony by 2–14 days depending on the year. While late hatch decreases fitness through reductions in fecundity, a shift of 5 days too early can result in mortality of 90 % or more, suggesting that an increasingly premature hatch relative to bud break is non-trivial. Winter moth has sufficient genetic variability that selection should act to push hatch time closer to bud break (van Asch et al. 2010), although this does not appear to have happened naturally thus far. Although the effects of climatic change on *O. brumata* phenology and synchronicity with host plants have not been studied in North America where it is invasive, its extensive use of multiple tree genera in the northeastern United States could buffer it from any deleterious consequences of climatic change.

Considerably less is known about the effects of climate change on phenology in other invasive forest insects, even for those that have been extensively studied. For example, while various models (Régnière et al. 2009; Williams and Liebhold 1995b; see Sect. 4.1) have suggested that the geographic range of *L. dispar* in North America will expand northward under various warming projections, the potential for asynchrony with host tree species has not been explored. The responses of trees to warming at higher latitudes may differ from the temperature response of *L. dispar* egg hatch, thus increasing the risk of phenological mismatch. Although *L. dispar* is sensitive to tree phenology (Hunter 1993; Hunter and Elkinton 2000; Stoyenoff et al. 1994), it has life-history attributes that could mitigate many of the most deleterious effects of asynchrony. Similar to *O. brumata*, *L. dispar* larvae feed on a wide-variety of woody plants, which ensures that at least spatially, some hosts will be available to neonates. Furthermore, the temporal distribution of egg hatch, both within a single egg mass and among egg masses in a population, spans extended periods (Gray et al. 2007; Hunter 1993), which also increases the likelihood that the highly mobile larvae will encounter phenologically suitable hosts.

The phenological relationship between insects and plants do not occur in a vacuum; rather, it is a template upon which other environmental factors also enhance or attenuate the effects of asynchrony. Temperature and climatic shifts also occur in concert with rising levels of CO_2, which could increase or decrease quality of plant tissue for herbivores depending on species and functional feeding guild (Cornellisen 2011; Stiling and Cornelissen 2007). Other environmental feedbacks and covariates associated with climate change could further confound any analysis. Based on limited studies to date, it seems unlikely that phenological asynchrony will be of significant long-term consequence for many native insect herbivores. Even less confidence can be attached to predictions about non-native species. As many forest insect invasions are initiated from genetically-limited founder populations from only portions of their native range, it is unclear if responses to climatic shifts will differ from that seen in populations of native species. However, many successful invasive species are habitat or host generalists and may express considerable phenotypic plasticity while other species, despite apparently limited genetic diversity, have nonetheless rapidly adapted to climatic variability in recipient regions (Gomi et al. 2009).

The potential for climatic disruption or alteration of coupled species relationships also applies to higher trophic levels, which often exert considerable top-down regulation on herbivore populations. The relative synchrony between natural enemies and their prey could be maintained under climatic change if the organisms respond similarly to the same variable or to variables that remain highly correlated. The decoupling of such relationships could occur as divergent responses to the same or to correlated variables, although few predator-prey interactions and even fewer multi-trophic systems have been examined in detail. A recent study documented considerable species turnover in samples of subarctic parasitoid communities when compared to historical data sets from the same localities, with patterns suggesting a link to climate warming (Fernández-Triana et al. 2011). In a different study, a meta-analytical approach suggested that an amplification of climatic variability was negatively correlated with parasitism of tree-feeding Lepidoptera, particularly for specialist hymenopteran parasitoids, which were disproportionately affected relative to tachinid parasitoids with broader host ranges (Stireman et al. 2005). Thus, at least in the short-term, the influence of parasitoids on the population dynamics of their prey could be reduced, which has important ramifications for outbreak species whether native or introduced. However, selection has favored herbivore life-history strategies that maximize temporal enemy-free or enemy-reduced space, and differential responses to climate variables could also force greater overlap between some herbivores and their natural enemies (Hance et al. 2007).

There is evidence of decoupled predator-prey relationships due to climatic shifts in a number of insectivorous birds in Europe. The fitness of these birds is greatest when the timing of reproduction corresponds with a peak in biomass of primarily lepidopteran caterpillars (Both et al. 2006; Visser et al. 1998, 2006). Warmer springs have shifted this peak earlier, and higher temperatures have compressed the period of abundance as larvae complete their development more rapidly. Although reproductive activity of birds has also advanced, it has not done so at the same rate as caterpillar biomass peak, and this relationship appears to becoming increasingly asynchronous. The implications for invasive forest insects are unclear, but since birds are important predators of many herbivorous insects, especially in low-density insect populations (Holmes et al. 1979; Marquis and Whelan 1994; Parry et al. 1997), a diminution of their capacities would likely benefit both native and non-native insects alike. However, there remains much uncertainty in the degree and importance of divergence of bird and insect prey phenology due to climate change; for example, a long time series data set in England suggests that the great tit, an important predator of *O. brumata* larvae, has been able to track the shift in spring phenology of its prey item over time (Charmantier et al. 2008).

The success of many invasive insects in forests owes at least in part to an incomplete or missing natural enemy component, and thus the potential effects of climate change on this trophic level could be largely moot for some species. However, for invasive species held in check by classical biological control introductions, climatic shifts potentially could alter these important interactions, leading to a resurgence of previously suppressed populations or hamper efforts to develop new biological control programs. Indeed, it has long been recognized that classic biological control could be vulnerable to climate change (Cannon 1998) because the interactions between introduced enemies and their non-native prey could be inherently more susceptible to decoupling than those interactions involving native species. Many introductions of control agents are initiated with relatively low genetic diversity, which potentially limits the adaptive response to changing climate. Second, specific biotypes of natural enemies, particularly parasitoids, are often selected to match current climatic conditions in a given region (Robertson et al. 2009); these may or may not be suitable for future climatic envelopes. Conversely, some biological control organisms that are currently climatically limited in parts of an invader's range may become more effective under warming scenarios (Siegert et al. 2009; Zalucki and van Klinken 2006).

Many of the predictions concerning the decoupling of insect herbivore-host or predator-prey interactions are overly general or simplistic because we lack the necessary knowledge to make these predictions in all but a few systems. The effects of climate change on decoupling interactions involving non-native species are even more difficult to generalize as the relationships are often novel and are occurring in environments different from those where the species evolved. For example, in North America, the non-native tachinid *Cyzenis albicans* (Fallén) is regarded as the most effective regulator of invasive *O. brumata* populations, but this parasitoid is a trivial source of mortality in native populations (Roland and Embree 1995; Varley et al. 1973). Thus, it may be difficult to generalize the effects of climate from donor to recipient regions, even for well-studied systems.

5 Economic Ramifications of Invasions in the Face of Climate Change

Despite the challenges associated with predicting the ecological consequences of climate-mediated effects on biological invasions, it is arguably even more difficult to quantify the economic costs due to all of these interacting forces. After all, reliable estimates of the economic costs due to specifically non-native forest insects alone are largely lacking (Aukema et al. 2011). Even though these costs are challenging to estimate, they are not always difficult to envision. For example, increases in the availability of suitable habitat due to changing climates facilitating invasions into new areas could in turn increase the costs associated with its management (Cannon 1998; Kiritani 2006). Similarly, increases in abundance, and outbreak intensity and frequency due to climate warming is likely to lead to increased management costs (Hellmann et al. 2008; Rosenzweig et al. 2001; Waring et al. 2009). Costs could also include the increase in the energy footprint of food and fiber production systems due to this increased need for pest control measures (Gandhi and Herms 2010; Pimentel 2002).

Other potential consequences, however, can be complex and involve a cascading array of effects across one or more trophic levels. One such effect of climate change, and specifically the role of increased concentrations of carbon dioxide and ozone in the atmosphere, is the potential change in host plant nutritional quality. For example, plants grown under high levels of carbon dioxide can cause changes in the carbon-to-nitrogen ratio of plant tissues (Hamilton et al. 2005); consequently, herbivores feeding on such plants could eat more leaf matter to compensate for the reduced nutritional quality of their host plants (Coviella and Trumble 1999; Dermody et al. 2008; but see Kopper and Lindroth 2003). Increases in herbivory due to changes in concentrations of atmospheric gases, coupled with increases in herbivore abundance, insect developmental rates, and voltinism owing to increases in surface temperatures (Bale et al. 2002; Chen et al. 2011; Tobin et al. 2008; Yamamura and Kiritani 1998), could have dramatic implications to pest management practices and the costs required to achieve pest control. Furthermore, a need to increase pest control tactics, specifically the use of chemical insecticides, could also intensify the inimical effects to non-target species (Pimentel et al. 1980) as well as select for resistance in the target species (Roush and McKenzie 1987).

Because of the potential for climate change to decouple interactions between natural enemies and their prey (Simon et al. 2002; Stireman et al. 2005), the use of biological control as a management tactic against non-native forest pests could be rendered less effective. In particular, classical biological control has received renewed interest in combating non-native insect pests (Hajek et al. 2007; Hajek and Tobin 2010; Hoddle 2004), and increased scrutiny is given to the specificity of introduced agents to avoid the historical blunders from the import of generalist natural enemies (Elkinton et al. 2006; Simberloff and Stiling 1996; Strong and Pemberton 2000). Because of the need for specificity in selecting a natural enemy for introduction, changes in climate – even if subtle – could influence aspects of these interspecific interactions, and the suitable range of one species could be affected by climate differently than the other. For example, the parasitic nematode *Deladenus siricidicola* is an effective biological control agent of the wood wasp *S. noctilio* in Argentina (Corley et al. 2007) and Australia (Neumann and Minko 1981). As a nematode, it is likely more sensitive to changes in moisture conditions, which are predicted to be affected by changes in climate (Rosenzweig et al. 2001), than its insect host. Indeed, the observed geographic variation in the effectiveness of *D. siricidicola* as a biological control agent of *S. noctilio* could be due to variation in climate among regions (Slippers et al. 2012b). Although additional and specific forest insect examples are still rare, climate and projections in climate will likely need to be considered when evaluating the short- and long-term efficacy of an introduced natural enemy (Zalucki and van Klinken 2006).

6 Conclusions

In addition to many of the "known unknowns" described above, a final consideration in the context of climate change and its effect on forest insect invasions is the proverbial "unknown unknowns". The dynamics of forest insects and their interactions with associated pathogens and natural enemies, together with interactions with host species, can be difficult to predict when species are introduced to a new area. Indeed, although many biological and ecological aspects are often highlighted as important when considering the invasion potential of a species and in formulating risk assessments (Liebhold and Tobin 2008; Lockwood et al. 2007; Worner and

Gevrey 2006), developing a general paradigm of species invasiveness with broad application has proved challenging (Hulme 2003; Lonsdale 1999; Rejmánek and Richardson 1996). Coupling the uncertainty of biological invasions with the complexity of climate change and its variable effect on individual species and to those communities in which they interact complicates this challenge even further. Innocuous species today could quite possibly become quite invasive under future climatic conditions, whether in their native range, an introduced habitat, or both. Greater attention should be given to this complexity through examinations of landscape-level climatic changes and its combined effect on ecosystem inhabitants.

Acknowledgments We are grateful to Andrew Liebhold and Trevor Fenning for helpful comments on an earlier draft of this chapter.

References

Altermatt F (2010) Climatic warming increases voltinism in European butterflies and moths. Proc R Soc Biol Sci Ser B 277:1281–1287

Ammunet T, Heisswolf A, Klemola N, Klemola T (2010) Expansion of the winter moth outbreak range: no restrictive effects of competition with the resident autumnal moth. Ecol Entomol 35:45–52

Arnaldo PS, Oliveira I, Santos J, Leite S (2011) Climate change and forest plagues: the case of the pine processionary moth in Northeastern Portugal. For Syst 20:508–515

Aukema BH, Carroll AL, Zhu J, Raffa KF, Sickley TA, Taylor SW (2006) Landscape level analysis of mountain pine beetle in British Columbia, Canada: spatiotemporal development and spatial synchrony within the present outbreak. Ecography 29:427–441

Aukema JE, McCullough DG, Von Holle B, Liebhold AM, Britton K, Frankel SJ (2010) Historical accumulation of nonindigenous forest pests in the continental US. Bioscience 60:886–897

Aukema JE, Leung B, Kovacs K, Chivers C, Britton KO, Englin J, Franke SJ, Haight RG, Holmes TP, Liebhold AM, McCullough DG, Von Holle B (2011) Economic impacts of non-native forest insects in the continental United States. PLoS One 6:e24587

Ayres MP, MacLean SF (1987) Development of birch leaves and the growth energetics of *Epirrita autumnata* (Geometridae). Ecology 68:558–568

Ayres MP, Scriber JM (1994) Local adaptation to regional climates in *Papilio canadensis* (Lepidoptera: Papilionidae). Ecol Monogr 64:465–482

Balanya J, Oller JM, Huey RB, Gilchrist GW, Serra L (2006) Global genetic tracks global climate warming in *Drosophila subobscura*. Science 313:1773–1775

Bale JS, Hayward SAL (2010) Insect overwintering in a changing climate. J Exp Biol 213:980–994

Bale JS, Masters GJ, Hodkinson ID, Awmack C, Bezemer TM, Brown VK, Butterfield J, Buse A, Coulson JC, Farrar J, Good JEG, Harrington R, Hartley S, Jones TH, Lindroth RL, Press MC, Symrnioudis I, Watt AD, Whittaker JB (2002) Herbivory in global climate change research: direct effects of rising temperature on insect herbivores. Glob Chang Biol 8:1–16

Barbosa P, Schultz JC (eds) (1987) Insect outbreaks. Academic Press, San Diego

Bartell SM, Nair SK (2003) Establishment risks for invasive species. Risk Anal 24:833–845

Battisti A, Stastny M, Netherer S, Robinet C, Schopf A, Roques A, Larsson S (2005) Expansion of geographic range in the pine processionary moth caused by increased winter temperatures. Ecol Appl 15:2084–2096

Beaumont LJ, Gallagher RV, Thuiller W, Downey P, Leishman MR, Hughes L (2009) Different climatic envelopes among invasive populations may lead to underestimations of current and future biological invasions. Divers Distrib 15:409–420

Benrey B, Denno RF (1997) The slow-growth-high-mortality hypothesis: a test using the cabbage butterfly. Ecology 78:987–999

Berg EE, Henry JD, Fastie CL, DeVoider AD, Matsuoka SM (2006) Spruce beetle outbreaks on the Kenai Peninsula, Alaska, and Kluane National Park and Reserve, Yukon Territory: relationship to summer temperatures and regional differences in disturbance. For Ecol Manag 227:219–232

Bjørnstad O, Ims RA, Lambin X (1999) Spatial population dynamics: analyzing patterns and processes of population synchrony. Trends Ecol Evol 11:427–431

Bjørnstad ON, Peltonen M, Liebhold AM, Baltensweiler W (2002) Waves of larch budmoth outbreaks in the European alps. Science 298:1020–1023

Blanckenhorn WU, Demont M (2004) Bergmann and converse Bergmann latitudinal clines in arthropods: two ends of a continuum? Integr Comp Biol 44:413–424

Both C, Bouwhuis S, Lessells CM, Visser ME (2006) Climate change and population declines in long distance migratory bird. Nature 441:81–83

Brasier CM (2008) The biosecurity threat to the UK and global environment from international trade in plants. Plant Path 57:792–808

Brockerhoff EG, Bain J, Kimberley M, Knížek M (2006a) Interception frequency of exotic bark and ambrosia beetles (Coleoptera: Scolytinae) and relationship with establishment in New Zealand and worldwide. Can J For Res 36:289–298

Brockerhoff EG, Liebhold AM, Jactel H (2006b) The ecology of forest insect invasions and advances in their management. Can J For Res 36:263–268

Buse A, Good JEG (1996) Synchronization of larval emergence in winter moth (*Operophtera brumata* L.) and budburst in pedunculate oak (*Quercus robur* L.) under simulated climate change. Ecol Entomol 21:335–343

Cannon RJC (1998) The implications of predicted climate change for insect pests in the UK, with emphasis on non-indigenous species. Glob Chang Biol 4:785–796

Cerezke HF (1995) Egg gallery, brood production, and adult characteristics of mountain pine beetle, *Dendroctonus ponderosae* Hopkins (Coleoptera, Scolytidae), in three pine hosts. Can Entomol 127:955–965

Charmantier A, McCleery RH, Cole LR, Perrins C, Kruuk LEB, Sheldon BC (2008) Adaptive phenotypic plasticity in response to climate change in a wild bird population. Science 320:800–803

Chen S, Fleischer SJ, Tobin PC, Saunders MC (2011) Projecting insect voltinism under high and low greenhouse gas emission conditions. Environ Entomol 40:505–515

Corley JC, Villacide JM, Bruzzone OA (2007) Spatial dynamics of a *Sirex noctilio* woodwasp population within a pine plantation in Patagonia, Argentina. Entomol Exp Appl 125:231–236

Cornellisen T (2011) Climate change and its effects on terrestrial insects and herbivory patterns. Neotrop Entomol 40:155–163

Coviella CE, Trumble JT (1999) Effects of elevated atmospheric carbon dioxide on insect-plant interactions. Conserv Biol 13:700–712

Coyle DR, Nebeker TE, Hart ER, Mattson WJ (2005) Biology and management of insect pests in North American intensively managed hardwood forest systems. Annu Rev Entomol 50:1–29

Crosby AW (1986) Ecological imperialism: the biological expansion of Europe, 900–1900. Cambridge University Press, Cambridge, UK

Cudmore TJ, Björklund N, Carroll AL, Lindgren BS (2010) Climate change and range expansion of an aggressive bark beetle: evidence of higher beetle reproduction in naïve host tree populations. J Appl Ecol 47:1036–1043

Cullingham CI, Cooke JEK, Dang S, Davis CS, Cooke BJ, Coltman DW (2011) Mountain pine beetle host-range expansion threatens the boreal forest. Mol Ecol 20:2157–2171

Dale VH, Joyce LA, Mcnulty S, Neilson RP, Ayres MP, Flannigan MD, Hanson PJ, Irland LC, Lugo AE, Peterson CJ, Simberloff D, Swanson FJ, Stocks BJ, Wotton BM (2001) Climate change and forest disturbances. Bioscience 51:723–734

Danks HV (2006) Insect adapatations to cold and changing environments. Can Entomol 138:1–23

Davidson J, Wickland AC, Patterson HA, Falk KR, Rizzo DM (2005) Transmission of *Phytophthora ramorum* in mixed-evergreen forest in California. Phytopathology 95:587–596

de la Giroday H-MC, Carroll AL, Lindgren BS, Aukema BH (2011) Incoming! Association of landscape features with dispersing mountain pine beetle populations during a range expansion event in western Canada. Landsc Ecol 26:1097–1110

de la Giroday H-MC, Carroll AL, Aukema BH (2012) Breach of the northern Rocky Mountain geoclimatic barrier: initiation of range expansion by the mountain pine beetle. J Biogeogr 39:1112–1123

Dermody O, O'Neill B, Zangerl A, Berenbaum M, DeLucia EH (2008) Effects of elevated CO_2 and O_3 on leaf damage and insect abundance in a soybean agroecosystem. Arthropod Plant Interact 2:125–135

Deutsch CA, Tewksbury JJ, Huey RB, Sheldon KS, Ghalambor CK, Haak DC, Martin PR (2008) Impacts of climate warming on terrestrial ectotherms across latitude. Proc Natl Acad Sci U S A 105:6668–6672

Dewar RC, Watt AD (1992) Predicted changes in the synchrony of larval emergence and budburst under climatic warming. Oecologia 89:557–559

di Castri F (1989) History of biological invasions with special emphasis on the old world. In: Drake JA, Mooney HA, di Castri F et al (eds) Biological invasions: a global perspective. Wiley, New York, pp 1–30

Dixon AFG (1976) Timing of egg hatch and viability of the Sycamore aphid, *Drepanosiphum platanoides* (Schr.), at bud burst of Sycamore, *Acer platanus* L. J Anim Ecol 45:593–603

Donnelly A, Caffarra A, O'Neill BF (2011) A review of climate-driven mismatches between interdependent phenophases in terrestrial and aquatic ecosystems. Int J Biometeorol 55:805–817

Elkinton JS, Liebhold AM (1990) Population dynamics of gypsy moth in North America. Annu Rev Entomol 35:571–596

Elkinton JS, Parry D, Boettner GH (2006) Implicating an introduced generalist parasitoid in the invasive browntail moth's enigmatic demise. Ecology 87:2664–2672

Elkinton JS, Boettner GH, Sremac M, Gwiazdowski R, Hunkins RR, Callahan J, Scheufele SB, Donahue CP, Porter AH, Khrimian A, Whited BM, Campbell NK (2010) Survey for winter moth (Lepidoptera: Geometridae) in northeastern North America with pheromone-baited traps and hybridization with the native Bruce spanworm (Lepidoptera: Geometridae). Ann Entomol Soc Am 103:135–145

Engel K, Tollrian R, Jeschke JM (2011) Integrating biological invasions, climate change and phenotypic plasticity. Comm Integr Biol 4:247–250

Esper J, Büntgen U, Frank DC, Nievergelt D, Liebhold A (2007) 1200 years of regular outbreaks in alpine insects. Proc R Soc Biol Sci Ser B 274:671–679

Evangelista PH, Kumar S, Stohlgren TJ, Young NE (2011) Assessing forest vulnerability and the potential distribution of pine beetles under current and future climate scenarios in the Interior West of the US. For Ecol Manag 262:307–316

Faccoli M (2009) Effect of weather on *Ips typographus* (Coleoptera Curculionidae) phenology, voltinism, and associated spruce mortality in the southeastern Alps. Environ Entomol 38:307–316

Feeny PP (1970) Seasonal changes in oak leaf tannins and nutrients as a cause of spring feeding by winter moth caterpillars. Ecology 51:565–581

Fernández-Triana J, Smith MA, Boudreault C, Goulet H, Hebert PDN, Smith AC, Roughley R (2011) A poorly known high-latitude parasitoid wasp community: unexpected diversity and dramatic changes through time. PLoS One 6:e23719

Forkner RE, Marquis RJ, Lill JT, Corff JL (2008) Timing is everything? Phenological synchrony and population variability in leaf-chewing herbivores of *Quercus*. Ecol Entomol 33:276–285

Fox CW, Waddell KJ, Groeters FR, Mousseau TA (1997) Variation in budbreak phenology affects the distribution of a leaf-mining beetle (*Brachys tessellatus*) on turkey oak (*Quercus laevis*). Ecoscience 4:480–489

Frost CJ, Hunter MD (2004) Insect canopy herbivory and frass deposition affect soil nutrient dynamics and export in oak mesocosms. Ecology 85:3335–3347

Gandhi JKJ, Herms DA (2010) Direct and indirect effects of alien insect herbivores on ecological processes and interactions in forests of eastern North America. Biol Invasions 12:389–405

Gilbert M, Grégoire J-C, Freise JF, Heitland W (2004) Long-distance dispersal and human population density allow the prediction of invasive patterns in the horse chestnut leafminer *Cameraria ohridella*. J Anim Ecol 73:459–468

Gilchrist GW, Huey RB, Balanya J, Pascual M, Serra L (2001) A time series of evolution in action: a latitudinal cline in wing size in South American *Drosophila subobscura*. Evolution 58:768–780

Gomi T (2007) Seasonal adaptations of the fall webworm *Hyphantria cunea* (Drury) (Lepidoptera: Arctiidae) following its invasion of Japan. Ecol Res 22:855–861

Gomi T, Takeda M (1996) Changes in life-history traits in the Fall Webworm within half a century of introduction to Japan. Funct Ecol 10:384–389

Gomi T, Adachi K, Shimizu A, Tanimoto K, Kawabata E, Takeda M (2009) Northerly shift in voltinism watershed in *Hyphantria cunea* (Drury) (Lepidoptera: Arctiidae) along the Japan Sea coast: evidence of global warming? Appl Entomol Zool 44:357–362

Gray DR (2004) The gypsy moth life stage model: landscape-wide estimates of gypsy moth establishment using a multi-generational phenology model. Ecol Model 176:155–171

Gray DR, Tanner JA, Logan JA, Munson AS (2007) Using sterile gypsy moth eggs as a survey and experimental tool in the field: a comparison of hatching patterns. Ann Entomol Soc Am 100:439–443

Hagen SB, Jepsen JU, Ims RA, Yoccoz NG (2007) Shifting altitudinal distribution of outbreak zones of winter moth *Operophtera brumata* in sub-arctic birch forest: a response to recent climate warming? Ecography 30:299–307

Hajek AE, Tobin PC (2010) Micro-managing arthropod invasions: eradication and control of invasive arthropods with microbes. Biol Invasions 12:2895–2912

Hajek AE, McManus ML, Delalibera I Jr (2007) A review of introductions of pathogens and nematodes for classical biological control of insects and mites. Biol Control 41:1–13

Hamilton JG, Dermody O, Aldea M, Zangerl AR, Rogers A, Berenbaum MR, DeLucia EH (2005) Anthropogenic changes in tropospheric composition increase susceptibility of soybean to insect herbivory. Environ Entomol 34:479–485

Hance T, Van Baaren J, Vernon P, Boivin G (2007) Impact of extreme temperatures on parasitoids in a climate change perspective. Annu Rev Entomol 52:107–126

Harrington R, Woiwood I, Sparks T (1999) Climate change and trophic interactions. Trends Ecol Evol 14:146–150

Harvey GT (1983) A geographic cline in egg weights in *Choristoneura fumiferana* (Lepidoptera: Tortricidae) and its significance to population dynamics. Can Entomol 115:1103–1108

Hassall C, Thompson DJ, French GC, Harvey IF (2007) Historical changes in the phenology of British Odonata are related to climate. Glob Chang Biol 13:933–941

Hayes KR, Barry SC (2008) Are there any consistent predictors of invasion success? Biol Invasions 10:483–506

Hayhoe K, Wake CP, Huntington TG, Luo L, Schwartz M, Sheffield J, Wood E, Anderson B, Bradbury J, DeGaetano A, Troy T, Wolfe D (2007) Past and future changes in climate and hydrological indicators in the U.S. Northeast. Clim Dynam 28:381–407

Haynes KJ, Liebhold AM, Fearer TM, Wang G, Norman GW, Johnson DM (2009) Spatial synchrony propagates through a forest food web via consumer–resource interactions. Ecology 90:2974–2983

Hellmann JJ, Byers JE, Bierwagen BG, Dukes JS (2008) Five potential consequences of climate change for invasive species. Conserv Biol 22:534–543

Hengeveld R (1989) Dynamics of biological invasions. Chapman and Hall, London

Hlasny T, Turcani M (2009) Insect pests as climate change driven disturbances in forest ecosystems. In: Strelcová K, Matyas C, Kleidon A et al (eds) Bioclimatology and natural hazards. Springer, Dordrecht, pp 165–177

Hoddle MS (2004) Restoring balance: using exotic species to control invasive exotic species. Conserv Biol 18:38–49

Holmes RT, Schultz JC, Nothnagle P (1979) Bird predation on forest insects: an exclosure experiment. Science 206:462–463

Holmes TP, Aukema JE, Von Holle B, Liebhold A, Sills E (2009) Economic impacts of invasive species in forests past, present, and future. Ann NY Acad Sci 1162:18–38

Hu J, Angeli S, Schuetz S, Luo Y, Hajek AE (2009) Ecology and management of exotic and endemic Asian longhorned beetle *Anoplophora glabripennis*. Agr For Entomol 11:359–375

Hulme PE (2003) Biological invasions: winning the science battles but losing the conservation war? Oryx 37:178–193

Hulme PE, Bacher S, Kenis M, Klotz S, Kühn I, Minchin D, Nentwig W, Olenin S, Panov V, Pergl J, Pyšek P, Roques A, Sol D, Solarz W, Vilà M (2008) Grasping at the routes of biological invasions: a framework for integrating pathways into policy. J Appl Ecol 45:403–414

Hunter AF (1991) Traits that distinguish outbreaking and non-outbreaking Macrolepidoptera feeding on northern hardwood trees. Oikos 60:275–282

Hunter AF (1993) Gypsy moth population sizes and the window of opportunity in the spring. Oikos 68:531–538

Hunter AF (1995) Ecology, life-history, and phylogeny of outbreak and non-outbreak species. In: Cappuccino N, Price PW (eds) Population dynamics: new approaches and synthesis. Academic Press, New York, pp 41–64

Hunter AF, Elkinton JS (2000) Effects of synchrony with host plant on populations of a spring-feeding Lepidopteran. Ecology 81:1248–1261

Intergovernmental Panel on Climate Change (2007) The physical science basis. Working group I. Contribution to the fourth assessment report of the IPCC. Cambridge University Press, Cambridge, UK

Isard SA, Gage SH, Comtois P, Russo JM (2005) Principles of the atmospheric pathway for invasive species applied to soybean rust. Bioscience 55:851–861

Jepsen JU, Hagen SB, Ims RA, Yoccoz NG (2008) Climate change and outbreaks of the geometrids *Operophtera brumata* and *Epirrita autumnata* in subarctic birch forest: evidence of a recent outbreak range expansion. J Anim Ecol 77:257–264

Jepsen JU, Kapari L, Hagen SB, Schott T, Vindstad OPL, Nilssen AC, Ims RA (2011) Rapid northwards expansion of a forest insect pest attributed to spring phenology matching with sub-Arctic birch. Glob Chang Biol 17:2071–2083

Johnson DM, Liebhold AM, Bjørnstad ON, McManus ML (2005) Circumpolar variation in periodicity and synchrony among gypsy moth populations. J Anim Ecol 74:882–892

Jones BC, Despland E (2006) Effects of synchronization with host plant phenology occur early in the larval development of a spring folivore. Can J Zool 84:628–633

Jonsson AM, Appelberg G, Harding S, Barring L (2009) Spatio-temporal impact of climate change on the activity and voltinism of the spruce bark beetle, *Ips typographus*. Glob Chang Biol 15:486–499

Jonsson AM, Harding S, Krokene P, Lange H, Lindelow A, Okland B, Ravn HP, Schroeder LM (2011) Modelling the potential impact of global warming on *Ips typographus* voltinism and reproductive diapause. Clim Chang 109:695–718

Kausrud K, Økland B, Skarpaas O, Grégoire J-C, Erbilgin N, Stenseth NC (2012) Population dynamics in changing environments: the case of an eruptive forest pest species. Biol Rev 87:34–51

Keane RM, Crawley MJ (2002) Exotic plant invasions and the enemy release hypothesis. Trends Ecol Evol 17:164–170

Kerdelhué C, Zane L, Simonato M, Salvato P, Rousselet J, Roques A, Battisti A (2009) Quaternary history and contemporary patterns in a currently expanding species. BMC Evol Biol 9:220

Kiritani K (2006) Predicting impacts of global warming on population dynamics and distribution of arthropods in Japan. Popul Ecol 48:5–12

Kopper BJ, Lindroth RL (2003) Effects of elevated carbon dioxide and ozone on the phytochemistry of aspen and performance of an herbivore. Oecologia 134:95–103

Kovacs KF, Haight RF, McCullough DG, Mercader RJ, Siegert NW, Liebhold AM (2010) Cost of potential emerald ash borer damage in U.S. communities, 2009–2019. Ecol Econ 69:569–578

Kunkel KE, Huang H-C, Liang X-Z, Lin J-T, Wuebbles D, Tao Z, Williams A, Caughey M, Zhu J, Hayhoe K (2008) Sensitivity of future ozone concentrations in the northeast USA to regional climate change. Mitig Adapt Strateg Glob Chang 13:5–6

Kurz WA, Dymond CC, Stinson G, Rampley GJ, Neilson ET, Carroll AL, Ebata T, Safranyik L (2008) Mountain pine beetle and forest carbon feedback to climate change. Nature 452:987–990

Liebhold AM, Tobin PC (2008) Population ecology of insect invasions and their management. Annu Rev Entomol 53:387–408

Liebhold AM, MacDonald WL, Bergdahl D, Mastro V (1995) Invasion by exotic forest pests: a threat to forest ecosystems. For Sci Monogr 30:1–49

Liebhold AM, Brockerhoff EG, Garrett LJ, Parke JL, Britton KO (2012) Live plant imports: the major pathway for forest insect and pathogen invasions of the United States. Front Ecol Environ 10:135–143

Lockwood JL, Hoopes M, Marchetti M (2007) Invasion ecology. Blackwell Publishing Ltd., Malden

Logan JA, Powell JA (2001) Ghost forests, global warming, and the mountain pine beetle (Coleoptera: Scolytidae). Am Entomol 47:160–173

Logan JA, Régnière J, Powell JA (2003) Assessing the impacts of global warming on forest pest dynamics. Front Ecol Environ 1:130–137

Logan JA, Macfarlane WW, Willcox L (2010) Whitebark pine vulnerability to climate-driven mountain pine beetle disturbance in the Greater Yellowstone Ecosystem. Ecol Appl 20:895–902

Lonsdale WM (1999) Global patterns of plant invasions and the concept of invasibility. Ecology 80:1522–1536

Lovett GM, Christenson LM, Groffman PM, Jones CG, Hart JE, Mitchell MJ (2002) Insect defoliation and nitrogen cycling in forests. Bioscience 52:335–341

Mann ME, Bradley RS, Hughes MK (1998) Global-scale temperature patterns and climate forcing over the past six centuries. Nature 392:779–787

Marquis RJ, Whelan CJ (1994) Insectivorous birds increase growth of white oak through consumption of leaf-chewing insects. Ecology 75:2007–2014

Martel J, Kause A (2002) The phenological window of opportunity for early-season birch sawflies. Ecol Entomol 27:302–307

Martel J, Hanhimaki S, Kause A, Haukioja E (2001) Diversity of birch sawfly responses to seasonally atypical diets. Entomol Exp Appl 100:301–309

Masters GJ, Brown VK, Clarke IP, Whittaker JB, Hollier JA (1998) Direct and indirect effects of climate change on insect herbivores: Auchenorrhyncha (Homoptera). Ecol Entomol 23:45–52

Menendez R, Gonzalez-Meglias A, Lewis OT, Shaw MR, Thomas CD (2008) Escape from natural enemies during climate-driven range expansion: a case study. Ecol Entomol 33:413–421

Menzel A, Sparks TH, Estrella N, Koch E, Aasa A, Ahas R, Alm-Kübler K, Bissolli P, Braslavská OG, Briede A, Chmielewski FM, Crepinsek Z, Curnel Y, Dahl Å, Defila C, Donnelly A, Filella Y, Jatczak K, Måge F, Mestre A, Nordli Ø, Peñuelas J, Pirinen P, Remišová V, Scheifinger H, Striz M, Susnik A, Van Vliet AJH, Wielgolaski F-E, Zach S, Zust A (2006) European phenological response to climate change matches the warming pattern. Glob Change Biol 12:1969–1976

Neumann FG, Minko G (1981) The *Sirex* woodwasp in Australian radiata pine plantations. Aust For 44:46–63

Nielsen DG (1989) Exploiting natural resistance as a management tactic for landscape plants. Fla Entomol 72:413–418

Parmesan C (2006) Ecological and evolutionary responses to recent climate change. Annu Rev Ecol Evol Syst 37:637–669

Parry D, Spence JR, Volney WJA (1997) The response of natural enemies to experimentally increased populations of forest tent caterpillar. Ecol Entomol 22:97–108

Parry D, Spence JR, Volney WJA (1998) Bud break phenology and natural enemies mediate survival of early instar forest tent caterpillar (Lepidoptera: Lasiocampidae). Environ Entomol 27:1368–1374

Parry D, Goyer RA, Lenhard GJ (2001) Macrogeographic clines in fecundity, reproductive allocation, and offspring size of the forest tent caterpillar *Malacosoma disstria*. Ecol Entomol
26:281–291

Payette S, Bhiry N, Delwaide A, Simard M (2000) Origin of the lichen woodland at its southern
range limit in eastern Canada: the catastrophic impact of insect defoliators and fire on the
spruce-moss forest. Can J For Res 30:288–305

Peltonen M, Liebhold AM, Bjørnstad ON, Williams DW (2002) Spatial synchrony in forest insect
outbreaks: roles of regional stochasticity and dispersal. Ecology 83:3120–3129

Pimentel D (ed) (2002) Biological invasions. Economic and environmental costs of alien plant,
animal, and microbe species. CRC Press, Boca Raton

Pimentel D, Andow D, Dyson-Hudson R, Gallahan D, Jacobson S, Irish M, Kroop S, Moss A,
Schreiner I, Shepard M, Thompson T, Vinzant B (1980) Environmental and social costs of
pesticides: a preliminary assessment. Oikos 34:126–140

Pimentel D, Zuniga R, Morrison D (2005) Update on the environmental and economic costs associated with alien invasive species in the United States. Ecol Econ 52:273–288

Poland TM, McCullough DG (2006) Emerald ash borer: invasion of the urban forest and the threat
to North America's ash resource. J For 104:118–124

Pöyry J, Leinonen R, Söderman G, Nieminen M, Heikkinen RK (2011) Climate-induced increase
of moth multivoltinism in boreal regions. Glob Ecol Biogeogr 20:289–298

Raimondo S, Liebhold AM, Strazanac J, Butler L (2004) Population synchrony within and among
Lepidoptera species in relation to weather, phylogeny, and larval phenology. Ecol Entomol
29:96–105

Régnière J, Nealis V, Porter K (2009) Climate suitability and management of the gypsy moth
invasion into Canada. Biol Invasions 11:135–148

Rehfeldt GE, Crookston NL, Warwell MV, Evans JS (2006) Empirical analyses of plant-climate
relationships for the western United States. Int J Plant Sci 167:1123–1150

Reichard SH, White P (2001) Horticulture as a pathway of invasive plant introductions in the
United States. Bioscience 51:103–113

Rejmánek M, Richardson DM (1996) What attributes make some plant species more invasive?
Ecology 77:1655–1661

Robertson C, Nelson TA, Jelinski DE, Wulder MA, Boots B (2009) Spatial-temporal analysis
of species range expansion: the case of the mountain pine beetle, *Dendroctonus ponderosae*.
J Biogeogr 36:1446–1458

Robinet C, Roques A (2010) Direct impacts of recent climate warming on insect populations.
Integr Zool 5:132–142

Robinet C, Baier P, Pennerstorfer J, Schopf A, Roques A (2007) Modelling the effects of climate
change on the potential feeding activity of *Thaumetopoea pityocampa* (Den. & Schiff.) (Lep.,
Notodontidae) in France. Glob Ecol Biogeogr 16:460–471

Roland J, Embree DG (1995) Biological control of winter moth. Annu Rev Entomol 40:475–492

Root TL, Price JT, Hall KR, Schneider SH, Rosenzweig C, Pounds JA (2003) Fingerprints of
global warming on animals and plants. Nature 421:57–60

Rosenzweig C, Iglesius A, Yang XB, Epstein PR, Chivian E (2001) Climate change and extreme
weather events: implications for food production, plant diseases, and pests. Glob Chang Hum
Health 2:90–104

Rouault G, Candau J-N, Lieutier F, Nageleisen L-M, Martin J-C, Warzée N (2006) Effects of
drought and heat on forest insect populations in relation to the 2003 drought in Western Europe.
Ann For Sci 63:613–624

Roush RT, McKenzie JA (1987) Ecological genetics of insecticide and acaricide resistance. Annu
Rev Entomol 32:361–380

Roy DB, Sparks T (2000) Phenology of British butterflies and climate change. Glob Chang Biol
6:407–416

Royama T (1984) Population dynamics of the spruce budworm, *Choristoneura fumiferana*. Ecol
Monogr 54:429–492

Safranyik L, Shrimpton DM, Whitney HS (1975) An interpretation of the interaction between lodgepole pine, the mountain pine beetle and its associated blue stain fungi in Western Canada. In: Baumgartner DM (ed) Management of lodgepole pine ecosystems symposium proceedings. Washington State University Cooperative Extension Service, Pullman, pp 406–428

Safranyik L, Carroll AL, Régnière J, Langor DW, Riel WG, Shore TL, Peter B, Cooke BJ, Nealis VG, Taylor SW (2010) Potential for range expansion of mountain pine beetle into the boreal forest of North America. Can Entomol 142:415–442

Sambaraju K, Carroll AL, Zhu J, Stahl K, Moore RD, Aukema BH (2012) Climate change could alter the distribution of mountain pine beetle outbreaks in western Canada. Ecography 35:211–223

Sherriff RL, Berg EE, Miller AE (2011) Climate variability and spruce beetle (*Dendroctonus rufipennis*) outbreaks in south-central and southwest Alaska. Ecology 92:1459–1470

Shigesada N, Kawasaki K (1997) Biological invasions: theory and practice. Oxford University Press, New York

Siegert NW, McCullough DG, Venette RC, Hajek AE, Andresen JA (2009) Assessing the climatic potential for epizootics of the gypsy moth fungal pathogen *Entomophaga maimaiga* in the North Central United States. Can J For Res 39:1958–1970

Simberloff D, Gibbons L (2004) Now you see them, now you don't! Population crashes of established introduced species. Biol Invasions 6:161–172

Simberloff D, Rejmánek M (eds) (2011) Encyclopedia of biological invasions. University of California Press, Berkeley

Simberloff D, Stiling P (1996) How risky is biological control? Ecology 77:1965–1974

Simon RB, Thomas CD, Bale JS (2002) The influence of thermal ecology on the distribution of three nymphalid butterflies. J Appl Ecol 39:43–55

Singer MC, Parmesan C (2010) Phenological asynchrony between herbivorous insects and their hosts: signal of climate change or pre-existing adaptive strategy? Philos Trans R Soc Lond B Biol Sci 365:3161–3176

Slippers B, de Groot P, Wingfield MJ (eds) (2012a) The *Sirex* woodwasp and its fungal symbiont: research and management of a worldwide invasive pest. Springer, Dordrecht

Slippers B, Hurley BP, Mlonyeni XO, de Groot P, Wingfield MJ (2012b) Factors affecting the efficacy of *Deladenus siricidicola* in biological control systems. In: Slippers B, de Groot P, Wingfield MJ (eds) The *Sirex* woodwasp and its fungal symbiont: research and management of a worldwide invasive pest. Springer, Dordrecht, pp 119–133

Smith MT, Turgeon JT, De Groot P, Gasman B (2009) Asian longhorned beetle *Anoplophora glabripennis* (Motschulsky): lessons learned and opportunities to improve the process of eradication and management. Am Entomol 55:21–25

Stange EE, Ayres MP, Bess JA (2011) Concordant population dynamics of Lepidoptera herbivores in a forest ecosystem. Ecography 34:772–779

Stefanescu C, Penuelas J, Filella I (2003) Effects of climatic change on the phenology of butterflies in the northwest Mediterranean Basin. Glob Chang Biol 9:1494–1506

Steinbauer MJ, Kriticos DJ, Lukacs Z, Clarke AR (2004) Modeling a forest lepidopteran: phenological plasticity determines voltinism, which influences population dynamics. For Ecol Manag 198:117–131

Stenseth NC, Mysterud A (2002) Climate, changing phenology, and other life history and traits: nonlinearity and match- mismatch to the environment. Proc Natl Acad Sci U S A 99:13379–13381

Stephens PA, Sutherland WJ, Freckleton RP (1999) What is the Allee effect? Oikos 87:185–190

Stiling P, Cornelissen T (2007) How does elevated carbon dioxide (CO_2) affect plant-herbivore interactions? A field experiment and meta-analysis of CO_2 mediated changes on plant chemistry and herbivore performance. Glob Chang Biol 13:1823–1842

Stireman JO, Dyer LA, Janzen DH, Singer MS, Lill JT, Marquis RJ, Ricklefs RE, Gentry GL, Hallwachs W, Coley PD, Barone JA, Greeney HF, Connahs H, Barbosa P, Morais HC, Diniz IR (2005) Climatic unpredictability and parasitism of caterpillars: implications of global warming. Proc Natl Acad Sci U S A 102:17384–17387

Stoyenoff JL, Witter JA, Montgomery ME, Chilcote CA (1994) Effects of host switching on gypsy moth (*Lymantria dispar* (L.)) under field conditions. Oecologia 97:143–157

Strong DR, Pemberton RW (2000) Biological control of invading species—risk and reform. Science 288:1969–1970

Sun JH, Miao ZW, Zhang Z, Zhang ZN, Gillette NE (2004) Red turpentine beetle, *Dendroctonus valens* LeConte (Coleoptera: Scolytidae), response to host semiochemicals in China. Environ Entomol 33:206–212

Tauber MJ, Tauber CA (1976) Insect seasonality: diapause maintenance, termination, and postdiapause development. Annu Rev Entomol 21:81–107

Taylor CM, Hastings A (2005) Allee effects in biological invasions. Ecol Lett 8:895–908

Taylor CM, Davis HG, Civille JC, Grevstad FS, Hastings A (2004) Consequences of an Allee effect on the invasion of a Pacific estuary by *Spartina alterniflora*. Ecology 85:3254–3266

Thuiller W, Richardson DM, Pyšek P, Midgley GF, Hughes GO, Rouget M (2005) Niche-based modelling as a tool for predicting the risk of alien plant invasions at a global scale. Glob Chang Biol 11:2234–2250

Tobin PC, Nagarkatti S, Loeb G, Saunders MC (2008) Historical and projected interactions between climate change and voltinism in a multivoltine insect species. Glob Chang Biol 14:951–957

Tobin PC, Berec L, Liebhold AM (2011) Exploiting Allee effects for managing biological invasions. Ecol Lett 14:615–624

Tobin PC, Bai BB, Eggen DA, Leonard DS (2012) The ecology, geopolitics, and economics of managing *Lymantria dispar* (L.) in the United States. Int J Pest Manag 58:195–210

Torchin ME, Lafferty KD, Dobson AP, McKenzie VJ, Kuris AM (2003) Introduced species and their missing parasites. Nature 421:628–630

van Asch M, Visser ME (2007) Phenology of forest caterpillars and their host trees: the importance of synchrony. Annu Rev Entomol 52:37–55

van Asch M, Julkunen-Tiito R, Visser ME (2010) Maternal effects in an insect herbivore as a mechanism to adapt to host plant phenology. Funct Ecol 24:1103–1109

Vanhanen H, Veleli TO, Paivinen S, Kellomaki S, Niemela P (2007) Climate change and range shifts in two insect defoliators: Gypsy moth and nun moth – a model study. Silva Fenn 41:621–638

Varley GC, Gradwell GR, Hassell MP (1973) Insect population ecology. An analytical approach. Blackwell Scientific, Oxford, UK

Veteli TO, Lahtinen A, Repo T, Niemela P, Varama M (2005) Geographic variation in winter freezing susceptibility in the eggs of the European pine sawfly (*Neodiprion sertifer*). Agr Forest Entomol 7:115–120

Visser ME, Holleman LJM (2001) Warmer springs disrupt the synchrony of oak and winter moth phenology. Proc R Soc Biol Sci Ser B 268:289–294

Visser ME, van Noordwijk AJ, Tinbergen JM, Lessells CM (1998) Warmer springs lead to mistimed reproduction in great tits (*Parus major*). Proc R Soc Biol Sci Ser B 265:1867–1870

Visser ME, Holleman LJM, Gienapp P (2006) Shifts in caterpillar biomass phenology due to climate change and its impact on the breeding biology of an insectivorous bird. Oecologia 147:164–172

Visser ME, Caro SP, van Oers K, Schaper SV, Helm B (2010) Phenology, seasonal timing and circannual rhythms: towards a unified framework. Philos Trans R Soc Lond B Biol Sci 365:3113–3127

Volney WJA, Fleming RA (2000) Climate change and impacts of boreal forest insects. Agric Ecosyst Environ 82:283–294

Walther G-R, Roques A, Hulme PE, Sykes MT, Pyšek P, Kühn I, Zobel M, Bacher S, Botta-Dukát Z, Bugmann H, Czúcz B, Dauber J, Hickler T, Jarošík V, Kenis M, Klotz S, Minchin D, Moora M, Nentwig W, Ott J, Panov VE, Reineking B, Robinet C, Semenchenko V, Solarz W, Thuiller W, Vilà M, Vohland K, Settele J (2009) Alien species in a warmer world: risks and opportunities. Trends Ecol Evol 24:686–693

Waring KM, Reboletti DM, Mork LA, Huang CH, Hofstetter RW, Garcia AM, Fule PZ, Davis TS (2009) Modeling the impacts of two bark beetle species under a warming climate in the Southwestern USA: ecological and economic consequences. Environ Manag 44:824–835

Werner RA, Holsten EH, Matsuoka SM, Burnside RE (2006) Spruce beetles and forest ecosystems in south-central Alaska: a review of 30 years of research. For Ecol Manag 227:195–206

Williams DW, Liebhold AM (1995a) Herbivorous insects and global change: potential changes in the spatial distribution of forest defoliator outbreaks. J Biogeogr 22:665–671

Williams DW, Liebhold AM (1995b) Influence of weather on the synchrony of gypsy moth (Lepidoptera: Lymantriidae) outbreaks in New England. Environ Entomol 24:987–995

Wingfield MJ, Bernard Slippers B, Wingfield BD (2010) Novel associations between pathogens, insects and tree species threaten world forests. N Z J For Sci 40:S95–S103

Wolda K (1988) Insect seasonality: why? Annu Rev Ecol Syst 19:1–18

Work TT, McCullough DG (2000) Lepidopteran communities in two forest ecosystems during the first gypsy moth outbreaks in northern Michigan. Environ Entomol 29:884–900

Worner SP, Gevrey M (2006) Modelling global insect pest species assemblages to determine risk of invasion. J Appl Ecol 43:858–867

Yamamura K, Kiritani K (1998) A simple method to estimate the potential increase in the number of generations under global warming in temperate zones. Appl Entomol Zool 33:289–298

Yan Z, Sun J, Don O, Zhang Z (2005) The red turpentine beetle, *Dendroctonus valens* LeConte (Scolytidae): an exotic invasive pest of pine in China. Biodivers Conserv 14:1735–1760

Yukawa J, Akimoto K (2006) Influence of synchronization between adult emergence and host plant phenology on the population density of *Pseudasphondylia neolitseae* (Diptera: Cecidomyiidae) inducing leaf galls on *Neolitsea sericea* (Lauraceae). Popul Ecol 48:13–21

Zalucki MP, van Klinken RD (2006) Predicting population dynamics of weed biological control agents: science or gazing into crystal balls? Aust J Entomol 45:331–344

Zalucki MP, Clarke AR, Malcolm SB (2002) Ecology and behavior of first instar larval Lepidoptera. Annu Rev Entomol 47:361–393

Part IV
The Economics of Forestry

How Cost-Effective Is Forestry for Climate Change Mitigation?

Gregory Valatin and Colin Price

Abstract Cost-effectiveness analysis is important in focusing policies on minimising the costs of meeting climate change mitigation targets and other policy goals. This chapter provides a review of previous cost-effectiveness estimates of forestry options and underlying approaches, focusing especially upon UK studies, and setting the estimates in the context of those for other mitigation measures. Methodological issues such as discounting affecting estimates are discussed and existing evidence gaps highlighted.

For the UK, research gaps include evidence on impacts of afforestation on forest soil carbon balance, on comprehensive GHG balances for forest stands, on carbon stock changes during early tree growth and once stands reach maturity, and carbon substitution (or displacement) benefits. Better evidence is also needed on opportunity costs and on leakage effects.

Existing evidence indicates that forestry options are generally cost-effective compared with a range of alternatives. Whether this conclusion holds in particular cases will vary between projects and regions, as well as being dependent upon the approach adopted. To the extent that cost-effectiveness estimates depend upon the methodology adopted and benchmark used, future comparisons could benefit from greater methodological transparency and consistency.

Not only may forestry options be relatively cost-effective but, given the challenging task of reaching current targets, they are likely to be critical if existing international objectives on climate change mitigation are to be met.

G. Valatin (✉)
Centre for Ecosystems, Society and Biosecurity, Forest Research, Northern Research Station, FR-NRS & Bangor University, Roslin EH25 9SY, Midlothian, Scotland
e-mail: gregory.valatin@forestry.gsi.gov.uk

C. Price
90 Farrar Road, LL57 2DU Bangor, Wales, UK
e-mail: c.price@bangor.ac.uk

T. Fenning (ed.), *Challenges and Opportunities for the World's Forests in the 21st Century*, Forestry Sciences 81, DOI 10.1007/978-94-007-7076-8_13, © Crown Employees UK 2014

1 Background

Cost-effectiveness analysis is important in focusing policies on minimising the costs of meeting climate change mitigation targets and other policy goals. However, underpinning assumptions and approaches to estimating the cost-effectiveness of forestry measures vary. This chapter provides a review of previous cost-effectiveness estimates of forestry options and underlying approaches.

Before focusing on cost-effectiveness issues, the remainder of Sect. 1 provides a summary of background information on global carbon balances and climate change, and on the global and UK potential for forests to contribute to climate change mitigation. Section 2 discusses a range of methodological issues that affect cost-effectiveness estimates. Section 3 discusses the range of cost-effectiveness estimates made to date, focusing primarily upon those for UK forestry, and sets these estimates in the context of carbon prices derived in other ways. The final section offers some tentative conclusions and highlights existing research gaps.

1.1 Global Carbon Balances and Climate Change

Evidence from ice core data indicates that the current concentration of atmospheric carbon dioxide (CO_2) is unprecedented in the past 800,000 years (Lüthi et al. 2008), with data from boron-isotope ratios in ancient planktonic shells suggesting that it is likely to be at its highest level for about 23 million years (Pearson and Palmer 2000; IPCC 2001, Fig. 3.2e, p. 201). Anthropogenic carbon emissions rose by 70 % between 1970 and 2004, from 29 to 49 thousand million tonnes of carbon dioxide equivalent ($GtCO_2e$) per year (IPCC 2007a), with global emissions rising by 3 % a year since 2000 (Peters et al. 2013). The current atmospheric concentration of CO_2 of over 390 parts per million (ppm) (Arvizu et al. 2011), which is around two-fifths higher than the pre-industrial level of about 280 ppm, is currently rising at an annual rate around 2 ppm (IPCC 2007a; GCP 2012; CO_2Now 2013).

As atmospheric CO_2 concentrations have increased over the past 150 years, the mean global temperature has risen. In the absence of new policy action, annual world greenhouse gas (GHG) emissions could rise by a further 70 % by 2050, and lead to a rise of 4 °C, or possibly 6 °C, above the pre-industrial global mean temperatures by the end of the century (OECD 2009), with greater temperature rises likely in some regions, including the Arctic (IPCC 2007a, Fig. 3.2, p. 46). Likely adverse impacts associated with exceeding a 1.5 to 2.5 °C temperature increase include increased risk of extinction of around 20 % to 30 % of plant and animal species, with many millions more people expected to be at risk of floods due to sea level rise by the 2080s (IPCC 2007a). Warming could lead to positive feedbacks that magnify temperature changes. These could include potential dieback of Amazon rainforest if warming exceeds 3 °C (see Lenton et al. (2008) and discussion in Dresner et al. (2007)). Thawing of the permafrost and subsequent soil decomposition could lead to the further release of up to 380 $GtCO_2e$ under a high warming

(7.5 °C increase) scenario by the end of the century (Schuur et al. 2011). Recent evidence shows that warming of the Arctic is occurring faster than had been predicted, with sea level rising more rapidly than expected (Le Page 2012).

In order to prevent 'dangerous climate change', international agreements reached at Cancun (UNFCCC 2011, paragraph 4) and under the Copenhagen Accord (UNFCCC 2010, paragraphs 2 and 12) call for limiting the average global temperature rise to no more than 2 °C above pre-industrial levels, with consideration of adopting a limit of 1.5 °C. To be confident of limiting the mean global temperature rise to between 2 °C and 2.4 °C is thought to require stabilisation of atmospheric GHG concentrations in the 445 to 490 ppm range, with reductions in annual global carbon emissions occurring no later than 2015, and emissions 50 to 85 % below 2000 levels by 2050 (Arvizu et al. 2011). However, some scientists have argued that even the existing GHG atmospheric concentration, which, including the effect of other GHGs, is equivalent to around 430 ppm CO_2e (Trumper et al. 2009), is too high for the temperature rise to stay below the 2 °C threshold. Ramanathan and Feng (2008), for example, argue that the increase in atmospheric GHGs since pre-industrial times to date probably commits the world to a warming of 2.4 °C (1.4 to 4.3 °C) above the pre-industrial level during the current century – although some underpinning assumptions have been argued to be over-pessimistic (e.g. Schellnhuber 2008). Hansen et al. (2008) also recommend a rapid reduction from the current concentration by around 10 % to no higher than 350 ppm of CO_2. The difficulties of achieving such a target are discussed in the final section.

1.2 Global Potential of Forestry for Climate Change Mitigation

Historically, forests in pre-agricultural times are thought to have covered around 5,700 million hectares globally, and to have stored around 1,200 thousand million tonnes of carbon (GtC) in total, including 500 GtC in living biomass and 700 GtC in soil organic matter (Mahli et al. 2002). Forests currently cover about 4,000 million hectares and, excluding woodlands under 0.5 ha, or primarily within agricultural or urban land uses, are estimated to store around 650 GtC, including around 290 GtC both in forest biomass and in soils, and 70 GtC in deadwood and litter (FAO 2010, Table 2.21). While comparisons are sensitive to definitional issues such as the depth of soil carbon covered, the latter estimates imply that the amount currently stored is of a similar order of magnitude to the total amount of carbon now in the earth's atmosphere: this is currently around 800 GtC (Lorenz and Lal 2010; Riebeek 2011).

Forestry has a potentially very significant contribution to make globally and might contribute two-thirds of the total climate change mitigation potential of land management activities (Mahli et al. 2002). There are two principal ways in which it can contribute.

Firstly, deforestation is a major source of GHG emissions. This is the reason, for instance, that forestry was the third largest source of global emissions in 2004, accounting for around 17 % of the total in that year (IPCC 2007a, Fig. 2.1, p. 36). It is also the reason that it has contributed an estimated 45 % of the total increase in atmospheric CO_2 since 1850 (Mahli et al. 2002). In the absence of mitigation efforts, deforestation could result in an increase of 30 ppm in atmospheric CO_2 by 2100, making stabilisation of atmospheric GHG concentrations at a level that avoids the worst effects of climate change highly unlikely (Eliasch Review 2008). Reducing emissions from deforestation and forest degradation (REDD) is therefore a very important climate change mitigation activity if the international community's current climate stabilisation aspirations are to be met, especially in countries where the level of annual deforestation is high.

Secondly, afforestation and reforestation activities can make significant contributions to sequestering atmospheric carbon, as well as providing a renewable source of energy and materials to substitute for use of fossil fuels and more fossil-carbon-intensive materials. By itself, carbon sequestration by forests is best viewed as a component of mitigation strategies – however, it is far from sufficient to sequester total emissions from burning fossil fuels. Under business-as-usual scenarios global emissions from burning fossil fuels may be of the order of 1,800 to 2,100 GtC over the twenty-first century, exceeding the maximum potential human-induced forest carbon sink by a factor of 5 to 10 (Mahli et al. 2002).

1.3   UK Potential of Forestry for Climate Change Mitigation

In total, UK forests are estimated to store around 162 million tonnes of carbon (MtC) in tree biomass, with a further 46 MtC estimated to be stored in forest litter and the top organic (F) layer of forest soils. Including soil carbon to a depth of 1 m, UK forests are estimated to store a total of 878 MtC (Morison et al. 2012, Table 2.1).

Britain cannot become carbon-neutral through domestic woodland creation alone (Broadmeadow and Matthews 2003). Nonetheless, considerable scope exists for increasing forest cover from the current low base to raise the contribution of British forests to climate change mitigation.

The 2010 UK Greenhouse Gas Inventory (Brown et al. 2012, Tables ES2.1 and ES2.2, pp. 10–11) indicates total net UK emissions of 590 million tonnes of CO_2 equivalent ($MtCO_2e$) in 2010, including the effect of carbon sequestration due to afforestation since 1990 and management of existing forests, which *removed* an estimated 3.6 $MtCO_2e$. Including areas afforested during the period 1921–1990, the contribution of UK woodlands to climate change mitigation is much greater than the level counted towards meeting the UK target under the Kyoto Protocol. (Changes associated with planting forests up to 1990 are not accounted for under the Protocol as they are treated as part of the baseline.) Estimates from the same model used for the UK Greenhouse Gas Inventory show total net carbon sequestration by UK woodlands rising from 2.4 $MtCO_2$ in 1945 to a peak of 16.3 $MtCO_2$ in 2004, before falling to 12.9 $MtCO_2$ in 2009 (Valatin and Starling 2011). Uncertainty remains over the precise magnitude of the UK forest sink, however, with estimates of current net uptake (after taking account of removals of around 6.5 $MtCO_2$ due to harvesting) ranging between 9 and 15 $MtCO_2$ (Morison et al. 2012).

There are currently 3 million hectares of woodland in the UK, accounting for around 12 % of the UK's total land area, a proportion far below the average for the EU as a whole of 37 % (FAO 2010). The impact of expanding UK woodland cover by about a third to 16 % by increasing woodland creation to 23,000 ha per year (an extra 14,840 ha per year above the current level) was considered by Read et al. (2009). Based upon a scenario which involves creating a mix of high-yielding short rotation forestry, broadleaf and conventionally managed coniferous woodlands, and underpinning assumptions (e.g. yield classes), this was estimated to increase the net carbon sequestration by UK forests planted since 1990 to over 10 $MtCO_2e$ (Read et al. 2009). Including carbon substitution benefits, total abatement was estimated to rise to 15 $MtCO_2e$ by the mid 2050s. This is equivalent to about one tenth of the total UK GHG emissions at that time if current emissions reduction commitments are achieved (Read et al. 2009), although approaches to ensuring that associated carbon sequestration benefits are maintained in perpetuity were not discussed.

2 Techniques and Issues in Cost-Effectiveness Assessment

Cost-effectiveness is an economic efficiency measure, in general terms evaluating the cost *per unit of achievement* of the desired objective. It is often employed when a target level of achievement is *given*, and what is sought is the best way of achieving it. It may be contrasted with optimality measures, which seek to achieve the *most desirable* level of an objective, given some trade-off between achieving it and the cost of doing so. In the context of climate change, optimality may be considered to entail the best balance between on the one hand the benefits that accompany continuing generation of GHGs (i.e. normal economic activity), and on the other the costs of the consequent climate change. However, where climate change impacts are highly uncertain, optimality may be indeterminate. (See also discussion of intergenerational and other ethical issues in Sect. 2.8 below.) Cost-effectiveness analysis is most appropriate when a target (e.g. limiting global average temperature rise to no more than 2 °C above pre-industrial levels) has been agreed and where the issue addressed is how to ensure the target is met at least total cost.

The appropriate cost-effectiveness measure for CO_2 mitigation is in essence very simple. The extra cost (financial or social according to context) of deploying the CO_2 mitigation measure – compared with the cost of *not* deploying it, is divided by the extra reduction in the atmospheric carbon level achieved by deploying the measure – compared with the level of carbon reduction ensuing if the measure is *not* deployed. In general terms, this is expressed as:

$$\frac{[\text{Net cost of the measure}] - [\text{Net cost of 'do-nothing' measure}]}{[\text{Carbon reduction of the measure}] - [\text{Carbon reduction of 'do-nothing'}]}$$

For non-forestry options there may be a one-off cost to achieve a single pulse of mitigation. This may become more complicated if there is a cost of maintaining sequestration, as for example in geo-engineering or carbon capture and storage options where there is a requirement for on-going monitoring and maintenance of the system. Carbon-fixing agricultural practices may also produce a one-hit outcome contemporary with cost, as when a less fuel-intensive practice is used, or when an annual crop is harvested as bioenergy.

In assessing forestry, problems include the time profiles not only of costs, but also those of carbon fluxes. For afforestation options there is year-on-year sequestration whose rate may not reach a maximum for many decades, especially in temperate or boreal conditions. For commercial regimes, there is also a discrete series of removals from the crop. Similarly, for such options as adopting reduced-impact logging in place of conventional logging, there is an immediate differential in the carbon removed from the forest, then a long period over which carbon biomass is re-established, but not necessarily at the same rate or to the same carbon stock with the two logging systems (Healey et al. 2000). To resolve these profile problems it is necessary to adopt some means of integrating fluxes over time, or defining a mean level of added sequestration. The two main approaches focus on respectively the fluxes and the stock of carbon sequestered. The flux approach may best be envisaged in terms of a price for the service of actively locking up carbon. It may be given by calculating the price for a unit of carbon's being locked up (i.e. for a carbon reduction) which would just suffice for the option to break even, including also as a debit any subsequent revolatilisation of the carbon by burning or decay. In cases where the discount rate is assumed constant, the derivation of that price is as follows; the issue of discounting cash flows and carbon fluxes is treated later.

For investment in forest carbon fixing to break even,

$$
\begin{aligned}
&\left[\{\text{difference of}\}\,\text{summed discounted cost}\right] = \\
&\left[\{\text{difference of}\}\,\text{summed discounted carbon credits}\right] = \\
&\sum_{t=0}^{t=T}[\text{carbon price}]\times[\{\text{difference of}\}\text{flux}]_t \div (1+[\text{discount rate}])^t = \\
&[\text{carbon price}]\times\sum_{t=0}^{t=T}[\{\text{difference of}\}\ \text{flux}]_t \div (1+[\text{discount rate}])^t \\
&\qquad\qquad\text{whence}\ [\text{carbon price}] = \\
&\frac{\left[\{\text{difference of}\}\ \text{summed discounted cost}\right]}{\{\text{difference of}\}\sum_{t=0}^{t=T}[\text{flux}]_t \div (1+[\text{discount rate}])^t}
\end{aligned}
$$

Where the discount rate changes through time, as under current UK Treasury advice, the term '$\div (1+[\text{discount rate}])^t$' is replaced by a discount factor compounded from the relevant discount rates for the relevant periods: e.g. for 50 years the discount factor is '$\div \{(1+3.5\ \%)^{30}\times(1+3\ \%)^{20}\}$'.

Such a price may properly be compared with mitigation costs calculated for non-forestry options if the carbon price is assumed constant over time. Where carbon prices are anticipated to change over time, comparisons based upon the above approach have either:

- To be confined to options that have the same profile of fluxes over the same time horizon; or
- To include some means of weighting fluxes according to relative price, most readily achieved by using a price-adjusted carbon discount rate.

By contrast, the stock approach considers the forest to be 'renting out' the service of maintaining carbon in a sequestered state for one time period, normally a year. The break-even cost of doing so is the annual cost of retaining an equilibrium condition in the forest stock. This equilibrium may represent a fully-developed [semi-]natural forest in which the composition of tree sizes remains the same from year to year; or the mean over a rotation period of the carbon stock in a growing forest; or, what is equivalent, the mean carbon stock averaged over all the age-classes in a normal forest.

To compare this rental value with the mitigation costs of non-forestry options it is necessary to render the latter as an annual cost of *maintaining* each tonne in a sequestered state. Or, where a constant discount rate is applied, the total discounted costs of *achieving* permanent sequestration of a tonne may be converted to an annuity, by multiplying them by whatever discount rate is deemed appropriate. Such an approach would be especially suitable in the context of steady-state economies – a popular concept in some circles.

Since they refer to different durations of carbon lock-up, the flux and stock approaches will not normally give similar prices per tonne. Where the time profiles of cost and of sequestration differ between options, they may not even give the same ranking of cost-effectiveness within a given set of options. More fundamentally, divergence can also arise where carbon storage in existing woodlands is treated as a benefit under the stock approach, but not under the flux one. Existing carbon storage is treated equally as a benefit under both approaches in the case of REDD projects.

2.1 Issues: Units

Much confusion has been caused in the past by failure to specify or recognise the unit of achievement. Measures commonly used include 'cost per tonne of carbon', 'cost per tonne of CO_2' (removed from or emitted to the atmosphere), and 'cost per tonne of CO_2 equivalent' (the last measure being used in making comparison with other greenhouse gases or with other mitigation measures). For example Ayers and Walter (1991) seem to have been early users of the $/tCO_2e$ measure, without making this clear, and found themselves in phantom conflict with other authors, who were using $/tC$. According to context, one or other of these may be considered the most appropriate, but it is of the first importance to ensure that figures drawn together from different sources all have the same basis, or are converted to so being. For example, because carbon constitutes about 12/44 of the mass of CO_2, cost given as 'per tonne CO_2' can be converted to 'per tonne carbon' by multiplying by 44/12. Since carbon is incorporated in different molecular structures during its transactions between earth, vegetation, atmosphere and ocean, there is something to be said for the 'per tonne carbon' measure.

Where there are other significant GHG fluxes, these can be converted to a 'per tonne of carbon dioxide equivalent' (tCO_2e) based upon their global warming potential compared to the 'radiative forcing' (measurable in terms of the increase in equilibrium temperature caused) of emitting a tonne of carbon dioxide. This entails complex modelling of, among other things, 'natural' uptake of atmospheric CO_2 into oceans and terrestrial ecosystems, particularly boreal forests, as well as parallel modelling for the options considered as alternatives. Global warming potential is defined as an index, usually computed as the cumulative radiative forcing over an arbitrary 100 years, compared to emitting a unit of carbon dioxide. Over this time-frame, the other GHGs have higher (up to 23,900 times higher – in the case of sulphur hexafluoride) global warming potentials than carbon dioxide, molecule for molecule (Brown et al. 2012, p. 39). The current preference of the UK government for using tCO_2e (e.g. HM Treasury and DECC 2012) arguably reflects best the primary concern with the impact of changes in atmospheric GHG balances and with using a metric that facilitates comparisons between sectors.

For climate impacts other than through reducing GHG concentrations, the appropriate cost-effectiveness measure would also be cost per unit of reduced radiative forcing. Forestry examples include impacts of afforestation on solar radiation reflectivity (the 'albedo effect') and of increased release of water vapour from forests ('evapotranspiration') on cloud cover and associated reflection of solar radiation, as well as in reducing surface temperatures. Jarvis et al. (2009) suggest that UK afforestion in general has little effect on albedo because forests' solar reflectivity is similar to that of previous vegetation, but precise calculations have not so far been done for UK conditions. Over a typical rotation increased solar radiation absorption by conifers compared with grassland may reduce the climate change mitigation benefits from carbon sequestration and substitution by 20 to 35 %, depending on conditions (James Morison, personal communication). Jarvis et al. also argue that any impact on cloud cover is likely to be small as existing UK weather patterns are generally determined at much larger scales, over the Atlantic Ocean, Europe and Russia.

2.2 Elements in Forest Sequestration

Carbon is locked up in the chemical components of trees, litter, soil, and wood products (especially those that are durable). Physical and economic valuations have differed in their focus on these, some concentrating on the trees themselves (Price and Willis 1993); others on soil (Cannell et al. 1993); others on wood products (Price and Willis 2011); while yet others have tried to incorporate all elements (Brainard et al. 2009).

The rate of accumulation of carbon in trees has long been a subject of physical study, modelling and economic analysis (Price 1990; Dewar and Cannell 1992; Olschewski and Benitez 2010). With appropriate conversion factors, tabular or parameterised yield models can give the required data: development of models that incorporate expected impacts of climatic changes on yields being a current research frontier.

Leaf and branch litter is a significant store of sequestered carbon, and a forest soil may contain more carbon per unit ground area than even the mature forest trees. It is an important matter, therefore, whether and how the silviculture affects the soil carbon stock (Jandl et al. 2007). This remains an ongoing area of research. In the UK, for example, there is currently considered to be insufficient data to quantify with confidence changes in soil carbon associated with afforestation (Morison et al. 2012).

After harvesting, sequestered carbon may remain in forest products for periods as short as a few months (paper) or as long as millennia (structural timbers). For example, millions of tCO_2e may be locked up in the roofs and fitments of Britain's medieval churches.

Forest operations consume fossil fuel and emit carbon, with most emissions within the forest occurring during harvesting, and subsequent haulage to the primary processor representing the largest source of emissions overall. However, fossil fuel usage during forestry operations (road building, ground preparation, thinning, harvesting and timber haulage) is relatively minor compared with the level of net carbon uptake by forests. In the UK these forestry operations have recently been estimated (Morison et al 2012) to result in total annual emissions of 0.22 $MtCO_2$ (a level around 1 to 2 % of net carbon uptake). This includes total emissions from harvesting of 0.07 $MtCO_2$ (a level under 1 % of net uptake).

Since all these represent components in the overall forestry carbon and GHG balance, it should be beyond question that, where significant, all are included in evaluation of forestry's cost-effectiveness for mitigation. However, some stores such as in the biomass of trees are readily measured and predicted, while other factors such as soil processes or the product life span are less predictable. These are not just issues for scientists, but are reflected profoundly in the economic evaluation.

2.3 Additionality and Leakage

Because in the real world comparison must always be made with what would happen in the absence of the specified measure, there is a potential problem in specifying the counter-factual: in economics, it could be said, the most important question is 'what changes? what difference does it make if I do this, rather than not-doing this?' This raises questions of 'additionality' (what changes within the specified project boundary) and 'leakage' (what changes outside the project boundary).

Although approaches to additionality vary and its determination is imprecise to the extent that it is based upon comparisons of future hypothetical scenarios

(Valatin 2011a), the key issue is: what change in the GHG balance, over and above what would otherwise have existed, is the consequence of a particular mitigation activity? Richards and Stokes (2004) identified additionality and leakage as particularly important problems for forestry sequestration studies.

Within forestry, the 'do-nothing' option may not be carbon neutral. Land abandoned for agriculture may, in time, accommodate a natural succession of vegetation whose end-point is a mature forest, perhaps one capable of storing more carbon than a human-made forest on a commercial rotation. Asked what would happen in the absence of their enrichment planting in a cut-over tropical forest, one agency that was drawing down funds for enhanced carbon storage said that the forest would probably regrow naturally anyway. What, then, would change as a result of enrichment planting? In terms of the final carbon storage, possibly nothing, although speeding the carbon accumulation could accelerate mitigation benefits. Contrariwise, not maintaining tree cover on steep slopes may lead to erosion and loss of soil carbon.

Less obvious is the effect of adding wood products to a world market, compared with *not* supplying them. If the UK increases its output of construction-grade timber, is the consequence that more timber is used in buildings? If so, is that through buildings' being larger (and thus needing more heating), or through displacing other, fossil-carbon-intensive materials? Or would UK timber displace imports from Scandinavia, Russia or North America? In this case, would reduction in timber exports from those regions lead to a greater accumulation of carbon in less-managed forests, or a carbon-reducing conversion to agriculture as forestry became a less economically viable activity? Such effects on 'invisible stakeholders' have not customarily entered economic analysis (Price 1988, 2007), and they are seldom mentioned in publications, but they could affect significantly both financial and carbon accounts of forestry.

Evidence from US studies is reported to imply that 'leakage', can range from 5 % to 93 % of project abatement benefits depending upon the activity and region (Murray et al. 2004; van Kooten et al. 2012). A primary concern is that conservation of domestic forests will lead to increased timber harvesting and environmental degradation in other countries (i.e. indirect land use change), with a lesser concern being that it may result in use of more energy-intensive materials (Gorte 2009). However, research on quantification of international leakage effects appears sparse, with a recent review (Henders and Ostwald 2012) relating to REDD projects identifying just two items. Both involved modelling exercises based upon complex data inputs, and are not currently used in practice for forestry projects under the voluntary carbon market standards considered.

More recognised is that afforestation involves withdrawal of land from agriculture, which has its own effects on soil carbon (Moran et al. 2008); on operational fossil fuel use; and on the carbon transactions of affected food imports (Hockley and Edwards-Jones 2009). If lost food production is replaced by intensification of agriculture on other land, the ensuing fossil fuel use and consequent CO_2 emissions would need to be considered too.

Other mitigation measures also present an adjustment from a do-nothing position that itself has consequences for carbon storage, via the adjustments of production and consumption technologies that accompany changing market conditions.

At present it is believed that no adequate examination of these complex secondary effects has been concluded. Richards and Andersson (2001), for example, noted that estimating the off-site effects of individual carbon projects is an onerous task as it requires analysing shifts in supply functions for forest products, agricultural products and agricultural land. In many countries suitable general equilibrium models (or even the requisite time-series datasets to build such models) may not currently exist, or be considered too costly to develop.

Leakage due to the potential for afforestation to result in deforestation of other areas is not an issue within the UK owing to the existing regulatory requirements for an environmental impact assessment for deforestation over 1 ha (0.5 ha in sensitive areas), for re-stocking of areas felled, and for protection of biodiversity and semi-natural habitats. The approach to leakage adopted under the Woodland Carbon Code (Forestry Commission 2011b) developed for UK forest carbon projects includes not accounting for reductions in GHG emissions associated with the cessation of the previous (e.g. agricultural) land use. This allows for the potential intensification of activities (e.g. agriculture) elsewhere in the UK. However, GHG emissions associated with any resultant more intensive use of land under the same ownership or lesseeship have to be accounted for in calculating the net carbon sequestration of a project (Forestry Commission 2012). As it currently just covers afforestation, leakage associated with forest conservation projects (the main focus of US studies) is not an issue at present for UK projects under the Code.

2.4 Treatment of Volatilisation and Repeat Projects

At the end of a commercial forest rotation, and often at intermediate times, timber is removed. The issue here is how this removal is treated within carbon accounts.

- It could be debited from the forest account, and credited to the account of the recipient.
- It could be taken as instant loss to the global fixed carbon account.
- Progressive return of the carbon to the atmosphere as CO_2 could be profiled as a generalised volatilisation, according to some specific functional form (Brainard et al. 2009).
- The volatilisation process could be disaggregated so as to occur at various rates according to product category (Thompson and Matthews 1989).
- The forest could be taken to sequester the long-term average level of carbon under the existing management regime (e.g. perpetual series of commercial rotations, or biological maturity if no harvesting is envisaged), so that this level of carbon stays permanently sequestered. (The fiction is sometimes perpetrated that, once the forest has grown to commercial rotation, that level of carbon remains permanently sequestered, while successor rotations add perpetually to the sequestration – despite the fact that this clearly cannot be the case).

Although which treatment is most appropriate may depend upon the purpose of the analysis, the UK Woodland Carbon Code (Forestry Commission 2011a) assumes that the long-term average carbon stock is maintained. In effect, once the long-run average level is attained, this results in placing an equal value on the capture and release of carbon. However, as capture tends to precede release, where a positive discount rate is used (and the carbon value is not increasing over time), the 'discounted tonnes' of capture will exceed those of release. Although the overall effect may be small, to the extent that this positive balance of 'discounted tonnes' is considered an abatement benefit, not taking it into account may tend to result in the net cost of projects being over-estimated.

Volatilisation is not so much an issue for assessments in which forests' sequestration is rented for fixed periods, as is the case under temporary storage certificates (Olschewski and Benitez 2010).

Volatilisation also becomes much less of an issue with a high discount rate, because of its occurrence late in the cycle (the same reason that tends to make forestry unprofitable with high discount rates).

With sufficient lapse of time (many centuries), nearly all the carbon sequestered by a single cycle of commercial forestry returns to the atmosphere, because ultimately all wood products (including biomass) decay or are otherwise oxidised (Price and Willis 1993). A perpetual sequence of rotations, which is the ground of sustainable forest management and the base assumption of classical forest economics, repeats the fluxes of the first, endlessly. But clearly its sequestration is not cumulative, apart from any accumulation of carbon in soil and litter layers, or unless some means is found of permanently preserving the harvested timber, or except in relation to recurrent displacement of fossil-carbon-intensive materials.

2.5 Combustion and Structural Displacement

In addition to storing carbon directly, forest products may be beneficial in displacing fossil-carbon-intensive materials such as steel or concrete. Use as biofuel also displaces combustion of fossil fuels, but necessarily involves instant and complete revolatilisation of sequestered carbon.

Inclusion of such functions may dramatically increase the profitability of forestry (Price and Willis 2011). Obversely, they may substantially reduce cost/tCO_2e, in forestry options that involve such commercial removals.

This is an area where, for the UK at least, significant scope remains for improving existing estimates of the associated abatement benefits (Morison et al. 2012).

2.6 Net Costs

To speak of 'the climate change mitigation cost' of forestry options is to assume that forestry is an unprofitable investment, or that it would not in any case be undertaken for a range of purposes. Offsetting its costs are revenues from sale of products and, in a public context, the value of providing net non-market benefits. These should be, and in some cases have been, deducted from costs in deriving a supply price for carbon sequestration services. In some circumstances a negative net cost will arise.

Mitigation cost is a concept of practical significance only in relation to additional forestry options, ones that would not be undertaken in the absence of carbon benefits. However, even where costs are negative (i.e. a project would have been expected to go ahead in the absence of carbon benefits), the mitigation cost may be of policy interest in comparing costs of different measures and developing marginal abatement cost curves.

Cost estimates may be based upon the costs to the private sector of implementing measures, or the social costs to the economy as a whole. The latter may extend to considering transaction and policy implementation costs, and ancillary costs and benefits, including life-cycle analysis of effects in related sectors.

2.7 Opportunity Costs

Agricultural opportunity costs may constitute the largest element of the cost of woodland creation measures. However, these vary widely. In some cases (e.g. where the most marginal land is used) the opportunity costs of converting farmland to woodland may be minimal, or even negative where environmental impacts associated with existing agricultural practices (Spencer et al. 2008) are accounted for, or where farming's profitability is only achieved through subsidy. (For example, a recent survey of UK agriculture (Defra et al. 2010, Table 2.5, p. 9) reports that almost a quarter of farms (22.1 %) had a net farm income below zero). This is an area where, for the UK at least, work to improve upon previous estimates is needed, including how statutory requirements for woodland cover to be subsequently retained in perpetuity affect the value of land converted from agricultural or other uses (Valatin 2012).

2.8 Discounting and Time Horizon

As will rapidly become apparent, the effect of discounting is not only of great importance in relation to climate change: it is also one of the most contested areas in natural resource economics. The authors are not necessarily in agreement over all the issues, and where this is so we have tried to make that plain.

The Case for Discounting

The general position of economists, and that of the UK Government, is that cash flows (of costs and benefits measured in current prices) should be discounted. The main justifications have been as follows:

- Financial resources can be invested to yield net revenues. Thus they have an opportunity cost in reduction of what other benefits can be generated in future, if cash flows are expended early or are received late. Alternatively, the later that costs occur, the smaller the sum that needs to be invested presently in order to provide compensation for future costs, since the period of growth of the compensation funds will be longer.

- People have an innate *time preference*, for early rather than late consumption, and a democratic government should respect that wish.
- Assumed future growth of income and consumption per capita entails diminishing marginal utility – a reduced significance of additional units of future consumption, or a lower opportunity cost of resources diverted from consumption in order to deal with environmental and social problems.
- The possibility exists that devastating events will eliminate, or radically and unpredictably alter, future returns (HM Treasury 2003), or result in human extinction (Lowe 2008).

In addition, concerns about the potential for exceeding critical tipping points combined with uncertainty about precise thresholds could be viewed as providing a reason for prioritising early abatement, either by valuing it more highly than later abatement (because of the longer period of ensuing benefit), or by discounting later abatement (Valatin 2011b).

The Case for Not Discounting Carbon Fluxes

Not discounting carbon fluxes generally implies that a tonne of carbon sequestered at the end of a 100-year rotation is as important as one sequestered immediately. Such an approach would be consistent with an intergenerational justice argument, that the costs of climate change to future generations, howsoever or whensoever caused or mitigated, ought to be treated at parity with costs to the present generation.

However, if issues of intergenerational equity forbid the discounting of carbon fluxes, why should they not also forbid discounting wood fluxes, or for that matter cash flows? If discounting *is* justified for benefits and costs generally, morally relevant differences should be shown why it should *not* be applied to carbon.

From a rights-based perspective, Spash (1994) notes that harms inflicted on future generations due to continuing GHG emissions are in no way balanced out by benefits enjoyed by the current generation in their use of fossil fuels (see also Spash (2002)). To the extent that preventing the most significant avoidable future harms is considered a moral imperative, this could be viewed as providing an ethical basis for not discounting either the causes *or* the effects of climate change.

Similarly, drawing upon Principle 1 of the UN Conference on the Human Environment 1972 Stockholm Declaration and a 2008 UN Human Rights Council resolution, Caney (2010) argues that persons have a human right to a healthy environment and that climate change poses a far-reaching threat to the enjoyment of this right by current and future generations. But note that similar arguments could apply to other actions which cause significant avoidable harm to future generations. It might be argued that in some circumstances 'avoidable harm' could include failing to provide wood and other natural resources that might be vital to future well-being.

Furthermore, inconsistencies could arise if equity/rights-based perspectives (not discounting the causes or effects of climate change), are combined with a preference for early abatement so that critical thresholds are not exceeded. This would imply the adoption of a different approach to GHG emissions (not discounting) from that for abatement (discounting), with increasing weight placed on emissions relative to abatement over time.

Discounting and Not-Discounting Under the UK Government Approach

In former times the governmental view was that 'environmental costs and benefits should be discounted just like any other costs and benefits' (Department of the Environment 1991). The Stern Report, commissioned by the UK Government and pilloried by some economists for the *low* discount rate used, did in fact discount the value of *effects* of climate change on human welfare.

However, the present UK government approach to cost-effectiveness (HM Treasury and DECC 2012) is that carbon fluxes themselves (the *cause* of climate change) are not discounted. Whether the *value* of carbon fluxes should or should not be discounted depends, according to this approach, upon the focus of the analysis. Two broad 'sector' categories are distinguished.

Emissions associated with industrial sources subject to emissions reduction targets under the EU emissions trading scheme are categorised as occurring in the 'traded' sector. These include carbon emissions from energy (combustion installations over 20 MWth, mineral oil refineries, coke ovens), ferrous metals production and processing, building materials production (cement, glass and ceramics), pulp, paper and board manufacture, and (from 2012) civil aviation. (Inclusion of flights to and from countries outside the scheme has been delayed, however, while coverage of the scheme is to be extended to petrochemicals, ammonia and aluminium industries in 2013.) Sources in this 'traded sector' are currently responsible for almost half of total EU CO_2 emissions and around 40 % of total EU GHG emissions (European Commission 2012).

Emissions and abatement from sources which are not covered by the EU ETS are categorised as occurring in the 'non-traded' sector. These include sequestration in forests and many of the carbon displacement benefits associated with use of wood products.

As separate targets (and markets with different prices) exist for the two, carbon in each is *treated* essentially as a separate commodity, although of course the physical transactions take place with a common atmospheric pool of carbon. Carbon prices in the two sectors are currently projected to converge and equalise in 2030 as a functioning global carbon market is established (HM Treasury and DECC 2012, p.13).

Discounting Abatement Benefits in Traded and Non-Traded Sectors

Although neither sectoral perspective involves discounting carbon fluxes per se, each does include the present value of any consequent fluxes in the other sector (which is computed by applying discounting).

Analyses from a forestry perspective, as part of the 'non-traded' sector, include the (discounted) present value of consequent fluxes in the 'traded' sector, such as those arising by substitution for fossil fuels in large-scale electricity generation. Contrariwise, analyses from the 'traded' sector's perspective include the present value of any forest fluxes – for example those of forests planted to yield those products.

Apart from the different treatment of carbon benefits, all project cash flows *are* discounted. Thus these would be the same, whichever sectoral perspective was taken.

Although their discounted value is taken into consideration, fluxes themselves in the other sector are not accounted for as part of the aggregate abatement associated with a project, and may thereby lose some of their significance.

Implications of the Discounting Protocol

To the extent that the time profiles of fluxes in the two sectors differ, discounting in computing the present value of fluxes in the other sector will have a quantitatively different effect. Consequently, if it were the case that the level of abatement in both sectors was the same, the calculated cost-effectiveness of carbon mitigation in the overall project would be likely to appear different, depending on which sector perspective the analysis adopts.

In general the two sectoral perspectives are not viewed as alternatives, however: the perspective that should be adopted is that of the sector in which most of the mitigation benefits arise. A non-traded sector perspective is appropriate for most UK forestry projects as most of the carbon benefits occur in this sector.

Given that separate targets exist for the 'traded' and 'non-traded' sectors, from an institutional perspective it may seem logical that the present value of fluxes in the 'other' sector should be included in this way in analysis performed within one sector. From a global perspective, and given that CO_2 fluxes interact with the same atmosphere irrespective of the institutional source or sink, it may seem strange that different approaches to valuation of carbon fluxes within the two sectors are used and different social values of carbon applied. However, as social values applied in valuing carbon in the regulated (i.e. traded) sector are related to market prices applying within the EU ETS, it does not appear entirely surprising that those applying outside this compliance market have a different basis, even if the current ratio of the values in the two sectors (around 1:10 for the central estimates in 2013–2014) is remarkable.

It could be argued that not discounting the *causes* of climate change (carbon emissions) or the causes of mitigation (carbon sequestration) shows inconsistency in application of discounting. On the other hand, this may appear unimportant to the extent that the primary purpose of the protocol is to determine whether measures are cost-effective, and to allow comparisons of the cost-effectiveness of different options, rather than focusing upon levels of abatement per se.

Declining Discount Rates

To the extent that equity/rights-based perspectives are consistent with discounting at all, they are arguably more consistent with using declining discount rates than with using the initial discount rate in perpetuity. Use of declining discount rates is currently the approach recommended for UK policy appraisal in the Treasury Green Book (HM Treasury 2003), based upon uncertainty about future values of time preference (Lowe 2008). For a discussion of the use of declining discount rates for policy, see OXERA (2002), Hepburn and Koundouri (2007), and Gerlagh and Liski (2012). For critiques of the approach, see Price (2005, 2010, 2011).

Again, however, it could be argued that if moral imperatives favour future reduction in discount rate, they would even more favour not discounting at all.

Adapting Conventional Discounting: An Alternative Approach

A more conventional economic perspective is that the argument of diminishing marginal utility applies in principle – though not to an equal extent – to all things that may be enjoyed, suffered, compensated for or mitigated by the deployment of investment funds or material resources. This includes what is required to defend against the consequences of climate change. It is important to note, however, even within this perspective, that not all environmental values experience diminishing marginal utility – at all or at the same rate. The proper approach is not to discount carbon *fluxes* differentially, but to discount the *effects* of those fluxes differentially, according to the expected and various influence of diminishing marginal utility. For example, some biodiversity values may not be susceptible to diminishing marginal utility, whereas products based on technological advance may have very rapidly diminishing marginal utility (Price 1993, Chaps. 16–18). Under some scenarios marginal utility may diminish: under others (e.g. catastrophic disruption of the world economy) it may increase, the appropriate approach then being to take a mean of outcomes, weighted by their probabilities (Price 1997).

The human extinction/catastrophe argument may be considered a valid reason for discounting future effects, and has long been discussed in the literature (Price 1973; Dasgupta and Heal 1979). However, the inclusion of this rationale for discounting risks the promotion of a self-fulfilling prophecy. Discounting for the uncertainty that surrounds climate change reduces the weight given to the future costs of climate change, and so, perversely, increases the value ascribed to the most risky strategy, business-as-usual.

As for the time preference argument, it has long been regarded by economic philosophers as arising 'merely from weakness of the imagination' (Ramsey 1928), and indeed as representing a misinterpretation of what it is that people prefer (Price 1993, Chap. 7). It has no relevance to the value of the future *to* future generations, whether that is from the causes or the effects of carbon fluxes, or from any other environmental values, or from wood, or from any other material values.

Thus diminishing marginal utility remains as the 'respectable case for discounting': the return on investment funds is dealt with in other, more appropriate, ways of giving an opportunity cost (Price 2003).

The UK Government Approach in Practice

Although on initial inspection the present UK government approach (HM Treasury and DECC 2012) might be interpreted to mean that a tonne of carbon sequestered is equally important irrespective of when it occurs, such an interpretation would be misleading for two reasons.

Firstly, some discounting may take place within forestry analyses (generally, in calculating the present value of any fluxes in the traded sector, as discussed in section "Discounting Abatement Benefits in Traded and Non-Traded Sectors").

Secondly, although the cost-effectiveness estimate is derived without discounting forestry carbon fluxes themselves, discounting is applied to the value with which this estimate is compared, as follows. To determine whether forestry is an attractive option, the cost-effectiveness estimate is compared with the social value of a $1\text{-tCO}_2\text{e}$ reduction in emissions (abatement) in terms of its contribution to meeting UK climate change mitigation targets. This comparator is taken as a weighted mean of the calculated social values of carbon at each of the times when abatement occurs, each of these being discounted according to the Treasury Green Book protocol, and is computed as:

$$\sum_{t=0}^{t=T} [\{\text{abatement}_t \div \text{lifetime abatement}\} \times \{\text{social value of carbon}_t \times \text{discount factor}_t\}]$$

Thus the timing of the abatement, while not affecting the cost-effectiveness estimate itself, does affect the value with which it is compared.

In judging whether a measure is cost-effective, the HM Treasury and DECC (2012) approach gives similar results to discounting the GHG savings in the sector and then comparing the estimate with a cost comparator computed as an undiscounted (abatement-weighted) social value of carbon (i.e. using the above formula but omitting the '$\times$ discount factor$_t$' term). Despite similarities, hypothetical examples can be constructed to show that the two methods do not invariably give the same result (for further discussion see Valatin 2012, p. 4).

Whether lack of discounting the carbon fluxes (combined with the employment of discounting in computing the cost-comparator used to judge cost-effectiveness) permits an appropriate comparison of abatement levels remains a matter of debate between the authors. For one, it provides a transparent comparison uncomplicated by discounting or other subsequent potential transformations of fluxes, that could usefully supplement alternative perspectives. For the other, it obscures the following fact: to the extent – and *only* to the extent – that delay in the *effects* of climate change justifies discounting, then delay in the *causes* of climate change (GHG fluxes), equally results in less importance for those causes. A consistent appraisal should explicitly reflect this by discounting of fluxes, as deemed appropriate.

The authors are agreed that numerically the two approaches will yield similar results, given consistent assumptions about discounting. The implication is that the comparator does, *in its effect*, simulate the discounting of carbon fluxes (later fluxes have less influence on the comparator), whatever might be said about ethical considerations.

Price Change and Discount Rate Adjustment

From the early days of climate change economics, it has been argued (Nordhaus 1991; Adger and Fankhauser 1993) that the economic impact of climate change will grow in line with gross world product (GWP): for example, because with advancing agricultural technology greater food production would be lost when a given area of farm land is inundated by sea-level rise; and because people will have larger houses to which air conditioning needs to be applied. If this is so, it is, arguably, correct to adjust the carbon price directly, rather than to subsume it in adjustment of the discount rate. This makes it possible to incorporate a number of factors that might be considered to affect the carbon price.

However, in the absence of such explicit adjustment, it is better to adjust the discount rate for carbon downwards by the margin of the best-guessed rate of carbon price increase, than to assume that the carbon price will remain constant. The practical advantage of adjusting the discount rate rather than increasing the carbon price over time – as done under the present UK government approach (HM Treasury and DECC 2012) – is that it simplifies calculations, and makes it easier to achieve consistency through the phases of evaluation (see Price in review).

Increasing carbon prices – and, equally, reduced carbon discount rates – can provide an incentive to delay abatement (Sohngen and Sedjo 2006; Murray et al. 2009). In fact a combination of low or zero discounting with a price rise for carbon makes it possible to generate a negative carbon account for a single-cycle forest sequestration and revolatilisation option (Price 2012). However, experience suggests that such cases do not arise for afforestation projects *where current UK government guidelines* (HM Treasury and DECC 2012) *are followed*. This is probably because discounted social values of carbon decline continuously after about the first 40 years, while future values for the 'non-traded' sector are generally below the initial value (see Valatin 2011b, Table IV and discussion of the evolution of the UK approach to determining the social value of carbon).

In Summary

The discount rate debate has been extremely long-running (back to the time of Moses) and wide-ranging (embracing everything from trivial pleasures to human life itself). The reader is referred to numerous reviews for further argumentation (e.g. Lind 1982; Broome 1992; Price 1993, 2006; Portney and Weyant 1999).

As a consequence of the variation in abatement and cost profiles over time, cost-effectiveness estimates are sensitive to the time horizon and base year focused upon. In this sense, abatement cost estimates provide only a snapshot of cost-effectiveness at a specific point in time over a particular time-horizon.

2.9   Sensitivity to Assumptions

The following table illustrates how the assumptions made concerning the above issues could significantly affect the cost derived. It is based on a spreadsheet model of forest stand growth and utilisation which includes both carbon fluxes and cash flows. Net cost is calculated according to normal net discounted cash flow procedures over a perpetual series of rotations. Carbon fluxes are included or not, and discounted or not, according to various protocols discussed above.

2.10   Risk

Forestry's long production cycle, as well as the indefinitely prolonged residence of some CO_2 fluxes into or from the atmosphere, make calculations concerning climate change mitigation susceptible to great problems of prediction. Fire, storms, attack by pests and pathogens, human incursion and revisions of governmental or landowner policy, all compromise the certainty of carbon storage, as much as that of timber production values. No-one can guarantee that carbon locked up by forests will remain so for ever, even if this is the plan.

Thus, in addition to the effect of discounting, future carbon values are reduced by these and other threats.

Risk has sometimes been treated, especially in financial markets, by adding a premium to the discount rate. In relation to physical threats to forests' survival, such a treatment is widely regarded as crude at best and at worst (in relation to future costs) systematically perverse (Price 1993, Chap. 11). Technically, risk is distinguished from uncertainty in that the probability distribution of possible outcomes is known. If this is the case, then the appropriate treatment is to take a range of possible outcomes and to combine their probability-weighted values in a mean expected value. This approach can be applied as readily to carbon flux figures as to the cash flow profiles associated with forestry options. Where future probability distributions are uncertain an alternative far simpler, if ad hoc, approach followed under several voluntary carbon standards (Valatin 2011b, Table 5, p. 14), including the Woodland Carbon Code, is to reduce the anticipated future abatement by a risk factor based upon past experience and expert judgement.

2.11 Forestry Options: Do They Offer a Limited Stock of Solutions?

While there is, world-wide, an area of land estimated at around 1,500 million hectares that might be afforested, even a massive afforestation programme would only sequester at most a few decades' emissions at current levels (see Mahli et al. 2002). That is a much shorter period than the limit imposed by availability of fossil fuels. Thus carbon sequestration in forest biomass at maximum represents only a medium-term solution to the problems of accumulating atmospheric CO_2 and climate change. Accumulation in forest soils may continue for longer, but itself is likely to rise to an asymptotic limit. By contrast the abatement benefits of biomass energy and structural displacement are cumulative over successive cycles of harvesting and regrowth, so have a role to play in climate change abatement over the longer term.

Moreover, as an afforestation strategy proceeded, it is likely that progressively more costly options would have to be adopted. Thus invitations to offset emissions (for example by paying for some afforestation, as offered by airlines) are accompanied by figures that omit the long-term costs entailed for later offsetters, because of the earlier withdrawal of the cheapest options (cf. Price (1984) on the cost of depleting mineral resources).

If the carbon fluxes do not include revolatilisation, then the costs of climate change mitigation must include those of perpetuating forest cover, or of achieving a different, permanent solution to the climate change problem.

3 Reviews of Previous Studies

3.1 UK Studies

The earliest published UK study of climate change mitigation cost may be that of Price (1990), who compared the cost of mitigating CO_2 concentration by growing biomass for displacement use in power generation, with that of using trees to sequester CO_2 emissions from fossil-fuel-based generation. It introduced the discounted net cost per discounted flux unit approach and applied it to a forest plantation of typical species and productivity for the UK. Its assumptions were very basic: a uniform rate of sequestration was used during growth and, for the CO_2 sequestering option, the carbon in timber was assumed to be permanently fixed. No revenues from sale of timber were included. Applying a 7 % discount rate to a representative north-temperate zone afforestation scheme gave a cost per tonne *coal* of £356 for growing wood fuel and of £76 for sequestering the CO_2 emitted by burning a tonne of coal.

The approach of Price and Willis (1993) refined this technique, deriving carbon fixing profiles from yield models, and carbon volatilisation from decay rates specific to wood product groups. Some illustrative results from the approach appear as the

Table 1 Some possible approaches to deriving marginal abatement costs in forestry

Discount carbon[a]	Use irregular carbon uptake profile[b]	Include all future revolatilisation[c]	Include displacement of fossil-carbon - intensive materials and energy production	Cost of CO_2 abatement[d] £/tonne
No	Immaterial	No	No	£2,913/463 tonnes = 6.3
Yes	No	No	No	£2,913 annualised/9.28 tonnes/year = 13.4[e]
Yes	Yes	No	No	18.0[f]
Yes	Yes	Yes	No	31.4[f]
Yes	Yes	Instant volatilisation	No	38.5[f]
Yes	Yes	Yes	Yes	19.4[f]

[a]Where carbon fluxes are discounted, an illustrative 3.5 % rate is used
[b]Irregular carbon uptake is according to the Forestry Commission yield model for thinned Sitka spruce yield class 12 (Edwards and Christie 1981)
[c]Volatilisation is at rates given by Thompson and Matthews (1989)
[d]In the absence of agreed figures, a notional figure only is used for the effect of displacement, it being taken to have similar magnitude to the direct lock-up of carbon in the timber
[e]An annuity (calculated at 3.5 %), equivalent to £2,913 over the rotation is divided by the mean annual CO_2 fixed (i.e. 9.28 tCO_2)
[f]Breakeven price

bottom four rows of Table 1. The paper also estimated the area of forest that would need to be planted (2.3 ha) to mitigate the CO_2 emissions associated with an international forestry conference.

A recent review of three studies estimating the climate change mitigation cost-effectiveness of UK forestry measures (Radov et al. 2007; Moran et al. 2008; ADAS forthcoming – results from the latter are also published in Matthews and Broadmeadow 2009) illustrates differences of approach in some recent studies (Valatin 2012). Estimates from the three studies, together with some of the under-pinning assumptions, are summarised in Table 2.

Note that the Radov et al. (2007) estimates and some of those in ADAS (forth-coming) are of the same order of magnitude as those calculated by the discounted carbon flux approach of Price and Willis (1993) shown in the bottom four rows of Table 1. However, differences in method between all the studies preclude over-arching conclusions from them about the relative cost-effectiveness of different forestry measures.

More recently, reporting large variations in land values between regions and grades of agricultural land, Nijnik et al. (2013) illustrate how abatement cost estimates tend to increase on higher quality agricultural land. Cost estimates reported for Scotland ranged from £4/tCO_2 (£15/tC) for Sitka spruce yield class 16 planted on poor quality uncultivated agricultural land previously used for livestock, to £21/tCO_2 (£76/tC) where prime arable land instead is used (Nijnik et al. 2013, Fig. 2, p. 39). Due to regional differences in opportunity costs and in timber prices, estimates also vary across the UK. Estimates for Sitka spruce yield class 12 planted on 'grade 3'

Table 2 Cost-effectiveness of UK forestry measures

	Radov et al. (2007)	Moran et al. (2008)	ADAS (forthcoming)
Time period(s) covered	(i) 2009–2012 (ii) 2009–2017 (iii) 2009–2022	to 2022	(i) to 2022 (ii) to 2050
Baseline land use	Arable	Sheep	Rough grazing/uncultivated
Carbon pools covered	T, S	T, L and S	T, L, S and HWP
Carbon benefits covered	Seq	(a) Seq (b) SeqSbm (c) SeqSbf	(a) Seq (b) SeqSbm(m) (c) SeqSbm(h)
Tree species and yield class options considered	2	1	14
Opportunity cost (£/ha/year)	£120 to £148[a]	£141	£50 to £350
Loss in land value (£/ha)	£2,500 to £7,500[a]	Not included separately	Not included separately
Establishment cost(s) (£/ha)	£1,250 to £3,000	£1,250	£1,310 to £5,400
Timber price profile	n.a.	2.5 % annual increase	2 % annual increase
Discount rate applied	7 %	3.5 %	3.5 %
Woodland creation cost-effectiveness (£/tCO$_2$e)	~£20 to ~£40	(a) −£7 (b) −£2 (c) −£6	(a) −£61 to £103 (b) −£61 to £73
Forestry management cost-effectiveness (£/tCO$_2$e)	Not considered	(a) £1 (b) £12[b]	(a) −£52[c]

Notes: Carbon pools: T: Tree; L: litter; S: Soil; HWP: harvested wood products; Carbon benefits: Seq carbon sequestration; SeqSbm carbon sequestration and materials substitution; (m) 'medium' materials substitution; (h) 'high' materials substitution benefits; SeqSbf carbon sequestration and fossil fuel substitution benefits in energy generation; Seqd carbon sequestration and displacement (including carbon storage in harvested wood products and fossil fuel substitution benefits in materials and energy generation)
[a]There may be an element of double-counting here (an issue not discussed in Radov et al. 2007).
[b]Assumes shortened rotation length (59 years to 49 years)
[c]Assumes increased management of currently under-managed woodland; Cost-effectiveness not estimated for medium substitution benefits or carbon sequestration alone due to apparent negative abatement potential

livestock land are reported to range from £7/tCO$_2$ (£27/tC) in Scotland to £17/tCO$_2$ (£65/tC) in south-east England (Nijnik et al. 2013, Fig. 3, p. 39). In general abatement costs are argued to be highest where land prices are greatest due to the stronger effect of land price differentials than of timber price ones (land prices and timber prices both tending to be higher in England than Scotland). The analysis does not account for differences in carbon displacement or ancillary (e.g. recreation and amenity) benefits, however: these could affect the ranking of options.

In each case the estimates in these studies generally suggest that forestry measures are cost-effective relative to social values of carbon recommended by the UK government for policy appraisal based upon the cost of meeting national abatement targets (Valatin 2011b). These social values include a central estimate for 2012 of £58/tCO$_2$e for 'non-traded' sectors (not part of the EU emissions trading scheme) at 2012 prices, rising over time to a peak of £334 per tCO$_2$e in 2077, declining thereafter (HM Treasury and DECC 2012, supporting Table 3). However, direct comparison of estimates in the table above is hampered by differing approaches, lack of clarity about the precise methodology in some cases, and the fact that the options are not generally alternatives for particular areas of land (Valatin 2012).

None of the three studies include ancillary benefits. More recent studies that have made climate change cost-effectiveness estimates while embracing ancillary benefits include Nisbet et al. (2011) and Valatin and Saraev (2012). The first of these focuses primarily upon benefits of woodland planting for flood risk reduction in the catchment upstream of Pickering, Yorkshire, but also covers habitat creation and erosion prevention benefits, while the second includes health and amenity benefits associated with woodland planting in Wales. As expected, inclusion of ancillary benefits improves the estimated cost-effectiveness of the forestry options. Although coverage of ancillary benefits is partial and differs, the studies otherwise adopt a similar approach based closely upon that recommended in UK government guidelines (DECC and HM Treasury 2011) and the approach to accounting for non-permanence adopted under the Woodland Carbon Code (Forestry Commission 2011a). Nisbet et al. (2011) report indicative cost-effectiveness estimates ranging from −£62/tCO$_2$ to £3/tCO$_2$, while Valatin and Saraev (2012) report estimates ranging from −£37/tCO$_2$ to £13/tCO$_2$. In both studies, the woodland creation options considered are judged highly cost-effective as climate change mitigation measures under the DECC and HM Treasury (2011) approach (although this is not the primary purpose of woodland creation in the first case).

By contrast to these results, evaluations which include non-market disbenefits, such as the effect on the landscape of large-scale clear-felling at the end of commercial rotations, would increase the associated social cost/tCO$_2$e. Potential lost hydroelectricity generation through afforestation may be of particular concern to the extent that its consequences include more electricity generation using fossil fuels (Barrow et al. 1986). The severity of impact on HEP, however, may be reduced by application of current practices and guidelines on forestry and watercourses (Nisbet 2005; Forestry Commission 2011b, requirement 74, p. 40).

3.2   International Studies of Forestry Mitigation Costs

The very low cost, equivalent to about $2/tCO$_2$, for the forest-based carbon sequestering option given by Sedjo and Solomon (1989) may result from the non-discounting of carbon fluxes, and a low opportunity cost of land, issues which also affect the $3/tCO$_2$ estimate of Sedjo and Ley (1997), and remain to this day.

An early study of change *within* forestry suggested a cost of $4 to $7/tCO$_2$ (depending on discount rate) for a limited modification of regional silviculture to enhance carbon fixing (Hoen and Solberg, 1994).

Healey et al. (2000) calculated the break-even price for a tonne of carbon flux, for a project defined as converting an area of conventional logging to one of reduced impact logging (RIL). The study included some ancillary benefits of RIL, to biodiversity and water quality: these were deducted from the net cost of conversion to RIL in deriving the break-even price. This ranged from $0 to $12/tCO_2$.

Richards and Stokes (2004) give a range of $10 to $150 per tonne of carbon (equivalent to $3 to $41/tCO_2$) but note the difficulties of making comparisons between studies, because of the different units, approaches and assumptions used, as we have discussed above.

Many more recent international studies suggest that forestry options are relatively inexpensive. Stern (2006), for example, notes that a substantial body of evidence suggests that preventing further deforestation would be relatively cheap, while Sohngen (2009) argues that forestry options could halve the total cost of abatement required to meet the 2 °C threshold target. Costs vary greatly between settings, with a range of $3 to $280/tCO_2$ given by van Kooten and Sohngen (2007). Estimates from 'bottom-up' studies suggest that forestry can offer abatement of around 6 $GtCO_2$e/year in 2030 at a cost of less than $100/tCO_2$e, just over half of this at under $50/tCO_2$e (IPCC 2007a, Fig. 4.2, p. 59).

To a considerable extent, this view – that forestry offers relatively inexpensive mitigation – is vindicated by comparison with other figures for carbon price mentioned below.

3.3 Other Approaches to Carbon Pricing

Comparative studies are a focus for government evaluations of marginal abatement costs. The relative cost-effectiveness of the many potential forestry options should be set in a context of other means of mitigating climate change, and the economic appraisal thereof. An early classification of economic approaches (Price and Willis 1993) recognised eight general methods for pricing carbon (including via cost of carbon sequestration by woodlands and other ecosystems). Even at this time a huge range of prices was quoted (Table 3).

Some of these alternative approaches to pricing carbon, with further illustrative figures where they have been found, are discussed below.

Other Means of Reducing CO_2 Concentrations

Alternative mitigation options under review include geo-engineering, which is defined by IPCC (2012, p. 2) as 'a broad set of methods and technologies that aim to deliberately alter the climate system in order to alleviate the impacts of climate change'. They also include means of physically storing CO_2 out of the atmosphere–ocean system, reducing the CO_2 intensity of energy production, and reducing energy consumption. IPCC (2007b, pp. 78–79) were cautious about the scope and cost of

Table 3 Methods of pricing CO_2: an early survey

Flux pricing method	Time-scale	Example of exponents at that time	Cost/tCO_2e
1. Constraint on growth of CO_2 emissions bottom-up	Phased	National Academy of Sciences (1991)	£0 to £65
2. Constraint on growth of CO_2 emissions top-down	Phased	Jorgensen and Wilcoxen (1990)	£1 to £8
3. Extra cost of low carbon fuel	Instant	Price (1990)	£97
4. Extra cost of low carbon fuel: delayed and discounted	Future	Anderson (1991)	£7
5. Cost of sequestering carbon (as discussed above)	Prolonged?	Sedjo and Solomon (1989)	£2
6. Cost of altering radiative balance	Prolonged?	National Academy of Sciences (1991)	Trivial?
7. Lost production, damage cost and defensive spending	Perpetual	Nordhaus (1991), Cline (1992)	5p to £25
8. Carbon tax to achieve target	Undefined	Cline (1992)	£18 to £49

Note: these prices were compiled in 1993, and reflating them to 2012 prices would imply that they can probably be more or less doubled now. For purposes of comparison, prices originally given in £/tC have been converted to £/tCO_2e

some of these options: 'geo-engineering solutions to the enhanced greenhouse effect have been proposed. However, options to remove CO_2 directly from the air, for example, by iron fertilization of the oceans, or to block sunlight, remain largely speculative and may have a risk of unknown side effects. ... Detailed cost estimates for these options have not been published and they are without a clear institutional framework for implementation.' Schellnhuber (2011) notes that carbon sequestration through industrial 'air capture' could well cost of the order of $1,000/t$CO_2$.

Sequestering by Other Means

Other biological processes of sequestering carbon apart from by growing trees, such as photosynthesis by phytoplankton, seaweed and other types of marine algae, or accumulation of peat or biochar, may also add to carbon stocks. Intervention to accelerate the processes is more problematic, and little attention seems to have been given to providing a comprehensive account of their potential to enhance climate change mitigation on a significant scale. Some indicative estimates are available. For example, Moxey (2011) suggests indicative costs of restoring degraded peatlands through grip blocking in order to restore them to CO_2 sinks from their current position as CO_2 sources (resulting from previously being drained, etc.) may typically be around £13/tCO_2 (see also Artz et al. 2012). However, the authors are not aware of any large-scale and well-agreed costings of these sequestration strategies (and would welcome information on such costings). This may be partly due to the present scarcity of evidence on associated carbon fluxes, although these are the focus of ongoing research – including some related to interests in the UK (e.g. Duke et al. 2012) – in developing a Peatland Carbon Code. (For a review of current evidence on carbon fluxes and GHG emissions associated with UK peatlands, for example, see Worrall et al. (2011)).

IPCC (2005) give a range of costs for carbon capture and storage of CO_2 emissions, particularly from power plant, from \$0 to \$270/tCO_2. Pöyry Energy Consulting (2007) estimate that in the UK some carbon capture and storage at power stations could be achieved at a cost below £25/tCO_2, but note that there is limited scope at this price. Storage in aquifers and oil-fields seems also susceptible to the kind of long-term risk that attends forestry options.

Biogeological processes have of course been responsible for reducing atmospheric CO_2 to a level at which present terrestrial life is possible, by incorporation in limestone and other carbonate rocks derived from animal skeletons and through chemical precipitation. However, these processes have taken hundreds of millions of years to sequester the existing amount of carbon. While human action to accelerate the processes is conceivable, it may seem hardly conceivable that this could take place on a short enough time scale to meet present targets. Schuiling (2012) does argue that this approach is feasible by spreading crushed olivine (involving enhanced weathering of crushed magnesium silicates) at a modest cost of around \$10/tCO_2. However, Schellnhuber (2012) advises caution about such 'silver bullet' solutions, noting their possible externalities (e.g. river acidification) and the associated cost.

This is not to say that none of these strategies could play a part in long-term solutions to climate change problems, but that in the present state of knowledge they cannot be relied upon to supply the needed quick solution, much less at a known, agreed and reasonable cost.

Reduced CO_2 Intensity in Energy, Materials and Services Production

Existing final products and services might be made available to consumers through using more energy-efficient technologies, generating less CO_2 per unit of product. At the production end, there have been thermal efficiency increases for conventional thermal power generation, and in iron and steel making and other metallurgical processes. However, these gains are partly offset by the reduced quality of available fossil energy resources, such as shale oil: this entails heavier CO_2 overheads resulting from fossil fuel use in exploration and exploitation. There are thermodynamic limitations on how energy-efficient processes can become, and some necessarily entail a given quantum of CO_2 release that can only be avoided by reliable carbon capture and storage. Although until recently this was thought to be the case for cement production from limestone ($CaCO_3$), Berger (2012) reports that a process is currently being developed that does not give rise to carbon dioxide emissions – instead producing lime, graphite and oxygen. However, this uses solar energy that could otherwise displace fossil-fuel-based energy production, and thereby affects CO_2 emissions indirectly. Lighter, more material-parsimonious structures also offer further savings.

At the consumer end, there are more energy-efficient cars, electrical appliances and light bulbs; and, more radically, change of transport mode from private to public with its further advantages of reducing congestion.

As well as by saving energy input, CO_2 emissions can be reduced by varying the mix of energy-generation technologies including renewables such as photo-voltaics, wind, hydroelectric power and waves. (Tidal energy, often included as a renewable, is technically a depleting stock resource, derived from the rotational energy of the Earth. However, rough estimation shows that the rotational energy of the earth is equivalent to hundreds of millions of years of current global energy consumption – it's amazing how much energy can be stored in a large, rapidly rotating flywheel!) Nuclear power, like all forms of electricity generation, entails GHG emissions during construction and operating phases, and also in uranium mining. Concerns about safety, augmented by major releases of radioactive pollution and the loss of life in accidents over the past 25 years, caused fresh capacity to be dropped from the future energy portfolio of some countries (e.g. Germany and, at least initially, Japan). A perception that these problems are not serious compared with those of climate change has led some to favour its reintroduction, but the concerns themselves generally remain unalleviated.

For an illustrative cost of renewables, consider a photo-voltaic system with an ascribed economic life of 40 years. Using installation cost, estimated generation figures and CO_2 saving for the system supplied by EvoEnergy (personal communication), and current prices of imported and exported electricity, the costs are:

- With 10 % discount rates, £576/tCO_2
- With 3 % discount rates, £147/tCO_2
- With 3 % rate for cash and 1 % for carbon, £105/tCO_2
- With 1 % rates, £51/tCO_2

At low discount rates, cost is sensitive to project life. With 55-year life and 1 % rates, the project breaks even and abatement is "free".

It would be expected that, with scale economies, a commercial installation using this technology would also provide free abatement, vindicating Anderson's speculation in 1991 that costs of low-carbon energy would fall dramatically.

These and technologies like them are in the end the ones that must be deployed to deal with the twin problems of CO_2 accumulation and depletion of fossil energy. Realistically, afforestation, reforestation and other biological CO_2 mitigation options do not provide more than a medium-term, 'holding' solution, to give time while these technologies' competence is evolved to a low enough cost, on a sufficient scale. In the meantime, each tonne of carbon prevented from entering the atmosphere as CO_2 has a cost which can be measured in essentially the same format (calculating £/tCO_2e) that would be used in appraising the cost-effectiveness of physical and biological systems for carbon sequestration. However, where a measure increases consumers' disposable income because it reduces their energy bills, only GHG savings net of the direct 'rebound effect' (associated increases in consumption of the main energy service in question) are accounted for under current UK government guidelines in valuing changes in energy use (HM Treasury and DECC 2012, pp. 17–19).

The premium willingly paid for current modes of transport and for current modes of production and consumption might be taken as reflecting their value over-and-above that of the low CO_2 modes. But such premia also reflect the inertia of technology and consumption patterns: the exigencies of climate change might prove a valuable incentive to adopt technologies that might be superior, irrespective of their impact on climate. The classic case, originally driven by energy conservation considerations, is the low-energy light bulb, which provided at first a win–win saving in energy and in electricity bills, and now offers a mandatory win–win–win–win change, with saved CO_2 emissions and reduced annual household investment cost added to the initially proclaimed benefits.

Low-energy refrigerators and tumble-dryers also offer rational investments in energy-saving to those already owning a model with high energy consumption, or choosing which model to buy, with CO_2 reduction benefits as a bonus. That these options are not taken up by consumers may reflect both a lack of knowledge of the benefits and a high discount on future energy savings (Gateley 1980).

Reduced Consumption

Energy economy can be achieved just by reducing what we consume: fewer holidays abroad, fewer hours of television, more fastidious switching off of lights and appliances not in use. In conventional economic terms, the cost of lost consumption would be considered as the price of energy, being a measure of consumers' willingness to pay for its use (no matter how wasteful the use might be). The view of alternative economics might be that such reduced material consumption might in fact enrich lives and health, by forcing us back on human relationships and physical activity as a means of finding satisfaction.

A naive estimate of what would be implied for the value of lost consumption could be given by dividing GDP by current emissions, resulting in a loss of more than £2,500/tCO_2! However, this takes no account of a selective effect whereby the least valued consumption/tCO_2 would be sacrificed first. It also takes no account of ways in which energy consumption can be reduced without reducing consumption of final goods and services (see discussion above). More sophisticated approaches illustrated on the top two lines of Table 3 consider how the economy can best adjust to constrained CO_2 emissions, respectively by top-down economic optimisation and by bottom-up 'engineering' approaches. (For a discussion of implications for cost-effectiveness estimates of adopting a bottom-up 'engineering', a sectoral optimisation, or an econometric modelling approach, see Dempsey et al. (2010)).

Other Means of Mitigating Climate Change

Among the options that seek to mitigate the effect of climate change other-than-by slowing or reversing CO_2 accumulation in the atmosphere, there is altering the radiative balance by circling the earth with a belt of 'smart mirrors' launched by rocket.

This would replicate the effect of industrial smoke and volcanic particulates in the upper atmosphere, which reflect back solar radiation, and have been responsible for the so-called 'global dimming effect', as a result of which it is estimated that the globe has warmed by 0.6 °C less than the present levels of GHGs would indicate (Hansen et al. 2005). NASA (quoted in Cline (1992)) once reported the cost of launching smart mirrors as 'trivial', but perhaps that should be interpreted in the context of an agency whose annual budget has been about $18 billion in recent years (Office of Management and Budget various years), and does not account for potential negative impacts.

More recently, injection of sulphate particles into the stratosphere has become the focus of such discussions. Crutzen (2006) suggests that the cost of 'injections to counteract effects of doubling CO_2 concentrations would be $25 to 50 billion a year.' Keith et al. (2010) claim that 'This is over 100 times cheaper than producing the same temperature change by reducing CO_2 emissions.' While implicitly less costly than CO_2 mitigation measures, no equivalent cost/tCO_2 was given. But problems of delivery, side-effects, and the short duration of the effect have been raised by Robock et al. (2009). Furthermore, IPCC (2012, p. 5) note that existing studies of the costs of solar radiation management methods are 'limited primarily to implementation (direct) costs, and even then there is limited literature for even the most prominent techniques; indirect costs and possible impacts are poorly explored, particularly in relative comparisons against ongoing climate change.' (See also discussion in Royal Society (2009) and Schellnhuber (2011)).

Business-As-Usual: The Social Cost of Climate Change

The laissez-faire option (method 7 in Table 3) was formerly favoured as the complete solution by climate change sceptics (Bate and Morris 1994). Many economists (e.g. Nordhaus 1993, 2007) consider it as one of a suite of options to be evaluated. It entails accepting the consequences – if any – of business-as-usual conduct of the world economy. If there are adverse effects – which the great majority of scientists now consider highly likely – then those can be subject to economic valuation. In a rational appraisal, the question is whether the cost of those adverse effects is greater than the cost of mitigating them, and if so how far the mitigation measures should go. In recent years, authoritative commentators have come to markedly different conclusions on this matter (Stern 2006; Nordhaus 2007). Note also that ethical issues discussed in section "The Case for Not Discounting Carbon Fluxes" imply that trade-offs are not always considered to apply (e.g. to the extent that preferences for avoiding significant harms are lexicographic).

The economic cost of *not* mitigating the effects of CO_2 accumulation, consequent climate change and sea-level rise can be interpreted through: the forgone benefits of a less productive global economy; the disbenefits such as poorer health; the costs of defensive measures such as sea-wall and flood defences and temperature amelioration and health protection; and those of 'retreat' from lands which it is deemed too expensive to defend.

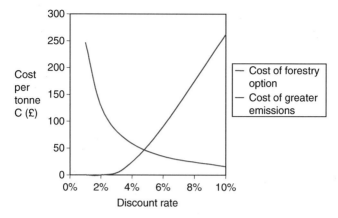

Fig. 1 Cost of mitigation versus cost of business-as-usual. Notes: The cost is per tC, not per tCO_2; The indicative cost of greater emissions has been calculated under reasonable assumptions about the relationships among CO_2 concentrations, rates of CO_2 uptake, temperature change, thermal inertia and economic damage. While sensitive to these assumptions, the relationship between discount rate and cost per tonne of carbon always shows this rising path. Source: Presentations made by Colin Price at various seminars in 2010

Some of the many problems of evaluating the costs of continuing emissions are:

- Rates of uptake and equilibrium levels in oceanic and terrestrial sinks;
- Relationships between CO_2 levels and temperature change;
- Rates of temperature adjustment given the thermal inertia imparted largely by the oceans;
- Mapping from temperature rise to economic damage;
- Nonlinearities in the above relationships; especially
- Threshold effects; and
- Positive feedbacks;
- Discounting, as discussed above.

All these mean that the effect of another tCO_2 does not have a simple, predictable cost. The uncertainty and the precautionary principle have led to reduced willingness to accept the business-as-usual approach, with costings increasingly erring on the side of caution.

Early figures quoted in the literature included \$5/tC (Nordhaus 1993) and \$20/tC (Fankhauser 1995). Clarkson and Deyes (2002) from an international survey report a range from \$9 to \$197/tC, equivalent to £2 to £34/tCO_2 at the present exchange rate of £1 = \$1.60, and give a preferred figure equivalent to £19/tCO_2. Stern (2006) focuses on costs of global scenarios, but uses a cost of \$85/tCO_2 under a business-as-usual scenario, acknowledging that this lies above typical figures in the literature, but pointing out that the figure falls well within the range of quoted figures, and that it takes account of risks explicitly.

Comparison of cost of mitigation, by a representative upland afforestation option in the UK, with the option of bearing climate change shows, as might be expected, acute sensitivity to discount rate, as Fig. 1 demonstrates. The higher the discount rate, the smaller the significance of long-drawn-out forest sequestration,

Table 4 Some prices in carbon markets

Source	Date	Price per tonne CO^2
Biocarbon Fund	2003	$3
Den Elzen and de Moor	2002	$4.5 to $5.5
International Emissions Trading Association	2003	$9.9 to $13.7
Grubb	2003	$9 to $22
PointCarbon	2008	$35

Source: Olschewski and Benitez (2010)

so the higher the cost per unit sequestered. On the other hand, the higher the discount rate, the smaller the significance of long-term climate change, so the *lower* the cost per unit emitted. Although the costs of climate change are also subject to significant uncertainty and their quantification is not uncontroversial (as noted above), it is noteworthy that Fig. 1 implies that the switching point between options occurs in the range of discount rates normally discussed in the academic literature and in commercial practice, but that the forestry option is cheaper than business-as-usual under the UK government's preferred range of discount rates of 3.5 % to 1 % (HM Treasury 2003).

Market Prices

Those with a marketist predisposition might point out that markets already exist for carbon, giving a price by an approach additional to those recorded in Table 3. However, prices fluctuate dramatically through time and even vary across markets at a point in time (Table 4).

Since this table was compiled, there has been some downward pressure on prices, partly in response to the onstreaming of many forestry projects 'undertaken under REDD. Nevertheless, Peters-Stanley and Hamilton (2012) report a price increase in voluntary markets. Prices in forest carbon markets are also reported to have risen from $3.8/tCO_2$ in 2008 to $5.5/tCO_2$ in 2011 (see Diaz et al. 2011).

The more fundamental objections to using 'market' prices are that:

- There are no 'natural' free-market prices for carbon sequestration or emissions. What prices exist, are constructed prices, derived in response to governmental and intergovernmental regulations, stipulations and moral suasion, and selling to consumers what may be misguided 'warm glows' (Price in review); and
- The level of market prices, based upon demand from a relatively small section of the economy, does not account sufficiently for damage costs to society of emissions.

Taxing to Achieve a Mitigation Target and Other Fiscal Measures

Road fuel tax in the UK amounts to around £250/tCO_2, though some such taxation has been in place since long before climate change became identified as a problem.

It has a mix of economic roles, in revenue raising, in funding transport infrastructure and in countering other externalities, such as those associated with congestion and road traffic accidents. Nowadays reduction of CO_2 emissions is a much-discussed and widely promulgated reason for such taxation. The absence of equivalent taxation of aviation fuel is seen by many as an anomaly that hampers achievement of emissions reductions, especially in an era of rapidly growing air travel.

The UK government presently offers a premium 'feed-in tariff' for solar-generated electricity, which might be taken to express its willingness to pay to meet an emissions reduction target. Each unit of electricity generated by thermal stations is associated with almost 0.5 kg CO_2 emission. The initial domestic feed-in tariff (FIT) of 43p per unit (kWh) therefore implied a valuation of abatement of around £800/tCO_2, and even the present values applying to new solar photovoltaic installed after August 2012, which range from 7p to 16p per unit (OFGEM 2012), imply a range of about £140/tCO_2 to £300/tCO_2. Rapid reduction in FIT suggests either that the falling cost of solar generation makes it easier to meet the target, or that the price is now seen as too generous.

Fiscal measures required to meet a standard may not always be an independent assessment of the price of CO_2 emissions, if the standards are set as a result of perceived costs of emissions, among other factors. Using these measures as a price then risks running into circular arguments.

4 Discussion and Conclusions

Climate change mitigation is considered by many governments and international agencies to constitute the greatest challenge currently facing humanity. The task is urgent if international aspirations to prevent 'dangerous' climate change are to be met, with the International Energy Agency's Chief Economist warning that'… the door to 2° – which is a must for a decent life – is closing forever'(IEA 2011; see also Peters et al. 2013).

Price Waterhouse Cooper (2012) have also concluded that the current global rate of CO_2 emissions reduction (0.7 % per year) is very far short of that required (5.1 % per year) to limit temperature change to 2 °C. They note too that reduction in US CO_2 emissions was partly achieved by switching to shale gas, and fear that the availability of this possibly short-term resource may reduce pressure to adopt renewable energy solutions. All this suggests an even greater urgency to adopt cost-effective options that can mitigate CO_2 levels in the medium term. Experience also suggests a case for front-loading emissions reductions rather than focusing upon meeting future targets (Latin 2012).

Forests potentially have a very important role to play globally in climate change mitigation. In the absence of mitigation efforts, current deforestation rates could make stabilisation of atmospheric GHG concentrations at a level that avoids the worst effects of climate change highly unlikely (Eliasch Review 2008). Afforestation and agroforestry options also provide a potential way of reducing existing atmospheric GHG concentrations from levels considered by some scientists to be already too high. If the challenging international target of limiting temperature increase due to anthropogenic causes to a maximum of 2 °C (or possibly 1.5 °C) is to be met, realising the potential of forests to increase carbon sequestration and also for increased substitution through use of wood products will be critical. Because CO_2 emissions persist in the atmosphere – some part of them indefinitely – merely *reducing emissions* does not reduce CO_2 levels in the atmosphere. Moreover, because of the long lags of adjustment processes, the need to reduce net emissions is immediate: by the time critical temperature changes are approached, it will be too late for mitigating action to prevent their being exceeded. In temperate forestry, there is the further consideration that many years may elapse before an afforestation project reaches its fastest sequestration rate.

Both the amount of abatement by forestry and the potential contribution of each type of option depend upon the level of incentives (e.g. carbon prices). A primary reason that deforestation and forest degradation are currently large net sources of carbon emissions is that traditionally there has been little incentive for landowners or forest users to account for non-market values, including the social value of carbon sequestration and storage. Opportunities for climate change mitigation by forestry are being lost in the existing institutional structure (Nabuurs et al. 2007), although the situation is slowly changing as payment-for-ecosystem-services schemes develop, including markets for forest carbon, often underpinned by regulatory change and new institutions. At international level, some progress has been made in agreeing financing mechanisms for reduced emissions from deforestation and forest degradation (REDD and REDD+) following agreement at Bali (UNFCCC 2008). National initiatives are also developing, such as the launch in 2011 of a Woodland Carbon Code by the Forestry Commission (2011a) to help underpin an emerging market for carbon sequestration by UK woodlands.

There remain research gaps in the supporting evidence too. As noted above, for the UK these include robust evidence on impacts of afforestation on forest soil carbon balance. Other gaps include a paucity of comprehensive GHG balances for UK forest stands, carbon stock changes during early tree growth and once stands reach maturity, and carbon substitution (or displacement) benefits (Morison et al. 2012). Research on albedo, evapotranspiration and other biophysical factors, including the impact of surface roughness on exchanges of energy and mass between the land surface and the atmosphere, remains at an early stage (see Anderson et al. 2011). Better evidence is also needed on opportunity costs and on leakage effects.

There remain, too, unresolved issues in the appropriate means of calculating a mitigation cost, associated with the long time-span of forestry.

However, unresolved issues and the multiplicity of economic and biophysical factors involved do not mean, as some argue (e.g. van Kooten et al. 2012), that the abatement benefits of forestry options are currently too uncertain to estimate.

Despite the lack of internationally accepted approaches at present, as noted above, available evidence indicates that forestry options are relatively cost-effective compared with a range of alternatives. Whether this conclusion holds in particular cases could be expected to vary between projects and regions, as well as being dependent upon the approach adopted. To the extent that the cost-effectiveness of forestry options depends upon the methodology adopted and benchmark used, future comparisons could benefit from greater methodological transparency and consistency. Lack of transparent methodology has been raised as a particular concern with some previous studies (e.g. see Kesicki 2011; Ekins et al. 2011).

Recent studies in the UK suggest that some forestry options are generally very cost-effective judged by current UK government benchmarks. Accelerating global emissions, impacts that are possibly more severe than anticipated, non-negligible probabilities of catastrophic impacts, and a desire for greater certainty that critical thresholds will not be exceeded, might lead to adoption of tighter abatement targets. If so, estimates of the social value of carbon would need to be revised upwards (Valatin 2011b). This would make forestry options even more cost-effective compared with the social value of carbon.

However, the main point to stress is not that forestry options are relatively cost-effective (although this is what the available evidence suggests), but that they will be critical if international objectives on climate change mitigation are to be met.

Reaching the climate change mitigation targets agreed at international level is challenging, but not impossible (Schellnhuber 2008; Den Elzen et al. 2010; Höhne et al. 2012; Peters et al. 2013), with a key issue being the cost entailed (e.g. Schellnhuber 2011). From a theoretical perspective, work on international environmental agreements (Valatin 2005) suggests that 'free rider' problems entailed in reaching a binding agreement may not be as significant a barrier as early work by economists and game theorists had suggested, with challenges relating more to coordination issues, including fairness and justice. (Subsequent work on minimum participation rules confirm the underlying results – see Carraro et al. (2009) and Weikard et al. (2009).) On a practical level, sources of disagreement include whether an agreement should be based upon equal per capita emissions (e.g. Chang 2012), or some other measure, perhaps related to needs, resources available, or historic responsibilities for causing the rises in atmospheric GHGs. They also relate to whether it should be based upon countries' 'carbon footprint' in terms of domestic consumption, or upon emissions due to domestic production (see Helm (2012a, b) who refers to evidence that emissions from production fell by 15 % in Europe over the period 1990–2005, but increased by 19 % in terms of consumption). Beyond these matters for negotiation, scope exists too for action by individual countries, that could potentially lead to the targets' being met, even in the absence of a strong global agreement. For example, Helm et al. (2012) argue that introduction of border carbon adjustments would provide incentives for subsequent adoption of similar measures by other countries, along similar lines to the most likely outcome of the current dispute between the EU and other countries over inclusion of aviation under the EU ETS.

Acknowledgements Thanks to James Morison, Chris Quine, Richard Haw, Trevor Fenning and Andrew Moxey for comments on earlier drafts.

References

ADAS (forthcoming) Analysis of policy instruments for reducing greenhouse gas emissions from agriculture, forestry and land management – forestry options. Report to Forestry Commission England, ADAS, Abingdon, Oxfordshire. (Final draft versions dated June 2009 and July 2011)

Adger N, Fankhauser S (1993) Economic analysis of the greenhouse effect: optimal abatement level and strategies for mitigation. Int J Environ Pollut 3:104–119

Anderson D (1991) The forestry industry and the greenhouse effect. Scottish Forestry Trust and Forestry Commission, Edinburgh

Anderson R, Canadell J, Randerson J, Jackson R, Hungate B, Baldocchi D, Ban-Weiss G, Bonan G, Caldiera K, Cao L, Diffenbaugh N, Gurney K, Kueppers L, Law B, Luyssaert S, O'Halloran T (2011) Biophysical considerations in forestry for climate protection. Front Ecol Environ 9(3):174–182

Artz R, Donnelly D, Cuthbert A, Evans C, Smart S, Reed M, Kenter J, Clark J (2012) Restoration of lowland raised bogs in Scotland: emissions savings and the implications of a changing climate on lowland raised bog condition. IUCN/James Hutton Institute/ Scottish Wildlife Trust/Esmée Fairburn Foundation

Arvizu D, Bruckner T, Chum H, Edenhofer O, Estefen S, Faaij A, Fischedick M, Hansen G, Hiriart G, Hohmeyer O, Hollands K, Huckerby J, Kadner S, Killingtveit Å, Kumar A, Lewis A, Lucon O, Matschoss P, Maurice L, Mirza M, Mitchell C, Moomaw W, Moreira J, Nilsson L, Nyboer J, Pichs-Madruga R, Sathaye J, Sawin J, Schaeffer R, Schei T, Schlömer S, Seyboth K, Sims R, Sinden G, Sokona Y, von Stechow C, Steckel J, Verbruggen A, Wiser R, Yamba F, Zwickel T (2011) Technical summary. In: Edenhofer O, Pichs-Madruga R, Sokona Y, Seyboth K, Matschoss P, Kadner S, Zwickel T, Eickemeier P, Hansen G, Schlomer S and von Stechow C (eds) (2012) Renewable energy sources and climate change mitigation: special report of the Intergovernmental Panel on Climate Change. Cambridge University Press. http://www.ipcc.ch/pdf/special-reports/srren/SRREN_Full_Report.pdf. Accessed 5 Mar 2013

Ayers R, Walter J (1991) The greenhouse effect: damages, costs and abatement. Environ Resour Econ 1:237–270

Barrow P, Hinsley AP, Price C (1986) The effect of afforestation on hydroelectricity generation: a quantitative assessment. Land Use Policy 3:141–151

Bate R, Morris J (1994) Global warming: apocalypse or hot air? Institute of Economic Affairs, London

Berger M (2012) Solar-powered cement production without carbon dioxide emissions. Nanowerk. http://www.nanowerk.com/spotlight/spotid=24883.php. Accessed 12 Nov 2012

Brainard J, Bateman IJ, Lovett AA (2009) The social value of carbon sequestered in Great Britain's woodlands. Ecol Econ 68:1257–1267

Broadmeadow M, Matthews R (2003) Forests, carbon and climate change: the UK contribution. Information Note 48, Forestry Commission

Broome J (1992) Counting the cost of global warming. White Horse Press, Cambridge

Brown K, Cardenas L, MacCarthy J, Murrells T, Pang Y, Passant N, Thistlethwaite G, Thomson A, Webb N, with contributions from Dore C, Gilhespy S, Goodwin J, Hallsworth S, Hobson M, Manning A, Martinez C, Matthews R, Misselbrook T, Thomas J, Walker C, Walker H, Watterson JD (2012) UK greenhouse gas inventory, 1990–2010: Annual Report for Submission under the Framework Convention on Climate Change, AEA Technology report to DECC. http://naei.defra.gov.uk/reports.php?list=GHG

Caney S (2010) Human rights, climate change and discounting. In: O'Brien K, St Clair AL, Kristoffersen B (eds) Climate change, ethics and human security. Cambridge University Press, Cambridge, UK, pp 113–130, Chapter 7

Cannell MGR, Dewar RC, Pyatt DG (1993) Conifer plantations on drained peatlands in Britain: a net gain or loss of carbon? Forestry 66:353–369

Carraro C, Macrchiori C, Oreffice S (2009) Endogenous minimum participation in international environmental treaties. Environ Resource Econ 42:411–425

Chang SJ (2012) Solving the problem of carbon dioxide emissions. Paper presented at the IUFRO conference on new frontiers in forest economics, Zurich, June, 26th–30th. http://neffe.ch/index.php?id=48#c205. Accessed 15 Nov 2012

Clarkson R, Deyes K (2002) Estimating the social cost of carbon emissions. Government Economic Service Working Paper 140. HM Treasury/Defra, London

Cline WR (1992) The economics of global warming. Institute for International Economics, Washington, DC

CO_2Now (2013) Annual data: atmospheric CO_2. CO_2now.com website, http://co2now.org/. Accessed 31 Jan 2013

Crutzen PJ (2006) Albedo enhancement by stratospheric sulfur injections: a contribution to resolve a policy dilemma? Clim Change 77:211–220

Dasgupta PS, Heal G (1979) Economic theory and exhaustible resources. Cambridge University Press, Cambridge, UK

DECC and HM Treasury (2011) Valuation of energy use and greenhouse gas emissions for appraisal and evaluation. Department of Energy and Climate Change, London, http://www.decc.gov.uk/en/content/cms/statistics/analysts_group/analysts_group.aspx

DEFRA, DARDNI, RERAD, DRAH (2010) Agriculture in the United Kingdom. Department for the Environment, Food and Rural Affairs, London

Dempsey J, Plantinga A, Alig R (2010) What explains differences in the costs of carbon sequestration in forests? A review of alternative cost estimation methodologies. Chapter 4. In: Alig R (ed) Economic modeling of effects of climate change on the forest sector and mitigation options: a compendium of briefing papers. Report PNW-GTR-833, pacific North West Research Station. US Department of Agriculture, Corvallis, pp 87–108

Den Elzen M, Hare W, Höhne N, Levin K, Lowe J, Riahi K, Rogelj J, Sawin E, Taylor C, van Vuuren D, Ward M, Bosetti V, Chen C, Dellink R, Fenhann J, Gesteira C, Hanaoka T, Kainuma M, Kejun J, Massetti E, Matthews B, Olausson C, O'Neill B, Ranger N, Wagner F, Xiusheng Z (2010) The emissions gap report: are the Copenhagen pledges sufficient to limit global warming to 2°C or 1.5°C? UN Environmental Programme, Nairobi, www.unep.org/publications/ebooks/emissionsgapreport/. Accessed 5 Mar 2013

Department of the Environment (1991) Policy appraisal and the environment. HMSO, London

Dewar RC, Cannell MGR (1992) Carbon sequestration in the trees, products, litter and soils of forest plantations: an analysis using UK examples. Tree Physiol 11:49–71

Diaz D, Hamilton K, Peters-Stanley M (2011) State of the forest carbon markets 2011: from canopy to currency. Forest Trends, Washington, DC, http://www.forest-trends.org/publication_details.php?publicationID=2963

Dresner S, Ekins P, McGeevor K, Tomei J (2007) Forest and climate change: global understandings and possible responses. Chapter 6 in Freer-Smith P, Broadmeadow M, Lynch J (eds) Forests and Climate Change, Forest Research/CABI

Duke G, Dickie I, Juniper T, ten Kate K, Pieterse M, Rafiq M, Rayment M, Smith S, Voulvoulis N (2012) Opportunities for UK business that value and/or protect Nature's services; elaboration of proposals for potential business opportunities. Attachment 1 to Final Report to the Ecosystem Markets Task Force and Valuing Nature Network. GHK, London, http://uncsd.iisd.org/news/defra-reports-on-business-opportunities-in-protecting-nature%E2%80%99s-services/

Edwards PN, Christie JM (1981) Yield models for forest management. Forestry Commission Booklet 48. HMSO, Edinburgh, London

Ekins P, Kesicki F, Smith A (2011) Marginal abatement cost curves: a call for caution. Report to Greenpeace UK by UCL Energy Institute, University College London, London

Eliasch Review (2008) Climate change: financing Global Forests, independent report to government, commissioned by the Prime Minister and prepared by Johan Eliasch with the support of the Office of Climate Change, London

European Commission (2012) Climate action: emissions trading system (EU ETS). European commission. http://ec.europa.eu/clima/policies/ets/index_en.htm. Accessed 12 June 2012

FAO (2010) Global Forest Reseources Assessment 2010. Forestry Paper 163, UN Food and Agriculture Organization, Rome. http://www.fao.org/forestry/fra/fra2010/en/.Accessed 18 Dec 2012

Forestry Commission (2011a) The woodland carbon code. Version 1.3. Forestry Commission, Edinburgh

Forestry Commission (2011b) Forests and water UK forestry standard guidelines, 5th edn. Forestry Commission, Edinburgh

Forestry Commission (2012) Carbon leakage. Woodland carbon code section 3.3. Forestry Commission, Edinburgh, http://www.forestry.gov.uk/forestry/infd-8hngvh. Accessed 7 Nov 2012

Fankhauser S (1995) Valuing climate change. Earthscan, London

Gateley D (1980) Individual discount rates and the purchase and utilization of energy-using durables: comment. Bell J Econ 11:373–374

GCP (2012) Carbon budget and trends 2012. http://www.globalcarbonproject.org/carbonbudget/12/hl-full.htm. Accessed 31 Jan 2013

Gerlagh R, Liski M (2012) Carbon prices for the next thousand years. CESifo Working Paper 3855. http://re3.feem.it/getpage.aspx?id=4941

Gorte R (2009) Carbon sequestration in forests. Congressional research Service, Washington, DC, www.fas.org/sgp/crs/misc/RL31432.pdf

Hansen J, Nazarenko L, Ruedy R, Sato M, Willis J, Del Genio A, Koch D, Lacis A, Lo K, Menon S, Novakov T, Perlwitz J, Russell G, Schmidt GA, Tausnev N (2005) Earth's energy imbalance: confirmation and implications. Science 308:1431–1435

Hansen J, Sato M, Kharecha P, Beerling D, Berner R, Masson-Delmotte V, Pagani M, Raymo M, Royer DL, Zachos JC (2008) Target atmospheric CO_2: where should humanity aim? Open Atmos Sci J 2:217–231, http://pubs.giss.nasa.gov/cgi-bin/abstract.cgi?id=ha00410c

Healey JR, Price C, Tay J (2000) The cost of carbon retention by reduced impact logging. For Ecol Manage 139:237–255

Helm D (2012a) The carbon crunch: how we are getting climate change wrong – and how to fix it. Yale University Press, London

Helm D (2012b) Forget the Kyoto accord and tax carbon consumption. Environment 360. http://www.dieterhelm.co.uk/sites/default/files/YALEEnv360081112.pdf. Accessed 5 Mar 2013

Helm D, Hepburn C, Ruta G (2012) Trade, climate change and the political game theory of border carbon adjustments. Working Paper 80, Centre for climate change economics and policy, and Grantham research institute on climate change and the environment, Leeds/London. http://www2.lse.ac.uk/GranthamInstitute/publications/WorkingPapers/Abstracts/80-89/Trade-climate-change-border-carbon-adjustments-abstract.aspx. Accessed 15 Nov 2012

Henders S, Ostwald M (2012) Forest carbon leakage quantification methods and their suitability for assessing leakage in REDD. Forests 3:33–58, http://www.mdpi.com/1999-4907/3/1/33

Hepburn C, Koundouri P (2007) Recent advances in discounting: implications for forest economics. J For Econ 13:169–189

HM Treasury and DECC (2012) Valuation of energy use and Greenhouse Gas (GHG) emissions: supplementary guidance to the HM treasury green book on appraisal and evaluation in central government. HM Treasury and Department of Energy and Climate Change, London, https://whitehall-admin.production.alphagov.co.uk/government/uploads/system/uploads/attachment_data/file/68947/supplementary.docx. Accessed 8 Mar 2013

Hockley N, Edwards-Jones G (2009) Growing food and fibre in a constrained world. In: Hemery G (ed). Timber, mutton or fuel? Debating the economics of land use and forestry. Proceedings of the ICF National Conference 2009. Cardiff

Hoen HF, Solberg B (1994) Potential and economic efficiency of carbon sequestration in forest biomass through silvicultural management. For Sci 40:429–451

Höhne N, Kejun J, Rogelj J, Segafredo L, Seroa da Motta R, Shukla PR, Angelsen A, Blok K, Chen C, Cruz R, Dalkmann H, Dellink R, den Elzen M, Egan C, Façanha C, Gesteira C, Graham P, Hargrave J, Hanaoka T, Karpay D, Levin K, Lowe J, Luderer G, Montzka S, Olivier J, Pestiaux J, Sawin E, Schaeffer M, Schaeffer R, Taylor C, Ton M, Ürge-Vorsatz D, van Vuuren D, Wagner F, Wunder S, and Xiusheng Z (2012) The emissions gap report 2012. UN Environmental Programme, Nairobi. http://www.unep.org/pdf/2012gapreport.pdf. Accessed 5 Mar 2013

IEA (2011) World Energy Outlook, press and media, quotes. International Energy Authority. http://www.iea.org/publications/worldenergyoutlook/pressmedia/quotes/5/. Accessed 13 Nov 2012

IPCC (2001) Climate Change 2001: the scientific basis. Contribution of Working Group I to the Third Assessment Report of the Intergovernmental Panel on Climate Change. In: Houghton J, Ding Y, Griggs D, Noguer M, van der Linden P, Dai X, Maskell K, Johnson C (eds) Cambridge University Press, Cambridge, UK, http://www.grida.no/publications/other/ipcc_tar/. Accessed 23 Jan.2013

IPCC (2005) Carbon dioxide capture and storage. Cambridge University Press, Cambridge, UK

IPCC (2007a) Climate change 2007: synthesis report. Fourth Assessment Report of the Intergovernmental Panel on Climate Change. Cambridge University Press, Cambridge, UK

IPCC (2007b) Climate change 2007: mitigation. Contribution of Working Group III to the Fourth Assessment Report of the Intergovernmental Panel on Climate Change. Cambridge University Press, Cambridge, UK

IPCC (2012) Meeting report of the intergovernmental panel on climate change expert meeting on Geoengineering. In: Edenhofer O, Pichs-Madruga R, Sokona Y, Field C, Barros V, Stocker TF, Dahe Q, Minx J, Mach K, Plattner G-K, Schlömer S, Hansen G, Mastrandrea M (eds) IPCC Working Group III technical support unit, Potsdam Institute for Climate Impact Research, Potsdam, Germany. http://www.ipcc-wg3.de/meetings/expert-meetings-and-workshops/em-geoengineering. Accessed 7 Nov.2012

Jandl R, Lindner M, Vesterdal L, Bauwens B, Baritze R, Hagedorn F, Johnson DW, Minkkinen K, Byrne KA (2007) How strongly can forest management influence soil carbon sequestration? Geoderma 137:253–268

Jarvis P, Clement R, Grace J, Smith K (2009) The role of forests in the capture and exchange of energy and greenhouse gases. Chapter 3. In: Read D, Freer-Smith P, Morrison J, Hanley N, West C, Snowdon P (eds) Combating climate change – a role for UK forests. An assessment of the potential of the UK's trees and woodlands to mitigate and adapt to climate change. The Stationery Office, Edinburgh, pp 21–49

Jorgensen DW, Wilcoxen PJ (1990) The cost of controlling US carbon dioxide emissions. Quoted in Cline (1992)

Keith DW, Parson E, Morgan MG (2010) Research on global sun block needed now. Nature 463:426–427

Kesicki F (2011) Marginal abatement cost curves for policy-making – expert-based vs model-derived curves. UCL Energy Institute, University College London. www.homepages.ucl.ac.uk/~ucft347/Kesicki_MACC.pdf. Accessed 5 Mar 2013

Latin H (2012) Climate change policy failures: why conventional mitigation approaches cannot succeed. World Scientific, London

Le Page M (2012) Global warming/global warning. New Sci 216(2891):34–39

Lenton T, Held H, Kreigler E, Hall J, Lucht W, Rahmstorf S, Schellnhuber H (2008) Tipping elements in the Earth's climate system. Natl Acad Sci USA 105(6):1786–1793

Lind RC (ed) (1982) Discounting for time and risk in energy policy. Johns Hopkins University Press, Baltimore

Lorenz K, Lal R (2010) Carbon sequestration in forest ecosystems. Springer, London

Lowe J (2008) Intergenerational wealth transfers and social discounting: supplementary Green Book guidance. HM Treasury, London, http://www.hm-treasury.gov.uk/d/4(5).pdf. Accessed 17 Jan 2013

Lüthi D, Le Floch M, Bereiter B, Blunier T, Barnola J-M, Siegenthaler U, Raynaud D, Jouzel J, Fischer H, Kawamura K, Stocker T (2008) High-resolution carbon dioxide concentration record 650,000-800,000 years before present. Nature 453:379–382, http://www.nature.com/nature/journal/v453/n7193/pdf/nature06949.pdf. Accessed 5 Mar 2013

Mahli Y, Meir P, Brown S (2002) Forests, carbon and global climate. Philo Trans R Soc Lond 360:1567–1591

Matthews RW, Broadmeadow MSJ (2009) The Potential of UK forestry to Contribute to Government's Emissions Reduction Commitments. In: Read DJ, Freer-Smith PH, Morison JIL, Hanley N, West CC, Snowdon P (eds) In Combating climate change – a role for UK forests – an assessment of the potential of the UK's trees and woodlands to mitigate and adapt to climate change. The Stationery Office, Edinburgh, pp 139–161

Moran D, Macleod M, Wall E, Eory V, Pajot G, Matthews R, Mcvittie A, Barnes A, Rees B, Moxey A, Williams A, Smith P (2008) UK marginal abatement cost curves for agriculture and land use, land-use change and forestry sectors out to 2022, with qualitative analysis of options to 2050. Final Report to the Committee on Climate Change, London

Morison JIL, Matthews R, Miller G, Perks M, Randle T, Vanguelova E, White M, Yamulki S (2012) Understanding the carbon and greenhouse gas balance of UK forests. Report for Forestry Commission, Forest Research

Moxey A (2011) Illustrative economics of peatland restoration. Report to IUCN UK Peatland Programme. www.iucn-uk-peatlandprogramme.org/scientificreviews. Accessed 5 Mar 2013

Murray B, McCarl B, Lee H-C (2004) Estimating leakage from forest carbon sequestration programs. Land Econ 80(1):109–124

Murray B, Lubowski R, Sohngen B (2009) Including international forest carbon incentives in climate policy: understanding the economics. Nicholas Institute, Duke University, USA, www.nicholas.duke.edu/institute/carbon.economy.06.09.pdf. Accessed 5 Mar 2013

Nabuurs G, Masera O, Andrasko K, Beneez-Ponce P, Boer R, Dutschke M, Elsiddig E, Ford-Robertson J, Frumhoff P, Karjalainen T, Krankina O, Kurz W, Matsumoto M, Oyhantcabal W, Ravindranath N, Sanz Sanchez M, Zhang X (2007) Forestry. In climate change 2007: mitigation. Contribution of Working Group III to the Fourth Assessment Report of the Intergovernmental Panel on Climate Change, Cambridge University Press

National Academy of Sciences (1991) Policy implications of greenhouse warming. National Academy Press, Washington, DC

Nijnik M, Pajot G, Moffat A, Slee B (2013) An economic analysis of the establishment of forest plantations in the United Kingdom to mitigate climate change. For Policy Econ 26:34–42

Nisbet T (2005) Water use by trees. Information Note, Forestry Commission, Edinburgh, http://www.forestry.gov.uk/forestry/infd-8bvgx9

Nisbet T, Marrington S, Thomas H, Broadmeadow S, Valatin G (2011) Slowing the flow at pickering, final report to Defra of FCERM multi-objective flood management demonstration project RMP5455. Forest Research, Farnham, http://www.forestry.gov.uk/fr/INFD-7ZUCQY#final1. Accessed 5 Mar 2013

Nordhaus W (1991) To slow or not to slow: the economics of the greenhouse effect. Econ J 101:920–937

Nordhaus WD (1993) Rolling the 'DICE': an optimal transition path for controlling greenhouse gases. Resour Energy Econ 15:27–50

Nordhaus WD (2007) A review of the Stern review on the economics of climate change. J Econ Lit 45:686–702

OECD (2009) The economics of climate change mitigation: policies and options for Global action beyond 2012. OECD, Paris, http://www.oecd.org/document/32/0,3746,en_2649_34361_41951200_1_1_1,00.html

Office of Management and Budget (various years) Budget of the United States Government. Office of Management and Budget, Washington, DC

OFGEM (2012) Feed-in Tariff Payment Rate Table for Photovoltaic Eligible Installations for FIT Year 3 (2012/13). Gas and Electricity Markets Authority (OFGEM), London. http://www.ofgem.gov.uk/Sustainability/Environment/fits/tariff-tables/Pages/index.aspx. Accessed 10 Jan 2013

Olschewski R, Benitez PC (2010) Optimizing joint production of timber and carbon sequestration of afforestation projects. J For Econ 16:1–10

OXERA (2002) A social time preference rate for use in long-term discounting. OXERA Consulting/Office of the Deputy Prime Minister, Department for Transport, Department for the Environment and Rural Affairs, London

Pearson P, Palmer M (2000) Atmospheric carbon dioxide concentrations over the past 60 million years. Nature 406:695–699

Peters G, Andrew R, Boden T, Canadell J, Ciais P, Le Quéré C, Marland G, Raupach M, Wilson C (2013) The challenge to keep global warming below 2°C. Nat Clim Chang 3:4–7

Peters-Stanley M, Hamilton K (2012) Developing dimension: state of the voluntary carbon markets 2012. Forest Trends, Washington, DC, http://www.forest-trends.org/publication_details.php?publicationID=3164, Accessed 5 Mar 2013

Portney PR, Weyant JP (eds) (1999) Discounting and intergenerational equity. Resources for the Future, Washington, DC

Pöyry Energy Consulting (2007) Analysis of carbon capture and storage cost–supply curves for the UK. Pöyry Energy, Oxford

Price C (1973) To the future: with indifference or concern? The social discount rate and its implications in land use. J Agric Econ 24:393–398

Price C (1984) Project appraisal and planning for over-developed countries. Environ Manage 8:221–242

Price C (1988) Does social cost-benefit analysis measure overall utility change? Econ Lett 26:357–361

Price C (1990) The allowable burn effect: a new escape from the bogey of compound interest. For Chron 66:572–578

Price C (1993) Time, discounting and value. Blackwell, Oxford, Also freely available in electronic format from c.price@bangor.ac.uk

Price C (1997) Analysis of time profiles of climate change. In: Adger WN, Pettenella D, Whitby M (eds) Climate change mitigation and European land use policies. CAB International, Wallingford, pp 71–87

Price C (2003) Diminishing marginal utility: the respectable case for discounting? Int J Sustain Dev 6:117–132

Price C (2005) How sustainable is discounting? In: Kant S, Berry AL (eds) Sustainability, economics and natural resources: economics of sustainable forest management. Springer, Amsterdam, pp 106–135

Price C (2006) Economics of sustainable development: reconciling diverse intertemporal perspectives. In Encyclopaedia of Life Support Systems (electronic book, no page numbers). EoLSS Publishers, Ramsey

Price C (2007) Sustainable forest management, pecuniary externalities and invisible stakeholders. For Policy Econ 9:751–762

Price C (2010) Low discount rates and insignificant environmental values. Ecol Econ 69:1895–1903

Price C (2011) Optimal rotation with declining discount rate. J For Econ 17:307–318

Price C (2012) Rising carbon flux price and the paradoxes of forest-based reduction of atmospheric carbon stock: an extended summary. Scand For Econ 44:240. Full manuscript available from the author

Price C (in review) A nightmare waiting in the wings: the social cost of carbon, forest economics and declining discount rates. Manuscript submitted to Journal of Forest Economics. Available from the author

Price Waterhouse Cooper (2012) Too late for two degrees? Low carbon economy index 2012. http://www.pwc.com/en_GX/gx/low-carbon-economy-index/assets/pwc-low-carbon-economy-index-2012.pdf. Accessed 26 Nov 2012

Price C, Willis R (1993) Time, discounting and the valuation of forestry's carbon fluxes. Commonw For Rev 72:265–271

Price C, Willis R (2011) The multiple effects of carbon values on optimal rotation. J For Econ 17:298–306

Radov D, Klevnas P, Skurray J, Harris D, Chambers B, Chadwick D, Dyer R, Nagler D (2007) Market mechanisms for reducing GHG emissions from agriculture, forestry and land management. NERA Economic Consulting, Defra, London

Ramanathan V, Feng Y (2008) On avoiding dangerous anthropogenic interference with the climate system: formidable challenges ahead. Natl Acad Sci USA 105(38):14245–14250

Ramsey FP (1928) A mathematical theory of saving. Econ J 38:543–559

Read DJ, Freer-Smith PH, Morrison JIL, Hanley N, West CC, Snowdon P (eds) (2009) Combating climate change – a role for UK forests. An assessment of the potential of the UK's trees and woodlands to mitigate and adapt to climate change. The Stationery Office, Edinburgh, http://www.tsoshop.co.uk/gempdf/Climate_Change_Main_Report.pdf. Accessed 24 Jan 2013

Richards K, Andersson A (2001) The leaky sink: persistent obstacles to a forest carbon sequestration program based upon individual projects. Clim Policy 1:41–54

Richards KR, Stokes C (2004) A review of forest carbon sequestration cost studies: a dozen years of research. Clim Change 63:1–48

Riebeek H (2011) The carbon cycle. Earth Observatory, NASA. http://earthobservatory.nasa.gov/Features/CarbonCycle/. Accessed 7 Nov 2012

Robock A, Marquardt A, Kravitz B, Stenchikov G (2009) Benefits, risks, and costs of stratospheric geoengineering. Geophys Res Lett 36, L19703

Royal Society (2009) Geoengineering the climate: science, governance and uncertainty. Royal Society, London, http://royalsociety.org/policy/publications/2009/geoengineering-climate/. Accessed 5 Mar 2013

Sedjo RA, Ley E (1997) The potential role of large-scale forestry in Argentina. In: Adger WN, Pettenella D, Whitby M (eds) Climate change mitigation and European land use policies. CAB International, Wallingford, pp 255–268

Sedjo RA, Solomon AM (1989) Climate and forests. In: Rosenberg NJ et al (eds) Greenhouse warming: abatement and adaptation. Resources for the Future, Washington, DC, pp 105–119

Schellnhuber H (2008) Global warming: stop worrying, start panicking? Natl Acad Sci USA 105(38):14239–14240

Schellnhuber H (2011) Geoengineering: the good, the MAD, and the sensible. Natl Acad Sci USA 108(51):20277–20278

Schellnhuber HJ (2012) Reply to Schuiling: last things last. Natl Acad Sci USA 109(20):E1211. http://www.pnas.org/content/109/20/E1211.full.pdf+html. Accessed 7 Nov 2013

Schuiling R (2012) Capturing CO_2 from the air. Natl Acad Sci USA 109(20):14239–14240

Schuur E, Abbott B (2011) High risk of permafrost thaw. Nature 480:32–33

Sohngen B (2009) An analysis of forestry carbon sequestration as a response to climate change. Copenhagen Consensus Center, Copenhagen, http://fixtheclimate.com/component-1/the-solutions-new-research/forestry

Sohngen B, Sedjo R (2006) Carbon sequestration costs in global forests. Energy J, Special Issue 27:109–126

Spash C (1994) Double CO_2 and beyond: benefits, costs and compensation. Ecol Econ 10:27–36

Spash C (2002) Greenhouse economics: value and ethics. Routledge, London

Spencer I, Bann C, Moran D, Mcvittie A, Lawrence K, Caldwell V (2008) A framework for environmental accounts for agriculture. Report SFS0601 by Jacobs with Scottish Agricultural College and Cranfield University to DEFRA Welsh Assemby Government, Scottish Government, and DARD (N Ireland), London

Stern N (2006) The economics of climate change, HM Treasury/Cambridge University Press. http://www.hm-treasury.gov.uk/stern_review_report.htm. Accessed 5 Mar 2013

Thompson D, Matthews R (1989) CO_2 in trees and timber lowers greenhouse effect. Forestry and British Timber October:19–24

Treasury HM (2003) The green book: appraisal and evaluation in central government. The Stationery Office, London

Trumper K, Bertzky M, Dickson B, van der Heijden G, Jenkins M, Manning P (June 2009) The natural fix? The role of ecosystems in climate mitigation. A UNEP rapid response assessment. United Nations Environment Programme, UNEPWCMC, Cambridge. http://www.grida.no/publications/rr/natural-fix/. Accessed 5 Mar 2013

UNFCCC (2008) Report of the conference of the parties on its thirteenth session, held in Bali from 3 to 15 December 2007. Part two: action taken by the Conference of the Parties at its thirteenth session. United Nations Convention on Climate Change, Decision 2/CP.13, FCCC/CP/2007/6/Add.1. http://unfccc.int/resource/docs/2007/cop13/eng/06a01.pdf. Accessed 5 Mar 2013

UNFCCC (2010) Report of the conference of the parties on its fifteenth session, held in Copenhagen from 7 to 19 December 2009. Part two: action taken by the conference of the parties at its fifteenth session. United Nations Convention on Climate Change, FCCC/CP/2009/11/Add.1 http://unfccc.int/resource/docs/2009/cop15/eng/11a01.pdf. Accessed 5 Mar 2013

UNFCCC (2011) Report of the conference of the parties on its sixteenth session, held in Cancun from 29 November to 10 December 2010. Part two: action taken by the conference of the parties

at its sixteenth session. United Nations Convention on Climate Change, FCCC/CP/2010/7/
Add.1. http://unfccc.int/resource/docs/2010/cop16/eng/07a01.pdf. Accessed 5 Mar 2013

Valatin G (2005) Justice, human security and the environment, PhD Dissertation, Economics
Department, University of Siena

Valatin G (2011a) Forests and carbon: a review of additionality. Research Report. Forestry
Commission, Edinburgh, http://www.forestry.gov.uk/fr/INFD-7WUEAN. Accessed 5 Mar 2013

Valatin G (2011b) Forests and carbon: valuation, discounting and risk management. Research
Report, Forestry Commission, Edinburgh, http://www.forestry.gov.uk/fr/INFD-7WUEAN.
Accessed 5 Mar 2013

Valatin G (2012) Marginal abatement cost curves for UK forestry. Research Report 19, Forestry
Commission, Edinburgh, http://www.forestry.gov.uk/PDF/FCRP019.pdf/$FILE/FCRP019.
pdf. Accessed 5 Mar 2013

Valatin G, Saraev V ((2012) Natural environment framework: woodland creation case study.
Report to Forestry Commission Wales, Forest Research, Edinburgh, http://www.forestry.gov.
uk/fr/INFD-8YAECD. Accessed 5 Mar 2013

Valatin G, Starling J (2011) Valuation of ecosystem services provided by UK woodlands, appendix
to chapter 22 of UK National Ecosystem Assessment. UNEP-WCMC/Defra, London

van Kooten GC, Sohngen B (2007) Economics of forest ecosystem carbon sinks: a review. Int Rev
Environ Resour Econ 1:237–269

van Kooten GC, Johnston C, Xu Z (2012) Economics of forest carbon sequestration. Working paper
2-12-4, Resource Economics and Policy Analysis Research Group. University of Victoria, Canada

Weikard H-P, Wangler L, Freytag A. (2009). Minimum participation rules with heterogeneous countries.
Jena Economic Research Papers, zs.thulb.uni-jena.de/receive/jportal_jparticle_00156984

Worrall F, Chapman P, Holden J, Evans C, Artz R, Smith P, Grayson R (2011) A review of current
evidence on carbon fluxes and greenhouse gas emissions from UK peatlands. Joint Nature
Conservation Committee, Peterborough, www.jncc.defra.gov.uk/pdf/jncc442_webFinal.pdf.
Accessed 5 Mar 2013

Additionality of Climate Change Mitigation Activities

Gregory Valatin

Abstract Although widely considered to be a core aspect of quality assurance of climate change mitigation activities, additionality remains a source of much controversy in relation to carbon accounting and carbon markets. This chapter illuminates the multi-faceted nature of the concept and develops a taxonomy of different forms. It provides an overview of how additionality is currently applied in relation to both compliance and voluntary carbon markets, including tests used and underlying evidence base requirements. This draws upon and updates an earlier review commissioned to help inform development of a Woodland Carbon Code designed to underpin climate change mitigation activities in the UK by the forest sector. Sources of uncertainty and trade-offs in practical application of the concept are highlighted, and potential perverse incentives explored.

1 Introduction

The concept of additionality is used to distinguish the net benefits associated with an activity or project by comparison with what would have happened in the absence of the intervention (HM Treasury 2003). In a climate change mitigation context, additionality is generally used to mean net greenhouse gas (GHG) emissions savings or sequestration benefits in excess of those that would have arisen anyway in the absence of a given activity or project (i.e. compared to a 'baseline').

Along with issues of permanence, leakage and displacement, additionality is widely considered a core aspect of quality assurance of climate change mitigation activities. Lack of additionality implies there are no GHG abatement benefits over and above those that would have arisen anyway. Credits issued for benefits which

G. Valatin (✉)
Forest Research, Northern Research Station Roslin, Midlothian EH25 9SY, Scotland, UK
e-mail: gregory.valatin@forestry.gov.uk

T. Fenning (ed.), *Challenges and Opportunities for the World's Forests in the 21st Century*, Forestry Sciences 81, DOI 10.1007/978-94-007-7076-8_14,

are not additional but are used as offsets result in an overall increase in GHG emissions. They would not provide net abatement benefits to those who purchase them, and would undermine wider climate change mitigation efforts. The requirement for climate change benefits to be 'additional' is reflected at international level in Articles 3.4, 6.1, and 12.5 of the Kyoto Protocol.

Additionality remains a source of much controversy in relation both to carbon accounting and carbon markets. This relates in part to the hypothetical nature of counterfactuals (i.e. identification of what would have otherwise occurred) upon which baselines and additionality determination are based (e.g. Schneider 2007; McCully 2008; Wara and Victor 2008; Shapiro 2010). It also relates to concerns that additionality criteria can provide perverse incentives to invest in relatively high-cost projects that offer comparatively few climate benefits (e.g. Bode and Michaelowa 2003), or even in projects that increase GHG emissions (Mukerjee 2009; Calel 2011). To the extent that the concept encompasses a range of wider environmental, institutional and social considerations that have no direct connection to GHG balances per se but relate to broader criteria for judging the value of abatement activities, the scope of additionality (i.e. which of the wider aspects are included) can also be controversial.

1.1 Structure

Focusing upon project additionality, this article explores the multi-faceted nature of this concept (for discussion of international finance additionality, see: Brown et al. 2010). It builds upon a review (Valatin 2011) commissioned to help inform development of the Woodland Carbon Code (Forestry Commission 2011b) designed to help underpin climate change mitigation activities by the UK forestry sector.

Drawing upon existing protocols, the next three sections develop a taxonomy of different forms of environmental additionality, legal, regulatory and institutional additionality, and financial and investment additionality. These forms are treated as distinct for the purposes of developing the taxonomy, although some are closely linked, and in practice distinctions are sometimes blurred and different forms combined within a single test.

The article then provides an overview of how additionality is currently applied in practice in both compliance and voluntary carbon markets, including explicit tests and underlying evidence requirements for forestry projects. This is based upon reviewing material on approaches to additionality published on the websites of the different carbon standards without prejudging coverage of the concept.

The article then considers how some additionality tests can give rise to perverse incentives. Such concerns have arisen in a wider context, especially in relation to Certified Emissions Reduction credits (CERs) issued for trifluoromethane (HCF-23) destruction projects implemented under the Clean Development Mechanism (CDM) of the Kyoto Protocol (Rajan 2011; Schwank 2004; Schneider 2011a, b). Analysis of how perverse incentives could potentially arise in relation to woodland carbon projects draws upon a modified version of the additionality game developed by Calel (2011) in the context of HCF-23 projects. A final section offers some conclusions. Readers preferring to skip details of the taxonomy or current approaches can find summary information on these in Table 7, and in Tables 2 and 3, respectively.

2 Types of Additionality

2.1 Environmental Additionality

Environmental aspects are fundamental to determining the additionality of GHG abatement projects. At least six forms can be distinguished.

GHG Additionality

A key component in quantifying abatement savings, GHG additionality relates to comparisons with a baseline level of emissions. The net impact of a project on GHG balances fundamentally depends upon the breadth of carbon pools and other GHG fluxes taken into account in determining the baseline, emissions and sequestration. In some cases a narrow focus may be taken on carbon fluxes associated with above-ground vegetation only. In others, account may also be taken of below-ground carbon pools including soils, storage in harvested wood products, and displacement effects of using wood products instead of fossil fuel and more fossil fuel intensive materials. Covering a wider range of impacts can be expected to improve the quality of additionality assessments providing reliable quantification methods are available, but also increase the cost.

The time horizon used to judge GHG additionality can also be important, as well as any explicit or implicit weighting system used in comparing GHG emissions and savings at different points in time. Where projects increase household incomes, 'rebound effects' of GHG emissions associated with higher incomes and energy use may also be taken into account. In some cases, the baseline may take account of alternative investment options, rather than purely projections for the project area in the absence of the project going ahead.

The net impact also depends upon the method used to determine the baseline. This may be established using a project-by-project ('bottom-up') approach, or using a standardised or benchmarked ('top-down') approach. Better able to take account of specific project attributes and site conditions, a project-by-project approach can be more precise. By contrast, a standardised or benchmarked approach has the advantage of reducing project-specific transactions costs (Bloomgarden and Trexler 2008), with project activities simply considered additional in some cases if they are of a particular type included on a 'positive list' (Peters-Stanley 2012).

Unit Additionality

To prevent what are considered 'business as usual' activities being credited and to promote efficient resource use, projects may only be considered additional where their output is associated with emissions per unit output below a specified benchmark level unrelated to the project baseline. For example, emissions may also be required to be below the average per unit output in the sector to be considered additional where a project-by-project approach to setting the baseline is used. Similarly, forestry activities may only be considered additional if GHG savings per unit of wood production, or GHG savings per unit of land area (specified independently of the project baseline) are above a particular level.

Project Additionality

The nature of the counterfactual assumed for the baseline is fundamental to additionality determination. In cases of avoided deforestation or forest degradation, activities are generally deemed additional only where these areas would otherwise have been expected to be deforested or degraded. In some cases afforestation, reforestation and forest management activities may similarly only be deemed additional where forests are considered unable to establish themselves in the absence of the project (e.g. through natural regeneration), or woody biomass is not expected to increase in the absence of the project. To prevent carbon markets providing perverse incentives (e.g. for prior deforestation or degradation to subsequently claim credits for project activities), projects on some types of land, such as areas subject to recent anthropogenic clearance of trees, may be excluded.

Intent Additionality

Abatement may have to be shown to have been an original objective of the project for associated GHG benefits to be considered additional. An aspect that Costa et al. (2000) term "intent" (or "program") additionality, the aim is to exclude projects for which GHG benefits are purely coincidental.

Tree Additionality

In some cases, the number of trees planted may have to exceed the number removed by a particular margin. For example, a 'no net loss' criteria may be applied in order for GHG benefits to be considered additional.

Ecological Additionality

To be considered additional, in some cases projects have to show positive net ecological impacts (e.g. enhance habitats, flora and fauna, and biodiversity), use 'native' species, increase resilience to climate change, or provide evidence that no genetically modified species are used. These elements could potentially be subdivided into associated sub-categories (e.g. 'habitat additionality', 'flora additionality' 'fauna additionality', 'biodiversity additionality', 'native additionality', 'resilience additionality', 'naturalness additionality', etc.). Projects may have to adopt practices considered ecologically beneficial (e.g. natural regeneration) and avoid those considered harmful (e.g. broadcast fertilisation). They may also have to pass an environmental impact assessment.

2.2   Legal, Regulatory and Institutional Additionality

Different forms of legal, regulatory and institutional additionality are apparent at the level of the individual enterprise, government and carbon certificate purchaser. At least 12 forms can be distinguished.

Compliance Additionality

At the level of individual enterprises, only GHG benefits that exceed those associated with meeting statutory minimum standards are generally considered additional, with the chosen baseline in part based upon regulatory requirements. This form is sometimes termed 'regulatory additionality' (e.g. Reynolds 2008). However, for clarity different terminology is adopted here as other elements of additionality are also associated with regulations.

Incentive Additionality

Changes may not be mandatory, but GHG savings may nonetheless be expected to accrue as a consequence of incentives provided by existing regulatory frameworks (e.g. woodland grant schemes) or GHG-related environmental credit (e.g. renewable energy certificate) schemes. Only benefits exceeding those expected under these incentive structures may be considered additional. To prevent over-reliance on public funding, projects may also be required to have a minimum proportion of private or voluntary sector finance to be considered additional (an aspect closely related to financial additionality – see section "Financial Additionality" below).

Threshold Additionality

In some cases a limit may be applied to the GHG savings considered additional. The cap may depend upon the type of project and be specified in terms of maximum production, aggregate abatement or abatement per unit of production. It may be introduced to limit the size of projects due to wider environmental (or other) considerations.

Norm Additionality

A project may have to comply with wider voluntary industry standards, codes or good practice benchmarks for GHG benefits to be considered additional. These may cover a range of socioeconomic (e.g. employment, income generation and poverty alleviation) and community engagement issues. They can also relate more broadly to sustainable forest management issues, such as watershed and soil erosion protection. For example, carbon savings from woodland creation in the UK may be considered additional only where a project conforms with various good practice guidelines under the UK Forestry Standard (Forestry Commission 2011a).

This form of additionality could potentially also be further subdivided into subcategories covering particular types of norms (e.g. 'employment additionality', 'income additionality' 'poverty additionality', 'engagement additionality', 'watershed additionality', 'erosion additionality', etc.).

Technological Additionality

To promote production efficiency, GHG savings may only be considered additional if they result from the application of a specific type or category of technology that differs from "business as usual" (and thus from that generally assumed under the baseline). To the extent that it involves lower monitoring costs, technological additionality may be viewed as a useful proxy for unit additionality in some cases.

Barrier Additionality

A variety of legal, social, technological, ecological or financial barriers may exist that normally prevent particular climate mitigation activities being undertaken. For example, these may relate to weakness of existing land tenure or property rights, to soil degradation, or unfavourable climatic conditions. Projects may only be considered additional if they can be shown to overcome existing barriers.

Practice Additionality

In some cases, only activities which are not common practice in the area in which they are located may be considered additional. For example, a woodland creation project may only be considered additional if similar projects without carbon funding are not already undertaken in a similar geographical and regulatory environment.

Reporting Additionality

Principles applied to GHG accounting and national reporting of emissions reductions at government level may affect which project level savings are considered additional. For example, additionality is not an issue in relation to afforestation, reafforestation and deforestation activities under Article 3.3 of the Kyoto Protocol, but it can be with respect to forest management measures under Article 3.4, and is a requirement in trades between countries under Article 6 (IPCC 2000, 5.7, Tables 5–10).

Institutional Additionality

From the perspective of a private or voluntary sector prospective buyer, GHG benefits may only be considered additional where activities are entirely independent of, rather than part of, meeting national targets, or where they exceed those envisaged under these targets. Carbon credits that are used to meet government targets (e.g. included in national GHG inventories reported under the Kyoto Protocol), or included under binding cap-and trade schemes, and sold in voluntary carbon markets give rise to potential double-counting. In the absence of mechanisms to retire carbon certificates sold on voluntary markets, lack of this form of additionality applies to many climate change mitigation activities within countries such as the UK that have legally binding emissions reduction commitments under the Kyoto Protocol (Kollmuss 2007). It can also exert a significant influence on voluntary carbon markets. The precipitous drop in voluntary carbon certificates sold in the EU from 2.3 $mtCO_2e$ in 2007 to 0.2 $mtCO_2e$ in 2008, for instance, is reported to have been due in part to double-counting concerns related to the associated reductions also being covered by national reporting under the Kyoto Protocol (Hamilton et al. 2009).

Date Additionality

GHG benefits may only be considered additional and credited if they occur after a specified date, or if they are associated with specific activities or projects that commenced after (or, in a few cases, before) a particular date. To the extent that projects commencing prior to a particular date are initiated for reasons entirely unrelated to climate change mitigation, for example, only accounting for abatement from projects initiated subsequently may be viewed as important to help underpin intent additionality.

Term Additionality

Related partly to practical project implementation issues such as time horizons for monitoring and verification, and level of commitment to land use change required of landowners, in some cases GHG benefits may be considered additional only if they occur within a particular time-frame. For example, abatement may be counted only if coming from a project with a time horizon above a specified minimum duration.

To ensure additionality criteria are periodically re-assessed, the maximum period that GHG benefits are considered additional may also be capped.

Jurisdiction Additionality

Although often more associated with defining coverage than additionality per se, only if GHG benefits arise within a particular geographical area (e.g. specific countries), or in some cases, if activities involve particular communities or social groups, are GHG benefits considered additional.

2.3   Financial and Investment Additionality

Financial and investment aspects of additionality are closely related. In some cases, they are viewed as the key determinants of additionality and merged into a single test. Merger (2008, p. 19) takes this approach, for example, stating that "a project must provide evidence that without the additional financial means from the sale of CO_2 certificates the project cannot be implemented." However, at least five aspects can be distinguished.

Financial Additionality

Evidence often has to be provided that a project would not have been financed without revenues from the sale of carbon certificates for GHG benefits to be considered additional. Activities that would have been financed anyway (e.g. with international development assistance) are not considered additional (cf. Au Yong 2009).

To the extent that availability of finance is a potential barrier to project implementation, financial additionality can be considered a type of barrier additionality. Financial additionality in this sense is quite different from meanings currently in international climate negotiations which focus on aggregate levels of finance (see: Bode and Michaelowa 2003, p.507; Brown et al. 2010).

Viability Additionality

To be considered additional, in some instances developers have to demonstrate that a project would not be economically viable without revenues from the sale of carbon certificates. In cases where projects are only financed if expected to be economically viable, this form of additionality is a prerequisite for financial additionality. It can also be considered a form of barrier additionality.

Investment Additionality

A project may have to demonstrate that it would not be the most financially attractive option without revenues from the sale of carbon certificates in order for GHG benefits to be considered additional. This does not necessarily imply either that a project would not be financially viable without revenues from the sale of carbon certificates (viability additionality), or that it would not have been financed (financial additionality).

Sales Additionality

Where a project commences before the date of registration under a particular standard or mechanism, often it may only be considered additional if the income from the sale of carbon credits was a decisive factor in the original decision to proceed. In many instances both investment additionality and financial additionality could be expected to be prerequisites for sales additionality. (An exception could arise if income from the sale of carbon certificates is a decisive factor despite a project not being expected to be economically viable).

Closely related to intent additionality (which is generally a prerequisite), the purpose of sales additionality is similar in aiming to exclude projects for which carbon revenues are purely coincidental.

Gaming Additionality

Some GHG emissions may be generated by activities primarily in order to obtain carbon credits for subsequent abatement (see subsequent Sect. 4 on "Perverse Incentives"). Where this occurs, abatement is not generally considered additional.

3   Tests Applied to Forestry Projects

This section provides a review of the coverage of existing additionality protocols, focusing upon the aspects covered by tests applied to forestry projects or credits, and comparing approaches used and associated evidence requirements. Tests applied under the Kyoto Protocol Clean Development Mechanism (CDM), Air Resources Board (ARB) protocols applying to the California cap-and-trade scheme, and six voluntary market standards, including the recently launched Woodland Carbon Code (Forestry Commission 2011b) are focused upon. Those voluntary market standards (including Greenhouse Friendly, VER+, and the Voluntary Offset Standard) currently based entirely, or almost entirely, upon CDM or other UNFCCC methodologies are excluded from separate consideration to avoid repetition. Similarly, as those used under the Kyoto Protocol Joint Implementation can follow the same methodologies used for CDM projects (see http://www.jirulebook. org/5091), these are also excluded from separate consideration below.

This expands comparisons in Valatin (2012) to include ARB, and serves to further illustrate the developmental stage of carbon market standards. Summary information for each of the standards examined is given in Table 1. This includes

Table 1 Carbon standards and associated certificate and project types

Mechanism or standard	Type of certificate	Timing of certificate issue	Name of unit(s)	Type of projects
American Carbon Registry Forest Carbon Project Standard (v2.1 Nov 2010)	Voluntary	Ex-post	Emission Reduction Ton (ERT)	A/R, IFM, REDD
California Air Resources Board (Oct 2011)	Compliance	Ex-post	ARB offset credit	R, IFM, AC, UF
Clean Development Mechanism	Compliance	Ex-post	Temporary Certified certified Emission emission Reduction (tCER); Long-term Certified Emission Reduction (lCER)	A/R
CarbonFix (v3.2 Dec 2011)	Voluntary	Ex-ante and ex-post	CO$_2$-certificate	A/R
Green-e (v2.0 June 2011)	Unrestricted	Ex-post	Unrestricted	F+
Plan Vivo (Oct 2008)	Voluntary	Ex-ante	Plan Vivo certificate	D
Verified Carbon Standard (v3.1 July 2011)	Voluntary	Ex-post	Verified Credit Unit (VCU)	ARR, APD, AUFDD, IFM, REDD
Woodland Carbon Code (v1.3 July 2011/ March 2012 additionality protocol)	UK (non-tradable internationally)	Ex-post	Unrestricted	A

Project types: A afforestation, R reforestation, A/R afforestation and reforestation, ARR afforestation, reforestation and revegetation, AC avoided conversion, APD avoiding planned deforestation, AUFDD avoiding unplanned frontier deforestation and degradation, REDD reduced emissions from deforestation and degradation, D developing country community forestry, IFM improved forest management, UF urban forestry, F+ forestry and other project types
See the following websites: http://americancarbonregistry.org/, http://www.arb.ca.gov/cc/capandtrade/offsets/offsets.htm, http://cdm.unfccc.int/, http://www.carbonfix.info/CarbonFix-Standard.html, http://www.green-e.org/, http://www.planvivo.org, http://www.v-c-s.org, http://www.forestry.gov.uk/carboncode

information on the version of each of the voluntary carbon market standards focused upon, and whether carbon certificates are issued ex-post (i.e. after), or ex-ante (i.e. before) monitoring and verification. (Note that even where carbon certificates are issued ex-post, arrangements also often allow them to be secured by advance payment prior to their issue – a practice termed 'forward crediting'). Web links for material drawn upon for comparisons – including for the tables that follow – are listed in the notes under Table 1.

Based upon reviewing published information on their websites, types of additionality tests applied explicitly as part of additionality protocols are summarised in Table 2. Tests applying to small- and large-scale projects are distinguished in the case of CDM forestry as separate protocols apply. Note, however, that the snapshot provided may be an incomplete guide to existing practice to the extent that this also depends upon informal norms and conventions (for example, if developers of small-scale CDM forestry projects also apply tests required for large-scale projects, or included under non-binding best practice guidance).

Focusing upon explicit tests is informative in identifying and comparing aspects covered by additionality protocols under different standards. This only provides a partial picture, however, as similar ('implicit' additionality) tests exist in some cases under general eligibility requirements or other sections of the standards rather than as part of an additionality protocol. Furthermore, some protocols note that certain types of projects are assumed to pass a particular additionality test (e.g. whether it is a common practice, or not). For the purposes of wider comparisons of the scope of different standards, this is termed 'presumed additionality' and is also included in the more comprehensive comparison of additionality tests provided in Table 3. This shows, for example, that GHG additionality (quantifying net GHG benefits compared to a baseline) underpins all the standards, although it is only focused upon explicitly in demonstrating additionality in all cases under the California Air Resources Board and for small-scale CDM projects.

Marked differences in the range of additionality tests exist between standards (Tables 2 and 3). However, these appear largely unrelated to whether ex-post and ex-ante certificates are issued (Table 1). Furthermore, there is flexibility under some standards for the project developer to choose how to demonstrate additionality, with specific combinations of the tests treated as alternatives in some cases.

Similarities and differences in the specific types of additionality tests applied are considered next.

3.1　Environmental Additionality Tests

GHG Additionality Tests

The GHG additionality test specified under the CDM for small-scale Afforestation and Reforestation (A/R) projects involves demonstrating that within the project boundary the net GHG removals by sinks are increased above the sum of changes in

Table 2 Explicit additionality tests applied to forestry projects

Category	Additionality test	American Carbon Registry Forest Carbon Project Standard	California Air Resources Board	Clean development mechanism: Small-scale	Large-scale	CarbonFix	Green-e	Plan Vivo	Verified Carbon Standard	Woodland Carbon Code
Environmental	GHG	~	√	√		~	~			
	Unit	~							~	
	Project	√	~		~	√			~	
	Intent	~	~	~			~			
	Tree		~							
	Ecological									
Legal, regulatory and institutional	Compliance	√	√		√	~	√	√	√	√
	Incentive			~						√
	Norm					√				
	Technological									
	Barrier	~		~		~	~		~	√
	Practice	~			~		~		~	
	Institutional				~		~		~	
	Date	√	~		√	~	√	√	~	
	Term									
	Jurisdiction									
Financial and investment	Financial	~			√	~	~		~	~
	Viability	~			√	~	~			
	Investment		~		~	~	~			√
	Sales	~				~	~			
	Gaming									

Note: √ denotes test applied in all cases; ~ denotes applies in some cases

Table 3 Explicit and implicit additionality tests applied to forestry projects

Category	Additionality test	American Carbon Registry Forest Carbon Project Standard	California Air Resources Board	Clean development mechanism: Small-scale	Clean development mechanism: Large-scale	CarbonFix	Green-e	Plan Vivo	Verified Carbon Standard	Woodland Carbon Code
Environmental	GHG	~‡	√	√	¶	~‡	¶	¶	¶	¶
	Unit	~			†	†	~		~	
	Project	√	~¶	¶	¶	√/¶	~		~	
	Intent	~		~	~		~		†	
	Tree		~							
Legal, regulatory and institutional	Ecological	¶	†¶		√	¶	¶	¶	¶	¶
	Compliance	√	√ρ			~	√	√	√	√
	Incentive									√
	Norm	¶		¶	¶	√¶		¶	¶	¶
	Technological				¶	~		√		√
	Barrier	~	ρ	~	~		~			
	Practice	~			~	~	~		~	
	Institutional		~			~	¶	¶	¶	
	Date	√	~	¶	√	~	√	¶	¶	¶
	Term	¶	¶	¶	¶	¶	¶	¶	¶	¶
	Jurisdiction	¶	¶	¶	¶		¶	¶	¶	¶
Financial and investment	Financial	~			√		~		~	~
	Viability	~	~		√	~	~			
	Investment				√	~	~			√
	Sales	~			~	~	~		†	
	Gaming								¶	

Notes: Explicit additionality test: √ applied in all cases; ~ applied in some cases. Implicit additionality test: ¶ applied in all cases; ‡ applied in cases an explicit test does not apply; † applied in some cases. Presumed additionality test applied in some cases: ρ. In some cases more than one test apply

carbon stocks in the carbon pools that would otherwise have occurred. Where used as part of the additionality protocol, the test under CarbonFix is similar, being based upon the latest CDM additionality tool. If project proponents provide relevant evidence indicating that no significant changes in the carbon stocks within the project boundary would occur in the absence of the project, existing carbon stocks prior to implementation of the project are considered as the baseline (and assumed constant throughout the crediting period). If significant changes in carbon stocks within the project boundary are expected in the absence of the project, an approved simplified baseline and monitoring methodology for small-scale A/R projects must be used. The carbon pools covered under these vary between different project categories and types of location as illustrated in Table 4.

Although not explicit additionality tests (except where the small-scale forestry CDM protocol is adopted under CarbonFix), it is notable that carbon pools covered in setting project baselines under the voluntary carbon standards also differ. This is illustrated in Table 5, in which, the two standards which do not specify carbon pools covered explicitly, are excluded. (Under Green-e validation, monitoring and verification standards must be "explicit, transparent and credible"; while under Plan Vivo carbon accounting has to be based upon "best available evidence"). The approach under the Woodland Carbon Code differs from most other standards in using standardised baselines (based upon 'carbon look-up tables'), rather than project-specific baselines.

Coverage of carbon pools and other GHG fluxes also varies between project types for large-scale forestry CDM projects (see: Valatin 2011, Table 6, p. 21).

Framing of the GHG additionality test under the ARB is similar to that for large-scale forestry CDM projects in terms of the focus upon GHG reductions or removal enhancements in excess of those expected under business-as-usual activities. However, in practice no GHGs apart from CO_2 are covered under the protocol for US forest projects (see: ARB 2011a, section 5) at present. In principle they are covered where 'significant' under the protocol for urban forest projects, but as no guidance on how they are to be estimated is provided (see: ARB 2011b, sections 4 and 5), in practice this protocol focuses upon carbon fluxes.

Unit Additionality Tests

Indicators used differ between the standards. Under Green-e it has to be shown that GHG emissions are reduced below levels of technologies commonly used to produce the same products/services. The Verified Carbon Standard (VCS) test involves demonstrating that carbon sequestration per unit of output by the project is above (or GHG emissions generated below) the benchmark level approved for the product, service, sector or industry. (Although a similar approach to Green-e is also specified, this appears less relevant to woodland projects). The test under the American Carbon Registry Forest Carbon Project Standard (ACRFCPS) focuses on demonstrating that an activity exceeds a performance standard benchmark representing typical forest management of the forest type and region in which the project takes

Table 4 Coverage of small-scale CDM A/R baseline methodologies

Project location (prior land use)/ type	CDM reference	Carbon pools					
		Above-ground tree biomass	Above-ground woody perennials biomass	Below-ground grassland biomass	Below-ground tree biomass	Below-ground woody perennials biomass	Soil organic carbon
Grasslands or croplands	AR-AMS0001 †	✓	✓	✓	✓	✓	
Settlements	AR-AMS0002 ‡	✓			✓		
Wetlands	AR-AMS0003	✓			✓		
Agroforestry	AR-AMS0004 ζ	✓			✓		✓
Low quality lands	AR-AMS0005 ξ	✓			✓		✓
Silvopasture	AR-AMS0006 Ψ	✓			✓		✓

Notes: † version 05 (EB 42); ‡ version 02 (EB 42); ¶ version 01; ζ version 02 (EB 47); ξ 02 (EB 46); Ψ 01 (EB 47)
Source: http://cdm.unfccc.int/methodologies/SSCmethodologies/SSCAR/approved.html

Table 5 Carbon pools and other GHGs covered under voluntary carbon standards

			California Air Resources Board					Verified Carbon Standard							
Pool			American Carbon Registry Forest Carbon Project Standard	Reforestation	Improved forest management	Avoided conversion	Urban forestry	Carbon Fix	Afforestation, reforestation and revegetation	Conversion to reduced impact logging (RIL) with minimal impact on timber	Conversion to RIL with over 25 % reduction in timber extracted, or from logged to protected forests	Extended rotation length/ conversion of low productive forests to productive forests	Conversion of forest to non-forest annual crop or pasture	Conversion of forest to perennial tree crop (e.g. oil palm, bananas, fruit trees, spice trees, tea shrubs)	Woodland Carbon Code
Above ground biomass	Tree		√	√	√	√			√	√	√	√	√	√	√
	Non-tree	Woody	√	√				√	△					√	√
		Non-woody						√							
Deadwood	Standing		√	√	√	√				√	√				
	All														
Below ground biomass	Tree														√
	Non-tree	Woody						√							√
		Non-woody						√							
	All								△	~	~	~	~	~	√
Tree biomass							√								
Litter				√					△						
Soil			~	√	√	~			△	~	~	~	△	△	√
Site preparation	Biological emissions			~	√										
	Fuel emissions			√											

Project activity	Source	(1)	(2)
Tree planting/care	Fuel emissions	√	
Woodland management			√
Clearing forest land outside project area	Biological emissions	√	√
Changes in wood harvesting outside project area	Biological emissions	~	~
Harvested wood products	In-use	√	√
	In landfills	~	~
	Decomposition	√	√
	All		
Unspecified/other		√	~

Notes: √ denotes covered in all cases; ∆ denotes has to be included where project activities may significantly reduce pool; ~ denotes covered in some cases; (1) All significant changes in carbon pools/GHG sources with exception of litter, and emissions from removal of herbaceous vegetation, fertiliser application, and of nitrous oxide (N_2O) from litter and fine root decomposition; (2) 0.5 % of future CO_2 fixation deducted to cover within project fossil fuel use (e.g. by machines and flights). Where fertiliser is used, 0.005 tCO_2 per kg of nitrogen is deducted. Any biomass burned in land preparation is assumed to add 10 % to baseline emissions to cover N_2O and CH_4 emissions

place. The benchmark in this case may be based upon net sequestration and emissions rates, or upon emissions per unit of output (e.g. of harvested wood products) along similar lines to standards applying in other sectors.

Labels used differ. Tests under the ACRFCPS and the VCS are termed a 'performance benchmark', while that under Green-e (in contrast to terminology adopted in this article) is termed a 'common practice' test.

Project Additionality Tests

A land eligibility test applying to (A/R) projects is included under the ACRFCPS which requires that none of the land was subject to anthropogenic clearing of native ecosystems within 10 years of the project start date. Where loss of cover occurred due to natural disturbance, it has to be demonstrated that there is no natural recovery. The test under CarbonFix similarly applies to woodland creation, with evidence required that there would be no increase in woody biomass on the area to be planted in the absence of the project (and where this does not hold, any increase has to be accounted for in establishing the baseline).

Under the ARB, official documentation has to be provided for avoided conversion projects demonstrating that the anticipated land use conversion is legally permissible and evidence provided that the project area is suitable for conversion. Where conversion to commercial, residential or agricultural land is anticipated, the project area must have a slope of not more than 40 %. Where conversion to agricultural land is anticipated, evidence has to be provided that the soil is suitable and water available for the expected agricultural land use, while where conversion to mining is anticipated, evidence has to be provided of the extent and amount of mineral resources within the project area.

General requirements of additionality demonstration and baseline determination under VCS mention deforestation and degradation rates among factors that can require assessment across a given geographical area.

Intent Additionality Tests

The form of intent additionality tests used for projects commencing prior to registration is similar under different standards. A project starting before November 1997 may be approved under the ACRFCPS if documentation is provided to show that GHG mitigation was an objective from the inception of the project (and approved methodologies can also require documentation to demonstrate that GHG mitigation was originally a primary objective). Similarly, credits from a project starting prior to 2000 may be deemed additional under Green-e if it is demonstrated that they are for activities initiated in part for the purpose of reducing or displacing GHG emissions.

Under the CDM, those projects commencing before 2 August 2008 and prior to publication of a project design document have to demonstrate that the CDM was seriously considered at the outset. Evidence has to be shown that CDM benefits were a decisive factor, and that continuing actions were made in parallel with implementation to secure CDM status, with these commencing no more than 3 years after the start of the project. For projects starting subsequently, within the first 6 months developers have to inform a designated national authority in the host country and the UNFCCC secretariat of their intention to seek CDM status.

Tree Additionality Tests

Under the ARB, urban forest projects undertaken by municipalities and on educational campuses must include planting at least as many trees as are removed, and exceed the business-as-usual net tree gain (the annual number of trees planted minus the annual number removed). This can be estimated for a single year, or series of years, over the previous 5 years, and the 5-year moving average annual net tree gain reported on an ongoing basis, with no carbon benefits associated with trees planted in any year in which the net tree gain is negative considered additional.

3.2 Legal, Regulatory and Institutional Additionality Tests

Compliance Additionality Tests

The structure of most of the regulatory compliance tests is fairly uniform across different standards. However, there are differences in the breadth of coverage and evidence requirements.

The test specified under Plan Vivo is relatively narrow in only considering compliance with legal requirements. Tests under the CDM for large-scale projects, the VCS and (where used) the CarbonFix standard, are slightly broader in considering both legal and regulatory requirements.

By comparison, tests under the ACRFCPS, the ARB, Green-e, and under the Woodland Carbon Code are broader still. The ARB protocol for U.S. forest projects takes into consideration federal, state and local laws, regulation and ordinance, as well as legally binding mandates including forest management plans, conservation easements and deed restrictions (except where enacted in support of the project). The Green-e standard considers compliance with public policy, regulations, legal mandate and guidance, including those not specifically related to GHG emissions. The ACRFCPS takes account of existing laws, regulations, statutes, legal rulings, and regulatory frameworks directly or indirectly affecting GHG emissions from a project or its baseline, including mandatory forest management/forest practice rules. Similarly, the Woodland Carbon Code considers laws, statutes, regulations, court orders, environmental management agreements, planning decisions and other legally binding agreements requiring woodland creation, or implementation of measures that would achieve equivalent levels of sequestration or other

GHG reductions. This excludes compensatory planting required to replace areas of woodland that are felled for purposes such as development, or restoration of open habitats, for example.

Exclusions apply in some cases. Under the VCS recent legal requirements, regulatory frameworks and policies that provide comparative advantages to some technologies do not need to be taken into account in some cases.

Incentive Additionality Tests

Projects that receive grant aid from a government funded initiative for woodland planting, establishment or management are only eligible under the Woodland Carbon Code if a minimum proportion of funding is from other sources. At present (as of April 2012) other sources of finance have to provide at least 15 % of the total planting, establishment and forest management costs over the life-time of the project, with woodland grant payments covering at most 85 % of the costs, but these limits are likely to change in future.

Norm Additionality Tests

The breadth of tests applied varies under different standards. Under Green-e, compliance with industry standards generally, including those not specifically related to GHG emissions, has to be demonstrated. (This forms part of a combined legal/regulatory/institutional test.) Under CarbonFix evidence has to be provided that a project contributes more to sustainable development in the short-, medium- and long-term than the most likely without-project scenario. Types of evidence required are unspecified.

Technological Additionality Tests

A similar core element is used under the different standards. The test under Green-e involves showing that the technology used is near the top of the standard's list of technologies on net GHG emission rates for similar technologies and practices producing similar products or services. The test under the VCS involves showing that a project uses less emissions-intensive technology than a business-as-usual option, and that it meets specific technology and performance criteria. It must also result in a minimum level of GHG savings (e.g. related to market penetration). Although a technology test could be applied to elements of woodland projects such as GHG emissions associated with machinery and chemical use in planting, establishment or forest management activities, associated documentation makes no mention of its applicability to such activities. It is therefore unclear to what extent the test is used at present for woodland projects or is confined to GHG emissions reduction projects in other sectors.

Barrier Additionality Tests

The breadth of barriers considered in distinguishing projects that would otherwise not be implemented differs between standards. Explicit coverage of different types of barrier is summarised in Table 6.

Although some overlap in classifications of different types exists between standards, on the whole, voluntary market standards explicitly account for a narrower range of potential barriers than the tests under the CDM. A notable exception is the lack of requirement under the test for small-scale CDM forestry projects (where used) to consider economic/financial barriers. The latter, which in some cases encompass financial and investment additionality criteria (see below), are explicitly covered under barrier tests under the ACRFCPS, Plan Vivo, VCS and the Woodland Carbon Code (as well as the test applying to large-scale CDM forestry projects). The focus of the test under Plan Vivo is broader than those under the other standards in taking account of barriers in the absence of broader project development (in addition to those in the absence of carbon finance).

The evidence required is specified explicitly only in the case of large-scale CDM forestry projects and can be of a variety of kinds. This may include legislation, environmental resource management norms or rules, statistical or market data, sectoral studies, minutes from Board meetings, correspondence, feasibility studies, financial or budgetary information, and documents written by independent experts.

Practice Additionality Tests

The test for large-scale A/R projects under the CDM (and CarbonFix in cases where a test is applied) initially focuses on whether similar activity not registered under the CDM already occurs within the geographical area of the proposed project. If so, developers have to demonstrate essential distinctions for a project to be considered additional. These may include a fundamental and verifiable change in circumstances since similar activities were implemented (e.g. due to the end of promotional policies, or the existence of barriers). Similar activities are defined as those of similar scale, taking place in the relevant geographical area and a similar environment – including with respect to the regulatory framework (UNFCCC 2007). Tests under the VCS are also similar in requiring that the project type without carbon finance is not common practice in the sector/ region, or where it is, identifying barriers faced compared with existing projects.

By contrast, the ACRFCPS test appears more nuanced. It involves demonstrating that the project activity exceeds the common practice of similar landowners managing similar forests in the region (e.g. by comparing forest management plans). This allows for potential differences in forest management approaches between different types of landowner.

Table 6 Coverage of barrier tests under different standards

Types of barriers	American Carbon Registry Forest Carbon Project Standard	Clean development mechanism		Green-e	Plan Vivo	Verified Carbon Standard (VCS)	Woodland Carbon Code
		Small-scale	Large-scale				
Capacity				✓			
Cultural					✓		
Ecological/environmental	✓	✓					✓
Financial	✓	✓	✓		✓	✓	✓
Investment		✓				✓	
Infrastructure			✓	✓			
Institutional	✓	✓	✓	✓		✓	
Legal					✓		
Local				✓			
Organisational							
Prevailing practice		✓	✓				
Property rights			✓				
Social		✓	✓	✓	✓		
Supply				✓			✓
Technological	✓	✓	✓		✓	✓	✓
Tradition		✓	✓		✓		

Date Additionality Tests

Tests under the CDM and the Green-e standard employ the same cut-off date of 1 January 2000 (as does CarbonFix in cases where a test is applied). Under the ACRFCPS a forestry project must have started after 1 November 1997. Under the ARB urban forest protocol, projects have to commence after December 31 2006 unless otherwise stipulated in the applicable Compliance Offset Protocol approved by the Board, or an early action offset project.

There is flexibility in some cases under ACRFCPS and Green-e, however, as projects commencing before these dates may be approved if an intent additionality test is passed, or (under Green-e) if a sales additionality test is passed.

3.3 Investment and Financial Additionality Tests

Viability and investment additionality test elements under some voluntary carbon standards are essentially indivisible. Under Plan Vivo, for example, it has to be shown that the project or activity could not have happened in the absence of carbon finance, with no particular methodology specified. Similarly, the test under the VCS (specified as a potential element of an investment barrier assessment) considers whether investment return constraints exist that can be overcome by carbon revenues.

Viability Additionality Tests

Tests under Green-e and ACRFCPS focus upon showing that the project would produce an unacceptably low rate of return in the absence of carbon funding. Similarly, the test used under the Woodland Carbon Code requires evidence that the project would not have been viable and therefore would not have gone ahead without carbon finance, taking account of grants available for UK woodland creation. Evidence such as net present value and internal rate of return calculations is required under the ACRFCPS, and under the Woodland Carbon Code, with a full financial analysis of expected costs and revenues over the lifetime of the project required under the latter.

The protocol for large-scale forestry projects under the CDM is more involved than those under other standards (apart from CarbonFix where the same test is used). It involves determining whether in the absence of carbon finance through the CDM the project would be less economically or financially attractive than continuation of the existing land use. A simple cost analysis test is applied to projects in cases with no financial benefits apart from CDM-related income. This involves documenting incomes and costs associated with the project and comparing these with those associated with continuation of the current land use. For projects which also generate non CDM-related income, investment comparison analysis is used to determine whether the project has a lower return based upon the financial indicator considered most suitable for the project type and context. This may be the internal rate of return on the project or on the equity, the net present value, payback period, or cost-benefit ratio. Sensitivity analysis is then used to identify whether this conclusion is robust.

Investment Additionality Tests

The tests for large-scale forestry projects under the CDM compare project returns with those for other alternative land use change scenarios (as opposed to continuation of the existing land use). Apart from this different focus, the tests are the same as for viability additionality under the CDM described in the previous subsection, with one exception. As an alternative to investment comparison analysis, benchmark analysis can be used based upon one of the same indicators, or upon the required rate of return on investment or equity. The indicator excluding carbon revenues is compared with the benchmark value and, if lower, sensitivity analysis is then used to identify whether this conclusion is robust to reasonable variations in critical assumptions. The CDM approach can also be used under CarbonFix.

For avoided conversion projects under the ARB, the fair market value of the land under the anticipated alternative use has to be at least 40 % greater than the current value under forest. Under Green-e, if credits come from a project that produces goods or services apart from GHG emissions reductions, it must be demonstrated that it is not the least-cost option to produce these.

Financial Additionality Tests

Tests under the ACRFCPS and the VCS are similar. Specified as an element of a financial barrier assessment, the test under the ACRFCPS focuses on demonstrating the existence of barriers such as limited access to capital in the absence of carbon revenues. Specified as a potential element of an investment barrier assessment, the test under the VCS considers whether there are capital constraints that can be overcome by carbon revenues. Where included, the test under the WCC is part of a barrier assessment, with supporting evidence (e.g. from a bank) required.

Sales Additionality Tests

The focus and form of tests used differs. Where used, the test under the CDM appears most demanding in terms of evidence requirements.

A project starting prior to 2000 may be deemed additional under Green-e if it is demonstrated that it was partly induced by the existence or anticipation of the voluntary carbon market. Similarly, for activities commencing before project registration, the developers of large-scale CDM projects have to provide documentary evidence that income from sale of CERs was seriously considered in the decision to proceed. It is preferred that legal, documentary or corporate evidence that had been available to third parties at, or prior to, the start of the project is provided. The CDM approach can also be used under CarbonFix.

Relating to whether carbon revenues are crucial in the decision to proceed, the test under the ACRFCPS aims to assess whether carbon market incentives are a key element in overcoming technological barriers. Similarly, one of three alternative financial tests applied under Green-e aims to determine whether emission reduction funding is essential for the project to move forward.

4 Perverse Incentives

Perverse incentives induce negative unintended consequences stemming from the characteristics or manner in which a mechanism is introduced. As a primary focus of concern about the additionality of carbon credits, it is useful to consider evidence in relation to HFC-23 destruction projects and the relationship to underlying incentive structures.

4.1 HFC-23 Destruction Projects

A byproduct of hydrocarbonflurocarbon-22 (HCFC-22) production, HCF-23 is a very potent GHG with a global warming potential (GWP) of 14,800 times that of carbon dioxide over a 100-year period (Forster et al. 2007, Table 2.14). With the cost of destroying HCF-23 a fraction of the market value of associated carbon reduction credits (typical marginal abatement costs are below $1/tCO_2e$) and the credits worth several times the HCFC-22 produced, companies are reported to make huge profits from HCF-23 destruction projects (EIA 2010; Scolnick 2010). Although HCFC-22 is an ozone-depleting substance and production for use as a refrigerant or other emissive purposes (e.g. air conditioning) is regulated at international level under the Montreal protocol, production for feedstock purposes (e.g. for production of polytetrafluoroethylene) is not. As a consequence companies have been able to establish new HCFC-22 production facilities and, until recently, claim carbon credits for destroying the HCF-23 produced. HCF-23 destruction projects account for the majority of CERs issued to date. They also account for the majority of offsets purchased by companies within the European Union Emissions Trading Scheme (EIA 2010) to date, although from 2013 their use will no longer be permitted. A comprehensive review of monitoring reports for HCF-23 projects approved under the CDM (Schneider 2011a) found that HCFC-22 production plants produced more HCF-23 when the destruction of this GHG could be credited than in other periods, suggesting that the CDM also provides perverse incentives for existing plants to generate more HFC-23.

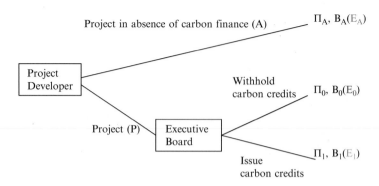

Fig. 1 Additionality game: The Project Developer chooses at the initial node (left) project P, or alternative A. If P is chosen, the Executive Board chooses at the second node whether to issue or withhold carbon credits. Any issue of credits is contingent on the Project Developer adopting emissions reduction technology. The payoffs to the Project Developer and Executive Board based upon the associated level of emissions are given at the game's terminal nodes (right). Solving by backward induction, the Executive Board chooses to issue carbon credits rather than withholding them if the Project Developer chooses project P as this leads to reduced emissions for the project. Anticipating the Executive Board will issue credits, the Project Developer chooses project P. The subgame-perfect Nash equilibrium is given by the pair of strategies {Undertake project P, Issue credits}, with the payoff to the Project Developer and associated emissions of {Π_1, E_1}

4.2   Underlying Incentive Structures

Project developers face incentives to implement projects which increase emissions if they expect to obtain revenues from the sale of carbon certificates for subsequently reducing emissions and this increases their net returns. The underlying incentive structure can be illustrated using a simplified version of the two-stage additionality game developed by Calel (2011) that is shown in Fig. 1. The three potential outcomes have payoffs to the project developer ranked $\Pi_1 > \Pi_A > \Pi_0$ and associated GHG emissions ranked $E_0 > E_1 > E_A$. In the first stage of the game the project developer has to decide whether to initiate project P, or alternative A (representing the baseline – possibly not undertaking a project at all). If P is chosen, the project may be issued with carbon credits in the second stage of the game by the CDM executive board (EB) providing the project developer adopts technology that leads to lower GHG emissions (E_1) than those (E_0) expected to arise for project P otherwise. The second stage of the game (which only occurs if P is chosen) involves the EB deciding whether to issue the project developer with carbon credits for the project. As GHG emissions are lower under project P if credits are issued than if they are withheld ($E_1 < E_0$), the EB's best response is to issue credits if the project developer chooses P. Anticipating that the EB will issue credits if P and associated GHG abatement technology is adopted, the project developer undertakes the project as, with the carbon credits, it yields a higher payoff than alternative A ($\Pi_1 > \Pi_A$). GHG emissions increase ($E_1 > E_A$) as a consequence of project P being undertaken.

Despite this increase in emissions, the outcome {undertake P, issue credits} is the unique equilibrium (sub-game perfect solution) of the one-shot game. For repeated games, the outcome also depends on other factors such as discount rates. However, repetition is argued to be unlikely to change the equilibrium, as in each game the best response of the EB to a project developer's choice of P is to issue credits (Calel 2011). Based upon a preference for low emissions, 'payoffs' to the EB are shown in Fig. 1 ranked $B(E_A) > B(E_1) > B(E_0)$.

4.3 Incentive Structures and Forestry

Similar perverse incentives to the structure shown in Fig. 1 could apply to reduced degradation and deforestation (REDD) projects if there is no project additionality test precluding land being deforested or degraded in order to claim carbon credits for subsequent reforestation or restoration. However, it is highly unlikely analogous cases could arise in countries such as the UK where the current regulatory framework includes requirements for replanting woodland after felling.

It is also difficult to envisage cases in a UK forestry context directly analogous in magnitude of effects to the situation arising from perverse incentives to produce HFC-23 in order to obtain carbon credits for its subsequent destruction. Nonetheless, a modified version of the additionality game which is applicable to UK woodland creation projects is shown in Fig. 2.

The 'Woodland Additionality' game posits a project developer facing the choice between two woodland creation options. GHG savings in each case are assumed independent of whether certificates are issued, with higher savings (S) associated with project H than project L ($S_H > S_L$). Two cases could plausibly lead the project developer to choose L in preference to H. In both cases the project developer anticipates that the EB views L as providing additional benefits and will issue carbon certificates if chosen, reflected in an implicit ordering of EB payoffs $B_1(S_L) > B_0(S_L)$.

First, if the project developer anticipates carbon certificates will not be issued for H (a similar case to the game in Fig. 1), reflected by an ordering of EB payoffs $B_0(S_H) > B_1(S_H)$, L is preferred if the expected return including revenue from carbon certificates is higher (i.e. if $\Pi_{L1} > \Pi_{H0}$). Despite the lower GHG savings (S_L), choosing L is rational for the project developer as it offers higher returns. Such a situation could arise if the application of additionality criteria (e.g. an investment additionality test) precludes claiming carbon certificates for H but not project L, for example. This would be consistent with the observation (e.g. Bode and Michaelowa 2003) that application of additionality criteria can result in incentives to invest in projects that are relatively high cost and offer relatively low GHG savings.

Second, if the project developer anticipates carbon certificates will be issued for H, reflected by an ordering of EB payoffs $B_1(S_H) > B_0(S_H)$, L is preferred if it offers a higher return taking account of the level of carbon certificates (i.e. $\Pi_{L1} > \Pi_{H1}$). Such a situation could arise where the baseline used for determining additionality and subsequent crediting is not comprehensive. If carbon displacement benefits of using

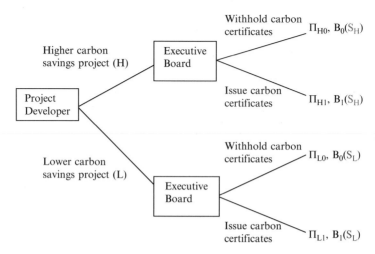

Fig. 2 Woodland additionality game: The Project Developer chooses at the initial node (left) between project H with high GHG savings, or project L with low GHG savings. Once this choice has been made, the Executive Board chooses at the next node whether to issue or withhold carbon certificates depending whether the project provides additional GHG savings compared to the baseline. The payoffs to the Project Developer and Executive Board are given at the game's terminal nodes (right). The payoff to the Executive Board is based upon the level of GHG savings it considers additional. Where the Executive Board is expected to issue carbon certificates only if project L is chosen, the Project Developer chooses project L if the payoff with the certificates is greater than from project H without certificates. Where the Executive Board is expected to issue carbon certificates for either project, the Project Developer chooses project L if the payoff with certificates is greater than from project H with certificates. In both cases the subgame-perfect Nash equilibrium is given by the pair of strategies {Undertake project L, Issue credits}, with a payoff to the Project Developer and associated GHG savings of $\{\Pi_{L1}, S_L\}$

timber instead of more fossil-fuel intensive materials are accounted for in estimating overall benefits, for example, but not in issuing carbon certificates, this might lead to projects being undertaken with high sequestration benefits, but relatively low overall GHG abatement benefits (S_L).

4.4 Potential Implications of Perverse Incentives for UK Woodland Projects

The woodland additionality game helps clarify how perverse incentives could arise due either to incomplete coverage of project baselines, or associated with use of certain types of additionality tests (e.g. financial, viability, or investment additionality tests). The extent to which perverse incentives affect UK woodland carbon projects in practice is unclear, however.

Perverse incentives arise in the case of HFC23 destruction projects in part due to the high revenues from carbon credits compared to the low abatement costs (Schneider 2011a). Involving principally carbon dioxide (which as noted above has a far lower GWP than HFC23), similar disparities between abatement costs and carbon revenues are unlikely at present for woodland carbon projects in the UK. Forestry returns in the UK have traditionally been modest (Valatin and Starling 2010), with emerging prices for forestry carbon apparently far below social values recommended for use in public policy appraisals (DECC 2010). Furthermore, one of the remedies suggested (issuing only a proportion of carbon certificates) is already a feature of the approach under the Woodland Carbon Code (Forestry Commission 2011b) – albeit adopted for different (permanence risk management) reasons.

More complete coverage of project baselines could help reduce risks of perverse incentives affecting woodland carbon projects in the UK. The focus on sequestration, for example, could be expected to favour longer rotations even if shorter rotations offer higher overall GHG benefits once displacement is accounted for. (Examples of impacts on optimal rotation length of the inclusion or exclusion of carbon sequestration and displacement benefits can be found in Price and Willis (2011)). However, greater information requirements, complexities and uncertainties, as well as potential impacts on the volume of abatement undertaken if costs to project developers increase, are also relevant in considering the potential for extending coverage of the UK Woodland Carbon Code or other standards. Changes in ancillary societal benefits associated with any shift in average harvesting age could be a further important consideration to the extent that taking these into account provides a more comprehensive perspective in considering potential impacts.

In some cases perverse incentives may be reduced by using a further (e.g. gaming additionality or date additionality) test. Willis et al. (2012), for example, reports introduction of new rules under the CDM which stipulate that HCFC-22 production plants have to have operated for at least 3 years between 2000 and 2004 and be running in 2005 in order for abatement by HCF-23 destruction to be credited.

Even where a particular test creates perverse incentives, this does not necessarily imply it should not be used. Bloomgarden and Trexler (2008) note in relation to tests based upon a hypothetical counterfactual that some fraction of non-additional reductions will always pass, while some fraction of truly additional reductions will always fail, and the challenge is to find an acceptable balance. The situation in considering use of an additionality test associated with perverse incentives is similar, but includes a further element. The challenge is to seek an acceptable balance between the fall in non-additional reductions due to use of the test on the one hand, and the increase in GHG emissions due to perverse incentives together with the fall in truly additional abatement due to application of the test on the other.

5   Summary and Conclusions

As demonstrated above, additionality is a multi-faceted concept. At least six forms of environmental additionality, 12 of legal, regulatory and institutional additionality, and five of financial and investment additionality, can be distinguished. These are summarised in Table 7 and cover key aspects under existing carbon standards and additionality protocols.

Table 7 Forms of additionality

Type	Description
Environmental	
GHG	Positive overall impact on GHG balances (net carbon benefit of activity or project)
Unit	Emissions per unit output below specified level (or possibly GHG savings per unit area above a threshold level)
Project	Afforestation and reforestation: forests unable to establish themselves in the absence of planned activities or project;
	Avoided deforestation or forest degradation: forests would have been deforested or degraded in the absence of the project
Intent	GHG abatement a decisive factor in decision to proceed
Tree	Positive impact on the total number of trees
Ecological	Positive net impacts on habitats, species and biodiversity
Legal, regulatory, Institutional	
Compliance	Exceeds statutory requirements
Incentive	Exceeds benefits associated with incentives provided by regulatory framework
Threshold	Does not exceed maximum GHG savings counted as additional
Norm	Meets voluntary industry standards, or good practice benchmarks
Technological	Application of specific technology.
Barrier	Overcomes implementation barrier
Practice	Not common practice
Reporting	National GHG accounting and reporting additionality rules
Institutional	Independent of statutory emissions reduction targets
Date	Activities occur after (or in some cases before) particular date
Term	Abatement arises within a specified time-scale
Jurisdiction	Activities in particular location, or undertaken by specific communities or social groups
Financial and investment	
Financial	Would not be financed without sale of carbon certificates
Viability	Not financially viable without sale of carbon certificates
Investment	Not most attractive option without sale of carbon certificates
Sales	Income from the sale of carbon credits a decisive factor in decision to proceed
Gaming	GHG emissions not generated for the purpose of subsequent abatement to claim carbon credits

Aspects covered by additionality tests vary between mechanisms and standards. By far the most prevalent of the tests incorporated in existing additionality protocols under the CDM and the six voluntary carbon standards considered are compliance and barrier tests (see Table 2).

Differences in types of tests appear unrelated to whether carbon certificates are issued ex-post or ex-ante. This is illustrated most clearly by the same additionality protocol being applied in both cases under CarbonFix (see also comparisons in Tables 2 and 3). However, coverage, methodology and evidence requirements of specific types of tests vary between standards.

Trade-offs exist between the rigour and cost of additionality tests. Differences in tests applied partly relate to these trade-offs. This is illustrated by the different tests applied to large-scale and small-scale projects under the CDM, as well as the different approaches to establishing baselines.

These trade-offs can affect both the quality and number of projects approved. In a context of asymmetric information between sellers and buyers of carbon certificates, buyers are likely to have greater confidence the more rigorous the additionality tests applied at project level. However, more rigorous tests will generally involve higher transactions costs for project developers and could reduce the number of projects seeking certification. To the extent that higher transaction costs lead to some projects that would pass more rigorous tests not being put forward, it will tend to reduce the overall climate change mitigation benefits obtained. This suggests a delicate balance between underpinning the quality of carbon certificates and market confidence on one hand, and maximising the expected overall climate change mitigation benefits on the other. Akin to Heisenberg's uncertainty principle (the impossibility of exactly measuring both the position and velocity of an object simultaneously) perhaps, applying more rigorous additionality tests could serve to reduce net GHG savings in some instances.

Inclusion of tests for aspects of additionality of most interest to buyers affects demand. In the absence of perfectly elastic supply this influences the price of certificates in carbon markets and thus incentives for future development of projects. To the extent that it is a concern to prospective purchasers, the lack of a mechanism to ensure institutional additionality could be a disincentive to developing woodland carbon projects in countries such as the UK and an impediment to climate change mitigation by the forest sector. From a public perspective, however, the trade-off between securing extra GHG abatement and the extra cost of implementing and administering such a mechanism is not necessarily straightforward. There may be increased costs of reaching national abatement targets if some woodland projects are no longer counted and other public incentives are needed. Furthermore, the extent to which these costs divert public expenditure from alternative uses resulting in a reduction in GHG abatement elsewhere in the economy is also relevant. The current government approach in the UK is to view woodland projects as contributing to meeting national GHG targets rather than as providing carbon credits that can be traded as offsets.

In considering which approach to apply, developers of carbon standards and public authorities could benefit from greater clarity about distinctions between the different aspects and the potential for perverse incentives associated with some tests. Precisely specifying which tests for additionality and associated institutional arrangements are best is not easy from first principles without detailed consideration of the associated costs and benefits. However, even where such information exists or can be reliably estimated, it may not help much in specifying criteria such as cut-off dates, or thresholds for inclusion or exclusion of projects. To the extent that the concept of additionality is open to interpretation and based upon comparison with a hypothetical scenario, its determination is necessarily imprecise and is likely to remain controversial, even where comparatively stringent tests are applied.

Acknowledgements Thanks to Pat Snowdon, Chris Quine and Trevor Fenning for their comments, to Christine Cahalan and three anonymous reviewers.

References

ARB (2011a) Compliance offset protocol U.S. forest projects. Air Resources Board, California Environmental Protection Agency, October, www.arb.ca.gov/regact/2010/capandtrade10/copusforest.pdf

ARB (2011b) Compliance offset protocol urban forest projects. Air Resources Board, California Environmental Protection Agency, October, www.arb.ca.gov/regact/2010/capandtrade10/copurbanforestfin.pdf

Au Yong HW (2009) Investment additionality in the CDM. Technical Paper Econometrica Press, Edinburgh

Bloomgarden E, Trexler M (2008) Another look at additionality. Environ Finance May 17

Bode S, Michaelowa A (2003) Avoiding perverse effects of baseline and investment additionality determination in the case of renewable energy projects. Energ Policy 31:505–517

Brown J, Bird N, Schalatek L (2010) Climate finance additionality: emerging definitions and their implications. Climate Finance Policy Brief No. 2, Overseas Development Institute

Calel R (2011) Perverse incentives under the CDM: a comment. Working Paper 63 Centre for Climate Change Economics and Policy, and Grantham Research Institute on Climate Change and the Environment, LSE. July

Costa PM, Stuart M, Pinard M, Phillips G (2000) Elements of a certification system for forestry-based carbon offset projects. Mitig Adapt Strat Glob Chang 5:39–50

DECC (2010) Valuation of energy use and greenhouse gas emissions for appraisal and evaluation: supplement to HM treasury's green book. Department of Energy and Climate Change, London

EIA (2010) HFC-23 offsets in the context of the EU emissions trading scheme. Policy Briefing, Environmental Investigation Agency/CDM Watch, July

Forestry Commission (2011a) The UK forestry standard: the Government's approach to sustainable Forestry. Forestry Commission, Edinburgh

Forestry Commission (2011b) The Woodland carbon code. Version 1.3. Forestry Commission, Edinburgh

Forster P, Ramaswamy V, Artaxo P, Berntsen T, Betts R, Fahey DW (2007) Changes in atmospheric constituents and in radiative forcing. In: Solomon S, Qin D, Manning M, Chen Z, Marquis M, Averyt KB (eds) Climate change 2007: the physical science basis contribution of working group I to the fourth assessment report of the intergovernmental panel on climate change. Cambridge University Press, Cambridge

Hamilton K, Sjardin M, Shapiro A, Marcello T (2009) Fortifying the foundation: state of voluntary carbon markets 2009. New Carbon Finance/Ecosystem Marketplace , New York/Washington, DC

HM Treasury (2003) The green book: appraisal and evaluation in central government. TSO, London

IPCC (2000) Land use, land-use change, and forestry. Intergovernmental panel on climate change. Cambridge University Press, Cambridge, UK

Kollmuss A (2007) Carbon offsets 101: a primer on the hottest – and trickiest – topic in climate change. World Watch Mag 20 July/August

McCully P (2008) The great carbon offset swindle: how carbon credits are gutting the Kyoto Protocol and why they must be scrapped. In: Pottinger L (ed) Bad deal for the planet: why carbon offsets aren't working… And how to create a fair global climate accord. Dams, Rivers and People Report 2008 International Rivers Berkeley, pp 2–14

Merger E (2008) Forestry carbon standards 2008 and the state of climate forestation projects. Carbon Positive, Athens

Mukerjee M (2009) A mechanism of Hot Air. Sci Am 4 June, pp 9–10

Peters-Stanley M (2012) Bringing it home: taking stock of government engagement with the voluntary carbon market. Ecosystem Marketplace, Forest Trends, Washington, DC

Price C, Willis R (2011) The multiple effects of carbon values on optimal rotation. J Forest Econ 17:298–306

Rajan SC (2011) Vested or public interest? the case of India. In global corruption report: climate change. Transparency International Earthscan, London, pp 57–62

Reynolds B (2008) Do we need financial additionality? Environ Finance 36 March

Schneider L (2007) Is the CDM fulfilling its environmental and sustainable development objectives? An evaluation of the CDM and options for improvement. Institute for Applied Ecology, Berlin

Schneider RL (2011a) Perverse incentives under the CDM: an evaluation of HFC-23 destruction projects. Clim Policy 11:851–864

Schneider L (2011b) The trade-offs of trade: realities and risks of global carbon markets. In global corruption report: climate change. Transparency International Earthscan, London, pp 131–143

Schwank O (2004) Concerns about CDM projects based on decomposition of HFC-23 emissions from 22 HCFC production sites. INFRAS, Zurich

Scolnick T (2010) Carbon market distortions and diminishing environmental returns: the clean development mechanism and China. Briefing note 2010-24, Pacific Institute for Climate Solutions/ISIS, Saunder School of Business. University of British Columbia, Vancouver

Shapiro M (2010) Conning the climate: inside the carbon-trading shell game. Harper's Mag. Feb, pp 31–39

UNFCCC (2007) Tool for the demonstration and assessment of additionality in A/R CDM Project Activities (Version 02), CDM Executive Board Report EB 35, Annex 17, http://cdm.unfccc.int/methodologies/ARmethodologies/approved_ar.html

Valatin G (2011) Forests and carbon: a review of additionality. Research Report Forestry Commission, Edinburgh

Valatin G (2012) Additionality and climate change mitigation by the UK forest sector. Forestry 85(4):445–462

Valatin G, Starling J (2010) Valuation of ecosystem services provided by UK woodlands, appendix to Chapter 22 of UK National Ecosystem Assessment. UNEP-WCMC/Defra, London

Wara M, Victor DG (2008) A realistic policy on carbon offsets. Working Paper 74, Program on Energy and Sustainable Development, Freeman Spogli Institute, University of Stanford California

Willis K, Ozdemiroglu E, Campbell D (2012) Environmental economics and policy. J Environ Econ Policy 1:1–4

Different Economic Approaches in Forest Management

David B. South

Abstract The management of forests is directly related to the landowner's objective, and to some extent, the type of economic system employed by the landowner. The system used to manage government forests is typically different than that used by industrial landowners. Likewise, the system used by large estate owners may differ from that used by industry. Therefore, the management of forests is directly affected by the individual's attitude toward money.

In this chapter, I describe seven approaches to the management of forests. The "slash and burn" approach is taken when forests are considered a liability. The "cut-out and get-out" approach is taken when the resource is exploited for short-term economic gains. A "stay and plant" system involves taking a long-term view of economic investments. Some "estate-owners" manage their forests using an "annual spreadsheet" approach (where only annual costs and income is considered). "Urban development" may occur where the "land expectation value" is lower than the "real-estate" value of the land. Government forests that are managed as "wilderness" are typically left alone with no timber harvesting or road building. Timber harvest are permitted on some "managed forests" that are owned by governments or non-government organizations. Additional management approaches exist but most of the world's forests generally fall into these seven categories.

1 Introduction

For millennia, humans have been altering forests using different objectives. Some clear forests in order to make room for food production, some clear forests while harvesting firewood and others attempt to "preserve" forests for future generations.

D.B. South (✉)
School of Forestry and Wildlife Sciences, Auburn University, Alabama, USA
e-mail: southdb@auburn.edu

T. Fenning (ed.), *Challenges and Opportunities for the World's Forests in the 21st Century*, Forestry Sciences 81, DOI 10.1007/978-94-007-7076-8_15, © Springer Science+Business Media Dordrecht 2014

Some people plant trees on grasslands (or deforested areas) in hopes of increasing their economic welfare. Others might plant trees in hopes of affecting the climate. In each of these cases, an economic justification might be used to explain the selected management option.

This chapter involves describing seven categories of forests (Table 1) in an attempt to explain why all forests are not managed the same. These categories exist because the views individuals have about economics have a direct impact on the management of forests. Some individuals value a quick return on investment, some are willing to accept low rates of return, others speculate that the market price of bare-land will increase, while others place a value on forests that far exceed the current monetary value. In this chapter I will refrain from delving into the equations used by forest economists. Instead, I will simply point out that some base management decision on economic theory while others do not. Finally, I reject the hypothesis that the likelihood of a sustainable and vibrant forest industry (i.e. providing many jobs) is independent of a nation's wealth.

2 Categories of Forests

2.1 Burn the Forest

Burning the forest has been a practice used by humans for thousands of years. In tropical regions where the climate might not have a dry season, the trees and brush are cut down by hand (e.g. slashed with machetes) to increase flammability. This practice is commonly referred to as "slash and burn" agriculture. One reason for burning the trees is to increase the production of food. When human populations grow, the need for food increases and, in some regions, this will increase the amount of "slash and burn" agriculture. The deforestation that results is largely due to an increase in either farmland or grazing land. In many cases (especially in aboriginal cultures), the forest is not burned for monetary reasons. Native people who choose to burn the forest often do not base their decision on the output from an economic spreadsheet. They may do so because they wish to provide more food for their families.

There are many examples of cultures burning the forest, especially when the forest is not "owned" by individuals. One example is the use of fire by natives. Halkett (1991) said that the Polynesians of New Zealand burnt almost one third of the native forests of New Zealand. Pyne (1982) said that "So extensive were the cumulative effects of these modifications that it may be said that the general consequence of the Indian occupation of the New World was to replace forested land with grassland or savannah, or, where the forest persisted, to open it up and free it from underbrush. Most of the impenetrable woods encountered by explorers were in bogs or swamps from which fire was excluded; naturally drained landscape was nearly everywhere burned." There were various reasons for burning the forests. Sometimes it was to reduce hunger by helping in the hunting of flightless birds (Guild and Dudfiel 2009) and sometimes it was to increase the carrying capacity of the land (i.e. number of ungulates per ha) and thereby reduce hunger.

Table 1 Large forests (e.g. >1,000 ha) are managed under different systems. The following lists seven types of systems but others also exist

	Burn forests	Cut-out and get-out	Stay and plant	Estate-owners	Urban develop	Managed GO or NGO forest	GO wilderness
Currency	Food	$	$	$	$	$	$
Private ownership	No	No/Yes	Yes	Yes	Yes	No	No/Yes
Trees have value as	Negative value	Timber	Timber	Timber, wildlife, recreation	Shade, beauty	Water, timber wildlife, recreation	Water, beauty, wildlife, climate
Rotation	None	None	Short	Medium	None	Long	–
Economic system	Barter	Cashflow	Faustmann	Cashflow	Cashflow Faustmann	Budget	Budget
Harvest trees for $	No	Yes	Yes	Yes	Perhaps	Perhaps	No
Pay taxes	No/Yes	Yes	Yes	Yes	Yes	No	No

Fig. 1 Painting by Eero Erik Nikolai Järnefelt entitled "Under the Yoke (burning the brushwood) is in the Finnish National Gallery (Source: http://commons.wikimedia.org/wiki/File:Raatajat_rah-analaiset.JPG)

A second example is from Scandinavia where there was a tradition of turning wilderness into gardens (Pyne 1995). The practice of Svedjebruk involved ring-barking or felling of trees, allowing the brush to dry, and then burning the forest (Fig. 1). The burned area was seeded in the spring and the rye was harvested in the autumn. Turnips or cabbage would be planted in the following 2 years. This would be followed by grazing and eventually the clearing would be abandoned. If the area returned to a forest following natural regeneration, the cycle would be repeated. When the forests were expansive and the population was low, Svedjeburk was encouraged. However, as the forests declined and population pressures increased (due to an increase in the production of food), limitations were placed on burning the woods. For example, during the first half of the seventeenth century, the King mandated that forests in Varmland not be burned and converted to rye fields.

In both of these examples from the past, the dense forests had a negative value since time was required to convert the forest to grassland or farmland. Also, those who were burning the forest did not hold title to the land. In North America, the natives did not pay taxes but in Varmland, the "Forest Finns" were taxed by the King. Any excess food obtained from the grassland or rye fields could have been used to barter for other goods.

2.2 Cut-Out and Get-Out

The term "cut-out and get-out" is used to describe the practice of purchasing forest-land, processing the large trees into lumber, exporting lumber to cities, selling the deforested land, and moving on to another forested area. Managing the land on a continuing basis was not done because this would involve a reduction in cash flow and possibly accepting a lower return on investment. In some cases, the long-term view of the owner was no more than 8 years (Jones 2005). Although purchasing forestland and selling the cutover land was common in some regions, it is not a requirement. In some cases, the timber was owned by the government and some-times the rulers profited from selling cutting rights to the forests.

The primary economic incentive was to make the highest profit, minimize expen-ditures, minimize property taxes, in a relatively short period of time (e.g. 8 years). As a result, investments in regenerating the forest were not made. The objective was to take a relatively inexpensive resource, increase the added value by converting logs into lumber and enjoy short-term profits.

2.3 Stay and Plant

In the early twentieth century, some timber barons (e.g. Goodyear and Weyerhaeuser) decided to keep their forestland (as opposed to selling their cutover land). They believed that long-term profits could be made by managing forests on a "perpetual" basis, where the area harvested annually was limited by the harvested rotation age (i.e. if the rotation age was 50 years, then 2 % of the forest would be cut down each year). Sawmills would not have to be moved and the industry could provide a steady source of jobs for the region. In areas where natural regeneration was diffi-cult, tree planting techniques were developed. However, tree planting involved a long-term investment. The long-term nature seemed to require a different eco-nomic approach than the short-term objectives employed by those who "cut-out and get-out."

2.4 Estate Owners

An "Estate Owner" is someone who owns large tracks of forests such as Royalty, land barons, and other wealthy individuals who own more than 2,000 ha of forests. These individuals typically understand the "time value of money" and many expect their forests to increase their net wealth over time. Many hire managers to help ensure the forest remains economically productive.

Some large estates are managed using the single tree selection method of harvesting (McIntyre et al. 2008; Neel 2010). Others estates are managed using a

Table 2 An example of the "Estate Owner's Method" spreadsheet listing annual costs and revenue. This estate has a 5,000 ha plantation and is managed as a "fully regulated" forest on a 50-year rotation. All previous management costs are "sunk" and no assumptions about future wood prices are required. A "land expectation value" is not calculated using this method (but in this case it would be $1,030/ha using a 6 % real discount rate)

	Hectares treated each year	Per ha (cost) or revenue	Annual cash-flow
Site preparation and planting	100	($500)	($50,000)
Forester costs	–	($10)	($50,000)
Total costs			($100,000)
Thinning (age 15 year)	100	$715	$71,500
Thinning (age 30 year)	100	$358.5	$358,500
Thinning (age 40 year)	100	$201.7	$201,700
Harvest (age 50 year)	100	$958.4	$958,400
Total revenue			$1,565,000
Net profit for year 2020			$1,466,000
Profit/ha in year 2020			$293
Benefit/cost ratio			15.6/1

"fully regulated" forest (a.k.a. normal forest) where the forest contains an even distribution of age classes, so that it would be capable of yielding the same volume of timber every year in perpetuity (Helms 1998; Tahvonen and Viitala 2006). Regardless of the management approach, many estate owners will examine the income and expenses on an annual basis. If income from harvests and processing exceeded expenses, the estate owner would be pleased. However, if expenses exceeded income for the year, the landowner would ask the forester why the expenses were so high or the harvests so low. The next year, the process would begin anew with an annual accounting of costs and income. All costs and income from the previous year would be ignored (or sunk).

This method of accounting has been referred to as the "Estate Owner's Method" (EOM) (South and Laband 2013). This approach to analyzing economic returns was used by the South African Timber Growers' (1993) in South Africa. Not all estate owners employ the EOM, but some use similar methods that only examine current year's costs and income. The EOM does not involve calculating a "land expectation value" (LEV), it does not involve long-term discounting, and typically does not include the cost of land in the calculation. An example of an EOM approach to forest economics is provided in Table 2.

2.5 Urban Development

Some wealthy individuals and institutions purchase farm and forestland as investments. For example, during the first decade of the twenty-first century, 30–40 %

of the farmland purchases in Iowa were made by investors. When investors purchase forestland, their primary goal is to sell the land in the future for a profit. Some investors will conduct a "residential thinning" in order to generate a partial payment for the property. Then after the thinning, the lots are surveyed and then sold for a housing subdivision. Some portions of the decline in total forestland (e.g. California, Florida, Virginia) are due to a conversion of a forested landscape into an urban one. Most land speculators have short-term objectives and therefore they do not use the LEV formula (Faustmann 1849) when making economic decisions about timberland.

In many places in North America, the LEV is much less than the price of forestland. Although the forest I own is worth more than $20,000/ha, the LEV (@ 5 %) according to Faustmann's formula is less than $1,000/ha. In many places in the USA, the LEV for a pine plantation is less than the land value. For example, the average price of pastureland in Florida was greater than $12,000/ha in 2010. Generally, most urban developers ignore the LEV calculation since the land value is often much higher than the LEV (when using current discount rates). Urban developers have no interest in managing the forest in perpetuity.

2.6   Managed GO or NGO Forests

Worldwide, government organizations (GO) control over 80 % of the forests (Agrawal et al. 2008). The management of GO forests cover the extremes; from promoting deforestation to the preservation of watersheds. In general, the attitude of government officials toward forests varies with the wealth of the nation. In relatively poor countries, the management of GO forests might (in some cases) result in a decline in forest area while wealthy countries may report an increase in forested area.

In relatively wealthy countries, GO forests are managed using various objectives. I will provide three examples near where I live. The first example is a 2,600 ha forest located in the foothills of the Blue Ridge Mountains in South Carolina. Each day the "Table Rock Reservoir Watershed" provides about 110 million liters of water to the city of Greenville. After filtering and treatment, this water is sold to consumers at about 2 cents per liter. Many of these consumers value clean water and some are even willing to pay more than $0.60 per liter for bottled water. Assuming untreated water has a value of 0.5 cent per liter, this forest supplies $550,000 of value per day to the city of Greenville (or $211/ha/day). Since this is a high-value, single-use forest, I am not allowed to either hike or camp in this publicly owned watershed. Timber harvests are also prohibited.

The second example is the Table Rock State Park where I am allowed to camp and hike. The objective of this 1,247 ha forest is primarily for recreation. There is a use fee (currently at $2/person) for visitors who wish to canoe on the lake, hike on the trails, picnic or fish for bass or bream. Timber harvests are not included in the management plan.

   The third example is the National Forests in the mountains of North Carolina where the US Forest Service manages forests for multiple objectives (including timber harvests). The total area of these forests amount to 507,700 ha from which about 0.1 % were harvested in 2010 (i.e. 608 ha). Revenues from timber harvesting (mainly hardwoods) that year were $1.228 million (or about $2,020/ha). Payments to county governments (in lieu of taxes) amounted to almost $6.6 million dollars (or $11/ha). The total revenues from all sources (timber, recreation, firewood, minerals, etc.) amount to almost $3.9 million (or $18.90/ha/year). If the total "appropriated" budget for the forest is $23.5 million then the net "cost" to the taxpayer is roughly $41/ha.

   Some forests are managed by non-government organizations (NGO) like the World Land Trust and The Nature Conservancy. NGO forests are managed under a variety of objectives, including objectives that involve the harvesting of trees. The Nature Conservancy (TNC) uses land acquisition as a principal tool of its conservation effort. For example, TNC in North Carolina purchased 2,060 ha area that is known as the Green River Game Land. The area (once owned by a power company and a timber company) is now managed for the public by the North Carolina Wildlife Resources Commission. The public can hunt, fish, canoe and hike in this area of the Appalachian Mountains. Worldwide, TNC has helped to protect more than 48 million hectares of land.

   The above examples illustrate only a few of the ways GO and NGO forests are managed. Even so, there is one common factor; all of the forests in the above examples operate under the constraints of a budget. For the most part, taxpayers pay to fund the budgets of GO forests (and even some timberland acquired by NGOs). Although there are some exceptions, in many cases the budget level assigned is not a function of timber sales (or revenue generated from the forest).

2.7   GO Wilderness

Many government organizations manage some forests as wilderness areas. In some countries, 4–5 % of the land is kept in a general natural condition. In the United States, four organizations (Forest Service, National Park Service, Bureau of Land Management, Fish and Wildlife Service) manage wilderness. Most of these areas have opportunities for a primitive and unconfined type of recreation, they are large enough to preserve and use as wilderness, and they likely contain ecological, geological, or other features of scientific, scenic, or historical value. Hunting is permitted in some wilderness areas but timber harvesting is prohibited.

   The cost to manage wilderness areas varies depending on size and budgets. For example, the Denali National Park has a budget of about $13 million and it contains 2.45 million hectares (only 868,700 ha are designated as wilderness). This equates to about $5.30 per hectare which is much lower than the average cost of $88/ha/year for all lands managed by the National Park Service (calculated using a budget of $3.16 billion/year).

3   Discussion

3.1   Economic Objectives Vary

The categories of forests listed in Table 1 are all influenced by underlying economic principles. Even so, those managing timberland differ in their economic approaches. Some view standing trees as a liability while others view them as an asset to be liquidated. Some say forests should not provide jobs that are based on making lumber, paper or bioenergy. Instead, they suggest future generations would benefit more if forests were simply left alone (and not sold to the highest bidder).

Forests are either managed for profit or for non-profit objectives. Forests managed for profit might be owned by private individuals, industrial companies and government-owned companies (e.g. City Forests at Dunedin, NZ). These forests may be further divided into two subgroups. One is the "Faustmann" group which is composed of those who use LEV or Net Present Value (NPV) calculations when selecting among several forest management options. The other group does not use either NPV or LEV and this group I will refer to as the "cashflow" group (see Table 1). Both groups hope to make a profit from managing timberland, but the "Faustmann" group does not "sink" costs incurred in the past.

Non-profit forests are generally those funded by either taxpayers or non-government organizations (NGOs). In many cases, the size of the allocated budget has no relationship to the productivity of the timberland. Management chiefs have to lobby politicians or CEOs for increases in the budget. In cases where the budget is relatively high, there often is no incentive to be efficient or cost effective with management practices. In fact, in some cases managers are penalized if they are frugal and spend less than their allocated budget. South and Laband (2013) provide a more detailed explanation of these two groups.

Economists are familiar with the terms "nominal dollar" and "real dollar." The "nominal" dollar refers to a value expressed in a given year or series of years. The "true" value changes over time since inflation often erodes the purchase power over time. For example, the value of one dollar spent in the year 1900 is greater than a dollar spent a century later. Comparing nominal dollars of different years is like comparing apples with oranges (i.e. we may not know if an increase in nominal dollars indicates a profit or loss). In contrast, the value of a "real" dollar is fixed since it is based on the purchase power for just a single calendar year (e.g. 1983). So for example, if $100 "real" dollars were spent in either 1983 or 2010, they would purchase the same amount of goods in either year. As a comparison, $100 nominal dollars in 1983 would purchase the same amount of goods as $222 "nominal" dollars in 2010. Therefore, if someone invested $100 in 1983 and achieved a 3 % "nominal" interest rate, they would have $222 in the year 2010. Unfortunately, this is equal to a 0 % "real" interest rate (which means the person could purchase the same amount of food in either 1983 (with $100) or 2010 (with $222). In this example, the 3 % value is equal to the rate of inflation.

3.2   Two Schools of Thought Regarding Land

There are two "theoretical" types of "dollars" that are found in the forestry literature. First, forest economists use theoretical dollars when using an equation to
calculate a "land expectation value" (LEV) or "bare land value" (Klemperer 2003).
This equation requires various assumptions about future prices and expenditures
and therefore the LEV is expressed in "expected" dollars. The second "theoretical"
dollar is an "ECO-dollar" that has been used to estimate the value of "ecosystem
services." Some suggest the value of the Earth's ecosystem exceeds $30 trillion
ECO-dollars per year (Costanza et al. 1997).

Many landowners place a higher value on nominal and real dollars than they do
on theoretical ECO- or LEV dollars. Most who practice "cut-out and get-out," and
most estate owners do not use LEV dollars when making management decisions.
Some who practice "stay and plant" or "urban development" do calculate an LEV
(either when deciding to purchase or sell land). For example, in cases where the
LEV is less than the market, some will choose to subdivide the forest and sell the
land for real-estate development.

There are some managers who place a higher value on ECO-dollars than they do
real dollars. These managers are often in charge of "non-profit" forests where the
budget originates from taxes or donated money. Some calculate the annual income
from their forest exceeds 1,000 Eco-dollars per ha (Agrawal et al. 2008; Moore et
al. 2011).

The price of land is often a factor when determining if forestry is an economical
venture. For some, a high price for bare-land (i.e. land without trees) is an impediment to the practice of forestry. For example, Mclaren (2005) used Faustmann's
formula to calculate an LEV of $978/ha for an average pine plantation (in 2005 US
dollars using a 6 % real discount rate). If the market price of land in 2005 was $978/
ha or less, then the one school of thought (i.e. School #1) says that establishing a
tree farm would be economical (and that buying land at this price would be wise).
In contrast, if the market price was $1,200/ha, those from School #1 would not rec-
ommend buying the land since the expected Return on Investment (ROI) would be
less than 6 % (real; before tax). Essentially, those from School #1 consider land cost
as an integral aspect when determining the economics of investing in a tree farm.

In contrast, there are others who consider purchase of land as a separate investment. Those from the second school of thought (i.e. School #2) would consider
investing in a tree farm even though the market price of land is higher than the LEV.
Some speculate that the market value of bare-land will increase over a 35-year time
horizon. Therefore, those from School #2 would not divest their forestry interests if
the market value of cutover-land exceeds their calculated LEV.

Figure 2 illustrates the primary difference between the two schools of thought.
If we examine the economics for a single tree farm, School #2 says that the for-
estry investment achieves the same ROI regardless of the market value of the land.
In contrast, School #1 assumes that for the tree farm to be profitable, the expected
ROI for the tree farm is a function of how much is paid for the bare land. In other

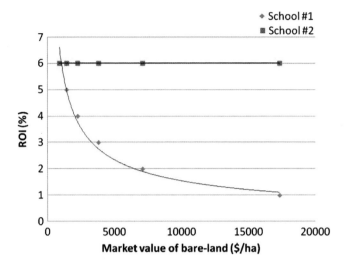

Fig. 2 An example of the relationship between the market value of land (and school of thought) on the Return on Investment (ROI) required for a tree farm to be "profitable." School #1 does not practice forestry if market value of bare-land is greater than the calculated Land Expectation Value (LEV). School #2 assumes the ROI from the tree farm is not affected by the market value of the bare land. School #2 views the land as a separate investment

words, those from School #1 limit the investment in tree farms to land that has a relatively low value. This might help to explain why some timberland firms sell their timberland in countries with high land values and then reinvest in countries where land prices are low.

3.3 Taxes and Forestry

The complexities of the tax system in the USA can help explain changes in ownership of timberland. Mangers of companies who originally adopted the "stay and plant" approach (Table 1) grew tired of the "double tax" system where the company paid taxes on profits and then shareholders paid taxes on their portion. As a result, most of the large "stay and plant" companies are gone. Their timberland is now mostly managed by timber Real Estate Investment Trusts (REIT), managed by timber investment organizations (TIMO) or non-government organizations (NGO). The TIMOs do not own the land but, instead, they acquire timberland for large institutional investors such as pension funds, University endowments, hedge funds, Foundations, and high net worth individuals (who own the lands). REITs and some TIMOs pay property taxes on timberland.

The next step in the evolution of timberland ownership involves selling timberland to tax-exempt NGOs. In some cases, the land is sold to an NGO for more than

$4,000/ha and sustainable forest management continues uninterrupted. This reduces the tax-burden for the previous landowner while it might increase the tax-burden for those who own adjacent timberland. Some local governments may want to increase taxes on private timberland in order to keep tax revenues at historical levels (or higher).

There are several advantages to selling land to a conservation organization or land trust. In some cases, those selling land can continue to use the land in the same manner as before the sale. In the USA, a landowner might achieve federal, state and local tax deductions from donating land to these groups.

For example, donors of land can claim a federal income tax deduction under section 170 of the U.S. Internal Revenue Code. In some states, a donation of land can also receive a tax credit or tax deduction (for state assessed taxes). At the county level, a wealthy individual owning 3,000 ha of forest might have to pay $30,000 per year in taxes while a conservation group would pay no taxes on the same property.

3.4   A Hypothetical Timeline

The following story involves the management of a hypothetical 1,000 ha forest over a 550-year period. In the year 1500, the forest was occupied by natives who periodically burn the forest in order to encourage the growth of grasses. This increased the carrying capacity for deer and increased the success rate of hunters. After a devastating epidemic of small-pox, the human population declined and natural regeneration increased the stocking rate in the forest. The land was eventually acquired in the year 1740 by a Scotsman of Royal Lineage (who traded two horses for the property).

The Scotsman harvested timber and wildlife and used the Estate-owner's method of accounting to determine annual profits (at that time, he did not know about Faustmann's equation). Eventually, his great-grand-daughter inherited the forest. In 1890, she sold the forest to a timber-baron who logged the forest, sold the lumber, and then stopped paying taxes on the eroded land. The new owner paid the back taxes and acquired title to the land in 1920. He had seen the success of tree planting in Germany and after using Faustmann's formula to calculate a LEV, he decided to invest in establishing a tree farm (by planting pine seedlings at a rate of 50 ha/year). By 1950, he had a "fully regulated" forest that was managed on a 25-year rotation.

By 1970, the market price of the timberland exceeded the LEV of the timberland (by 50 %). Although the profit amounted to $250/ha/year, the owner decided that his fully-regulated timberland was no longer profitable. He sold the land (at $2,000/ha) to a Real Estate developer who intended to subdivide the property into 200 5-ha plots.

During a round of golf, the developer mentioned to a wealthy media mogul that he just purchased a 1,000 ha forest that was adjacent to a 4,000 ha wilderness area (managed by a state government). The mogul expressed her concern that the eco-dollars generated by that forest outweighed any financial gains derived from fragmenting the forest with a subdivision. The developer asked "Exactly how much is an Eco-dollar worth?" She replied: "It is worth the same as a US dollar except that instead of declining in value over time (due to inflation), Eco-dollars increase in value (because with development, the total amount of wildlife habitat generally declines over time). Seeing an opportunity, the developer asked if the worth of ecosystem services his forest provides would be worth as much $2,000 Eco-dollars per hectares? After she said yes, he offered to sell the property to her for $4 million ($2 million for the market price and $2 million for ecosystem services). She declined saying the ecosystem services were not worth that much to her as an individual, but that it would be worth that much to society. After a few minutes of bartering they agreed to a price of $2.5 million. Both were pleased with the exchange.

By 1990, the mogul was still enjoying her timberland and some of the trees were now 55 years old. However, she was growing tired of spending money on road maintenance and taxes, so she decided to sell the property to a conservation group. The market value of the property had increase by 50 % in real dollars (i.e. quintupled in nominal dollars). She donated part of the property and sold the remainder to the conservation group. The deal allowed her to continue to use the property (as she had before) but she no longer had to pay property tax.

The transfer of the deed to the conservation group had a long-term impact on the sustainability of the forest. In contrast, the adjacent 4,000 ha GO wilderness was not as fortunate. The wilderness area was established in 1930 and was managed by the government as a wilderness for 90 years. However, the politicians practiced deficit spending and the Debt-to-GDP (Gross Domestic Product) ratio exceeded 200 %. This resulted in an economic downturn and newly elected government tried to find ways out of the financial hole. They decided to sell off some government resources to help pay off the debt. Since large coal deposits were discovered under the wilderness, a decision was made to create a government corporation to mine the coal. This decision was fought in the courts with arguments that the Eco-dollars generated by the wilderness far exceeded the need to reduce public debt. The judges ruled in favor of the government and the "wilderness" aspect ceased.

This hypothetical scenario (which is influenced by my regional experiences) is intended to illustrate how 1,000 ha can be managed under very different economic objectives. This example is just one of many possible sequences. It suggests that as the wealth of an area increases; the price of land increases (in real dollars), the LEV no longer exceeds the price of bare land, forest industry may move to less populated regions (where jobs are scarce), NGOs and wealthy individuals purchase timberland (once owned by the forest industry), and, in the public's eye, the relative importance of ecosystem services increases. This is supported by Von Thunen's theory (von Thünen 1826) that forestry is only worthwhile where land rent is low enough to justify growing wood for lumber and fuel (Hall 1966). If there is some truth to this relationship, then forest industry might exist for a longer time in developing countries where land prices are relatively low as compared to wealthy countries where land prices (and incomes) are relatively high.

4 Final Questions

What is your view of forest economics? If your survival depended on it, would you cut down your forest so that you could farm the land to keep your family fed? Would you ever consider buying 100 ha of forest for $500/ha if you knew the timber alone was worth $200,000? Would you harvest all the timber from 25 ha to payback the $50,000 loan? Have you ever invested $1,000 in tree planting with the expectation of a nominal 6 % return on investment in 20 years? If you inherited a 1,000 ha forest from your grandmother, would you consider all her previous expenditures before you decided on an optimal rotation age for management? If a real-estate developer offered you a price that was double the calculated LEV for the forest, would you sell the property? If not, why not? If you were an elected treasury official of a government, would you consider selling government managed pine plantations to investors from overseas? If you were the director of a national park, would you ever allow the drilling for oil in a wilderness area? Assuming you answered all these questions, were your replies influenced by your current level of personal wealth?

References

Agrawal A, Ashwini C, Hardin R (2008) Changing governance of the world's forests. Science 320:1460–1462

Costanza R, d'Arge R, de Groot R, Farber S, Grasso M, Hannon B, Limburg K, Naeem S, O'Neil RV, Paruelo J, Raskin RG, Sutton P, van den Belt M (1997) The value of the world's ecosystem services and natural capital. Nature 387:253–260

Faustmann M (1849) Calculation of the value which forestland and immature stands possess for forestry. J For Econ 1(1):7–44, 1995

Guild D, Dudfiel M (2009) A history of fire in the forest and rural landscape in New Zealand; part 1, Pre-Maori and Pre-European influence. N Z J For 54:34–38

Halkett J (1991) The native forests of New Zealand. GP Publications Ltd, Wellington, p 149

Hall P (ed) (1966) Von Thünen's isolated state. Pergamon Press, p 304

Helms JA (1998) The dictionary of forestry. The Society of American Foresters, Bethesda, p 210

Jones WP (2005) The tribe of black Ulysses: African American lumber workers in the Jim Crow South. The University of Illinois Press, p 256

Klemperer WD (2003) Forest resource economics and finance. McGraw-Hill, p 551

McIntyre RK, Jack SB, Mitchell RJ, Hiers JK, Neel W (2008) Multiple value management: the Stoddard-Neel approach to ecological forestry in longleaf pine grasslands. Joseph W. Jones Ecological Research Center, Newton, p 36, Online at: http://www.jonesctr.org/education_and_outreach/publications/EFpub.pdf

McLaren P (2005) Realistic alternatives to radiata pine in New Zealand – a critical review. N Z J For 50(1):3–10

Moore RT, Moore R, Williams T, Rodriguez E, Hepinstall-Cymerman J (2011) Quantifying the value of non-timber ecosystem services from Georgia's private forests. Georgia Forestry Foundation, Forsyth, Online at: http://www.warnell.uga.edu/news/wp-content/uploads/2011/02/Final-Report-1-24-11.pdf

Neel L (2010) The art of managing longleaf: a personal history of the Stoddard-Neel approach. University of Georgia Press, Athens, p 211

Pyne JS (1982) Fire in America: a cultural history of wildland and rural fire. Princeton University Press, Princeton, pp 66–122, See especially Chapter 2 "The Fire from Asia"

Pyne JS (1995) World fire: the culture of fire on earth. Henry Holt and Company, New York

South African Timber Growers' Association (1993) Annual report and financial statements for the year ended 31 March 1993. South African Timber Growers' Association, Pietermaritzburg, South Africa, 36 pp

South DB, Laband D (2013) The estate owner's approach to forest economics. In: Guldin JM (ed) Proceedings of the 15th biennial southern silvicultural research conference. Asheville, NC: USFS e-Gen. Tech. Rep. SRS-GTR-175, pp 169–174

Tahvonen O, Viitala E-J (2006) Does Faustmann rotation apply to fully regulated forests? For Sci 52(1):23–30

von Thünen JH (1826) Isolated State; an English edition of Der isolierte Staat. Translated by Carla M. Wartenberg. Pergamon Press [1966], Oxford, New York

Part V
Tree Breeding and Commercial Forestry

Eucalyptus Breeding for Clonal Forestry

Gabriel Dehon S.P. Rezende, Marcos Deon V. de Resende, and Teotônio F. de Assis

Abstract As global demand for wood increases, planted forests will also become increasingly important. Accepting and promoting them as the only way to address the wood scarcity problem and also to help suppress the demand for illegally logged timber from natural forests is a major issue globally. Eucalypt clonal forestry is proving to be an iconic alternative in this context, due to their fast growth, wood quality appropriate to many different uses, huge existing variability, and suitability to vegetative propagation. However, efficient breeding and deployment strategies are essential. The present chapter aims to present, based on the authors' practical experience, an overview on the most successful approaches that may be used during the different phases of eucalypt breeding programs for clonal forestry. Relevant topics covered are: identifying breeding objectives and related traits for the main eucalypt businesses worldwide; the major planted species and their value for different objectives; breeding strategies (recurrent selection methods, breeding cycle, etc.); recombination issues, such as effective population size, mating designs and controlled pollination methods; evaluation and selection procedures as applied to progeny and clonal trials; and deployment aspects, such as number of commercial clones, large scale vegetative propagation methods, and risk management.

G.D.S.P. Rezende (✉)
Forest Breeding and Biotechnology Manager, Fibria S. A., Rod. Gal. Euryale Jesus Zerbini, Km 84, SP 66, Jacareí, SP, Brazil, CEP 12340-010
e-mail: gabriel.rezende@fibria.com.br

M.D.V. de Resende
Embrapa – Brazilian Agricultural Research Corporation and Federal University of Viçosa, Viçosa, MG, Brazil

T.F. de Assis
Assistech Ltda, Belo Horizonte, MG, Brazil

T. Fenning (ed.), *Challenges and Opportunities for the World's Forests in the 21st Century*, Forestry Sciences 81, DOI 10.1007/978-94-007-7076-8_16,
© Springer Science+Business Media Dordrecht 2014

1 Introduction

The world population has recently reached seven billion people, and every day, everyone makes use of forest products at homes, offices and schools, but often don't realize how important they are. Forest products are present in buildings structural and decorative materials, furniture, printing and writing paper, toilet paper, steel products, pharmaceutical products, cosmetics and many other necessities of life.

Global demand for wood will increase, driven mainly by the following trends (FAO 2010, 2011):

- The world population is increasing, forecasts indicating that there will be nine billion people in 2050.
- Wood availability is decreasing. The global forest area and wood availability per capita has been reduced over time.
- Economic growth, especially in the emerging economies.
- Globalization of the forest products market. China, India and Brazil are now driving future forest products demand and investments.
- Climate change, with increasing temperatures and water scarcity trends.
- Environmental and energy policies, valuing the potential of wood as a renewable source, with positive environmental benefits.
- Demand for high quality value added and certified wood products.

In this scenario planted forests will become increasingly important. Accepting and promoting them as the only way to address the wood scarcity problem and to avoid illegal logging in natural forests is a major issue (Fenning and Gershenzon 2002). Expanding the range of commercial forests will be as important as increasing productivity in current plantation areas, in order to minimize competition for land, whether for agricultural, conservation or carbon capture needs, as well as for increasing the availability of wood while reducing its cost.

Eucalypt forests are an iconic alternative in this context, due to its astonishing fast growth (rotation ages ranging from 6 to 15 years), wood quality appropriate to many different uses, huge existing variability, and suitability to vegetative propagation. The global area of eucalypt plantations is already around 20 million hectares across all the inhabited continents, with large areas located in Brazil and India (more than 4 million hectares each) (FAO 2010, 2011).

According to the Brazilian Association of Planted Forests Producers (ABRAF 2011), in 2010 the harvest from Brazilian eucalypt forests was 112,955,222 m^3 of wood, with 48.5 % being used for pulp and paper production, 13.6 % for charcoal production mainly for the steel industry, 3.9 % for chipboard panels production, 3.1 % for added value solid wood products production (lumber for furniture, flooring, etc.) and 29.3 % for industrial firewood production. The estimated gross value of this forest production was ca. US$25,000,000,000.

In Brazil, nearly 50 % of current eucalypt forests are clonal. The development of clonal forestry for eucalypt in this country, including genetic and silvicultural improvements, is emblematic (Fig. 1) because it moved mean productivity from

Fig. 1 Eucalypt clonal forests, Veracel Company, Brazil

25–30 m³/ha/year to 35–45 m³/ha/year in the last 30 years. Productivity can exceed 60 m³/ha/year in some areas where proper conditions (soil, climate, genetic material and silvicultural practices) are combined. However, clonal eucalypt forests are also successful in many other countries, such as China, South Africa, Portugal, Spain, Chile, Argentina and Uruguay, and are mostly used for pulp and paper production.

Given further developments, eucalypt clonal forestry could go a long way to fulfill present and future global wood needs. In this context, specific breeding and deployment strategies will be required to optimize gains in productivity and wood quality. The present chapter aims to present the authors' practical experience concerning the most relevant aspects for the development and implementation of such strategies.

2 Breeding Objectives

Breeding objectives must be aligned with the long term business strategy. This assumption, which is obvious at first sight, sometimes creates a huge challenge for breeders, simply because they must ensure all internal and external clients will be satisfied when a specific clone is planted, harvested and converted into a final product. This is especially true for wood quality traits as they strongly impact the final product performance. The larger the number of traits involved, the smaller will be gain from selection in each one of them. That means the breeders must choose to work on relatively few individual wood quality traits, which explain the variability in the long term objectives for the most important product attributes.

The most relevant breeding objectives and related traits usually considered in eucalypt breeding programs worldwide, for the main eucalypt businesses (pulp and paper, charcoal for steel industry, biomass and solid wood production) are presented in Table 1. From this table some important observations can be made:

– Increased productivity is needed for any business. Forest productivity is the ulti-mate expression of adaptation. In highly productive areas (defined mostly by the absence of significant water stress), where there is broad adaptation of the breed-ing population, individual tree wood volume will play a major role as compared to survival. In poor or restrictive areas, especially those affected by drought or frost conditions, survival ability will be of increased importance.
– Increased tolerance to pests and diseases is a key determinant of forest productiv-ity and forest production stability. For each point in time there will be specific agents to be monitored that might damage clonal forest plantations.
– Increased rooting ability is crucial for eucalypt clonal forestry, as this trait is the most important factor impacting clonal plant production costs. However, the application of this selection criterion must be done with parsimony, as the acceptable rooting ability of a given clone must take into account the cost/benefit relationship defined also by the gains obtained from its use (e.g. pulp productivity). Sprouting ability is also important, both in the nursery and the field, especially where coppice (second or third rotation) is a common practice.
– Wood basic density is relevant for almost all end uses, as it impacts both productivity and product quality. Special attention must be given to this trait, in any situation.

In this context, understanding the genetic control and correlations between traits is very important as this will impact different choices and activities related to the breeding program. Some important lessons obtained so far are as follows (Araujo et al. 2012; Assis 2000; Assis and Resende 2011; Borralho et al. 1993, 2008, Costa e Silva et al. 2006, 2009; Downes et al 1997; Drew et al. 2009, 2011; Greaves and Borralho 1996; Potts 2004; Raymond and Apiolaza 2004; White et al. 2007):

– Heritabilities for wood quality traits are higher than for growth traits (usually above 0.4 and below 0.25, respectively).
– Genetic control of growth and wood quality traits is predominantly additive (determined by the alleles mean effects) but evidence exist that for growth traits non-additive effects (effects from interactions between alleles or dominance effects, and between gene loci or epistasis) are important too.

Table 1 Most relevant breeding objectives and related traits in eucalypt breeding programs for pulp and paper, charcoal for steel industry, biomass and solid wood production

Business	General objectives	Specific objectives	Related or complementary traits (selection criteria)
Pulp and paper	Increased pulp productivity (adt/ha/year)[a]	Increased forest productivity (m³/ha/year)	Increased tree volume (m³/tree)
			Increased survival (%)
			Increased tolerance to biotic (pests/diseases) and abiotic (drought/frost) stress
			Increased rooting ability (%)
		Reduced specific consumption (m³/adt)[a]	Increased basic density (kg/m³)
			Increased pulp yied (%)
			Reduced lignin content (%)
			Increased S:G lignin type ratio[a]
	Increased product quality (major examples)	Printing and writing paper: increased bulk, stiffness and opacity	Increased basic density (kg/m³)
			Increased number of fibers (million/g in pulp)
			Increased runkel ratio[b]
		Tissue paper: increased softness, bulk and tensile strength	Basic density upper limited to 600 kg/m³
			Increased number of fibers (million/g in pulp)
Charcoal	Increased charcoal productivity (t/ha/year)	Increased forest productivity (m³/ha/year)	Increased tree volume (m³/tree)
			Increased survival (%)
			Increased tolerance to biotic (pests/diseases) and abiotic (drought/frost) stress
			Increased rooting ability (%)
		Reduced specific consumption on carbonization (m³/t)	Increased basic density (kg/m³)
			Increase gravimetric yield (%)
	Increased product quality (major examples)	Increased mechanical resistance and reduced specific consumption on pig iron making (m³/t)	Increased basic density (kg/m³)
			Increased lignin content (%)
			Increased particle size

(continued)

Table 1 (continued)

Business	General objectives	Specific objectives	Related or complementary traits (selection criteria)
Biomass	Increased biomass productivity (t/ha/year)	Increased forest productivity (m³/ha/year)	Increased tree volume (m³/tree)
			Increased survival (%)
			Increased tolerance to biotic (pests/diseases) and abiotic (drought/frost) stress
			Increased rooting ability (%)
		Reduced specific consumption (m³/t)	Increased basic density (kg/m³)
			Increased drying capacity
	Increased product quality (major example)	Increased calorific power (Kcal/kg)	Increased basic density (kg/m³)
			Increased lignin content (%)
Solid wood	Increased lumber productivity (m³/ha/year)	Increased forest productivity (m³/ha/year)	Increased tree volume (m³/tree)
			Increased survival (%)
			Increased tolerance to biotic (pests/diseases) and abiotic (drought/frost) stress
			Increased rooting ability (%)
		Increased industrial recovery (%)	Reduced taper
			Increased straightness
			Reduced spiral grain
			Reduced end splitting
	Increased product quality (major examples)	Increased clear and clean wood	Increased natural debranching
			Light-colored wood
		Increased dimensional stability	Reduced anisotropic factor[c]
		Increased mechanical resistance	Increased basic density (kg/m³)

[a]Specific consumption is the amount of wood (m³) required for the production of one air dry ton of pulp (adt). S:G lignin type ratio means syringyl/guaiacyl ratio. High S:G ratios are advantageous for pulping process (cooking and bleaching) because S lignin type is more easily removed from wood

[b]Runkel index = (2 × cell wall thickness)/lumen diameter. High runkel indexes are advantageous for pulp and paper quality because it means the fibers are more consistent

[c]Anisotropic factor: relation between tangential contraction and radial contraction. Low anisotropic factors are advantageous for solid wood quality because they are associated with a low probability of splitting, cracking or warping

– Environmental correlations may occur between growth and wood quality traits, e.g. the same clone planted in both a more favorable adaptive region and in a less favorable one, will grow faster and present lower wood density and vice-versa, respectively.
– Genetic correlation exist between wood traits and between wood and final products traits.

– Within the same broad adaptation region, usually there is no significant genotype x environment interaction, either for volume or wood quality. This kind of interaction, however, is to be expected to some extent between regions with great differences in adaptive potential.
– Age effects occur both for growth and wood quality. Knowing general growing curves, at least for growth and basic density (basic density also increases with age), will help decisions in different phases of the breeding program.

3 Choosing Species

Eucalypt belong to the division *Angiospermae*, class *Dicotyledon*, order *Myrtales*, family *Myrtaceae*, genus *Eucalyptus* and *Corymbia*. There are around 900 species belonging to these genera (mostly *Eucalyptus*), originating from Australia, Papua New Guinea and parts of Indonesia. Thus, they naturally occur across a wide range of latitudes, meaning huge differentiation among species along with evolutionary processes and adaptation to very contrasting environmental conditions. Eldridge et al. (1993) and Boland et al. (2006) present detailed descriptions of eucalypt species and respective centers of origin. Recent reviews discuss several aspects of the biology and genetics of eucalypt (Grattapaglia et al. 2012; Myburg et al. 2007).

Despite the large number of existing species, less than 20 are commercially planted worldwide. *E. grandis*, *E. urophylla*, *E. globulus*, *E. camaldulensis*, *E. saligna*, *E. nitens*, *E. tereticornis* and *E. dunnii* are among the most important ones. Fonseca et al. (2010) present a more complete list of the main species planted around the world, as follows:

(a) Genus *Eucalyptus*

   – Subgenus *Symphyomyrthus*

     • Section *Transversaria*: *E. grandis*, *E. urophylla*, *E. saligna* and *E. pellita*.
     • Section *Maidenaria*: *E. globulus*, *E. nitens*, *E. dunnii*, *E. benthamii*, *E. viminalis* and *E. smithii*.
     • Section *Exsertaria*: *E. camaldulensis*, *E. tereticornis* and *E. brassiana*.

   – Subgenus *Idiogenes*: *E. cloeziana*.
   – Subgenus *Monocalyptus*: *E. pilularis*.

(b) Genus *Corymbia* (originally included in genus *Eucalyptus* as subgenus *Corymbia*): *C. citriodora*, *C. torelliana* and *C. maculata*.

All planted species have 2n = 22 chromosomes, predominantly allogamous (selfing may occur at rates between 10 % and 35 %), with hermaphrodite and protandric flowers. The most important natural pollination vectors are insects. Hybridization between species belonging to different subgenera is rare, but it does occur between species within the same subgenus and especially within the same section (Griffin et al. 1988; Assis 2000).

Table 2 Pulp productivity differences between E. "*urograndis*" and E. *globulus* (mean range values)

Species	Volume (m³/ha/year)	Specific consumption (m³/adt)	Pulp Productivity (adt/ha/year)
E. *globulus*	10–30	2.8–3.4	3–10
E. "*urograndis*"	30–50	3.4–4.2	7–14

The world-class benchmark for clonal forest productivity is the tropical inter-specific hybrid type between E. *grandis* and E. *urophylla* planted in Brazil and some other countries, usually known as E. "*urograndis*". This hybrid type was developed during the 1980s in Brazil and became a standard because it combines fast growth, increased tolerance to pests and diseases, excellent rooting ability, as well as wood quality suitable to different uses. E. "*urograndis*" clones are now widely planted in Brazil for pulp and paper, charcoal and solid wood production. This kind of material easily exceeds 40 m³/ha/year in traditional planting areas.

On the other hand, E. *globulus* clonal forests grown in temperate climates, as in Portugal, Spain and Chile, are benchmarks regarding wood quality for pulp and paper production. This species presents reduced specific consumption, derived from its high basic density and high pulp yield, low lignin content and better lignin quality. Moreover, it presents excellent fiber morphology (increased fibers/g and runkel index). The combination of these characteristics makes its paper the most appreciated in different segments of the world market.

The main differences between E. "*urograndis*" and E. *globulus* regarding pulp productivity are presented in Table 2. It is easily seen that the E. *globulus* advantage in specific consumption is not sufficient to offset E. "*urograndis*" increased volume, resulting in huge differences in pulp productivity ranges. Moreover, rotation age for E. "*urograndis*" is 6–7 years whereas it is around 12 for E. *globulus*. These differences have important impacts in business competitiveness as affected by wood costs.

Ranges within the above "species" are due to both genetic and environmental variability observed in respective traditional plantation areas. Thus, one logical question arises from this analysis: is it possible to combine E. *globulus* wood quality with tropical species growth rates into an inter-specific hybrid clone? Because of significant adaptive distances between temperate and tropical species, this possibility has rarely been realized and many abnormalities or poor performances have been observed in hybrid progenies (Potts and Dungey 2004). However, there is some evidence of success, especially in temperate/tropical transition areas (e.g. southern Brazil), where some outstanding individuals have been found, regardless of their family behavior. Once vegetative propagation is established, value from these "transgressing recombinants" can be captured in clonal commercial plantations (Assis 2000; Bison et al. 2007).

The current success of inter-specific hybrid clones, especially E. "*urograndis*" in Brazil, regardless of the genetic control behind it (complementarity of additive effects, heterosis or epistasis), in combination with the huge unexplored genetic variability available between and within different species in genus *Eucalyptus* and genus *Corymbia*, and the need for increasing production goals to fulfill world demands for wood, suggest that a multispecies hybrid development approach will be very attractive to breeding programs around the world. This approach, however, is unlikely to compete with programs dedicated to pure species in regions where the adaptation of these pure species is unquestionable and clear benefits exist from using them, e.g. the excellent wood quality obtained from highly productive *E. globulus* clonal plantations located in the maritime regions of northern Portugal and southern Chile.

A basic description of the main planted species potentialities, in accordance with the main objectives pointed out in Sect. 2, is presented in Table 3, subject to specific environmental effects. It is easily seen that countless combinations of species in different proportions might be used in a multispecies program, and could bring novel variation and complementary attributes to a synthetic population and to the specific clones within it. However a relatively limited number of core species must be carefully chosen, otherwise, the breeding program may become unmanageable.

4   Breeding Strategy

The vegetative propagation of outstanding individuals is a millenary technique extensively used in agriculture. Many important crops including potato, sugar-cane, banana, grape, etc., are produced by vegetative propagation. Its application to forests is not recent either (Zobel and Talbert 2003), but initially it was used only to preserve genotypes or to establish seed orchards, through grafting. The first attempts to root eucalypt cuttings were accomplished by French and Australians in northern Africa during the 1950s. Yet, the first commercial eucalypt clonal plantations were only established in Brazil during the 1980s, as previously mentioned.

The basic assumption underlying the use of clonal material is the possibility of fully transmitting to "offspring", by vegetative propagation, an outstanding individual genotype. This way, selection gains are maximized, once all kinds of genetic effects, additive (allelic mean effects) and non-additive (effects from interactions between alleles and between gene loci) are capitalized. Thus, any eucalypt commercial clone may be defined as a group of genetically identical plants, produced by vegetative propagation from the same common ancestor, which was selected for its all round outstanding performance.

However, the gains from clonal selection are a dead-end for an existing population at any given moment. Breeding programs, based on recurrent selection (recombination, evaluation and selection in successive generations), are required to allow the generation and deployment of new commercial clones. But breeding for clonal selection requires some extra effort, because individual performance evaluation in progeny trials is not robust enough to predict performance of the same individual as a clone over a range of different environmental and silvicultural conditions (Reis et al. 2011; White et al. 2007).

Thus, it is usually necessary to go through two evaluation and selection phases in each breeding cycle (initial evaluation/selection in progeny trials and final evaluation/selection in clonal trials), as illustrated in Fig. 2. This has a huge impact on the time

Table 3 Main *Eucalyptus* and *Corymbia* species potentialities, in accordance to Table 1. Species marked +, ++ or +++ are respectively good, very good and excellent

	Individual volume[a]	Drought tolerance[a]	Frost tolerance[a]	Disease tolerance[a]	Rooting ability	Basic density	Pulp yield[b]	Lignin content[b,c]	Fiber morphology[b]
Tropical species									
E. grandis	+++				++	+	++	++	++
E. urophylla	++	++		++	+++	++	++	++	+
E. saligna	++		+		++	+	++	++	+
E. pellita	+	++	+	+++	+++	+++			
E. robusta	+	++	+	+++	+++	+++			
E.camaldulensis		+++	+	++	+++	+++			
E. tereticornis		+++	+	++	++	+++			
E. brassiana		+++			+	+++			
E. cloeziana	+	+	+	+		+++	++	++	
C. citriodora	+	++		+		+++	++	++	
C. torelliana	+	+++	+++	++	++	+++	+	+	
Temperate species									
E. globulus	++	++	++		+	++	+++	+++	+++
E. nitens	+++		+++			+	+	++	+++
E. dunnii	++		++	+	+	++	++	++	++
E. benthamii	+++		+++	+	+	++	++	++	+++
E. viminalis	+		+++	++	+	++	+	+	+++
E. smithii	+		++	++		+++	+++	+++	+++

[a]Some other eucalypt species to mention due to evidence regarding disease, drought or frost tolerance are *E. resinifera, E. longirostrata, E. dalrympleana, E. rudis, E. badjensis, E. dorrigoensis, E.pilularis, E. paniculata, E. microcorys, E. cypellocarpa* and *C.maculata*
[b]Species marked according to suitability to pulp and paper production (e.g. low lignin content)
[c]Species presenting high lignin content are *E. camaldulensis, E. tereticornis, E. pellita, E. resinifera* and *E. paniculata* (*E. paniculata* presents very high density as well)

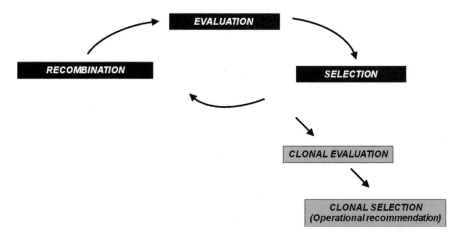

Fig. 2 Basic eucalypt breeding cycle (12–16 years from recombination to final recommendation)

required for clonal deployment. Even the most efficient programs, which make use of advanced techniques, such as flowering induction (see Sect. 5) and early selection (see Sect. 7), require about 12–16 years from initial recombination to clonal recommendation and deployment.

Some alternative approaches may enable short-cuts within this breeding cycle, but so far all of them also have limitations or still need to be proven operationally. The most relevant to mention are as follows:

– Simultaneous testing of families and clones: the vegetative propagation of seedlings as soon as they are obtained might avoid clonal tests at a later stage and save significant amounts of time. However, it is extremely difficult to multiply a representative minimum number of individuals from each progeny (see Sect. 6), so that a sufficient number of plants is available for all required trials. If a limited number of individuals from each cross are randomly selected, the variability may not be fully represented and some outstanding individuals will be lost. Moreover, only good rooters will pass on to the field trials, giving to rooting ability an excessive weight in the global selection process. Alternatively, if an effort is made to multiply a representative number of individuals from each progeny, costs rapidly increase and operational efficiency becomes a significant obstacle.
– Genome-wide selection: this relatively recent method (Meuwissen et al. 2001) relies on a genome-wide variant of marker-assisted selection (MAS) for quantitative traits. It is currently a hot topic in plant breeding and is being tested as a way to incorporate desirable alleles at many loci of small effect. In this approach, genomic breeding values (GBV) of individuals can be predicted in an experimental or "training" population, based on prediction models constructed from regressing phenotypes on whole-genome marker genotypes for several hundred to a few thousand individuals. These prediction models can then be used to estimate the GBVs of yet to be phenotyped individuals at early ages based on

marker data only (Grattapaglia and Resende 2011). Some results available for eucalypt suggest that the method might lead to elimination of progeny trials through genome based preliminary screening of seedlings, followed by cloning of only the top performing individuals for establishment in clonal field trials (Resende et al. 2012). In spite of this expectation, selection based on the phenotype evaluation is still unavoidable (Resende et al. 2011) and many standing issues still remain before the method can be widely adopted: will calibrations work both for small, short-term breeding populations and larger, long-term breeding ones? Will it work for both pure species and hybrid multi-species (multiple alleles) approaches? Will it be able to capture both additive and non-additive effects, allowing effective selection for parents and for clones? Will it be able to identify favorable transgressing recombinants, which might not be present in the training population? However, if this technology is successfully proven, it will be of great help to breeders.

– Clonal variety: this approach can be effective when there is great confidence about the performance of outstanding progenies or individual trees being tested, as compared to current commercial clones (controls). If so, clonal trials can be eliminated through the creation of a clonal variety composed of tens of outstanding individuals, which would be promptly propagated and deployed as a clonal mix into pilot plantations. From phenotypic evaluation of these pilot plantations and DNA fingerprinting of the best performers (Grattapaglia et al. 2004), purification of the clonal variety can be driven in successive steps, until an optimized mix composition of a few top clones is obtained, which can be individualized or not.

Regardless of the use of any above mentioned short-cuts, recombination of individuals selected in progeny trials according to their additive values must move forward at the same time as individuals selected from the same progeny trials according to their total genetic values are cloned for establishment in clonal field trials. This will make the breeding cycle a bit faster. Furthermore, it is mandatory to carry out parallel breeding sub-cycles (with sub-populations), every 3-4 years, to ensure periodic clonal portfolio renewal.

As previously indicated, one of the main assumptions justifying the use of clones is the maximization of selection gains by capturing all kinds of genetic effects (additive and non-additive). Although additive effects are the most important in determining total genetic variability available for growth related traits, current evidence suggests that non-additive effects also play an important role, both in tropical and temperate species (Araújo et al. 2012; Bouvet et al. 2009; Li et al. 2007; Rezende and Resende 2000). Reported non-additive genetic variance estimates for growth reach up to 80 % of additive genetic variance estimates, hence nearly doubling the total genetic variance available. More consistent information on the magnitude of heterosis and epistasis in eucalypt species and inter-specific hybrids is required, but this expected substantial amount of non-additive variance gives support to the continuing use of clonal propagation as a preferable deployment strategy for eucalypt.

There are many different efficient recurrent selection methods described in the literature (Comstock 1996), but in theory, reciprocal recurrent selection (RRS) methods and respective variants should work better for inter-specific programs and also for pure species with contrasting populations, because they allow capture of non-additive effects to some extent from one cycle to another. However, RRS methods are more time and money consuming, because they require recombination of individuals from pure species (or pure contrasting populations) based on respective "hybrid" progenies performance. As saving time and money is vital for any breeding program, and this becomes even more important in forest breeding due to the long cycles involved, in general terms the most effective approach suggested for eucalypt breeding is using simple recurrent selection for both pure species or multispecies programs (synthetic pure populations or synthetic hybrid populations, respectively) (Kerr et al. 2004; Resende and Assis 2008). In synthetic hybrid multispecies programs, the different types of hybrids available will be crossed each other, but pure species trees may be added to the group to improve specific attributes of the desired mix. As a consequence, parallel recurrent selection of pure species involved in the multispecies programs is desirable, although as a secondary priority.

It is important to draw attention to the fact that neglecting the capture of non-additive effects by choosing the simple recurrent selection method, does not mean neglecting the capture of non-additive effects during the clonal selection phase, as will be pointed out in Sect. 7.

5 Recombination

Combining short and long term breeding strategies is strongly recommended, especially for programs that are just starting, in order to provide results as soon as possible and address business needs and pressures. Short term programs involve narrowly based populations composed by elite individuals promptly available. If no information exists on additive and non-additive merit of available elite trees, recombining the best existing commercial clones is an interesting approach.

On the other hand, long term recombination must ensure maximum exploration of the population's improvement potential, which depends upon keeping an appropriate effective population size (number of unrelated individuals that effectively contribute with alleles to the breeding population along cycles). This will avoid favorable allele loss, in balance with the application of selection intensities high enough to provide satisfactory genetic gains. In general terms, regardless of the heritability and family structure concerned, keeping the effective population size between 30 and 50 along the breeding cycles will ensure the maintenance of adequate levels of genetic variability and consequent application of strong selection intensities (Namkoong et al. 1988; Resende and Barbosa 2005). But once a parent's gene pool is sufficiently sampled in progeny trials (see Sect. 6), it is not worth keeping it or the remaining seeds of its progenies for future breeding activities, as adopting this conservative practice reduces the breeding program overall efficiency and increases associated costs.

Recombination generates a new set of progenies. Mating designs used in this phase will determine family structure of progeny tests to be evaluated in the ongoing breeding cycle. Family structure is important because it will provide relevant information for decision support, namely the relatedness between individuals. This information is particularly important for estimating genetic parameters (e.g. additive and non-additive variances, heritabilities, etc.), for parent selection (keeping effective population size), and for potential clone selection (minimizing genetic vulnerability).

There are basically three types of family structures: half-sibs (HS), full-sibs (FS) and self-sibs (SS). The amount of additive and non-additive variance expressed between and within these types of families is obviously different. For example, comparisons between HS families provide a good estimate of the additive variance or general combining ability, and comparisons between FS families provide good estimates of the specific combining ability of parents, which has a non-additive variance expression component (Lynch and Walsh 1997; White et al. 2007).

The mating designs for developing recurrent selection programs and clonal selection should be analyzed in terms of four purposes: (i) efficiency in estimating the general combining ability of the parents; (ii) efficiency in identifying superior families; (iii) efficiency in selecting clones; (iv) the possibility of evaluating a large number of parents, which is desirable for enabling high selection intensity whilst maintaining adequate effective population size for future selection.

The main available designs are: polimix (half-sib families); single pair (full-sib families); disconnected factorial (half-sib families and full-sib families, simultaneously); partial diallel (half-sib families and full-sib families, simultaneously). The relative efficiency of the different designs for satisfying each above mentioned purpose was studied by Resende and Barbosa (2005), and the major conclusions were as follows:

– Polimix is the best design for estimation of general combining ability of parents (purpose i), as it maximizes efficiency in terms of heritability and accuracy. Disconnected factorial designs become almost as efficient as polimix when three or four crosses by parent are accomplishable.
– A single pair disconnected factorial or a partial diallel design with three or four crosses per parent is the best option for identifying superior families and selecting new clones (purposes ii and iii).
– The possibility of evaluating a large number of parents (purpose iv) should be considered by fixing the maximum number of crosses that can be afforded to be generated and evaluated in the field. For example, by limiting to 200 the total number of crosses or families that one can generate and evaluate in the field, the following number of parents can be used: 400 for single pair mating, 200 for polimix and 50 for disconnected factorial or partial diallel designs with four crosses by parent. Considering the recombination of the 30 best parents, it can be seen that selection intensities (in terms of the selected proportion) in the disconnected factorial and partial diallel designs are very high (60 %) providing low genetic gains.

It can be observed that none of the designs fulfill all the purposes in the most efficient way. For this reason, when good estimates of population genetic parameters are already available, i.e. in mature programs, it may be a good option to use polimix approaches, because of the significant gain in time and labor that is obtained. To complement this design, specific single pair crosses can be made emphasizing the best parents, i.e. the ones with highest breeding values, aiming at minimizing potential polimix disadvantages related to genetic sampling and selective accuracy of families and clones.

One major challenge faced by breeders during the recombination phase is getting the parents to flower as early and evenly as possible to enable the controlled pollinations. Significant advances have been achieved in recent decades, reducing the time required for flowering from the normal 4–8 years to 2–4 years (Fonseca et al. 2010; Griffin et al. 1993; Hasan and Reid 1995; Moncur and Hasan 1994). The methods combine grafting, water/nutrition management and hormonal treatments, both in the field and greenhouse (Fig. 3). But these methods are still very species and genotype dependent and further research into the physiology of flowering in eucalypt is required, as improvements in this area offer large opportunities for reducing breeding cycles.

Important advances have also been achieved in controlled pollination methods. Because eucalypt flowers are hermaphrodite (male and female structures are present in the same flower) and protrandric (the male structures mature prior to the female) complex manipulation (emasculation/isolation/pollination/isolation) is required for traditional controlled pollination. These operations are time consuming and negatively impact time, labor and seed productivity. However, in the late 1990s an important advance was made with the OSP (One Stop Pollination) technique, developed for *E. globulus* in Chile (Harbard et al. 2000). This method makes it possible to combine emasculation and pollination in a single operation, by cutting the female stylus tip just after emasculation and immediately applying pollen on top of it (Fig. 4). Some years later, further advances were achieved by Assis et al. (2005), who developed AIP (Artificially Induced Protogyny), with the purpose of avoiding emasculation for improved operational efficiency. The method consists of making the stylus receptive to external pollen prior to flower opening, by cutting the flower bud tip in the pre-anthesis phase (Fig. 4).

Eucalypt species present huge variation regarding flower size and this must be taken into account in inter-specific crosses. When crossing large-flowered species (e.g. *E. globulus*) with small-flowered ones (e.g. *E. nitens*), usually the former must be used as male parents because otherwise pollen tubes will not be robust enough to fertilize ovules.

At this point a brief discussion about the potential for genetic transformation is required. For breeders, genetic transformation is often regarded as no more than an alternative recombination system for "adding" a specific trait or trait expression controlled by one or few genes which cannot normally be found within the breeding population, into top performing clones. The reason behind this view is that the overall performance of a commercial clone in relation to adaptability (individual growth, survival) and product suitability (wood quality), is determined by thousands of

Fig. 3 Eucalypt
recombination phase in
greenhouse

Fig. 4 OSP (*left*) and AIP (*right*) controlled pollination techniques

genes and their interactions, which were naturally arranged in a very specific way to
provide an overall favorable phenotype. Therefore, based on quantitative genetics
principles and the most recent studies on QTL mapping and genomic selection efficiency in trees (Grattapaglia and Kirst 2008; Grattapaglia et al. 2009; Grattapaglia
and Resende 2011; Resende et al. 2012), a single gene effect most certainly will not
turn a bad performer into a good one. Detecting genes capable of providing a substantial

phenotypic benefit has often been frustrating, with many reported QTLs proving to have overestimated or inconsistent effects.

Moreover, as previously noted, recombination is only one in many phases of a breeding cycle. Using genetic transformation as an alternative recombination method for eucalypt is itself a challenging approach because significant technical bottlenecks related to genotype dependent regeneration/transformation protocols and expression stability persist, and the most amenable clones might not necessarily be the best performers available. Even if success is obtained in generating many transgenic lines of a specific clone, subsequent phases of field evaluation and selection are required. These field tests and eventual deployment of a transgenic clone will certainly take much more money and time than usual, due to regulatory, biosafety, intellectual property, certification and public acceptance issues involved.

However, genetic modification of trees is taking place in many countries. This effort will hopefully bring about knowledge and progress in sustainable forest production, especially for qualitative traits. But transgenic eucalypt clones are unlikely to be deployed commercially in much of the world for some more years, largely due to their cost and regulatory issues, greatly limiting the benefits that this technology is supposed to be capable of providing, as detailed in the related chapters of this book.

6   Evaluation

As previously mentioned, usually there are two evaluation steps in each breeding cycle, namely progeny trials and clonal trials (Fig. 2). In progeny trials seedlings obtained from the recombination phase are directly established in field trials. At this stage each tree represents a single genotype in the field. In clonal trials, the best trees identified in progeny trials are cloned and rooted cuttings are established in field trials over different locations. The objective of progeny and clonal trials is providing reliable data for the subsequent selection phases. In this context, it is important to take the following aspects into account:

– Locations: the number of locations where field trials must be established is as large as the number of different environmental conditions that have to be covered for the projected geographical extent of the breeding programs' ambitions. The larger this coverage is, the more contrasting will be the environmental parameters which in turn will cause complex type genotype × environment (G × E) interactions, with significant changes in progeny or clone ranking over locations. In the initial cycles, the identification of these contrasting areas depends upon soil (depth, stoniness, texture, fertility, etc.) and climate (rainfall, temperature, etc.) characterized as impacting forest performance. However, in more advanced cycles, the number of locations will be defined by a thorough analysis of G × E interactions over locations where the initial tests were established. At this stage, it is usually possible to group regions originally classified as contrasting based on soil and climate information, but which in fact, are similar from a G×E perspective (Rezende and Resende 2001). Nevertheless, it is advisable for clonal trials and especially when only two or three contrasting grouped regions exist, to have an extra test in each macro-region, to ensure against the loss of genetic material and information if a natural or operational disaster occurs, such as fire, wind or mistaken harvest.

– Experimental procedures: accurate evaluation of the genetic merit of an individual tree, a family or a clone requires the application of appropriate experimental procedures which maximize environmental control and genetic material representativeness. In this regard, the most relevant recommendations to breeders are as follows:

• Incomplete block designs (lattices) are recommended for eucalypt breeding field trials because large numbers of progenies or clones are usually tested (White et al. 2007).

• Many plot sizes have been studied in forest species, but the current consensus is that the use of Single Tree Plots (STP) is the best approach, both for progeny and clonal trials, as it provides higher experimental precision (larger number of repetitions and smaller and more homogeneous blocks) (White et al. 2007). Nevertheless, STP requires additional care in tests planning, establishing, controlling and taking measurements.

• Accuracy is also affected by the total number of individuals per family in progeny trials, and by the number of ramets per clone in clonal trials, at each location and among all locations. These numbers may vary depending on the heritability of the trait and the genetic structure of the material under evaluation (HS families, FS families, S1 families, clones). The studies carried out by Resende and Barbosa (2005) suggest that acceptable accuracy and variability levels can be achieved in progeny trials regardless the trait heritability and the family structure, by using a minimum of 100 individuals per progeny along all locations and a minimum of 30 individuals per progeny in each location. The genetic representativeness or effective size of a family is relevant to maximize selection accuracy of potential clones in progeny trials. The effective size (Ne) of a FS and a HS family is respectively given by $Ne = (2n)/(n+1)$ and $Ne = (4n)/n+3$, where n is the number of individuals per family. Simulations showed that n = 100 provide 99 % and 97 % of representativeness of a FS family and a HS family respectively. So, including more individuals than this will barely add to the variability of the sample, as it is already sufficient to include the best individuals of FS and even HS families. In the clonal trials, acceptable accuracy levels are achieved by using between 20 and 30 ramets per clone per location.

• Controls must be included in all field trials. One or two stable clones from the first recommendations shall be used as "permanent controls" in every progeny or clonal trial along cycles, allowing comparisons between materials established in different locations, ages or cycles and also the estimation of cumulative gains over time. In addition, current commercial clones should be used as controls in every planned progeny and clonal trial, to provide a reliable estimate of the candidates merit as compared to the best current planting material.

- Silvicultural practices as applied to field trials, including soil preparation, fertilization and weed control, must be similar to those used in large scale operational forestry, to ensure the suitability of the recommended clones to the standard conditions. Nevertheless, special attention should be given to environmental control, so that experimental precision is not compromised by environmental variation inside blocks or repetitions.

– Measurements: phenotyping progeny and clonal trials properly is a crucial step in any breeding program. Intensity, timing and type of measurements will largely depend upon the resources available. A sequential approach which privileges growth traits has been unavoidable especially due to the lack of non-destructive, efficient, cheap and large scale evaluation technologies for traits related to wood quality, rooting, and tolerance to pests, diseases and abiotic stress. Survival should be evaluated in the very early stages of field trials and for all individuals. Growth traits (diameter, height, health/physiological condition) should be evaluated at least from near to half the rotation age, allowing resources to be saved as well as early selection. Non-destructive large scale wood quality evaluation should, as far as possible, be applied at the same intensity as it is for growth traits, but this is difficult in practice because only a limited number of technologies are currently available for that purpose. Pilodyn has been used traditionally for basic density evaluation, but it presents low accuracy and its efficiency is limited to ranking purposes. Near-infrared spectroscopy (NIRS) from wood increment cores has been used successfully for wood chemical evaluation, including pulp yield and lignin content, but it requires significant investment in calibration models and its efficiency is also limited to ranking purposes (Apiolaza 2009; Downes et al 1997; Raymond and Schimleck 2002; Schimleck et al. 2006). More sophisticated methods such as *Silviscan*™, combine image and x-ray analysis, and provide accurate and detailed wood characterization (Buksnowitz et al. 2008; Raymond 2002; Wu et al. 2009; Wynne and Nelson 2006; Yang et al. 2006), but are very expensive for evaluating hundreds of trees every year, as required in regular breeding programs. Thus, a final complete and rigorous assessment of wood volume (including taper and straightness), wood quality, sprouting and rooting ability, nutritional efficiency, tolerance to pests and diseases, drought and frost, both in the field and laboratory, can only be accomplished for a limited number of pre-selected clones (20–50) in the final stages of clonal trials.

7   Selection

Breeding for clonal deployment involves three selection approaches (Fig. 2): selecting parents for recombination based on progeny trials evaluation; selecting candidate clones for clonal trials based on progeny trials evaluation; and selecting operational clones based on clonal trials evaluation. In every situation the most effective selection procedure is estimating the genetic value of individuals via BLUP (Best Linear

Unbiased Prediction). This statistical procedure maximizes selection accuracy by better separating environmental effects from genetic values (Henderson 1984). Furthermore, the method does not require balanced data, it accounts for trees genealogy (allowing prediction of additive and non-additive value), and combines all the information available including different locations, years, mating designs, field designs and breeding generations.

The whole process of genetic evaluation for selection purposes involves the use of the so called mixed model methodology in which the fitted model encompasses random effects such as genetic values and fixed effects such as locations and years. Treating genetic values as random effects leads to a better prediction of the genetic value of the candidates, which turn out to be both unbiased and with minimum prediction error variance (consequently more accurate). Besides correcting for environmental effects, the methodology takes into account the quantity and distribution of the information associated with each individual and also the heritability of the trait. As such, the procedure enables the comparison of individuals across time and space, enabling genetic gain to be maximized by the selection process.

This approach assumes knowledge of many genetic parameters as input data. As genetic parameters are not really known, they must be estimated by the most precise method available, which is the Restricted Maximum Likelihood (REML) procedure (Patterson and Thompson 1971). The REML method is superior to traditional Analysis of Variance (ANOVA) when the data is unbalanced and the designs are non-orthogonal. It takes into account the genetic relationship matrix to specify all the possible genetic relationships in the data set, and these relationships are then used to produce unbiased precise estimates of variance components. According to Searle et al. (1992) ANOVA estimates of variance components are unbiased only under the following conditions: data are balanced, meaning that there is 100 % survival and all families are planted in equal numbers of blocks at all test sites; parents are unselected, non-related and come from the same generation; and parents are inter-mated in a single, structured mating design (such as a factorial or diallel) to produce a single type of collateral relatives such as full-sib families.

By using REML estimates when fitting the mixed model, the so called empirical BLUP predictions are produced. Frequently, the REML variance components estimation and the BLUP prediction are performed simultaneously by the REML/BLUP procedure which corrects for environmental effects, estimates genetic parameters and produces the individual BLUP predictions at the same time. Different software, such as ASREML (Gilmour et al. 2002), can be used to perform this kind of analysis. Given the complexity involved, careful consistency analysis of data and results is recommended to software adoption and use at operational scales.

The BLUP approach is at least equal and often superior to other selection methods such as simple phenotypic selection (mass selection), between and within family selection and combined family – within family selection. The results diverge as the heritability of the trait under selection diminishes due to poor experimental precision, and the data becomes unbalanced. With BLUP there is a tendency to choose as winners the better tested candidates, while the opposite is true for classical methods. In other words BLUP penalizes the least tested genetic materials, which is a desirable feature. For mass selection, each tree's phenotypic measurements are the only data used to predict the tree's genetic value for each trait, while BLUP entails all the information genetically linked to each selection candidate. However, uniform stands with low levels of environmental noise and high levels of trait expression give higher heritabilities, and, therefore, the genetic values are less regressed for these stands than others. In these situations, simpler selection methods, such as mass selection tends to be as efficient as BLUP selection (Reis et al. 2011; White et al. 2007).

Parental selection for recombination must be based on estimated additive genetic values while clonal selection must be based on estimated total genetic (or genotypic) values. The merits diverge and so do the rankings from these predictions, as the degree of allelic dominance and epistasis increases. Thus, for traits with some level of dominance such as growth and survival, top parents may not be top clones and vice-versa. Even if some high positive ranking correlations are observed for the whole population under evaluation, important changes may occur in the best elite material, which will impact the final program results. Moreover, the strength of the G×E interactions differs for the additive and total genetic values, the latter effect being the stronger (Araújo et al. 2012; Rezende and Resende 2001). This will be particularly relevant to the final selection of clones for operational deployment. Such features emphasize the need for thorough planning of the linear model to enable the BLUP prediction for the both additive and total genetic values.

While many traits are taken into account when selecting genetic material (Table 1), selection can be applied according to three different approaches: selection indices, tandem selection and independent culling levels (White et al. 2007). In practice combining selection indices and independent culling levels is often necessary for estimating the ultimate value of candidates (e.g. pulp productivity, as an index estimated from volume, basic density and pulp yield values, combined with minimum 70 % rooting). As previously mentioned, due to practical issues related to evaluation costs and time, a sequential approach is frequently used, meaning only a smaller group of pre-selected clones for volume and survival are evaluated/selected for wood quality traits, rooting ability, etc. Selection indices which consider the most relevant traits as affected by their economic importance, heritability and genetic correlations are useful when the relative economic importance of the measured traits can be accurately evaluated, and when these relative weights are expected to be stable across the time period that the selected genotypes are being deployed and bred. In general, they become more useful when there is no expected constraint to the long term wood supply, making the economic value of the wood more important than its existence in terms of productivity.

If breeding values are predicted for each trait separately (i.e. a separate BLUP analysis for each measured trait), then selection indices can be easily formed to combine data from all the traits under analysis into a single value for each candidate. The validity of this multiple-step approach rests on a property that the BLUP of any linear combination of traits is equal to that linear combination of the BLUP predicted values of the individual traits (White and Hodge 1989). So, for selection indices that are linear functions, e.g. pulp productivity, the indices values provide the best unbiased predictions of that linear combination of traits.

Indices combining more traits are capable of identifying candidates that are above average for that combination of traits, but not outstanding for any one. Therefore, when selecting parents for recombination it is important to include some trees that are outstanding for individual traits, even if their indices values are smaller than other candidates. This allows the capture of superior alleles in future breeding cycles, especially by crossing selections that are truly excellent for different sets of traits.

Some other issues to be considered while using BLUP are:

- Application requires compatible hardware processing capacity.
- It is important that an experienced person conducts the analysis, making sure data is clean and consistent prior to analysis.
- The quality of predictions can be improved by reducing the experimental error through environmental co-variables and spatial analysis. Incorporation of coordinates concerning the position of each experimental data point into mixed model analyses enables spatial analysis, and has the potential to reduce experimental error and increase heritability and gain from selection. However, spatial analysis is not currently widely used in forest genetic data analysis, as long as the trials are usually established according to adequate experimental designs. In such cases there is no advantage from spatial analysis as the models tend to be over-fitted. Examples of complex models including spatial analysis in forest trees are reported by Resende and Thompson (2004).
- Taking into account any missing plots around each tree may also help reducing experimental error. Such numbers can be fitted as co-variables simultaneously to the REML estimation and BLUP prediction. Competition effects at the environmental and genetic levels can be also fitted to improve the accuracy of the REML/BLUP procedure (Resende et al. 2005).

After predicting the additive genetic values of candidate parents (recombination) some restrictions may still be necessary regarding the family contribution to the selected group of individuals. To do this it is necessary to determine the amount of relatedness to be permitted in the selected group. The concept of effective population size is important in this connection (see recommendations in Sects. 5 and 6). For selecting the candidate clones to be tested in clonal trials, the relatedness issue is less relevant but not negligible. Some care must be taken to avoid a high probability of deploying too many related materials as operational clones (see Sect. 8).

In any situation, an early selection approach, as applied both to progeny and clonal trials, is strongly recommended to reduce the generation time and maximize genetic gain per unit time (Borralho et al 1992; Osorio 1999; Rezende et al. 1994; Stackpole et al. 2010). Despite the lack of consistent scientific information on this matter, selection near to half the rotation age is an efficient and safe approach for the most important eucalypt commercial species and traits (White et al. 2007). Practical experience has shown that potential loss of some superior individuals may occur, but this negative impact is clearly offset by gains per unit time. However, it is always advisable to confirm the performance of selected clones at rotation age and also after harvest, in coppice trials.

8 Deployment

As previously mentioned, the use of clones is justified by the possibility of fully maintaining all kinds of genetic effects expressed in the founding ancestor, both additive and non-additive, in the plantation material. In eucalypt non-additive genetic effects for growth traits seems to be expressive, hence maximizing the total genetic variance available (Araújo et al. 2012; Bouvet et al. 2009; Rezende and Resende 2000; White et al. 2007). This expected importance of non-additive variance supports the use of clonal propagation as the preferred means of deploying improved eucalypt. In other words, gains associated with clonal forestry would be expected to be larger than if seedlings were used from an orchard, both based on the same original genotypes. Moreover, favorable "transgressive segregants" or "correlation breakers" trees, which might be neglected in a seedling population can be immortalized as clones.

So long as non-additive genetic effects are not being captured in the seeds from a seed orchard, but outstanding improved seed varieties are available, even better plantations can be established with the best clones selected from such varieties. Furthermore, there is evidence from a number of leading forestry companies that on average, commercial clonal plantations have historically provided at least 25 % higher realized gain in volume as compared to seedling plantations from the same breeding population, managed in the same locations and with the same silvicultural care. The use of clones also enables greater homogeneity of the wood products, which is no small advantage, and can capture the special attributes of a single tree, such as tolerance to diseases, drought or frost. Thus, in practical terms, clonal plantations present better productivity and uniformity, also resulting in improved efficiency in forest management.

Clonal forests may not represent the best option in every situation however. Improved seedlings may be a better option for plantations to be established in locations presenting specific edapho-climatic conditions, such as regular water deficits above the normal adaptation limits of species, very shallow or swampy soils, etc., for which suitable clones have not yet been sufficiently evaluated nor accurately selected. Therefore, when there is no clear evidence of adaptive superiority of available clones, it is usually a safer option to use batches of improved seed, provided that they were broadly developed for the target region. This is because the genetic variability and buffering ability of the seed population will help minimize any potential losses caused by biotic or abiotic stresses.

One frequent question breeders need to answer is "what is the ideal number of clones to be used in operational nurseries?" This number should not be as large as to reduce genetic gains or to cause operational constraints in the nursery. Nor yet should the number be as small as to increase the risk of genetic vulnerability or the risk of overall bad performance originated in mistaken selection.

Experience has shown that nurseries should operate with five to ten operational clones. These clones should be unrelated, although some slight relatedness between two or maximum three clones out of ten is often acceptable or unavoidable.

Stable clones are preferred because they allow gains in the operational efficiency of the production nursery due to the larger scale of their production and, if chosen judiciously, they can represent a sufficiently generic option for use in most situations. In short, they make the nursery managers life easier. Yet, given that $G \times E$ interaction can be expected to some extent (Sect. 7), clear good performers under specific environmental conditions should not be discarded, at the risk of losing potential gains in the shorter or longer term.

Another important issue is the timing of introducing any new clones into operational nursery production. Recommending new clones every year can be disruptive to the nursery planning and routine. It is important to allow time for learning about the propagation performance of new clones and to find solutions for overcoming any difficulties that might be encountered, such as water management, nutrition and disease control.

Breeders would rather recommend two to five top candidates for pre-commercial pilot plantations (at least 50 ha per year) every 3 or 4 years. This way periodic clonal portfolio renewal is built in accordance to the sub-cycles planned in the breeding strategy (Sect. 4). A good general approach, at any given moment, is to have nearly half the operational clones with a proven track record in commercial plantations accounting for 70–80 % of nursery production. The other half should consist of newly recommended clones, accounting for 20–30 % of nursery production.

Within the 3 or 4 years following the recommendation of a new set of clones, their operational performance shall be monitored in pilot plantations and in the nursery. Some of them will be confirmed as operational clones and will have their production expanded, replacing obsolete material. Some others will fall however, but when this happens other sets will be undergoing introduction to the nursery. Breeders must be deeply involved in this post recommendation monitoring phase, working very closely with operational foresters.

The procedures suggested above usually provide long term sustainable gains, rendering the large scale clonal production of tree material in nurseries and forests feasible, and also avoiding the risks of genetic vulnerability. Genetic vulnerability, especially to pests and diseases, is often assumed as a disadvantage of large scale clonal forests, but this potential problem is mostly overestimated. Factors that mitigate these risks are as follows:

- By recommending two to five unrelated top new clones every 3 or 4 years, a full replacement of the clonal nursery portfolio will happen every 10–15 years. This means that it is unlikely that any one specific clone will be in use in a specific area for more than two operational rotations.
- Simulations with the number of clones needed for managing risks in clonal forestry carried out by Bishir and Roberds (1999), suggested that the level of risk is unlikely to change significantly after the number of clones used in the whole plantation area exceeds 30–40. If about ten clones are constantly used in nurseries and these are completely replaced every 10–15 years, 30 or 40 different clones will have contributed to the total planted area after 30 or 40 years, which seems to be acceptable (one different clone per year on average).

Furthermore, this arrangement provides that any damage will not be significant if any one clone proves to be problematic, because its contribution to the planted area will never be predominant. Yet, problematic clones can be rapidly taken out of production as soon as difficulties are recognized and many alternatives will be immediately available.

- Local plantations are planned so as to use many different operational clones. A plantation will usually have many blocks, each one ranging from 5 to 50 ha, at a rate of only one clone per block, but having many blocks with different clones in the same area. This mosaic structure further reduces risks.

- Susceptibility to pests and diseases is determined more by species and provenances than by clones (e.g. Gonipterus beetle and *E. globulus*). Moreover, some pests/diseases are opportunistic and occur due to poor adaptation of planting material to environmental conditions or suboptimal silvicultural practices employed.

- Breeding programs are usually established with broadly based populations providing high levels of genetic diversity (including tolerance to biotic and abiotic stresses) and, as stated in Sect. 7, specific screening for pest and disease tolerance should occur in the final selection steps, prior to operational recommendation. In fact, it is important to understand that most often clonal deployment is the solution for overcoming problems with pests and diseases because it allows large scale production of tolerant individual trees. Overcoming the notorious problems caused by canker disease in Brazil during the 70s was a landmark example and irrefutable evidence to this end (Alfenas et al. 1983, 2009; van Heerden et al. 2005).

- In the last decades, severe genetic vulnerability problems related to the use of clones in forestry or agriculture were not widely reported. It should be remembered that cultivars belonging to autogamous (self-pollinating) species such as rice, soybeans, common beans and wheat, behave just like clones, since all the plants of a given cultivar are genetically identical. The same happens with simple hybrid maize cultivars which originate from pure homozygous lines. Yet, every year millions of hectares of a restricted number of cultivars of these crop species are planted around the world.

Large scale propagation methods have evolved impressively since commercial plantations were first established in the mid 1980s in Brazil. The first operational method was based on the use of macro-cuttings (two-leafed cuttings, 8–12 cm long) from sprouts collected from the stumps of early harvested commercial plantations or clonal banks. Because rooting ability declines with ageing, these commercial plantations were usually harvested prior to half the rotation age. These methods required large areas for producing the necessary numbers of macro-cuttings (usually 1 ha of clonal bank for 100 ha of planned forest plantations). Macro-cuttings also required hormonal treatment (IBA) for rooting induction.

In the late 1980s and early 1990s macro-cutting origin moved from clonal banks to clonal gardens, which were designed and managed with the specific objective of producing cuttings. In this system spacing is reduced and intensive management

Fig. 5 *E. globulus* clonal garden and macro-rooted-cutting, Viveiros Aliança, Portucel Group, Portugal

(topping, pruning, nutrition, etc.) is used to retard plant maturation allowing cuttings production from more juvenile material. Better rooting performance was achieved with this method, but the cutting still requires IBA treatment. Moreover, the area required for cutting production was reduced by 90 %. This approach is still used successfully in different countries, with different eucalypt species (Fig. 5).

Although clonal gardens brought important benefits, they have some well known limitations: silvicultural practices are not homogeneous and are strongly affected by climatic conditions; the success rate with macro-cuttings is very genotype dependent; and the architecture of their root system is often not so good.

During the 1990s in Brazil, Assis (2011) developed a new propagation method aimed at minimizing these problems. The method is based on the use of mini-cuttings (two or four leafed cuttings, 4–8 cm long) from sprouts collected in mini-stumps managed in mini-clonal gardens (Fig. 6). This system, although more labor intensive, is currently the mostly widely used in the world, especially for tropical species. Its main advantages are:

- Mini-clonal gardens are much smaller. They are established under homogeneous and protected conditions, allowing more precise control of water, temperature and nutrients, which positively affect the cuttings productivity as well as rooting ability and health.
- Mini-cuttings are much more juvenile, which promotes significant gains in rooting ability (often eliminating the need for IBA treatments), reduced nursery life-cycle and more balanced root system quality.

Choosing between macro and mini-cuttings is not always easy. The advantages of mini-cuttings might suggest this as the natural option. Usually it is, but macro-cuttings or mixed systems (macro + mini) can be justified in some specific situations, when species are poor rooters, or when there is a significant interaction between clones and cutting type, or when the demand for plants is limited. In summary, one size does not fit all and a common sense approach always helps.

Micro-propagation is another system with huge potential, due to the reinforced juvenility of the material produced via tissue culture. Yet, no large scale feasible method is currently available for eucalypt due to economic constraints, despite being used for the maintenance of germplasm and the rescue of rare or valuable genotypes for incorporation into the breeding programs. While the relevance of micro-propagation for large scale production of eucalypt plants is still unproved, it is advisable, whenever possible, to establish macro or mini-clonal gardens using micro-propagated plants. Practical evidence exists that rooted-cuttings produced from micro-propagated stumps produce plants with better juvenile attributes and performance, including rooting ability, precocity and plant quality. Being so, starting tissue cultures of new clones when they are just recommended for pilot operational plantations may be a good strategy because it allows the delivery of very juvenile material for macro or mini-clonal gardens establishment at a later stage, when some of them are definitely chosen for large scale production.

Regardless of the system in place, the physiology of clonal propagation is very complex. The operational efficiency of any eucalypt clonal nursery is highly sensitive to numerous variables, including physiological aspects (genotype rooting ability, cutting juvenility as affected by the maturation of the original stump, cutting size and nutritional status, C-effects such as topophysis, cyclophysis and periphysis, etc.) (White et al. 2007), substrate quality, container type and quality, water quality and management, fertilization, climatic control, sanitary control, operator effects, infra-structure, automation, etc. These variables will interact one with each other in multiple combinations throughout the plant production process (cutting production and collection, planting, rooting, elongation, acclimation, quality control, expedition), thus affecting the final nursery success rates. Much practical knowledge about

Fig. 6 *E. "urograndis"* mini-clonal garden and mini-rooted-cuttings, Brazil

these matters has been generated in the last decades and much more is expected in the forthcoming years. Nevertheless, such a complex theme cannot be more than superficially addressed in this article. Further information can be found in Alfenas et al. (2009).

Finally, it is very important to keep track of the genetic identity of operational clones. Contamination may occur at various stages of the process, from the development phase to gradual mixing in nursery operations, oversight of which requires regular sampling, followed by DNA fingerprinting using molecular markers where available (Grattapaglia and Kirst 2008). This kind of analysis can be easily and cheaply carried out these days, bringing important quality control to this long term, technically intensive and usually highly profitable integrated forestry investment chain.

Acknowledgements The authors are very grateful to Dr. Magno Antônio Patto Ramalho (UFLA - Federal University of Lavras, Brazil) and Dr. Dario Grattapaglia (Embrapa - Brazilian Agricultural Research Corporation) for their helpful comments on this manuscript.

The authors also acknowledge the Companies Veracel (Brazil) and Portucel Group (Portugal), for making available Figs. 1 and 5 respectively.

References

ABRAF (2011) Anuário estatístico da ABRAF 2011, ano base 2010. ABRAF, Brasília, Brazil, p 130

Alfenas AC, Jeng R, Hubbes M (1983) Virulence of *Cryphonectria cubensis* on *Eucalyptus* species differing in resistance. Eur J For Pathol 13:197–205

Alfenas AC, Zauza EAV, Mafia RG, Assis TF (2009) Clonagem e doenças do eucalipto, 2nd edn. UFV, Brazil, p 500

Apiolaza LA (2009) Very early selection for solid wood quality: screening for early winners. Ann For Sci 66:6

Araujo JA, Borralho NMG, Dehon G (2012) The importance and type of non-additive genetic effects for growth in *Eucalyptus globulus*. Tree Genet Genomes 8:327–337

Assis TF (2000) Production and use of *Eucalyptus* hybrids for industrial purposes. In: Dungey HS, Dieters MJ. Nikles DG (eds) Hybrid breeding and genetics of forest trees: proceedings of QFRI/CRCSPF symposium, Australia, pp 63–74

Assis TF (2011) Hybrids and mini-cutting: a powerful combination that has revolutionized the *Eucalyptus* clonal forestry. BMC Proc 5:I18

Assis TF, Resende MDV (2011) Genetic improvement of forest tree species. Crop Breed Appl Biotechnol 11:44–49

Assis TF, Warburton P, Harwood C (2005) Artificially induced protogyny: an advance in the controlled pollination of *Eucalyptus*. Aust For 68:27–33

Bishir J, Roberds JH (1999) On numbers of clones needed for managing risks in clonal forestry. For Genet 6(3):149–155

Bison O, Ramalho MAP, Rezende GDSP, Aguiar AM, Resende MDV (2007) Combining ability of elite clones of *Eucalyptus grandis* and *Eucalyptus urophylla* with *Eucalyptus globulus*. Genet Mol Biol 30:417–422

Boland DJ, Brooker MIH, Chippendale GM, Hall N, Hyland BPM, Johnston RD, Kleinig DA, McDonald MW, Turner JD (2006) Forest trees of Australia, 5th edn. CSIRO, Australia, 736 pp

Borralho NMG, Almeida MH, Potts BM (2008) O melhoramento do eucalipto em Portugal. In: Alves AM, Pereira JS, Silva JMN (eds) Impactes ambientais do eucaliptal em Portugal. ISAPress, Portugal, pp 61–110

Borralho NMG, Cotterill PP, Kanowski PJ (1992) Genetic control of growth of *Eucalyptus globulus* in Portugal. II. Efficiencies of early selection. Silvae Genet 41(2):70–77

Borralho NMG, Cotterill PP, Kanowski PJ (1993) Breeding objectives for pulp production of *Eucalyptus globulus* under different industrial cost structures. Can J For Res 23:648–656

Bouvet JM, Saya A, Vigneron PH (2009) Trends in additive, dominance and environmental effects with age for growth traits in *Eucalyptus* hybrid populations. Euphytica 165:35–54

Buksnowitz C, Muller U, Evans R, Teischinger A, Grabner M (2008) The potential of SilviScan's X-ray diffractometry method for the rapid assessment of spiral grain in softwood, evaluated by goniometric measurements. Wood Sci Technol 42:95–102

Comstock RE (1996) Quantitative genetics with special reference to plant and animal breeding. Iowa State University Press, USA, p 421

Costa e Silva J, Borralho N, Araújo J, Vaillancourt R, Potts B (2009) Genetic parameters for growth, wood density and pulp yield in *Eucalyptus globulus*. Tree Genet Genomes 5:291–305

Costa e Silva J, Potts B, Dutkowski GW (2006) Genotype by environment interaction for growth of *Eucalyptus globulus* in Australia. Tree Genet Genomes 2:61–75

Downes GM, Hudson IL, Raymond CA, Dean GH, Michell AJ, Schimleck LR, Evans R, Muneri A (1997) Sampling plantation eucalypts for wood and fibre properties. CSIRO, Australia, 144 pp

Drew DM, Downes GM, Evans R (2011) Short-term growth responses and associated wood density fluctuations in variously irrigated *Eucalyptus globulus*. Trees 25:153–161

Drew DM, Downes GM, O'Grady AP, Read J, Worledge D (2009) High resolution temporal variation in wood properties in irrigated and non-irrigated *Eucalyptus globulus*. Ann For Sci 66:406

Eldridge K, Davidson J, Hardwood C, van Wyk G (1993) Eucalypt domestication and breeding. Clarendon, UK, p 288

FAO (2010) Global forest resources assessment, 2010 – main report. FAO, Rome, Italy, p 378

FAO (2011) State of the world's forests, 2011. FAO, Rome, Italy, p 179

Fenning TM, Gershenzon J (2002) Where will the wood come from? Plantation forestry and a role for biotechnology. Trends Biotechnol 20(7):291–296

Fonseca SM, Resende MDV, Alfenas AC, Guimarães LMS, Assis TF, Grattapaglia D (2010) Manual prático de melhoramento genetico de eucalipto. UFV, Brazil, 192 pp

Gilmour AR, Cullis BR, Welham SJ, Thompson R (2002) ASReml reference manual. Release 1.0. 2 ed. Harpenden: Biomathematics and Statistics Department – Rothamsted Research, UK, p 187

Grattapaglia D, Kirst M (2008) *Eucalyptus* applied genomics: from gene sequences to breeding tools. New Phytol 179:911–929

Grattapaglia D, Plomion C, Kirst M, Sederoff RR (2009) Genomics of growth traits in forest trees. Curr Opin Plant Biol 12:148–156

Grattapaglia D, Resende MDR (2011) Genomic selection in forest tree breeding. Tree Genet Genomes 7:241–255

Grattapaglia D, Ribeiro VJ, Rezende GDSP (2004) Retrospective selection of elite parent trees using paternity testing with microsatellite markers: an alternative short term breeding tactic for *Eucalyptus*. Theor Appl Genet 109(1):192–199

Grattapaglia D, Vaillancourt R, Shepherd M, Thumma B, Foley W, Külheim C, Potts B, Myburg A (2012) Progress in Myrtaceae genetics and genomics: *Eucalyptus* as the pivotal genus. Tree Genet Genomes 1–46. doi:10.1007

Greaves BL, Borralho NMG (1996) The influence of basic density and pulp yield on the cost of eucalypt Kraft pulping: a theoretical model for tree breeding. Appita J 49:90–95

Griffin AR, Burgess IP, Wolf L (1988) Patterns of natural and manipulated hybridisation in the genus *Eucalyptus* L'Herit – a review. Aust J Bot 36:41–66

Griffin AR, Whiteman P, Rudge T, Burgess IP, Moncur M (1993) Effect of paclobutrazol on flower-Bud production and vegetative growth in 2 species of *Eucalyptus*. Can J For Res 23:640–647

Harbard JL, Griffin R, Espejo JE, Centurion C, Russel J (2000) "One stop pollination": a new technology developed by Shell Forestry technology unit. In: Dungey HS, Dieters MJ. Nikles DG (eds) Hybrid Breeding and Genetics of Forest Trees: Proceedings of QFRI/CRCSPF Symposium, Department of Primary Industries, Brisbane, Australia, pp 430–434

Hasan O, Reid JB (1995) Reduction of generation time in *Eucalyptus globulus*. Plant Growth Regul 17:53–60

Henderson CR (1984) Aplications of linear models in animal breeding. University of Guelph, Canada, p 462

Kerr RJ, Dieters MJ, Tier B (2004) Simulation of the comparative gains from four hybrid tree breeding strategies. Can J For Res 34(1):209–220

Li Y, Dutkowski GW, Apiolaza LA, Pilbeam D, Costa e Silva J, Potts BM (2007) The genetic architecture of a *Eucalyptus globulus* full-sib breeding population in Australia. For Genet 12(3):167–179

Lynch M, Walsh B (1997) Genetics and analysis of quantitative traits. Sinauer Associates, Sunderland, MA, USA, p 980

Meuwissen TH, Hayes BJ, Goddard ME (2001) Prediction of total genetic value using genome-wide dense marker maps. Genetics 157:1819–1829

Moncur MW, Hasan O (1994) Floral induction in *Eucalyptus nitens*. Tree Physiol 14:1303–1312

Myburg AA, Potts BM, Marques CM, Kirst M, Gion JM, Grattapaglia D, Grima-Pettenati J (2007) *Eucalyptus*. In: Genome mapping and molecular breeding in plants. Springer, USA, pp 115–160

Namkoong G, Kang HC, Brouard JS (1988) Tree breeding: principles and strategies. Springer, USA, p 180

Osorio LF (1999) Estimation of genetic parameters, optimal test designs and prediction of the genetic merit of clonal and seedling material of *Eucalyptus grandis*. School of Forest Resources and Conservation, University of Florida, Gainesville

Patterson HD, Thompson R (1971) Recovery of inter-block information when block sizes are unequal. Biometrika 58:545–554

Potts BM (2004) Genetic improvement of eucalypts. In: Burley J, Evans J, Youngquist JA (eds) Encyclopedia of forest science. Elsevier Science, UK, pp 1480–1490

Potts BM, Dungey HS (2004) Interspecific hybridization of *Eucalyptus*: key issues for breeders and geneticists. New For 27(2):115–138

Raymond CA (2002) Genetics of *Eucalyptus* wood properties. Ann For Sci 59:525–553

Raymond CA, Apiolaza LA (2004) Incorporating wood quality and deployment traits in *Eucalyptus globulus* and *Eucalyptus nitens*. In: Walter C, Carson M (eds) Plantation forest biotechnology for the 21st century. Research Signpost, Kerala, India, pp 87–99

Raymond CA, Schimleck LR (2002) Development of near infrared reflectance analysis calibrations for estimating genetic parameters for cellulose content in *Eucalyptus globulus*. Can J For Res 32:170–176

Reis CAF, Gonçalves FMA, Rosse LN, Costa RRGF, Ramalho MAP (2011) Correspondence between performance of *Eucalyptus* spp. Trees selected from family and clonal tests. Genet Mol Res 10(2):1172–1179

Resende KFM, Santos FMC, Dias MAD, Ramalho MAP (2011) Implication of the changing concept of genes on plant breeder's work. Crop Breed Appl Biotechnol 11(4):345–351

Resende MDV, Assis TF (2008) Seleção recorrente recíproca entre populações sintéticas multi-espécies (SRR-PSME) de eucalipto. Pesqui Florest Bras 57:57–60

Resende MDV, Barbosa MHP (2005) Melhoramento genético de plantas de propagação assexuada. Embrapa Florestas, Brazil, p 130

Resende MDV, Resende MFR, Sansaloni CP, Petroli CD, Missiaggia AA, Aguiar AM, Abad JM, Takahashi EK, Rosado AM, Faria DA, Pappas GJ, Kilian A, Grattapaglia D (2012) Genomic selection for growth and wood quality in *Eucalyptus*: capturing the missing heritability and accelerating breeding for complex traits in forest trees. New Phytol 194:116–128

Resende MDV, Striger JK, Cullis BC, Thompson R (2005) Joint modeling of competition and spatial variability in forest field trials. Braz J Math Stat 23(2):7–22

Resende MDV, Thompson R (2004) Factor analytic multiplicative mixed models in the analysis of multiple experiments. Braz J Math Stat 22(2):31–52

Rezende GDSP, de Bertolucci F LG, Ramalho MAP (1994) Eficiência da seleção precoce na recomendação de clones de eucalipto avaliados no Norte do espírito Santo e sul da Bahia. Cerne 1(1):45–50

Rezende GDSP, Resende MDV (2000) Dominance effects in *Eucalyptus grandis*, *Eucalyptus urophylla* and hybrids. In: Dungey HS, Dieters MJ, Nikles DG (eds) Hybrid Breeding and Genetics of Forest Trees: Proceedings of QFRI/CRCSPF Symposium, Department of Primary Industries, Brisbane, Australia, pp 93–100

Rezende GDSP, Resende MDV (2001) Genotypic evaluation and genotype x environment interaction in *Eucalyptus* clones selection at Aracruz Celulose S.A., Brazil. In: Developing the Eucalypt of the Future: Proceedings of Iufro International Symposium, Instituto Forestal, Valdivia, Chile, pp 69–81.

Schimleck LR, Rezende GDSP, Demuner BJ, Downes GM (2006) Estimation of whole-tree wood quality traits using near infrared spectra from increment cores. Appita J 59(3):231–236

Searle SR, Casella G, McCulloch CE (1992) Variance components. Wiley, USA, p 528

Stackpole DJ, Vaillancourt RE, Aguigar M, Potts BM (2010) Age trends in genetic parameters for growth and wood density in *Eucalyptus globulus*. Tree Genet Genomes 6:179–193

van Heerden SW, Amerson HV, Preisig O, Wingfield BD, Wingfield MJ (2005) Relative pathogenicity of *cryphonectria cubensis* on *Eucalyptus* clones differing in their resistance to C-cubensis. Plant Dis 89:659–662

White TL, Adams WT, Neale DB (2007) Forest genetics. CABI, USA, p 682

White TL, Hodge G (1989) Predicting breeding values with application in forest tree improvement. Kluwer, UK, p 367

Wu Y, Wang SQ, Zhou DG, Xing C, Zhang Y (2009) Use of nanoindentation and silviscan to determine the mechanical properties of 10 hardwood species. Wood Fiber Sci 41:64–73

Wynne RH, Nelson RF (2006) SilviScan special issue – lidar applications in forest assessment and inventory - foreword. Photogramm Eng Remote Sens 72:1337–1338

Yang JL, Bailleres H, Evans R, Downes G (2006) Evaluating growth strain of *Eucalyptus globulus* labill. From SilviScan measurements. Holzforschung 60:574–579

Zobel B, Talbert J (2003) Applied forest tree improvement, 3rd edn. Blackburn Press, Caldwell, NJ, USA, p 505

Conifer Somatic Embryogenesis and Multi-Varietal Forestry

Yill-Sung Park

Abstract The global forestry sector, managing both natural forests and commercial plantations, is faced with many future challenges, including increased production of wood with desirable attributes, changing to new forest products, adaption to climate change, forest protection, and species conservation and restoration. To meet these challenges, a forest management system should be sufficiently flexible. Such flexibility is offered by the use of emerging tree biotechnologies, such as somatic embryogenesis (SE) and cryopreservation. SE is a tissue culture technique whereby genetically identical trees can be mass produced. Through the implementation of industrial multi-varietal forestry (MVF; the use of tested high-value tree varieties in plantations), it offers a new paradigm in tree breeding and deployment that is more flexible than the current seed orchard system. In addition to gaining economic benefits from MVF, SE enables research to elucidate genetic response to environmental factors, diseases, and insects and provides a tool for species conservation and restoration.

1 Introduction

Tree improvement efforts around the world in the past 50 years have contributed greatly to the productivity and quality of plantation forestry. Increased productivity is delivered through a breeding scheme based on the seed orchard, and this will continue to be the primary means of achieving genetic improvement in the near future. Seed orchard-based tree breeding schemes typically produce about 10 % volume increase per generation (Tosh 2012). Although conventional tree breeding

Y.-S. Park (✉)
Natural Resources Canada, Canadian Forest Service – Canadian Wood Fibre Centre,
PO Box 4000, Fredericton, NB E3B 5P7, Canada
e-mail: Yillsung.Park@nrcan.gc.ca

T. Fenning (ed.), *Challenges and Opportunities for the World's Forests*
in the 21st Century, Forestry Sciences 81, DOI 10.1007/978-94-007-7076-8_17,
© Crown Employee Canada 2014

provides a substantial increase in productivity, plantation forestry in the future will face new challenges: productivity will be pushed to even higher levels; breeding goals are changing as we search for new products; projected climate change scenarios cast uncertainty about tree adaptation as well as resistance to new pests; and tree breeders are expected to contribute to conservation and restoration of threatened tree species. Thus, tree breeders are required to develop "flexible" breeding and deployment systems to meet these challenges. Recent advances in tree biotechnology for several conifer species have enabled the development of more flexible tree breeding and deployment strategies than conventional seed orchard approaches can offer. In particular, somatic embryogenesis (SE) and cryopreservation offer the implementation of multi-varietal forestry (MVF), which is defined as the deployment of genetically tested tree varieties in plantations. It is also known as clonal forestry, but MVF is considered to be a more descriptive term when applied to industrial plantation forestry (Park 2004). Generally, a clone refers to any genotype with its genetic copies or ramets, whereas *a variety* refers to a clone that is selected or bred for certain attributes (and has field trial data to show to what extent these attributes has been achieved). Despite its many benefits, MVF has not been practiced in conifers due to the inability to produce the same genotypes consistently over time. With the application of SE and cryopreservation, it is now possible to produce the same tested genotypes consistently over time, which is analogous to the production of agronomic and horticultural varieties. The possibility of developing value-added tree varieties for plantation forestry offers a new paradigm in tree breeding and deployment that is more flexible than seed orchards. In addition, the use of other biotechnology tools such as molecular markers can improve the efficiency of such a strategy (Park and El-Kassaby 2006). Development and industrial implementation of a MVF strategy using conifer SE are discussed in this chapter. Application of SE in forest management and research is also discussed.

2 Conifer Somatic Embryogenesis

Simply, SE is a cloning technique based on tissue culture whereby genetically identical copies of a genotype are produced in unlimited numbers. SE in conifers was first reported in 1985 in Norway spruce (*Picea abies*, Hakman et al. 1985; Chalupa 1985), European larch (*Larix decidua*, Nagmani and Bonga 1985), and sugar pine (*Pinus lambertiana*, Gupta and Durzan 1986). Since then, SE has been widely available in many coniferous species, although there are still varying degrees of difficulty in obtaining SE. For several economically important conifers, however, SE is sufficiently refined to the point that it can be implemented in industrial production.

Conifer propagation by SE is accomplished in four stages: initiation and proliferation of embryogenic tissue; maturation of somatic embryos; germination of somatic embryos; and greenhouse/nursery culture (Fig. 1). In general, the initiation of SE is most efficiently obtained by using immature zygotic embryos as the starting material; however, in many spruce species, SE has been obtained from mature or

Process of somatic embryogenesis in multi-varietal forestry

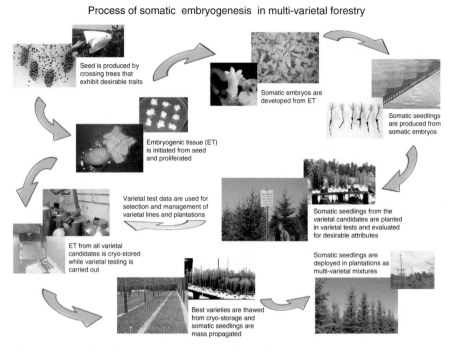

Seed is produced by crossing trees that exhibit desirable traits

Somatic embryos are developed from ET

Somatic seedlings are produced from somatic embryos

Embryogenic tissue (ET) is initiated from seed and proliferated

Varietal test data are used for selection and management of varietal lines and plantations

Somatic seedlings from the varietal candidates are planted in varietal tests and evaluated for desirable attributes

ET from all varietal candidates is cryo-stored while varietal testing is carried out

Somatic seedlings are deployed in plantations as multi-varietal mixtures

Best varieties are thawed from cryo-storage and somatic seedlings are mass propagated

Fig. 1 Process of conifer somatic embryogenesis in multi-varietal forestry

stored seeds in relatively high frequencies. Although SE has been achieved from bud explants of 10-year-old SE-derived trees, the frequency of such initiation is still low at this time (Klimaszewska et al. 2011). Thus, an early stage of zygotic embryo explants such as megagametophytes containing zygotic embryos is preferred, particularly for pine species.

Several formulations of initiation media are used, for example, DCR (Gupta and Durzan 1986), MS (Murashige and Skoog 1962), LV (Litvay et al. 1985), and their modifications. The initiation media typically contain plant growth regulators (PGR), most frequently auxin (2,4-D) and cytokinin (BA). The explants are placed on the initiation medium, which is usually solidified with agar or gellan gum. Responsive explants typically initiate embryonal mass (EM) in 4–6 weeks. Once EM is obtained, it can be proliferated in liquid suspension culture or in a bioreactor. EM is continuously proliferated as long as fresh medium is supplied. The proliferation medium is essentially the same as initiation medium or a slightly modified formulation for further growth. The initiation and proliferation of EM are carried out in darkness at 23–25 °C. Initiation of SE is affected by the level of PGR, genotype, and the stage of zygotic embryo development (particularly for pines).

Maturation of somatic embryos is achieved by withdrawing the auxin and cytokinin and adding abscisic acid (ABA), and the inorganic and organic medium composition is usually the same as for initiation. A critical factor that promotes development of a large number of mature somatic embryos is the restriction of water availability by physical means, or use of osmotic means, or combination of both. Accordingly, control of gel strength, polyethylene glycol, and a combination of the two are frequently used. To improve mature embryo production, the EM is first suspended in PGR-free liquid medium before culture on a filter disk placed on the maturation medium. Improved somatic embryo quality is an important factor because the embryos must be converted to vigorously growing plants.

Germination of somatic embryos is usually carried out on semi-solid medium without PGR. It has been found that it is beneficial to culture the somatic embryos in darkness for first 7 days before exposure to light. This usually ensures the elongation of hypocotyls. At present, the germinants are transplanted into the substrate used in the greenhouse and cultured in the greenhouse or in the nursery. However, it is also possible to germinate somatic embryos directly in micro-plugs filled with vermiculite saturated in liquid media. Initially, the somatic plantlets require immediate fertilization and high relative humidity, often requiring a fine misting system in the greenhouse. Thereafter, fertilization and pesticide application are the same as the regular schedule.

3 Cryopreservation and Thawing

The most important advantage of plant propagation by SE that the EM can be stored at ultra-low temperatures (−140°C to −196 °C) without changing genetic make-up or losing viability. This allows for the development of high-value tree varieties usually involved in lengthy field tests, while the corresponding varietal lines are cryogenically stored. Once those varietal lines with desirable attributes have been identified, the corresponding EM can be thawed and mass produced for deployment. Thus, cryopreservation ensures the production of the same varietal lines consistently over time.

Since the first publication on cryopreservation of white spruce (Kartha 1985), the protocol has been modified for use with several conifer species. The current protocol entails suspending 2 g (fresh mass) of EM in 7 ml of liquid maintenance medium supplemented with 0.4 M sorbitol for 18–24 h, and then, just before freezing, cold DMSO solution is added to the cell suspension on ice. The cell suspension is kept on ice for 1–2 h and then dispensed to cryo-vials. The cryo-vials are placed in alcohol-insulated containers ("Frosty" Nalgene™ container) that are pre-cooled for 2 h at −80 °C. The container with cryo-vials is then placed in a freezer at −80 °C for 1–2 h, during which time slow cooling of the cell suspension takes place (approximately −1 °C/min). Subsequently, the container is plunged in a liquid nitrogen cryo-tank and stored.

For thawing and regrowth of EM, the cryo-vials are rapidly thawed in water bath at 37 °C for 1–2 min, and the cell suspension is poured over a filter-paper disk placed on a thick pad of sterile blotting paper, allowing the storage solution to drain off. The top-most filter paper with cells is transferred to semi-solid initiation medium. Culture growth typically occurs in 1–2 weeks.

The recovery rate of cryopreserved genotypes has been high at about 95 % for many spruce and pine species; however, some species may require special treatment, such as a nurse culture (Hargreaves et al. 2002). It was generally observed that successful cryogenic storage is somewhat dependent on EM vigor. At this time, no published reports of the long-term recovery rate of cryopreserved EM are available; however, it was observed that cryostorage of five lines of *P. strobus* up to 7 years did not show any adverse effect on the recovery of cultures (K. Klimaszewska, personal communication). Park et al. (1998a) compared the genetic stability of a set of embryogenic lines after 3 and 4 years of cryopreservation, respectively, for in vitro and ex vitro characters and found highly consistent results between two thawing dates. Cryopreservation maintains juvenility and minimizes undesirable genetic change caused by prolonged subculture because ultra-low temperatures stop cellular metabolic functions (Kartha 1985). Therefore, it is prudent to use cryogenic storage as a means of minimizing any potential genetic change. There is, however, a possibility that the initial freezing may cause alterations. Therefore, it is recommended to cryopreserve EM lines first, then thaw a part of stored EM and propagate candidate lines for field testing and subsequent deployment.

4 Breeding and Variety Deployment

Among the many applications of SE, the most important application is its use in MVF. There are many advantages of MVF (Libby and Rauter 1984) including: (1) the capture of much greater genetic gain than is possible through conventional seed orchard breeding by exploiting both additive and non-additive genetic variation (Park et al. 1998b); (2) the flexibility to rapidly deploy suitable varieties in line with changing product goals, climate, or environment; and (3) the ability to design and manage genetic gain and plantation diversity according to need.

It is likely that MVF will be practiced on high productivity sites with intensive forest management in connection with long-term tree improvement programs, where seed orchards produce improved seeds for general reforestation. Therefore, with an existing long-term program, MVF may be used as a complementary strategy. The complementary function of MVF in connection with traditional seed orchard breeding is schematically illustrated in Fig. 2. Typical clonal seed orchard (CSO) breeding uses some form of recurrent selection and maintains a breeding population (BP) consisting of genetically selected individuals. Grafts of these are planted in seed orchards for the production of improved seeds. To obtain genetic improvement for the next generation, the individuals in the BP are mated, often by positive assortative mating (PAM), to produce material for next-generation selection and genetic testing and so form the next-generation breeding population. Based on the genetic testing, the initial CSO can be rogued for further improvement. Seeds from the seed orchards are genetically improved as they capture the additive genetic variability among the parents and are well suited for extensive reforestation. Complementary MVF may begin with the elite individuals selected from the BP. The elite parents

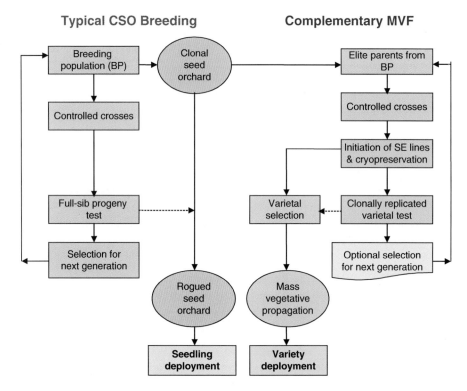

Fig. 2 Schematic illustration of a typical seed orchard breeding with complementary multi-varietal forestry

can be control pollinated (CP), open pollinated (OP), or supplementary mass pollinated (SMP) to produce an offspring population. The resulting seeds from these crosses are subjected to SE initiation and, once SE is initiated, the embryogenic tissue (ET) lines are cryogenically stored. Parts of the ET from each line are then thawed, propagated, and planted in a clonally replicated varietal test (CVT). Based on the CVT, high-value embryogenic varieties are retrieved from cryostorage and mass propagated for planting on the productive sites. The deployment of tested varieties offers much greater genetic gain than CSO seeds as it captures both additive and non-additive genetic variation.

MVF is adaptable to new and small-scale breeding programs. For example, a new program may be initiated by collecting OP seeds from phenotypically selected trees from wild stands, followed by the development of embryogenic varieties by SE. For shorter term, small-scale programs, a breeder may opt out of establishing a seed orchard. In other words, the cost for seed orchard establishment, management, cone collection, and seed processing are replaced by that of cryogenic storage. Assembly of a next-generation breeding population may, normally, not be carried out, but an optional selection may be made. In this case, a breeder may have to rely on the use of molecular markers to control and monitor relatedness within the selected population.

5 Varietal Testing

The main advantage of MVF is the ability to obtain greater genetic gain by using all available genetic variances including non-additive variances (Park 2002). This benefit can be realized only through varietal selection based on varietal testing (VT), which is focused on selection and deployment of suitable varieties rather than genetic testing aimed at estimating genetic parameters. Thus, genetic gain calculated from selected varieties of the VT is the actual realized gain, as the same varieties will be deployed in the plantations.

In many established long-term breeding programs, the BP is often structured into sublines, which are groups of 10–20 parents within which breeding is carried out. The top tier of these sublines is referred to "nucleus" or elite subline, which contains the best parents of the BP. This strategy is designed to deliver fast genetic gain from elite crosses (Coterill et al. 1989). Thus, the development of varietal lines can begin by making elite crosses with the nucleus subline. If the breeding program is new, it can be started with open-pollinated seeds of phenotypic selections from wild stands.

In seed orchard breeding, the selected parents from these sublines are planted in a seed orchard and allowed to produce wind-pollinated seeds; however, the strategy is often inefficient due to pollen contamination, asynchronous flowering leading to unequal parental contribution, etc. In MVF, the development of varietal lines requires controlled crossing of the best parents and the propagation of several individuals within each cross by SE. For example, with a set of 20 parents, a breeder can produce up to 380 possible crosses without selfing, but, these are unrealistic numbers. In reality, the number of crosses to make and the number of individuals to propagate within each cross depend on the availability of resources and logistics. Typically in eastern Canada, VT is established using 200–300 varietal lines developed from 20 to 30 crosses, e.g., 200 candidate varieties may consist of 10 genotypes each from 20 crosses. A series of tests are planted in multiple years to ensure that large numbers of candidate varietal lines are available for selection, and the tests are conducted at multiple locations.

Differential SE success rate could have an impact on capturing potential genetic gain for MVF; however, a simulation study demonstrated that embryogenic propensity among families had no significant impact on gain (Lstibůrek et al. 2006). It also indicated that, even though there is reduced variation among families due to differential SE success, the variation within family is not changed. Thus, it is reasonable to expect much of the genetic gain will be derived from within-family variation. In establishing VT, one would generate more embryogenic candidate lines within a family than by using more families with smaller number of individuals within a family.

However, a balancing act is necessary to obtain optimal gains from both among- and within-family selection as well as for diversity concerns.

The VT is a key part of MVF because it will provide appropriate data for varietal selection and for the management of plantations. The flexibility of MVF is primarily derived from VT because it is intended that VT will continuously provide relevant data on growth, quality, insect and disease resistance, adaptation to changing environments and climate, and other traits throughout the rotation age and beyond. The final assessments of candidate varieties in the VT may take a long time; however, whenever the relevant and updated data become available, the breeders are offered the flexibility to rapidly adapt to the change by simply thawing and deploying appropriate varieties from the cryopreservation as environmental sensitivities emerge or goals are changed. Thus, it is important to establish VT with large numbers of candidate varieties across a wide range of sites. This may require planting tests in several successive years.

The use of clonal replicates, as in VT, is essential for disease and insect resistance screening experiments, especially when the objective of the experiment is to develop resistant varieties. Challenge tests based on a family or provenance level lack the precision to identify resistant individual genotypes because these tests include several different individuals comprising family or provenance, and thus, they only provide the resistance levels at the family or provenance average. The removal of the "genotype effect," which is achieved by the use of clonal replicates that provide genetic uniformity of candidate varieties in the challenge test, is essential for capturing individual genotypes that are truly resistant. Similar to VT, clonally replicated genetic testing (CRT) has been used in tree breeding experiments designed to obtain additional genetic parameters. For example, using control-pollinated families, Mullin and Park (1992) further demonstrated the partitioning of epistatic variance. Thus, CRT provides additional genetic information for breeding programs and offers efficiency and precision of parameter estimation for quantitative traits (Foster and Shaw 1988).

6 Balancing Genetic Gain and Diversity

In breeding, an increased genetic gain is usually achieved through a reduction in genetic variability. The major concern in deployment of cloned varieties is that a narrow genetic base may make clonal varieties vulnerable to disease and insects, and this may lead to plantation failure. For known diseases and insects, MVF has an advantage because resistant varieties may be developed or identified, but, for unknown or introduced diseases and insects, protection against them is less predictable despite the large genetic variability with a tree species, as entire tree species can on occasion succumb to novel pests or diseases (as discussed in various other chapters). It is difficult, if not impossible, to design a protection scheme against unknown diseases and insects. However, it is generally assumed that the more varieties are deployed in a plantation, the lower the risk will be. The increased number of

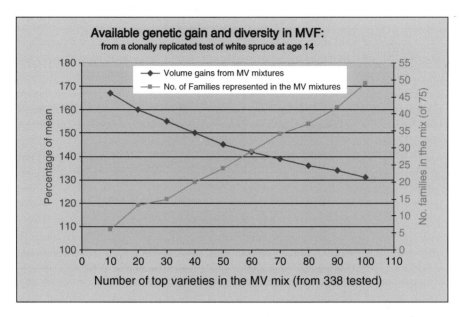

Fig. 3 Available genetic gains as percentage of test mean for multi-varietal mixes (—◆—) and corresponding diversity of MV mixes represented by the number of families in the mix (—■—)

varieties in the plantation will reduce genetic gain. Therefore, it is necessary to balance genetic gain and diversity. Lindgren (1993) discussed some basic principles for determining an appropriate number of varieties in a plantation: (1) if the species being deployed is short lived or short rotation, a lower number of varieties may be used because the exposure to potential risks is short; (2) a lower number of varieties may be used if plantation management is intensive and includes pest management; and (3) the more well known a variety, the more acceptable is its extensive use.

Once an appropriate number of varieties is selected, they may deployed in varietal blocks or random mixtures (Libby 1982). In eastern Canada, however, an approach called "desired gain and diversity" is used. This approach uses the best available VT data to select varieties while considering a predetermined level of genetic gain or diversity. For example, the available genetic gain and management of plantation diversity is illustrated in Fig. 3. The data shown in the graph were obtained from a 14-year-old clonally replicated test of white spruce (*Picea glauca*) that included 338 candidate varieties derived from 75 full-sib crosses using a disconnected diallel mating. In the graph, if we take the top ten varieties of the 338 in the test, the genetic gain for volume is 68 % better than the average of all varieties in the test. Even when we take the 100 best, the genetic gain is 30 % better than the average. This is still a large genetic gain compared with that obtained by seed orchard breeding. Incidentally, the average (100 %) represents an idealized seed orchard gain without inefficiencies. The number of families represented in a varietal mix may be considered as an indicator of diversity. If we take the ten best varieties, this varietal mix is represented by only seven of the 75 full-sib families in the test. If we take the top 100, i.e., taking 30 % above the seed orchard gain, the varietal mix is represented by 48 of 75. In other words, we can balance a desired level of genetic gain and diversity based on the VT test results. For example, it is likely that a large number of varieties may be included in the MV mix at an early stage of VT; however, when the test is mature or varietal characterization is sufficient, a smaller number of elite lines may be used. Therefore, determining an appropriate number of varieties is a dynamic process based on best available data at a given time.

Another approach to manage genetic diversity in plantations is to mix in selected varietal lines with lower-cost seed orchard seedlings. This will increase the initial plantation diversity. Typically, in eastern Canada, about 40 % of plantation basal area will be commercially thinned at about half rotation age, leaving superior quality trees for final harvesting. Thus, it is likely that most final crop trees would consist of the tested varietal lines, and this reduces exposure to potential risks for the crop trees. The thinning will favor the best phenotypes regardless of genetic origin, and exceptional trees originating from seed orchard seeds will also become crop trees to fill gaps created when some varieties or ramets perform poorly and are removed during thinning. The highest quality trees, regardless of genetic origin, will remain in the plantations. This approach can be combined with the 'desired gain and diversity' strategy. Therefore, the diversity of plantations is dynamically managed spatially and over time, and selection of varieties is continuously revised based on VT throughout the testing period and as new varieties are introduced at each breeding cycle, resulting in different compositions of varietal mix.

7 Industrial Production

Industrial production of varieties by SE is at an early stage of development, and it is known that several companies and organizations in Canada, Chile, France, New Zealand, USA, and the UK are actively involved in development. Most conifer SE applications have been developed using the Petri dish-based system and in vitro germination. These approaches are suitable for establishing VT and small-scale commercial production; however, they are viewed as more expensive for commercial production than seedlings. Improving the efficiency of SE for a mass propagation system that is amenable to automation is an important task for commercialization. Some promising developments for an efficient SE system include rapid proliferation of embryogenic tissue in liquid culture and photoautotrophic micro-propagation techniques (Kozai et al. 2005) applicable to SE, and a computer-aided robotic system for a transplant system (Find 2009). Despite the lack of reliable automated systems for SE production, the cost is similar to commercial rooting of cuttings production. Although planting stock production by SE is preferred, it is still complex and relatively expensive due to a lack of automation. Alternatively, mass propagation of varietal lines can be achieved by the use of serial rooting of cuttings, whereby a few juvenile plants of varietal lines thawed from cryogenic storage are used as donor plants (Park et al. 1998b). Mass vegetative propagation by rooting of cuttings from juvenile donor plants is readily available for several conifers species, especially for spruce species. In this case, SE and cryopreservation, in conjunction with VT, are used as a tool for developing varietal lines. In addition to the lack of an automated SE system, another factor impeding industrial application is a shortage of the tested embryogenic varieties to deploy. Therefore, any breeding program that includes future MVF must start establishing well-designed VT now, because there will be at least about 5 years of lag time between the establishment of VT and the first deployment of MVF. Current productivity and quality improvement in MVF is achieved by careful exploitation of existing natural genetic variability; however, MVF is likely to be the delivery mechanism for value-added tree biotechnology products in the future.

8 Hybrid Varieties

In agriculture, hybridization usually refers to crossing of different strains (or homo-zygous lines) within a species, but in forestry, it refers to crossing between different species or distinctly different races within a species. The benefits of hybridization in forestry include the capture of hybrid vigor in growth traits and the combination of desirable traits. An example of hybrid vigor is demonstrated by the interspecific crosses between Japanese (*Larix kaempferi*) and European larches; an example of combining desirable traits is demonstrated by pitch (*P. rigida*) and loblolly (*P. taeda*) pine hybrids, which are successfully used in Korea to take advantage of the cold tolerance and fast growth in the respective species. Despite the huge poten-tial benefits, hybridization in conifers has rarely been used as a modern breeding method, partly due to the labor intensiveness of hybrid seed production by mass controlled pollination and the inefficiencies of seed production in a bi-species orchard designed to produce hybrid seed. SE is an ideal tool for developing hybrid varieties as it can be used for mass production of hybrid plants from a relatively small number of interspecific crosses. Also, the use of SE enables the selection of elite individuals within the crosses, further improving hybrid characteristics. The development of blister-rust-resistant white pine varieties through interspecific hybridization between *P. strobus* and *P. wallichiana* (and backcrossing to *P. strobus*) is underway in Ontario, Canada, which adopted the use of SE for variety development (Lu 2008). The deployment considerations, including selection for variety mix, would be similar to those for MVF, as discussed earlier.

9 Other Applications of Somatic Embryogenesis

9.1 Embryo Rescue

Somatic embryogenesis can be used in embryo rescue. Crossing between distantly related species can be a useful technique for creating unusual genotypes, but such crosses usually result in abortion. Embryo rescue has been carried out with a large number of species, including tree species. A couple of recent examples of the latter are citrus triploid hybrids (Aleza et al. 2010) and banana (Uma et al. 2011) but, to our knowledge, embryo rescue has never been attempted with conifers. It is generally believed that embryo abortion is due to a genetic incompatibility between the developing embryo and the maternal tissue surrounding it. Even though embryos may not develop fully within the megagametophyte tissue with the seed, these embryos can develop normally if they are removed from seeds before abortion occurs and can then be grown to maturity in vitro, or immature embryos may be induced to initiate SE. The latter is a possibility because SE often initiates during the initial cleavage polyembryony stage of the zygotic embryo. Thus, embryo rescue has the capacity to create genotypes that are not possible to obtain through breeding. Subsequent mass propagation of a rescued embryo can be highly effective if propagated by SE.

9.2 Species Conservation and Restoration

Somatic embryogenesis combined with cryopreservation can have an important impact on species conservation and restoration. For example, whitebark pine (*P. albicaulis*) is a keystone species growing in the subalpine regions of Alberta and British Columbia and is threatened in its natural range. No seeds are available for reforestation as they are a food source for birds and animals, and its wingless seed is dispersed only by birds. The serious threats come from white pine blister rust (*Cronartium ribicola*), white pine weevil (*Pissodes strobi*), mountain pine beetle (*Dendroctonus ponderosae*), prolonged fire suppression preventing natural regeneration, and projected climate change. In this case, SE can be used as an alternative means of producing planting stock as the seed availability is very limited. Through SE, valuable genotypes can be cryogenically stored over the long term, and restoration of the sites with better adapted and pest-resistant genotypes can be accomplished at a later date. This is a strategy adopted by the Alberta Sustainable Resource Development in Canada (Park et al. 2010). Similar efforts are also in progress for limber pine (*P. flexilis*) in Alberta.

9.3 Genetic Engineering

Genetic transformation allows integration of valuable genes or traits that are absent from even the elite breeding material into selected genotypes. For example, genetic engineering for pest resistance, or genes providing tolerance to salt, heavy metal, or drought in trees, can be extremely valuable. Obtaining disease and insect resistance by traditional selection and breeding is very difficult for forest trees, primarily due to the long generation time and difficulty in finding resistant genotypes in the wild. Merkle et al. (2007) examined the possibilities of restoring American chestnut (*Castanea dentata*) devastated by the chestnut blight fungus (*Cryphonectria parasitica*) which was accidentally introduced into North America from China early in the 20th century, through introducing antifungal genes using embryogenic culture. SE is the primary enabling technology for both transformation and subsequent propagation. Among the available transgenic technologies, transformation of embryogenic culture is the most common one. It is generally achieved through co-cultivation with *Agrobacterium* carrying the transgene. The advantages of using SE in genetic engineering are: the process can be carried out in a strictly confined environment; the transformed and non-transformed cells can be separated by inclusion of an antibiotic resistance gene; and, as SE in most species starts from single cells, one can avoid ending up with chimera, i.e., individuals with both transformed and non-transformed cells. However, the stability and containment of transgenes are important issues that have to be dealt with before genetically modified trees are deployed. Engineering of sexual sterility genes into SEs is a means of achieving containment, but it is not yet widely available. Due to a potentially adverse environmental impact and bio-safety issues, most jurisdictions around the world regulate testing and deployment of transgenic trees (Trontin et al. 2007; and also see the chapter of Häggman et al. in this book for an overview of this area).

9.4 *Epigenetic Memory*

It is presently generally accepted that epigenetics refers to changes in phenotype or gene expression caused by mechanisms other than changes in the underlying DNA sequence (Meehan et al. 2005). Epigenetic memory effects in Norway spruce during zygotic embryogenesis and seed maturation have been reported by Johnsen et al. (2005), who found that adaptive traits in progenies were influenced by the temperature that the maternal trees were exposed to during seed development. von Aderkas et al. (2007), working with interior spruce (*P. glauca* x *P. engelmannii*) SE, reported that somatic embryos, matured at a lower temperature (5 °C), showed significantly higher cold tolerance than those matured at 20 °C. However, it is not known whether such epigenetic memory effects will continue during the seedling and adult tree stages; however, Johnsen et al. (2005) found that epigenetic memory effects persist for many years in the filial generation. This is an important research area for plant adaptation especially under climate change scenarios. SE can be used for studying epigenetic memory effects as well as embryo development.

10 Concluding Remarks

Somatic embryogenesis is the first conifer biotechnology to be applied in tree improvement for the implementation of MVF and has opened new commercial opportunities for the forest industry. Several companies around the world are in the process of implementing MVF to take advantage of the increased productivity, production of wood with desired attributes, and flexibility in meeting future demands. Currently, the system of producing plants by SE is slightly more expensive than the traditional seedling production system due to the lack of an automated production system, but this might change as SE is highly amenable to automation. In addition to the economic benefits of implementing MVF, SE can be used as an important research tool for conifer development, ecophysiology, pest resistance, functional genomics, among other things.

References

Aleza P, Juarez J, Cuenca J, Ollitrault P, Navarro L (2010) Recovery of citrus triploid hybrids by embryo rescue and flow cytometry from 2x x 2x sexual hybridization and its application to extensive breeding programs. Plant Cell Rep 29:1023–1034

Chalupa V (1985) Somatic embryogenesis and plant regeneration from cultured immature and mature embryos of *Picea abies* (L.) Karst. Commun Inst For Cech 14:57–63

Coterill P, Dean C, Cameron J, Brindbergs M (1989) Nucleus breeding: a new strategy for rapid improvement under clonal forestry. In: Proc. IUFRO Conf.: Breeding tropical trees: population structure and genetic improvement strategies in clonal and seedling forestry. Pataya, Thailand, pp 39–51. Nov. 1988

Find J (2009) Integration of biotechnology, visualisation technology and robot technology for automated mass production of elite trees. Abstract 2009 IUFRO Tree Biotechnology Conference, June 28–July 2, 2009. Whisler, BC, Canada

Foster GS, Shaw DV (1988) Using clonal replicates to explore genetic variation in a perennial plant species. Theor Appl Genet 76:788–794

Gupta PK, Durzan D (1986) Somatic embryogenesis from callus of mature sugar pine embryos. Bio/Technol 4:643–645

Hakman I, Fowke LC, von Arnold S, Eriksson T (1985) The development of somatic embryogenesis in tissue culture initiated from immature embryos of *Picea abies* (Norway spruce). Plant Sci 38:53–59

Hargreaves CL, Grace LJ, Holden DG (2002) Nurse culture for efficient recovery of cryopreserved *Pinus radiata* D. Don embryogenic cell lines. Plant Cell Rep 21:40–45

Johnsen O, Daehlen OG, Ostreng G, Skropa T (2005) Daylength and temperature during seed production interactively affect adaptive performance of *Picea abies* progenies. New Phytol 168:589–596

Kartha KK (1985) Meristem culture and germplasm preservation. In: Kartha KK (ed) Cryopreservation of plant cells and organs. CRC Press, Florida, pp 115–134

Klimaszewska K, Overton C, Stewart D, Rutledge RG (2011) Initiation of somatic embryos and regeneration of plants from primordial shoots of 10-year-old somatic white spruce and expression profiles of 11 genes followed during the tissue culture process. Planta 233:635–647

Kozai T, Afreen F, Zobayed SMA (eds) (2005) Photoautotrophic (sugar-free medium) micropropagation as a new micropropagation and transplant production system. Springer, Dordrecht, p 316

Libby WJ (1982) What is a safe number of clones per plantation? In: Heybrook HM, Stephan BR, von Weissenberg K (eds) Resistance to disease and pests in forest trees. Pudoc, Wageningen, pp 342–360

Libby WJ, Rauter RM (1984) Advantages of clonal forestry. For Chron 60:145–149

Lindgren D (1993) The population biology of clonal deployment. In: Ahuja MR, Libby WJ (eds) Clonal forestry I: genetics and biotechnology. Springer, Berlin, pp 34–49

Litvay JD, Verma DC, Johnson MA (1985) Influence of loblolly pine (*Pinus taeda* L.) culture medium and its components on growth and somatic embryogenesis of the wild carrot (*Daucus carota* L.). Plant Cell Rep 4:325–328

Lstibůrek M, Mullin TJ, El-Kassaby YA (2006) The impact of differential success of somatic embryogenesis on the outcome of clonal forestry programs I. Initial comparison under multi-trait selection. Can J For Res 36:1376–1384

Lu P (2008) Breeding eastern white pine for blister rust resistance through interspecific hybridization and back crossing: a review of progress in Ontario (extended abstract). In: Proc. breeding and genetic resources of five-needle pines, Yangyang, Korea, pp 81–82

Meehan RR, Dunican DS, Ruzov A, Pennings S (2005) Epigenetic silencing in embryogenesis. Exp Cell Res 309:241–249

Merkle SA, Andrade GM, Nairin CJ, Powell WA, Maynard CA (2007) Restoration of threatened species: a noble cause for transgenic trees. Tree Genet Gen 3:111–118

Mullin TJ, Park YS (1992) Estimating genetic gains from alternative breeding strategies for clonal forestry. Can J For Res 22:14–23

Murashige T, Skoog F (1962) A revised medium for rapid growth and bioassays with tobacco cultures. Physiol Plantar 15:473–497

Nagmani R, Bonga JM (1985) Embryogenesis in subcultured callus of *Larix decidua*. Can J For Res 15:1088–1091

Park YS (2002) Implementation of conifer somatic embryogenesis in clonal forestry: technical requirements and deployment considerations. Ann For Sci 59:651–656

Park YS (2004) Commercial implementation of multi-varietal forestry using conifer somatic embryogenesis. In: Proc. IUFRO joint Conf. of Div. 2 Forest genetics and tree breeding in the age of genomics: progress and future. Charleston, SC, p 139. Nov. 1–5, 2004

Park YS, El-Kassaby YA (2006) New breeding and deployment strategy using conifer somatic embryogenesis and pedigree reconstruction. In: Proc. of IUFRO Div. 2 joint Conf.: Low input breeding and conservation of forest genetic resources. Antalya, Turkey, pp 194–195. Oct 9–13, (Abstract)

Park YS, Barrett JD, Bonga JM (1998a) Application of somatic embryogenesis in high-value clonal forestry: deployment, genetic control, and stability of cryopreserved clones. In Vitro Cell Dev Biol Plant 34:231–239, Plant3

Park YS, Bonga JM, Mullin TJ (1998b) Clonal forestry. In: Mandal AK, Gibson GL (eds) Forest genetics and tree breeding. CBS Publishers and Distributors, New Delhi, pp 143–167

Park YS, Barnhardt L, Klimaszewska K, Dhir N, MacEacheron I (2010) Somatic embryogenesis in whitebark pine (*Pinus albicaulis*) and its implication for genetic resource conservation and restoration. In: Proc. of IUFRO conference on Somatic embryogenesis of trees. Suwon, Korea. August 18–21, 2010

Tosh K (2012) New Brunswick tree improvement council. http://nbforestry.com/?section=13&sub section=66. Accessed Aug 2012

Trontin JF, Walter C, Klimaszewska K, Park YS, Lelu-Walter MA (2007) Recent progress on transformation of four *Pinus* species. Invited review. Transgen Plant J 1(2):314–329

Uma S, Lakshmi S, Saraswathi MS, Akbar A, Mustaffa MM (2011) Embryo rescue and plant regeneration in banana (*Musa* spp.). Plant Cell Tiss Org Cult 105:105–111

von Aderkas P, Kong L, Hawkins B, Rohr R (2007) Effects of non-freezing low temperatures on quality and cold tolerance of mature somatic embryos of interior spruce (*Picea glauca* (Moench) Voss x *P. Engelmanni* Parry ex. Engelm.). Propag Ornamental Plants 7:112–121

Modern Advances in Tree Breeding

Yousry A. El-Kassaby, Fikret Isik, and Ross W. Whetten

Abstract Traditional tree improvement programs are long-term endeavours requiring extensive resources. They require establishing mating designs, installing progeny tests on multiple sites to evaluate parents and their offspring over large geographic areas, monitoring those tests over extended periods of time, and eventual analysis of measurements to assess economic traits. Most tree breeding programs follow the classical recurrent selection scheme, resulting in the generation of multiple breeding and production populations. This process, while successful in attaining appreciable gains, remained static for a long time. The availability of plentiful, reliable, and most of all increasingly affordable genetic markers brought about drastic changes to present-day breeding methods. In this chapter, we focus on four significant genetic marker-dependent approaches with significant potential to directly or indirectly change contemporary tree breeding methods. These include pedigree reconstruction, pedigree-free models, association genetics, and genomic selection.

1 Introduction

Tree breeding programs are resource- and time-dependent endeavours. The selection and testing phases are often conducted over vast geographic areas with large trials, requiring frequent and long-time monitoring and assessment. The lowest-intensity

Y.A. El-Kassaby (✉)
Department of Forest Sciences, Faculty of Forestry,
University of British Columbia, Vancouver, BC V6T 1Z4, Canada
e-mail: y.el-kassaby@ubc.ca

F. Isik • R.W. Whetten
Department of Forestry and Environmental Resources,
North Carolina State University, Campus Box 8002, Raleigh, NC 27695, USA
e-mail: fisik@ncsu.edu; Fikret_Isik@ncsu.edu; ross_whetten@ncsu.edu

T. Fenning (ed.), *Challenges and Opportunities for the World's Forests*
in the 21st Century, Forestry Sciences 81, DOI 10.1007/978-94-007-7076-8_18,
© Springer Science+Business Media Dordrecht 2014

approach to tree improvement is a reciprocal transplanting-like approach known as provenance testing (Callaham 1964) for the identification of superior seed sources for reforestation. Provenance testing allowed evaluating several seed sources originating from multiple locations within the species' natural range through their field-testing over potential target planting areas. This process aided in identifying superior seed sources and their adaptability for the safe transfer of their seed to the new planting sites (Rehfeldt 1983). Provenance testing focused on acquiring precise knowledge of the seed sources and their performance over testing sites (Konig 2005). This process is a simple population improvement method, as the pedigree or genealogy of the tested material is often unknown. The main achievement of provenance testing is the delineation of areas for safe seed transfer, known as seed zones (Campbell 1986).

The first and simplest pedigree-known testing utilized wind-pollinated/open-pollinated families (also known as half-sib families because their offspring share the seed donors' genotype). Wind-pollinated testing, as a partial pedigree method, permits within and among family selection, thus it is expected to yield greater gains than provenance testing. The New Zealand radiata pine tree improvement program is the most notable program for adopting this approach (Burdon and Shelbourne 1971). The main attractive feature of this method is its simplicity and suitability for testing large number of families; however, it is often considered as a spring-board to full pedigree testing (Jayawickrama and Carson 2000). It should be stated that wind-pollinated testing is fraught with assumptions that cannot be either tested or fulfilled, and often leads to inaccuracies in estimates of individual breeding values (Namkoong 1966).

The utilization of a full pedigree (i.e., individuals with known genealogy) is the most common testing mode in tree breeding programs (White et al. 2007). The formation of a structured pedigree, created through the implementation of a mating design of controlled pollinations, provides greater control of the genealogy and the eventual accurate estimation of genetic parameters such as trait heritabilities and parent and offspring breeding values (Namkoong et al. 1988). It should be stated that the successful completion of structured pedigree is an elaborate process requiring time and substantial painstaking effort. The recurrent selection scheme is the most common breeding framework used when full pedigree is used (Allard 1960).

2 Pedigree Reconstruction

Structured pedigree designs (full- and half-sib families) constitute the backbone for most tree breeding programs, resulting in impressive gains and better management of inbreeding and genetic diversity (White et al. 2007). Lambeth et al. (2001) introduced an idea of polymix breeding and pedigree reconstruction. El-Kassaby and Lstibůrek (2009) further implemented this idea via the posterior analysis of naturally-occurred crosses among a group of parents. They coined the method "Breeding without Breeding (BwB)" and proposed the utilization of molecular markers, SSRs in this case, and pedigree reconstruction models (see Jones and

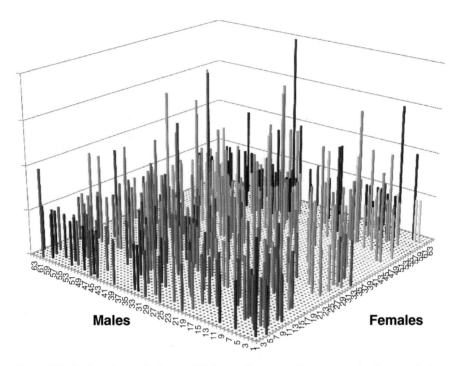

Fig. 1 Distribution of posteriorly assembled naturally-occurred crosses among 63-parent lodge-pole pine seed orchard revealed by full pedigree reconstruction of bulk offspring (i.e., unknown maternal and paternal parentage) using DNA microsatellite markers (nine nuclear and six chloroplast loci) and pedigree reconstruction (El-Kassaby, unpublished)

Ardren 2003 for review) to by-pass the costly and time consuming breeding phase. The disconnected partial diallel mating scheme is often employed to create the structured pedigree for generating the offspring needed for testing (Namkoong et al. 1988). The BwB concept is illustrated using bulk seed sample from a 63-parent lodgepole pine seed orchard (El-Kassaby, unpublished), and can be compared with the disconnected partial-diallel design. With this number of parents and the implementation of a six-parent scheme, 153 full-sib families are expected to be generated (seven 6-parent and three 7-parent partial diallel units). However; when pedigree reconstruction was implemented, a total of 446 full-sib families were assembled without making any controlled crosses (Fig. 1). The resulting mating is far more efficient as many more crosses were created as compared to the classical disconnected partial diallel.

Furthermore, El-Kassaby et al. (2011) extended the BwB concept and increased the method's efficiency through the application of two distinct steps: (1) the use of simplified half-sib progeny testing with large sample size per parent and (2) restricting offspring sampling for DNA fingerprinting and pedigree reconstruction to a random sample of offspring from a subset of parents rather than the entire parental population. The use of half-sib families in testing is expected to simplify the

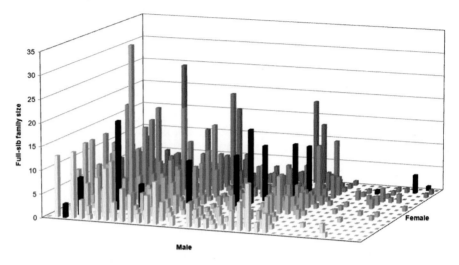

Fig. 2 Pedigree reconstruction from natural mating produced from seed collected from 15 seed donors growing in a 41-parent western larch seed orchard showing the formation of full-sib families nested within the maternal and paternal half-sib families with selfing presented as *black bars* (After El-Kassaby et al. (2011))

progeny test design as compared to multiple full-sib families. A random sample of offspring from a subset of seed parents is expected to capture most of the un-sampled parents as fathers (i.e., paternal half- and full-sib families) and therefore their breeding values can be estimated. Finally, the inclusion of all the offspring phenotypic information from both full- and half-sib families is expected to increase the estimated genetic parameters' precision; however, it should be stated that the breeding value of the half-sib individuals will be estimated with lesser precision as compared to those of full-sib families. El-Kassaby et al. (2011) empirically tested this concept and assessed offspring generated from only 15 seed-donors (i.e., half-sib families) out of a 41-parent western larch seed orchard. In this experiment, each half-sib family was represented by 400 seedlings bringing the total experiment sample size to $N \approx 6,000$. They randomly sampled 1,500 individuals, irrespective of their half-sib family designation, for DNA fingerprinting and pedigree reconstruction. As expected, an unbalanced mating structure was produced reflecting variation in parental reproductive output (Fig. 2).

It is interesting to note that the assembled matings produced offspring sired by all 41 parents in the orchard, indicating that the pedigree reconstruction successfully captured the un-sampled parents as pollen donors even when the offspring sampling was restricted to 15 seed-donors only. The most interesting observation from the data analyses is the congruence between height breeding values from the combined analysis (1,500 FS + 4,500 HS) and that based on the conventional full-sib families alone (1,500 individuals). This was observed for both parents and offspring (Fig. 3). The great advantage of the FS and HS combined analysis is the role

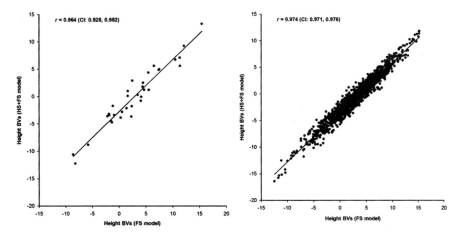

Fig. 3 Scatter plot of predicted breeding values for parents (*left*) and offspring (*right*) from the incomplete (combined HS + FS) and complete (FS) pedigree models. Pearson correlation (r) is in the left corner of each graph (After El-Kassaby et al. (2011))

played by the 1,500 FS individuals in linking the remaining 4,500 HS to the paternal and maternal parents and their half- and full-sib families (Fig. 3). Furthermore, El-Kassaby et al. (2011) demonstrated that individuals' breeding values precision did not change drastically if the random sampling of individuals for fingerprinting and pedigree reconstruction was reduced to approximately one third (i.e., less fingerprinting efforts).

Pedigree reconstruction is an effective method in situations where the posterior determination of offspring genealogy is needed or for species that do not lend themselves to controlled pollination. Using pedigree reconstruction for trees from plantation blocks that originated from seed orchards or breeding arboreta can instantaneously convert them to progeny test trials (Hansen and McKinney, 2010). While this approach requires good GIS tracking of plantations polygons over the landscape (see Ding et al. 2012), it also requires rigorous spatial analysis to account for site heterogeneity (see Cappa et al. 2011).

3 Pedigree-Free Models

Fundamentally, Breeding without Breeding is anchored to the utilization of pedigree reconstruction to assemble half- and full-sib families needed for conducting standard intra-class correlation analyses for estimating quantitative genetics parameters such as traits' heritabilities and parental and offspring breeding values (Falconer and Mackay 1996). In situations where pedigree reconstruction is not feasible, molecular genetic markers offer an alternative approach for estimating quantitative genetic parameters. Molecular markers can be used to estimate "marker-based pairwise relationships" among any group of individuals irrespective of their genealogy, based on the assumption that markers identical by state are also identical by descent (Li et al.1993; Queller and Goodnight 1989; Lynch and Ritland 1999; Wang 2002). The use of "marker-based pairwise relationship" created an opportunity to studying domesticated and undomesticated species in experimental or natural setting with and without the availability of pedigree, thus permitting the estimation of genetic parameters in an unstructured population. Efficient methods have been developed for the use of high-density marker information for a group of individuals to estimate their realized relationship matrix (vanRaden 2008). This matrix is used in place of the classical pedigree-based numerator relationship matrix required in quantitative genetics analyses. This approach allows estimating quantitative genetic parameters such as narrow sense heritability and breeding values using the genomic best linear unbiased prediction method, as described in more detail below (Zapata-Valenzuela et al. 2011; El-Kassaby et al. 2012; Porth et al. 2012).

The realized relationship matrix was successfully used to estimate narrow sense heritability, breeding value and genetic and phenotypic correlations in an unstructured black cottonwood population (El-Kassaby et al. 2012; Porth et al. 2012). More interesting is the study of Klápště et al. (2013) in which a pedigree-free model was compared to a marker-based pairwise relationship model. Surprisingly, Pearson's product moment and Spearman's rank correlations between western larch offspring breeding values produced from the two approaches were highly significant, indicating that the generated DNA-based pair-wise relationship matrix is indeed a valid substitute for the classical pedigree matrix (Fig. 4). This approach was further extended to accommodate a mixture of information generated from both genetic markers and conventional pedigree by Korecký et al. (2013). This approach is unique as the combination of historical and contemporary co-ancestry generated by the genetic markers and pedigree, respectively, could not be attained by either approach individually. Thus, combining both data sets is expected to improve the accuracy of the estimated genetic parameters as the often ignored Mendelian sampling term in structured pedigree is precisely accounted for when molecular markers are used.

The availability of molecular markers is expected to effectively increase breeding efficiency. The use of densely well dispersed SNP data to estimate the realized relationship among individuals is expected to result in a greater kinship resolution and offers an opportunity improvement to classical breeding efforts.

4   Marker-Trait Association

The availability of cost-effective molecular genetic marker systems opens the door to analysis of the genetic basis of phenotypic traits measured in breeding populations. Classical quantitative genetics approaches, whether based on provenance, pedigree, or realized relationship matrices, are based on the 'infinitesimal model' proposed by

FS

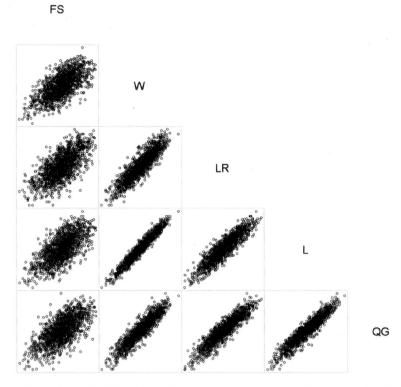

Fig. 4 Correlations of individuals' breeding values produced from pedigree-based full-sib (FS) and four molecular genetic markers-based pairwise relationship estimation methods (W: Wang (2002); LR: Lynch and Ritland (1999); L: Li et al. (1993); QG: Queller and Goodnight (1989)) (After Klápště et al. (2013))

Fisher (1918). Fisher's model reconciled the disparate views of geneticists who studied quantitative traits that show continuous variation, and geneticists who studied discrete characters controlled by single genes, by hypothesizing that continuous variation is the cumulative effect of many different genes, each with a small and approximately equal additive effect on the phenotype. This model has been extremely useful for close to a century, and recent publications have reviewed the substantial body of evidence supporting the main features of the model (Hill et al. 2008; Stranger et al. 2011). This model has important implications for efforts to understand the molecular genetic mechanisms that underlie phenotypic variation in forest tree breeding programs, and for breeders interested in accurately predicting genetic merit of individuals based on genotype information.

The analytical approach called "association genetics" was described over 15 years ago (Lander and Schork 1994; Risch and Merikangas 1996) as an alternative to family-based linkage mapping approaches to characterize the genetic basis of human disorders. Much more work has been done using association genetics in the field of human biomedical genetics than in any other area, and much has been learned about the strengths and weaknesses of the approach (reviewed by Stranger et al. 2011; Rowe and Tenesa 2012). Neale and Savolainen (2004) reviewed key requirements for association genetics, and proposed that populations of conifers (and by extension, other wind-pollinated forest tree species) would be suitable experimental materials for association genetics. Applications of association genetics in tree breeding were described by White et al. (2007, pp. 543–547) and Wilcox et al. (2007); a brief overview will be given to set the stage for discussion of the current status.

The fundamental concept in association genetics is to test for a statistical association between the allelic state at a genetic marker locus in an individual and the phenotype of that individual, for many individuals in a population. The value of such associations is that they can help to identify the molecular basis for phenotypic variation, which in turn may provide molecular markers useful for marker-assisted breeding (Neale and Savolainen 2004). The power to detect associations is a function of several parameters, including the presence of population structure (Neale and Savolainen 2004), the extent of linkage disequilibrium in the test population, the size of the test population, and the proportion of phenotypic variation accounted for by each causative genetic variant involved in the phenotype of interest. The genetic variants tested for association with phenotype may be in known genes that are believed to play a role in controlling the phenotype under study (the 'candidate gene' approach), or they may be chosen on the basis of the allele frequencies in the population and distribution in the genome (the 'genome-wide' approach). As with any statistical testing procedure, if multiple tests of the same hypothesis are conducted, false positive (Type I) errors are likely unless the significance threshold is corrected for the number of tests made. Risch and Merikangas (1996) proposed a threshold of 5×10^{-8} for genome-wide significance in an experiment testing associations of one million single-nucleotide polymorphism (SNP) loci in the human genome; more recent publications have refined this estimate slightly for different sets of human SNP loci (Li et al. 2012). Linkage disequilibrium (LD), the non-random association between allelic states at different loci, affects the independence of multiple tests, and so correction for multiple testing should take into account patterns of LD among the loci analyzed.

An early study of linkage disequilibrium in Douglas fir, based on a relatively small sample of 18 genes from 32 haploid megagametophyte samples, concluded that each gene contained 2–3 independent "haploblocks" of genetic variation, and 4–5 SNP loci per gene would be required to adequately sample the genetic variation in each gene (Krutovsky and Neale 2005). This study focused on transcribed regions, because relatively few resources were available at the time for analysis of non-transcribed regions of genomic DNA in any conifer species. The majority of SNPs identified as significantly associated with target traits in human GWA studies are in non-coding sequences (45 % in introns and 43 % in intergenic regions; Hindorff et al. 2009), suggesting that efforts to model the genetic variation underlying phenotypic variation must include analysis of non-coding genomic DNA sequences. Fortunately, reference genome sequencing projects are now underway for loblolly pine, white spruce, and Norway spruce (searchable abstracts available on-line at https://pag.confex.com/pag/xx/webprogram/start.html), and reference genome sequences are already available for poplar (Tuskan et al. 2006) and eucalyptus (available on-line at http://phytozome.net/), so genomic sequence information will be more readily available for future efforts to model genetic variation.

Determination of the appropriate sample size and number of genetic loci to test in order to achieve a specific level of power in an association study requires evaluation of several population parameters that affect power (Ball 2005; Spencer et al. 2009). The magnitude of the genetic effect of a locus, the frequency in the population of the allele that causes an effect, and the extent of LD between the causative allele and nearby genetic markers (e.g. SNPs) are some of these parameters. Association studies in humans primarily focus on disease-related phenotypes, and the magnitude of the genetic effect is often expressed as a ratio of the likelihood of disease occurrence in a heterozygous individual to the likelihood of disease in an individual homozygous for the most common allele (genotypic risk ratio, Risch and Merikangas 1996, or relative risk per allele, Spencer et al. 2009). The structure of linkage disequilibrium in the human genome is complex enough that simulation is the most general approach to modeling the dependence of experimental power on sample size, relative risk, and allele frequency (Spencer et al. 2009). Such simulations indicate that power is lower for lower risk allele frequencies, for lower risk per allele, and for lower numbers of genetic variant loci tested; for a relative risk per allele of 1.5, an array that assays one million SNP loci provides only about 50 % power in a sample size of 5,000 when the risk allele frequency is less than 10 % (Spencer et al. 2009). A relative risk per allele of 1.5 is roughly equivalent to accounting for 5 % of phenotypic variation, although that equivalence is affected by allele frequency in the population; relatively few loci detected to date in human genome-wide association studies have effects that large (Stranger et al. 2011). This suggests that association genetics studies will not be powerful enough to detect individual genes that account for a significant proportion of phenotypic variation in complex traits in forest trees, if the infinitesimal model is accurate. Some traits of interest to tree breeding programs, such as resistance to fusiform rust disease in *Pinus taeda*, are controlled by individual genes with major effects (Wilcox et al. 1996); association genetics approaches are well-suited to analysis of such traits.

Height growth is an important phenotype in many tree breeding programs, so results of association genetics analysis of height in humans are of interest. Yang et al. (2010) reported that joint analysis of all SNPs as random effects in a mixed linear model that incorporated relationship information derived from marker genotypes explained almost half the genetic variation in height in a sample human population of less than 4,000 individuals, although all 180 loci identified by meta-analysis of association studies in a combined population of 183,727 individuals (Lango Allen et al. 2010) together explained about 14 % of the genetic variation in height. The difference between the analytical approaches taken by these two groups is that Yang et al. focused their attention on creating a predictive model, without concern for identifying specific loci, while Lango Allen et al. followed a more classical association approach using rigorous statistical methods to reduce the likelihood of false positive results and identify loci and pathways mechanistically related to height growth. Many of the loci identified by Lango Allen et al. can be grouped into biological pathways with recognized effects on growth and development, and in many cases, multiple genetic variants were identified per gene (Lango Allen et al. 2010). This phenomenon, referred to as allelic heterogeneity, reduces power in association analyses, because the same phenotype can be due to multiple different genetic variants, even at the same functional gene. Occurrence of multiple genetic variants within genes that affect the same phenotype creates the possibility for epistatic interactions; epistatic interactions within genes or between tightly-linked genes can result in differences between the heritability estimated from closely-related individuals versus distantly-related individuals (Haig 2011; Würschum et al. 2012; Zuk et al. 2012). The approach of analyzing association genetics data by grouping variants into functional genes, organizing genes into pathways, and integrating genetic pathways with gene expression data may provide additional power for understanding phenotypic variation, if modeling approaches that can take pathway structure and gene expression patterns into account can be developed (Cookson et al. 2009; Bennett et al. 2012; Kreimer et al. 2012; O'Hagan et al. 2012). Another approach, similar to that used by Yang et al. (2010), is to incorporate all SNP loci as random effects in the association analysis; this approach has been reported to overcome disadvantages of both traditional linkage analysis and association analysis methods in livestock (Kemper et al. 2012). This type of analysis has much in common with genomic selection, discussed later in the chapter.

Allele frequency of the minor allele at biallelic SNP loci has a major impact on the power of association genetics studies (Spencer et al. 2009; Stranger et al. 2011). Most SNP loci in a sample of over 3,000 SNPs assayed in over 900 loblolly pine trees had minor allele frequencies of less than 15 % (Eckert et al. 2010). Such low minor allele frequencies in samples of unrelated populations contributes to a requirement for extremely large sample sizes to achieve significance in traditional association genetics studies; only alleles with relatively large effects can be detected unless sample sizes exceed 5,000 and marker allele frequency is close to causative variant allele frequency (Ball 2005; Stranger et al. 2011). Structured populations descended from a smaller number of parents can reduce this problem by increasing the frequency of rare alleles that occur in that sample of parents. This strategy has been used to develop the maize Nested Association Mapping (NAM) population (Yu et al. 2008; McMullen et al. 2009), and methods to deal with the population structure that arises in populations produced from mating designs have also been developed (Yu et al. 2006). The combined use of the NAM population and a more typical association population of 282 inbred lines allowed identification of several SNPs that affect maize kernel composition (Cook et al. 2012). Similar strategies may become feasible in forest tree breeding programs, once reference genome sequences are available and haplotype information can be readily developed for the parents of elite breeding populations.

Understanding of molecular mechanisms underlying phenotypic variation is not the primary objective of breeding programs – instead, the objective is to create models of genetic variation in breeding populations that have predictive power to identify individuals of high genetic merit. Studies that increase understanding molecular mechanisms can contribute to development of predictive genetic models in the long term, while studies that focus on developing models of inheritance of complex traits in breeding populations have more immediate value in the short term. Understanding molecular mechanisms can be challenging in human biomedical genetics (Peters and Musunuru 2012), and will be even more challenging for most trees of interest to breeding programs. The association genetics approach can contribute fundamental understanding of mechanisms underlying traits controlled by relatively small numbers of genes, but traits controlled by many genes of equal and small effects will be very expensive to analyze using this method.

5   Genomic Selection

5.1   Background

Many traits of interest to breeders are polygenic, being controlled by many genes each with small effect (Hill et al. 2008). These small-effect genes are crucial for the success of complex trait improvement (Crosbie et al. 2003). For many decades plant and animal breeders relied on phenotype and resemblance among relatives to capture genetic variance explained by these small effect genes. The methods used to improve complex traits were 'black box' as breeders did not know the underlying genetic architecture of complex traits, such as the number of genes controlling the trait and their location in the genome. Tree breeders have adopted these methods since 1950s. The success in improvement of tree characteristics has been relatively modest because breeding-testing-selection cycles for forest trees take many years to complete and tree breeding is logistically complex. Breeders have long looked to molecular markers to overcome challenges and improve the efficiency of selection (Neale and Savolainen 2004).

Beginning in late 1970s quantitative trait loci (QTL) mapping and later candidate gene approaches have been explored as tools to explain gene architecture of complex traits. The idea was that if alleles with large effects on the trait are traced (oligogenic model) with the markers, they could be used for selection of superior genotypes in breeding populations. This concept is called marker aided selection (MAS). However, QTL mapping and candidate gene approaches have had limited use to improve quantitative traits in most plant and animal breeding programs. Major reasons include the cost of producing large number of markers, and the observation that most quantitative traits are controlled by many QTLs, each with small effect, as predicted by the infinitesimal model. Individual QTLs often explained only a small percent (<5 %) of total variance and marker-trait associations discovered in individual families were not repeatable across the population (Goddard and Hayes 2009; Neale 2007).

QTL mapping experiments have been useful in discovering the genetic architecture of quantitative traits important in agricultural and forestry, but the focus is on identifying genetic loci associated with phenotypes. In breeding, on the contrary, the emphasis is on predicting genetic merit of individuals or lines rather than on discovering individual genes. A good predictor of genetic merit does not have to identify the underlying genes (Goddard and Hayes 2009). What is needed is a large number of markers to populate the genome and to explore the LD between these markers and the many QTL with small effect. This approach is called genomic selection (GS) or genome-wide selection. Since the introduction of the concept by Meuwissen et al. (2001), GS has shifted the paradigm, driven by the increased efficiency in DNA sequencing technologies and computing power.

GS contrasts greatly with traditional MAS, because in GS there is no defined subset of significant markers used for selection. Instead, GS jointly analyzes all markers in a population, attempting to explain the total genetic variance with dense genome-wide marker coverage through summing marker effects to predict breeding values of individuals (Meuwissen et al. 2001). The idea is that if we populate the genome with high-density markers, we can capture the LD between markers or marker haplotypes and causal polymorphism. Such association would be consistent across different families (Meuwissen et al. 2001). With advancement in DNA sequencing technologies and efficiency in genotyping, GS has become a reality in dairy cattle breeding (Goddard and Hayes 2009). Many livestock breeding programs now routinely apply GS to market bulls (Hayes et al. 2009). Genomic selection processes start from a training population. Candidates to establish a next cycle of breeding are selected through GS. The training can be performed iteratively as new phenotype and marker data accumulate (Heffner et al. 2011).

5.2 Empirical Examples from Forest Trees

Forest tree breeding programs are still at the first stage of breeding-testing and selection cycles with little genetic difference from natural populations. If successful, the impact of genomic selection on forest tree breeding could be far greater than for other crops or animal breeding programs. A few early empirical studies on genomic selection in forest trees are encouraging. For example, in a cloned loblolly pine breeding population, accuracies of GS varied between 0.55 and 0.88, matching those achieved by conventional phenotypic selection (Resende et al. 2012). Similarly in the same species, Isik et al. (2011) reported genomic estimated breeding values with reliability as high as breeding values based on resemblance among relatives and phenotypic data. These studies estimated the individual marker effect and summed up the coefficients to estimate genomic estimated breeding values of trees.

Alternatively a smaller subset of markers can be used to estimate realized genomic relationships using frequency of alleles shared by individuals (Legarra and Misztal 2008). Then, the additive genetic relationship matrix derived from pedigree is substituted by the genomic relationship matrix to predict genomic estimated breeding values. Genomic BLUP (GBLUP) could be a powerful tool for forest tree breeding programs. Such models can capture the Mendelian segregation effect in

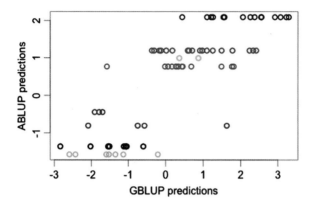

Fig. 5 Predicted breeding values of loblolly pine clones based on pedigree (*y-axis*) and genomic BLUP (*x-axis*) for eight crosses. Each cross is designated with a different color. In the absence of phenotype, the expected breeding value of sibs would be the same, which is the mid-parent value (ABLUP). However, DNA markers can capture Mendelian sampling effect within each cross as shown here, and thus, sibs can be ranked and selected without progeny testing (Zapata-Valenzuela et al. 2011)

full-sib families, which was not the case using the average additive genetic relationships. For example, Zapata-Valenzuela et al. (2011) showed that accuracies of genomic estimated breeding values using GBLUP were comparable to traditional pedigree-based BLUP methods. In the same study, breeding values of a training population were estimated using GBLUP and classical BLUP (Henderson 1984). In the absence of phenotype, sibs from a cross had the same mid-parent breeding values when classical BLUP was used (Fig. 5). However, genomic relationship matrices based on SNP markers allowed prediction of different genetic values for sibs from a single cross.

5.3 Statistical Machinery

Classical linear mixed models are not efficient to handle large number of markers as predictors because the number of predictors (p) is larger than the number of data points (n) to explain variance in the phenotype. Such large *p* and small *n* effect causes lack of degrees of freedom. Statistical analysis of large number of markers has been a very active area of research in recent years, and many statistical methods have been proposed in the literature (Gianola et al. 2009). The effect of markers or haplotypes can be estimated by simultaneously including all markers in a model, but the challenge is to estimate the variances of marker effects. The best linear unbiased prediction (BLUP) method and ridge regression approaches have been proposed to estimate individual marker effects (Meuwissen et al. 2001; Whittaker et al. 2000). These methods make the assumption that markers are sampled from a population with expectation $N \sim \left(0, \sigma_g^2\right)$ and each marker explain the same (σ_g^2 / n)

amount of genetic variance. Rather than categorizing markers as either significant or as having no effect, ridge regression and BLUP shrink all marker effects toward zero (Meuwissen et al. 2001). This is not a realistic assumption because regardless of association of markers with the trait loci, all the markers are shrunk towards the mean at the same level. Bayesian methods have a natural way of taking into account uncertainty about all unknowns in a model (e.g., Gianola et al. 2009) and, when coupled with the power and flexibility of Markov Chain Monte Carlo, Bayesian methods can be applied to almost any parametric statistical model. Meuwissen et al. (2001) introduced BayesA and BayesB and compared them with BLUP method in their original paper on GS. In BayesA, all the markers explain a fraction of genetic variance and the variance explained by each marker can vary based on the scaled inverted chi-square distribution as prior. Method BayesB corrects the shortcoming of BayesA by shrinking a high proportion (π) of markers to zero. Bayes C, Cπ, and D and Dπ were introduced to address the undesirable effect of priors on estimations observed for BayesA and BayesB. Habier et al. (2011) concluded that accuracies of the alternative Bayesian methods were similar and none of them outperformed all others across all traits and training data sizes. The choice of statistical methods for GS is sometimes is a matter of practicality, time and ease of application. Examples on empirical and simulated data suggest that Bayesian approaches are efficient to increase accuracy of predictions but the increase is usually minimal unless a large fraction of genetic variance in the trait in question is controlled by a few loci.

5.4   Challenges of GS in Forest Tree Breeding

Despite advances in the efficiency of genotyping technologies, genotyping is still costly for forest trees. For example, the Illumina SNP genotyping platform costs about $150 per sample for loblolly pine as of 2012, though the cost is decreasing. Several labs in the USA and other countries are working on alternative genotyping technologies, such as genotyping by sequencing (Baird et al. 2008; Elshire et al. 2011; Peterson et al. 2012; Poland et al. 2012; Truong et al. 2012), and we expect that the cost of genotyping could be less than $50 as of 2013.

GS has been successful in cattle breeding because the number of founders in these populations is relatively small (<30) and the LD between markers and trait loci are large, thanks to deep pedigree in the populations and small effective population size. Tree breeding populations still are at their infancy. The pedigree structures are still shallow with very low linkage disequilibrium (Neale and Savolainen 2004). Marker-trait phase detected in one generation may not hold in a subsequent generation because of meiotic recombination. For GS to be successful, well-structured populations (small effective population size, multiple generations) are needed.

Conifers are major targets of breeding programs in the northern hemisphere, and they have large and complex genomes. GS require dense coverage of whole genome to trace many QTLs associated with phenotype. Many more markers

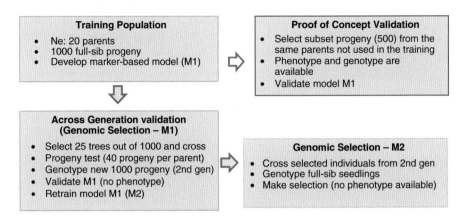

Fig. 6 Genomic selection process for an elite breeding population. A marker-based prediction model is retrained across multiple generations. Such process would make the model more powerful for genomic estimated breeding values to trace LD of markers and QTLs

might be needed to populate genome of conifers. Grattapaglia and Resende (2011) suggested that 20 markers/cM are needed for an effective population size of greater than 30.

Forest trees have some advantages in implementation of GS. A large population can be put together easily. Each family can be represented by large number of progeny (several hundreds) with little investment and time. Phenotyping can be quite accurate thanks to efficient experimental designs and cloning of individuals.

An example GS plan has been proposed for a loblolly pine breeding population within the North Carolina State University Tree Improvement Program in the USA (Fig. 6). In the diagram given in Fig. 6, the process starts with creating a training population with an effective population size (Ne) smaller than 50 parents. In this example, 20 parents are used. Relatedness among the 20 founders is desirable, because that will make the marker-based model more powerful to predict GEBV by tracing historical LD in the population. From full-sib crosses of 20 parents, about 1,000 individuals can be genotyped. This progeny population is field-tested and breeding values are obtained Deregressed breeding values of 1,000 individuals or phenotypic values adjusted for fixed effects can be obtained to use as new 'phenotype' for development of a marker-based model (M1).

There are different methods to validate the predictive ability of markers. An additional 500 progeny from the same crosses (with known phenotype and genotypes) can be used as a validation population. Alternatively, random sampling of a small subset of progeny or selection of subset of progeny within each full-sib family can be used to validate model M1. This step is a proof of concept to show that the model has predictive power, and is not necessarily an application of GS. In order to utilize the benefit of GS approaches, we need to breed the selected individuals from the training population, obtain seeds, and use M1 to make selection decisions. This can be called 'across generation' GS application. The M1 model can be retrained when more genotypic and phenotypic data become available as breeding progresses (M2). GS training models would have more reliability as new data are included and can be used for multiple generations.

6 Conclusions

The availability of cost-effective genetic markers in forest tree species is expanding rapidly due to advances in DNA sequencing technology and investment in determining the reference genome sequences for several commercially-important species of forest trees. These resources are likely to fundamentally change the way tree breeding programs characterize genetic variation in their breeding populations, and several research groups are actively working to develop methods for applications of these tools in practical breeding programs. Molecular markers are already useful tools for population management applications such as validation of crosses, pedigree reconstruction, and unambiguous identification of clones. Association genetics results have already been reported for several traits in various species of forest trees, and application of these results in practical breeding programs may follow soon. Development of more sophisticated analytical methods capable of integrating the analysis of genetic variation detected by SNP assays with variation in gene expression patterns, metabolite levels, and phenotypic measurements may provide new tools capable of more accurate prediction of genetic value based on molecular assays. Predictive modeling of genetic value is the central objective of genomic selection methods, which have shown considerable promise in livestock and crop species that have appropriate patterns of LD in breeding populations. Forest tree breeding populations are likely to have very different patterns of LD than livestock or crop species, and new approaches to genomic selection may be required in order for this method to reach its full potential in applied tree breeding programs.

References

Allard RW (1960) Principles of plant breeding. Wiley, New Year
Baird NA, Etter PD, Atwood TS, Currey MC, Shiver AL, Lewis ZA, Selker EU, Cresko WA, Johnson EA (2008) Rapid SNP discovery and genetic mapping using sequenced RAD markers. PLoS One 3:e3376
Ball RD (2005) Experimental designs for reliable detection of linkage disequilibrium in unstructured random population association studies. Genetics 170:859–873
Bennett BD, Xiong Q, Mukherjee S, Furey TS (2012) A predictive framework for integrating disparate genomic data types using sample-specific gene set enrichment analysis and multi-task learning. PLoS One 7(9):e44635
Burdon RD, Shelbourne CJA (1971) Breeding populations for recurrent selection: conflicts and possible solutions. N Z J For Sci 1:174–193
Callaham RZ (1964) Provenance research: investigation of genetic diversity associated with geography. Unasylva 18:40–50

Campbell RK (1986) Mapped genetic variation of Douglas-fir to guide seed transfer in southwest Oregon. Silvae Genet 35:85–95

Cappa EP, Lstiburek M, Yanchuk AD, El-Kassaby YA (2011) Two-dimensional penalized splines via Gibbs sampling to account for spatial variability in forest genetic trials with small amount of information available. Silvae Genet 60:25–35

Cook JP, McMullen MD, Holland JB, Tian F, Bradbury P, Ross-Ibarra J, Buckler ES, Flint-Garcia SA (2012) Genetic architecture of maize kernel composition in the nested association mapping and inbred association panels. Plant Physiol 158:824–834

Cookson W, Liang L, Abecasis G, Moffatt M, Lathrop M (2009) Mapping complex disease traits with global gene expression. Nat Rev Genet 10(3):184–194

Crosbie, TM, Eathington, SR, Johnson, GR, Edwards, M, Reiter, R, Stark, S et al (2003) Plant breeding: past, present, and future. In: Lamkey, KR, Lee, M (eds) Plant breeding: the Arnel R. Hallauer International Symposium, Mexico City. 17–23 Aug 2003. Blackwell, Oxford, UK, pp 1–50

Ding C, McAuley L, Meitner MJ, El-Kassaby YA (2012) Evaluating interior spruce seed deployment with GIS-based modeling using British Columbia's Prince George seed planning zone as a model. Silvae Genet 61:271–279

Eckert AJ, van Heerwaarden J, Wegrzyn JL, Nelson CD, Ross-Ibarra J, González-Martínez SC, Neale DB (2010) Patterns of population structure and environmental associations to aridity across the range of loblolly pine (*Pinus taeda* L., Pinaceae). Genetics 185:969–982

El-Kassaby YA, Lstibůrek M (2009) Breeding without breeding. Genet Res 91:111–120

El-Kassaby YA, Cappa EP, Liewlaksaneeyanawin C, Klápště J, Lstiburek M (2011) Breeding without breeding: is a complete pedigree necessary for efficient breeding? PLoS One 6:e25737

El-Kassaby YA, Klápště J, Guy RD (2012) Breeding without Breeding: selection using the genomic best linear unbiased predictor method (GBLUP). New For 43:631–637. doi:10.1007/s11056-012-9338-4

Elshire RJ, Glaubitz JC, Sun Q, Poland JA, Kawamoto K, Buckler ES, Mitchell SE (2011) A robust, simple genotyping-by-sequencing (GBS) approach for high diversity species. PLoS One 6:e19379

Falconer DS, Mackay TFC (1996) Introduction to quantitative genetics. Longman, New York

Fisher RA (1918) The correlation between relatives on the supposition of Mendelian inheritance. Trans Roy Soc Edinb 52:399–433

Gianola D, de los Campos G, Hill WG, Manfredi E, Fernando R (2009) Additive genetic variability and the Bayesian alphabet. Genetics 183:347–363

Goddard ME, Hayes BJ (2009) Mapping genes for complex traits in domestic animals and their use inbreeding programmes. Nat Rev Genet 10:381–391

Grattapaglia D, Resende MDV (2011) Genomic selection in forest tree breeding. Tree Genet Genomes 7:241–255

Habier D, Fernando RL, Kizilkaya K, Garrick DJ (2011) Extension of the Bayesian alphabet for genomic selection. BMC Bioinforma 12:186

Haig D (2011) Does heritability hide in epistasis between linked SNPs? Eur J Hum Genet 19:123

Hansen OK, McKinney LV (2010) Establishment of a quasi-field trial in *Abies nordmanniana* – test of a new approach to forest tree breeding. Tree Genet Genomes 6:345–355

Hayes BJ, Bowman PJ, Chamberlain AJ, Goddard ME (2009) Genomic selection in dairy cattle: progress and challenges. J Dairy Sci 92:433–443

Heffner EL, Jannink J-L, Sorrells ME (2011) Genomic selection accuracy using multifamily prediction models in a wheat breeding program. The Plant Genome 4:65–75

Henderson CR (1984) Applications of linear models in animal breeding. University of Guelph, Ontario

Hill WG, Goddard ME, Visscher PM (2008) Data and theory point to mainly additive genetic variance for complex traits. PLoS Genet 4:e1000008

Hindorff LA, Sethupathy P, Junkins HA, Ramos EM, Mehta JP, Collins FS, Manolio TA (2009) Potential etiologic and functional implications of genome-wide association loci for human diseases and traits. Proc Natl Acad Sci USA 106:9362–9367

Isik F, Whetten R, Zapata-Valenzuela J, Ogut F, McKeand S (2011) Genomic selection in loblolly pine – from lab to field. From IUFRO tree biotechnology conference 2011: from genomes to integration and delivery. BMC Proc 5(Suppl 7):I8

Jayawickrama KJS, Carson MJ (2000) A breeding strategy for the New Zealand radiata pine breeding cooperative. Silvae Genet 49:82–90

Jones AG, Ardren WR (2003) Methods of parentage analysis in natural populations. Mol Ecol 12:2511–2523

Kemper KE, Daetwyler HD, Visscher PM, Goddard ME (2012) Comparing linkage and association analyses in sheep points to a better way of doing GWAS. Genet Res (Camb) 94:191–203

Klápště J, Lstibůrek M and El-Kassaby YA (2013) Estimates of genetic parameters and breeding values from western larch open-pollinated families using marker-based relationship. Tree Genet Genome (in press)

Konig AO (2005) Provenance research: evaluating the spatial pattern of genetic variation. In: Geburek TH, Turok J (eds) Conservation and management of forest genetic resources in Europe. Arbora Publishers, Zvolen, pp 275–333

Korecký J, Klápště J, Lstibůrek M, Kobliha J, Nelson CD El-Kassaby YA (2013) Comparison of genetic parameters from marker-based relationship, sibship, and combined models in Scots pine multi-site open-pollinated tests. Tree Genet Genomes. doi:10.1007/s11295-013-0630-z

Kreimer A, Litvin O, Hao K, Molony C, Pe'er D, Pe'er I (2012) Inference of modules associated to eQTLs. Nucleic Acids Res 40:e98

Krutovsky KV, Neale DB (2005) Nucleotide diversity and linkage disequilibrium in cold-hardiness and wood quality-traits candidate genes in Douglas-fir. Genetics 171:2029–2041

Lambeth C, Lee B-C, O'Malley D, Wheeler N (2001) Polymix breeding with parental analysis of progeny: an alternative to full-sib breeding and testing. Theor Appl Genet 103:930–943

Lander ES, Schork NJ (1994) Genetic dissection of complex traits. Science 265(5181): 2037–2048

Lango Allen H et al (2010) Hundreds of variants clustered in genomic loci and biological pathways affect human height. Nature 467(7317):832–838, 292 additional co-authors

Legarra A, Misztal I (2008) Technical note: computing strategies in genome-wide selection. J Dairy Sci 91:360–366

Li CC, Weeks DE, Chakravarti A (1993) Similarity of DNA fingerprints due to chance and relatedness. Hum Hered 43:45–52

Li MX, Yeung JM, Cherny SS, Sham PC (2012) Evaluating the effective numbers of independent tests and significant p-value thresholds in commercial genotyping arrays and public imputation reference datasets. Hum Genet 131(5):747–756

Lynch M, Ritland K (1999) Estimation of pairwise relatedness with molecular markers. Genetics 152:1753–1766

McMullen MD et al (2009) Genetic properties of the maize nested association mapping population. Science 325(5941):737–740, 31 additional co-authors

Meuwissen THE, Goddard ME, Hayes BJ (2001) Prediction of total genetic value using genome-wide dense marker maps. Genetics 157:1819–1829

Namkoong G (1966) Inbreeding effects on estimation of genetic additive variance. For Sci 12:8–13

Namkoong G, Kang HC, Brouard JS (1988) Tree breeding: principles and strategies. Springer, New York, Monograph, Theor Appl Genet 11

Neale D (2007) Genomics to tree breeding and forest health. Curr Opin Genet Dev 17:539–544

Neale D, Savolainen O (2004) Association genetics of complex traits in conifers. Trends Plant Sci 9(7):325–330, ISSN 1360–1385, 07/2004

O'Hagan S, Knowles J, Kell DB (2012) Exploiting genomic knowledge in optimising molecular breeding programmes: algorithms from evolutionary computing. PLoS One 7:e48862

Peters DT, Musunuru K (2012) Functional evaluation of genetic variation in complex human traits. Hum Mol Genet. doi:10.1093/hmg/dds363

Peterson BK, Weber JN, Kay EH, Fisher HS, Hoekstra HE (2012) Double digest RADseq: an inexpensive method for de novo SNP discovery and genotyping in model and non-model species. PLoS One 7:e37135

Poland JA, Brown PJ, Sorrells ME, Jannink J-L (2012) Development of high-density genetic maps for barley and wheat using a novel two-enzyme genotyping-by-sequencing approach. PLoS One 7:e32253

Porth I, Klápště J, Skyba O, Lai BSK, Geraldes A, Muchero W, Tuskan GA, Douglas CJ, El-Kassaby YA, Mansfield SD (2012) Populus trichocarpa cell wall chemistry and ultrastructure trait variation, genetic control, and genetic correlations. New Phytol 197:777–790

Queller DC, Goodnight KF (1989) Estimating relatedness using genetic markers. Evolution 43:258–275

Rehfeldt GE (1983) Seed transfer guidelines for Douglas-fir in Central Idaho. U. S. For Serv Res Note INT-337

Resende MFR, Muñoz P, Acosta JJ, Peter GF, Davis JM, Grattapaglia D, Resende MDV, Kirst M (2012) Accelerating the domestication of trees using genomic selection: accuracy of prediction models across ages and environments. New Phytol 193:617–624

Risch N, Merikangas K (1996) The future of genetic studies of complex human diseases. Science 273(5281):1516–1517

Rowe SJ, Tenesa A (2012) Human complex trait genetics: lifting the lid of the genomics toolbox – from pathways to prediction. Curr Genom 13:213–224

Spencer CC, Su Z, Donnelly P, Marchini J (2009) Designing genome-wide association studies: sample size, power, imputation, and the choice of genotyping chip. PLoS Genet 5:e1000477

Stranger BE, Stahl EA, Raj T (2011) Progress and promise of genome-wide association studies for human complex trait genetics. Genetics 187:367–383

Truong HT, Ramos AM, Yalcin F, de Ruiter M, van der Poel HJA, Huvenaars KHJ, Hogers RCJ, van Enckevort LJG, Janssen A, van Orsouw NJ, van Eijk MJT (2012) Sequence-based genotyping for marker discovery and co-dominant scoring in germplasm and populations. PLoS One 7:e37565

Tuskan GA et al (2006) The genome of black cottonwood, *Populus trichocarpa* (Torr. & Gray). Science 313(5793):1596–1604, 109 additional co-authors

vanRaden PM (2008) Efficient methods to compute genomic predictions. J Dairy Sci 91:4414–4423

Wang JL (2002) An estimator for pairwise relatedness using molecular markers. Genetics 160:1203–1215

White TL, Adams WT, Neale DB (2007) Forest genetics. CABI Publishing, Cambridge, MA

Whittaker JC, Thompson R, Denham MC (2000) Marker-assisted selection using ridge regression. Genet Res 75:249–252

Wilcox PL, Amerson HV, Kuhlman EG, Liu B-H, O'Malley DM, Sederoff RR (1996) Detection of a major gene for resistance to fusiform rust disease in loblolly pine by genomic mapping. Proc Natl Acad Sci USA 93:3859–3864

Wilcox PL, Echt CE, Burdon RD (2007) Gene-assisted selection: applications of association genetics for forest tree breeding (Ch 10). In: Oraguzie NC, Rikkerink EHA, Gardiner SE, de Silva HN (eds) Association mapping in plants. Springer, New York, p 278

Würschum T, Maurer HP, Dreyer F, Reif JC (2012) Effect of inter- and intragenic epistasis on the heritability of oil content in rapeseed (*Brassica napus* L.). Theor Appl Genet 126:435–441. doi:10.1007/s00122-012-1991-7

Yang J, Benyamin B, McEvoy BP, Gordon S, Henders AK, Nyholt DR, Madden PA, Heath AC, Martin NG, Montgomery GW, Goddard ME, Visscher PM (2010) Common SNPs explain a large proportion of the heritability for human height. Nat Genet 42:565–569

Yu J, Pressoir G, Briggs WH, Vroh Bi I, Yamasaki M, Doebley JF, McMullen MD, Gaut BS, Nielsen DM, Holland JB, Kresovich S, Buckler ES (2006) A unified mixed-model method for association mapping that accounts for multiple levels of relatedness. Nat Genet 38:203–208

Yu J, Holland JB, McMullen MD, Buckler ES (2008) Genetic design and statistical power of nested association mapping in maize. Genetics 178:539–551

Zapata-Valenzuela J, Isik F, Maltecca C, Wegryzn J, Neale D, McKeand S, Whetten R (2011) BMC Proc 5(Suppl 7):P60

Zuk O, Hechter E, Sunyaev SR, Lander ES (2012) The mystery of missing heritability: genetic interactions create phantom heritability. Proc Natl Acad Sci USA 109:1193–1198

A 'Reality Check' in the Management of Tree Breeding Programmes

Heidi S. Dungey, Alvin D. Yanchuk, and Richard D. Burdon

Abstract Modern tree breeding is now at least 80 years old, and has many successes to report, notably with loblolly pine since the 1920s and with *Pinus radiata* in New Zealand since the 1950s. Nevertheless, tree breeders are facing various difficult issues, some longstanding, and some new, which will not disappear. These include: meeting the requirement to preserve genetic variation while capturing genetic gain; accelerating the capture of genetic gain through both enhanced selection technology and rapid-deployment propagation systems; coping with changes in perceptions of appropriate breeding goals driven by science, biotic influences, changes in forestry ownerships, and market shifts; harnessing DNA technology with what are still essentially wild organisms; doing so without incurring undue opportunity costs with respect to genetic gains available from conventional breeding; managing cultural differences between classical breeders and those practising the new genomic sciences; and coping with institutional and funding changes.

The New Zealand *Pinus radiata* breeding programme is used as a case study to examine these issues closely and highlight opportunities for the continued development of well-thought-out and successful breeding programmes.

H.S. Dungey (✉) • R.D. Burdon
Scion Research, 49 Sala Street, Rotorua, New Zealand
e-mail: Heidi.Dungey@scionresearch.com

A.D. Yanchuk
Scion Research, 49 Sala Street, Rotorua, New Zealand

British Columbia Ministry of Forests, Lands and Natural Resource Operations, Victoria, BC, Canada

T. Fenning (ed.), *Challenges and Opportunities for the World's Forests in the 21st Century*, Forestry Sciences 81, DOI 10.1007/978-94-007-7076-8_19, © Springer Science+Business Media Dordrecht 2014

461

1 Introduction

Modern tree breeding is now at least 80 years old, with many successes to report, including with loblolly pine since the 1920s (Lambeth et al. 2005) and with *Pinus radiata* in New Zealand (NZ) since the 1950s (Burdon 2008; Burdon et al. 2008; Dungey et al. 2009; Jayawickrama 2001). The New Zealand programme is a case study for closer examination of difficult issues now facing tree breeders, issues that breeders need to constantly review in their own programmes, including:

- The need to preserve genetic variation.
- Determining the best population sizes.
- Reviewing and updating breeding goals.
- Accelerating delivery of genetic gain.
- Maximising delivery of genetic gain through propagation; and
- Coping with institutional and funding changes.

All these points are, of course, related. Here we review these issues and the key points and opportunities for the continued development of thoughtful and successful breeding programmes.

2 The Need to Preserve Genetic Variation

Several early forest genetics investigations started with systematic collections planted in provenance and progeny tests (e.g. in New Zealand, (Burdon 1992; Burdon et al. 2008; Shelbourne et al. 2004)). These 'core' collections came from native populations and/or land-race 'plus-tree' selections and formed the basis for most modern breeding populations and/or improved commercial tree stocks (White et al. 2007). Many of these field tests remain, but they are ageing, and will be lost if the genetic variation is not properly secured. This is even more important where the tree breeding populations now represent the entire evolutionary future of the species (El-Kassaby and Ritland 1996; Stoehr et al. 2008; Yanchuk 2001). We need to get it right, as future 'corrections' will be much more difficult than with most other organisms, largely because of the long generation intervals of trees, the need to improve multiple traits, and likely minimal appetite to search for alternatives in material from less-improved, previous generations.

2.1 The Pervasive Trade-Off

Within a breeding population there is always a trade-off between maintaining genetic diversity and level of genetic gain (Gea et al. 1997; Lindgren et al. 1996; Yanchuk 2001; Zheng et al. 1997). Consistent delivery of genetic gain across multiple generations requires diversity (White et al. 2007), but maintaining too much diversity will slow down the breeding and selection process. As such, breeders are often interested in knowing the effective population size (N_e) of their breeding population(s). Recommended values of N_e typically range from 20 to 80, depending on the traits, species of interest, and an expected number of generations that the population could be bred for (Danusevicius and Lindgren 2005; White et al. 2007; Yanchuk 2001).

Trade-offs and gains have been well studied (e.g., Andersson et al. 1998; Danusevicius and Lindgren 2002a, b, 2005; Fernández and Toro 2001; Hallander and Waldmann 2009; Kang et al. 2007; Kang et al. 2001; Kerr et al. 1998; Lstibůrek et al. 2004a, b; Wei and Lindgren 2006). Although some opinions are conflicting on the matter, and depending on the species and traits, the use of positive assortative mating in a single population appears to be an advantage (Rosvall and Mullin 2003; Rosvall et al. 2003), and cloning the elite (a sub-population comprising the best individuals) or whole breeding populations should deliver more gain (Shelbourne et al. 2007). The development of selection algorithms to minimize coancesty build up have and will continue to be very valuable (Hallander and Waldmann 2009; Lindgren and Mullin 1997; Zheng et al. 1997). These theoretical algorithms, to optimise combinations of gains and levels of diversity, are particularly useful where selecting over multiple generations.

More recently, the rate at which inbreeding accumulates per generation in tree breeding populations is seen as crucial. Traditional F (inbreeding) values (Falconer and Mackay 1996) and/or inbreeding effective population size (Caballero 1994), used in animal breeding programs, appear more meaningful statistics of expected loss of genetic variation over time. These measures give some guide to the per-generation fixation of deleterious alleles. As such, population sizes need to be assessed and managed in this context, so the acceptable per-generation trade-offs between the level of gain and rate of inbreeding are set and implemented. Multiple traits complicate this situation further. Regardless, it is difficult insist on breeding populations of substantially over 100–120 parents, as the rate of inbreeding must be managed to deliver enough genetic gain per generation to make programs worthwhile (Yanchuk and Sanchez 2011).

'Tracking' of genetic diversity in relation to the original or wild or progenitor populations can also be an important question, particularly where breeding and re-planting is undertaken within the natural range, e.g. with Douglas-fir (El-Kassaby and Benowicz 2000; Hubert and Bastien 1999; Krakowski and Stoehr 2009; Lipow et al. 2003; Yanchuk 2001). In the mid-1990s isozyme or allozyme studies were one of the few ways to look at 'loss of genetic diversity', e.g. in interior spruce in British Columbia (Stoehr and El-Kassaby 1997), and results of these studies more or less fitted expectations from genetic sampling theory. Conservation of low-frequency alleles is important as a form of 'genetic insurance', but breeding populations do not seem to be the most appropriate place to try and conserve these (Yanchuk 2001). One alternative is to measure genetic distance, and preferentially select diverse parents (geographically and genetically) to form the next generation, as was suggested by (Zhang et al. 2009) in *Pinus massoniana*. Again, this option still faces the conflict between diversity and gain, not to mention adaptation of diverse provenances. Yet another alternative, which may sit better in programmes focussed on gain delivery, is to seek fresh diversity that may arise within a huge 'seedling' population from broadly-based seed orchards and crosses, where new desirable mutations can be captured, and some rarer alleles conserved (Burdon 1997). Setting the size of the breeding population, in practice, depends on the gain required to justify the expense of tree breeding, the available genetic material, and the budget available.

2.2   Diversity and Risk

Forests with reduced genetic diversity incur some elevated risk, which has been reviewed for *Pinus radiata* in New Zealand (Aimers-Halliday and Burdon 2003; Burdon 2001a, b, 2010a,b; Burdon and Aimers-Halliday 2003) and for loblolly pine (*P. taeda*) (Lambeth et al. 2005). However, risks can arise that have no direct relationship to within-species diversity. Some of the worst outbreaks of pests and diseases in forestry are largely attributed to the increase in winter temperatures due to climate change (Bentz et al. 2010; Carroll et al. 2004; Cudmore et al. 2010; Macias Fauria and Johnson 2009). Outbreaks of mountain pine beetle (*Dendroctonus ponderosae*) (Coleoptera: Scolytidae) on lodgepole pine (*P. contorta*) were not foreseen, were in a native population with a large and natural genetic base, and yet still caused widespread devastation.

Breeders often attempt to maintain genetic diversity through establishing archives or large block plantings in order to capture rare alleles that cannot be reliably retained within an intensive breeding programme (Burdon 2010a; Yanchuk 2001). In practice, while it is essential to have some gene conservation and/or archives to at least mitigate the impacts of any serious pest or disease incursions, it may not be enough.

3   Reviewing and Updating Breeding Goals

Questions are always arising on how much genetic gain is being attained, what are the 'real' traits (e.g., measures of final stand value, rather than individual-tree performance) that should be focused on, and what future investment is justifiable. The global shift away from the vertically integrated forestry companies has further complicated tree breeding, as investors see little return on investing in up-front development costs of better germplasm for markets they cannot directly see.

In order to ensure gain delivery, most tree breeding programmes consider some suite of 'key' traits over the longer-term. Focus and gains can, however, be diluted by shifting objectives and expectations over the short term (e.g. within a decade). These distractions can include immediate biological imperatives (e.g. a new pest or disease), shorter-term industry planning horizons, and shifts in the prices driven by developments in processing technologies and target markets. Short-term distractions have sometimes led to serious problems for tree breeders. For instance, improved growth has always been a major component of most programs, reflecting the simple economics of more volume and larger piece sizes, with the hope of more value; but it was well known early on that that wood quality would likely be lowered without strong culling against low-wood-quality genotypes. In the short term, things would look good on the yield tables, but once the logs started coming in and going out to end users, the bad news started coming back. So the mistaken input of the previous generations of forest managers, pressing for strong selection primarily on growth, is generally worn by the current generation of breeders.

In New Zealand dramatic genetic improvement has been achieved in *P. radiata* for tree form, along with substantial improvement for growth and health (Burdon et al. 2008). Yet for various reasons there is now a strong focus on genetic improvement of wood properties (Burdon 2010b). Internationally, that is also occurring with other species, notably loblolly pine (e.g. Eckard et al. (2010)) and Douglas-fir (e.g. Jayawickrama et al. (2011)).

Gains required to justify the expense of tree breeding are linked strongly to the rotation age, the volume and value of product produced, the discount rate used, and the profit margin required. Where all traits are very favourably correlated genetically, gains are typically not diluted too badly by selecting for multiple traits, but if traits are adversely correlated, simultaneous gains can be drastically restricted. Adding additional traits to cull undesirable properties, even without adverse genetic correlations, makes the job more difficult as many more genotypes simply need to be tested to find an adequate number of parents for the breeding or the production populations. Shelbourne et al. (1989) cited an unpublished study showing that the investment in breeding and associated research to produce a specific level of improvement in *P. radiata* would provide a benefit:cost ratio of 46:1– the greater the investment, the greater the benefit! Other studies, however, have implied that the gains are perhaps closer to a 9:1 ratio (A. Katz pers. comm.), but the benefits are clearly there.

Quantifying realised genetic gain is essential to maintain funding support from both the forest industry and from funding agencies. Moreover, knowing what gain is being achieved in what traits is helpful for reviewing breeding goals. Proving effectiveness of genetic improvement is traditionally through realised genetic gain trials in large-block plots e.g. Stoehr et al. (2010). Although it can be difficult to get significant differences between different treatments (i.e., among genetic entries with different levels of predicted gain), due to the number of replicates required (up to six or more is ideal), a good compromise is to lower the replication of the large blocks to say three, but also include several replicates of row-plots of the same material (Stoehr et al. 2010). Plots of the improved material can also be deployed into commercial plantings, and compared with the performance of other material in the surrounding forest. Although not as statistically robust for individual comparisons on individual sites, this approach will suit a regression analysis if sufficient data on sufficient plots are monitored over time. Data from large-block genetic gain trials have formed the basis of empirical modelling for forest growth and yield prediction in New Zealand (Carson et al. 1999) and have fulfilled more than one purpose.

One thing is apparent, hedging your bets, changing your mind within a breeding cycle and attempting to cover all bases in a breeding programme will not deliver the products that users will support in the long run. A strategic approach is essential. Yet tree breeding strategy must address not only long critical paths in breeding operations, but also great market and biological uncertainties. We recommend updating a breeding strategy only when new technologies are available or when something changes in the operating environment that warrants an adjustment. A good example of this is perhaps when genetic marker technologies become economically available. Implementing these should be very valuable, especially if they describe a large proportion of the genetic variation or can change the way we develop and use pedigree information.

4   Accelerating Delivery of Genetic Gain

4.1   The Basics

A lot of basic information is required for optimal management and improvement practices for forest tree species, including methods for grafting, vegetative reproduction, seed orchard management/production, mating systems in orchards, nursery raising practices, establishment practices and early performance in the field, to name only a few. These are sometimes neglected as key underpinning technologies for delivering on genetic gain expectations. All systems, with experienced staff in all steps of the process, are essential to delivering gain efficiently. Additional steps can be taken, which include understanding optimal times for selection, and introduction of molecular biology tools. These are described below.

4.2   Early Selection

It is always desirable to shorten a breeding cycle to deliver more gains faster to forest growers. Quantitatively, early selection is a form of indirect selection (Burdon 1989a). Assessments on traits at early ages are used as proxies for measurements on breeding-goal traits at harvest ages. Success depends on good genetic correlations between the early and late assessments and good additive genetic variation of both. In *P. radiata* there has been considerable research to determine if early selection is possible. Quantitative genetics studies have shown that a good selection age for this species in New Zealand is around age eight (King and Burdon 1991). This is when genetic correlations between early-age measurements and later-age measurements become acceptable, usually around 0.8. More recent studies on wood quality have concurred, despite indicating that the heritability of some wood properties can be quite low before ring seven from the pith at breast height (Dungey et al. 2006).

Wood density is the most promising trait for successful early selection, and is now an important component in selection in *P. radiata*. However, various authors have noted that wood density alone is not a good predictor of the performance of corewood (so-called 'juvenile' wood), the problem zone approximating to the first ten rings from the pith in *P. radiata* (Apiolaza 2009; Burdon et al. 2004a, b; Cown et al. 2004). The microfibril angle (MfA) in this corewood zone can be very high; density can look acceptable, but the stiffness of the corewood is not well predicted by either trait individually (e.g. Dungey et al. 2006; Kumar et al. 2006). These properties mean that early selection for wood stiffness is not ideal using just MfA or density. Rapid screening around age 8–10 for *P. radiata* using acoustic measurements as surrogates for wood stiffness should give more progress (Apiolaza 2009; Dungey et al. 2006).

Success in selecting early for desirable traits may also be hampered by a mindset. Traditionally we seek to breed for a later-age trait, and measure it using a surrogate. What if we were to step back and think about what we have learned in wood quality, physiology and forest health? In terms of wood density and stiffness, we know in most tree species that outer wood (so-called mature wood) from older trees is not the problem, so why do we worry about age-age correlations in this instance? Breeding for thresholds in problem areas (e.g. corewood) may be more appropriate (Apiolaza et al. 2000). A much harder problem, however, is quantifying the trade-off between density and stem volume at the level of the crop at harvest age (Burdon 2010b), which has ramifications for setting breeding goals as well as for selection.

Recent examples of early selection success include studies by Luo et al. (2010) that showed that selection for volume at age two or three would work well in *Eucalyptus grandis*, with a rotation age of 5 years. Age eight has worked well for selecting *P. radiata* for growth and form and wood density in New Zealand (Dungey et al. 2006; King and Burdon 1991). Ye, T.Z. and K. Jayawickrama (pers. comm. 2012); Jayawickrama et al. (2011) have shown in Douglas-fir that selection age could be brought forward at least 1–9 years, from 10 years (family selection for height) (Johnson et al. 1997), based on extensive correlation analyses that showed acceptable correlations were obtained at age 9 from planting. However, when selecting for volume at age 50, selection was optimised at age 10 years, and DBH and volume at 12 and 11 years respectively. Gains per generation can be sacrificed in order to maximise gain per unit time.

4.3 Molecular Genetics and Breeding

A pervasive issue with using molecular biology in breeding forest trees is disconnection between the science and operational application. Very often, molecular geneticists believe that their technology can just be handed over to the tree breeder. The technology can be developed without operational breeding populations being properly considered, and often application of techniques is effectively limited to small, closed populations (cf Burdon and Wilcox 2011). The challenge is to develop these technologies with breeders, within an operational breeding context. Only if molecular techniques (e.g. QTLs, other markers, SNPs) can show clear applications, capturing gain that is otherwise unobtainable, or is cheaper and/or quicker to obtain the same gain than using field-based phenotyping, is when the technologies should be pursued. There is always the reciprocity with the technologies that need to be explored, whereby accurate and effective molecular technologies rely on high-quality phenotyping, usually in replicated trials across environments. In other words, pursuing molecular marker tool development may require more phenotyping of better quality than has been typical in a tree-breeding operation. There will always be a place for molecular genetics for the pursuit of knowledge, but if a real impact is to be made, it must be targeted, with a much more collaborative relationship with quantitative geneticists, breeders and the breeding programme.

The development of molecular tools for early selection has been, nevertheless, extremely exciting for breeders. Using genetic markers can be expected to increase the efficiency of phenotypic selection for low-heritability traits if particular chromosome sites exert large effects (Lande and Thompson 1990). The comparison of phenotypic selection with that using markers and phenotype showed that the latter could provide higher average gain compared to the former, for a trait heritability of 0.25 (Kumar and Garrick 2001). In reality, however, marker-associated effects are much harder to find for traits of lower heritability (e.g. growth and form; Burdon and Wilcox 2011; Lande and Thompson 1990), typically explaining only a very small proportion of the phenotypic variance in forest trees (Burdon and Wilcox 2011), and so can account for relatively little genetic variation. Compared with crop species where a substantial amount of the technology has been developed, most tree species are highly heterozygous, lack appreciable linkage disequilibrium except within families or among populations, and do not usually have ideal test populations and/or inbred lines available for rapid application. For this reason, there are no documented instances in forest tree breeding where markers have actually been applied in real populations and have delivered gain to the forest. Furthermore, the community of forest molecular geneticists globally are largely losing faith in this technology to deliver meaningful differences in the production populations, with a few exceptions, forest health and some wood properties. Resistance and/or tolerance genes are much more likely to be genes of large effect, explaining a relatively large amount of genetic variation, e.g. fusiform rust resistance in loblolly pine (Burdon and Wilcox 2011; Wilcox et al. 1996).

Molecular genetics research that has focussed on providing bio-markers for early selection is not a new science, but as with many developing areas in science, promises and expectations of important practical applications are usually followed by warranted criticisms when they do not deliver. Quantitative trait loci (QTL) mapping was once the 'hot idea', and it would surely improve and be more useful with better genetic markers (e.g., microsatellite markers/SSRs and even RFLP's). The switch to genome-wide marker-assisted selection (GWAS) and even gene-assisted selection, where markers and/or genes were tagged or linked to important breeding traits, has also not yet succeeded in delivering usable tools in forest tree breeding

(Burdon and Wilcox 2011). With these 'gene-based' technology approaches, it is unlikely they will have much impact in tree breeding unless we find genes of large effect.

While substantial efforts are still being made to improve trait prediction with GWAS approaches, a major move has been toward developing high-density markers across whole genome. Whole-genome sequences or 'shot gun sequences' are used to develop genomic selection tools, and this technology has raised the greatest hopes. A dense SNP (single nucleotide polymorphism) marker 'chip', with 1,000's of random SNP variants across the genome, is used to predict genomic breeding values, rather than breeding values based on field phenotypic data in replicated trials (Zapata-Valenzuela and Zaror 2011). Once these are discovered and validated to phenotypic traits in a particular population, the chip can be used in populations with the same or similar genetic structure. Grattapaglia and Kirst (2008) review the development of the technique in forest trees. The exciting point for breeders is that if the marker resource is adequate they can be 'trained' to new populations and to multiple traits. For the first time, it appears that a flexible approach may be evolving. However, the phenotypic data still needs to be of high quality, genotyping costs have to be relatively low, and as such will not come cheaply or quickly.

Genomic selection (GS) was aggressively pursued in in animal breeding, (e.g. Eisen 2007; Goddard and Hayes 2007; Togashi et al. 2011) and more recently in crop genetics (e.g. Thomson et al. 2012). In New Zealand, the importance of applying within a closed population emerged when SNP estimates from Holstein-Friesian dairy cattle data did not produce accurate breeding values in Jersey data, and vice versa (Harris et al. 2009). Although there have been some challenges so far for GS, it is rapidly progressing world-wide (Dekkers 2012; Duchemin et al. 2012; Haile-Mariam et al. 2012; Lillehammer et al. 2011; Sattler 2009; Weigel et al. 2012). The application of GS in forestry faces some special challenges, but as with developments in forest tree breeding we can learn from the early application in other species to optimise our research investment.

A current dichotomy between molecular and quantitative geneticists is seen in the move away from nucleus or 'elite'-based breeding populations, and this will have implications for GS (Dungey et al. 2009; Jayawickrama and Carson 2000; White et al. 1999). The principle of nucleus or 'elite' breeds is to achieve faster gain in a few specific traits. Multiple elite populations can deliver more specific gains, but also dilute overall progress in the breeding population. Movement away from closed elite or nucleus-based populations is largely based on work by Lstibůrek et al. (2004a); (Lstibůrek et al. 2004b) showing that an open-nucleus breeding strategy is more likely to deliver gain. Elites, on the other hand, can be the type of closed population where the application of marker-aided selection and/or GS should be possible.

Fingerprinting using markers, usually SSRs, has had immediate applications in tree breeding. Quality control is a big issue in seed orchards, archives and crossing orchards: genetic 'fingerprinting' enables mistakes in identification of progeny or grafts to be identified so that any dilution in gain can be eliminated. Fingerprinting can also be applied to identify how well the orchard is inter-pollinating, and how accurate the estimates of genetic gain for the orchard are. In some instances, fingerprinting can also serve to identify and/or confirm an individual's pedigree, i.e. who was the Mum and who was the Dad. Pedigree analysis was proposed by Lambeth et al. (2001) and has since been tested in a number of species, e.g., *Cryptomeria japonica, Populus,* Douglas-fir, larch and *Pinus radiata,* (El-Kassaby et al. 2010; Funda et al. 2008; Kumar et al. 2007; Moriguchi et al. 2009; Wheeler et al. 2006; Emily Telfer, pers. comm. 2012). Such applications of molecular techniques may be used for reconstructing the pedigree for an entire field trial, before analysis and ranking of the genotypes, recently described as 'breeding without breeding' by El-Kassaby and Lstibůrek (2009).

All these options promise advantages as tools for the tree breeder. But, as always, these technologies come with a cost. Phenotyping large numbers of individuals, necessary to obtain gain, costs money, and phenotyping costs, if greater than \$10–\$20 per tree, in large programs are typically not affordable. Additional costs from molecular tools without proven application and gain delivery will prevent rapid uptake of technology. This reflects the limitations of the forest industry: the crop is long-term, cash flow is usually limited, and high internal rates of return are required from many investors with relatively high discount rates (currently around 7–8 % in New Zealand). Risk versus gain is of course the final factor that has to be considered.

Finally, use of molecular technology for selection, while it stands to save costs eventually, will actually increase costs in the short term. In other words, using such technologies in tree breeding helps intensify domestication, but domestication entails accepting increased costs in order to obtain better returns.

5   Maximising Delivery of Genetic Gain Through Propagation

Clonal uniformity is an aspiration in a forestry production system. Foresters like to see plantations that are uniform, but what are the real advantages? In South Africa, the advantages are clear in clonal eucalypts (Clarke 2000). Clones with known qualities are deployed across sites where their performance is relatively reliable and predictable. These clones can then to be marketed to the log buyer for their known properties. To justify the investment in characterising clones and matching them to sites there must be a premium. Actually, forest growers may need such investment simply to remain competitive in the fibre marketplace.

Clonal forestry can be very important to a profitable and productive forestry operation. The primary advantage is that it can deliver the best material relatively quickly, resulting in higher operational gain, i.e. gains actually in the forest (Balocchi 1997; Wright et al. 1996). Importantly, however, clonal forestry is a means to deploy high-gain material from a robust breeding population, not a means of increasing gain on its own (King and Johnson 1989). Another advantage of clones lies in their uniformity within a stand, compared with the variation expected from seed lots involving several crosses, or even pair-crosses. Uniformity also requires little or no non-genetic variation (Burdon 1989b), which depends on everything else in the production system is being place, including effective clonal propagation, silviculture, genetic improvement, and nutrition. An understanding of wood quality and of the forest resource, and excellent establishment practices, are vital for lucrative matching of clonal stock to sites. If everything is right, substantial financial and operational gains can be made (Wright and Rosales 2000).

Examples of success include the canker-resistant hybrid eucalypt clones in Brazil (Comério and Iannelli 1997; Ferreira and Santos 1997) and eucalypt clones for pulp South Africa (Arbuthnot 2000; Clarke 2000; Duncan et al. 2000). Clones work well for these programmes because investment, persistence and faith in the ultimate success of these programs, as well as distinct biological advantages of the eucalypt hybrids involved (e.g. resistance to canker in Brazil, ability to propagate these sub-tropical species), have set them up to succeed. Short rotation ages also make economic and biological risk versus gain tradeoffs much easier to accommodate in a forestry operation.

Propagation, however, sometimes fails. Why? The research and small-scale systems very often succeed (Aimers-Halliday et al. 2003; Hargreaves et al. 2009, 2011; Menzies et al. 2000), but in scaling up to operational levels some attention to detail is lost. In *P. radiata* successful stool-bed systems have been available for some time (e.g. (Menzies et al. 1988), but the stool-bed management and lack of understanding of physiological age can often cause poor rooting, poor root quality and poor cutting performance (Aimers-Halliday et al. 2003; Menzies et al. 2000, 2001). Many species are notoriously difficult to propagate, e.g. temperate eucalypts and stringy barks (sub-genus *Symphyomyrtus*; Brooker (2000)) are very difficult, although not impossible. Some loss of genotype capture arises at all stages of propagation, including tissue culture, involving all the steps in the somatic embryogenesis process (Hargreaves et al. 2009) and in nursery stool bed systems (Aimers-Halliday and Burdon 2003). Loss of genotypes can erode genetic gain and/or genetic diversity, effectively adding ability to propagate to the list of selection traits. The consequent erosion of genetic gain can be especially severe if ability to propagate is in genetic trade-off with breeding-goal traits (Haines and Woolaston 1991).

The trick will be investing to address the key road-blocks, and a closer involvement of the operational nurseryman or tissue culture specialist, known simply as tech transfer. Why does this not always happen? It seems there is a gap in investment, the science usually attracts its investment, and the operational side usually attracts investment from within the company; however, there is usually very little attention paid by government agencies or industry themselves, in the investment for interaction between science teams and companies, to ensure operational implementation.

Success in propagation involves investment in appropriate infrastructure, choosing the right people to run the programs, and continued investment to overcome hurdles in scaling-up from initial research. The right, highly-skilled nurseryman knows how to delay maturation in the stool beds and how to keep tissue-culture plants alive and vigorous (Zhou et al. 1998). Yet there are instances where clonal forestry should not be pursued. These include severely degraded or nutrient-poor sites where genetic gains are not expressed. If the product the trees are being grown for does not require specific qualities or uniformity, then cloning may not be important. If the clones do not significantly exceed the population average, then investing in their deployment will not be justified by the expected returns. Managing risks of existing or new pests and diseases with clones will also be important. Unless clones confer additional advantages in terms of resistance or tolerance, then it may not be worth addressing challenges of achieving appropriate diversity in a clonal mix to provide a buffer against local incursions from pests or diseases.

6   Coping with Institutional and Funding Changes

With the increasingly complex issues that most tree breeding programs are now addressing, we definitely need programs equipped with a greater breadth of scientific skills and expertise. Our quantitative genetics training from the traditional texts like Becker (1985), Falconer (1960); Falconer and Mackay (1996) remain the 'backbone' of all programs, but gain was probably 'lost' in the early days of these programs, without techniques that are now routinely available. Statistical algorithms, computer power, computer simulations, experimental designs, etc., are vastly better than only a few decades ago, and all have contributed to improved genetic evaluations in breeding programs.

These developments in quantitative genetics have had an almost immediate effect on improving efficiency of estimating heritabilities, genetic correlations and breeding values estimates. However, and as mentioned previously, the new 'big sciences' (e.g., the recent 'omics research areas) promise large benefits but they also provide significant distractions from the core and established breeding programs, not to mention the an erosion of the education and training opportunities in traditional quantitative genetics (Knight 2003). Besides this generally creating confusion on where research dollars should be allocated, the net result has been a pervasive reduction in tree breeders' ability to focus on the development and delivery of genetic gain through the normal recurrent selection processes. To be sure, these new 'breakthrough' technologies have to be developed, and will be important once their appropriate roles are found in tree improvement, but much genetic gain has surely been lost for want of a better balance of investments in technologies for when and where they could and would be appropriate (Namkoong et al. 2005).

While we are all now familiar with and can see the proper places for many of these technological developments, such as with tissue culture, transgenics and genomics, all three actually provide additional challenges and require careful thinking about the applications. These techniques can actually magnify the problems of managing genetic diversity and long-term population improvement, but they can also be used to solve the problems. In the future breeders, academics, and those developing the higher technologies must work closer together to pinpoint how and where the technology can truly support the accumulation and delivery of genetic gain and yet better genetic resource management.

As noted above, the list of questions about the continued benefits of tree improvement, faced by forest managers, industry executives, government funding agencies, continues to evolve. Along with this evolution of the direction of breeding programs has been a global quest of governments and industry to make both academic and non-academic research more accountable for their research investments with respect to delivering 'practical outcomes'. Despite this narrow focus on research applications, we have notoriously little ability to predict or know when a basic scientific finding that appears largely academic may lead to plugging an important 'gap' in an applied area. Regardless, governments and their partnered funding agencies often impose many business-type measures and administrative systems on research agencies, and researchers themselves, to ensure that "research results are delivered". Overseeing research projects, programs and infrastructure investments, are numerous review committees, boards, councils, etc., which are intended to keep the science focussed on priorities of the day, but now create a debilitating drain on scientists, and worse, on the much need scientific exploration process (e.g., basic funding which allows for a sort of a 'free discovery' time, is generally lacking now). Once the laborious administrative procedures are developed to 'streamline reporting', further time commitments are in practice imposed on the researcher by requiring quarterly or even monthly progress reports for systems staffed with people who are ill-equipped to judge research 'progress' and to appreciate the day-to-day difficulties of doing original research. Typically, other researchers are then asked to provide reviews, which further tax the whole system.

Forestry, and many forest genetics research programs around the world have unfortunately fallen into this system. Needless to say, this additional burden of non-technical meetings, and bureaucratic report writing take a much time from the actual business of tree breeding, and ultimately impede actual genetic progress. A better balance is needed not only between genetic gain and genetic diversity targets in breeding population, but a better balance between funding and administrative agencies and the science and technical providers and managers is required. In other words, technical progress is thus double-edged for tree breeding, in that the advances in information technology can become instruments of counterproductive management systems.

7 Conclusions

Short-term industrial planning horizons, fickle markets and uncertain comparative values of different traits, and a relatively large diversion of resources over the last decade to molecular genetics research, have made tree breeding and other genetic resource management a very complex business now, and probably have had a substantial impact on our broader genetic development and management plans and activities. What is needed now is for the technologies to work together rather than compete.

References

Aimers-Halliday J, Burdon RD (2003) Risk management for clonal forestry with *Pinus radiata* – analysis and review. 2: technical and logistical problems and countermeasures. N Z J For Sci 33(2):181–204

Aimers-Halliday J, Menzies MI, Faulds T, Holden DG, Low CB, Dibley MJ (2003) Nursery systems to control maturation in *Pinus radiata* cuttings, comparing hedging and serial propagation. N Z J For Sci 33(2):135–155

Andersson EW, Spanos KA, Mullin TJ, Lindgren D (1998) Phenotypic selection compared to restricted combined index selection for many generations. Silva Fenn 32(2):111–120

Apiolaza LA (2009) Very early selection for solid wood quality: screening for early winners. Ann For Sci 66(6):601–610

Apiolaza LA, Garrick DJ, Burdon RD (2000) Optimising early selection using longitudinal data. Silvae Genet 49(4–5):195–200

Arbuthnot A (2000) Clonal testing of *Eucalyptus* at Mondi Kraft, Richards Bay. In: Forest genetics for the next millenium. IUFRO Working party 2.08.01. Tropical species breeding and genetic resources. Institute for Commonwealth Forestry Research, Durban, pp 61–64

Balocchi CE (1997) Realised versus operational gain: the role of propagation strategies. In: Burdon RD, Moore JM (eds) IUFRO working party S2.02.19 '97 Genetics of radiata pine. FRI Bulletin No 203. New Zealand Forest Research Institute, Rotorua, pp 253–255

Becker WA (1985) Manual of quantitative genetics, 4th edn. McNaughton & Gunn, Ann Arbor

Bentz BJ, Rgnire J, Fettig CJ, Hansen EM, Hayes JL, Hicke JA, Kelsey RG, Negrn JF, Seybold SJ et al (2010) Climate change and bark beetles of the western United States and Canada: direct and indirect effects. Bioscience 60(8):602–613

Brooker MIH (2000) A new classification of the genus *Eucalyptus* L'Her. (Myrtaceae). Aust Syst Bot 13(1):79–148

Burdon RD (1989a) Early selection in tree breeding: principles for applying index selection and inferring input parameters. Can J Forest Res 19:499–504

Burdon RD (1989b) Some economic issues in clonal forestry. In: Miller JT(ed) FRI/NZFP Forests Ltd Clonal Forestry Workshop. FRI Bulletin No. 150. New Zealand Forest Research Institute, Rotorua, pp 53–54

Burdon RD (ed Miller, JT) (1992) Introduced forest trees in New Zealand: recognition, role, and seed source. 12. Radiata pine *Pinus radiata* D. Don. FRI Bulletin – N Z J For Sci, Forest Research Institute 124(12):59

Burdon RD (1997) Genetic diversity for the future: conservation or creation and capture? In: Burdon RD, Moore JM (eds) IUFRO working party S2.02.19 '97 Genetics of radiata pine. FRI Bulletin No 203. New Zealand Forestry Research Institute, Rotorua, pp 238–247

Burdon RD (2001a) Genetic aspects of risk – species diversification, genetic management and genetic engineering. N Z J For 45(4):20–25

Burdon RD (2001b) Genetic diversity and disease resistance: some considerations for research, breeding, and deployment. Can J For Res 31(4):596–606

Burdon RD, Aimers-Halliday J (2003) Risk management for clonal forestry with *Pinus radiata* – analysis and review. 1: strategic issues and risk spread. N Z J For Sci 33(2):156–180

Burdon R, Walker J, Megraw B, Evans R, Cown D (2004a) Juvenile wood (*sensu novo*) in pine: Conflicts and possible opportunities for growing, processing and utilisation. N Z J For 49(3):24–31

Burdon RD, Kibblewhite PR, Walker JCF, Megraw RA, Evans R, Cown DJ (2004b) Juvenile versus mature wood: a new concept, orthogonal to corewood versus outerwood, with special reference to *Pinus radiata* and *P. taeda*. For Sci 50(4):399–415

Burdon RD (2008) Breeding radiata pine – future technical challenges. N Z J For 53(1):3–5

Burdon RD (2010a) Conservation and management of potentially resistant tree germplasm: a key but easily neglected part of a robust biosecurity strategy. N Z J For Sci 40:115–122

Burdon RD (2010b) Wood properties and genetic improvement of radiata pine. N Z J For 55(2):22–27

Burdon RD, Wilcox PL (2011) Integration of molecular markers in breeding. In: Plomion C, Bousquet J, Kole C (eds) Genetics, genomics and breeding of conifers. CRC Press, St Helier, pp 276–322

Burdon RD, Carson MJ, Shelbourne CJA (2008) Achievements in forest tree genetic improvement in Australia and New Zealand 10: *Pinus radiata* in New Zealand. Aust For 71(4):263–279

Caballero A (1994) Developments in the prediction of effective population size. Heredity 73(6):657–679

Carroll AL, Taylor S, Régnière J, Safranyik L (2004) Effects of climate change on range expansion by the mountain pine beetle in British Columbia (2004) M. In: Shore TL, Brooks JE, Stone JE (eds) Proceedings of the mountain pine beetle symposium: challenges and solutions, 30–31 Oct 2013, Kelowna BC Canada. Canadian Forest Service, Pacific Forestry Center, Victoria BC. Information Report BC-X-399, pp 223–232

Carson SD, Garcia O, Hayes JD (1999) Realized gain and prediction of yield with genetically improved *Pinus radiata* in New Zealand. For Sci 45(2):186–200

Clarke C (2000) Are *Eucalyptus* clones advantageous for the pulp mill? In: Forest genetics for the next millenium. IUFRO Working party 2.08.01. Tropical species breeding and genetic resources, Institute for Commonwealth Forestry Research, Durban, pp 45–48

Comério XA, Iannelli CM (1997) Eficiéncia da estaquia, da microestaquia e da micropropagação na clonagem de *Eucalyptus* spp/Efficiency of the cutting, microcutting and micropropagation processes in the cloning of *Eucalyptus* spp. *In* Proceedings of the IUFRO conference on silviculture and improvement of eucalypts. v2: biotechnology applied to genetic improvement of tree species. Embrapa, Colombo, pp 40–45

Cown DJ, Ball RD, Riddell MJC (2004) Wood density and microfibril angle in 10 *Pinus radiata* clones: distribution and influence on product performance. N Z J For Sci 34(3):293–315

Cudmore TJ, Björklund N, Carroll AL, Staffan Lindgren B (2010) Climate change and range expansion of an aggressive bark beetle: evidence of higher beetle reproduction in naïve host tree populations. J App Ecol 47(5):1036–1043

Danusevicius D, Lindgren D (2002a) Efficiency of selection based on phenotype, clone and progeny testing in long-term breeding. Silvae Genet 51(1):19–26

Danusevicius D, Lindgren D (2002b) Two-stage selection strategies in tree breeding considering gain, diversity, time and cost. For Genet 9(2):145–157

Danusevicius D, Lindgren D (2005) Optimization of breeding population size for long-term breeding. Scand J For Res 20(1):18–25

Dekkers JCM (2012) Application of genomics tools to animal breeding. Curr Genomics 13(3):207–212

Duchemin SI, Colombani C, Legarra A, Baloche G, Larroque H, Astruc JM, Barillet F, Robert-Granié C, Manfredi E et al (2012) Genomic selection in the French Lacaune dairy sheep breed. J Dairy Sci 95(5):2723–2733

Duncan EA, van Deventer F, Kietzka JE, Lindley RC, Denison N (2000) The applied subtropical *Eucalyptus* clonal programme in Mondi forests, Zululand coastal region. In: Forest genetics for the next millenium. IUFRO Working party 2.08.01. Tropical species breeding and genetic resources. Institute for Commonwealth Forestry Research, Durban, pp 95–97

Dungey HS, Matheson AC, Kain D, Evans R (2006) Genetics of wood stiffness and its component traits in *Pinus radiata*. Can J For Res 36(5):1165–1178

Dungey HS, Brawner JT, Burger F, Carson M, Henson M, Jefferson P, Matheson AC et al (2009) A new breeding strategy for *Pinus radiata* in New Zealand and New South Wales. Silvae Genet 58(1–2):28–38

Eckard JT, Isik F, Bullock B, Li B, Gumpertz M (2010) Selection efficiency for solid wood traits in *Pinus taeda* using time-of-flight acoustic and micro-drill resistance methods. For Sci 56(3):233–241

Eisen EJ (2007) Animal breeding: what does the future hold? Asian-Australas J Anim Sci 20(3):453–460

El-Kassaby YA, Benowicz A (2000) Effects of commercial thinning on genetic, plant species and structural diversity in second growth Douglas-fir (*Pseudotsuga menziesii* (Mirb.) Franco) stands. For Genet 7(3):193–203

El-Kassaby YA, Lstibůrek M (2009) Breeding without breeding. Genet Res 91(2):111–120

El-Kassaby YA, Ritland K (1996) Impact of selection and breeding on the genetic diversity in Douglas-fir. Biodivers Conserv 5(6):795–813

El-Kassaby YA, Funda T, Lai BSK (2010) Female reproductive success variation in a *Pseudotsuga menziesii* seed orchard as revealed by pedigree reconstruction from a bulk seed collection. J Hered 101(2):164–168

Falconer DS (1960) Introduction to quantitative genetics. Longman Scientific & Technical, Essex

Falconer DS, Mackay TFC (1996) Introduction to quantitative genetics. Addison Wesley Longman Ltd, Essex

Fernández J, Toro MA (2001) Controlling genetic variability by mathematical programming in a selection scheme on an open-pollinated population in *Eucalyptus globulus*. Theor Appl Genet 102(6–7):1056–1064

Ferreira M, Santos PETd (1997) Melhoramiento genético florestal dos *Eucalyptus* no Brasil–Breve histórico e perspectivas/Genetic improvement of *Eucalyptus* in Brazil– Brief review and perspectives. In: Proceedings of the IUFRO conference on silviculture and improvement of eucalypts v1:Tree improvement strategies. Embrapa, Colombo, pp 14–34

Funda T, Chen CC, Liewlaksaneeyanawin C, Kenawy AMA, El-Kassaby YA (2008) Pedigree and mating system analyses in a western larch (*Larix occidentalis* Nutt.) experimental population. Ann For Sci 65(7):705. doi:10.1051/forest:2008055

Gea LD, Lindgren D, Shelbourne CJA, Mullin T (1997) Complementing inbreeding coefficient information with status number: implications for structuring breeding populations. N Z J For Sci 27(3):255–271

Goddard ME, Hayes BJ (2007) Genomic selection. J Anim Breed Genet 124(6):323–330

Grattapaglia D, Kirst M (2008) *Eucalyptus* applied genomics: from gene sequences to breeding tools. New Phytol 179(4):911–929

Haile-Mariam M, Nieuwhof GJ, Beard KT, Konstatinov KV, Hayes BJ (2012) Comparison of heritabilities of dairy traits in Australian Holstein-Friesian cattle from genomic and pedigree data and implications for genomic evaluations. J Anim Breed Genet. doi: 10.1111/j.1439-0388.2012.01001.x

Haines RJ, Woolaston RR (1991) The influence of reproductive traits on the capture of genetic gain. Can J For Res 21(2):272–275

Hallander J, Waldmann P (2009) Optimum contribution selection in large general tree breeding populations with an application to Scots pine. Theor Appl Genet 118(6):1133–1142

Hargreaves CL, Reeves CB, Find JI, Gough K, Josekutty P, Skudder DB, van der Maas SA, Sigley MR, Menzies MI, Low CB, Mullin TJ (2009) Improving initiation, genotype capture, and family representation in somatic embryogenesis of *Pinus radiata* by a combination of zygotic embryo maturity, media, and explant preparation. Can J For Res 39(8):1566–1574

Hargreaves CL, Reeves CB, Find JI, Gough K, Menzies MI, Low CB, Mullin TJ (2011) Overcoming the challenges of family and genotype representation and early cell line proliferation in somatic embryogenesis from control-pollinated seeds of *Pinus radiata*. N Z J For Sci 41:97–114

Harris BL, Johnson DL, Spelman RJ (2009) Genomic selection in New Zealand and the implications for national genetic evaluation. In: Sattler JD (ed) 36th ICAR Session. ICAR, Via G. Tomassetti 3, 1/A, 00161 Rome, pp 325–330

Hubert C, Bastien C (1999) Genetic benefits, economic risks, ecological risks: how are they connected? [Gain génétique, risque économique, risque écologique: Quels liens ?]. Revue Forestière Française 51(4):496–510

Jayawickrama KJS (2001) Genetic parameter estimates for radiata pine in New Zealand and New South Wales: a synthesis of results. Silvae Genet 50(2):45–53

Jayawickrama KJS, Carson MJ (2000) A breeding strategy for the New Zealand radiata pine breeding cooperative. Silvae Genet 49(2):82–90

Jayawickrama KJS, Ye TZ, Howe GT (2011) Heritabilities, intertrait genetic gorrelations, GxE interaction and predicted genetic gains for acoustic velocity in mid-rotation coastal Douglas fir. Silvae Genet 60(1):8–18

Johnson GR, Sniezko RA, Mandel NL (1997) Age trends in Douglas-fir genetic parameters and implications for optimum selection age. Silvae Genet 46(6):349–358

Kang KS, Lindgren D, Mullin TJ (2001) Prediction of genetic gain and gene diversity in seed orchard crops under alternative management strategies. Theor Appl Genet 103(6–7):1099–1107

Kang KS, Cheon BH, Han SU, Kim CS, Choi WY (2007) Genetic gain and diversity under different selection methods in a breeding seed orchard of *Quercus serrata*. Silvae Genet 56(6):277–281

Kerr RJ, Goddard ME, Jarvis SF (1998) Maximising genetic response in tree breeding with constraints on group coancestry. Silvae Genet 47(2–3):165–173

King JN, Burdon RD (1991) Time trends in inheritance and projected efficiencies of early selection in a large 17-yr-old progeny test of *Pinus radiata*. Can J For Res 21(8):1200–1207

King JN, Johnson, GR (1989) Computer simulation of clonal selection gains. In: Miller JT (ed) FRI/NZFP Forests Ltd Clonal Forestry Workshop. FRI Bulletin No. 150. New Zealand Forest Research Institute, Rotorua, pp 51–52

Knight J (2003) A dying breed. Nature 421(6923):568–570

Krakowski J, Stoehr MU (2009) Coastal douglas-fir provenance variation: patterns and predictions for British Columbia seed transfer. Ann For Sci 66(8):811, p811-811p810

Kumar S, Dungey HS, Matheson AC (2006) Genetic parameters and strategies for genetic improvement of stiffness in radiata pine. Silvae Genet 55(2):77–84

Kumar S, Garrick DJ (2001) Genetic responses to within-family selection using molecular markers in some radiata pine breeding schemes. Can J For Res 31:779–785

Kumar S, Gerber S, Richardson TE, Gea L (2007) Testing for unequal paternal contributions using nuclear and chloroplast SSR markers in polycross families of radiata pine. Tree Genet Genomes 3(3):207–214

Lambeth C, Lee BC, O'Malley D, Wheeler N (2001) Polymix breeding with parental analysis of progeny: an alternative to full-sib breeding and testing. Theor Appl Genet 103(6–7): 930–943

Lambeth C, McKeand S, Rousseau R, Schmidtling R (2005) Planting nonlocal seed sources of loblolly pine – managing benefits and risks. South J Appl For 29(2):96–104

Lande R, Thompson R (1990) Efficiency of marker-assisted selection in the improvement of quantitative traits. Genetics 124(3):743–756

Lillehammer M, Meuwissen THE, Sonesson AK (2011) A comparison of dairy cattle breeding designs that use genomic selection. J Dairy Sci 94(1):493–500

Lindgren D, Mullin TJ (1997) Balancing gain and relatedness in selection. Silvae Genet 46(2–3):124–129

Lindgren D, Gea L, Jefferson P (1996) Loss of genetic diversity monitored by status number. Silvae Genet 45(1):52–58

Lipow SR, Johnson GR, St. Clair JB, Jayawickrama KJ (2003) The role of tree improvement programs for *ex situ* gene conservation of coastal Douglas-fir in the Pacific Northwest. For Genet 10(2):111–120

Lstibůrek M, Mullin TJ, Lindgren D, Rosvall O (2004a) Open-nucleus breeding strategies compared with population-wide positive assortative mating: I. Equal distribution of testing effort. Theor Appl Genet 109(6):1196–1203

Lstibůrek M, Mullin TJ, Lindgren D, Rosvall O (2004b) Open-nucleus breeding strategies compared with population-wide positive assortative mating: II. Unequal distribution of testing effort. Theor Appl Genet 109(6):1169–1177

Luo J, Zhou G, Wu B, Chen D, Cao J, Lu W, Pegg RE, Arnold RJ et al (2010) Genetic variation and age-age correlations of *Eucalyptus grandis* at Dongmen Forest Farm in southern China. Aust For 73(2):67–80

Macias Fauria M, Johnson EA (2009) Large-scale climatic patterns and area affected by mountain pine beetle in British Columbia, Canada. J Geophys Res 114(G1):G01012. doi:10.1029/200 8JG000760

Menzies MI, Aimers JP, Whitehouse, LJ (1988) Workshop on growing radiata pine from cuttings. FRI Bulletin – New Zealand Forest Service, Forest Research Institute 135

Menzies MI, Dibley MJ, Faulds T, Aimer-Halliday J, Holden DG (2000) Morphological markers of physiological age for *Pinus radiata*. N Z J For Sci 30(3):359–364

Menzies MI, Holden DG, Klomp BK (2001) Recent trends in nursery practice in New Zealand. New For 22(1–2):3–17

Moriguchi Y, Ishiduka D, Kaneko T, Itoo S, Taira H, Tsumura Y (2009) The contribution of pollen germination rates to uneven paternity among poly crosses of *Cryptomeria japonica*. Silvae Genet 58(3):139–144

Namkoong G, Lewontin RC, Yanchuk AD (2005) Plant genetic resource management: the next investments in quantitative and qualitative genetics. Genet Resour Crop Evol 51(8):853–862

Rosvall O, Mullin TJ (2003) Positive assortative mating with selection restrictions on group coancestry enhances gain while conserving genetic diversity in long-term forest tree breeding. Theor Appl Genet 107(4):629–642

Rosvall O, Mullin TJ, Lindgren D (2003) Controlling parent contributions during positive assortative mating and selection increases gain in long-term forest tree breeding. For Genet 10(1):35–53

Sattler JD (2009) ICAR technical series No 13. Identification, breeding, production, health and recording of farm animals. In: Sattler JD (ed) 36th biennial ICAR Session. ICAR, Via G. Tomassetti 3, 1/A, 00161 Rome, pp 325–330

Shelbourne CJA, Carson MJ, Wilcox MD (1989) New techniques in the genetic improvement of radiata pine. Commonw For Rev 68(3):191–201

Shelbourne CJA, Low CB, Gea LD, Knowles RL (2004) Achievements in forest tree genetic improvement in Australia and New Zealand 5: genetic improvement of Douglas-fir in New Zealand. Aust For 70:28–32

Shelbourne CJA, Kumar S, Burdon RD, Gea LD, Dungey HS (2007) Deterministic simulation of gains for seedling and cloned main and elite breeding populations of *Pinus radiata* and implications for strategy. Silvae Genet 56(6):259–270

Stoehr MU, El-Kassaby YA (1997) Levels of genetic diversity at different stages of the domestication cycle of interior spruce in British Columbia. Theor Appl Genet 94(1):83–90

Stoehr M, Yanchuk A, Xie CY, Sanchez L (2008) Gain and diversity in advanced generation coastal Douglas-fir selections for seed production populations. Tree Genet Genomes 4(2):193–200

Stoehr M, Bird K, Nigh G, Woods J, Yanchuk A (2010) Realized genetic gains in coastal Douglas-fir in British Columbia: implications for growth and yield projections. Silvae Genet 59(5):223–233

Thomson MJ, Zhao K, Wright M, McNally KL, Rey J, Tung C, Reynolds A, Scheffler B, Eizenga G, McClung A, Kim H, Ismail AM, de Ocampo M, Mojica C, Reveche MY, Dilla-Ermita CJ, Mauleon R, Leung H, Bustamante C, McCouch SR (2012) High-throughput single nucleotide polymorphism genotyping for breeding applications in rice using the BeadXpress platform. Mol Breed 29(4):875–886

Togashi K, Lin CY, Yamazaki T (2011) The efficiency of genome-wide selection for genetic improvement of net merit. J Anim Sci 89(10):2972–2980

Wei RP, Lindgren D (2006) Stepwise penalty index selection from populations with a hierarchical structure. Silvae Genet 55(2):62–70

Weigel KA, Hoffman PC, Herring W, Lawlor TJ (2012) Potential gains in lifetime net merit from genomic testing of cows, heifers, and calves on commercial dairy farms. J Dairy Sci 95(4):2215–2225

Wheeler N, Payne P, Hipkins V, Saich R, Kenny S, Tuskan G (2006) Polymix breeding with paternity analysis in *Populus*: a test for differential reproductive success (DRS) among pollen donors. Tree Genet Genomes 2(1):53–60

White TL, Matheson AC, Cotterill PP, Johnson RG, Rout AF, Boomsma DB (1999) A nucleus breeding plan for radiata pine in Australia. Silvae Genet 48(3–4):122–133

White TL, Adams WT, Neale DB (2007) Forest genetics. CABI Publishing, CAB International, Oxfordshire

Wilcox PL, Amerson HV, Kuhlman EG, Liu BH, O'Malley DM, Sederoff RR (1996) Detection of a major gene for resistance to fusiform rust disease in loblolly pine by genomic mapping. Proc Natl Acad Sci U S A 93(9):3859–3864

Wright JA, Rosales L (2000) Clonal eucalypt forestry in Venezuela. In: 2000 pulping/process and product quality conference. TAPPI, pp 783–787. http://www.scopus.com/inward/record.url?eid=2-s2.0-0142218968&partnerID=40&md5=48ace2ef8728cdb589742c9a5a3c4648

Wright JA, Osorio LF, Dvorak WS (1996) Realised and predicted genetic gain in the *Pinus patula* breeding program of Smurfit Carton de Colombia. S Afr For J 175(1):19–22

Yanchuk AD (2001) A quantitative framework for breeding and conservation of forest tree genetic resources in British Columbia. Can J For Res 31(4):566–576

Yanchuk AD, Sanchez L (2011) Multivariate selection under adverse genetic correlations: impacts of population sizes and selection strategies on gains and coancestry in forest tree breeding. Tree Genet Genomes 7(6):1169–1183

Zapata-Valenzuela J, Zaror RH (2011) Accelerated forest genetic breeding using genomic selection. (Mejoramiento genético forestal acelerado mediante selección genómica). Bosque 32(3):209–213

Zhang Y, Chu DY, Jin GQ, Zhou ZC, Qin GF (2009) Molecular characterization of elite genotypes within a first generation breeding population of *Pinus massoniana* using ISSR. For Res 22(6):772–778

Zheng YQ, Lindgren D, Rosvall O, Westin J (1997) Combining genetic gain and diversity by considering average coancestry in clonal selection of Norway spruce. Theor Appl Genet 95(8):1312–1319

Zhou T, Zhou J, Shelbourne CJA (1998) Clonal selection, propagation, and maintenance of juvenility of Chinese fir, and afforestation with monoclonal blocks. N Z J For Sci 28(3):275–292

Seed Orchards and Aspects on Supporting Tree Breeding

Dag Lindgren

Abstract The main quantitative output and work horse for plant breeding today and accumulated efforts since half a century is seed orchards. This document do not cover deployment of vegetatively propagated material or control crosses (less than 2 % of number of Swedish forest plants produced) or GMO or genomic based selections (zero plant production) and focus mainly on Sweden. Seed orchards established now are mainly with tested grafted clones. With Norway spruce, where vegetative propagation of young plants is easy, current testing and seed orchard deployment is based on testing clonal performance. For Scots pine it is progeny tested clones, but progeny testing is a painfully slow and seemingly inefficient procedure. The number of clones is typically 20 or slightly more when clones are unrelated. It is more efficient to deploy clones in different proportions and it is not economic to strive for equal proportions. Pollen contamination is an important aspect of seed orchards, a practical remedy has not been found. However, seed orchard crops from genetically young seed orchards with 100 % contamination are still better than crops from mature but genetically outdated alternatives. Earlier deployed clones were unrelated, but it seems to become inefficient avoiding related clones after the first generations. Genetic thinning is rare and difficult to defend from a gain point of view, but selective harvesting becomes increasingly common. The breeding population (typically 1,000) is shared in compartments (typically 50) and seed orchards draw on several compartments. Probably both seed orchards and breeding would benefit from a larger "breeding population". There would be advantages if breeding efforts and seed orchard establishment could be better synchronized. New cohorts of recently selected clones should be deployed to pine seed orchards more often, they tend to be genetically worn out and expensive to harvest.

D. Lindgren (✉)
Swedish University of Agricultural Sciences, Umeå, Sweden
e-mail: Dag.Lindgren@slu.se

T. Fenning (ed.), *Challenges and Opportunities for the World's Forests*
in the 21st Century, Forestry Sciences 81, DOI 10.1007/978-94-007-7076-8_20,
© Springer Science+Business Media Dordrecht 2014

Projections of the impact of seed orchards on the national forest harvest almost a century ahead in Sweden is 10 % assuming no technology change. The possible "ecological risks" with the seed orchard technology seem a small addition to that of plantation forestry.

1 Introduction

Seed orchards are still the major quantitative output and work horse for forest tree breeding. That is in spite of a more than half a century of history in support industrial conifer forestry from economically important programs and tremendous efforts to create more efficient outlets for the breeding effort. Seed orchards themselves have developed surprisingly little. Seed yield is typically less than a percent of what a grain farmer could produce. Less than a few percent of Swedish forest plant production uses clonal propagation techniques or controlled crosses. Even in conditions with long rotation times, the investment in seed orchards seems worthwhile. Swedish forestry supports seed orchard programs covering the full seed need. Some general references about seed orchards are Kang (2001), Prescher (2007) and Lindgren (2008).

2 Seed Orchards Have Advantages Besides Gain from Tree Breeding

Seed orchards generally bring together genotypes from different stands, which are not related and thus inbreeding and inbreeding depression tends to be lower than for stand seeds. Very probably there is a stimulating effect from heterosis. Seed orchards are a more reliable seed supply, in particular if origins from other areas or countries perform better than the areas intended for reforestation by the seed orchard.

Good seeds. Seed orchards are established in good climatic environment and suitable ground conditions. They are closely managed. Cone collection and seed storage is kept under good control. The seeds will usually be heavier, better filled and healthier, germinate faster and more uniform compared to stand seeds. The nursery crop will also develop faster and be more uniform. That results in plant crops which develop faster and more predictably, and are therefore easier to manage and more uniform. This results in a better tree crop which performs better both in growth and survival and gives a more uniform forest plantation. The physiological superiority of the seeds leads often to a stand that performs some percent better, besides the genetic advantages. If used for direct seeding the higher quality of the seeds will produce a better result with fewer seeds.

The Swedish Scots pine seed orchard Västerhus. Photo: Dag Lindgren

Generally seed orchards are pruned to limit their height, control crowns and sometimes to get cones down on ground. Thus cone harvest costs can be kept low. The seed supply will also often be larger, more reliable, more uniform and more predictable.

Domestication. Most modern seed orchards are based on selections based on results from experimental plantations. That means that trees are selected in – and often based on results from – environments, which are similar to what will be used in practical forestry. This in itself has an improvement effect.

2.1 Flexibility

The client area and genetic output of a seed orchard is not engraved in stone once the establishment decision is made. If the environment changes, perhaps because of a warmer climate, the target area of the seed orchard can be modified. The genetic characteristics of the crop will change when the pollen production raises and pollen contamination decreases, thus a young orchard may be used for a different target area than a mature one. Genetic and selective harvesting thinning can improve the genetic quality of the crop. Supplementary pollination can improve the genetic quality of seed orchard crop and artificial crosses can readily be made in a seed orchard.

For some cases it may be beneficial to amplify the best part of the harvest by clonal propagation. Old seed orchards should be replaced with newer genetically improved material. But seed demand, the need for back-ups and often the need for ground for establishing new seed orchards regulate when old seed orchards are finally retired.

2.2   Seedling Seed Orchards

Most seed orchards are clonal, but seedling seed orchards constitute an alternative. Selfing will be less common than in clonal seed orchards, as there are no genetically identical replications. But inbreeding caused by milder relatedness will be more common, and this may be more harmful as it is more likely to result in marginally handicapped trees, which may ultimately cause larger production losses in the mature forest, as selfed genotypes seldom survive to maturity in a forest. Seeds collected following open pollination in selected trees are often used for seedling seed orchards. They do not have selected fathers, in a forest the pollen parents are unselected trees and even in a seed orchard many of the fathers are not from selected trees (contamination), this reduce the gain compared to grafts of selected trees. But seedling seed orchards can be based on controlled cross of selected trees, when it does not matter for gain if selections or their progeny are placed in a seed orchard. Seedling seed orchards can be combined with a tree improvement programme, so the orchard can fill several purposes. The seed, and in particular the pollen production, is delayed as compared to a grafted seed orchard with mature trees. This disadvantage may look greater than it is in the long run and in advanced generations trees will be less mature at selection and when the difference is probably small.

2.3   Protected Seed Production

For some species and circumstances seeds are produced under protected conditions like plastic tents. This offer advantages, including eliminating pollen contamination, but has not been wide-spread for most major species.

2.4   Pollen Contamination

Fertilizing pollen from outside the seed orchard is often an important factor. General and efficient remedies have been difficult to find where this contamination is large. Reducing or eliminating such contamination will probably get a higher priority, as long-term breeding raises the potential gain. Often seed orchards operators have to live with and adapt to pollen contamination: Deliberately choose locations where the expected contamination has a reasonable genetics; reduce the contamination with available methods even if they have limited effects; adjust the target area to predicted contamination; and welcome the positive effects of contamination (more seeds, more diversity and less inbreeding)!

2.5   Monoclonal Harvests

Clonal seed orchards may be harvested by clone. More uniform seeds are obtained, which is an advantage for plant production. Genetic differences between clones can be utilized. Genetically better clones can be used where the superiority results in the highest advantages, probably where site index is high. Seed price may be related to genetic quality. Lower ranking clones may be kept in storage as reserves. Still lower ranking clones may not be harvested years when the seed need is limited. Harvests from low ranking clones may be used for direct seeding. When a seed orchard becomes genetically outdated, seeds harvested from the best clones may still be competitive.

2.6   Genetic Diversity

Genetic diversity usually results in higher and safer biological production, but uniformity is sometimes preferred because a more uniform product; faster deployment of gain; and easier management. Genetic diversity is often demanded by legislation and it is desirable for public, costumer and authority acceptance or it may be the policy of the owner. Often diversity is expressed just as a clone number, it is more informative with some variant of "effective number". Extra diversity in a seed orchard offers options for later genetic thinning and selective harvesting. Sometimes seed orchards are blamed for their ecological impact, but plantation forestry may be the proper target for such criticism rather than tree improvement itself.

2.7   Genetic Diversity in Seed Orchards and Supporting Breeding

The reason for long term tree breeding is to support future deployment of genetically improved material to forestry. For most situations that means seed orchards, or at least that the option remains open. Genetic diversity in the breeding stock is the raw material of the breeder and a valuable resource which should be managed with care in a sustainable way and not exhausted. The genetic diversity and relatedness in the breeding stock should allow the best genetic material to be creamed for deployment to seed orchards with a high degree of improvement and sufficient diversity. To be able to make efficient selections with limited relatedness many generations ahead, it is important that the breeding population is large enough to limit genetic drift and that build-up of relatedness within the breeding population is minimal. A breeding program may appear more efficient in the short time if it sacrifices diversity for gain, but much of that extra gain may be lost later, as the gain in the final step to deployment as seed production will be reduced.

Both in seed orchards and in the supporting breeding programme it is beneficial if deployment from the recruitment population utilizes somewhat more of the genetically better genomes and founder genomes than is done by truncation (either select or reject). It is not efficient or practical to overemphasize equal utilization of clones or genomes, but better to use them in a more gradual way.

The most attractive share of the recruitment population for seed orchard deployment will become more related as generations pass. It is possible to arrange the breeding stock so for example twenty unrelated selections will remain available (sublining), but this does not look like good breeding economy. It seems more optimal to use few sublines and allow some slight inbreeding by deploying relatives. To keep such inbreeding low is an argument in favor of starting long term breeding efforts with many genotypes and restricting the loss of diversity in long term breeding.

Some current typical values for the major conifers in Sweden are given: Typical young seed orchards recruited from the breeding population have around 20 clones deployed in slightly different proportions. The Swedish national breeding metapopulation comprises slightly more than 1,000 genotypes structured in subpopulations of 50 each. Selection has a large within family component. The subpopulations have slightly different target areas and characteristics. A seed orchard is typically recruited from around four adjacent subpopulations. It has been projected that seed orchards will raise the national annual forest harvest in Sweden by about 10 % till year 2100.

2.8 Clonal Testing

Long term breeding is theoretically most efficiently performed by using clonal copies rather than being guided by progeny performance, but of course information from relatives should also be used when primary testing is done using clones. Tested clones may be used as parent trees in seed orchards. Alternatively, superior tested clones in seed orchards may be mated and the seeds (or embryos) clonally propagated, in that way the latest information from clone testing may reach the forest much faster than would otherwise be the case. Such approaches are utilized with Norway spruce in Sweden.

2.9 Seed Orchard Rotation

It is generally inefficient to use a seed orchard for a very long time. One reason is that the cost of harvesting the seeds and the effort required often increases steeply with the height of the trees, which increases as the trees age, of course. Harvest costs are often the dominating factor in the cost of the seeds, usually more than the capital costs. Another reason is that a seed orchard gets genetically outdated, as newly improved breeding stock becomes available. Normally a seed orchard program should be supported by a long term breeding program, and the potential to harvest gain from that breeding program increases by time. The age of some Swedish seed orchards seem to indicate that optimally short rotation has not been planned for. Seed orchard establishment should interface with breeding efforts so the breeding stock can be efficiently creamed at the right moment. Replacement of retired seed orchards serves to introduce new clones as parents to forests, and thus contribute to genetic diversity on the regional level.

It takes considerable time for pollen production to rise in a seed orchard, but if the breeding stock is good and the location has a genetically reasonable pollen cloud, seeds from a young seed orchard may still be genetically superior to seeds that can be obtained from alternative seed sources. It may be a good idea to maintain the mature seed orchard for some years extra in case the performance of the new orchards should fall below prediction or the demand for seed increases. The seeds harvested from the best clones from the overly mature seed orchard might also still be genetically compatible with harvests from a young seed orchard.

2.10   Seed Orchards and the Future of Mankind

Forests create the very materials that Man needs from air, water and sunshine and recycle it back to air and water after use. Seed orchard forestry is not mining, but sustainable and renewable; and actually better than that. Seed orchards improve the green production apparatus each time they are harvested and create additional resources in an environmental friendly way. Seed orchards form the foundation for much of the future physical and ecological realities in our countries and are part of what we give to the people coming after us, and so may be seen as a moral activity.

Seed orchards are used for land managed as production or cultivated forest by planting or direct seeding, but good seed orchards raise the production on such land, and thus give more room for other land uses including "conservation" needs. Better economic output from cultivated forests will also offer forest operators more room for using land for other purposes.

Seed orchards is a safe and mature technology (although still with room for improvement) and is able to benefit from the development of improved technologies like genomics, flower stimulation, exotics and GMO, but are not dependent on them. Clonal propagation or crossing technology may replace seed orchards, geneticists certainly see advantages in that. But most of these high expectations have failed in the past, and a cautious attitude seems advisable.

References

Kang, KS (2001) Genetic gain and gene diversity of seed orchard crops. Acta Universitatis Agriculturae Sueciae. Silvestria 187 75pp+11 chapters

Lindgren, D (ed) (2008) Proceedings of a seed orchard conference, Umeå, Sweden, 26–28 September 2007. ISBN: 978-91-85911-28-8. 256 pp

Prescher, F (2007) Seed orchards – genetic considerations on function, management and seed procurement. Acta Universitatis Agriculturae Sueciae 2007:75. ISBN: 978-91-576-7374-9

Part VI
Biotechnological Approaches

Biosafety Considerations in the Context of Deployment of GE Trees

Hely Häggman, Suvi Sutela, Christian Walter, and Matthias Fladung

Abstract According to FAO world population will increase from the current seven to nine billion by 2050. This combined with ongoing climate change will lead to increased demands for land resources for food and feed production and subsequently to changes in land use from forestry to agricultural purposes. With an increased awareness of the importance of native forests for the world's climate, harvesting from these forests is expected to decrease and conversions of native forest land to agricultural land may also become undesirable. Another factor for consideration will be an increased demand on land resources for the production of biofuels and bioproducts. In future, forests will be planted for raw materials for the pulp and paper industry and fiber production along with providing these new resources. In this scenario, productivity, in particular that of planted forests needs to be increased significantly, while at the same time native forests must be protected from further exploitation. Genetic engineering offers a potential to significantly and in relatively short time frames increase volume and quality of forest-based raw materials, thus, providing options to reduce world wide consumption of petrochemicals and increase the use of sustainable resources. Since forests can be grown on marginal lands, competition with land resources suitable for agricultural production can be avoided. At the same time, the increased productivity from bioengineered forests will provide an option to protect native forests.

H. Häggman (✉) • S. Sutela
Department of Biology, University of Oulu, P.O. Box 3000, FI-90014 Oulu, Finland
e-mail: Hely.Haggman@oulu.fi

C. Walter
The New Zealand Forest Research Institute Ltd, 49 Sala Street, Rotorua, New Zealand
e-mail: Christian.Walter@scionresearch.com

M. Fladung
Thünen-Institute of Forest Genetics, D-22927 Grosshansdorf, Germany
e-mail: matthias.fladung@ti.bund.de

T. Fenning (ed.), *Challenges and Opportunities for the World's Forests in the 21st Century*, Forestry Sciences 81, DOI 10.1007/978-94-007-7076-8_21, © Springer Science+Business Media Dordrecht 2014

The biosafety of genetically modified (GM) forests is an important consideration given the benefits expected from these plantations and the scale to which this will need to happen to have a meaningful impact. We assess the current scientific knowledge around the environmental safety of GM trees and discuss this in the context of expected environmental and economic benefits, and in the context of risk associated with accepted conventional tree breeding and forestry practices.

1 Introduction

According to the Food and Agriculture Organisation FAO report (2010) of Global Forest Resource Assessment, global forests cover 4 billion hectares i.e. 3 % of the total habitable land area. The forest area is, however, not equally distributed but five countries including the Russian Federation, Brazil, Canada, the United States of America and China account for more than half of the total forest area. The rate of deforestation and in particular the conversion of tropical forests to agricultural land, shows signs of decreasing, but is still alarmingly high. Nevertheless, afforestation and natural expansion of forests in some countries and regions have significantly reduced the net loss of forest area (i.e. from −8.3 million hectares per year between 1990 and 2000 to −5.2 million between 2000 and 2010) at the global level. Outbreaks of forest insect pests damage some 35 million hectares of forest annually, primarily in the temperate and boreal zone. At present, legally established protected areas cover around 13 % of the world's forest. 30 % i.e. 1.2 billion hectares of the world's forests are primarily used for production of wood and non-wood forest products (FAO 2010).

One main driver of future forestry is the increasing population which recently (August 2013) exceeded seven billion and it is expected to exceed nine billion by 2050. Increasing population numbers in combination with accelerated climate change including weather extremes, floods, droughts etc. (Nellemann et al. 2009) are predicted to increase the need for more food production. The current estimates of FAO's high level expert forum "How to Feed the World in 2050" in Rome (2009) (http://www.fao.org/wsfs/forum2050/wsfs-forum/en/) indicated that food production must increase by 70 % in the next 40 years and if more land is not available, 370 million people could be facing famine by 2050. This will also push new demands for increases in plantation and productivity for future forestry. This scenario was already realized by Fenning and Gershenzon (2002) and they highlighted the role of plantation forestry on a long-term basis to limit the harvest pressures on natural forests. For improved yield of the plantations, the authors suggest biotechnology tools to be applied to tree-improvement processes in order to save time.

The total land area designated primarily for wood production has decreased by more than 50 million hectares during the last 20 years and the area designated for multiple uses has increased by 10 million hectares during the same period. Simultaneously, 330 million hectares globally are designated for soil and water conservation (protective functions) and the management of forests for social, cultural and recreational functions is increasing. These trends indicate that the future forest area will have to serve multiple purposes including conservation, recreation and wood production. There is a definite need for efficient timber production using marginal land which is not suitable for economically viable agricultural uses. This is a major driver for applying insights and technologies from quantitative and molecular tree breeding programs. This will have to include the deployment of bioengineered planting stock in effectively managed plantation forests to increase productivity, while at the same time native forests must be protected from further exploitation.

The term "bioeconomy" refers to the sustainable production and conversion of biomass into a range of food, health, fibre, industrial products and energy (EFI, The European Forest Institute 2011). At present, world economies are faced by the urgent need to find alternatives to the dwindling fossil fuel resources (especially oil and gas). The more support that is directed at the use of renewable energy or the higher the price of oil or CO_2, the higher a demand for wood energy or wood fibers and other bio-products (bio-chemicals, bio-plastics, bio-fuels, food additives etc. derived from sustainable forestry) will be (FPAC, The Forest Products Association of Canada 2011). Bioeconomy is a wide concept including environmental and economic aspects as well as biotechnology derived research and technology aspects and it promotes both products and businesses based on renewable biomass.

In the above presented scenario, the traditional pulp and paper industry and fibre production will still have an important role. Furthermore, a bioeconomy concept will provide multiple new production possibilities which will most certainly transform the whole forestry sector. Towards these goals, the productivity, in particular that of plantation forests urgently needs to be increased due to the restricted land area available. In this context, genetic engineering, amongst other tools, offers a potential to significantly and in relatively short time frames increase volume, quality and chemical composition of forest-based raw materials, thus, providing options to reduce world-wide consumption of petrochemicals and increase the use of sustainable resources (Fenning et al. 2008). In the present review, our aim is to consider biosafety issues in the context of deployment of genetically engineered (GE) forest trees in plantation forestry.

2 Contributions of Modern Biotechnology to Future Tree Breeding and Forestry

At present, the applications of modern biotechnology to tree breeding and forestry are diverse including quantitative and molecular breeding, clonal propagation, cryopreservation and genetic engineering (Fig. 1). The advent and rapid development of next generation sequencing technology has a key role providing sequence information for several tree species, which have extraordinarily high genetic diversity as compared to crop plants. The very first reference genome of a forest tree species, that of the black cottonwood (*Populus trichocarpa* Torr. & Gray), was published in 2006 (Tuskan et al. 2006). Today, genomic or organellar sequences of several other tree species are available, including the first drafts of Norway spruce (*Picea abies* L. Karst) (Nystedt et al. 2013) and white spruce [*Picea glauca* (Moench) Voss] (Birol et al. 2013) genome sequences. Sequence information is an invaluable

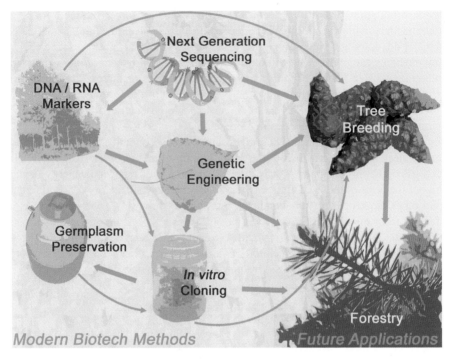

Fig. 1 The applications of modern biotechnology to tree breeding and forestry. The main contributors of biotechnology include increasing sequence information by next generation sequencing enabling development of specific markers and genetic engineering possibilities. Important biotech elements also include germplasm preservation by cryopreservation and *in vitro* cloning techniques which enable the use and multiplication of specific, potentially tailor-made, cultivars in future tree breeding or in forestry applications

resource for modern biotechnology applications and an important prerequisite to develop tools for genetic improvement of production populations for plantation forestry. Sequence information is also seen as supporting stewardship of natural populations as emphasized by Neale and Kremer (2011).

Most of the important traits of forest trees such as growth and adaptation, are quantitatively inherited i.e. regulated by several genes. An individual gene controlling some genetic variance of a phenotypic trait is regarded as a quantitative trait locus (QTL). Forest trees have long generation intervals and life cycles and most of them are also obligate outcrossers which makes quantitative tree breeding efforts very slow. In comparison to highly domesticated agricultural crops, tree breeding efforts have only been underway for a very short time frame and a very few generations and therefore forest trees are considered much less domesticated than crop plants. Seed produced by breeding programs is therefore genetically highly diverse, which provides significant opportunities for further tree improvement through quantitative or molecular breeding.

Since the late 1980s, DNA-based molecular markers have been identified as potential tools to enhance plant breeding. The first DNA based molecular marker tool, restriction fragment length polymorphism (RFLP), allowed the production of moderately dense genetic maps for breeding purposes. The second wave of genetic markers were all based on polymerase chain reaction (PCR), the most common ones being random amplified polymorphic DNA (RAPD), amplified fragment length polymorphism (AFLP), inter simple sequence repeats (ISSRs), and simple sequence repeats (SSRs) (Semagn et al. 2006). Both RFLPs and especially PCR-based marker systems allowed the scanning of genomes and the mapping of QTLs. However, these markers could not overcome the low level of linkage disequilibrium (LD) and recombination events with each generation in tree breeding populations. Thus, despite of the increasing number of markers, QTLs could not efficiently be applied to marker assisted selection (MAS) or brought to application in tree breeding as pointed out by Neale (2011).

Single nucleotide polymorphism (SNP) i.e. single nucleotide variation at a unique physical location within a locus among different individuals, has been described in the 1980s (reviewed by Brookes 1999), but their large scale discovery and efficient genotyping became available with the recent availability of high throughput second and third generation sequencing platforms (Glenn 2011). Of late, association genetics studies have been successfully applied to find candidate gene SNPs associated with a broad range of quantitative traits such as growth, stress, wood properties and disease resistance of forest trees (Neale and Savolainen 2004; González-Martínez et al. 2007; Dillon et al. 2010; Quesada et al. 2010; Thavamanikumar et al. 2011).

Clonal propagation of forest trees has been considered an important tool in traditional breeding to realize additional gains due to the potential exploitation of non-additive genetic variation, to improve forests by using a genetically improved planting stock and to increase its homogeneity. Clonal material has also been used to compensate for potential shortages of improved seed from seed orchards. Recently *in vitro* propagation techniques such as somatic embryogenesis (especially in conifers) and organogenesis (specifically in hardwoods) have also been emphasized when the germplasm conservation (as reviewed by Häggman et al. 2008; Lambardi et al. 2008) or production of transgenic forest trees (as reviewed by Flachowsky et al. 2009; Kole and Hall 2008) has been considered.

The principal method for genetic conservation of widely distributed and wind-pollinated species is the establishment of gene-reserve forests where a single stand contains a considerable proportion of the genetic diversity within a species. This kind of *in situ* conservation enables continuous evolution of the species. However, due to climate change and other potential disturbances, backup methods including *ex situ* conservation via cryopreservation have also been widely used. Furthermore, the first global assessment of the conservation status of trees worldwide indicated that around 8,000 tree species are threatened by extinction (Oldfield et al. 1998) indicating an urgent need for germplasm conservation.

Genetic engineering efforts with forest tree species started around 25 years ago and have since become an important addition to the toolbox of modern biotechnology and tree improvement. The first GE plant species was *Populus alba* × *grandidentata* genetically transformed using *Agrobacterium tumefaciens* and neomycin phosphotransferase (*nptII*) as a marker gene and the bacterial 5-enolpyruvyl-shikimate-3-phosphate (EPSP) synthase gene (*aroA*) conferring herbicide tolerance (Fillatti et al. 1987). Since then the transformation protocols, including *Agrobacterium*-mediated and biolistic transformation using a variety of tissue types, have been developed and/or improved for several coniferous (Henderson and Walter 2006) and deciduous tree species (as reviewed by Kole and Hall 2008; Busov et al. 2010). Moreover, several new genetic transformation applications already exist for model plant species and will certainly be included in the methodological toolbox of forest trees. These applications include usage of artificial miRNAs and overexpression of siRNAs which are capable of down regulating endogenous gene function, virus induced gene silencing (VIGS) which utilizes the advantage of natural posttranscriptional silencing (PTGS) mechanisms used for viral defence (reviewed by Busov et al. 2010), and targeted mutagenesis with zinc finger nucleases (Zhang et al. 2010).

In comparison with traditional tree breeding approaches, modern gene transfer techniques are very precise and involve very limited numbers of genes. The change introduced to the resulting organism will only be in a very specific location as compared to traditional methods where the whole genome is affected. Recent research has confirmed that the amount of change induced by the same trait is significantly higher in an organism produced by traditional breeding as compared to genetic modification (Batista et al. 2008). Also, significant change induced by traditional methods becomes obvious when considering the breeding history of corn or carrot for instance in which the recent cultivars are so different from the wild types, that it is not immediately obvious that they are the same species.

GE crops have a 17 years track record of commercial deployment with a cumulative area of more than 1.5 billion hectares planted by 2011, with a total area of over 170 million hectares planted in 2012 (James 2012). This makes GE crops the agricultural technology with the highest ever adoption rate, with farmers from 28 countries as well as global economists and researchers, reporting multiple economic and environmental benefits. At the same time, numerous scientific studies have failed to find any incidence of harm. In the forestry sector, it is now 26 years since the first transgenic tree *Populus* was reported (Fillatti et al. 1987), and since then over 700 field tests with transgenic trees have not identified any generic risk to the environment (Walter et al. 2010). However, there is still much debate and controversial views are frequently offered regarding genetic engineering technology. In the next section we will examine risks discussed and evaluate them in the context of accepted practise in forestry.

3   The Perceived Risks Associated with New Biotech Approaches

3.1   Gene Dispersal and Related Consequences

Major concerns have been raised regarding the potential risks of introducing GE trees into the natural environment, in particular the risks related to the transfer of recombinant DNA into the gene pool of a given species via vertical or horizontal gene transfer (Strauss et al. 1995; Rottmann et al. 2000). In GE field trials, however, no indication of risk has been found either for the environment or human health (Walter et al. 2010).

Vertical Gene Transfer

In particular, the vertical gene transfer (VGT) defined as the sexual transfer of genes via pollen between plants of the same or related species, and is one of the main concerns that has been raised with respect to commercial use of transgenic plants and therefore part of intensive discussions. Risks have been asserted drawing horror scenarios that "aggressive" and "non-native" genes when released with GE trees may contaminate natural gene pools and thus threaten the stability of natural ecosystems. However, the question arises as to whether a transgene present in the genome of pollen grains of flowering GE trees is able to establish for long term in the gene pool of a given species. Further, we need to consider whether this new gene actually presents a threat to the host organism or the ecosystem. In practice, the effect of the new gene may be negative, positive or neutral. It is also instructive to consider that natural gene flow exists and may lead to the transfer of the very same genes or combination of genes.

To answer this question for poplar, DiFazio (2002) studied gene flow and its implications for transgenic risk assessment in a poplar plantation by a modelling approach. He clearly showed that gene flow is related to long- rather than short-distance transfer of pollen and competitiveness, with the degree of fertility/sterility of transgenic trees being critical to the level of gene flow that might occur. High fertility leads to a very high degree of gene flow, however, when fertility is reduced, gene flow is dramatically slowed or even eliminated.

The STEVE model (DiFazio 2002) evaluates three different scenarios, gene flow of (a) neutral transgenes, (b) herbicide resistance transgenes, and (c) insect resistance transgenes. For a neutral transgene, the model predicted that after 50 years the number of GE trees forming as part of wild poplar populations would be less than 1 %. The herbicide resistance gene behaved similarly to the neutral gene in respect to gene flow and establishment in the gene pool as long as the corresponding herbicide was not applied. However, introgression of the herbicide transgene in the natural gene pool was enhanced when the GE trees were cultivated in close relationship to agricultural fields where the herbicide was applied for weed control. Selective forces were dictating whether the transgene is retained within the population or not. Unsurprisingly therefore, the situation was different for the insect resistance transgene. Dependent on insect pressure, this gene conferred a high degree of insect resistance leading to undamaged trees in GE tree plantations, as compared to non-resistant wild trees. Following introgression of the insect resistance transgene into the natural gene pool, this gene may lead to a substantial growth benefit in natural ecosystems if it is under a sufficiently high insect pressure to influence the survival of the affected trees. Taken together, gene flow from GE trees to wild populations seems low under existing conditions. Therefore, absolute sterility of GE trees is probably scientifically not essential in order to avoid gene introgression into natural tree populations, however, it may be socially desirable.

Another study into outcrossing rates has been performed by Fladung (unpublished) by using microsatellite analyses in non-transgenic aspens grown in artificial planted stands in the Arboretum of the Institute of Forest Genetics in Grosshansdorf, Germany. In 2003, seeds of two female aspen trees grown in these stands were collected, sown and genotyped. Further, 47 male (putative father) trees also growing in these stands were genotyped and their positions determined. Also weather data including temperature and other meteorological parameters (humidity, radiation, rainfall, wind strength, sunshine duration) were collected for the year 2003, as well as data on pollen emission during the same year were considered. A correlation between the concentration of poplar pollen in the air and temperature in 2003 was found.

Almost 200 descendants per female aspen mother tree were genotyped. Parental analyses showed that:

(a) Flight direction of reproductively effective poplar pollen seemed not to be dependent on wind or other climatic factors: in one mother tree, the two most likely putative father trees grew north of the mother tree approximately 25 m away. In the second mother tree, the two most likely putative father trees were located 50 m east and 40 m north-west, respectively.

(b) A large proportion of about 60–70 % of reproductively effective pollen probably originated from poplar trees growing less than 100 m away from the mother tree. Considering that poplar is wind-pollinated and thus the pollen can be transported (at least theoretically) long distances, the result was surprising.

(c) On the other hand it can be concluded, that 30–40 % of reproductively effective pollen originated from poplar trees grown outside of the stands investigated. This result suggests that about one-third of reproductively active poplar pollen traveled more than 100 m.

(d) The vast majority of the putative father trees are located close to both mother trees (within a radius of only 30–60 m).

Horizontal Gene Transfer

Exchange of genes between sexually non-compatible organisms is called "Horizontal Gene Transfer" (HGT). This mechanism is evolutionary common and mainly found in micro-organisms, but also known to occur between bacteria and plants (Brown 2003). Here, the *Agrobacterium*-based natural gene transfer system is one of the best characterised examples of HGT (Chilton et al. 1977; Schell et al. 1979). Other than for *Agrobacterium* and related bacteria, the possibility of a HGT is potentially higher when there are tight associations between plants and other organisms (mainly bacteria, fungi and viruses).

One famous example of such a tight association is the symbiosis of many trees with ectomycorrhizal (ECM) fungi. These fungi are tightly associated with the roots of the plants, thus a HGT event is theoretically more likely in such mycorrhizal symbiosis. The possibility of HGT from transgenic poplar trees grown in the field under natural environmental conditions to ECM fungi was intensively studied by Nehls et al. (2006). The transgenic trees were transformed with the *bar* gene from *Streptomyces hygroscopicus* encoding a gene conferring resistance against the herbicide BASTA, cloned downstream of a fungus-specific promoter. The usage of a fungus- specific promoter prevents the transcription of the *bar* gene in the trees thus leaving the trees BASTA-sensitive. If the construct is transferred to ECM fungus by HGT, however, the gene would be activated and the fungus would show BASTA-resistance. Thus, thousands of root-mycorrhiza associations could easily be screened by simply transferring them onto BASTA containing media. Using this method, about 100,000 mycorrhizas were isolated from transgenic tree roots, and putative BASTA-tolerant fungi were further analysed by molecular methods. No indication for a HGT from transgenic aspen to ECM fungi was found, neither under optimized laboratory nor under field conditions (Nehls et al. 2006). In another study, the transgene specific PCR product was not obtained from soil samples collected from field sites under hybrid triploid poplars indicating that little if any HGT was occuring (Zhang et al. 2005b). These findings were supported by Keese (2008), who also found that although HGT from plants to other eukaryotes (or prokaryotes) exists, the frequency is extremely low, being in most cases lower than background rates.

The possibility of HGT has also been studied between *Agrobacterium* and en-dophytic bacteria. A large endophyte community can be found in poplar clones grown in the field as well as under aseptic conditions *in vitro* (Ulrich et al. 2008b). One endophyte genus, *Paenibacillus* was found to be predominantly present in most of the tissue cultures of woody plants (Ulrich et al. 2008a). Even with HGT being frequently observed between different bacterial communities, no indication of HGT could be detected *in vitro* between *Agrobacterium* and these endophytic bacteria (http://www.gmo-safety.eu/database/1011.transgenic-poplars-horizontal-gene-transfer-agrobacteria-endophytic-bacteria-possible.html).

Again, for considerations of real risk, we need to keep in mind that the insect resistance genes used in transgenic plants, for example, were originally derived from naturally occurring bacteria (e.g., *Bacillus thuringiensis*) and they are therefore abundantly present in almost all natural environments. Considering the long periods that plant ecosystems have been exposed to such bacteria, we may conclude that such gene transfer may have already happened (horizontally at first, but also vertically subsequent to horizontal gene transfer) at various times and locations, however with negligible impact. Alternatively, this may actually indicate that such gene transfer is infinitesimally small and can be neglected in our risk considerations.

Nevertheless, it is widely accepted that genes can naturally pass higher plant species boundaries. The significance of a low HGT ratio for genetic novelty and adaptation during evolution of eukaryotes has been pointed out by Bock (2009). He stated that HGT can expand the species gene pool that offers the opportunity to take advantage of additional gene resources, possibly leading to increased fitness values. An example was given in a recent study by Christin et al. (2012), where a minimum of four independent lateral gene transfer events were described for the grass lineage

Alloteropsis. This grass group contains mainly C3 but also five C4 species, where the latter ones differ in C4 anatomy and biochemistry to each other. The authors could show that for optimisation of the C4 pathway periodically genomic units containing genes for already fully adapted C4 function were laterally transferred several times over the last 10 million years. Also a transfer of larger portions of the chloroplast genome or even entire chloroplasts without sexual reproduction via HGT has quite recently been discovered (Stegemann et al. 2012).

Spread of Antibiotic Resistance Genes

Very often, genes encoding for antibiotic or herbicide resistances are being used as selectable markers in genetic engineering techniques. These markers help to select those cells that have incorporated the foreign genes within their genome early in the engineering process. Subsequently, these genes have no further use, but they are continuously being expressed during the whole life period of the transgenic plant.

In case that the transgenic plants are processed to food, the spread of antibiotic resistance genes to humans, animals or pathogenic organisms could theoretically be harmful if somehow being expressed in the new host, resulting in reduction of effectiveness of the antibiotics. However, with exception of kanamycin, antibiotic resistance genes used in genetic engineering are not related to those antibiotics being applied to humans. Kanamycin is the only significant antibiotic used for human health, however, the *nptII* gene encoding for kanamycin resistance is known to be ubiquitously present in soils around the world. Further, as shown in the previous paragraph, the unmediated transfer of genes from plants to bacteria is highly unlikely. Thus, the risk by resistance genes seems more to be a political than scientific argument.

Evolution of Resistances

One important goal in genetic engineering of trees is the introduction of herbicide tolerance. The purpose of this modification is to allow the establishment of tree plantations in weed infested areas and to support tree growth during the first years of establishment in the forest. Also, herbicide tolerance could have an economic advantage in tree nurseries, where weeds can become a major problem. It has been argued that the application of herbicides in forestry can have negative effects on neighboring ecosystems through wind drift and also that the use of herbicides may promote the evolution of tolerances or resistances in wild populations. It is often argued that this may happen by repeated treatments with the same herbicide or by gene flow from herbicide tolerant plants to wild relatives. Plants from these populations may then acquire a weedy potential (see below) to become invasive. Once the use of a herbicide has selected for resistant individuals, continued use of the same herbicide will select for resistant individuals and discriminate against susceptible ones.

However, in plantation forestry the evolution of resistances is very unlikely because herbicides will only be used in the year when planting of the trees is planned in order to clear the field of weeds. In the first year after establishment of the plantation, herbicides are sometimes used to control weed growth, but generally no further use of herbicides is needed in the plantation once the trees have grown for the first year. It is also important to consider that the transfer of the herbicide resistance trait from a GE tree to an unmodified plant species via HGT is highly unlikely, and this has never been demonstrated even in ideal laboratory conditions. Further, if such a transfer happened, the receiving organism can only become a weed if there is continued application of the same herbicide over a longer period of time. We should also be mindful of the fact that herbicide resistance is in fact a natural trait that was originally derived from naturally occurring microorganisms which can be found in soils in most parts of the world. If HGT would be an issue and genes could be transferred to plants via this mechanism in an efficient way, we would already see a number of herbicide resistant plants of a variety of species. Millions of years of evolution could have already made this happen. The fact that we do not detect this, is another strong argument against the HGT hypothesis.

A similar chain of argument has been developed against the use of Bt-poplar. These plants have been transformed with *cry* genes encoding the *Bacillus thuringiensis* (Bt) toxin making the transgenic plants resistant to insect attack. At present, Bt-transgenic poplar are commercially grown only in China (Ewald et al. 2006) in an area of more than 300–500 ha (Walter et al. 2010). Concerns have been raised that insects might adapt to the Bt-toxin because life cycles of insects are much shorter than the ones of trees. Once evolved, highly mobile Bt-resistant or -tolerant insects may move to other tree stands where Bt-toxin is sprayed classically to control pest. However, such evolution of resistance in insects can also happen through the conventional application of Bt-sprays, for example in organic agriculture.

A number of studies are already available discussing this aspect of risk in context. An example of resistance formation has been reported for Bt-maize (Gassmann et al. 2011). In the US, where annual damage caused by the root borer is more than 1 billion US $, beetle-resistant Bt-maize varieties were successfully applied. Starting from 2009, however, farmers observed symptoms typical of rootworm damage on the maize plants on four fields that were actually planted with western corn rootworm-resistant maize. Entomological studies on all four fields clearly showed that the rootworms were less sensitive against the Bt-protein than normal, however, the insects did not demonstrate complete resistance. Following interviews with the farmers it was found that the fields had been planted for at least for three consecutive years and in one case for 6 years using the same Bt-maize variety. Gassmann et al. (2011) assumed that the repeated cultivation of the same Bt-maize variety has led to resistance formation against the Bt-protein and to spread of Bt-resistant insects. Farmers using GE Bt-crops are advised to use non-GE varieties intermittently or to create "refugia" of non-modified crops to reduce the probability of Bt-resistance development. Plantations with both GE Bt and non-transgenic trees was found to protect non-transgenic trees against insect herbivory (Hu et al. 2001). Thus, the plantation forestry with both non-transgenic and transgenic trees could also prevent the selection pressure against Bt-resistant insect pests.

The examples described need to be evaluated in the context of millions of hectares of Bt-varieties planted from 1996 to 2011, without any development of insects resistant to Bt. This confirms the safety of this modification. Further, while single *Bt*-genes may in very rare circumstances lead to resistance development; this becomes very different when various insect resistance genes are used ("gene stacking"). Here it was calculated that the probability of development of insect resistance with stacked genes is infinitesimally small. For Bt-transgenic trees no information on resistance formation is available up to now. In China where more than 300–500 ha are cultivated with Bt-poplars (Walter et al. 2010), no example of insect outbreak has been reported.

Production and Distribution of Toxins or Allergens

Naturally, many organisms are known to produce toxic substances or human allergens. In plants, many substances such as terpenes, waxes and alkaloids play an important role in defense reactions or are involved in responses against pathogen attack. These substances are produced in secondary metabolic pathways, but in rare cases, plants contain incomplete pathways which do not produce toxic substances. The transfer of genes by vertical or horizontal gene transfer could complete these inactive pathways and potentially lead to new toxic substances. However, due to the fact that HGT indeed exists but occurs at a very low level, the question still remains whether this risk is only of hypothetical nature or a true one.

Also when used in phytoremediation, such as the detoxification of heavy metals or organic compounds from soils by using plants, the transgenic trees involved were considered as being potentially harmful to plants, animals or humans. The purpose of creating such trees is to remove hazardous substances from the soil and concentrate them in the plant tissue, for subsequent safe storage or conversion into non-toxic forms. However, it has been suggested that if these substances do not accumulate in the desired organs or are released back to the environment, there may be a risk that the food chain will be re-contaminated with the toxic substances. In the case of long-living forest trees, however, these risks will surely be much less than the original problems. Further, the exposure to risk in this case is not represented by the transgenic plant, but by the toxic substances that were released into the environment in the first place. The purpose of those transgenic trees is to remedy the problem, not to make it worse. It should also be pointed out that such risks are not created by the genetic modification as such, but by the original pollution events and/or the improper use of the modified organisms. Improper use of a certain crop can always create extra risks, as exemplified by the consumption of peanuts or peanut containing products by people who are allergic to them.

3.2   Physical Spread of GE Trees

In addition to vertical and horizontal gene flow, the "physical" spread of GE trees is sometimes seen as an issue, in particular with field testing of transgenic trees for risk assessment. The spread of plants may happen via vegetative means, e.g., root suckering and this is known for a number of tree species (Fladung et al. 2003). Root suckers are formed by the lateral root system via adventitious buds. When large numbers of suckers are formed by a single tree, a high number of clonal individuals will be formed surrounding the "mother" tree. This became an issue during the first field trial with transgenic trees in Germany in 1996 and effective strategies were needed to eliminate root suckers, in order to fully contain the GE trees within the test area.

Vegetative Spread

The first experiments with the controlled release of genetically transformed trees in Germany was initiated in 1996 for a period of 5 years to investigate the performance of GE trees under natural environmental conditions including ecological studies. Eight transgenic aspen lines carrying either the 35S::*rolC* or the rbcS::*rolC* gene construct, and three control lines were transferred to the field at the age of 1 year (Fladung et al. 1996; Fladung and Kumar 2002). In 1999, at an age of 4 years, one small GE aspen plant identified as a root sucker was observed, followed by an increasing number of root suckers derived from transgenic and non-transgenic trees in 2000 and 2001 (Fladung et al. 2003). In 2001, the last year of the field trial, 15 root suckers were detected outside the field margins. This was possible because root length of individual trees was found to be up to 10 m (M. Kaldorf, personal communication).

The phenotype of all suckers was wild type or rbcS::*rolC*-like. In total, 234 root suckers were harvested in 2000 and 2001 and analysed by PCR for their transgenic status. More than half of the roots suckers investigated showed the presence of the *rolC* gene. However, no biosafety problems occurred because the suckers were eliminated and the area was carefully observed before the trial ended in autumn 2001. We concluded that in addition to generative propagation, the vegetative dispersal capacity of transgenic perennial plants needs to be considered in risk assessment studies. These studies are particularly important in case GE trees are ever planted outside a controlled environment, deliberately or otherwise. Here, we investigate whether in the event of root suckering risks arise only by "physical" presence of the GE tree in the environment, however, it can be an issue in case of possible weediness or change of fitness characteristics of GE trees.

Weediness

Following the transfer of genes and their random and unpredictable integration somewhere in the host genome, researchers can identify where that location is and select only those genetic engineering events where the integration is not in a gene region or other active region of the genome that may have an influence on the fitness of the respective organism. Where these checks are not performed, however, a change in the fitness of the GE trees may occur and this needs to be evaluated. When the overall fitness of an organism in its natural environment is reduced by any human intervention, the probability of long term survival of this organism (for example a GE tree) in the natural environment is likely to be significantly reduced. For example, GE plants carrying the *rolC* gene from *Agrobacterium rhizogenes* and therefore displaying slow growth and a dwarfy phenotype (Fladung 1990; Fladung et al. 1996) will be selected against in the natural environment and are unlikely to spread (Hancock 2003).

Alternatively, increased fitness may result in more invasiveness, thus leading to increased weediness of an organism. However, the definition for a "weed" is that a plant is undesirable and unattractive at a given place, is highly durable combined with high potential of generative and/or vegetative dispersal. Further, crop plants may also be considered as "weeds", such as when rape plants re-appear from seeds spilt from the previous year's crop and negatively affect subsequent crops grown there, such as wheat or barley. These examples show that weediness is a common phenomenon and not *per se* related to genetic engineering. Weediness is more related to culture of a given plant species rather than to the transfer of isolated genes used for genetic transformation. We must also keep in mind that weediness can result from any modification that humans have performed on the natural environment and continue to do when applying a variety of crop breeding techniques. These may not involve GE plants, but still have the potential to increase or decrease the fitness of the affected organisms. To assess whether a new organism may become a weed, the characteristics of that organism should be the focus of research, not the way it was produced. This concept is reflected for example in the Canadian regulation governing the release of new plant varieties, where the new trait is the focus of attention, not the method the trait was created by.

Similarly as shown by Modi et al. (1992) for *Escherichia coli*, a small genetic change may theoretically increase fitness in plants. In this bacterial example, movement of a transposon from a plasmid into the bacterial chromosome was shown to increase relative fitness of the bacteria by approximately 6–7 %. In plants, the potential impacts of individual transgenes should be determined by evaluating their phenotypic effects (Hancock 2003). Londo et al. (2010) showed with glyphosate-resistant canola (*Brassica napus* L.) and hybrids that increased plant fitness could be observed for transgenic genotypes as a result of sub-lethal glyphosate treatments in mesocosm units. This indicates a potential persistence of glyphosate resistance transgenes in weedy plant communities at areas nearby crop fields (roadsides, field edges), which are within the range of possible glyphosate-drift. However, it remains to be shown how important small changes in the fitness of GE plants are in the context of their survival under changing environments.

In plants, transgenes improving tolerance to challenging biotic or abiotic stress factors may increase fitness. For example, modifications with the *cry* gene discussed above, which encodes the Bt-toxin leading to insect resistance in transgenic plants (Genissel et al. 2000) may have a positive effect on fitness in certain conditions. However, this gene only leads to higher fitness values when feeding pressure by insects is high. In absence of such insects, no advantage will occur. Both examples underline the importance of the consideration of genotype by environment interactions in risk assessment studies, a consideration that must be taken into account for all modifications of the natural environment, not just GE.

In summary, when considering the potential for contaminating the gene pool of native tree species, all prospective "novel" tree clones along with exotic or ornamental tree species as well as the various spreading mechanisms (vertical, horizontal, vegetative) should be considered. From a scientific point of view, "novel" should not just mean GE trees, but also those produced by conventional breeding technologies including those advanced technologies such as forced hybridization and embryo-rescue. Further, the introduction of exotic tree species into native environments may contain large genetic risks. This is due to the introduction of 20,000–30,000 exotic genes contained within the new tree species. At the same time, a very large number of genes from accompanying fungal and bacterial organisms are introduced. All of these exotic genes are subject to the same vertical and horizontal transfer mechanisms, and sometimes introduced foreign species have become invasive because they have a higher fitness in their new environment, although none of the examples known to date have been attributed to GE plants (Hoenicka and Fladung 2006a). This illustrates again that genetic engineering and its risk potential must be evaluated in the context of accepted practice. Most importantly, due to the very low number of novel genes involved and the relative precision of genetic engineering technology, the magnitude of risk is almost certainly significantly lower than that associated with common forestry practices.

4   History of GE Crops

In the context of the potential deployment of GE trees in plantation forestry and the risk evaluation required for regulatory approval, it is instructive to briefly consider the 17 year history of commercial planting of GE food crops. Agricultural deployment of GE plants started in 1996 and in the last 17 years more than 1.5 billion hectares have cumulatively been planted by farmers in the developing and developed world, totalling 28 countries. In 2012 alone, over 170 million hectares were planted (James 2012) and the major crops were soybean, cotton and maize, mainly modified with herbicide and/or insect resistance. During this period, numerous studies were published on the environmental safety and economic benefit of these crops (Pray et al. 2002; Raney 2006; Wu 2008; Gómez-Barbero et al. 2008; Qaim 2009; Carpenter 2010; Brookes and Barfoot 2011). What are the major lessons that can be learned?

First of all, GE crops are safe for the environment. While initially there was significant concern from environmental groups and scientists, intensive research has failed to identify or substantiate any harm to the environment or humans (references above). Conversely, these studies have actually shown conclusively, that GE crops have led to significant environmental benefit. This is substantiated by the use of less environmentally damaging herbicides (Kleter et al. 2007), the preservation of topsoil through adoption of no-till agriculture associated with GE crops (Fawcett and Towery 2002), the significant reduction of pesticide applications by farmers using Bt crops and even unexpected positive side effects such as the significant reduction of pest pressure in areas where Bt crops were used, and even benefited farmers who did not use such crops (Hutchison et al. 2010).

Secondly, GE crops have had large economic benefits. This is realised through increased crop production, reduced application of expensive agrochemicals and reduced number of trips that farmers need to make to the field to manage their crops, the latter also reducing the impact on the environment by reduced fuel use and carbon dioxide emission (Gonzalvez 2004; Raney 2006; Qaim 2009; Brookes and Barfoot 2011).

Thirdly, GE crops have social benefit in creating healthier agricultural environments and thereby increased farmer health (reduced application of harmful chemicals) and in reducing the overall carbon footprint of agriculture worldwide (Brookes and Barfoot 2011). Application of genetic engineering has also been a significant benefit to public health (Barros et al. 2009) and farm animals, such as through improved nutritional characteristics (Dawe et al. 2002). Further, the use of GE crops has improved the incomes of farmers particularly in the developing world and therefore their options to provide better schooling to their children for instance. It is in fact interesting to note that the largest number of farmers applying GE crops are resource poor farmers in developing countries and they are realising the biggest benefits (James 2012).

Finally we need to note that unlike any other technology whether related to food production or other human endeavour (except perhaps medicines), the application of genetic engineering technology is highly regulated and deployment is costly, to the extent that only large agricultural companies can afford to bring such crops to market. This has concentrated the efforts in this area on a few crops which have large areas of cultivation, whereas crops of smaller importance have been neglected.

The case can be made that whereas initially caution was justified and strict regulation necessary in making sure that genetic engineering technology was properly assessed with regards to risks, now that we are able to derive knowledge and expertise from 17 years of commercial application, such strict regulation is neither necessary nor warranted. We are no longer in a situation where the argument can be made that not enough is known about the risks and benefits of GE crops to deregulate their use, which is a fact that regulatory bodies worldwide should take note of and act accordingly, to the benefit of all of mankind. Most importantly, regulatory efforts should focus on whether a new trait has been introduced and evaluate any risk related to this trait rather than the method which was used to create it.

5   The Performance of the GE Forest Trees in Field Trials

The confined field trials have been organised to reveal the performance of the GE trees, whether for growth, development, or the stability of introduced trait and any unintended effects to their natural habitats, and to ascertain the various interactions of tree species and the complexity of forest ecosystems. The traits tested by commercial companies and the public sector with field experiments include wood properties, resistance to insects, herbicides and diseases, abiotic stress resistance and plant growth vigour (Robischon 2006; Walter et al. 2010). Published biosafety assessments on the field trials of GE forest trees considering growth improvement, abiotic stress tolerance and herbicide-resistance are still few, however these traits have been shown to remain stable under natural environmental conditions (Meilan et al. 2000, 2002a, b; Jing et al. 2004; Li et al. 2008b; Zawaski et al. 2011). Next, we will review the current knowledge of the performance of GE forest trees in field trials focusing on selectable marker genes, disease and insect resistance, and wood properties.

5.1 Marker Genes

Selectable marker genes are generally needed in the early phases of most of the genetic transformation protocols in order to provide the identification of transformed cells. Usually, selectable marker genes are antibiotic resistance genes such as *nptII*, which confers resistance in the transformed tissues to the antibiotic kanamycin. In field trials marker genes have also been used to determine the stability and the expression levels of the transgene, and to unravel unintended environmental effects of the transgene or transformation process. In general transgene expression levels have been found to be stable, with the exception of one or few transgenic lines, as in studies conducted with marker genes *rolC* (Kumar and Fladung 2001a), *uidA* (Hawkins et al. 2003), the green fluorescent protein encoding gene *gfp* (together with *bar*) (Li et al. 2009), and *nptII* (Walter 2004). In addition, the stability of gene expression suppression induced by RNAi has shown to be stable in a 2-year field experiment with *gfp/bar* hybrid poplars (Li et al. 2008a). Furthermore, no unintended effects have been detected in the arbuscular or ectomycorrhizal status of transgenic aspens in studies of Kaldorf et al. (2002) and Stefani et al. (2009).

Field studies with GE radiata pine (*Pinus radiata* D. Don) in New Zealand focused on the continued expression of the *nptII* gene in transgenic lines that were created using biolistic bombardment (Walter et al. 1998). GE lines had a very high transgene copy number (sometimes many hundred copies) and fragmented integrations were also found. Therefore, high levels of silencing of gene expression were expected to occur over a period of a few years in the field. However, while silencing was observed in some transgenic lines early in embryogenesis and plant development, most of those GE lines included in the field retained their original expression level up to an age of 8 years, when the research was completed (Walter and Fenning 2004).

In the same field trial, Schnitzler et al. (2010) studied the influence of *nptII* expression on invertebrate populations living in close association with the GE trees in the field trial. The trees also had genes related to reproductive development, including *leafy*, *apetala*, and *constans*, however their expression levels were very low.

Based on invertebrate samples taken during the 3 last years of the 5-year long field experiment, the heterologous *nptII* expression did neither affect the abundance of invertebrate species, their richness, diversity, nor composition. Similarly, two independently transformed *nptII* and *leafy* expressing radiata pine lines did not have any significant impact on microbial communities, even though some variation between transgenic and non-transgenic radiata pines was detected in bacterial and fungal communities at some sampling dates (Lottmann et al. 2010). The differences detected were smaller than seasonal variations in bacterial and fungal communities and were probably not significant.

5.2 Disease Resistance

Only few reports have been published on field trials assessing the environmental effects of GE trees with improved disease resistance, even though enhanced pathogen resistance has been achieved using various approaches (reviewed in Powell et al. 2006). In two field trials completed with transgenic silver birches (*Betula pendula* Roth) expressing the sugar beet chitinase IV gene (*chi*IV) (Pasonen et al. 2004, 2008, 2009; Vauramo et al. 2006; Vihervuori et al. 2008), and American elms (*Ulmus americana* L.) (Newhouse et al. 2007) producing the synthetic ESF39A peptide, no unintended effects related to the genetic modification were detected. No variation in the fungal biomass nor diversity was detected in white spruces [Picea glauca (Moench) Voss] expressing the endochitinase gene (*ech42*) of *Trichoderma harzianum* (Lamarche et al. 2011), nor in the ectendomycorrhizal interaction in glasshouse experiments (Stefani et al. 2010).

The insect herbivore community (Vihervuori et al. 2008) and the ECM fungi colonization (Pasonen et al. 2009) was determined with a field trial of three growing seasons established with 18-month-old saplings of 15 *chi*IV silver birch lines. The *chi*IV gene was found to be stably expressed over a period of 3 years (Pasonen et al. 2004). The transgene did not have any effect on the resistance of trees to *Pyrenopeziza betulicola* (leaf spot disease), but it improved the resistance of *chi*IV lines to *Melampsoridium betulinum* (birch rust). The insect densities indicated that the leaves of *chi*IV lines were more susceptible to aphids, and also suffered higher levels of leaf damage by leaf miners (Vihervuori et al. 2008). The composition of leaf damage types however, was associated with the genotype of the tree rather than transgenic status. Similarly, cluster analysis of the root-associated fungal species identified using the ITS1 region revealed that the genotype had a greater impact on the structure of the ECM fungal community than the transgenic status of the trees, and that the ECM colonization structure could not be related to *chi*IV expression or endochitinase activity (Pasonen et al. 2009). In addition, leaves were collected from field grown silver birches during the second growing season and used in a decomposition experiment. The leaf samples were buried in the soil in litter bags for a period of 8 and 11 months (Vauramo et al. 2006). The fungal and total microbial biomass did not vary between the litter samples of GE and non-transgenic silver birches.

In addition, the *chi*IV lines and non-transgenic clones did not differ in the leaf decomposition process, addressed by studying the litter mass loss, or in the mean number of nematodes with the exception of one *chi*IV line (Vauramo et al. 2006).

Mycorrhizal status was determined for transgenic American elms expressing the synthetic antimicrobial peptide ESF39A after 3 months in a field trial (Newhouse et al. 2007). The ESF39A was targeted to *Ophiostoma novo-ulmi* fungi causing the Dutch-elm disease. No variation was found in the colonization rate of arbuscular mycorrhizal between the GE and non-transgenic lines.

5.3 Insect Resistance

The δ-endotoxin producing genes (*cry1Ab, cry1Ac, cry1Ac646, cry1B* and *cry3A*) of the soil bacterium *B. thuringiensis* have been introduced into eucalyptus, birch, larch, pine, poplar and spruce species (reviewed in Balestrazzi et al. 2006), to reduce damage caused by coleopteran, dipteran and lepidopteran insects (Schnepf et al. 1998). Transgenic trees have been tested in contained field trials and found to have reduced feeding damage by pest species (Meilan et al. 2000; Hu et al. 2001, 2007, cited in Zhang et al. 2011a; Lachance et al. 2007), or lower number of target insect individuals or pupae (Hu et al. 2001; Gao et al. 2003, cited in Ewald et al. 2006; Hu et al. 2007, cited in Zhang et al. 2011a, b). Similarly, transgenic radiata pine trees expressing *cry1Ac* were tested in laboratory experiments with needles fed to lepidopteran insects, which proved toxic to the insects (Grace et al. 2005). Genes targeted against insect herbivores, have been stably expressed in field grown trees transformed with *cry* genes (Yang et al. 2003; Wang et al. 2004, cited in Ewald et al. 2006; Lachance et al. 2007; Zhang et al. 2011a, b), proteinase inhibitor oryzacystatin I gene (*OC-I*) (Hou et al. 2009; Zhang et al. 2011b), and cowpea trypsin inhibitor (*CpTI*) (Zhang et al. 2005a). Black poplar (*Popolus nigra* L.) expressing *cry1Ac* and the hybrid poplar line 741 [*P. alba* L. × (*P. davidiana* Dode + *P. simonii* Carr.) × *P. tomentosa* Carr.] expressing the modified *cry1Ac* gene and the proteinase inhibitor gene (*API*) have been released commercially in China, and the influence of transgenes to the insect herbivores have been monitored in various studies (Gao et al. 2003, cited in Ewald et al. 2006; Zhang et al. 2004, cited in Ewald et al. 2006; Gao et al. 2006, cited in Ewald et al. 2006; Yao et al. 2006, cited in Zhang et al. 2011a; Hu et al. 2007, cited in Zhang et al. 2011a; Jiang et al. 2009, cited in Zhang et al. 2011a).

No differences were found between the *cry*1Ac and non-transgenic black poplars in the soil bacteria, actinomycetes or mould (Hu et al. 2004, cited in Ewald et al. 2006), however, the *cry1Ac* expressing plants curiously were found to have a shift in the dominating insect and spider communities, and, moreover, differences were also detected in the insect community structure between a plot of pure transgenic black poplars and a 1:1 mixture of transgenic and control poplars (Zhang et al. 2004, cited in Ewald et al. 2006). In addition, the variety, number, and ratio of para-sitic natural enemies of insects were elevated in the Bt producing black poplar plantations in the study of Hu et al. (2007, cited in Zhang et al. 2011a). This indicates that the genetic modification can potentially increase overall insect biodiversity.

The *cry1Ac* expressing hybrid poplar line 741 was also found to have increased arthropod species abundance and species richness, and the evenness of insect pest sub-communities (Gao et al. 2003, cited in Ewald et al. 2006). Furthermore, Gao et al. (2006, cited in Zhang et al. 2011a) observed variation in the ecological niches of some of the arthropod species between the 741 line and non-transgenic hybrid poplar. The non-target Asian lady beetles (*Harmonia axyridis* Pallas) had a shorter pupal period when fed with non-target sucking pest, poplar aphid (*Chaitophorus populeti* Panzer), grown on the line 741. No differences were detected between the non-transgenic and line 741 in the survival, body mass, eclosion, or sex ratio of the Asian lady beetle (Yao et al. 2006, cited in Zhang et al. 2011a). The individual numbers of Asian lady beetles were, however, shown to differ in another study between the field grown non-transgenic and 741 line (Jiang et al. 2009, cited in Zhang et al. 2011a). Also the number of crab spiders (*Misumenops tricuspidatus* Fabricius) and *Vulgichneumon leucaniae* (Uchida) paratisoids differed between the 741 and non-transgenic line (Jiang et al. 2009, cited in Zhang et al. 2011a).

Soil microorganisms were monitored in a 2-year field trial from the rhizosphere of *P. alba*×*P. glandulosa* Uyeki lines expressing *cry3A*, and also with stacked resistance lines expressing both *cry3A* and *OC-I* (Hou et al. 2009). Some transgenic poplar lines differed from the non-transgenic lines in respect to the community of soil microorganisms, but the observed differences were not related to the transgene type or to the number of transgenes. The same poplar lines were used in the studies of Zhang et al. (2011a, b) in which the composition of arthropod communities and their dominance were monitored over three growing seasons. As might be expected, the number of defoliator species on a double coleopteran-resistance line was lower than on the non-transgenic line, however, some changes were also observed in the communities of non-target arthropods (Zhang et al. 2011a). Differences between *cry3A* and non-transgenic lines were found with regards to guild dominance, family composition and family dominance, but in general, the arthropod communities were similar (Zhang et al. 2011b). However, the authors hypothesized that the observed differences between the poplar lines in the level of arthropod families were not a direct consequence of the leaf toxicity, but a result of a shift in the arthropod community, caused by the decreased population of the target coleopteran species (Zhang et al. 2011b).

The effect of stacked genes introduced into hybrid poplar (*Populus* × *euramericana*'Guariento') on the insect community was studied in a field trial established in 2006 (Su et al. 2011). The transgenic D5-20 and D5-21 lines expressing *cry3A* and *OC-I* in addition to three other genes introduced to improve drought tolerance and waterlogging-induced hypoxia were found to have a lower dominance of coleopteran target species. The authors stated that genetic engineering did not have a significant effect on the arthropod community, even though the number sucking pests was shown to be increased on the D5-21 line (Su et al. 2011).

The *cry3Aa* gene was shown to have an effect to the insect communities of aquatic stream in the study of Axelsson et al. (2011a), where the litter of *cry3Aa* hybrid poplars (*P. tremula×P. tremuloides*) had a 25–33 % increase in average abundance of aquatic insects. In this context it should be however noted, that when used as sprayable pesticide, the Bt itself can also accumulate in aquatic systems and affect the insect communities. Yet, Axelsson et al. (2011b) showed that the *cry3Aa* hybrid poplars lines had altered phytochemistry and non-target herbivores (*Deroceras reticulatum* and *D. agreste*) had increased preference to the *cry3Aa* hybrid poplars based on the bioassays.

No difference was detected in the endophyte frequency or distribution between non-transgenic white spruces or white spruce lines expressing *cryIAb* together with *uidA* and *nptII*, nor with white spruce lines expressing the marker genes *uidA* and *nptII* (Stefani and Bérube 2006). However, a shift in the rhizosphere microbial community was detected between the non-transgenic white spruce and *cryIAb/uidA/nptII* or *uidA/nptII* white spruce lines (LeBlanc et al. 2007). Moreover, the microbial community determined from samples collected after 3 years in the field also differed with respect to the two transgenic lines, *uidA/nptII* and *cryIAb/uidA/nptII*. The authors concluded that the transgenes were responsible for the observed differences, acknowledging that more than one sampling date and site would have been needed to understand the mechanisms involved in the microbial community structure shifts (LeBlanc et al. 2007). In another study on *cryIAb/uidA/nptII* white spruces, however, the N_2-fixing diazotroph communities were similar in the rhizosphere of non-transgenic and transgenic white spruce lines collected after 4-year field experiment (Lamarche and Hamelin 2007). Moreover, the only difference detected in the diazotroph community was between the natural white spruce stand and the white spruce plantation (Lamarche and Hamelin 2007).

Although all these studies show that the Bt producing trees may influence the species richness and the insect community structure directly and indirectly, the principle effect is a reduction in the number of target coleopteran, dipteran and lepidopteran individuals surviving on the trees. This is the expected and intended effect of the genetic modification. Indirect effects may arise as a result of these changes in the target organisms, causing changes in the species richness of other species and/or the dynamics of insect population as suggested by Zhang et al. (2011a). Furthermore, the changes in the arthropod communities may arise from the changed phenolic profiles of GE tree leaves as observed in the study of Axelsson et al. (2011b). However, the extent of the changes to the non-target organisms is difficult to evaluate, and it may be positive or negative. Moreover, in these studies the alternative to Bt transgenic trees, is the spraying of Bt which was not included as a control to the experiments. To evaluate the risk specifically associated to Bt transgenic trees, their effect need to be compared with the effect of the alternative technology, which is spraying of affected plants with Bt spray.

5.4 Wood Properties

Wood properties are mainly determined by the chemical composition and structure of the secondary cell wall of wood (xylem) cells. Alterations in cell wall chemistry have been achieved by modulating the expression of a variety of genes encoding enzymes forming part of the monolignol biosynthetic pathway (reviewed in Simmons et al. 2010). Results published on the performance of GE trees modified for their wood properties and tested in field trials include hybrid white poplars (*P. tremula*×*P. alba*) with altered expression of cinnamyl alcohol dehydrogenase (*CAD*) (Pilate et al. 2002), caffeate/5-hydroxyferulate O-methyltransferase (*COMT*) (Pilate et al. 2002), cinnamoyl-CoA reductase (*CCR*) (Leplé et al. 2007), and 4-coumarate:coenzyme A ligase (*4CL*) (Kitin et al. 2010; Voelker et al. 2010, 2011a, b). The *CCR* expressing hybrid white poplar lines grown for 8 years in the field showed stable reduction of *CCR* transcripts, reduced lignin and hemicellulose, increased cellulose, and reduced growth (Leplé et al. 2007). Similarly, in the 2-year field trial with saplings of 14 *4CL*-downregulated poplar lines, large scale reductions in lignin content caused reduced growth (Voelker et al. 2010). Moreover, the *4CL* lines with the most drastic reduction in the lignin content also showed physiological abnormalities in their wood structure (Kitin et al. 2010; Voelker et al. 2010), along with reductions in wood strength and stiffness (Voelker et al. 2011b). Furthermore, tyloses and deposition of phenolics were observed in the xylem wood vessels, which resulted in reduced transport efficiency of water (Kitin et al. 2010), increasing the susceptibility of the low lignin *4CL* lines to drought (Voelker et al. 2011a).

The poplar lines with *CAD* and *COMT* modifications grown on two separate trial sites for 4 years experienced similar pressure by insects, fungi and bacteria as compared to the non-transgenic hybrid poplars. Additionally, the properties of soils did not differ between samples taken under the non-transgenic and *CAD* and *COMT* lines (Pilate et al. 2002). Moreover, the change in the CAD and COMT enzyme activities and in the lignin composition remained stable during the experiment. The quality and decomposition of the leaf litter of the *CAD* and *COMT* lines, and the assemblages of aquatic insects colonizing the litter in three natural streams was monitored in the study of Axelsson et al. 2010. The litter decomposition was different between *CAD* expressing and non-transgenic line, and some differences were also found between hybrid poplars in the concentration of phenolics and carbon, but no major differences were found in the composition of the insect communities among litter of GE or non-transgenic hybrid poplars.

Tilston et al. (2004) used the trunks of *CAD* and *COMT* white poplars lines remaining from the field trials described in Pilate et al. (2002) to assess their decomposition in soil. The authors reported that the results were inconsistent, and that no significant effects on decomposition could be correlated to the phenotypic variations between trees and variations in soil. In addition to the field studies assessing the unintended effects of lignin modified trees in field trials (Pilate et al. 2002; Tilston et al. 2004; Axelsson et al. 2010), glasshouse and *in vitro* experiments have been conducted to assess the development, performance or feeding preferences of insect herbivores (Tiimonen et al. 2005; Brodeur-Campbell et al. 2006), as well as ECM symbiosis (Seppänen et al. 2007; Tiimonen et al. 2008; Sutela et al. 2009), and soil microbe profiles (Bradley et al. 2007) with no substantial effects reported.

5.5 General Remarks

When considering the longevity of trees, field trials with GE trees need to take place for long period of time and should ideally include both the juvenile and mature stages of the trees. Although past and present field trials have usually not covered the full life span of the trees, a large body of useful data has nevertheless still been generated. For example it has been clearly demonstrated with a number of trials that gene transformation usually leads to stable expression of the transgenes with few if any unintended side effects.

Studies focusing on the impact of transgenic tree lines on non-target organisms under natural environmental conditions have included a range of organisms such as fungi, bacteria, the microfauna of rhizosphere, and arthropods (insects, spiders). So far, unintended changes have only been detected in studies evaluating population dynamics of arthropods on poplar lines and the rhizosphere microbial community of white spruces producing the δ-endotoxin of *B. thuringiensis* or marker genes *uidA* and *nptII*. It is interesting to note here that these effects may be positive or negative with regards to the community under investigation, although it should be emphasized that these changes have been minor in comparison to the effects of other standard forestry practises and could not be directly linked to the genetic engineering events themselves.

In addition, no HGT has been detected in field trials, and the transgene expression was found to remain stable under natural environment in the trees which were previously found to behave consistently under *in vitro* and greenhouse conditions (reviewed in Ahuja 2009; Walter et al. 2010; Fladung et al. 2013). Accordingly, it should be possible to eliminate most or all of the pleiotropic effects related to the performance of GE tree lines as observed in some field trials, at the early phases of the assessment of GE lines. This is an important finding and needs to be viewed in the context of accepted breeding practice where it is not necessarily easy to understand all changes introduced and/or to eliminate all undesired combinations.

6 Mitigation of Risk

6.1 Vertical Gene Transfer

As shown by DiFazio (2002), gene flow can dramatically be slowed down or even eliminated when trees have reduced levels of fertility. An early study by Strauss et al. (1995) describes the possibilities for establishing reproductive sterility in poplar to reduce or even avoid the gene flow of transgenes into non-transgenic relatives, by the expedient of including in any genetic transformation for commercial purposes of genes to induce sterility, at least for the male flowers. A similar approach was discussed by Höfig et al. (2006). However, the evaluation of the successful induction of sterility in transgenic forest tree species has been impaired by the long vegetative periods of trees. In other studies, early flowering transgenic poplar lines were used to evaluate several sterility constructs (Weigel and Nilsson 1995; Hoenicka and Fladung 2006b; Hoenicka et al. 2012). This strategy allows the evaluation of sterility strategies in transgenic trees 2 years after transformation instead of 5–10 years otherwise.

As described by Hoenicka and Fladung (2006b), sterility could be induced in early flowering poplar lines. Two different sterility constructs, TA29:: *Barnase* and C-GPDHC:: *Vst1* were evaluated in their study. The induction of sterility was clearly demonstrated for poplar plants transformed with the construct C-GPDHC:: *Vst1*, as the flowers formed but were found to be sterile. However, transgenic aspen carrying the TA29:: *Barnase* did not form flowers at all and the induction of absolute sterility in these trees by this gene construct could not be fully verified. Due to possible 'leaky' expression, this construct might disturb not only the flower organ development but also the development of other plant organs at a very early stage.

Another possibility to avoid VGT is either to shorten the rotation cycle that GE trees remain in the vegetative growth phase or prolongation of the vegetative period or even complete suppression of flower formation, allowing only vegetative growth of the GE trees. Genes can be utilised that delay the onset of flowering. The Flowering Locus C (FLC) for example is a repressor of flowering, thus overexpression of this gene leads to late flowering (Michaels and Amasino 1999).

6.2   Expression Stability

The integration of the T-DNA into the plant genome is a random event, however, chromatin accessibility may limit integration sites and has a major influence on the integration (Kumar and Fladung 2001b). The existing literature suggests that we are just beginning to understand the complex mechanism of transgene integration in plants. For targeted transgene transfer, there are many hurdles to place a new gene into a desired region of host genome. This may however not be such a big problem, because the exact location of a transgene can actually be determined rather easily after transformation. Using a PCR approach and data from sequencing project, it is possible to determine where the transgene actually is located and predictions can be made as to whether problems may follow from this later on in the life of the tree. Undesirable integrations can be eliminated at a very early stage and only those transclones that have integration in desired areas, can be propagated.

There may be several approaches which could lead to targeted integration. In comparison to plants homologous recombination is more frequent in bacteria and unicellular eukaryotes such as yeast. In contrast to illegitimate recombination, end products of transgene integration are predictable in homologous recombination. This precision has been exploited for the modification of chromosomal genes (gene targeting) in a number of biological systems. However, the efficiency of gene targeting is still unsatisfactory in higher plants due to the low frequency of homologous recombination.

Site-specific recombination systems have been shown to be effective tools for eliminating antibiotic marker genes which are needed for the transformation process, but which are no longer needed during later stages of development, and also for targeting transgenes to a desired locus (Kumar and Fladung 2001b). It has already been demonstrated that cre/*lox* from bacteria and FLP/*FRT*-recombination systems from yeast do function in poplar (Fladung et al. 2005). Subsequent "proof-of-concept" experiments using the FLP/*FRT*-recombination system from yeast, were able to demonstrate that a recognition target (or reporter gene) could be precisely placed into a target region of the host genome with simultaneous removal of the antibiotic selection marker (Fladung et al. 2010).

7 Conclusion

Considering the future global scenarios including continued increases in population numbers and challenges expected due to climate change, rising demands on land resources for food and feed are highly likely. Further, only limited areas will be available for forestry purposes and this restriction will be even more serious due to the additional expectations on what forests need to produce besides pulp and paper and timber. With an increased awareness of the importance of native forests for the world's climate, and so harvesting from these forests is expected to decrease and conversions of native forest land to agricultural land is also likely to become undesirable. In the changing world, the traditional pulp and paper industry and fibre production will still have important roles, but the bioeconomy concept will provide several new production possibilities. To satisfy these demands and to ensure sustainable production of raw materials and energy, the productivity of plantations and other managed forests urgently needs to be increased and forest trees may have to become more suitable for marginal lands that are unsuitable for agricultural purposes. So far plantation forestry has been used in the global scale in many variations including multiple or single forest tree species, genotypes/seed from tree breeding programs and clonally propagated superior genotypes. Production efficiencies are already much higher in plantation forests as compared to natural forests and consequently, plantation forestry with good management practices has turned out to be the future choice for many production purposes.

When considering the deployment of GE forest trees they are well suited to the already well developed plantation forestry practices and management procedures. Based on research evidence we can state that conventionally bred trees have much larger changes in their genomes as compared to transgenic trees modified for a specific trait, where the change is in fact very small. Due to the regulatory focusing on genetic engineering, we do know a lot more about risk associated with GE trees as compared to risk related to conventionally bred trees and therefore the risk evaluation may be complicated. However, for studying the behavior of transgenic trees under natural environments, the potential risks of transgenic trees in general must be evaluated in science-based field trials before their wider use. Such field testing of transgenic trees is a prerequisite for decision-making prior to the deployment of transgenic trees in commercial plantations. During the last 20 years several field trials have been established worldwide and these studies have failed to demonstrate any risk to the environment or human health. Nevertheless, after the first generation commercial transgenic crops (including herbicide or insect resistance or both), the second-generation transgenic crops (including stresses such as drought, cold, salt, heat, flood, increased yield, lower nutrient requirements, or increased tolerance to diseases and pathogens) are already being tested in field experiments around the world. Next generation traits are also considered in forestry and these traits need to be field tested before commercial deployment.

One of the major concerns about GE trees relates to the transfer of recombinant DNA into the gene pool of other plants in the vicinity of the same or different species via vertical or horizontal gene transfer. One of the strategies proposed to avoid vertical gene transfer might be the production of sterile trees. This approach has already been successfully tested using early-flowering genotypes of poplar and birch. If successful, this strategy would not only be suitable for reducing or eliminating the spread of GE material, but also to prevent plantation trees from becoming a pest in areas where they might have a tendency to outperform native plants. This situation exists for example in the South Island of New Zealand where Douglas-fir (*Pseudotsuga* sp.) is being planted as a preferred plantation tree for its resistance to colder temperatures. It can however become a weed spreading across the land, which is undesirable and very costly to control. A sterility approach with Douglas-fir could prevent this uncontrolled spread from happening, and provide an opportunity to have highly productive Douglas-fir plantations without a negative impact on the surrounding environment.

Legislation addressing the deliberate release or marketing of GE crops is being applied all over the world, but many differences exist between countries. In addition to the differing legislation, genome/gene sequences may be regulated by patents which complicates the situation and can cause difficulties not only for the scientific research, but also for the marketing of specific products. A good example of regulatory hurdles providing significant barriers to the humanitarian use of GE crops is the case of golden rice where there is a very lengthy and highly political process that has so far prevented the use of this plant for the health benefit of millions of people in countries where rice is the staple food

GE forest trees have been intensively tested for almost 20 years in field trials with no indications of any risk to the environment or human health having been detected. Many applications have shown to be appropriate for commercialisation in plantation forestry and GE papaya and poplar are already in commercial use. In several cases deregulation has been proposed (as in the case of papaya).

In most cases, the regulation of GE crops and trees is based on scientific knowledge and data that was available 15–20 years ago, when our expertise around genetic engineering and its potential risks was still at its infancy. It is now no longer reasonable to argue that not enough is known about the risks of this technology and an overwhelming majority of scientists and the major science organisations worldwide have stated that genetic engineering is at least as safe, if not more so, than conventional plant breeding technology. Regulatory bodies, politicians and NGOs urgently need to take note and speed up the deployment of GE trees to ensure that we can apply the sustainable and highly productive solutions to address climate change and rising population numbers that they offer.

The EU-COST (European Cooperation in Science and Technology) Action FP0905 (2010–2014), entitled "Biosafety of forest transgenic trees: improving the scientific basis for safe tree development and implementation of EU policy directives", focuses on key aspects related to the biosafety of field trials of commercially released genetically modified trees (GMTs). The main objective of the COST Action FP0905 is evaluating and substantiating scientific data relevant to the biosafety of GM trees (Fladung et al. 2012). The information obtained will strengthen the scientific bases of policy directives of the European Union (EU) related to the potential release of GMTs for commercial cultivation in Europe. A wide exchange of scientific knowledge has been initiated and it provides a unique opportunity to develop a common scientific baseline for biosafety research and development of engineered trees.

References

Ahuja MR (2009) Transgene stability and dispersal in forest trees. Trees Struct Funct 23:1125–1135

Axelsson EP, Hjältén J, LeRoy CJ, Julkunen-Tiitto R, Wennström A, Pilate G (2010) Can leaf litter from genetically modified trees affect aquatic ecosystems? Ecosystems 13:1049–1059

Axelsson EP, Hjältén J, LeRoy CJ, Whitham TG, Julkunen-Tiitto R, Wennström A (2011a) Leaf litter from insect-resistant transgenic trees causes changes in aquatic insect community composition. J Appl Ecol 48:1472–1479

Axelsson EP, Hjältén J, Whitham TG, Julkunen-Tiitto R, Pilate G, Wennström A (2011b) Leaf ontology interacts with Bt modification to affect innate resistance in GM aspens. Chemoecology 21:161–169

Balestrazzi A, Allegro G, Confalonieri M (2006) Genetically modified trees expressing genes for insect pest resistance. In: Fladung M, Ewald E (eds) Tree transgenesis – recent developments. Springer, Berlin

Barros G, Magnoli C, Reynoso MM, Ramirez ML, Farnochi MC, Torres A, Dalcero M, Sequeira J, Rubinstein C, Chulze S (2009) Fungal and mycotoxin contamination in Bt maize and non-Bt maize grown in Argentina. World Mycotoxin J 2:53–60

Batista R, Saibo N, Lourenço T, Oliveira MM (2008) Microarray analyses reveal that plant mutagenesis may induce more transcriptomic changes than transgene insertion. Proc Natl Acad Sci USA 105:3640–3645

Birol I, Raymond A, Jackman SD et al (2013) Assembling the 20Gb white spruce (Picea glauca) genome from whole-genome shotgun sequencing data. Bioinformatics. doi:10.1093/bioinformatics/btt178

Bock R (2009) The give-and-take of DNA: horizontal gene transfer in plants. Trends Plant Sci 15:11–22

Bradley KL, Hancock JE, Giardina CP, Pregitzer KS (2007) Soil microbial community responses to altered lignin biosynthesis in *Populus tremuloides* vary among three distinct soils. Plant Soil 294:185–201

Brodeur-Campbell SE, Vucetich JA, Richter DL, Waite TA, Rosemier JN, Tsai CJ (2006) Insect herbivory on low-lignin transgenic aspen. Environ Entomol 35:1696–1701

Brookes A (1999) The essence of SNPs. Gene 234:177–186

Brookes G, Barfoot P (2011) GM crops: global socio-economic and environmental impacts 1996–2009. PG Economics Ltd, UK

Brown JR (2003) Ancient horizontal gene transfer. Nat Rev Genet 4:121–132

Busov VB, Strauss SH, Pilate G (2010) Transformation as a tool for genetic analysis in *Populus*. Genetics and genomics of Populus. Springer, New York

Carpenter JE (2010) Peer-reviewed surveys indicate positive impact of commercialized GM crops. Nat Biotechnol 28:319–321

Chilton MD, Drummond MH, Merlo DJ, Sciaky D, Montoya AL, Gordon MP, Nester EW (1977) Stable incorporation of plasmid DNA into higher plant cells: the molecular basis of crown gall tumorigenesis. Cell 11:263–271

Christin P-A, Edwards EJ, Besnard G, Boxall SF, Gregory R, Kellogg EA, Hartwell J, Osborne CP (2012) Adaptive evolution of C4 photosynthesis through recurrent lateral gene transfer. Curr Biol. doi:10.1016/j.cub.2012.01.054

Dawe D, Robertson R, Unnevehr L (2002) Golden rice: what role could it play in alleviation of Vitamin A deficiency? Food Policy 27:541–560

DiFazio SP (2002) Measuring and modeling gene flow from hybrid poplar plantations: implications for transgenic risk assessment. Dissertation, Oregon State University

Dillon SK, Nolan M, Li W, Bell C, Wu HX, Southerton SG (2010) Allelic variation in cell wall candidate genes affecting solid wood properties in natural populations and land races of *Pinus radiata*. Genetics 185:1477–1487

EFI (2011) The white paper, European bioeconomy in 2030 – delivering sustainable growth by addressing the grand societal challenges. published in Brussels at the final event of the BECOTEPS project. http://www.efi.int/portal/news___events/press_releases/?id=306

Ewald D, Hu J, Yang M (2006) Transgenic forest trees in China. Tree transgenesis – recent developments. In: Fladung M, Ewald E (eds) Tree transgenesis – recent developments. Springer, Berlin

FAO (2010) Global forest resources assessment 2010. FAO Forestry Paper 163, Rome http://www.fao.org/docrep/013/i1757e/i1757e.pdf

Fawcett R, Towery D (2002) Conservation tillage and plant biotechnology: how new technologies can improve the environment by reducing the need to plow. CTIC. http://www.whybiotech.com/resources/tps/ConservationTillageandPlantBiotechnology.pdf

Fenning TM, Gershenzon J (2002) Where will the wood come from? Plantation forests and the role of biotechnology. Trends Biotechnol 20:291–296

Fenning TM, Walter C, Gartland KMA (2008) Forest biotech and climate change. Nat Biotechnol 26:615–617

Fillatti JJ, Sellmer J, McCown B, Haissig B, Comai L (1987) *Agrobacterium* mediated transformation and regeneration of *Populus*. Mol Gen Genet 206:192–199

Flachowsky H, Hanke V-M, Peil A, Strauss SH, Fladung M (2009) A review on transgenic approaches to accelerate breeding of woody plants. Plant Breed 128:217–226

Fladung M (1990) Transformation of diploid and tetraploid potato clones with the *rolC* gene of *Agrobacterium rhizogenes* and characterization of transgenic plants. Plant Breed 104:295–304

Fladung M, Kumar S (2002) Gene stability in transgenic aspen-*Populus*. III. T-DNA repeats influence transgene expression differentially among different transgenic lines. Plant Biol 4:329–338

Fladung M, Muhs HJ, Ahuja MR (1996) Morphological changes observed in transgenic *Populus* carrying the *rolC* gene from *Agrobacterium rhizogenes*. Silvae Genet 45:349–354

Fladung M, Nowitzki O, Ziegenhagen B, Kumar S (2003) Vegetative and generative dispersal capacity of field released transgenic aspen trees. Trees Struct Funct 17:412–416

Fladung M, Nowitzki O, Kumar S, Hoenicka H (2005) The site-specific recombination systems Cre-*lox* and FLP-*FRT* are functionally active in poplar. For Genet 12:121–130

Fladung M, Schenk TMH, Polak O, Becker D (2010) Elimination of marker genes and targeted integration via FLP/FRT-recombination system from yeast in hybrid aspen (*Populus tremula* L. × *P. tremuloides* Michx.). Tree Genet Genomes 6:205–217

Fladung M, Altsaar I, Bartsch D, Baucher M, Boscaleri F, Gallardo F, Häggman H, Hoenicka H, Nielsen K, Paffetti D, Séguin A, Stotzky G, Vettori C (2012) European discussion forum on transgenic tree biosafety. Nat Biotechnol 30:37–38

Fladung M, Hoenicka H, Ahuja MR (2013) Genomic stability and long-term transgene expression in poplar. Transg Res. doi:10.1007/s11248-013-9719-2

FPAC (2011) The new face of Canadian forest industry. The emerging bio-revolution. The Bio-pathways project. http://www.fpac.ca/publications/BIOPATHWAYS%20II%20web.pdf

Gao BJ, Zhang F, Hou DY, Wu BJ, Zhang SP, Zhao XL (2003) Structure of arthropod community in stands of transgenic hybrid poplar 741. J Beijing For Univ 25:62–64 (in Chinese with an English abstract)

Gao B, Gao S, Liu J, Jian W (2006) Variation of nutritional structure and ecological niche of arthropod community in plantation of transgenic insect-resistance hybrid poplar 741. J Beijing For Univ 25:3499–3507 (in Chinese with an English abstract)

Gassmann AJ, Petzold-Maxwell JL, Keweshan RS, Dunbar MW (2011) Field-evolved resistance to Bt maize by Western Corn Rootworm. PLoS One 6:e22629. doi:10.1371/journal.pone.0022629

Genissel A, Viard F, Bourguet D (2000) Population genetics of *Chrysomela tremulae*: a first step towards management of transgenic *Bacillus thuringiensis* poplars *Populus tremula* × *P. tremuloides*. Hereditas 133:85–93

Glenn TC (2011) Field guide to next-generation DNA sequencers. Mol Ecol Resour 11:759–769

Gómez-Barbero M, Berbel J, Rodríguez-Cerezo E (2008) Bt corn in Spain – the performance of the EU's first GM crop. Nat Biotechnol 26:384–386

González-Martínez SC, Wheeler NC, Ersoz E, Nelson CD, Neale DB (2007) Association genetics in *Pinus taeda* L I. Wood property traits. Genetics 175:399–409

Gonzalvez D (2004) Transgenic papaya in Hawaii and beyond. AgBioForum 7:36–40

Grace LJ, Charity JA, Greham B, Kay N, Walter C (2005) Insect resistant transgenic *Pinus radiata*. Plant Cell Rep 24:103–111

Häggman H, Rusanen M, Jokipii S (2008) Cryopreservation of in vitro tissues of deciduous forest trees. In: Reed BM (ed) Plant cryopreservation: a practical quide. Springer, New York

Hancock JF (2003) A framework for assessing the risk of transgenic crops. Bioscience 53:512–519

Hawkins S, Leplé J, Cornu D, Jouanin L, Pilate G (2003) Stability of transgene expression in poplar: a model forest tree species. Ann For Sci 60:427–438

Henderson AR, Walter C (2006) Genetic engineering in conifer plantation forestry. Silvae Genet 55:253–262

Hoenicka H, Fladung M (2006a) Biosafety in *Populus spp.* and other forest trees: from non-native species to taxa derived from traditional breeding and genetic engineering. Trees 20:131–144

Hoenicka H, Fladung M (2006b) Faster evaluation of sterility strategies in transgenic early flowering poplar. Silvae Genet 55:241–292

Hoenicka H, Lehnhardt D, Polak O, Fladung M (2012) Early flowering and genetic containment studies in transgenic poplar. iForest 5:138–146. doi:10.3832/ifor0621-005

Höfig KP, Möller R, Donaldson L, Putterill J, Walter C (2006) Towards male sterility in *Pinus radiata* – a stilbene synthase approach to genetically engineer nuclear male sterility. Plant Biotechnol J 4:333–343

Hou Y, Su X, Jiao R, Huang Q, Chu Y (2009) Effects of transgenic *Populus alba* × *P. glandulosa* on soil microorganism. Sci Silvae Sinicae 45:148–152 (in Chinese with an English abstract)

Hu JJ, Tian YC, Han YF, Li L, Zhang BE (2001) Field evaluation of insect-resistant transgenic *Populus nigra* trees. Euphytica 121:123–127

Hu JJ, Zhang Y, Lu MZ, Zhang J, Zhang S (2004) Transgene stability of transgenic *Populus nigra* and its effects on soil microorganism. Sci Silvae Sinicae 40:105–109 (in Chinese with an English abstract)

Hu JJ, Li SM, Lu MZ, Li JX, Li KH, Sun XQ, Zhao ZY (2007) Stability of insect-resistance of Bt transformed *Populus nigra* plantation and its effects on the natural enemies of insects. J For Res 20:656–659 (in Chinese with an English abstract)

Hutchison WS, Burkness EC, Mitchell PD, Moon RD, Leslie TW, Fleischer SJ, Abrahamson M, Hamilton KL, Steffey KL, Gray ME, Helmich RL, Kaster LV, Hunt TE, Wright RJ, Pecinovsky K, Rabaey TL, Flood BR, Raun ES (2010) Areawide suppression of European corn borer with Bt maize reaps savings to non-Bt maize growers. Science 330:222–225

James C (2012) Global status of commercialized biotech/GM crops: 2012. ISAAA Brief No. 44. ISAAA, Ithaca, NY

Jiang WH, Liu JX, Zhang F, Gai BJ (2009) Population dynamic of target pest, non-target pests and major natural enemy in transgenic hybrid poplar 741. J Shandong Agric Univ (Nat Sci) 40:195–199 (in Chinese with an English abstract)

Jing Z, Gallardo F, Pascual M, Sampalo R, Romero J, de Navarra A, Canovás F (2004) Improved growth in a field trial of transgenic hybrid poplar overexpressing glutamine synthetase. New Phytol 164:137–145

Kaldorf M, Fladung M, Muhs H, Buscot F (2002) Mycorrhizal colonization of transgenic aspen in a field trial. Planta 214:653–660

Keese P (2008) Risks from GMOs due to horizontal gene transfer. Environ Biosafety Res 7:123–149

Kitin P, Voelker S, Meinzer F, Beeckman H, Strauss S, Lachenbruch B (2010) Tyloses and Phenolic deposits in xylem vessels impede water transport in low-lignin transgenic poplars: a study by cryo-flourescence microscopy. Plant Physiol 154:887–898

Kleter GA, Bhula R, Bodnaruk K, Carazo E, Felsot AS, Harris CA, Katayama A, Kuiper HA, Racke KD, Rubin B, Shevah Y, Stephenson GR, Tanaka K, Unsworith J, Wauchope RD, Wong S-S (2007) Altered pesticide use on transgenic crops and the associated genral impact from an environmental perspective. Pest Manag Sci 63:1107–1115

Kole C, Hall TC (eds) (2008) Compendium of transgenic crop plants: transgenic forest tree species. Wiley. doi:10.1002/9781405181099.k0407

Kumar S, Fladung M (2001a) Gene stability in transgenic aspen (*Populus*) II. Molecular characterization of variable expression of transgene in wild and hybrid aspen. Planta 213:731–740

Kumar S, Fladung M (2001b) Controlling transgene integration in plants. Trends Plant Sci 6:55–159

Lachance D, Hamel L-P, Pelletier F, Valéro J, Bernier-Cardou M, Chapman K, van Frankenhuyzen K, Séguin A (2007) Expression of a *Bacillus thuringiensis cry1Ab* gene in transgenic white spruce and its efficacy against the spruce budworm (*Choristoneura fumiferana*). Tree Genet Genomes 3:153–167

Lamarche J, Hamelin RC (2007) No evidence of an impact on the rhizosphere diazotroph community by the expression of *Bacillus thuringiensis* Cry1Ab toxin by Bt white spruce. Appl Environ Microbiol 73:6577–6583

Lamarche J, Stefani FOP, Séguin A, Hamelin RC (2011) Impact of endochitinase-transformed white spruce on soil fungal communities under greenhouse conditions. FEMS Microbiol Ecol 76:199–208

Lambardi M, Aylin Ozudogru E, Benelli C (2008) Cryopreservation of embryogenic cultures. In: Reed BM (ed) Plant cryopreservation: a practical quide. Springer, New York

LeBlanc PM, Hamelin RC, Filion M (2007) Alteration of soil rhizosphere communities following genetic transformation of white spruce. Appl Environ Microbiol 73:4128–4134

Leplé J, Dauwe R, Morreel K, Storme V, Lapierre C, Pollet B, Naumann A, Kang K-Y et al (2007) Downregulation of cinnamoyl-coenzyme a Reductase in poplar: multiple-level phenotyping reveals effects on cell wall polymer metabolism and structure. Plant Cell 19:3669–3691

Li J, Brunner AM, Meilan R, Strauss SH (2009) Stability of transgenes in trees: expression of two reporter genes in poplar over three field seasons. Tree Physiol 29:299–312

Li J, Brunner AM, Shevchenko O, Meilan R, Ma C, Skinner JS, Strauss SH (2008a) Efficient and stable transgene suppression via RNAi in field-grown poplars. Transgenic Res 17:679–694

Li J, Meilan R, Ma C, Barish M, Strauss SH (2008b) Stability of herbicide resistance over 8 years of coppice in field-grown, genetically engineered poplars. West J Appl For 23:89–93

Londo JP, Bautista NS, Sagers CL, Lee EH, Watrud LS (2010) Glyphosate drift promotes changes in fitness and transgene gene flow in canola (*Brassica napus*) and hybrids. Ann Bot 106:957–965

Lottmann J, O'Callaghan BD, Walter C (2010) Bacterial and fungal communities in the rhizosphere of field-grown genetically modified pine trees (*Pinus radiata* D.). Environ Biosafety Res 9:25–40

Meilan R, Auerbach D, Ma C, DiFazio S, Strauss S (2002a) Stability of herbicide resistance and *GUS* expression in transgenic hybrid poplars (*Populus* sp.) during four years of field trials and vegetative propagation. Hort Sci 37:277–280

Meilan R, Han K-H, Ma C, DiFazio SP, Eaton JA, Hoien EA, Stanton BJ, Crockett RP, Taylor ML, James RR, Skinner JS, Jounin L, Pilate G, Strauss SH (2002b) The *CP4* transgene provides high levels of tolerance to Roundup® herbicide in field-grown hybrid poplars. Can J For Res Rev Can Rech For 32:967–976

Meilan R, Ma C, Cheng S, Eaton J, Miller L, Crockett R, DiFazio S, James R, Strauss S (2000) High levels of Roundup® and leaf beetle resistance in genetically engineered hybrid cottonwoods. In: Blatner KA, Johnson JJ (eds) Hybrid poplars in the pacific northwest: culture, commerce and capability. Washington State University Cooperative Extension, Pullman

Michaels SD, Amasino RM (1999) Flowering locus C encodes a novel MADS domain protein that acts as a repressor of flowering. Plant Cell 11:949–956

Modi RI, Castilla LH, Puskas-Rozsa S, Helling RB, Adams J (1992) Genetic changes accompanying increased fitness in evolving populations of *Escherichia coli*. Genetics 130:241–249

Neale D (2011) Genomics-based breeding in forest trees: are we there yet? BMC Proc 5:14. doi:10.1186/1753-6561-5-S7-I4

Neale DB, Kremer A (2011) Forest tree genomics: growing resources and applications. Nat Rev Genet 12:111–122

Neale D, Savolainen O (2004) Association genetics of complex traits in conifers. Trends Plant Sci 9:325–330

Nehls U, Zhang C, Tarkka M, Hampp R, Fladung M (2006) Investigation of horizontal gene transfer from transgenic aspen to ectomycorrhizal fungi. In: Fladung M, Ewald E (eds) Tree transgenesis – recent developments. Springer, Berlin

Nellemann C, MacDevette M, Manders T, Eickhout B, Svihus B, Prins AG, Kaltenborn BP (eds) (2009) The environmental food crisis – The environment's role in averting future food crises. A UNEP rapid response assessment. United Nations Environment Programme, GRID-Arendal. http://www.grida.no/files/publications/FoodCrisis_lores.pdf

Newhouse AE, Schrodt F, Liang H, Maynard CA, Powell WA (2007) Transgenic American elm shows reduced Dutch elm disease symptoms and normal mycorrhizal colonization. Plant Cell Rep 26:977–987

Nystedt B, Street NR, Wetterbom A et al (2013) The Norway spruce genome sequence and conifer genome evolution. Nature. doi:10.1038/nature12211

Oldfield S, Lusty C, MacKinven A (1998) The world list of threatened trees. Word Conservation Press, Cambridge

Pasonen H-L, Lu J, Niskanen A-M, Seppänen S-K, Rytkönen A, Raunio J, Pappinen A, Kasanen R, Timonen S (2009) Effects of sugar beet chitinase IV on root-associated fungal community of transgenic silver birch in a field trial. Planta 230:973–983

Pasonen H-L, Seppänen S-K, Degefu Y, Rytkönen A, von Weissenberg K, Pappinen A (2004) Field performance of chitinase transgenic silver birches (*Betula pendula*): resistance to fungal diseases. Theor Appl Genet 109:562–570

Pasonen H-L, Vihervuori L, Seppänen S-K, Lyytikäinen-Saarenmaa P, Ylioja T, von Weissenberg K, Pappinen A (2008) Field performance of chitinase transgenic silver birch (*Betula pendula* Roth): growth and adaptive traits. Trees Struct Funct 22:413–421

Pilate G, Guiney E, Holt K et al (2002) Field and pulping performances of transgenic trees with altered lignification. Nat Biotechnol 20:607–612

Powell WA, Maynard CA, Boyle B, Séquin A (2006) Fungal and bacterial resistance in transgenic trees. In: Fladung M, Ewald E (eds) Tree transgenesis – recent developments. Springer, Berlin

Pray CE, Huang J, Hu R, Rozelle S (2002) Five years of Bt cotton in China – the benefits continue. Plant J 31:423–430

Qaim M (2009) The economics of genetically modified crops. Annu Rev Resour Econ 1:665–693

Quesada T, Gopal V, Cumbie WP, Eckert AJ, Wegrzyn JL, Neale DB, Goldfarb B, Huber DA, Casella G, Davis JM (2010) Association mapping of quantitative disease resistance in a natural population of loblolly pine (*Pinus taeda* L.). Genetics 186:677–686

Raney T (2006) Economic impact of transgenic crops in developing countries. Curr Opin Plant Biol 17:174–178

Robischon M (2006) Field trials with transgenic trees – state of the art and developments. In: Fladung M, Ewald E (eds) Tree transgenesis – recent developments. Springer, Berlin

Rottmann WH, Meilan R, Sheppard LA, Brunner AM, Skinner JS, Ma C, Cheng S, Jouanin L, Pilate G, Strauss SH (2000) Diverse effects of overexpression of LEAFY and PTLF, a poplar (*Populus*) homolog of LEAFY/FLORICAULA, in transgenic poplar and *Arabidopsis*. Plant J 22:235–245

Schell J, Van Montagu M, De Beuckeleer M, De Block M, Depicker A, De Wilde M, Engler G, Genetello C, Hernalsteens JP, Holsters M, Seurinck J, Silva B, Van Vliet F, Villarroel R (1979) Interactions and DNA transfer between *Agrobacterium tumefaciens*, the Ti-plasmid and the plant host. Proc R Soc Lond B Biol Sci 204:251–266

Schnepf E, Crickmore N, Van Rie J, Lereclus D, Baum J, Feitelson J, Zeigler D, Dean D (1998) *Bacillus thuringiensis* and its pesticidal crystal proteins. Microbiol Mol Biol Rev 62:775–806

Schnitzler F, Burgess EPJ, Kean AM, Philip BA, Barraclough EI, Malone LA, Walter C (2010) No unintended impacts of transgenic pine (*Pinus radiata*) trees on above ground invertebrate communities. Environ Entomol 39:1359–1368

Semagn K, Bjornstad Å, Ndjiondjop MN (2006) An overview of molecular marker methods for plants. Afr J Biotechnol 5:2540–2568

Seppänen S-K, Pasonen H-L, Vauramo S, Vahala J, Toikka M, Kilpeläinen I, Setälä H, Teeri TH, Timonen S, Pappinen A (2007) Decomposition of the leaf litter and mycorrhiza forming ability of silver birch with a genetically modified lignin biosynthesis pathway. App Soil Ecol 36:100–106

Simmons BA, Logué D, Ralph J (2010) Advances in modifying lignin for enhanced biofuel production. Curr Opin Plant Biol 13:313–320

Stefani FOP, Bérube JA (2006) Evaluation of foliar fungal endophyte incidence in field-grown transgenic Bt white spruce trees. Can J Bot 84:1573–1580

Stefani FOP, Moncalvo J, Séguin A, Bérubé JA, Hamelin RC (2009) Impact of an 8-year-old transgenic poplar plantation on the ectomycorrhizal fungal community. Appl Environ Microbiol 75:7527–7536

Stefani FOP, Tanguay P, Pelletier G, Piche Y, Hamelin RC (2010) Impact of endochitinase-transformed white spruce on soil fungal biomass and ectendomycorrhizal symbiosis. Appl Environ Microbiol 76:2607–2614

Stegemann S, Keuthe M, Greiner S, Bock R (2012) Horizontal transfer of chloroplast genomes between plant species. Proc Natl Acad Sci U S A 109:2434–2438

Strauss SH, Rottmann WH, Brunner AM, Sheppard LA (1995) Genetic engineering of reproductive sterility in forest trees. Mol Breed 1:5–26

Su X, Chu Y, Li H, Hou Y, Zhang B, Huang Q, Hu Z, Huang R, Tian Y (2011) Expression of multiple resistance genes enhances tolerance to environmental stressors in transgenic poplar (*Populus × euramericana* 'Guariento'). PLoS One 6:e24614. doi:10.1371/journal.pone.0024614

Sutela S, Niemi K, Edesi J, Laakso T, Saranpää P, Vuosku J, Mäkelä R, Tiimonen H, Chiang VL, Koskimäki J, Suorsa M, Julkunen-Tiitto R, Häggman H (2009) Phenolic compounds in ectomycorrhizal interaction of lignin modified silver birch. BMC Plant Biol 9:124. doi:10.1186/1471-2229-9-124

Thavamanikumar S, Tibbits J, McManus L, Ades P, Stackpole D, Hadjigol S, Vaillancourt R, Zhu P, Bossinger G (2011) Candidate gene-based association mapping of growth and wood quality traits in *Eucalyptus globulus* Labill. BMC Proc 5:O15. doi:10.1186/1753-6561-5-S7-O15

Tiimonen H, Aronen T, Laakso T, Saranpää P, Chiang V, Häggman H, Niemi K (2008) *Paxillus involutus* forms an ectomycorrhizal symbiosis and enhances survival of PtCOMT-modified *Betula pendula* in vitro. Silvae Genet 57:235–242

Tiimonen H, Aronen T, Laakso T, Saranpää P, Chiang VL, Ylioja T, Roininen H, Häggman H (2005) Does lignin modification affect feeding preference or growth performance of insect herbivores in transgenic silver birch (*Betula pendula* Roth)? Planta 222:699–708

Tilston EL, Halpin C, Hopkins DW (2004) Genetic modifications to lignin biosynthesis in field-grown poplar trees have inconsistent effects on the rate of woody trunk decomposition. Soil Biol Biochem 36:1903–1906

Tuskan GA, DiFazio S, Jansson S et al (2006) The genome of black cottonwood, *Populus trichocarpa* (Torr. & Gray). Science 313:1596–1604

Ulrich K, Stauber T, Ewald D (2008a) *Paenibacillus* – a predominant endophytic bacterium colonising tissue cultures of woody plants. Plant Cell Tiss Org 93:347–351

Ulrich K, Ulrich A, Ewald D (2008b) Diversity of endophytic bacterial communities in poplar grown under field conditions. FEMS Microbiol Ecol 63:169–180

Vauramo S, Pasonen H, Pappinen A, Setälä H (2006) Decomposition of leaf litter from chitinase transgenic silver birch (*Betula pendula*) and effects on decomposer populations in a field trial. Appl Soil Ecol 32:338–349

Vihervuori L, Pasonen H, Lyytikäinen-Saarenmaa P (2008) Density and composition of an insect population in a field trial of chitinase transgenic and wild-type silver birch (*Betula pendula*) clones. Environ Entomol 37:1582–1591

Voelker SL, Lachenbruch B, Meinzer FC et al (2010) Antisense down-regulation of *4CL* expression alters lignification, tree growth, and saccharification potential of field-grown poplar. Plant Physiol 154:874–886

Voelker SL, Lachenbruch B, Meinzer FC, Kitin P, Strauss SH (2011a) Transgenic poplars with reduced lignin show impaired xylem conductivity, growth efficiency and survival. Plant Cell Env 34:655–668

Voelker SL, Lachenbruch B, Meinzer FC, Strauss SH (2011b) Reduced wood stiffness and strength, and altered stem form, in young antisense *4CL* transgenic poplars with reduced lignin contents. New Phytol 189:1096–1109

Walter C (2004) Genetic engineering in conifer forestry: technical and social considerations. In Vitro Cell Dev Biol Plant 40:434–441

Walter C, Fenning T (2004) Deployment of genetically-engineered trees in plantation forestry – an issue of concern? The science and politics of genetically modified tree plantations. In: Walter C, Carson M (eds) Plantation forest biotechnology for the 21st century. Research Signpost, Trivandrum

Walter C, Boerjan W, Fladung M (2010) The 20-year environmental safety record of GM trees. Nat Biotechnol 28:656–658

Walter C, Grace LJ, Wagner A, White DWR, Walden AR, Donaldson SS, Hinton H, Gardner RC, Smith DR (1998) Stable transformation and regeneration of transgenic plants of *Pinus radiata* D. Don. Plant Cell Rep 17:460–468

Wang J, Zhang JG, Hu JJ, Zhang Z, Zhang SG (2004) Studies on safety assessment of transgenic Bt poplar. China Biotechnol 24:49–52 (in Chinese with an English abstract)

Weigel D, Nilsson O (1995) A developmental switch sufficient for flower initiation in diverse plants. Nature 377:495–500

Wu F (2008) Field evidence: Bt corn and mycotoxin reduction, ISB News Report Feb:1–4. http://www.isb.vt.edu/news/2008/feb08.pdf

Yang M, Lang H, Gao B, Wang J, Zheng J (2003) Insecticidal activity and transgene expression stability of transgenic hybrid poplar clone 741 carrying two insect-resistant genes. Silvae Genet 52:197–201

Yao L, Zhou GN, Feng ZH, Gao BJ, Yuan SL (2006) Survival and development immature *Harmonia axyridis* (Pallas) feeding on *Chaitophorus popleti* (Panzen) propagated on transgenic insect-resistance hybrid poplar 741. J Agric Univ Hebei 29:73–76

Zawaski C, Kadmiel M, Pickens J, Ma C, Strauss S, Busov V (2011) Repression of gibberellin biosynthesis or signaling produces striking alterations in poplar growth, morphology, and flowering. Planta 234:1285–1298

Zhang Z, Wang JH, Zhang JG, Zhang SG (2004) Effects of transgenic poplars to the structures of insect community. Sci Silvae Sinicae 40:84–89 (in Chinese with an English abstract)

Zhang Q, Zhang Z-Y, Lin S-Z, Lin Y-Z (2005a) Resistance of transgenic hybrid triplolds in *Populus tomentosa* Carr. Against 3 species of lepidopterans following two winter dormancies conferred by high level expression of cowpea trypsin inhibitor gene. Silvae Genet 54:108–116

Zhang Q, Zhang Z-Y, Lin S-Z, Lin Y-Z, Yang L (2005b) Assessment of rhizospheric microorganisms of transgenic *Populus tomentosa* with cowpea trypsin inhibitor (*CpTI*) gene. For Stud China 7:28–34

Zhang F, Maeder ML, Unger-Wallace E, Hoshaw JP, Reyon D, Christian M, Li X, Pierick CJ, Dobbd D, Peterson T, Joung JK, Voytas DF (2010) High frequency targeted mutagenesis in *Arabidopsis thaliana* using zinc finger nucleases. Proc Natl Acad Sci U S A 107:12028–12033

Zhang B, Chen M, Zhang X, Luan H, Diao S, Tian Y, Su X (2011a) Laboratory and field evaluation of the transgenic *Populus alba* × *Populus glandulosa* expressing double coleopteran-resistance genes. Tree Physiol 31:567–573

Zhang B, Chen M, Zhang X, Luan H, Tian Y, Su X (2011b) Expression of Bt-Cry3A in transgenic *Populus alba* × *P. glandulosa* and its effects on target and non-target pests and the arthropod community. Transgenic Res 20:523–532

Scientific Research Related to Genetically Modified Trees

Armand Séguin, Denis Lachance, Annabelle Déjardin, Jean-Charles Leplé, and Gilles Pilate

Abstract Over the last decade, we have witnessed impressive advances in tree molecular biology and the consolidation of tree genomics. We have essentially moved from a small portfolio of genes focusing on a specific genetic trait to large databases including thousands of genes and their respective expression profiles. In 2006, we saw the publication of the first genomic sequence of a tree, the model tree species *Populus trichocarpa*. Though, not surprisingly, much progress has been made with *Populus,* impressive research results have also been realized in more recalcitrant coniferous species such as pines and spruces.

Despite the rapid advances in tree genomics, tree genetic engineering (GE) proved to be a bottleneck requiring the development of whole-tree regeneration protocols using in vitro culture and an effective method of DNA transfer. The introduction of simple single gene traits such as insect resistance was the early target of tree genetic engineers. Today more tree species are compatible with GE and at a higher throughput, making functional genomics approaches possible to improve our understanding of gene functions.

In this chapter we will provide an historical overview of the advances made in GE of trees. We will also explore the various applications of tree GE to improving response to biotic and abiotic stresses, which is becoming more important in an ever-changing environment. Improvement of specific traits for tree domestication will also be covered. Lastly, we will briefly discuss issues related to the regulation of GM trees, particularly concerning genetic containment and environmental risk assessment.

A. Séguin (✉) • D. Lachance
Natural Resources Canada, Canadian Forest Service, Laurentian Forestry Centre,
1055 du P.E.P.S., P.O. Box 10380, Stn. Sainte-Foy, Québec, QC, Canada G1V 4C7
e-mail: armand.seguin@NRCan.gc.ca

A. Déjardin • J.-C. Leplé • G. Pilate
INRA, UR0588 Amélioration, Génétique et Physiologie Forestières,
CS 40001 Ardon, F-45075 Cedex 2, Orléans, France
e-mail: Gilles.pilate@orleans.inra.fr

T. Fenning (ed.), *Challenges and Opportunities for the World's Forests in the 21st Century*, Forestry Sciences 81, DOI 10.1007/978-94-007-7076-8_22,
© Crown Employee Canada 2014

1 Introduction

The first genetically modified (GM) organism described was the bacterium *Escherichia coli* in which a recombinant plasmid derived from two separate sources of DNA was expressed (Cohen et al. 1973). That accomplishment and the awareness of the potential related applications led to the holding of the Asilomar Conference on Recombinant DNA organized by the Nobel laureate Paul Berg. Held in 1975, the conference brought together prominent scientists from around the world as well as members of the press, government officials and lawyers in order to review the scientific progress in recombinant DNA research and to debate the appropriate measures to take in order to deal with the potential associated biohazards (Berg et al. 1975a, b). The conference was in response to a voluntary moratorium on recombinant DNA research imposed in 1973–1974 by the scientists themselves over safety concerns. The scientific community recognized the importance of developing the appropriate safeguards related to recombinant DNA research, particularly in the context of the health impact on laboratory staff and the general population as well as the effects on animal and plant species. A major outcome of the conference was the classification of experiments under four categories according to an estimated risk with an associated level of containment.

It took almost a decade following that first report on recombinant DNA work on *E. coli* before the stable transfer and long-term expression of a bacterial antibiotic resistance gene in plant cells would be reported. Almost simultaneously three separate research groups were able to demonstrate that *Agrobacterium tumefaciens* T-DNA vectors could be manipulated through recombinant DNA techniques and used to introduce and stably integrate a marker gene conferring antibiotic resistance into plant chromosomes (Bevan et al. 1983; Fraley et al. 1983; Herrera-Estrella et al. 1983). It was the start of a new era in plant biology with genetic engineering being used to obtain genetically modified plants for specific traits as well as to better understand the function of plant genes and their complex regulation. Despite rapid advances in plant genetic engineering and the use of improved promoters and in vitro culture systems, it took an additional 4 years before the first production of a transgenic tree with a silvicultural trait was reported (Fillatti et al. 1987). That initial report described the stable transformation of a poplar clone (NC-5339; *Populus alba x Populus grandidentata*) with genes encoding the antibiotic resistance gene neomycin phosphotransferase II (NPTII) and the bacterial EPSP (5-enolpyruvylshikimate 3-phosphate) synthase gene conferring resistance to the herbicide glyphosate.

In the early days of plant genetic transformation, very few genes, mostly of bacterial origin, were available. Herbicide resistant crops, primarily resistant to glyphosate, were amongst the first transgenic plants brought to market in 1996 and rapidly adopted by farmers (Dill 2005). Other strategies for herbicide resistance were developed, but insect resistance was also an undeniable major research goal (Castle et al. 2006). The insecticidal properties of *Bacillus thuringiensis* (*Bt*) spores containing crystal formulations were well established and used as a biological insecticide. The potent and specific insecticidal activity was attributed to the Cry δ–endotoxins and the availability of genes encoding these Cry proteins prompted the production of transgenic plants expressing these genes. Transgenic plants producing their own biological pesticide were more resistant to insects (Sanahuja et al. 2011). Early transformation successes were initially limited to a few crop species, with transformation of recalcitrant species being a major challenge (Vain 2007). Thanks to the improvement of specific procedures for recalcitrant species, a wide array of plant species has been amenable to genetic engineering even though this may concern a small number of genotypes for each species. Likewise, with the huge progress in genomics, the array of genes to be introduced is no longer a limitation to modifying/improving all kinds of agronomical or technological traits. Recent scientific breakthroughs in plant molecular biology and genomics such as the discovery of microRNA and complex gene regulation by transcription factors have made it possible to develop novel strategies to genetically modify plants to improve stress tolerance, pest and disease resistance as well as nutritional values, to name a few (Farre et al. 2010).

2   Genetic Modification for Tree Domestication

In comparison with crop plants that have been domesticated for centuries, tree improvement through conventional breeding programs was established only within the last century. Most tree improvement programs are in their third generation. Tree genetic improvement through breeding is a tedious task considering the biology of forest tree species. Trees are long lived and can take many years to reach sexual maturity, and traits such as growth, tree architecture, wood properties and pest resistance require more years to evaluate unless strong correlations can be established between juvenile and mature measurements. Despite these difficulties, outstanding advances have been made for several economically important species. Part V, "Tree Breeding and Commercial Forestry" covers some examples of this in detail. It is worth mentioning however that genomic developments in trees, in the context of tree domestication and adaptation, have been rapidly evolving (Boerjan 2005; Harfouche et al. 2011). Obviously, forest tree species will benefit in the near future from recent genomic-based breeding technologies (Neale 2007). Indeed, tree breeding programs for several forest tree species have taken advantage of the development of molecular markers and genetic maps on several occasions. Certainly, tree breeding also benefits from the rapid development of both tree genomics (Kumar and Fladung 2004; Neale and Kremer 2011) and efficient ways for clonal propagation (Merkle and Dean 2000). Elite genotypes and hybrids of important forest tree species such as *Populus* and *Eucalyptus* are easily multiplied by cuttings while conifers could also use somatic embryogenesis for rapid multiplication. Likewise, tree genetic engineering, which offers numerous benefits to ease forest tree domestication (Campbell et al. 2003), is now available for a number of tree species of high economic value (such as poplar, eucalyptus, pine and

spruce), making it possible in the near future to produce and multiply new genotypes with improved expression of the target gene(s) identified by molecular breeding (Harfouche et al. 2012). However, to fully conserve genetic gain, genetic
modification should at least be carried out at the end of the process on elite clones.
This is still a challenging issue for poplar since several commercially important
clones belong to hybrid poplar species, where reliable transformation protocols are
not yet available. However the situation is different with conifers as robust methods for in vitro culture using somatic embryogenesis have been developed for several commercially important species such as pines and spruces (Tang and Newton
2003; Henderson and Walter 2006).

In the drug development process, clinical trials are the ultimate step in validating
the expected benefits of a medication and also evaluating in a broader context
potential side effects and unexpected physiological interactions. In plant breeding,
field trials are somewhat similar to clinical trials and represent a critical step in
validating the success of the primary laboratory/greenhouse-based work. Genetic
gain and improved phenotypes following plant breeding are essentially evaluated in
natural field conditions, in order to submit them to a wide array of environmental cues.
For crop plant breeders, this step is essential as success in obtaining an improved
phenotype is based on parent lines bearing particular attributes and subsequent
progeny screening will reveal the best individuals.

2.1 Obtaining Genetically Transformed Material: A Real Bottleneck

Tobacco and petunia species were the preferred plant species in the early days of
molecular biology, as they were compatible with in vitro organogenesis and amenable to genetic transformation using *Agrobacterium tumefaciens*. Today recombinant technology approaches are applicable to a relatively wide spectrum of
vascular plants, from crops to ornamentals and forest tree species. In fact, species
from nearly all botanical groups have been successfully transformed, but often for
a given species, only a few selected genotypes are actually amenable to genetic
transformation. For example, GM poplars have been described in only a handful
of genotypes, chosen primarily for their ease of transformation, and in most cases
distantly related to homologated cultivars clonally planted for commercial production. Commercial application of GM technology would require the development of routine transformation protocols for commercial clones: for example in
poplar see Yevtushenko and Misra 2010. Tree genetic engineers are also faced
with long juvenile time periods making introgression from one transgenic line to
another a very long process relative to crop plants. In addition, several easy to
transform hybrid species exhibit limited or no possibility of crossing with desirable commercial genotypes.

In this chapter, we will not cover in much detail the developments related to in
vitro propagation and DNA transformation technologies but will instead give an

overview of these aspects, while reporting on specific cases of GM trees. Indeed, summarizing the considerable efforts realized by the scientific community in achieving stable transformation in forest tree species would be a significant task. To save space, we invite readers to consult the following reviews: Tang and Newton 2003 describe advances in the genetic transformation of coniferous species whereas Merkle and Nairn 2005 focus more on the hardwood species and Ye et al. 2011 on poplar. The review by Nehra et al. 2005 also provides an excellent summary of the technical strategies leading to the production of GM trees.

As previously mentioned, the first transgenic tree to be produced was a poplar genotype, owing to its amenability to in vitro culture through organogenesis and its susceptibility to *Agrobacterium.* The same could not be said for other species however. For a long time, gymnosperms were thought to be incompatible with *Agrobacterium*-based transformation procedures. Despite progress in conifer tissue culture through SE (e.g. initiation of embryogenic cultures, cryopreservation), success in genetic engineering of conifers initially relied on direct gene transfer technologies with microprojectile bombardment being considered the preferred method for loblolly pine (Walter et al. 1998) and spruce (Séguin et al. 1996). *Picea glauca* was the first conifer to be genetically modified with the *cry1* gene from *Bt* for insect resistance, a trait of commercial interest (Ellis et al. 1993). Though first reported in larch by Shin et al. 1994, it was only in the late 1990s that routine transformation protocols for several coniferous species using disarmed *A. tumefaciens* strains became available. Stable transformation of hybrid larch (Levée et al. 1997), spruces and pine (Levée et al. 1999; Wenck et al. 1999) was achieved using embryogenic cultures as starting material. Transformation by *Agrobacterium* has since become the method of choice for conifers owing to its relative ease of use and simpler genomic insertion patterns. In most of those early studies only marker genes such as GUS or GFP were tested. Specific silvicultural traits will be discussed in the next section. Reports on GM eucalyptus, a forest species of great economic importance, were initially absent but are now nearly routine for at least a few eucalyptus species. Despite its susceptibility to *Agrobacterium*, whole plant regeneration following transformation was a major challenge to overcome (Ho et al. 1998; Tournier et al. 2003). The following sections will highlight examples of traits engineered in woody species.

3 Exploring GM Technologies in Trees; Applied and Fundamental Research

3.1 Herbicide Resistance

Weed control has always been a major issue in agriculture with its associated costs and management practices. Herbicide-resistant crop plants could be obtained by the introduction of a single gene and enable the use of broad-spectrum herbicides for weed control, reducing the environmental impacts and enabling the use of reduced tillage practices. Herbicide resistance in forest trees offers similar advantages, although practices are different than for crop plants, with herbicide usage only at the initial stages of plantation establishment in order to limit weed competition until canopy closure. The first reported modification in trees by Fillatti et al. 1987 described the stable expression of a modified EPSP synthase (*aroA*) gene to reduce sensitivity to the herbicide glyphosate. Later reports in poplars described resistance to the herbicides phosphinothricin (De Block 1990) and chlorsulfuron (Brasileiro et al. 1992). In conifers, Shin et al. 1994 described glyphosate-resistant European larch using a construct with the *aroA* gene and confirmed the resistance with bioassays. Field trials were needed to fully validate the efficacy of herbicide resistance. Such field evaluations over an extended time period were conducted with transgenic poplars bearing herbicide resistance for glyphosate (Meilan et al. 2002) over 4 years and for glufosinate over 3 and 8 years respectively (Li et al. 2008, 2009). All these evaluations indicated stable levels of resistance over time, with no cases of resistance breakdown.

3.2   Biotic Stress

Transgenic crops for insect resistance, primarily using *Bt* endotoxin genes, have been at the forefront of GM plants together with herbicide resistance. Similar strategies were used for forest tree species resulting in the development of GM trees with increased resistance to certain insect pests. More than two decades ago a group at the University of Wisconsin-Madison reported increased resistance to two lepidopteran pests (forest tent caterpillar and gypsy moth) by means of feeding assays (McCown et al. 1991). One of the very first field trials of GM trees was performed with poplars expressing the *Bt* Cry1A insecticidal protein at the Agricultural Research Farm, University of Wisconsin in the USA (Kleiner et al. 1995). The goal of that study was to validate the efficacy, in open field conditions, of the *Bt* endotoxin in transgenic poplar for preventing gypsy moth and forest tent caterpillar defoliation. At the same time the authors wished to validate if transgene expression and the resulting insect resistance phenotype would persist after winter dormancy. In 1994, the year after plantation establishment, evaluation of insect resistance was performed using non-choice feeding assays. Forest tent caterpillar foliage consumption tests and larval weight gain measurements were performed in early June and late August and showed significant positive effects of the GM poplars as compared with the controls. Similar and conclusive data were also obtained in comparable assays with gypsy moth. Mortality was also shown to be higher when the forest tent caterpillar fed on *Bt* poplar material, but no significant difference was observed with the gypsy moth.

   Transgenic poplars were also tested for insect resistance in controlled field trial conditions in other locations in the world. A plantation using *Populus nigra* engineered with the *Bt* endotoxin (*cry*) gene was established in 1994 in northern

China: in that country, economic losses in forest plantations were mainly due to insect pests causing either reduced growth or tree death and an effective pest management approach was urgently needed (Ewald et al. 2006). Trees showing reduced susceptibility to two insect defoliators resulting in less leaf damage and lower pupae numbers were tested with success in this field trial (Hu et al. 2001). The success of this field evaluation of GM poplar led the way for several other initiatives in China using transgenic technologies in tree plantations (see the chapter by Meng-zhu Lu et al. for full details of the current state of the Chinese program). A research team at Oregon State University tested both herbicide and insect resistance in the northwestern United States in 1998 and observed very low feeding damage due to the cottonwood leaf beetle in transgenic *Populus* hybrids (Meilan et al. 2000). Additional trials with insect-resistant poplars including commercial releases have taken place in China and an increasing body of knowledge on the potential environmental impacts has been gathered (reviewed in Ewald et al. 2006). More recently, transgenic poplars bearing both *Bt* endotoxin and oryzacystatin specific for Coleoptera were evaluated in open field conditions as well as their potential effects on a non-targeted Lepidoptera (Zhang et al. 2011). Insect resistance using a *Bt* endotoxin gene was also achieved in eucalyptus (Harcourt et al. 2000).

As an alternative to the *Bt* endotoxin, a few research laboratories investigated the efficacy of various proteinase inhibitors that confer insect resistance in poplars (Leplé et al. 1995; Confalonieri et al. 2003) and also through manipulation of tryptophan decarboxylase (Gill et al. 2003). However, the deleterious effects of orizacystatin (a cystein proteinase inhibitor) against *Chrysomela tremulae* observed in the controlled conditions of the greenhouse (Leplé et al. 1995) were not confirmed in natural field conditions (Cornu et al., unpublished data). Engineered insect resistance has also been reported in conifers, including loblolly pine (Tang and Tian 2003), radiata pine (Grace et al. 2005) and white spruce (Ellis et al. 1993; Lachance et al. 2007). GM spruces engineered with the *cry1Ab* gene were tested for 5 years in Canada and showed increased resistance to defoliation by the spruce budworm (Lachance et al. 2007).

To date, increasing plant resistance to microbial pathogens using genetic engineering has been limited in agriculture with the GM crop market being centred on weed control and insect resistance. In fact none of the approaches based on expression of antimicrobial proteins (e.g. defensin, pathogen-related proteins) have been deployed commercially (Collinge et al. 2010). A small number of publications have reported enhanced microbial resistance by GM in forest tree species reviewed in Powell et al. 2006. In poplars, improved resistance to the fungal pathogen *Septoria musiva* by increasing production of oxalic acid (Liang et al. 2001) and to *Melampsora medusae* by expressing an exogenous endochitinase (Noël et al. 2005) have been reported. Greenhouse experiments with transgenic birch expressing a *chitinase IV* gene also demonstrated the efficacy of this approach in limiting damage caused by the fungal pathogen responsible for leaf spot disease (Pappinen et al. 2002). Transgenic birches engineered with a sugar beet chitinase were tested in field conditions and showed increased resistance to birch rust (*Melampsoridium betulinum*) (Pasonen et al. 2004). Black spruce was also engineered with a *Trichoderma* endochitinase and showed reduced disease symptoms to the root rot disease causal agent (Noël et al. 2005). Lastly it is important to mention that American elm has also been engineered to enhance Dutch elm disease resistance through targeted vascular expression of a recombinant antimicrobial peptide (Newhouse et al. 2007) and that fungal chestnut blight resistance could be improved by the introduction of an oxalate oxidase gene in American chestnut (Welch et al. 2007).

3.3 Abiotic Stress

Climate change could not only facilitate emergence of new forest pests, but also increase environmental stress resulting in extreme temperature variation and drought conditions. Fortunately a large body of information on gene function and interconnecting regulation networks is now available to better understand the physiological and molecular mechanisms underlying abiotic resistance in plants (Cramer et al. 2011). In addition, an increasing number of tree genes have been studied and clearly demonstrated to have several underlying genetic components that are very similar to those of crops and model plant systems with shared gene families and functions. A few transgenic approaches have been utilized successfully to increase environmental abiotic stress tolerance in trees. The manipulation of expression levels of genes encoding transcription factors (TF) involved in abiotic stress regulation has been quite successful in recent years. Over-expression of the C-repeat binding factor (CBF) from Arabidopsis significantly increases freezing tolerance in transgenic poplars (Benedict et al. 2006). Overexpression of ERF/AP2 in *Pinus strobus* was clearly demonstrated to result in a significant increase in salt stress, drought and freezing due to a change in polyamine biosynthesis (Tang et al. 2007). An ethylene responsive factor (ERF) like TF, when overexpressed in *Populus alba x berolinensis*, resulted in taller GM trees producing more dry biomass than controls under increased salt (NaCl) concentrations (Li et al. 2012). Foliar proline concentrations were also elevated in these transgenic poplars in response to salt treatments. Modification of other metabolite concentrations such as the osmoprotectant glycinebetaine through introduction of a choline oxidase gene from *Arthrobacter globiformis* was also shown to confer salt tolerance in transgenic eucalyptus (Yu et al. 2009).

3.4 Wood Properties

Wood is conventionally used as a resource for energy, timber and fibres. With the depletion of global stocks in fossil energy, wood has the potential to become a major source of lignocellulose-based biofuels. Wood is mostly made of fibrous cell walls, a multi-layer structure resulting from the assembly of different polymers, mainly cellulose, hemicelluloses and lignins, with traces of pectins and proteins (Déjardin et al. 2010). Wood properties strongly derive from the composition and structural arrangement of these polymers. Timber can be considered as a composite material, characterized by a network of long bundles of cellulose microfibrils, cemented in an amorphous matrix of hemicelluloses and lignins, thus preventing easy access to cellulose (Kerstens et al. 2001). Depending on wood uses, different wood traits need to be improved. For example, cellulose is the valuable biochemical compound for pulp or bioethanol production: wood should ideally have a high cellulose/lignin ratio, and be easily delignified. On the other hand, lignins and phenolics are important biochemical compounds for wood calorific properties.

The GM approach is an invaluable tool in helping to decipher the function of genes involved in the complex process of wood formation. As an example, GM trees altered for the expression of genes involved in monolignol biosynthesis provided unique information on the lignin biosynthetic pathway, and accurately complemented biochemical studies (Pilate et al. 2012; Wagner et al. 2012). These trees further demonstrated the high plasticity of the lignin polymer, which was already suspected from previous studies (Sederoff et al. 1999). Genetic engineering also opens an alternative route to classical breeding to create innovative wood material that may not necessarily exist in natural tree populations (Mellerowicz and Sundberg 2008; Pu et al. 2011). One future challenge is to produce wood with improved properties for biofuel production, with lignins easier to extract and/or higher cellulose content. Genes involved in lignin, cellulose and hemicellulose metabolism were therefore candidate targets for genetic engineering and these approaches have already shown promising results for potential applications in the field.

Modification of lignin biosynthesis. Lignin is the second major polymer after cellulose in wood. It is synthesized from the oxidative coupling of p-hydroxycinnamyl alcohol monomers, called monolignols, differing in the number of methoxylations in their aromatic cycle (Boerjan et al. 2003; Bonawitz and Chapple 2010). Angiosperm lignins mainly contain guaiacyl (G) and syringyl (S) units, derived respectively from coniferyl and sinapyl alcohols; in contrast, gymnosperm lignins contain mainly G units, with a small proportion of p-hydroxyphenyl (H) units, derived from p-coumaryl alcohol. The proportions between the different elementary units affects their linkage within the polymer, and therefore lignin structure and properties. For example, a high S/G ratio is favourable to an easier delignification. Most if not all of the genes involved in monolignol biosynthesis have been identified (Boerjan et al. 2003; Bonawitz and Chapple 2010). A major research effort was devoted to modifying the amount and composition of lignin in trees to reduce the ecological chemical footprint in the pulping process for paper production (Baucher et al. 2003). Genes encoding enzymes involved in lignin biosynthesis were functionally tested by mis-regulation approaches (either up- or down-regulation) and several studies demonstrated the possibility of modifying tree lignin using a transgenic approach. The first GM trees were obtained two decades ago; the most promising transgenic lines were also grown in field trial conditions for agronomical, technological and/or environmental evaluations (for recent a review, see Pilate et al. 2012). Modifying the expression levels of Cinnamate 4-hydroxylase (C4H),

Cinnamyl alcohol dehydrogenase (CAD), Cinnamoyl CoA reductase (CCR) and Ferulate 5-hydroxylase (F5H) genes in GM trees gave interesting results in terms of wood compositional properties.

A group of European laboratories joined efforts in performing a duplicate field evaluation of transgenic trees with modified lignin to improve pulping performance (Pilate et al. 2012). These experiments were set up at Jealott's Hill in the United Kingdom and Ardon in France and evaluated parameters such as growth, insect resistance and the potential risk of GM trees on the soil microbial communities. Also, this work validated the persistence over time of the expected lignin modification resulting from the down-regulation of the CAD and caffeate/5-hydroxyl-ferulate O-methyltransferase (COMT) genes in transgenic poplar lines. Wood from 4-year-old CAD-down-regulated transgenic lines showed improved chemical pulping performance. This gene modification had only a slight impact on total lignin content, but the structure of the lignin polymer was modified due to an increase in free phenolic groups (Lapierre et al. 1999), while the growth of a number of CAD transgenic lines was unaffected in field conditions. Other genes encoding enzymes of the lignin biosynthetic pathway were also targeted in an attempt to modify cell wall properties in trees. For instance, down-regulation by antisense inhibition of the gene encoding the 4-coumarate:coenzyme A ligase (4CL) in transgenic *Populus tremuloides* resulted in up to a 45 % reduction in lignin content and a 15 % increase in cellulose (Hu et al. 1999). These appealing results obtained in the greenhouse were however not confirmed in field conditions as *4CL* down-regulated poplars showed no more growth increases. In addition, the lines with a strong reduction in lignin content displayed physiological vulnerabilities (Voelker et al. 2010). It was also shown that wood from CCR-silenced poplars had reduced lignin and hemicellulose contents, combined with an increased proportion of cellulose: 5-year-old field-grown transgenic lines produced wood easier to delignify with less alkali, but unfortunately, the gene modification had a negative impact on growth performance (Leplé et al. 2007). Over-expression of F5H promoted high S/G ratio in the lignin polymer in poplar and eucalyptus (Franke et al. 2000; Huntley et al. 2003; Hinchee et al. 2009), which had a positive impact on wood chemical pulping performance. Two-year-old field-grown eucalyptus with C4H reduced activity produced wood with 20 % less lignin, without any negative impact on tree growth. Cell wall composition analyses showed a greater accessibility to cellulose for pulp or biofuel production with fewer chemicals (Hinchee et al. 2009). These examples illustrate the possibility of modifying both the content and structure of the lignin polymer, thus creating modified properties. However, it seems that more than a 20 % reduction in lignin content often resulted in negative effects on agronomical traits (Pilate et al. 2012). Taken as a whole, these studies also highlighted the importance of conducting field trials in evaluating the performance of GM tree varieties.

Modification of cellulose biosynthesis. Cellulose is the most abundant polymer in wood. Cellulose is a linear chain of β 1–4 linked glucose units. Although this molecule has a simpler structure than the lignin polymer, its synthesis is not yet completely understood. A model of cellulose biosynthesis has been proposed where the association of different types of cellulose synthases (CesA) arranged in a six-lobed rosette structure within the plasma membrane are necessary for the biosynthesis of cellulose microfibrils (Somerville 2006). Recently, an aspen secondary wall-associated cellulose synthase (PtdCesA8) was over-expressed in *P. tremuloides* in an attempt to increase the amount of cellulose in wood (Joshi et al. 2011). This unexpectedly resulted in the silencing of CesA8 in all of the transgenic lines. The silenced trees showed reduced growth with weeping-type branches. Wood contained fourfold less cellulose on a dry-weight basis, but more lignin and non-cellulosic polysaccharides. Wood displayed typical collapsed or 'irregular xylem' vessels with altered secondary cell walls. An alternative strategy to produce more cellulose in GM trees was to increase the availability of precursor substrates such as UDP-glucose. UDP-glucose can be produced by two different enzymes, UDP-glucose pyrophosphorylase (UGPase) and sucrose synthase (SuSy). In 4-month-old transgenic hybrid poplars (*Populus alba x P. grandidentata*), the over-production of *Acetobacter xylinum* UGPase resulted in a significant increase in cellulose content (6.6 %), as well as soluble sugars and starch, and a decrease in lignin content (12–21 %) (Coleman et al. 2007). However, the transgenic trees showed growth alterations, with a reduction in both stem height and diameter, limiting any potential application. When over-expressing the *Gossypium hirsutum* SuSy in 4-month-old GM poplars grown in a greenhouse, however, the amounts of soluble carbohydrates were significantly affected with an increase of 2–6 % in cell wall cellulose without any effects on lignin content and growth characteristics (Coleman et al. 2009).

Modification of hemicellulose biosynthesis. Hemicelluloses are the second most abundant structural carbohydrates in wood after cellulose. These polysaccharides include xyloglucans, xylans, mannans and glucomannans, and β-(1$\rightarrow$3,1$\rightarrow$4)-glucans. Their content and composition vary widely between species, tissues and cell wall types. In dicot secondary walls, glucuronoxylans are the most abundant and represent approximately 20–30 % of the dry weight whereas arabinoxylans are the most abundant in monocot primary walls and represent about 20 % of the dry weight (Scheller and Ulvskov 2010). Xylans and their biosynthesis pathways have been receiving increased attention as they may influence biomass properties, in particular regarding the conversion of biomass to biofuel (York and O'Neill 2008). Recently, it was shown that down-regulation of a glycosyltransferase belonging to the family GT47 (PoGT47C) in poplar resulted in a reduced glucuronoxylan content and an increased wood digestibility by cellulase (Lee et al. 2009). Two other glycosyltransferase genes identified in aspen (PoGT8D1 and PoGT8D2), when down-regulated in transgenic poplar, resulted in a 29–36 % reduction in stem wood xylan content translated into alteration of wood mechanical properties (Li et al. 2011). Some of these transgenic lines were further shown to have improved saccharification potential when no pre-treatment was included in the conversion process (Min et al. 2011). Likewise the constitutive expression of *Aspergillus aculeatus* xyloglucanase in *Populus alba* loosened the xyloglucan network and increased cellulose content (Park et al. 2004). Cellulose from this transgenic wood was more prone to saccharification and fermentation, likely due to its increased accessibility to hydrolytic enzymes (Kaida et al. 2009).

Unexpected effects on wood properties of other gene target modifications. Besides these approaches aimed at modifying specific secondary cell wall metabolisms, additional work has occasionally resulted in some unexpected modifications of wood characteristics. For example, hybrid poplar (*Populus tremula X P. alba*) over-expressing a pine cytosolic glutamine synthetase gene (GS1a) showed improved growth in a field trial (+ 41 % for stem height after 3 years, a higher number of leaves, increased leaf length in comparison with controls (Gallardo et al. 1999; Jing et al. 2004). Recently, the GM trees grown in the field were analyzed for their wood properties. The ectopic expression of GS1a was shown to cause alterations in wood properties and wood chemistry (Coleman et al. 2012). More intriguingly, the constitutive expression of the flowering promoter factor 1 gene (FPF1) in hybrid aspen affected cellulose and glucomannans, lignin deposition and therefore wood density, raising the question of the precise regulatory role of this gene in wood formation (Hoenicka et al. 2012).

Many of the possible ways of modifying wood properties using genetic engineering have yet to be explored. For example, little is known about the mechanisms responsible for the transport and delivery of matrix polymer precursors to the cell wall. Likewise, the functions of cell wall structural proteins and enzymes remains largely unknown: this may provide interesting approaches for the fine tuning of wood properties (Liu 2012). A recent major advance was certainly the discovery of the regulatory cascades of transcription factors involved in the differentiation of xylem and the formation of lignified secondary cell walls (Kubo et al. 2005; Demura and Fukuda 2007; Zhong and Ye 2007). This network appears less hierarchical than previously thought but much more inter-connected with putative feedback control loops, which makes the complete picture highly complex. Deciphering the equivalent regulatory networks in wood formation is still at an early stage and has mainly been scrutinized in poplar (Zhong et al. 2010, 2011; Du et al. 2011; Robischon et al. 2011) and eucalyptus (Goicoechea et al. 2005; Legay et al. 2010; Hussey et al. 2011). We may expect that this will lead to the identification of regulatory master switches that are likely to be targeted for genetic engineering in the near future.

4 Regulating GM Trees

As mentioned in the other chapters, the establishment of large-scale tree plantations has been very successful in supporting the forest industry in the context of the increasing commercial demand for forest products as a consequence of the growing human population. Moreover, in various parts of the world, tree plantations also offer better productivity on a smaller land area leading to increased conservation of natural forests (Fenning and Gershenzon 2002). Additionally, productive plantations will also benefit from biotechnological tools including clonal propagation and GM technologies (Walter and Fenning 2004). Overall, tree plantations are able to take advantage of several biotechnological opportunities to improve tree resistance to biotic and abiotic stresses that are increasing with climate change and movement of pests as a result of international trade. The technological progress on GM trees described in this chapter is providing new and important opportunities for improving forest productivity and the development of new resources, particularly in the context of biomass and bioenergy production as well as for the development of new biomaterials. GM technology will open up the development of new approaches that otherwise could not be obtained through exploitation of natural genetic variability. Such strategies may be compulsory to overcome the recalcitrance of the secondary cell wall of wood for bioethanol production, for instance. Overall the proper deployment of GM plantations in various parts of the world will need an appropriate regulatory framework allowing scientific research on GM trees with a proper assessment of the risk vs. benefits, taking into account both public perception and social benefits, before any potential future deployment of GM trees in commercial plantations (Walter et al. 2010).

4.1 Genetic Containment

Most field trials with GM trees worldwide (with the exception of a few cases in China) have been established for research purposes only, with the goal of gathering information on transgene expression and persistence and the effects of the introduced traits. Despite the fact that these trials essentially look like small tree plantations, they are considered to be confined, as they are under strict regulatory control, and established on designated areas for a pre-determined duration. Confined field trials with GM trees such as these are also essential in gathering scientific data regarding the environmental risks related to the deployment of a specific GM tree. These trials will make it possible to acquire knowledge about the efficiency of the modification with regard to tree performances and wood properties as well as any potential risks to the environment of GM trees, such as effects on non-targeted species, weediness, and effects on microflora. This type of research is needed to enable the formulation of recommendations and limitations related to the commercial deployment of GM trees. At the moment, regulations imposed on confined field trials with GM trees in most countries aim to prevent the dispersal of pollen, seed and vegetative propagules. Despite the long juvenile stages of most forest tree species, relative to the length of most trials, timely monitoring and removal of reproductive structures, if they occur, is required. In addition, dissemination through root suckers needs to be prevented. To achieve this, guard rows of non-transformed trees are planted around the trial site in combination with a buffer zone of several meters that is kept clear in order to monitor and eliminate the growth of any GM material.

Despite the fact that most trees take several years to flower, engineering reproductive sterility has been viewed as an effective way of confinement both for the trees in the confined field trials themselves as well as commercial release whenever the need for confinement of GM trees will be demonstrated, for instance when the genetic modification may provide a selective advantage of the GM material upon untransformed trees. However, understanding floral development in forest tree species at the genetic level is a challenging but prerequisite task before considering engineering reproductive sterility (reviewed in Lemmetyinen and Sopanen 2004; Ahuja 2009). As a result, an interesting portfolio of genes associated with reproductive development have been developed for commercially important forest tree species (Flachowsky et al. 2009). Various approaches have been used to engineer sterility in trees, including tissue-specific ablation using a flower promoter driving a phytotoxic gene, suppression of expression of key flower-related genes using an RNAi approach or expression of a modified gene resulting in a dominant negative mutation (reviewed in Brunner et al. 2007). To date, no fully effective methods have been developed, but even partial sterility could considerably reduce gene flow into wild trees and so reduce the risks of transgene escape to an acceptable risk level.

Obtaining gene flow data from GM trees into the natural forest and plant communities is an important step in the evaluation of the potential environmental effects of transgene dispersal. Several studies have been carried out in order to assess potential gene flow from GM tree plantations in comparison with non-GM material (see Ahuja 2009 and references therein). A recent study (DiFazio et al. 2012) on established hybrid poplar plantations in the US Pacific Northwest demonstrated that probable gene escape is likely possible spatially to a relatively high degree. However, in most cases gene flow occurs locally and management practices such as early harvesting and coppicing, or the use of confinement strategies based on floral sterility as mentioned above, could greatly attenuate transgene spread. Long-term stability of transgene expression is also an area of concern, not only for the proper maintenance of the engineered trait but also in the context of GM tree containment if it is also required for that (Ahuja 2009). For example, the loss or reduction of transgene expression in insect-resistant GM trees could result in a reduction of tree growth under epidemic conditions, but is unlikely to increase environmental risk. Accelerating the likelihood of insect resistance due to the lack of proper management practices (such as refugee zones to avoid selection of resistant insect populations) is not a direct impact of the use of GM trees, but should also be taken into consideration. However in the case of GM trees engineered for reproductive sterility for confinement purposes, loss of transgene expression could cause a reversion of the phenotype resulting in increased risk of dispersal of these GM trees. Several avenues of investigations have been explored to evaluate long-term expression in GM tree field trials (Ahuja 2010). Beyond the recent technological improvements in manipulating tree sterility and lowering weediness, environmental assessment for a specific engineered trait should be clearly put into the proper context and treated on a case-by-case basis. For instance, the establishment of a GM tree plantation in an area with a low predominance of related species would reduce the possibility for inter-breeding and consequently limit concerns about gene flow. Other biological factors also have to be taken into consideration, such as pollen viability (despite the fact that under some circumstances pollen can be transported long distances) and potential GM seed production and establishment on remote sites. Domesticated GM trees should also have a reduced ability to survive in the wild as they have a reduced fitness outside the protected environment of the plantation.

4.2 Environmental Risk Assessment Studies

There is a substantial body of literature dealing with the environmental risk assessment of GM trees and it would take too much space here to review it entirely. The reader is referred to the chapter by Häggman et al. on biosafety considerations related to GM trees. Recent reviews on GM trees particularly in the context of biomass production (Harfouche et al. 2011) and describing the current state of testing of GM trees worldwide (Robischon 2006) are also available. Nevertheless we provide here specific examples to illustrate how research on environmental assessment of GM trees using confined field trials is important, particularly in the context of regulatory initiatives and social acceptance. As mentioned earlier, trees are different from crop plants in terms of their longevity, their reproductive cycles and related dispersal distances, their level of domestication, and obviously their physical size. All these characteristics have to be taken into account during environmental risk assessment of GM trees. Thus, despite the extensive literature assessing the potential risks of GM crops (e.g. *Bt* insecticidal resistance), the existing research data cannot be simply transposed to GM trees with similar traits. Although the large body of information obtained from research on GM crops could be very informative to address issues related to the potential environmental risk of GM trees, it does not provide all the answers. In short the establishment of confined field trials with GM trees engineered with various traits in different genetic backgrounds and physical locations are needed to test the combination of the effects of a specific trait in a particular environment (e.g. gene x environment interactions). It is crucial to document potential unintended environmental effects related to the engineered trait and this is the goal of confined field trials using GM trees. Recent genomics tools also make it possible to conduct much broader analyses of the potential unintended effects of GM trees (e.g. associated microbial communities) as well as the effects of the transgene on the transcriptome. Moreover the latest developments on RNA interference and microRNA have opened up new avenues to precisely control the expression of selected genes in order to obtain a desirable phenotype.

In the early days of the development of commercial varieties of GM crops, the use of antibiotic resistance raised a great deal of concerns in the public, mainly in relation to the possible lateral gene transfer of antibiotic resistance from plant DNA to soil microorganisms. An evaluation of the potential risks of obtaining harmful new antibiotic-resistant bacterial strains was needed. In early 2000, despite the numerous publications that determined that there was little if any environmental risk from the persistence of DNA from transgenic crops in the environment or potential horizontal spread of antibiotic resistance genes, there were only very limited studies addressing similar issues with GM trees. A limited number of GM tree field trials had been established by that time, but the first one in Canada was established in 1997 in the province of Quebec near Quebec City for the express purpose of evaluating the potential risks of transgenic trees when established in field trials. Transgenic poplar with the β-glucuronidase reporter (GUS) gene under the control of a wound-inducible proteinase inhibitor gene promoter was used to monitor transgene expression and potential environmental risk studies of GM trees. The first paper in 2002 assessed the persistence of recombinant DNA (mainly the *nptII* gene encoding kanamycin antibiotic resistance) from decomposing leaves in the soil. Quantitative analysis of tree DNA stability demonstrated that recombinant gene markers were not detectable in decomposing leaves in the soil after 4 months, thus limiting the possibilities of lateral gene transfer to natural microbial communities (Hay et al. 2002).

In most cases the approach for testing the potential environmental risk of a GM tree is related to the trait used. For instance, GM trees engineered for increased resistance to fungal pathogens might affect their colonization with favourable mycorrhizal symbiotic fungi. Relatively small numbers of scientific papers have addressed these issues, but in general found little or no effect of the engineered traits on soil fungal communities and capability of forming symbiotic interaction with GE trees. Again, the details of these studies are systematically presented in the chapter by Häggman et al. Beyond any potential risk of the engineered proteins on the tree's associated organisms, complex tree-environment interactions regulated by complex signalling biochemical pathways could also potentially be altered by the engineered trait. In that context, a multifaceted approach for environmental risk assessment should be undertaken. It may sound unrealistic to question the potential risk of engineering of secondary cell wall composition (wood modification) in tree-environment interaction, but well-designed studies could provide enough basic information for rigorous risk assessments. As an example, it has been shown that the establishment of ectomycorrhizal symbiosis is dependent on specific plant phenolic compounds. However, a study on GM birch and poplars engineered for modified lignin showed no difference between the GM and non-GM trees in establishing ectomycorrhizal symbiotic interactions (Sutela et al. 2009).

5   Conclusion

The reasons for using GM forest trees includes faster incorporation of new genetic traits for tree domestication, or obtaining traits that are not accessible by traditional breeding approaches. In this chapter we have taken a historical point of view concerning the various technical developments related to production of GM trees and described the various traits obtained. GM trees are the product of a well understood and accurate technology that has been successfully applied to overcome specific silvicultural problems and has demonstrated clear environmental and economic benefits. The potential ecological and economic advantages of transgenic trees have been extensively discussed previously, and ensuring biosafety has been a foremost objective of the scientific community for several years already.

Nonetheless, the use of GM trees may lead to changes in forest management and a scientifically sound management plan for tree plantations should be applied to take into consideration all the potential risks and potential benefits for ecosystem processes. Situations where the extensive use of value-added GM trees may reduce diversity by using only a single or a few genotypes should also be avoided. A substantial amount of data on risk assessment in relation to the deployment of GM trees has been obtained during the last decade showing essentially the absence of any special environmental risk of these trees. Government bodies are responsible for the regulation of GM trees in the context of environmental release. Regulatory agencies worldwide should be up to date with the technological developments, building their regulations for GM trees based on scientific facts, and not on the vagaries of public opinion. As mentioned earlier, scientific data on key issues related to environmental risks could only be obtained with confined field trials and their establishment should be facilitated as much as possible.

A major obstacle for the deployment and acceptance of transgenic trees is the social perspective associated with it, despite positive scientific endeavours. One of the challenges facing the scientific community is to put forward specific examples that will impact public opinion that demonstrate clear environmental advantages, for instance. There are clear benefits for using GM approaches to tackle severe problems that affect trees species, such as elms that have been severely impacted by accidental introduction of the Dutch elm disease pathogen, or American chestnuts by chestnut blight (Merkle et al. 2007). Despite efforts to obtain more tolerant cultivars (American elm) and resistant species (American chestnut), the recent technological progress in the production of GM trees resistant to their pathogenic fungus also provides promising and complementary avenues. From the point of view of the general public, the use of GM technologies might be viewed as more acceptable in the context of efforts to restore iconic species such as the American chestnut.

Alternatively, improving trees for their capabilities in phytoremediation by genetic engineering could also represent a technological breakthrough with great environmental benefits. Phytoremediation is an approach that takes advantage of the capability of specific plants to remove and/or decontaminate harmful molecules present in contaminated soils that might be difficult by other means (Azzarello et al. 2011). A number of plants from different families are known for their tolerance and bioaccumulation (contaminant uptake and compartmentalization) properties and most of them are herbaceous species. In other cases, plants will remove these soil contaminants by biodegradation. Only a limited number of tree species are compatible with phytoremediation and most of the studies performed so far were done with willows and poplars (Azzarello et al. 2011). Improving the phytoremediation capabilities of these trees using genetic engineering has been shown to be possible. Improving the tolerance and biodegradation of specific volatile hydrocarbons such as trichloroethylene by the overexpression of a mammalian P450 2E1 enzyme has been obtained in transgenic poplars (Doty et al. 2007). Using a hydroponic setup, GM poplars showed an increased capability for removing and degrading several volatile hydrocarbons that are common and hazardous pollutants. Lastly, other traits such as drought resistance could also provide both economic and environmental advantages and other examples could be added to the list.

Considering the recent developments in genomics, it is an appropriate time to begin a dialogue among scientists, the public, and regulators for carefully integrating the use of GM trees into forestry. It is exactly a quarter of a century since the production of the first GM tree and close to two decades after the first field trial with GM trees. During that time, several studies were conducted on improving various components related to GM trees to make the technology more robust, precise and efficient. At the same time, a substantial amount of data from greenhouse and field trial experiments about the possible unintended impacts of the modified traits has been accumulated, and no damaging effects related to the GM trees have been found. It is time to make known and demonstrate the benefits of this technology in our changing world.

Acknowledgements The authors wish to thank Pamela Cheers for her editorial work. We apologize to all our colleagues whose scientific contributions could not be acknowledged in this chapter owing to space limitations. This work is supported by a grant from the Canadian Regulatory Systems for Biotechnology to AS and support from the EU-COST Action FP0905 entitled "Biosafety of forest transgenic trees: improving the scientific basis for safe tree development and implementation of EU policy directives".

References

Ahuja MR (2009) Transgene stability and dispersal in forest trees. Trees Struct Funct 23:1125–1135

Ahuja M (2010) Fate of transgenes in the forest tree genome. Tree Genet Genomes 7:221–230

Azzarello E, Pandolfi C, Pollastri S, Masi E, Mugnai S, Mancuso S (2011) The use of trees in phytoremediation. CAB Rev Perspect Agric Vet Sci Nutr Nat Resour 6(037). doi:10.1079/PAVSNNR20116037

Baucher M, Halpin C, Petit-Conil M, Boerjan W (2003) Lignin: genetic engineering and impact on pulping. Crit Rev Biochem Mol Biol 38:305–350

Benedict C, Skinner JS, Meng R, Chang Y, Bhalerao R, Huner NP, Finn CE, Chen TH, Hurry V (2006) The CBF1-dependent low temperature signalling pathway, regulon and increase in freeze tolerance are conserved in *Populus* spp. Plant Cell Environ 29:1259–1272

Berg P, Baltimore D, Brenner S, Roblin RO 3rd, Singer MF (1975a) Asilomar conference on recombinant DNA molecules. Science 188:991–994

Berg P, Baltimore D, Brenner S, Roblin RO, Singer MF (1975b) Summary statement of the Asilomar conference on recombinant DNA molecules. Proc Natl Acad Sci U S A 72:1981–1984

Bevan MW, Flavell RB, Chilton M-D (1983) A chimaeric antibiotic resistance gene as a selectable marker for plant cell transformation. Nature 304:184–187

Boerjan W (2005) Biotechnology and the domestication of forest trees. Curr Opin Biotechnol 16:159–166

Boerjan W, Ralph J, Baucher M (2003) Lignin biosynthesis. Annu Rev Plant Biol 54:519–546

Bonawitz ND, Chapple C (2010) The genetics of lignin biosynthesis: connecting genotype to phenotype. Annu Rev Genet 44:337–363

Brasileiro ACM, Tourneur C, Leple JC, Combes V, Jouanin L (1992) Expression of the mutant *Arabidopsis thaliana* acetolactate synthase gene confers chlorsulfuron resistance to transgenic poplar plants. Transgenic Res 1:133–141

Brunner AM, Li J, DiFazio SP, Shevchenko O, Montgomery BE, Mohamed R, Wei H, Ma C, Elias AA, VanWormer K, Strauss SH (2007) Genetic containment of forest plantations. Tree Genet Genomes 3:75–100

Campbell MM, Brunner AM, Jones HM, Strauss SH (2003) Forestry's fertile crescent: the application of biotechnology to forest trees. Plant Biotechnol J 1:141–154

Castle LA, Wu G, McElroy D (2006) Agricultural input traits: past, present and future. Curr Opin Biotechnol 17:105–112

Cohen SN, Chang AC, Boyer HW, Helling RB (1973) Construction of biologically functional bacterial plasmids in vitro. Proc Natl Acad Sci U S A 70:3240–3244

Coleman HD, Canam T, Kang KY, Ellis DD, Mansfield SD (2007) Over-expression of UDP-glucose pyrophosphorylase in hybrid poplar affects carbon allocation. J Exp Bot 58:4257–4268

Coleman HD, Yan J, Mansfield SD (2009) Sucrose synthase affects carbon partitioning to increase cellulose production and altered cell wall ultrastructure. Proc Natl Acad Sci U S A 106:13118–13123

Coleman HD, Canovas FM, Man H, Kirby EG, Mansfield SD (2012) Enhanced expression of glutamine synthetase (GS1a) confers altered fibre and wood chemistry in field grown hybrid poplar (*Populus tremula* X *alba*) (717-1B4). Plant Biotechnol J 10:883–889

Collinge DB, Jorgensen HJ, Lund OS, Lyngkjaer MF (2010) Engineering pathogen resistance in crop plants: current trends and future prospects. Annu Rev Phytopathol 48:269–291

Confalonieri M, Balestrazzi A, Bisoffi S, Carbonera D (2003) *In vitro* culture and genetic engineering of *Populus* spp.: synergy for forest tree improvement. Plant Cell Tiss Org Cult 72:109–138

Cramer GR, Urano K, Delrot S, Pezzotti M, Shinozaki K (2011) Effects of abiotic stress on plants: a systems biology perspective. BMC Plant Biol 11:163

De Block M (1990) Factors influencing the tissue culture and the *Agrobacterium tumefaciens*-mediated transformation of hybrid aspen and poplar clones. Plant Physiol 93:1110–1116

Déjardin A, Laurans F, Arnaud D, Breton C, Pilate G, Leplé JC (2010) Wood formation in angiosperms. Comptes Rendus Biologies 333:325–334

Demura T, Fukuda H (2007) Transcriptional regulation in wood formation. Trends Plant Sci 12:64–70

DiFazio SP, Leonardi S, Slavov GT, Garman SL, Adams WT, Strauss SH (2012) Gene flow and simulation of transgene dispersal from hybrid poplar plantations. New Phytol 193:903–915

Dill GM (2005) Glyphosate-resistant crops: history, status and future. Pest Manag Sci 61:219–224

Doty SL, James CA, Moore AL, Vajzovic A, Singleton GL, Ma C, Khan Z, Xin G, Kang JW, Park JY, Meilan R, Strauss SH, Wilkerson J, Farin F, Strand SE (2007) Enhanced phytoremediation of volatile environmental pollutants with transgenic trees. Proc Natl Acad Sci U S A 104:16816–16821

Ellis DD, McCabe DE, McInnis S, Ramachandran R, Russell DR, Wallace KM, Martinell BJ, Roberts DR, Raffa KF, McCown BH (1993) Stable transformation of *Picea glauca* by particle acceleration. Biotechnology 11:84–89

Ewald D, Hu J, Yang M (2006) Transgenic forest trees in China. In: Fladung M, Ewald D (eds) Tree transgenesis – recent developments. Springer, Berlin, pp 25–45

Farre G, Ramessar K, Twyman RM, Capell T, Christou P (2010) The humanitarian impact of plant biotechnology: recent breakthroughs vs bottlenecks for adoption. Curr Opin Plant Biol 13:219–225

Fenning TM, Gershenzon J (2002) Where will the wood come from? Plantation forests and the role of biotechnology. Trends Biotechnol 20:291–296

Fillatti JJ, Sellmer J, McCown B, Haissig B, Comai L (1987) *Agrobacterium*-mediated transformation and regeneration of *Populus*. Mol Gen Genet 206:192–199

Flachowsky H, Hanke MV, Peil A, Strauss SH, Fladung M (2009) A review on transgenic approaches to accelerate breeding of woody plants. Plant Breed 128:217–226

Fraley RT, Rogers SG, Horsch RB, Sanders PR, Flick JS, Adams SP, Bittner ML, Brand LA, Fink CL, Fry JS, Galluppi GR, Goldberg SB, Hoffmann NL, Woo SC (1983) Expression of bacterial genes in plant cells. Proc Natl Acad Sci U S A 80:4803–4807

Franke R, McMichael CM, Meyer K, Shirley AM, Cusumano JC, Chapple C (2000) Modified lignin in tobacco and poplar plants over-expressing the *arabidopsis* gene encoding ferulate 5-hydroxylase. Plant J 22:223–234

Gallardo F, Fu JM, Canton FR, Garcia-Gutierrez A, Canovas FM, Kirby EG (1999) Expression of a conifer glutamine synthetase gene in transgenic poplar. Planta 210:19–26

Gill RIS, Ellis BE, Isman MB (2003) Tryptamine-induced resistance in tryptophan decarbox-ylase transgenic poplar and tobacco plants against their specific herbivores. J Chem Ecol 29:779–793

Goicoechea M, Lacombe E, Legay S, Mihaljevic S, Rech P, Jauneau A, Lapierre C, Pollet B, Verhaegen D, Chaubet-Gigot N, Grima-Pettenati J (2005) *Eg*MYB2, a new transcriptional acti-vator from *Eucalyptus* xylem, regulates secondary cell wall formation and lignin biosynthesis. Plant J 43:553–567

Grace LJ, Charity JA, Gresham B, Kay N, Walter C (2005) Insect-resistant transgenic *Pinus radi-ata*. Plant Cell Rep 24:103–111

Harcourt RL, Kyozuka J, Floyd RB, Bateman KS, Tanaka H, Decroocq V, Llewellyn DJ, Zhu X, Peacock WJ, Dennis ES (2000) Insect- and herbicide-resistant transgenic eucalypts. Mol Breed 6:307–315

Harfouche A, Meilan R, Altman A (2011) Tree genetic engineering and applications to sustainable forestry and biomass production. Trends Biotechnol 29:9–17

Harfouche A, Meilan R, Kirst M, Morgante M, Boerjan W, Sabatti M, Scarascia MG (2012) Accelerating the domestication of forest trees in a changing world. Trends Plant Sci 17:64–72

Hay I, Morency M-J, Séguin A (2002) Assessing the persistence of DNA in decomposing leaves of genetically modified poplar trees. Can J For Res 32:977–982

Henderson AR, Walter C (2006) Genetic engineering in conifer plantation forestry. Silvae Genet 55:253–262

Herrera-Estrella L, De Block M, Messens E, Hernalsteens JP, Van Montagu M, Schell J (1983) Chimeric genes as dominant selectable markers in plant cells. EMBO J 2:987–995

Hinchee M, Rottmann W, Mullinax L, Zhang C, Chang S, Cunningham M, Pearson L, Nehra N (2009) Short-rotation woody crops for bioenergy and biofuels applications. In Vitro Cell Dev Biol Plant 45:619–629

Ho CK, Chang SH, Tsay JY, Tsai CJ, Chiang VL, Chen ZZ (1998) *Agrobacterium tumefaciens*-mediated transformation of *Eucalyptus camaldulensis* and production of transgenic plants. Plant Cell Rep 17:675–680

Hoenicka H, Lautner S, Klingberg A, Koch G, El-Sherif F, Lehnhardt D, Zhang B, Burgert I, Odermatt J, Melzer S, Fromm J, Fladung M (2012) Influence of over-expression of the flower-ing promoting factor 1 gene (FPF1) from Arabidopsis on wood formation in hybrid poplar (*Populus tremula* L. x *P. tremuloides* Michx.). Planta 235:359–373

Hu WJ, Harding SA, Lung J, Popko JL, Ralph J, Stokke DD, Tsai CJ, Chiang VL (1999) Repression of lignin biosynthesis promotes cellulose accumulation and growth in transgenic trees. Nat Biotechnol 17:808–812

Hu JJ, Tian YC, Han YF, Li L, Zhang BE (2001) Field evaluation of insect-resistant transgenic *Populus nigra* trees. Euphytica 121:123–127

Huntley SK, Ellis D, Gilbert M, Chapple C, Mansfield SD (2003) Significant increases in pulping efficiency in C4H-F5H-transformed poplars: improved chemical savings and reduced environ-mental toxins. J Agric Food Chem 51:6178–6183

Hussey SG, Mizrachi E, Spokevicius AV, Bossinger G, Berger DK, Myburg AA (2011) SND2, a NAC transcription factor gene, regulates genes involved in secondary cell wall development in Arabidopsis fibres and increases fibre cell area in eucalyptus. BMC Plant Biol 11:173

Jing ZP, Gallardo F, Pascual MB, Sampalo R, Romero J, De Navarra AT, Cánovas FM (2004) Improved growth in a field trial of transgenic hybrid poplar overexpressing glutamine synthe-tase. New Phytol 164:137–145

Joshi CP, Thammannagowda S, Fujino T, Gou JQ, Avci U, Haigler CH, McDonnell LM, Mansfield SD, Mengesha B, Carpita NC, Harris D, Debolt S, Peter GF (2011) Perturbation of wood cel-lulose synthesis causes pleiotropic effects in transgenic aspen. Mol Plant 4:331–345

Kaida R, Kaku T, Baba K, Oyadomari M, Watanabe T, Nishida K, Kanaya T, Shani Z, Shoseyov O, Hayashi T (2009) Loosening xyloglucan accelerates the enzymatic degradation of cellulose in wood. Mol Plant 2:904–909

Kerstens S, Decraemer WF, Verbelen JP (2001) Cell walls at the plant surface behave mechanically like fiber-reinforced composite materials. Plant Physiol 127:381–385

Kleiner KW, Ellis DD, McCown BH, Raffa KF (1995) Field evaluation of transgenic poplar expressing a *Bacillus thuringiensis crylA(a) d*-endotoxin gene against forest tent caterpillar (Lepidoptera: Lasiocampidae) and gypsy moth (Lepidoptera: Lymantriidae) following winter dormancy. Environ Entomol 24:1358–1364

Kubo M, Udagawa M, Nishikubo N, Horiguchi G, Yamaguchi M, Ito J, Mimura T, Fukuda H, Demura T (2005) Transcription switches for protoxylem and metaxylem vessel formation. Genes Dev 19:1855–1860

Kumar S, Fladung M (2004) Molecular genetics and breeding of forest trees. Haworth Press Inc., New York, 436 p

Lachance D, Hamel LP, Pelletier F, Valéro J, Bernier-Cardou M, Chapman K, Van Frankenhuyzen K, Séguin A (2007) Expression of a *Bacillus thuringiensis crylAb* gene in transgenic white spruce and its efficacy against the spruce budworm (*Choristoneura fumiferana*). Tree Genet Genomes 3:153–167

Lapierre C, Pollet B, Petit-Conil M, Toval G, Romero J, Pilate G, Leplé JC, Boerjan W, Ferret VV, De Nadai V, Jouanin L (1999) Structural alterations of lignins in transgenic poplars with depressed cinnamyl alcohol dehydrogenase or caffeic acid O-methyltransferase activity have an opposite impact on the efficiency of industrial kraft pulping. Plant Physiol 119:153–164

Lee C, Teng Q, Huang W, Zhong R, Ye Z-H (2009) Down-regulation of PoGT47C expression in poplar results in a reduced glucuronoxylan content and an increased wood digestibility by cellulase. Plant Cell Physiol 50:1075–1089

Legay S, Sivadon P, Blervacq AS, Pavy N, Baghdady A, Tremblay L, Levasseur C, Ladouce N, Lapierre C, Séguin A, Hawkins S, Mackay J, Grima-Pettenati J (2010) EgMYB1, an R2R3 MYB transcription factor from eucalyptus negatively regulates secondary cell wall formation in Arabidopsis and poplar. New Phytol 188:774–786

Lemmetyinen J, Sopanen T (2004) Modification of flowering in forest trees. In: Kumar S, Fladung M (eds) Molecular genetics and breeding of forest trees. Haworth, New York, pp 263–291

Leplé JC, Bonade-Bottino M, Augustin S, Pilate G, Le Tan VD, Delplanque A, Cornu D, Jouanin L (1995) Toxicity to *Chrysomela tremulae* (Coleoptera: Chrysomelidae) of transgenic poplars expressing a cysteine proteinase inhibitor. Mol Breed 1:319–328

Leplé JC, Dauwe R, Morreel K, Storme V, Lapierre C, Pollet B, Naumann A, Kang KY, Kim H, Ruel K, Lefèbvre A, Joseleau JP, Grima-Pettenati J, De Rycke R, Andersson-Gunnerås S, Erban A, Fehrle I, Petit-Conil M, Kopka J, Polle A, Messens E, Sundberg B, Mansfield SD, Ralph J, Pilate G, Boerjan W (2007) Downregulation of cinnamoyl-coenzyme A reductase in poplar: multiple-level phenotyping reveals effects on cell wall polymer metabolism and structure. Plant Cell 19:3669–3691

Levée V, Lelu M-A, Jouanin L, Cornu D, Pilate G (1997) *Agrobacterium tumefaciens*-mediated transformation of hybrid larch (*Larix kaempferi* x *L. decidua*) and transgenic plant regeneration. Plant Cell Rep 16:680–685

Levée V, Garin E, Klimaszewska K, Séguin A (1999) Stable genetic transformation of white pine (*Pinus strobus* L.) after cocultivation of embryogenic tissues with *Agrobacterium tumefaciens*. Mol Breed 5:429–440

Li J, Meilan R, Ma C, Barish M, Strauss SH (2008) Stability of herbicide resistance over 8 years of coppice in field-grown, genetically engineered poplars. West J Appl For 23:89–93

Li J, Brunner AM, Meilan R, Strauss SH (2009) Stability of transgenes in trees: expression of two reporter genes in poplar over three field seasons. Tree Physiol 29:299–312

Li Q, Min D, Wang JP-Y, Peszlen I, Horvath L, Horvath B, Nishimura Y, Jameel H, Chang H-M, Chiang VL (2011) Down-regulation of glycosyltransferase 8D genes in *Populus trichocarpa* caused reduced mechanical strength and xylan content in wood. Tree Physiol 31:226–236

Li D, Song S, Xia X, Yin W (2012) Two CBL genes from *Populus euphratica* confer multiple stress tolerance in transgenic triploid white poplar. Plant Cell Tiss Org Cult 109:477–489

Liang H, Maynard CA, Allen RD, Powell WA (2001) Increased *Septoria musiva* resistance in transgenic hybrid poplar leaves expressing a wheat oxalate oxidase gene. Plant Mol Biol 45:619–629

Liu CJ (2012) Deciphering the enigma of lignification: precursor transport, oxidation, and the topochemistry of lignin assembly. Mol Plant 5:304–317

McCown BH, McCabe DE, Russell DR, Robison DJ, Barton KA, Raffa KF (1991) Stable transformation of *Populus* and incorporation of pest resistance by electric discharge particle acceleration. Plant Cell Rep 9:590–594

Meilan R, Ma C, Cheng S, Eaton JA, Miller LK, Crocket RP, DiFazio SP, Strauss SH (2000) High levels of Roundup® and leaf beetle resistance in genetically engineered hybrid cottonwoods. In: Blattner KA, Johnson JD, Baumgartner DM (eds) Hybrid poplars in the Pacific Northwest: culture, commerce and capability. Washington State University, Pullman, pp 29–38

Meilan R, Han KH, Ma C, DiFazio SP, Eaton JA, Hoien EA, Stanton BJ, Crockett RP, Taylor ML, James RR, Skinner JS, Jouanin L, Pilate G, Strauss SH (2002) The CP4 transgene provides high levels of tolerance to Roundup® herbicide in field-grown hybrid poplars. Can J Forest Res 32:967–976

Mellerowicz EJ, Sundberg B (2008) Wood cell walls: biosynthesis, developmental dynamics and their implications for wood properties. Curr Opin Plant Biol 11:293–300

Merkle SA, Dean JFD (2000) Forest tree biotechnology. Curr Opin Biotechnol 11:298–302

Merkle SA, Nairn CJ (2005) Hardwood tree biotechnology. In Vitro Cell Dev Biol Plant 41:602–619

Merkle SA, Andrade GM, Nairn CJ, Powell WA, Maynard CA (2007) Restoration of threatened species: a noble cause for transgenic trees. Tree Genet Genomes 3:111–118

Min DY, Li QZ, Jameel H, Chiang V, Chang HM (2011) Comparison of pretreatment protocols for cellulase-mediated saccharification of wood derived from transgenic low-xylan lines of cottonwood (*P. Trichocarpa*). Biomass Bioenergy 35:3514–3521

Neale DB (2007) Genomics to tree breeding and forest health. Curr Opin Genet Dev 17:539–544

Neale DB, Kremer A (2011) Forest tree genomics: growing resources and applications. Nat Rev Genet 12:111–122

Nehra NS, Becwar MR, Rottmann WH, Pearson L, Chowdhury K, Chang S, Wilde HD, Kodrzycki RJ, Zhang C, Gause KC, Parks DW, Hinchee MA (2005) Forest biotechnology: innovative methods, emerging opportunities. In Vitro Cell Dev Biol Plant 41:701–717

Newhouse AE, Schrodt F, Liang H, Maynard CA, Powell WA (2007) Transgenic American elm shows reduced Dutch elm disease symptoms and normal mycorrhizal colonization. Plant Cell Rep 26:977–987

Noël A, Levasseur C, Le VQ, Séguin A (2005) Enhanced resistance to fungal pathogens in forest trees by genetic transformation of black spruce and hybrid poplar with a *Trichoderma harzianum* endochitinase gene. Physiol Mol Plant Pathol 67:92–99

Pappinen A, Degefu Y, Syrjälä L, Keinonen K, Von Weissenberg K (2002) Transgenic silver birch (*Betula pendula*) expressing sugarbeet chitinase 4 shows enhanced resistance to *Pyrenopeziza betulicola*. Plant Cell Rep 20:1046–1051

Park YW, Baba K, Furuta Y, Iida I, Sameshima K, Arai M, Hayashi T (2004) Enhancement of growth and cellulose accumulation by overexpression of xyloglucanase in poplar. FEBS Lett 564:183–187

Pasonen HL, Seppänen SK, Degefu Y, Rytkönen A, Von Weissenberg K, Pappinen A (2004) Field performance of chitinase transgenic silver birches (*Betula pendula*): resistance to fungal diseases. Theor Appl Genet 109:562–570

Pilate G, Déjardin A, Leplé JC (2012) Field trials with lignin-modified transgenic trees. In: Jouanin L, Lapierre C (eds) Advances in botanical research. Academic Press, Burlington, pp 1–36

Powell WA, Maynard CA, Boyle B, Séguin A (2006) Fungal and bacterial resistance in transgenic trees. In: Fladung M, Ewald D (eds) Tree transgenesis – recent developments. Springer, Berlin, pp 235–252

Pu YQ, Kosa M, Kalluri UC, Tuskan GA, Ragauskas AJ (2011) Challenges of the utilization of wood polymers: how can they be overcome? Appl Microbiol Biotechnol 91:1525–1536

Robischon M (2006) Field trials with transgenic trees – state of the art and developments. In: Fladung M, Ewald D (eds) Tree transgenesis – recent developments. Springer, Berlin, pp 3–23

Robischon M, Du J, Miura E, Groover A (2011) The *Populus* class III HD ZIP, popREVOLUTA, influences cambium initiation and patterning of woody stems. Plant Physiol 155:1214–1225

Sanahuja G, Banakar R, Twyman RM, Capell T, Christou P (2011) *Bacillus thuringiensis*: a century of research, development and commercial applications. Plant Biotechnol J 9:283–300

Scheller HV, Ulvskov P (2010) Hemicelluloses. Annu Rev Plant Biol 61:263–289

Sederoff RR, MacKay JJ, Ralph J, Hatfield RD (1999) Unexpected variation in lignin. Curr Opin Plant Biol 2:145–152

Séguin A, Lachance D, Charest PJ (1996) Transient gene expression and stable genetic transformation into conifer tissues by microprojectile bombardment. In: Lindsey K (ed) Plant molecular biology manual. Kluwer Academic Publishers, Dordrecht, pp 1–46

Shin DI, Podila GK, Huang Y, Karnosky DF, Huang YH (1994) Transgenic larch expressing genes for herbicide and insect resistance. Can J For Res 24:2059–2067

Somerville C (2006) Cellulose synthesis in higher plants. Annu Rev Plant Biol 22:53–78

Sutela S, Niemi K, Edesi J, Laakso T, Saranpää P, Vuosku J, Mäkelä R, Tiimonen H, Chiang VL, Koskimäki J, Suorsa M, Julkunen-Tiitto R, Häggman H (2009) Phenolic compounds in ectomycorrhizal interaction of lignin modified silver birch. BMC Plant Biol 9:124

Tang W, Newton RJ (2003) Genetic transformation of conifers and its application in forest biotechnology. Plant Cell Rep 22:1–15

Tang W, Tian Y (2003) Transgenic loblolly pine (*Pinus taeda* L.) plants expressing a modified δ-endotoxin gene of *Bacillus thuringiensis* with enhanced resistance to *Dendrolimus punctatus* Walker and *Crypyothelea formosicola* Staud. J Exp Bot 54:835–844

Tang W, Newton R, Li C, Charles T (2007) Enhanced stress tolerance in transgenic pine expressing the pepper *CaPF1* gene is associated with the polyamine biosynthesis. Plant Cell Rep 26:115–124

Tournier V, Grat S, Marque C, El Kayal W, Penchel R, de Andrade G, Boudet AM, Teulières C (2003) An efficient procedure to stably introduce genes into an economically important pulp tree (*Eucalyptus grandis* x *Eucalyptus urophylla*). Transgenic Res 12:403–411

Vain P (2007) Thirty years of plant transformation technology development. Plant Biotechnol J 5:221–229

Voelker SL, Lachenbruch B, Meinzer FC, Jourdes M, Ki C, Patten AM, Davin LB, Lewis NG, Tuskan GA, Gunter L, Decker SR, Selig MJ, Sykes R, Himmel ME, Kitin P, Shevchenko O, Strauss SH (2010) Antisense down-regulation of 4CL expression alters lignification, tree growth, and saccharification potential of field-grown poplar. Plant Physiol 154:874–886

Wagner A, Donaldson L, Ralph J (2012) Lignification and lignin manipulations in conifers. In: Jouanin L, Lapierre C (eds) Advances in botanical research. Academic Press, Burlington, pp 37–76

Walter C, Fenning T (2004) Deployment of genetically-engineered trees in plantation forestry – an issue of concern? The science and politics of genetically modified tree plantations. In: Walter C, Carson M (eds) Plantation forest biotechnology for the 21st century. Research Signpost, Kerala, pp 423–446

Walter C, Grace LJ, Wagner A, White DWR, Walden AR, Donaldson SS, Hinton H, Gardner RC, Smith DR (1998) Stable transformation and regeneration of transgenic plants of *Pinus radiata* D. Don. Plant Cell Rep 17:460–468

Walter C, Fladung M, Boerjan W (2010) The 20-year environmental safety record of GM trees. Nat Biotechnol 28:656–658

Welch AJ, Stipanovic AJ, Maynard CA, Powell WA (2007) The effects of oxalic acid on transgenic *Castanea dentata* callus tissue expressing oxalate oxidase. Plant Sci 172:488–496

Wenck AR, Quinn M, Whetten RW, Pullman G, Sederoff R (1999) High-efficiency Agrobacterium-mediated transformation of Norway spruce (*Picea abies*) and loblolly pine (*Pinus taeda*). Plant Mol Biol 39:407–416

Ye X, Busov V, Zhao N, Meilan R, McDonnell LM, Coleman HD, Mansfield SD, Chen F, Li Y, Cheng ZM (2011) Transgenic *Populus* trees for forest products, bioenergy, and functional genomics. Crit Rev Plant Sci 30:415–434

Yevtushenko DP, Misra S (2010) Efficient agrobacterium-mediated transformation of commercial hybrid poplar *Populus nigra* L. x *P. maximowiczii* A. Henry. Plant Cell Rep 29:211–221

York WS, O'Neill MA (2008) Biochemical control of xylan biosynthesis – which end is up? Curr Opin Plant Biol 11:258–265

Yu X, Kikuchi A, Matsunaga E, Morishita Y, Nanto K, Sakurai N, Suzuki H, Shibata D, Shimada T, Watanabe KN (2009) Establishment of the evaluation system of salt tolerance on transgenic woody plants in the special netted-house. Plant Biotechnol 26:135–141

Zhang B, Chen M, Zhang X, Luan H, Diao S, Tian Y, Su X (2011) Laboratory and field evaluation of the transgenic *Populus alba* x *Populus glandulosa* expressing double coleopteran-resistance genes. Tree Physiol 31:567–573

Zhong R, Ye Z-H (2007) Regulation of cell wall biosynthesis. Curr Opin Plant Biol 10:564–572

Zhong R, Lee C, Ye Z-H (2010) Functional characterization of poplar wood-associated NAC domain transcription factors. Plant Physiol 152:1044–1055

Zhong R, McCarthy RL, Lee C, Ye ZH (2011) Dissection of the transcriptional program regulating secondary wall biosynthesis during wood formation in poplar. Plant Physiol 157:1452–1468

Forest Biotechnology Futures

Kevan M.A. Gartland and Jill S. Gartland

Abstract The contribution of forest biotechnology to the future of our global forests is explored. Questions relating to how forest biotechnology can most appropriately fully contribute to achieving sustainable development goals in the light of REDD + and the United Nations Rio + 20 developments are raised. Opportunities for international support for preservation of old growth forest areas, whilst allowing for other areas to be semi-managed, or used for high density plantation forestry are discussed. The role of 'trees of technology' using modified practices or genetic components in tree breeding are described, with reference to future energy, pulp, food and construction uses. The crucial role of biotechnology in conserving forest biodiversity and uniquely valuable tree genotypes is evaluated. The current and future potential for leading edge biotechnological breakthroughs in manipulating rapid growth, extending geographical ranges, developmental control of flowering, carbohydrate commitment, 'omics technologies and resistance to biotic and abiotic stresses are explored. The ability of forest biotechnology to contribute to delivering economic, societal and environmental benefits globally is discussed.

1 Introduction

The 1995 Convention on Biodiversity defines biotechnology as 'any technological application that uses biological systems, living organisms, or derivatives thereof, to make or modify products or processes for specific use'. Any products or processes derived from trees or other forest ecosystem organisms may therefore usefully be considered as forest biotechnology. The longevity of trees, frequently including an extensive juvenile phase and their size relative to almost all other organisms has

K.M.A. Gartland (✉) • J.S. Gartland
Glasgow Caledonian University, Glasgow, Scotland, UK
e-mail: Kevan.Gartland@gcu.ac.uk

T. Fenning (ed.), *Challenges and Opportunities for the World's Forests in the 21st Century*, Forestry Sciences 81, DOI 10.1007/978-94-007-7076-8_23,
© Springer Science+Business Media Dordrecht 2014

made trees important features of humankind's development through a wide range of uses, from shade to fuel, from fibres to heritage conservation and from construction materials to pharmaceuticals. As yet, we are only really beginning to domesticate trees, a process which lags many hundreds of generations behind that of annual crop plants. The various forms of biotechnology provide a wide range of tools both to harness the genetic diversity of tree form and function for economic benefit and to conserve trees of significance, to protect biodiversity, for leisure, heritage or well-being purposes. Forests cover >30 % of the Earth's land area, providing in excess of 3.4 billion m^3 of harvested wood annually (Fenning and Gartland 2007). Few groups of plants provoke such strong emotions as trees, reflecting their importance to the development of humankind and our anthropogenic effects on the environment.

Whilst there is general agreement on the need to use resources in a more appropriate manner, balancing consumption with availability and to avoid prejudicing resource supplies for future generations, there are a wide range of definitions for sustainable development. Given the role of forests in dealing with climate adaptation, renewable energy supply demands, food, construction and shade needs, it is not surprising that applying the principles of sustainable development to the utilisation of our planet's forests provokes vigorous international debate. The classical and arguably most widely accepted definition of sustainable development, provided by Brundtland (United Nations 1987), requires consideration of economic, societal and environmental aspects of resource utilisation. Adopting this approach however, requires judgements to be made about the relative importance of these three types of imperative and the consequences of these judgements on resources available for future generations. The simplicity of the three-pronged Brundtland approach is however complicated by the plethora of alternative interpretations of what sustainable development might mean. So, for example, the UK Government currently adopts a set of guiding principles for sustainable development, promoting a goal of living within environmental limits in a just society, to be achieved by means of a sustainable economy, good governance and sound science. The UK Government approach appears to differ from that of Brundtland through greater emphasis on ensuring a just society, using sound science responsibly including specific reference to the controversial 'Precautionary Principle' and promoting good governance. The most important interpretative difference though, lies with how a sustainable economy is to be achieved, through the provision of an economy providing opportunities for all and in which environmental and social costs fall on those who impose them: the so-called 'Polluter Pays' principle, whilst incentivising efficient resource usage. Recent US Government contributions have adopted a contrasting approach, emphasising that economic growth although essential, should not be followed at all costs. The US State Department interpretation of sustainable development also includes consideration of the environment, natural resource usage, social dimensions, gender inequality and addressing poverty in striking a suitable balance. There are many more interpretations, but all share a recognition of the need to use natural resources sensibly, in a way that limits detriment to the needs of future generations. In the case of our global forest resources, achieving a sustainable balance between economic gains, conservation and preservation of natural resources and respect for the environmental beliefs and values of citizens remains an elusive goal. For forest biotechnology to contribute fully to achieving sustainable development in the decades ahead, key questions include:

- How can increasing demands for wood, forest products and biofuels be met without endangering future renewable resource supplies?
- How can native forest biodiversity be maintained without unnecessarily preventing economic growth?
- How can the contribution of forests to mitigate climate change effects be optimized, for a sustainable future?
- How do the needs of sustainably managed forests match up with, or indeed differ from the landscape-scale approach favoured by international bodies such as the United Nations Forum on Forests, after the Rio + 20 United Nations Conference on Sustainable Development?

Forest biotechnology can contribute to addressing these key questions in a variety of ways. Doing nothing to address the consequences of increasing global emissions and climate change, perhaps 20 % of which may be due to deforestation, surely risks large scale release of stored carbon from dead and dying trees due to global warming. Using biotechnological tools to characterise, breed and select for the ability to grow at increased carbon dioxide concentrations and surface temperatures is one way to maintain forest environments and the rich biodiversity associated with them. Forests may be able to provide up to 60 % of the mitigation capacity needed to overcome forecast increases in atmospheric carbon levels over the next 50 years. Applying biotechnological tools to the development of new biofuel production platforms can also deliver reduced dependence on fossil fuels. Trees can deliver highly efficient lignocellulosics energy conversion, at 2–15× the effectiveness of maize or sugarcane (Fenning et al. 2008). Efficient use of our forest resources for renewables productivity, whilst at the same time conserving forest biodiversity will require a three-tiered strategy to be adopted. Old growth forest areas, acting as major carbon sinks, will need to be left almost undisturbed, perhaps due to international conservation incentive payments. A second group of forests will be semi-managed, allowing for small scale extraction in developing countries for firewood and local construction needs. Both the preserved old growth and the semi-managed forests will serve as conservation havens for biodiversity which may one day be needed to mitigate climate change effects and at the same time act as current and future carbon stores. The costs of this approach should not however, be underestimated. Stern has suggested that paying countries to leave their forests untouched might cost $11 bn annually (Stern 2007). Devoting sufficient international resources and priority to these tiers of forest ecosystems would allow other large areas to be used for rapid growth, high productivity plantation forests. By careful selection of tree species and genotypes, radiata pine and eucalyptus plantations can already be 20- and 40-fold respectively more productive than unmanaged forests in southern latitudes. Biotechnological tools to extend the growing areas for such trees to more northerly latitudes will allow even further productivity increases to be obtained in the future. These tools will become even more powerful when allied to genes associated with faster growth, altered lignin content and modified carbohydrate commitment favouring vegetative rather than reproductive organ development. By perfecting and deploying such approaches, forest biotechnology can make a significant and positive contribution to the REDD + agenda, of 'reducing emissions from deforestation and degradation' through sustainable management of carbon stocks and the trees within our forests (Hofstad et al. 2009). To be truly effective, REDD + must play an important future role in international climate change negotiations, with polluting countries providing incentives for tropical and developing nations to maintain global carbon stores and refrain from unsustainable deforestation practices. Although influenced strongly by direct drivers such as agricultural expansion, wood extraction, energy policy and infrastructure development (Phelps et al. 2010), wider factors including property rights, governance principles and macroeconomic factors will also be significant, throughout this century.

2 Trees of Technology

Within the broad definition of forest biotechnology, three groups of tools are particularly relevant to the genetic improvement of trees:

- Conventional Breeding Technologies
- Genetic Modifications
- 'Omics Technology Applications

Conventional breeding technologies include vegetative propagation, grafting, rooting of cuttings, micropropagation, cryopreservation and somatic embryogenesis in laboratories, and marker aided selection. Each of these technologies allows for desired genotypes to be multiplied on a large scale. In the case of marker aided selection, genotypes with desired properties can be identified at an earlier stage of growth than would otherwise be possible. Such breeding technologies can deliver reductions in the length of the development cycle for elite tree genotypes. The second group of tools use genetic modification approaches to achieve reductions in development cycle length by specifically and precisely targeting control of genes inserted or already present in tree genomes. Typically, the insertion of a novel gene will be undertaken using the soil microbe *Agrobacterium* species, which naturally transfers genes into the nucleus of dicotyledonous plants, or using DNA-coated microprojectiles in the biolistic process. Whole genes, novel regulatory elements or parts of genes can be inserted using either approach. The amount, cellular location, timing, or developmental status during which gene expression takes place can be modified with increasing precision using these tools. As yet, the genetic modification approaches have usually transformed one or at most a few, genes into tree genomes. Releasing genetically modified trees into the environment frequently remains controversial, despite well established regulatory regimes and a history of safe use from field trials in many countries. This reflects in part, the potential of trees to emit pollen for extended periods of time and our limited abilities to prevent flowering and pollen formation. Whilst conifer pollen is known to potentially travel several hundred miles, less is known about the population dynamics of resultant distantly-pollinated trees. There are however, useful lessons to be learnt from experience with commercial scale release of genetically modified crop plants. The International Service for the Acquisition of Agri-Biotech Applications (ISAAA) reports that in 2011, 160 MHa of such crops were planted in 29 different countries, an increase of 8 % on 2010. Developing countries make up the majority of those choosing to adopt the products of genetic modification technologies, with planted areas increasing by 11 %, to 44 % of the globally planted areas. Brazil, India, China, Argentina and South Africa are all significant contributors to the increasing trend towards planting biotech crops. In the USA, 69 MHa have been planted, with a 90 % adoption rate being reached for four major crops maize, soya, wheat and alfalfa. Globally, 16.7 million farmers are now choosing to deploy genetically modified plants. Planting of genetically modified trees will grow in a similar fashion, especially as advances are made in limiting reproductive potential, since demand for wood and wood products continues to rise. The European Union which continues to apply a very narrow approach to the 'precautionary principle', has only very limited commercial scale plantings, despite insect resistant *Bacillus thuringiensis* 'Bt-maize' areas increasing by 25 % in 2011, in six countries. The politico-regulatory impasse for commercial scale environmental release applications in the European Union (EU) shows every sign of continuing to prevent final decisions being reached on deployment of many potential biotech crops. For this reason, the EU risks becoming a desert for most plant biotechnology applications. This is ironic, given that many European companies continue to develop such products for deployment out with EU borders.

Whilst there have been many hundreds of field scale release trials of genetically modified trees, with no significant environmental or human health impacts being reported, almost all of these trials have been concluded before flowering. Only two commercial scale releases have taken place. In Hawaii, the devastating losses to *Carica papaya* fruit yields due to papaya ring spot virus, which threatened to wipe out an important island industry worth $47 M, have been overcome using genetic modification technologies. Deliberate over-expression of a coat-protein gene from the pathogenic virus delivers a high degree of protection against viral symptoms. This allows fruit yields to be preserved, with the most widely planted transgenic variety 'Rainbow' yielding up to 313,000 lbs fruit/ha/year c.f. yields of only 13,000 lbs/ha/year from the virus-susceptible parent genotype 'Kapoho'. Within 3 years of deployment, Hawaiian papaya yields increased by 77 %, including exports to Canada. Obtaining regulatory approval for other markets proved more difficult however. From December, 2011, 'Rainbow' papaya became the first genetically modified direct to consumer product to be approved for import into Japan, a market worth up to $11 Million annually. This successful approach is being replicated and is awaiting regulatory approval in a range of other countries, including Thailand and Brazil. Extension of the available markets for genetically modified papaya, be it from Hawaii or elsewhere, is likely to follow in the coming years, adding economic sustainability to the environmental and societal benefits of this industry. Fruit trees are likely to provide further examples of disease resistance by genetic modification in the years ahead. US regulatory approvals for environmental, food and feed uses have been obtained for 'C5 Honeysweet' plum (*Prunus domestica*), constitutively expressing the plum pox potyvirus coat protein gene. Whilst not yet deployed on a commercial scale, this development highlights the importance of international collaborations in developing forest biotechnology products, as virus resistant C5 plum reached regulatory approval after extensive field trials in five different European and North American countries (Zagrai et al. 2008).

Chinese scientists have a more flexible environmental release regulatory regime than in many other countries. This has led to a wide range of genetically modified tree field trials having been undertaken over the past 15 years. These trials have frequently investigated resistance to insect pests through expression of Bt crystal toxin proteins or proteinase inhibitor genes and are described in the chapter by Meng-zhu Lu and colleagues herein. A second large group of field trials have been undertaken for resistance to environmental stresses such as salt tolerance or drought resistance. China has approved commercial scale growth of Bt-expressing black poplar (*Populus nigra*) for more than 10 years. In contrast to the Hawaiian papaya release, for protection against one specific viral pathogen and commercial scale fruit production, the Chinese black poplar commercial release will allow many significant insights to be gained into the effects or otherwise, of genetically modified trees on a forest ecosystem. This reflects the ability of some Bt constructs to affect more than one insect species. Whilst the utility of this approach increases potential applicability, this must be balanced with the possibility of unintended effects on non-target species. The scale of the Chinese black poplar trials both in area and in duration however, should allow such potential effects to be thoroughly investigated.

Opponents of the use of genetic modification technologies frequently cite the potential for pathogen resistance to break down with time as a reason for forestalling deployments. This has to some extent, proven to be the case with transgenic Bt cotton failing to control pink bollworm after extended use in some parts of Gujarat in India. Despite the likelihood that imported seeds have been used there inappropriately, scientists should be able to overcome such criticisms by the use of 'gene stacking' providing more than one resistance trait in a single plant genotype. Amongst the gene stacking approaches successfully being used to prevent resistance traits breaking down is Bollgard II cotton (MON 15985) expressing two different bollworm resistance genes, Cry 2Ab and Cry 1Ac. Ten years of use in Australia have shown <80 % reductions in pesticide use and a 10 % reduction in water requirements are obtainable with Bollgard II. As well as having these environmental advantages, this multiple resistance approach will remain popular in the years ahead since two separate resistance characteristics would have to break down for pest susceptibility to recur. The longevity and pollen spreading abilities of trees mean that future forest biotechnologists must frequently consider using gene stacking strategies. As yet there are only a few examples of designing in multiple traits into trees, but it is possible that, having seen how this can be done for annual crop plants, regulators may require it, if commercial scale deployment of transgenic trees is to become widespread. An alternative gene stacking strategy is being developed by ArborGen, in their recently approved permit application for large scale deployment of genetically modified eucalypts at 28 different locations, totalling 120 Ha, in seven areas of the South Eastern United States. This company, who can justifiably claim to be at the forefront of forest biotechnology developments, have invested heavily in attempting to extend the growing range of tropical eucalypts. Genetically modified traits being developed include male sterility, decreased lignin content, enhanced growth rates and freeze tolerance. Male sterility has been obtained through tapetal-specific PrMC2 promoted expression of the barnase gene which destroys RNA in a precise and tissue-limited manner to prevent pollen development. Barnase has been shown to be effective in maize, oilseed rape and chicory. Having amassed considerable biosafety data from several small scale trials where genetically modified male sterile *Eucalyptus grandis x E. urophylla* hybrids have been allowed to flower, ArborGen will seek to explore deployment of such eucalypt hybrids at larger scales in future. The US Department of Agriculture Animal and Plant Health Inspection Service (APHIS) has concluded that the barnase gene engineered into cold tolerant eucalypt hybrids is effective at preventing pollen formation. Effective freeze tolerance, to temperatures of -8 °C., due to over-expression of C-Repeat Binding Factors, could increase the areas of the South-Eastern United States suitable for growing high yielding tropical eucalypt hybrids from 6 to 23 MHa. Whilst each of the improved traits is likely to have commercial benefits, by stacking two or more such traits into the same genotype, very considerable pulping and biomass yield benefits can be gained. The stacked biotech products being developed include increased growth rates combined with freeze tolerance, and a combination of faster growth and improved pulping properties through reduced lignin content. By combining these traits with barnase male sterility a higher degree of biological containment is likely to be obtained. Freeze tolerant genotype biomass yields 15–28 % higher than conventional sub-tropical eucalypts appear possible, making a substantial contribution to American biomass requirements for future pulping and sustainable energy needs, in a country with an avowed intent to drastically reduce dependency on imported energy. Should the recently approved field releases prove successful and economically viable, there can be little doubt that a range of other genetically modified trees will be deployed. Recent changes to APHIS approval processes should enable faster decisions to be reached on new biotechnology products whilst maintaining appropriate environmental considerations.

Approaches to fast growing, high yielding eucalypts are also being pursued by FuturaGene in Brazil. This subsidiary of Suzano, the second largest producer of eucalyptus pulp and paper globally, has obtained consent from the Brazilian biosafety authorities for the fourth in a series of genetically modified eucalyptus field trials, investigating modified cell wall properties leading to faster growth and harvesting for biomass energy purposes. FuturaGene also has extensive commercial interests in China, the United States and Israel as well as a wide range of research and development partnerships with Universities and institutes to investigate the use of eucalypts and poplars as biomass and bioenergy crops. FuturaGene's genetically modified eucalypts grow faster than typical Brazilian eucalypts and are expected to yield up to 104 m3/ha/year of wood from plantings on the hundred hectare scale at five Brazilian locations. This represents a potential increase of >25 % over conventional genotypes. Eucalypts grown for energy crops represent a highly attractive investment proposition for the future, since they can be harvested at 1–1.5 years, as compared to 7 years for paper and pulp uses. The shorter timescale to potentially obtaining a return on investment this offers will remain attractive to investors given the continuing global economic uncertainty.

Manipulating the carbohydrate commitment of trees has long been a holy grail for forest biotechnologists. Altering the 25 % lignin and 70 % cellulose/hemicellulose content of wood is now possible, although there is much evidence that selection for high growth rates can be accompanied by reduced lignin content. Low lignin trees could be most suited to pulp, paper and biofuels processing. As our understanding of wood biochemistry and transcriptional control increases in the future, we will be able to manipulate wood structure and properties more precisely. Modelling studies suggest that the role of sucrose, from photosynthate, and phenylalanine, may be important in controlling carbon partitioning between lignin, cellulose and hemicelluloses, in secondary cell walls than has been previously thought (Novaes et al. 2010). Another area in which trees may prove productive biorefineries is in the production of bioplastics such as poly hydroxybutyrate (PHB) which has the potential to partially replace petroleum-derived plastics, is biodegradable and may make a positive contribution to achieving carbon balance. PHB yields of up to 2 % have been achieved in transgenic poplar expressing the *phbABC* genes from the bacterium *Ralstonia eutrophus* (Dalton et al. 2012). Targeting expression towards senescence may overcome a limitation of current approaches, where current approaches delivering higher PHB contents lead to chlorotic plants, as seen in *Arabidopsis* for example (Bohmert 2000).

The trend towards genetically modified trees for non-food commodity purposes becoming acceptable is likely to increase in the years ahead. Restricting potential for flowering or use of male sterile genotypes, as used in the ArborGen eucalypt trial plantings may also overcome citizens concerns regarding possible spread of genetically modified pollen over potentially long distances. One thing is certain however, that it is only currently possible to ascertain whether restricted gene flow traits such as these will remain stable and effective over long time periods by allowing their environmental release and carefully monitoring potential gene flow. This type of approach when combined with thoughtful mathematical modelling of release data obtained (DiFazio et al. 2012), at both local and landscape scales, will provide much valuable evidence for evaluating gene flow potential in the coming decade.

3   Biotechnology for Genetic Conservation

Biotechnology tools are being widely used in the genetic conservation of forest trees. Whilst this approach sometimes uses genetic modification, frequently it does not. This is particularly relevant for the propagation of threatened or elite germplasm materials. In vitro micropropagation will continue to be widely used to bulk up materials, although use of somatic embryogenesis technologies allows faster routes to large scale multiplication of such genotypes (Gatica-Arias et al. 2008). Since neither of these approaches is regarded by society as particularly controversial, expansion of their usage is likely to increase further. One promising avenue for future research into somatic embryogenesis is the identification and quantification of potential molecular markers for transitional competence to form large numbers of somatic embryos. A somatic embryogenesis receptor-like kinase (SERK) gene has been isolated from coconut (*Cocos nucifera*; Perez-Nunez et al. (2009) Typical SERK genes contain a signal peptide, a leucine zipper domain, a series of leucine rich-repeats, a distinctive serine-proline-proline domain, a single transmembrane domain, a complex kinase domain and a C-terminal region). Increased expression of the coconut SERK appears to be a potential marker for the ability of in vitro cultured cells to form somatic embryos. Studies with pineapple (*Ananas comosus*, Ma et al. 2012) have extended these observations, providing further evidence that SERKs are highly expressed during the transition of individual cells to competence for somatic embryo production. Similarly structured genes have also been found in citrus, papaya and grapevine genomes. Genes such as the pineapple (*AcSERK1*) or coconut (*CnSERK*) receptor-like kinases, could prove to be valuable markers for this transition competence in recalcitrant species or those under threat, where somatic embryo production has high economic or conservation value. Once cell cultures with high embryogenic potential have been identified, RNA analysis could routinely be used to check for retention of the competent state, potentially reducing downtime by giving early indicators of declining or lost embryogenic potential. When allied to proven cryopreservation protocols, including the monitoring of nuclear microsatellite sequence stability (Burg et al. 2007), enhanced somatic embryogenesis tools are likely to prove especially valuable in the conservation of rare or threatened tree genotypes, as well as for commercial multiplication purposes.

4   Pathogen Problems

Globally, our forests face an increasing range of stresses, be they biotic, such as attacks from pathogens with extended range due to climate change, or abiotic, due to extended drought periods, or increased nitrogen oxide pollutant levels. The biosecurity of forests is adversely affected by the transmission of pathogens between countries, often as a result of commodity trading. Introducing pathogens to new environments can frequently result in outbreaks of infection. The lack of previous exposure to a pathogen can result in new hosts having no resistance or defence mechanisms for introduced pathogens. This can destabilise an ecosystem, leading to dramatic effects on biodiversity at a local or larger level. Biotic threats to European forests include *Chalara fraxinea* affecting 90+ % of ash through fungal dieback, with perhaps only 1 % of resistant trees emerging and *Pseudomonas syringae* induced bleeding canker of European horse chestnuts (*Aesculus hippocastanum*), which may have been introduced from India, throughout North-West Europe. *Phytophthora* species will pose increasing problems for forestry, from *Ph. pseudosyringae* infected beech (*Nothofagus spp.*) and hornbeam, with up to 70 % dieback possible, to *Ph. ramorum* induced dieback of larch whilst also causing cankers on a range of other conifers in the UK, as well as birch. Many of the fungal or bacterial pathogen attacks occurring have been influenced by the global transportation of trees or forest products. Whilst this has been known to occur for at least 100 years, it is an increasing problem, particularly when allied to new host-pathogen interactions as a result of environmentally-induced range changes. *Ph. lateralis* for example, a highly damaging pathogen of Lawson cypress (*Chaemacyparis lawsoniana*) in the Pacific North-West of America, may have originated in Taiwan, where it causes minimal damage to native conifers, but has also been found causing dieback in France and Scotland. *Ph. ramorum* which is known to have at least 150 host species in Europe and North America, represents a particular problem for trees. Although likely to have originated in Asia, multiple European and North American genetic races of this pathogen have already developed. Infected rhododendrons, camellias or viburnums are likely to continue disseminating such pathogens, and infective *Ph. ramorum* spores have been found on at least one invasive species *Rhododendron ponticum*. Since 2009, large numbers of the commercially important Japanese larch (*Laryx kaempferi*) and larch hybrids have developed stem cankers and died in south-western and western parts of the British Isles, and on a single Sitka spruce (*Picea sitchensis*) tree in Ireland. Almost all of the infected areas have warm, moist climates and spread of this pathogen may be heavily influenced by environmental conditions, as outbreaks have not yet been reported in the colder, drier eastern UK. In contrast, American genetic races of *Ph. ramorum* have caused millions of losses due to 'Sudden Oak Death' amongst tan and native oaks in the Pacific north-western United States. As climate change is predicted to produce warmer, wetter summers for many European locations, such threats will increase in scale and magnitude of importance in coming decades.

In some circumstances, particular tree species can come under different threats at relatively close proximity. Fraser fir for example, has suffered large scale damage at elevated altitudes in the Appalachian Mountains of North Carolina, USA, due to the ravages of the balsam woolly adelgid. It is likely that this insect pathogen would not normally have reached such high elevations, but that climate change has allowed it to advance higher in recent years, causing up to 90 % losses and threatening the centre of biodiversity for this commercially important conifer. The consequences of balsam woolly adelgid attack for local ecosystems can be extremely severe, with 18 species of birds, invertebrates and plants from the spruce-fir ecosystem coming under threat on these mountain slopes (Gartland and Gartland 2004). At lower altitudes in North Carolina, where a lucrative 'Christmas Tree' industry has developed, the water-borne *Ph. cinnamomi* is causing root rot, with few symptoms evident until death is likely. As climate change continues to make greater differences to landscapes and habitats, new combinations of pathogens and potential hosts, or of existing pathogens extending their range, will arise more frequently. Such new pathogen

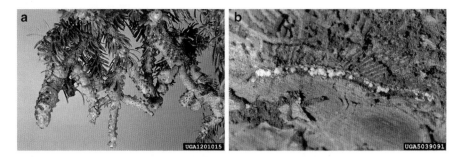

Fig. 1 (**a**) Balsam woolly adelgid damage on Frasier fir (Source: Wayne Brewer, Auburn University, Bugwood.org). (**b**) Coremial spore structures of Dutch elm disease fungus (*Ophiostoma ulmi*) in elm bark beetle breeding galleries (Source: Joseph O'Brien, USDA Forest Service, Bugwood.org)

attacks are an increasingly important threat to forest biosecurity. Unfortunately, no large scale approaches to dealing with them, once outbreaks occur, have been devised so far (Fig. 1).

Biotechnological approaches have begun to deliver new tools in combating threats to forest trees. For many of the threats posed to our forest trees by imported pathogens, other biotic or abiotic stresses such as climate change, biotechnology is probably the only effective way to develop novel solutions. This is especially so where tree biodiversity is under threat, or for individual trees of significance, sometimes known as 'heritage trees'. Initial progress is encouraging and in the years ahead, the number of examples where biotechnological tools can be used to overcome threats to valuable biodiversity and key germplasm will increase. Future deployment, on forest or landscape scales will be strongly influenced by the degree of public acceptability for such biotech trees. Dutch elm disease is caused by the fungus *Ophiostoma novo-ulmi*, transmitted by elm bark beetles (*Scolytus* spp.) has ravaged elms throughout the Northern Hemisphere, with so far limited outbreaks in New Zealand and Japan. Scientists have developed *Agrobacterium*-mediated gene transfer techniques for English elm (*Ulmus procera* Salisbury) and transferred a range of potential anti-fungal genes into the clonal SR4 genotype found in the southern British Isles (Gartland et al. 2005). These candidate genes appear to restrict fungal growth either by preventing spore germination or restricting hyphal growth. State University of New York (SUNY) scientists have used a similar delivery approach to successfully express a synthetic anti-microbial peptide ESF 39A, in American elm (*U. americana* L.). Promising results against the fungal pathogen *O. novo-ulmi*, have been obtained, whilst undertaking the first field trials of genetically modified American elms (Newhouse et al. 2009).

The American chestnut (*Castanea dentata*), has been devastated by the introduced chestnut blight fungal pathogen *Cryphonectria parasitica*, with >3.5 bn chestnuts destroyed in the last 100 years. Attempts to restore this dominate forest tree to the North American landscape, have utilised a range of biotechnological and conventional breeding tools. The American Chestnut Foundation have used conventional breeding approaches involving crosses with blight-resistant Chinese chestnut

(*C. mollissima*), followed by three rounds of backcrosses to the American parent and three rounds of intercrossing amongst the third backcross progeny has produced a B3F3 population, expected to be at least 15/16ths American chestnut genetically, with, as a result of fungal challenge and visual selection within each progeny generation, a high likelihood of chestnut blight resistance (Huckabee Smith 2012). In future years, 14,000 potentially blight resistant chestnuts will be planted on reclaimed mine lands, across five States, allowing for the first time, large scale multi-site trials to be conducted. A genomics based approach involving comparison of the transcriptomes of the American and Chinese chestnuts in response to blight infection, using ultra high throughput pyrosequencing has identified 28,890 and 40,039 unique genes 'unigenes' in American and Chinese chestnuts respectively (Barakat et al. 2009). The expressed sequence tags and unigenes obtained will provide an invaluable resource for the identification and understanding of genes involved in plant defence against biotic stresses. Several of the genes so far identified are associated with hypersensitive responses, physical blocking to prevent pathogen transport, as well as generic defence responses. Potential pathways to blight resistance and candidate genes for local and systemic defence responses will be investigated further in the years ahead (Hirsch 2012). This approach will generate additional understanding and benefits for many other tree-pathogen interactions based on specific hydrolases, altered patterns of lignin biosynthesis, signalling and cell death cascade pathways.

5 Applications of 'Omics Technologies

Tree genomes are considerably more complex than the 25,000 gene human genome, with conifers such as loblolly pine, having at least a seven-fold larger genome. The significant reductions already achieved in the speed and cost of sequencing complex genomes using 'Next Generation' sequencing for example, will undoubtedly continue in the years ahead, making it both faster and easier to obtain increasing amounts of genomic sequence data. Since poplar became the first tree genome to be sequenced, the range of genomics resources at or near to completion has steadily increased. Table 1 describes a selection of these genomes and some of the exciting and interesting questions which enhanced genomic knowledge will address in the future. The trend towards increasing numbers of finished tree genomes will undoubtedly accelerate (Wegrzyn et al. 2012). Whilst knowledge of vast amounts of genomic sequence information is useful, its value will be increased dramatically when allied to developing concepts in the regulation of gene expression, of growth and development in response to varying environmental conditions. Realising this potential value requires integrating findings from detailed investigation using the so-called 'omics technologies, including proteomics, epigenetics, epigenomics and ionomics (Baxter and Dilkes 2012) alongside the mere acquisition of genomic sequence information, with ecosystem level studies, especially where adaptation to changing environmental conditions may be important (Kremer et al. 2011).

Table 1 Tree genomics projects

Tree	Species	Indicative problem
Poplar	*Populus trichocarpa*	Gender determination
		Fate of transgenes model
		41,377 genes
Amborellid	*Amborella trichopoda*	Earliest diverging angiosperm still extant
		Primitive tree
		Organ differentiation
Apple	*Malus x domestica* 'golden delicious'	Fruit properties
		57,386 genes
Peach	*Prunus persica*	Selfing behaviour control
		27,852 genes
Pear	*Pyrus bretschneideri* cv. Dangshansuli	Fruit flesh quality
		Processing properties
Papaya	*Caricus papaya*	Fruit colour and yield control
		28,629 genes
Cocoa	*Theobroma cacao* 'Criollo'	Cocoa bean butter properties
		35,000+ genes
Grape	*Vitis vinifera* 'Pinot Noir'	Fruit processing properties
		26,346 genes
Eucalypts	*Eucalyptus grandis*	Pulping properties
		40,000+ genes
Sweet orange 'ridge pineapple'	*Citrus sinensis*	Juice properties
		25,376 genes
Clementine Mandarin	*Citrus elementina*	Taste properties
		25,385 genes

Proteomics allows the structure-function relationships of proteins to be related to genomic sequences and has revealed many insights into how gene expression products influence physical attributes, or phenotype. Biologists have for many decades wondered how clonal organisms can have different phenotypes in different environments and until recently, speculated widely on how environment interacts with genotype to produce heritable changes in phenotype, in what has become known as epigenetics (Bonetta 2008). Advances in 'omics technologies now allow epigenetic tools to be applied across whole genomes, in epigenomic investigations. As is so often the case, epigenomics is being led by biomedical science developments looking at how epigenetic factors can cause major human diseases (see Fig. 2). Many epigenomic tools are, however, equally applicable to tree genomes and an explosion of new findings and mechanistic models for how environment can heritably influence gene expression is just around the corner. The recent publication of an integrated encyclopaedia of DNA elements in the human genome (ENCODE Project Consortium 2012) is arguably the most significant advance since sequencing genomes became possible. The raft of 30 accompanying papers demonstrates that our previous understanding of gene expression being regulated by a small number of elements located at or close to exon-coding gene sequences,

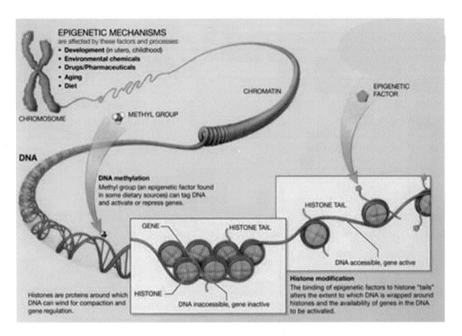

Fig. 2 Principles of epigenomics: millions of gene expression switches are influenced by DNA methylation and epigenetic factors binding to histones, altering gene activity (Source: Adapted from National Institutes of Health (USA) Epigenetics Portal)

is woefully simplistic. Since as many as four million individual switches may be involved in controlling the 25,000 genes and 147 cell types in the human genome over a typical 75 year lifespan, the plethora of switching events in tree gene expression over lifespans of hundreds and sometimes thousands of years, as in yew, for example, will be enormous. The chapter by Muchero and colleagues in this book discusses these issues in detail, building on the work of the poplar genome programme. Amongst the ways in which epigenetic control can take place which we are beginning to understand for trees are DNA methylation, histone modification and the role of non-coding RNAs. It is highly likely that the sites, timing and extent of cytosine methylation to form 5-methyl cytosine in trees will all exert crucial influences on transcription. Findings have suggested that a mixture of three distinct cytosine methylation contexts occur in poplar for example, with CG sites being the most highly methylated, followed by CHG sites and CHH sites, where H can be one of the bases adenosine, cytosine or thymine (Feng et al 2010). Methylation of cytosine can take place in coding regions, known as the body of a gene, or in promoter regions, regulating gene expression. Extensive cytosine methylation is usually thought to reduce transcriptional activity. Recent poplar epigenomic data suggests that gene body methylation reduces transcription to a greater extent than promoter sequence methylation (Vining et al. 2012). Whilst as yet, we do not have a comprehensive atlas of cytosine methylation patterns and effects on gene expression in trees, considerable variations seem certain to be found in different tree organs and under varying developmental stages or environmental stresses. Mapping such variations will reveal much more about the influence environmental conditions play on genes and development.

Histone modification patterns appear to be extremely complicated, with nucleosome histone N-terminal tails in chromatin likely to be tagged in a wide number of different ways (see Fig. 2). Exposed chromatin N-terminal histone tails may be methylated at lysine or arginine residues in different positions, sometimes multipally with up to three methylations taking place on one amino acid. Alternatively, exposed histone tail lysines may be acetylated, whilst serine or threonine may be phosphorylated, sumoylated or ubiquitinylated, and proline may be isomerised (Bonetta 2008). Each of these reactions are reversible, adding to the enormous complexity of transcriptional switching that now appears possible through modification of DNA or chromatin elements. Small RNAs provide another promising avenue for future gene expression studies in trees, ranging from epigenetic effects due to non-coding RNAs possibly directing DNA methylation, small interfering RNAs (siRNAs) acting to post-transcriptionally down regulate gene expression, to micro RNAs and the possibility of piRNAs interacting with Argonaute proteins (Ghildiyal and Zamore 2009). Human epigenomic studies have allowed encyclopaedic knowledge of such histone modifications to be mapped, leading to up and down regulation of gene expression in whole regions of chromosomes, due to altered chromatin folding (ENCODE Project Consortium 2012). Since >80 % of the human genome is now known to be involved in some form of RNA- or chromatin-associated event, it is inevitable that the number of molecular switches involved in tree gene expression, will greatly exceed the four million suggested for man. The larger genomes and extended lifespans of trees will utilise many more transcriptional switches across all stages of development and differentiation, whilst being exposed to a wider range of environmental stresses. An explosion of new knowledge and understanding will undoubtedly arise by comparing DNA and histone modification patterns with other plant and animal systems.

However the international regulatory regime changes in the years ahead (Viswanath et al. 2012; see also the chapter by Haggman et al. in this book), as long as oil prices remain high and demand for forest products continues to rise globally, forest biotechnology is assured of a bright and exciting future. The new insights to be gained into the control of growth, differentiation, carbohydrate partitioning, reproductive control and both food- and non-food uses of trees will all play major parts in determining the precise nature of the future of forest biotechnology (Troggio et al. 2012). The waves of new epigenomic data now beginning to emerge will undoubtedly provide increased understanding of the control of gene expression, environmental stress responses and development. When put together, all of these enhancements will ensure that forest biotechnology makes a significantly larger contribution to meeting our global economic, environmental and societal needs for many decades to come.

References

Barakat A, DiLoreto DS, Zhang Y, Smith C, Baier K, Powell WA, Wheeler N, Sederoff R, Carlson JE (2009) Comparison of the transcriptomes of American chestnut (*castanea dentata*) and Chinese chestnut (*castanea mollissima*) in response to the chestnut blight infection. BMC Plant Biol 9:51

Baxter I, Dilkes BP (2012) Elemental profiles reflect plant adaptations to the environment. Science 336:1661–1663

Bohmert K (2000) Transgenic Arabidopsis can accumulate polyhydroxybutyrate to up to 4% of their fresh weight. Planta 211:841–845

Bonetta L (2008) Epigenomics: detailed analysis. Nature 454:795–798

Burg K, Helmersson A, Bozhkow P, von Arnold S (2007) Developmental and genetic variation in nuclear microsatellite stability during somatic embryogenesis in pine. J Exp Bot 58:687–698

Dalton DA, Ma C, Murthy GS, Strauss SH (2012) Bioplastic production by transgenic poplar. Information Systems for Biotechnology News Report (Jan) 7–10

DiFazio SP, Leonardi S, Slavov GT, Garman SL, Adams WT, Strauss SH (2012) Gene flow and simulation of transgenic dispersal from hybrid poplar plantations. New Phytologist 193: 903–915

ENCODE Project Consortium (2012) An integrated encyclopaedia of DNA elements in the human genome. Nature 489:57–74

Feng S, Cokus SJ, Zhang X, Chen P-Y, Bostick M, Goll MG, Hetzel J, Jain J, Strauss SH, Halpern ME, Ukomadu C, Sadler KC, Pradhan S, Pellerini M, Jacobsen SE (2010) Conservation and divergence of methylation patterning in plants and animals. Proc Natl Acad Sci USA 107:8689–8694

Fenning TM, Gartland KMA (2007) Why forests matter. Prospect 133:17–18

Fenning TM, Walter C, Gartland KMA (2008) Forest biotech and climate change. Nat Biotechnol 26:615–617

Gartland KMA, Gartland JS (2004) Biotechnology applied to conservation, insects and diseases. In: Kellison R, McCord S, Gartland KMA (eds) Forest biotechnology in Latin America. Institute of Forest Biotechnology, North Carolina, pp 109–116

Gartland KMA, McHugh AT, Crow RM, Garg A, Gartland JS (2005) Biotechnological progress in dealing with Dutch Elm disease. In Vitro Cell Dev Biol Plant 41:364–367

Gatica-Arias AM, Arrieta-espinoza G, Espinoza-Esquivel AM (2008) Plant regeneration via somatic embryogenesis and optimisation of genetic transformation in coffee (*coffea Arabica* L.) cvs. Caturra and catuai. Electron J Biotechnol 11(1):9

Ghildiyal M, Zamore PD (2009) Small silencing RNAs: an expanding universe. Nat Rev Genet 10:94–108

Hirsch R (2012) Blight resistance: it's in the DNA. J Am Chestnut Found 26:20–22

Hofstad O, Kohlin G, Namaala J (2009) How can emissions from woodfuel be reduced? In: Angelsen A (ed) Realising REDD+: national strategy and policy options. CIFOR, Bogor

Huckabee Smith A (2012) Breeding for resistance: TACF and the Burnham hypothesis. J Am Chestnut Found 26:11–15

Kremer A, Vincenti B, Alia R, Burczyk J, Cavers S, Degen B, Finkeldey R, Fluch S, Gomory D, Gugerli F, Koelwijn HP, Koskela J, Lefevre F, Morgante M, Mueller-Starck G, Plomion C, Taylor G, Turok J, Savolainen O, Ziegenhagen B (2011) Forest ecosystem genomics and adaptation. Tree Genet Genomes 7:869–875

Ma J, He Y, Wu C, Liu H, Hu Z, Sun G (2012) Cloning and molecular characterization of a SERK gene transcriptionally induced during somatic embryogenesis *Ananas comusus* cv. Shenwan. Plant Mol Biol Report 30:195–203

Newhouse AE, Kaczmar NS, Powell WA, Maynard CA (2009) American elm. In: Kole C, Hall TC (eds) A compendium of transgenic plants. Blackwell, Oxford, pp 241–262

Novaes E, Kirst M, Chiang V, Winter-Sederoff H, Sederoff R (2010) Lignin and biomass: a negative correlation for wood formation and lignin content in trees. Plant Physiol 154:555–561

Perez-Nunez MT, Souza R, Saenz L, Chan JL, Zuniga-Aguillar JJ, Oropreza C (2009) Detection of a SERK-like gene in coconut and analysis of its expression during the formation of embryogenic callus and somatic embryos. Plant Cell Rep 28:11–19

Phelps J, Webb EL, Agrawal A (2010) Does REDD + threaten to recentralise forest governance? Science 328:312–313

Stern N (2007) The economics of climate change. Cambridge University Press, Cambridge

Troggio M, Gleave A, Salvi S, Chagne D, Cestaro A, Kumar S, Crowhurst RN, Gardiner SE (2012) Apple, from genome to breeding. Tree Genet Genomes 8:509–529

United Nations (1987) Report of the World Commission on Environment and Development. General Assembly Resolution 42/187, 11 December, 1987

Vining KJ, Pomraning KR, Wilhelm LJ, Priest HD, Pellegrini M, Mockler TC, Freitag M, Strauss SH (2012) Dynamic DNA cytosine methylation in the Populus trichocarpa genome: tissue-level variation and relationship to gene expression. Biomed Cent Genom 13:27

Viswanath V, Albrechtsen BR, Strauss SH (2012) Global regulatory burden for field testing of genetically modified trees. Tree Genet Genomes 8:221–226

Wegrzyn JL, Main D, Figueroa B, Choi M, Yu J, Neale DB, Jung S, Lee T, Stanton M, Zheng P, Ficklin S, Cho I, Peace C, Evans K, Volk G, Oraguzie N, Chen C, Olmstead M, Gmitter G Jr, Abbott AG (2012) Uniform standards for genome databases in forest and fruit trees. Tree Genet Genomes 8:549–557

Zagrai I, Capote N, Ravelonandro M, Cambra M, Zagrai L, Scorza R (2008) Plum pox virus silencing of C5 transgenic plums is stable under challenge inoculation with heterologous viruses. J Plant Pathol 90:63–71

Research and Application of Transgenic Poplar in China

Jianjun Hu, Lijuan Wang, Donghui Yan, and Meng-Zhu Lu

Abstract Poplar is a major tree species used in both ecological and economic plantations in China. Notably, a total of 7.04 million hectares of selected poplar clones were planted in China for commercial production and shelter belts during the past 20 years, this represents a significant 19 % of total tree plantations in China. However, these mono-clonal plantations are facing damage by insect pests which caused severe infestations resulting in significant loss, estimated at millions of US dollars annually. In addition, it is estimated that by the year 2020, China will increase tree plantation by 40 million hectares and timber storage by 1.3 billion cubic meters in 2020. In order to meet this challenging goal, the development of tree plantations in northern dry and semi-dry areas is accelerated, poplar as a fast-growing candidate needs improvement of stress tolerance. Unfortunately, poplar plantation clones under these challenges are not easily solved by conventionally breeding techniques like hybridization and selection.

Genetic engineering is one of the most powerful and promising technique to improve the tolerance of poplar clones being used in plantations. In order to develop poplars that were more tolerant to insect attack, GM/biotech poplars were developed in China. More specifically, *Populus nigra* clones (12, 172 and 153) were developed with *cry1Aa* and a hybrid white poplar, clone 741, was transformed with a fusion of *cry1Aa* and *API* (coding for a proteinase inhibitor from *Sagittaria sagittifolia* Under rigorous testing, the *Bt* poplar clones are exhibited a high level of resistance to leaf pests, resulting in a substantial 90 % reduction in leaf damage.

J. Hu • L. Wang • M.-Z. Lu (✉)
State Key Laboratory of Tree Genetics and Breeding, Chinese Academy
of Forestry, Wan Shou Shan, Beijing 100091, China
e-mail: lumz@caf.ac.cn

D. Yan
Institute of Forest Ecology and Protection, Chinese Academy
of Forestry, Wan Shou Shan, Beijing 100091, China

T. Fenning (ed.), *Challenges and Opportunities for the World's Forests*
in the 21st Century, Forestry Sciences 81, DOI 10.1007/978-94-007-7076-8_24,
© Springer Science+Business Media Dordrecht 2014

The two clones were first commercialized in 2001 and by the year 2011, the transgenic poplars occupied 490 ha compared with 453 ha in 2010. The transgenic poplar plantations have effectively inhibited the fast-spread of target insect pests and have significantly reduced the number of insecticide applications required. The performance of the *Bt* black poplar plantations are significantly better than the clones deployed locally.

Transgenic approaches have been also used in breeding trees for high tolerance to environmental stresses in China. Several transgenic poplars with tolerance of abiotic stresses has been reported, with two field testing cases showed a high salinity tolerance of transgenic poplar in marine salina. A hybrid poplar (*P. simoni* X *P. deltoids*) clone "Balizhuang" transformed with 1-p-mannitol dehydrogenase (*mtlD*) gene under 35S promoter were generated and the cuttings from this transgenic clone could survive in 0.4 % NaCl solution, and its seedlings gave a high survival rate and grew better in marine salina with 0.5 % salt concentration, comparing to non-transgenic clones. A much deployed hybrid clone (*Populus* X *euramericana*) 'Guariento' in plantation were transformed with multiple tolerance genes by biolistic bombardment and exhibited tolerance to drought and salinity, and field testing showed that the transgenic poplars perform well in harsh soil. Transgenic poplars aiming to change the wood property and the success in lowering the lignin content could improve the efficiency in pulping and saccharification of the wood. As of the end of 2010, 33 field trials had been approved and implemented featuring tolerance to insects, diseases, drought, and wood quality traits. The availability of commercial *Bt* poplar plantations has made it possible to empirically assess gene flow via pollen and seeds, and also for assessing the impact of *Bt* poplar on the insect community when intercropping with *Bt* cotton. The preliminary results showed generally no significant negative effects on the ecosystem.

The transgenic *Populus nigra* has also been used for hybridizing with non-transgenic *P. deltoides* to generate an insect resistant source in a breeding program designed to generate new hybrid clones. The 2-year old hybrids are grew in nursery and the insect bioassay showed nearly ten times in decrease of weight and 60–100 % of mortality of *Lymantria dispar* L. larvae after feeding with hybrid leaves for 2 weeks. The better clones will be farther selected for field testing and expected to be used in a larger region since the *P. deltoids* is more suitable to the south region comparing to the *P. nigra*.

The search for new genes or modification of available genes in further improvement of poplar clones with significant effects on their tolerance of biotic, abiotic stresses and quality of timber began about 10 years ago. Most studies are using transcript and protein profiling techniques like RNA-Seq, microarray and proteomic analysis during trait development in novel stress tolerance trees like *P. euphratica*, *Tamarix androssowii*, etc.. Several genes or gene clusters were used to further testing the usefulness of them in poplar breeding for stress tolerance. Modified available genes like *Bt* fused with activity enhancer or other genes to increase the stress tolerance are also in progress.

1 Overview of the Genus *Populus* in China

1.1 Natural Poplar Forest

China is one of the central habitats of poplars, where various poplar species are broadly distributed from 25° to 53° N, and from 76° across 134° E. Thus a large number of poplar species including 56 origins, 50 varieties and 35 endemic species can be found in this area. Some endemic species inhabit extreme environments such as *Populus kangdingens* at altitudes above 3,500 m in southwest of China, and *P. suaveolens* Fisch in areas near 72° N where temperature may drop to 69 °C below zero (Xu et al. 1988). Due to the resistance to saline, air drought, heat, light and sandstorm, *P. euphratica* thrives in harsh regions where salt content exceeds 2 %. *P. tomentosa*, another endemic species, has beautiful shapes and wood with good physical and mechanical properties (Wang et al. 2011).

Poplar species from all five sections can be found in China. Nine species and seven varieties from Section Leuce are present, of which five species are endemic, including *P. alba* L, *P. canescens* Smith, *P. tomentosa* Carr., *P.* × *pseudo-tomentosa* Wang et Tung, *P. davidiana* Dode, *P. tremula* L., *P. hopeiensis* Hu et Chow, *P. adenopoda* Maxim and *P. ningshanica* C. Wang et Tung, mainly distributed from the northeast and northwest to southwest China. Species from Section Tacamahaca form the largest section in China, including 37 species and 22 varieties, many of which are endemic species or unique native species, such as *P. ussuriensis*, *P. suaveoens*, *P. koreana*, *P. maximowiczii*, *P. cathayanna*, *P. simonii*, *P. szechuanica* and *P. yunnanensis* etc., distributed from the northeast to northwest, and from the southwest to eastern China (Zhao and Chen 1994). Three species and three varieties from Section Aigeiros, including *P. nigra* L., *P. jrtyschensis* CH. Y. Yang, and *P. afghanica* Schneid are centralized in Xinjiang Uygur Autonomous Region. Only two species from Section Turanga, *P. euphratica* and *P. pruinosa*, exist in China and 90 % of the natural forests of these species are distributed in the Tarim River Basin where the annual rainfall is only about 30–60 mm, while evaporation can be up to 3000 mm, and the temperature swings from 42 °C above zero to 30 °C below. Four species from Section Leucoides, including *P. lasiocarpa* and *P. wilsonii* Schneid, mainly inhabit the plateau areas of Hubei, Sichuan, Yunnan and Tibet.

1.2 Poplar Plantations

The development of forest plantations is the only way to compensate the shortage of forestry resources in China. With its capacity for fast-growth, poplars have become one of the most important groups of cultivated tree species. Currently, poplar plantations now occupy 7.04 million hectares, among which

4.5 million hectares are for timber production, with a total timber production volume of about 220 million cubic meters. In the 1950s, poplars were used mainly for shelter-forest plantation in China, such as the Shelter Forest System Program in Three-North Regions of China. In the 1960s and 1970s, poplars were mainly used for construction of farmland shelter-forest, and to contribute to the farmland combined ecological system, described as the "*Four Sides*" plantation approach (house side, village side, roadside and waterside). During 1980s and 1990s, fast-growing forests were generated and an intensive cultivation management was applied to these poplar plantations. Starting from 2010, a plan for an additional 9.2 million hectares of fast-growing forest was issued, which aims to provide 96.7 million cubic meters of wood including 7.32 million cubic meters of large diameter logs, 11.9 million tons of wood pulp and 13.15 million cubic meters of wooden materials and timber. In 2015, the total area of fast-growing forest is expected to reach 13.33 million hectares and provide 40 % of the domestic wood demand.

Presently, most poplar plantations are comprised of selected hybrid clones produced in China. From the late 1950s, a poplar breeding program was initiated and produced the first batch of new poplar cultivars such as *P. beijingensis* W. Y. Hsu, *P. cooperation*, *P. popularis*, *P. × xiaohei* T.S. Huang et Y. Ling etc. by hybridizing endemic species. The following projects on poplar breeding during 1980s to 1990s produced further cultivars including *P. deltoids* CL 'Zhonglin 46', *P. deltoids* CL 'Nankang', *P. × euramericana* cv. 'Nanlin', *P. deltoldes* Bartr. × *P. cathayana* Rehd. CI. 'xifeng', *P. deltoids* CL 'Danhong', *P. deltoides* CL 'Langfang' and *P. tomentosa* Carr. triploid clones by hybridization with both endemic and exotic poplar species. With the addition of new cultivars such as *P. deltoids* CL 'Zhonghuai' available from the beginning of this century, Chinese poplar plantations are being regularly updated with new clones.

At the same time, the introduction and domestication of poplar species from other countries is underway. Since the 1950s, more than 300 clones of *P. × euramericana*, *P. deltoids* and *P. nigra* were introduced from Italy, Germany, Romania, the former Soviet Union, the Netherlands, France and Japan. More than ten clones were selected for large-scale plantation after introduction and regional cultivation tests in China, including *P. × euramericana* Guinier cv. 'Sacrau-79', *P. × euramericana* Guinier cv. 'I-214', and *P. × euramericana* 'Robusta' in southern and northern China, *P. deltoides* Bartr. CL'Harvard (I-63/51)', *P. deltoides* Bartr. CL. 'Lux' I-69/55, and *P. euramericana* San Martino in the middle and downstream areas of Yangtze River. In the 1980s, 331 selected poplar clones which were introduced from 15 countries including France and Italy, led to the establishment of the first germplasm collection of species and clones in Section Aigeiros. Based on these resources, some excellent hybrid species such as *P. × euramericana* cv. 'Neva' and *P. × euramericana* cv. 'Guariento' were selected and planted on a large scale (Fig. 1). The total plantation area of these two clones has now reached 1.3 million hectares and produces 30 million cubic meters wood each year.

Fig. 1 Field performance of *P. × euramericana* cv. 'Neva' (Puyang, Henan province. By Jianjun Hu, 2008)

2 The Challenges to the Poplar Plantation in China

2.1 Biotic Stresses

Insects and diseases can cause severe damage to poplar plantations in China. In the past decade, the affected areas have kept increasing on a regular trend. In 2010, the area infected by insect and other diseases surpassed 0.93 million hectares with direct economic losses of two billion dollars. Principal factors contributing to occurrence of insects and diseases on poplars in China are associated with genetic uniformity of the trees since single clones are commonly used in plantations, degraded soils due to environmental stresses at the growing sites, as well as increasingly frequent abnormal weather conditions brought on by global climate change.

The major insect and pathogen pest species currently causing economic loss and ecological problems to China's poplars are mainly Lepidopteran and Coleopteran insects and ascomycete fungi. In detail, foliar insect pests often are found to be *Apocheima cinerarius*, *Clostera anachoreta*, *Clostera anastomosis*, *Lymantria dispar*, *Malacosoma neustria*, *Micromelalopha troglodyte*, *Stilpnotia candida* and *Stilpnotia salicis*. Foliar pathogens are mainly *Marssonina brunnea*, *Melampsora larici-populina* and *Coryneum populinum*. Insect pests on stem or branches are identified as *Anoplophora glabripennis*, *Apriona germari*, *Cryptorrhynchus lapathi*, *Paranthrene tabaniformis* and *Saperda populnea*. The diseases on stems and branches of poplars usually are *Botryosphaearia* and *Cytospora* cankers. There are frequent epidemics of these insect pests and diseases which cause various economic losses in vast regions mainly in northern China, but recently also in southern areas. The Asian Longhorned Beetle, *A. glabripennis*, is one of the most hazardous pests to poplars in the "Three North" shelterbelts project in northern China, while *A. germari* has caused significant damage to poplars mainly found in the Anhui, Shandong, Henan and Jiangsu provinces and *Batocera horsfields* (Hope) has affected trees along the Yangzi River in central China. The foliar insect pests *C. anachoreta* and *M. troglodyte* most seriously affect poplars in southern regions more than in northern regions, while *A. cinerarius* mainly causes damage to poplars in northwestern and northern regions. The diseases on stem, such as *Botryosphaeria* and *Cytospora* cankers, and diseases on leaves, such as black spot (*M. brunnea*) and grey spot (*C. populinum*), have a correlation with abiotic pressures due to climate or environmental changes. And previously innocuous exotic insects and fungi may suddenly become serious pests as the climate changes.

With the rapid social and economic development, China will face a growing demand for wood and fiber materials, and meeting the need for ecological reconstruction, environment improvement, as well as the need to respond to the pressures of climate change. Therefore, it is inevitable that the area of fast growing pure poplar plantations with a high rate of biomass production must be increased. To minimize the damage to the plantation, however, it is necessary to have an effective way to use clones with pest tolerance.

2.2 Abiotic Stresses

Nearly half of Chinese land has an arid or semi-arid climate, and there are 12 million hectares of it with saline soil, and desertification trends are increasing, with 3,460 km^2 of land area affected annually by soil erosion and desertification, with the loss of 50 million tons soil. The lost nutrient caused by the erosion is equivalent to 40 million tons of standard fertilizer. Besides, most regions of China frequently are impacted by early frost and cold weather in early spring. These abiotic stresses such as drought, saline and chilling severely limit the development of poplar plantation in a large-scale. And as global climate change intensifies, extreme weather and climate events, such as drought, flood, reduction in underground water, serious hazards of sand storms, are becoming more common. It is predicted that by 2050, China will lose 84 km^2 dry and 70 km^2 of land to desertification annually. If the current trend of climate change continues, it is predicted that the suitable distribution area for *Larix gmelinii* will be reduced by 58.1 % in 2020, 99.7 % in 2050 and will disappear from China entirely by 2100. Similarly, climate change will inevitably affect the poplar adaptability and ecological distribution, thus original poplar forests in some regions may be much reduced in area or even be completely lost. In such situations, there is a pressing need to apply biotechnological tools for improving poplar's tolerances to drought, high and low temperatures, and saline stresses, as well as wood property, to acquire faster growing and higher biomass.

3 Transgenic Poplar Research

3.1 Necessity of Application of Transgenic Poplar

In the past, great achievements have been obtained with poplar breeding in China. However, the breeding goal was merely focused on primary-growth traits, which led to the severe scarcity of new varieties appropriate for afforesting land with poor growing conditions and exposed to the increasing threats of diseases and pests, and instead resulted in the large-scale use in plantations of single fast-growing clones/varieties. Therefore, breeding new poplar varieties with disease-, pest-, drought- and salt-resistance remains to be solved. Conventional breeding techniques usually require a long term and high cost due to the long life cycle, the lack of genetic resources, incompatibility of distant inter-specific hybridizations. In 1980s, the emergence of genetic engineering technology brought unprecedented changes to genetic improvement of trees. Combination of genetic engineering technology and conventional breeding techniques can greatly reduce breeding cycle to accelerate breeding progress, and to expand breeding objectives to the enhancement of disease-, insect-, stress-resistance, improvement of wood quality and other economic traits. Since the initiation of research on transgenic poplar in 1989, remarkable achievements have been made in China. Field trials with dozens of transgenic poplar lines are under way for environmental release and plantation tests. Moreover, two of them have been already commercialized. Twenty years experience on transgenic poplar research shows not only the importance of genetic engineering technology for poplar breeding, but that it is also an important way to cope with the unpredictable environmental challenges.

3.2 Pest-Resistant

The generation of new pest-resistant transgenic poplar clones is one of the major goals in poplar breeding programs. A *Bacillus thuringiensis* (*Bt*) gene *Cry1A*, which is toxic to lepidopterans was introduced into the genome of *Populus nigra* in 1991, which marked the prelude of pest-resistant transgenic poplar research in China (Wu and Fan 1991). In 1993, enhanced pest-resistance was achieved with transgenic *P. nigra* by modification of the *Cry1A* sequence and by utilization of a transcriptional enhancer (Tian et al. 1993). Thereafter, many pest-resistant transgenic poplar programs have been performed, among which *Bt* genes are the most commonly used insecticidal protein genes, with *Cry1s* for lepidopterons and *Cry3s* for coleopterons. Presently, pest-resistant genes have been successfully applied to *P. nigra, P. deltoides, P. euramericana*. Pesticidal assay's in the laboratory indicated that these transgenic poplars demonstrated strong resistance to *Apcchimia cinerarius, Lymantria dispar* Linnaeus and *Clostera anachoreta* Fabricius (Tian et al.

1993; Chen et al. 1996). In 2005, a *Bt* strain Bt886 was isolated from dead *Tribolium macleby* lavae and showed strong toxicity to long-horned beetles such as *Anoplophora glabripennis* and *Apriona germari*. The corresponding *Bt* gene *Bt886cry3Aa* was cloned and proved to be useful in longhorn beetles control using a transgenic approach (Chen et al. 2005). *Cry3A* genes were introduced into the genome of *P. alba* × *P. glandulosa* which conferred resistance in this hybrid poplar to *Anoplophora glabripennis* Motsch and *Plagiodera versicolora* (Zhang et al. 2011). In order to further improve the toxicity of Bt886cry3Aa against longhorn beetles, an intestinal cellulase binding peptide was fused with Bt886cry3Aa, which almost tripled the toxicity of this protein (Guo et al. 2012). Furthermore, the toxicity of a cowpea trysin inhibitor gene *CpTI* to both lepidopterons and coleopterons was shown by introducing it into *P. tomentosa* Carr., *P. alba* Var. *pyramidalis* and *P. euramericana* CL 'Neva', where the growth and development of these *CpTI* transgenic poplars was not obviously affected (Liu et al. 1993; Zhuge et al. 2003). Recently, kunitz trypsin inhibitors (*KTIs*) were shown to be active in pest-resistance in poplars, and more importantly *KTIs* genes presented specific toxicity profiles that are of ongoing interest (Ma et al. 2011).

In addition to *Bt* genes and protease inhibitor genes, nerve toxin genes and anti-microbial peptide coding sequences were exploited for pest-resistant poplar research. An insect nerve toxin gene *AaIT* was introduced into the genome of *P. simonii* × *P. deltoides* and the transgenic plants showed strong insect-resistance with the lethality of 80 % to the first instar gypsy moth larvae (Wu et al. 2000). An anti-bacterial peptide coding sequence *LcI* was isolated and the peptide showed toxic activity to *A. glabripennis*. Then the *LcI* gene was introduced into the genome of *P. deltoids*, *P. nigra* and *P. euramericana* in order to generate new transgenic poplars with resistance to longhorn beetles (Chen et al. 1996).

The often low toxicity of certain genes and pest-tolerance has made researchers consider multiple gene transformations. Research on co-transformation of poplar with two genes has been reported. Transgenic poplar 741 (*P. alba* L. × (*P. davidiana* Dode+ *P. simonii* Carr.) × *P. tomentosa* Carr.) carrying a *Bt* gene and a protease inhibitor gene *API* showed higher toxicity to young larvae of *Lymantria dispar* Linnaeus, *Clostera anachoreta* Fabricius and *Hlyphantria cunea* and stronger inhibition on their adults than either transgenic poplars carrying the single gene (Yang et al. 2003, 2006). Co-transformation of *Bt* genes and other protease inhibitors or anti-microbial peptide coding sequences into *P. alba* × *P. glandulosa*, *P. tomentosa* Carr. and *P. euramericana* has also been reported to present synergistic effects to possibly delay the development of insect resistance (Li et al. 2000; Cao et al. 2010).

In relation to pathogen resistant research, canker and other fungal diseases cause major problems in poplar plantations. Pathogen resistant genes such as oxalate oxidase *OXO*, rabbit defensive gene *NP-1* and chitinase genes have been transformed into *P. tomentosa* Carr. and some species in *Aigeiros* section (Liang et al. 2001; Zhao et al. 1999). As a result, these transgenic poplars demonstrated substantially enhanced resistance to *Bacillus subtilis* and *Rhizoctotonia* (Zhang et al. 2003).

3.3 Drought, Salt and Cold-Tolerance

Breeding stress-resistant cultivars to expand or even to maintain the existing area of poplar plantations is one of the efficient ways to alleviate land-shortage conflicts between agriculture and forestry. Following the introduction of a *mtl*-D gene into *P. tremuloides,* it has been reported that the transgenic plants can tolerate 0.45 % of NaCl (Liu et al. 1995). Subsequently, transgenic *P. deltoids, P. tomentosa* Carr., *P. simonii* × *P. nigra, P. tremuloides* and the hybrid poplar 84K carrying drought-, cold- and salt-tolerant genes were successfully generated. Among these transgenic poplars, three categories of tolerant genes have been utilized for transformation. The first type includes genes involved in osmotic regulator biosynthesis, whose coding enzymes can effectively regulate osmotic potential to maintain a cell's normal turgor and metabolic functions, but without too much interference on macromolecular solute system, and thus being beneficial to cell membrane stability. The second type includes genes involved in signaling transduction that can switch on and off the expression of certain tolerance-related genes under stresses that may already be present, but insufficiently expressed. The third category includes drought, salt-related functional proteins that are highly hydrophilic and can absorb enough water to protect cells from the damage caused by drought and salt stresses. In addition, cold-tolerant transgenic research has been carried out to solve the freezing damage in winter or early spring in poplar plantation in China.

Most of the early transformed stress tolerant genes were osmotic regulator biosynthetic genes. As mentioned previously, after the introduction of a *mtl*-D gene into *P. tremuloides* it was reported that the transgenic plants could tolerate 0.45 % of NaCl (Liu et al. 1995). In 2000, a genetic transformation system optimized for *P. xiaozhannica* cv. 'Balizhuangyang' was established and the *E. coli mtl*-D gene transformed plants showed an obvious increase in salt-tolerance when growing on media containing 0.4–0.6 % of NaCl (Liu et al. 2000). Similarly, application of *mtl*-D gene in the production of transgenic *P. tremuloides*, poplar 84K and *P. tremuloides* also resulted in increased salt-tolerant plants (Sun et al. 2002). Utilization of *Bet-A* gene for salt tolerance has also been performed and the resultant plants showed a 27.1 % increase in betaine content and a decrease of 15.8 % in salt stress damage on average (Yang et al. 2001). Recently, an *Arabidopsis* gene *AtNHX1* was introduced into the genome of *P. euramericana* 'Neva' and the transgenic plants also demonstrated enhanced salt tolerance (Jiang et al. 2012).

Genes associated with signal transduction also regulate drought- and salt-tolerance gene expression under stress conditions, thus they are being widely used in drought and salt tolerance of transgenic poplars. Transformation of *PLDγ* antisense led to 0.7 % increase in NaCl tolerance in transgenic *P. tomentosa* Carr. (Liu et al. 2002). Introduction of a *PLDγ* antisense sequence into *P. deltoides* resulted in enhancement of both salt-tolerance and drought-tolerance (Zou et al. 2006). A dehydration responsive element gene *DREB1* was transformed into *P. euramericana cv.* Nanlin895, and salt-tolerance has been significantly enhanced in transgenic poplars surviving in 1/2 MS liquid nutrient media containing as much as

150 mmol/L of NaCl, where non-transgenic plants cannot survive at all (Yang et al. 2009). In addition, the drought-tolerance of these transgenic plants has also been increased as well. The water induced gene *BsPA*, betaine aldehyde dehydrogenase (*BADH*) and fructan sucrose transferase gene (*SacB*) were also used for salt- and drought-resistant research on poplars (Li et al. 2007). In recent years, researchers have cloned the dehydration associated genes *Pedhn*, ascorbate peroxidase gene (*PtAPX*) and leucine-rich receptor kinase gene (*PdERECTA*) from *P. euphratica*, *P. tomentosa* Carr. and *P. deltoides*, which give the opportunity to engage in drought- and salt-tolerant transgenic research by exploitation of poplar genes (Lu et al. 2009; Wang et al. 2011; Xing et al. 2011).

Polyunsaturated fatty acids are essential components of cell membrane in plants. The content of unsaturated fatty acids is relevant to cold- and salt-stresses. Therefore, fatty acid desaturases (*FADs*) have become target genes for cold-tolerance genetic engineering. *PtoFAD2* and *PtoFAD3* were isolated from *P. tomentosa* Carr. and over-expressing of these genes in hybrid poplar 84K caused a significant increase in survival rate under cold-stress conditions as compared to non-transformed controls (Zhou et al. 2010). These data strongly indicate that increasing unsaturated FAs in poplars is a highly feasible approach to breed cold tolerance trees, and thus to improve the establishment of poplar plantations in northern climes.

3.4 Wood Property

In China, poplar wood is used as the raw material for pulp-making. However, when utilizing the poplar wood as the fiber raw material, lignin is often removed via chemical approaches, which typically needs consumption of a large amount of chemicals thus leading to environmental pollution. Therefore, it is of great importance to genetically modify its lignin content so as to improve the yield and quality of paper and reduce the economical cost and pollution in paper-making procedure. It is quite difficult to effectively carry out lignin modifications using the traditional methods. In recent years, many researchers have attempted to modify the content or composition of lignin to improve the desired characters of the paper-making raw material. Many important enzymes involved in the process of lignin synthesis have been characterized and so they have become the target genes for the production of transgenic plants for the modifications of the lignin content and composition.

From the end of 1990s to the beginning of the twenty-first century, the metabolic pathway of monolignol biosynthesis has largely been elucidated, and a large suite of genes involved in regulation of lignin or cellulose biosynthesis have been cloned and characterized, such as phenylalanine ammonia-lyase (*PAL*), cinnamate-4-hydroxylase (*C4H*), 4-coumarate-CoA ligase (*4CL*), coumaroy-CoA-3-hydroxylase (*C3H*) and caffeoyl CoA *O*-methyltransferase (*CCoAOMT*) etc.. Typically, the lignin content in poplar wood and also the ratio of lignin and cellulose can be significantly altered or reduced by employing RNAi or anti-sense constructs to inhibit the expression of one or more of these targets. In China, a *CCoAOMT* gene was isolated from *P. trichocarpa*, and transformed into *P. tremula* × *P. alba* with an anti-sense construct using the *Agrobacterium*-mediated method. The lignin content of the 6 month-old transgenic plants was determined and they exhibited a significant decrease in Klason lignin content, which was 17.9 % lower as compared to the control plants (Wei et al. 2001). In addition, experiments using alkaline pulping methods have suggested that transgenic plants with reduced lignin content also had better pulp qualities (Zhao et al. 2005). The pulp characteristics of 3-year-old transgenic poplar, showed a decline of 13 % in lignin content of transgenic poplar had occurred and that phenyl-alcohol active chemicals were also reduced, resulting in a dramatic increase of yield and quality of pulp with reduced the alkaline use (Wei et al. 2008). Similarly, *4CL* gene was used to transform triploid *Populus tomentosa Carr*, and the lignin content demonstrated that inhibition of the expression of endogenous *4CL* gene dramatically lowered the lignin content of transgenic plants with a maximum decline by 41.73 % compared to the control plants (Jia et al. 2004).

In recent years, there have been some fundamental discoveries regarding the genetic modifications in the other aspects of the wood characteristics of poplar. For instance, the *PtCOBL4* gene from *Populous trichocarpa*, a member of COBRA gene family encoding glycosylphosphatidylinositol-anchored protein, regulates the accumulation of cellulose of the cell wall as well as the directed expansion of cell (Zhang et al. 2010). Using transcriptomic and proteomic approaches to explore and analyze the expression of genes involved in the development of secondary vascular system in poplar, allows researchers select and identify sets of key genes regulating formation of wood, facilitating the further practice to improve poplar wood characteristics through genetic engineering (Du et al. 2006; Wang et al. 2009).

3.5 Field Testing and Commercialization

Up to the year 2011, China has granted licenses for 33 field trials (Lu and Hu 2011). Among them, transgenic *P. nigra* 'Robasta' with *CryIA*, *P. alba* X *P. glands* with *BtCry3A* or *BtCry3A* fused with *OC-1*, *P. tomentosa* '741' with *Cry1Ac* fused with *AP1*, all conferring resistance against insect damage; Poplar clone 'Balizhuang' with *mtl-D*, *P. alba* X *P. glands* clone 84K with *SacB*, *P. alba×P. berolinensis* with *JERFs* (Jasmonate Ethylene Responsive Factor), for tolerance to abiotic stresses; Poplar plant (*P. tremula* × *P. alba*) with *GST* and hybrid poplar 84K with *vgb*, conferring changed growth and development; *P. tomentosa* plants transformed with *4CL* or *CCoAOMT* for lignin reduction; *P.* × *euramericana* 'Guariento' with multiple genes (*vgb+SacB+Cry3A+OC-1+JERF36*) for tolerance of abiotic and biotic stresses.

Two *Bt* poplar clones were commercialized in 2002 in China. One was *P. nigra* transformed with *cry1Aa* and three clones (12, 172 and 153) were selected for field-testing and clone 12 was commercialized and is used in poplar plantation in

Table 1 Comparisons between performance of *Bt* poplar clones and non-*Bt* clones in China

Sites	Poplar cultivars	DBH (cm)	Height (m)	Growing years	Area (hm²)
Huairou, Beijing	*Bt* poplar	28.2	29.1	10	40
	P. nigra				
	ZL46	25.4	28.8	10	
	P. euramericana 'Zhonglin46'				
Renqiu, Hebei	*Bt* poplar	20.8	19.1	8	33
	P. nigra				
	Chuangxin	18.6	19.7	8	
	P. deltoides cv 'Chuangxin'				

northern China. Another is a hybrid white poplar, clone 741, transformed with a fusion of *cry1Aa* and *API* (coding for a proteinase inhibitor from *Sagittaria sagittifolia*. By 2011, they occupied 490 ha. Under large scale field-testing, the *Bt* poplar clones exhibited a high level of resistance to leaf pests, resulting in a substantial 90 % reduction in leaf damage. The transgenic poplar plantations have effectively inhibited the fast-spread of target insect pests and have significantly reduced the number of insecticide applications required. Comparisons to non-transgenic poplar plantations, the *Bt* poplars required no insect pest control in the first 6 years, while the non-*Bt* plantations required 2–3 insecticide sprays per year. This is consistent with experimental data (Table 1) confirming that *Bt* clones performed better and grew faster than their conventional counterparts. For example, at 10 years old the tree trunk diameter was 28.2 cm for the *Bt* clone at the Beijing location versus 25.4 cm for the non- *Bt* clone "Zhonglin 46", and similarly a 20.9 cm diameter for the *Bt* clone at the Hebei location after 8 years, versus 18.6 cm for the non-*Bt* clone "*P. deltoides* cv Chuangxin". The transgenic *Populus nigra* has also been used for hybridizing with non-transgenic *P. deltoides* to generate an insect resistant source in a breeding program designed to generate new hybrid clones.

4 Biosafety Assessment of Transgenic Poplars

4.1 Biosafety of Transgenic Trees

The wide application of transgenic technology in genetics and breeding of poplar has greatly accelerated the progress on poplar breeding. However, with the development of genetically modified poplars, the issue of ecological safety issues caused by transgenic trees containing foreign genes has attracted increasing attention. The biosafety of transgenic poplar usually includes transgene stability, effect of transgenic poplar on target and non-target organisms, resistance of target organism to transgenic poplar, weedness, and the impact of possible horizontal gene transfer on the ecosystem.

4.2   Stability of Transgene

It's very important for assessing transgenic plants in the breeding process, whether the exogenous gene can be stably expressed in transgenic plant throughout the whole life circle, and this is also the foundation of safety evaluation of transgenic plants. The detection of *Bt* gene stability from the 7 year old *P. nigra* transformed with the *Bt* gene and located in Manasi Plain Forest Station, Xinjiang Uygur Autonomous Region, China, showed that the *Bt* gene was still present and stably expressed in transgenic plants and a ratio of 1 : 1 of *Bt* and non-*Bt* was observed in the progeny of crosses between the *Bt* poplars with a non-*Bt P. deltoids* clone, in accordance with the laws of Mendelian inheritance, suggesting the *Bt* gene is stably inserted into the genome (Hu et al. 2007). The researchers also studied the impact of transgenic hybrid poplar 741 on different insect communities, and the durability of the insect resistance (Yang et al. 2003; Liu et al. 2009). The increased larval mortality, decreased density of larvae populations, as well as the inhibition of the development of larvae observed over 3 years indicated that the resistance to the insect pests of transgenic poplar plantations was effective and sustainable.

4.3   Impact on Non-target Insects

The impact of transgenic plants on population of non-target organisms and biodiversity is an essential part of ecological risk assessment. Studies have shown that the variety, number and parasitic ratio of the natural enemies of the target insects in the *Bt P. nigra* plantations were all were higher than those in non-transgenic poplar plantations. Meanwhile, assessments of the inoculation rate of parasitic wasps on the insect pupae collected from transgenic and control poplar plantations showed that there was no significant difference in the eclusion rate and the quantity of parasitic wasps, and that all the newly emerged parasitic wasps were normal (Hu et al. 2004, 2007). Another study investigated the population changes of the target and non-target insects in 2–5 years old insect-resistant transgenic 741 poplar plantations, and found that there was a significant change in the classes and the vertical distribution of the arthropod community. In detail, the number of defoliator insects were as expected, significantly reduced in the transgenic 741 poplar plantations, while the species richness of arthropod community as a whole increased significantly. This indicated that there were negative effects of the transgenic poplars on the target insects and also on non-target herbivorous insects, which might be expected to lower the population trend index, while positive effects were found on the composition and occurrence of natural enemies and neutral arthropods. The spatial niche breadth of lepidopteran insects was narrowed, while the others were relatively widened. Predators and parasitoids had a narrow niche overlap with the lepidopteran pests, but had a wider niche overlap with the neutral species. This indicated that transgenic insect-resistance poplar had a positive effect on the natural enemies in utilization of the resources, and the stability of arthropod community could be increased in the forest where high-resistant and medium-resistant *Bt* and

Fig. 2 *Bt* poplar-cotton agroforestry eco-system (Manasi Plain Forest Station, Xinjiang Uygur Autonomous Region, China. By Jianjun Hu, July 2011)

non-*Bt* poplars were planted together (Gao et al. 2006a, b). The results of these study showed that transgenic poplar not only could effectively control the target pest population, but also improve the bio-diversity of insect communities.

Soil is essential for the circulation of nutrients and energy and the cycling of soil ecosystem, and potentially the microorganisms involved may be affected by the products generated from transgenic plants through the secretions of their root systems. Therefore, it is of great importance to ecologically evaluate the effects of transgenic plants on soil microorganism populations. Researchers in China in 2004 reported that there were no significant differences in the quantity of microorganisms, such as bacteria, actinomycete and fungi, among individual trees and the plantations of transgenic or non-transgenic *P. nigra*. Similar results about soil microbial community in the roots of *P. alba* × *P. glandulosa* were obtained in another study in 2009. An investigation of the distribution of *Bt* toxic protein in the soil under 4 year old transgenic 741 poplars showed that even though the *Bt* toxin could be found around the transgenic plants roots area, the concentration decreased exponentially away from the root surface and underneath the soil to the soil surface. There was no obvious correlation between the distributions of Bt toxic proteins and quantity of soil microbiology. It was also found that there were seasonal changes in soil fungi, bacteria, actinomycete quantity in the experimental field, but there were no significant differences between transgenic and non-transgenic plantations in any season.

The commercialized *Bt* transformed poplar and cotton were used to establish transgenic poplar-cotton ecosystems (Fig. 2), commonly used in agro-forestry, to study the effects on the arthropod community. In this eco-system, the effects of transgenic poplar and cotton on target, non-target insects and natural enemies of insects were studied. The 2-year experiments showed that the transgenic poplar-cotton eco-system had a significant effect on arthropod community structure and a strong inhibition to the target insect. Although this system is effective in controlling the target pests, the increase in the category and quantity of the non-target insects might stimulate them to become the main pests. Clearly, more studies were required in order to access the ecological consequence of the transgenic poplar-cotton eco-systems.

4.4 Transgene Escape

There is much concern that transgene flow to natural populations of related species might potentially change their genetic composition, since this may affect the genetic diversity of the natural populations important for the species. Gene flow occurs by the successful fertilization of flowers by pollen and germination of the seeds produced. Poplar is dioecious and a transgenic male plant could spread the transgene through it's pollen, while female plants through their seeds. It was found that the pollen from transgenic Bt Populus nigra planted in Manas, Xinjiang Uygur Autonomous Region, could pollinate non-transgenic female poplar trees 500 m away and produce transgenic seeds at the rate of 0.03 %, but no transgenic seeds were detected from non-transgenic females at 800 m in distance. On the other hand, transgenic seeds collected from transgenic female trees were tested for their germination rate in spring time in northern China and found that the seeds could not germinate without watering and lost their ability to germinate in 2 weeks in the field. The experiment was repeated and a similar result was obtained (Lu et al. 2006). This indicated that the transgene flow of transgenic poplars through pollens or seeds is low in northwestern China, where annual precipitation is 200–600 mm. However, care should be taken not to generalize from these results, considering the unique climate of the testing site and that long-term observation is required to robustly support them.

Horizontal transfer of transgene refers to the gene exchange between different species other than the gene flow between closely related organisms, which may enable the transgene to spread even more widely. Bacteria exist in very large numbers in nature and many have the competence to take up surrounding DNAs either in a free form or from other bacteria, giving rise to the possibility that the horizontal transfer may occur in the transgenic poplar plantation. If a plant derived transgene were taken up by a bacterium, then it might serve as the donor of transgene for other organisms. It was found that the Agrobacteria harboring construct of BtCry1Ac fused with API-A could stay with the infected triple-ploidy P. tomentosa materials and survive in the followed regeneration and planting procedures in green house conditions, as shown by the PCR products from plasmid DNA. In addition, the Agrobacteria could survive about 24 months during the culture of transgenic plants in bottles and could also be observed or detected in soil planted for nearly 1 month in the pots (Yang et al. 2006). These observations demonstrate the need for transgenic plants to be free of *Agrobacteria* before genetically modified plants are released into the environment.

In conclusion, to date, there have been no reports of significant negative environmental effects caused by the transgenic poplars. However, a long-term investigation are needed, to collect more data before a conclusive risk assessment can be made on the impact on human health and the environment that we live in after application of transgenic poplars on a large scale.

References

Cao CW, Liu GF, Wang ZY, Yan SC, Ma L, Yang CP (2010) Response of the gypsy moth *Lymantria dispar* to transgenic poplar *Populus simonii x P. nigra* expressing fusion protein gene of the spider insecticidal peptide and *Bt*-toxin C-peptide. J Insect Sci 10(200):1–13

Chen Y, Li L, Han YF (1996) Transformation of antibacterial gene *LcI* into poplar species. Forest Res 9(6):646–649 (in Chinese with an English abstract)

Chen J, Dai LY, Wang XP, Tian YC, Lu MZ (2005) A *cry3A* gene from *Bacillus thuringienesis* Bt886 coding the toxin against long-horned beetles. Appl Microbiol Biotechnol 67:351–356

Du J, Xie HL, Zhang DQ, He XQ, Wang MJ, Li YZ, Cui KM, Lu MZ (2006) Regeneration of secondary vascular system in poplar (*Populus tomentosa* carr.) as a novel system to investigate gene expression by a proteomic approach. Proteomics 6:881–895

Gao BJ, Liu JX, Gao SH, Jiang WH (2006a) Variation of nutritional structure and ecological niche of arthropod community in plantation of transgenic insect-resistance hybrid poplar 741. Acta Ecol Sin 26(10):3499–3507 (in Chinese with an English abstract)

Gao SH, Gao BJ, Liu JX, Guan HY, Jiang WH (2006b) Impacts of transgenic insect-resistance hybrid poplar 741 on the population dynamics of pests and natural enemies. Acta Ecol Sin 26(10):3491–3498 (in Chinese with an English abstract)

Guo CH, Zhao ST, Ma Y, Hu JJ, Han XJ, Chen J, Lu MZ (2012) *Bacillus thuringiensis* Cry3Aa fused to a cellulase-binding peptide shows increased toxicity against the long horned beetle. Appl Microbiol Biotechnol 93(3):1249–1256

Hu JJ, Zhang YZ, Lu MZ, Zhang JG, Zhang SG (2004) Transgene stability of transgenic *Populus nigra* and its effects on soil microorganism. Sci Silvae Sin 40(5):106–110 (in Chinese with an English abstract)

Hu JJ, Li SM, Lu MZ, Li JX, Li KH, Sun XQ, Zhao ZY (2007) Stability of insect-resistance of *Bt* transformed *Populus nigra* plantation and its effects on the natural enemies of insects. Forest Res 20(5):656–659 (in Chinese with an English abstract)

Jia CH, Zhao HY, Wang HZ, Song YR, Wei JH (2004) Obtaining the transgenic poplars with low lignin content through down-regulation of *4CL*. Chin Sci Bull 49(5):905–909 (in Chinese)

Jiang CQ, Zheng QS, Liu ZP, Xu WJ, Liu L, Zhao GM, Long XH (2012) Overexpression of *Arabidopsis thaliana* Na+/H+ antiporter gene enhanced salt resistance in transgenic poplar (*Populus × euramericana* 'Neva'). Trees 26:685–694

Li ML, Zhang H, Hu JJ, Han YF, Tian YC (2000) Study on insect resistant transgenic poplar plants containing both *Bt* and *PI* gene. Sci Silvae Sin 36(2):93–97 (in Chinese with an English abstract)

Li YL, Su XH, Zhang BY, Zhang ZY (2007) Molecular detection and drought tolerance of *SacB-transgenic Populus alba × P. Glandulosa*. J Beijing Forese Univ 29(2):1–6 (in Chinese with an English abstract)

Liang HY, Maynard CA, Allen RD, Powell WA (2001) Increased *Septoria musive* resistance in transgenic hybrid poplar leaves expressing a wheat oxalate oxidases gene. Plant Mol Biol 45:173–176

Liu CM, Zhu Z, Zhou ZL, Sun BL, Li XH (1993) cDNA cloning and expression of cowpea trypsin inhibitor in *Escherichia coli*. Chin J Biotechnol 9(2):152–157 (in Chinese with an English abstract)

Liu JJ, Peng XX, Wang HY, Mang KQ (1995) Cloning, sequencing and high level expression of *mtlD* gene and *gutD* gene from *Escherichia coli*. Chin J Biotechnol 11(2):157–161 (in Chinese with an English abstract)

Liu FH, Sun ZX, Cui DC, Du BX, Wang CR, Chou SY (2000) Cloning of *Escherichia coli. mtlD* gene and its expression in transgenic Balizhuangyang (*Populus*). Acta Genet Sin 27(5):428–433 (in Chinese with an English abstract)

Liu B, Li HS, Wang QH, Cui DC (2002) Transformation of *Populus tomentosa* with *anti-PLD* gene. Hereditas 24(1):40–44 (in Chinese with an English abstract)

Liu JX, Gao BJ, Zhang F, Jiang WH (2009) Life table of population of target insect in the stands of transgenic hybrid poplar 741. Sci Silvae Sin 45(5):102–108 (in Chinese with an English abstract)

Lu MZ, Hu JJ (2011) A brief overview of field testing and commercial application of transgenic trees in China. BMC Proceedings 5(Suppl 7):O63

Lu MZ, Chen XL, Hu JJ (2006) Empirical assessment of gene flow from transgenic poplar plantation. In: Proceedings of 9th international symposium on the biosafety of genetically modified organisms, Jeju Island, Korea

Lu H, Han RL, Jiang XN (2009) Heterologous expression and characterization of a proxidomal ascorbate peroxidase from *Populus tomentosa*. Mol Biol Rep 36:21–27

Ma Y, Zhao Q, Lu MZ, Wang JH (2011) Kunitz-type trypsin inhibitor gene family in *Arabidopsis* and *Populus trichocarpa* and its expression response to wounding and herbivore in *Populus nigra*. Tree Genet Genomes 7:431–444

Sun ZX, Yang HH, Cui DC, Zhao CZ, Zhao SP (2002) Analysis of salt resistance on the poplar transferred with salt tolerance gene. Chinese Journal of Biotechnology 18(4):481–485 (in Chinese with English abstract)

Tian YC, Li TY, Mang KQ, HanYF LL, Lu MZ, Dai LY, Han YN, Yan JJ (1993) Insect tolerance of transgenic *Populus nigra* plants transformed with *Bacillus thuringiensis* toxin gene. Chin J Biotechnol 9(4):291–297 (in Chinese with an English abstract)

Wang MJ, Qi XL, Zhao ST, Zhang SG, Lu MZ (2009) Dynamic changes in transcripts during regeneration of the secondary vascular system in *Populus tomentosa* Carr. Revealed by cDNA microarrays. BMC Genomics 10:215

Wang YZ, Wang HZ, Li RF, Ma Y, Wei JH (2011) Expression of a SK2-type dehydrin gene from *Populus euphratica* in a *Populus tremula* × *Populus alba* hybrid increased drought tolerance. Afr J Biotechnol 10(46):9225–9232

Wei JH, Zhao HY, Zhang JY, Liu HR, Song YR (2001) Cloning of cDNA encoding *CCoAOMT* from *Populus tomentosa* and down-regulation of lignin content in transgenic plant expressing antisense gene. Acta Bot Sin 43(11):1179–1183

Wei JH, Wang YZ, Wang HZ, Li RF, Lin N, Ma RC, Qu LQ, Song YR (2008) Pulping performance of transgenic poplar with depressed Caffeoyl-CoA O-methyltransferase. Chinese Science Bulletin 53:3553–3558

Wu NF, Fan YF (1991) Establishment of engineered poplars containing *Bacillus thuringiensis* δ-endotoxin gene. Chin Sci Bull 9:705–708 (in Chinese)

Wu NF, Sun Q, Yao B, Fan YL (2000) Insect-resistant transgenic poplar expressing *AaIT* gene. Chin J Biotechnol 16(2):129–133 (in Chinese with an English abstract)

Xing HT, Guo P, Xia XL, Yin WL (2011) PdERECTA, a leucine-rich repeat receptor-like kinase of poplar confers enhanced water use efficiency in *Arabidopsis*. Planta 234:229–241

Xu WY (1988) Poplar. Heilongjiang People's Publishing House, Harbin (in Chinese)

Yang CP, Liu GF, Liang HW, Zhang H (2001) Study on the transformation of *Populus simonii* × *P. nigra* with salt resistance gene *Bet-A*. Sci Silvae Sin 6:34–38 (in Chinese with an English abstract)

Yang CX, Li HG, Cheng Q, Chen Y (2009) Transformation of drought and salt resistant gene (DREBlC) in Populus× euramericana cv. Nanlin 895. Scientia Silvae Sinicae 18(2):17–21 (in Chinese with English abstract)

Yang MS, Lang HY, Gao BJ, Wang JM, Zheng JB (2003) Insecticidal activity and transgene expression stability of transgenic hybrid poplar clone 741 carrying two insect-resistant genes. Silvae Genet 52:5–6

Yang MS, Mi D, Ewald D, Wang Y, Liang HY, Zhen ZX (2006) Survival and escape of *Agrobacterium tumefaciens* in triploid hybrid lines of Chinese white poplar transformed with two insect-resistant genes. Acta Ecol Sin 26(11):3555–3561 (in Chinese with an English abstract)

Zhang BY, Chen M, Zhang XF, Luan HH, Tian YC, Su XH (2011) Expression of *Bt-Cry3A in transgenic Populus alba × P. glandulosa* and its effects on target and non-target pests and the arthropod community. Transgenic Res 20(3):523–32

Zhang XY, Lou YQ (2003) Major forest diseases and insect pests in China. China Forestry Publication House, Beijing (in Chinese)

Zhang DQ, Yang XH, Zhang ZY, Li BL (2010) Expression and nucleotide diversity of the poplar *COBL* gene. Tree Genet Genomes 6:331–344

Zhao TX, Chen ZS (1994) The poplar intensive cultivation in China. China Science and Technology, Beijing (in Chinese)

Zhao SM, Zu GC, Liu GQ, Huang MR, Xu JX, Sun YR (1999) Introduction of rabbit defensin *NP-1* gene into poplar (*P. tomentosa*) by *Agrobacterium*-mediated transformation. Acta Genet Sin 26(6):711–714 (in Chinese with an English abstract)

Zhao HY, Lu J, Lü SY, Zhou YH, Wei JH, Song YR, Wang T (2005) Isolation and functional characterization of a cinnamate 4-hydroxylase promoter from *Populus tomentosa*. Plant Sci 168(5):1157–1162

Zhou Z, Wang MJ, Hu JJ, Lu MZ, Wang JH (2010) Improve freezing tolerance in transgenic poplar by overexpressing a ω-3 fatty acid desaturase gene. Mol Breeding 25:571–579

Zhuge Q, Wang JC, Chen Y, Wu NF, Huang MR, Fan YL (2003) Transgenic plantlets of *Populus alba var. pyramidalis* by transformation with *CpTI* gene. Mol Plant Breeding 1(4):491–496 (in Chinese with an English abstract)

Zou WH, Zhao Q, Cui DC, Wang B (2006) Transformation of *Populus deltoides* with *anti-PLDγ* gene and chitinase gene. Sci Silvae Sin 42(1):37–42 (in Chinese with an English abstract)

Part VII
Genomic Studies with Forest Trees

Genome Resequencing in *Populus*: Revealing Large-Scale Genome Variation and Implications on Specialized-Trait Genomics

Wellington Muchero, Jessy Labbé, Priya Ranjan, Stephen DiFazio, and Gerald A. Tuskan

Abstract To date, *Populus* ranks among a few plant species with complete genome sequences and other highly developed genomic resources. With the first reference genome among all tree species, *Populus* has been adopted as a suitable model organism for genomic studies in trees. However, far from being just a model species, *Populus* is a key renewable resource that plays a significant role in providing raw materials for the biofuel and pulp and paper industries. Therefore, aside from leading frontiers of basic tree molecular biology and ecological research, *Populus* leads frontiers in addressing global economic challenges related to fuel and fiber production. The latter fact suggests that research aimed at improving quality and quantity of *Populus* as a raw material will likely drive the pursuit of more targeted and deeper research in order to unlock the economic potential tied in biological processes that drive this tree species. Advances in genome sequence-driven technologies, such as resequencing individual genotypes, which in turn facilitates large scale SNP discovery and identification of large scale polymorphisms are key determinants of future success in these initiatives. In this treatise we discuss implications of genome sequence-enable technologies on *Populus* genomic and genetic studies of complex and specialized traits.

W. Muchero (✉) • J. Labbé • G.A. Tuskan
Biosciences Division, Oak Ridge National Laboratory, 1 Bethel Valley Road,
Oak Ridge, TN 37831, USA
e-mail: mucherow@ornl.gov

P. Ranjan
University of Tennessee, Knoxville, TN 37916, USA

S. DiFazio
West Virginia University, Morgantown, VA 26506, USA

T. Fenning (ed.), *Challenges and Opportunities for the World's Forests
in the 21st Century*, Forestry Sciences 81, DOI 10.1007/978-94-007-7076-8_25,
© Springer Science+Business Media Dordrecht 2014

1   Introduction

The genus *Populus* is an economically important tree crop widely grown as feedstock for lignocellulosic biofuels and pulp and paper products, in part, for its rapid growth (Tuskan 1998; Yang et al. 2009) and ability to thrive on economically marginal lands that are not suitable for food crop production (Tuskan and Walsh 2001). Aside from its key importance as an industrial feedstock, *Populus* is also a biological model system for perennial tree crops due to its relatively compact genome, high level of interspecies diversity and ease of experimental manipulation compared to other tree genera (Tuskan et al. 2006). Given its central role in both economic and scientific realms, substantial investment has been made in developing genomic and genetic resources to facilitate experimental enquiry into the biology of adaptive traits in *Populus*. These resources include (1) the first reference genome of any tree species (Tuskan et al. 2006), over 900 resequenced *P. trichocarpa*, a repository of more than 48 million high-quality single nucleotide polymorphism (SNPs), Illumina Infinium SNP arrays with 34,131 and 5,390 SNP probes (Slavov et al. 2012; Geraldes et al. 2013), over 2,200 SSR markers with known positions on the genome assembly (Muchero et al. 2013). These cumulative resources have been used in wide-ranging studies ranging from population genetics studies to candidate gene identification using quantitative trait loci (QTL) mapping. The afore-mentioned 5,390 SNP array was recently applied to genotype three mapping pedigrees, 52–124, 545 and 54B, resulting in successful construction of higher density genetic maps with 3,500, 1,200 and 530 SNP markers, respectively. All together, these resources have been brought to bear on fundamental questions of *Populus* genome characteristics such as patterns of recombination across chromosomes, linkage disequilibrium and *P. trichocarpa* population structure along a Pacific Northwest cline (Slavov et al. 2012). In addition, the genomic resources and understanding of genome attributes are being used to identify genetic determinants as well as putative causal mutations for traits such as cell wall characteristics, biomass production, tolerance to heavy metals (Induri et al. 2012) and phenological traits using QTL and association mapping approaches.

The on-going resequencing initiative targeted at completion of 1,100 genomes is well underway with the fascinating possibility of saturating nucleotide polymorphisms discovery. In this regard, *Populus* is keeping pace with the *Arabidopsis* model system, which represents the most mature genomic resources of any plant species, where at least 80 ecotypes have been resequenced in the on-going 1001 genomes project (Cao et al. 2011). At present, over 900 *Populus* genotypes have been successfully resequenced. This collection of genomes will revolutionize approaches in identifying causal polymorphism on scales beyond individual SNPs (Abraham et al. 2012). Still, large-scale structural variation such as insertions/deletions (INDELs) and whole-gene deletions are not readily apparent based high-throughput genotyping technologies and hence, are typically under-represented in genetic mapping studies. Sequenced-based studies targeted at associating INDELs and whole-gene deletion polymorphisms to relevant economically important traits should, therefore, benefit greatly from current advances in *Populus* genomics.

It is our objective, in remainder of this chapter, to illustrate this potential by exploring the genomic distribution of INDELS using receptor kinases, which exhibit disproportionately higher levels in INDEL polymorphisms compared to other gene classes, as a proof-of-concept. In addition, we explore SNP marker coverage and segregation patterns in regions harboring this class of genes in genetic maps as a means of illustrating the poor representation resulting from technical challenges presented by INDELs in high-throughput genotyping assays.

2   Beyond Single Nucleotide Polymorphism

As stated above, one of the areas of *Populus* genomics that should receive significant benefit from whole-genome resequencing will be the identification of large-scale polymorphisms such as small INDELs and whole-gene deletions. Due to the fact that the majority of genotyping assays are designed to characterize difference in allelic forms that are expected to conform to expected segregation patterns, instances where genome segments are missing often result in "anomalous" segregation that are typically excluded from further analysis. In on-going analyses, we observed that this is especially true for genomic regions that are typified by long stretches of tandem repeats that in turn exhibit frequent events of insertions and or deletions. Receptor kinases possess these hallmark features and therefore, provide good targets for exploring limitations of current genotyping technologies in identification and mapping of INDELs. We demonstrate that in light of these limitations, the genetic basis of traits mediated by these polymorphisms may not be fully accounted for. Additionally, receptor kinases have been implicated in a variety of economically important biological process that require highly specific recognition between individual cells such as pollen recognition in self incompatibility (Iwano and Takayama 2012), between organisms as innate immune response to pathogen attack (Schwessinger and Ronald 2012), and/or in mycorrhizal formation (Kereszt and Kondorosi 2011). Specifically, D-mannose lectins are known to facilitate host recognition of bacterial, fungal, viral and insect pests or symbionts in plants based on highly specific binding of mannose moieties on microbial cell walls (Schwessinger and Ronald 2012). With abundant evidence linking receptor kinases such as lectins to innate immunity in both plants and animals, exploration of polymorphisms occurring among these key genes and their potential effects on *Populus*-microbe or *Populus*-insect interactions hold significant promise in unraveling the molecular genetics behind such processes.

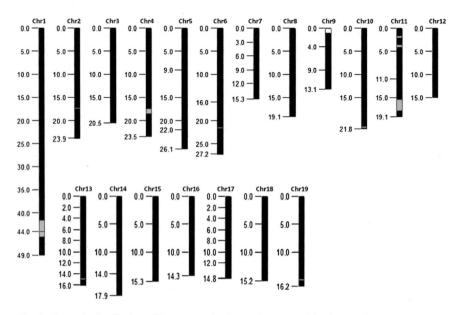

Fig. 1 Genomic distribution of D-mannose lectin paralogs (*green*) in the *Populus* genome

To that end, we analyzed the distribution of 96 paralogous D-mannose lectins within the *Populus* genome. This exercise revealed their presence on chromosomes I, II, IV, VI, X, XI, XIII and XIX with chromosomes I and XI accounting for 78 (81 %) of the D-mannose lectin genes in large tandem repeats (Fig. 1).

3   INDELs Based on Resequenced Genotypes

Based on the above observations, we focused attention on tandem repeats on chromosome I and XI in assaying structural variation in 45 *P. trichocarpa*, 2 *P. deltoides* and 1 *P. tremuloides* resequenced genomes. Variation among lectins within *P. trichocarpa* species revealed patterns of small to large-scale deletions involving intergenic, promoter and genic regions (Fig. 2). Notably, there was evidence of apparent whole-gene deletions when *P. trichocarpa* was compared to other *Populus* species, albeit based on limited alternate species genomes. In such cases, only SNP polymorphisms were found in within-*trichocarpa* comparisons, suggesting high levels of conservation of these genes in this species whereas no apparent homologs were found in either *P. deltoides* or *P. tremuloides*. These observations are in line with *Populus*-microbe interactions that exhibit within-species and between-species specificity. We hypothesize that significant advances in understanding these highly specific interaction can be made by focusing on the types of polymorphisms that occur at the species level.

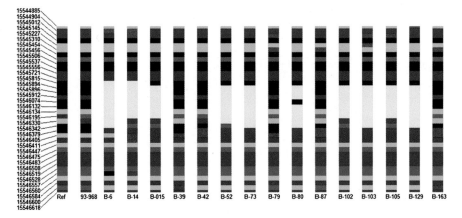

Fig. 2 INDEL polymorphisms (*yellow*) up to 564 bp in a region harboring D-mannose lectins on chromosome XI. Y-axis represents physical positions and the X-axis represents re-sequenced genotypes

4   Coverage in Genetic Maps

Tagu et al. (2001) reported significant segregation in *Populus* colonization by the symbiont *Laccaria bicolor* in F_1 progeny from a *P. trichocarpa* x *P. deltoides* inter-specific cross. Further, Labbé et al. (2011) suggested preferential colonization of *P. trichocarpa* compared to *P. deltoides* and mapped QTL associated with percent colonization in the F_1 pedigree family 54B. Interestingly, a major QTL appeared to co-locate with regions harboring receptor kinases on chromosome XI described above. Recognizing that most *Populus* genetic maps did not include deletion polymorphisms, we were interested in exploring how alternate markers would be distributed in that region. We analyzed three recently completed SNPs maps for the hybrid *Populus* families 52–124 (pseudo-backcross), 545 (F_1) and 54B (F_1) for marker coverage in these specific regions. We observed large gaps in all three maps with the 52–124 having the shortest gap of 584 KB in linkage group intervals corresponding to the genomic region harboring the receptor kinase tandem repeat. Figure 3 illustrates these gaps on chromosome XI for F_1 maps 545 and 54B, respectively. Interestingly, the apparent presence of INDELs was independently validated based on SNP genotyping with probes targeting SNPs in chromosome I and XI intervals. SNP segregation violating expected segregation patterns were observed in the F_1 progeny (Fig. 4c and d). In this regard, both parental genotypes appear to be homozygous (red and blue dots), as such, all F_1 progeny would be expected to be heterozygous (maroon dots in the center). However, both heterozygous and homo-zygous forms are observed for the progeny in the SNP assay. This segregation pattern could be explained if one parent had A- genotype, - representing a deletion; and the other parent having the BB genotype. The SNP assay recognizes a single signal 'A' hence the appearance of 'A-'as being homozygous. Similarly, progeny

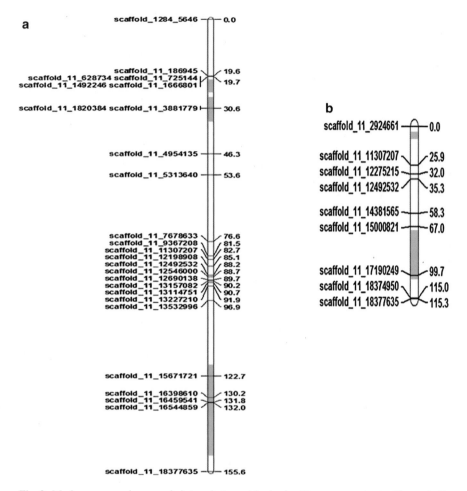

Fig. 3 Marker coverage in genomic intervals (*green*) harboring D-mannose receptor kinases in F_1 genetic maps (**a**) Family 545 and (**b**) Family 54B

that inherit the chromosome with the deletion will have genotype 'B-', which appears in the SNP assay to be homozygous BB, given the detection of only the B allele signal. Without prior knowledge of INDELs, these results would be treated as erroneous and typically excluded from map creation leading to gaps evident in the genetics maps illustrated in Fig. 3.

These observations highlight the challenges of identifying genomic regions that harbor genetic determinants of plant-microbe or plant-insect interactions, since resistance genes typically share the hallmark feature of occurring in tandem repeats with INDELs as the predominant form of functional polymorphism. As such, in work aimed at understanding plant-microbe interaction, inclusion of INDELs in genetic maps is paramount and its importance cannot be over-emphasized. Since the

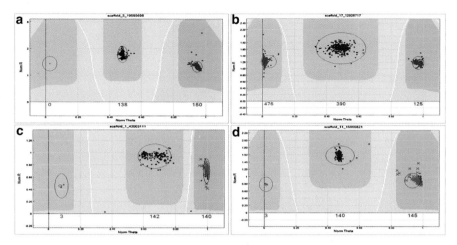

Fig. 4 Genome studio graphical representation of a SNP segregating normally in (**a**) a *Populus* pedigree and (**b**) a naturally varying collection of *P. trichocarpa*, unexpected SNP segregation in a F₁ *Populus* pedigree corresponding the region harboring D-mannose lectins on (**c**) chromosome I and (**d**) chromosome XI. GenomeStudio software recognizes progeny with unexpected genotypes (blue *x* instead of *dots*) based on pedigree information. Three samples on the *left* in panels c and d represent the replicate parent samples with the A-genotype

present discussion only considered D-mannose lectins, we believe this lack of adequate coverage is characteristic of other gene classes that have been shown to exhibit similar patterns of distribution and polymorphism content (Rodgers-Melnick et al. 2012).

5   *Populus* Phenomics

With the unprecedented amount of genotypic information becoming available for *Populus* through the resequencing initiative, the key limitation now remains access to high-quality phenotypic data. In an effort to bridge that gap, an international collaboration resulted in setting up of four common gardens with replicated 1,100 *P. trichocarpa* genotypes established to represent the genetic diversity in *P. trichocarpa* along the Pacific Northwest cline (Slavov et al. 2012). As a step toward association studies that include SNP and INDEL polymorphisms, we have established the presence of sufficient genetic variation in key traits such as susceptibility to key pathogens *Taphrina* spp. and *Venturia* spp. in common garden setting (Fig. 5). In addition, beneficial rhizosphere microbial interactions are being explored within the context of a U.S. Department of Energy project entitled Plant Microbe Interfaces at the Oak Ridge National Laboratory. This project seeks, in part, to establish the role that *Populus* genotype plays in shaping

Fig. 5 Disease symptoms showing evidence of genetic segregation in *P. trichocarpa* common gardens in the Pacific Northwest. (**a**) Leaf blistering caused by *Taphrina populina*, (**b**) shepherd's crook and (**c**) leaf necrosis caused by *Venturia*

microbial community composition in the rhizosphere. This work will undoubtedly benefit from the revelation of large structural polymorphisms in addition to individual SNPs as has been described above.

Acknowledgements This work was supported, in part, by the BioEnergy Science Center and by the Office of Science, Biological and Environmental Research (BER), as part of the Plant Microbe Interfaces Scientific Focus Area at Oak Ridge National Laboratory. Oak Ridge National Laboratory is managed by UT-Battelle, LLC, for the U.S. Dept. of Energy under contract DE-AC05-00OR22725. The funders had no role in study design, data collection and analysis, decision to publish, or preparation of the manuscript. No additional external funding received for this study.

References

Abraham P, Adams R, Giannone R, Kalluri U, Ranjan P, Erickson B, Shah M, Tuskan G, Hettich R (2012) Defining the boundaries and characterizing the landscape of functional genome expression in vascular tissues of *Populus* using shotgun proteomics. J Proteome Res 11(1):449–460

Cao J, Schneeberger K, Ossowski S, Günther T, Bender S et al (2011) Whole-genome sequencing of multiple Arabidopsis thaliana populations. Nat Genet 43:956–963

Geraldes A, DiFazio S, Slavov GT, Priya R, Muchero W et al (2013) A 34K SNP genotyping array for *Populus trichocarpa* (black cottonwood): design, application to the study of natural populations and transferability to other *Populus*. Molecular Ecology Resources 13(2):306–323

Induri BR, Ellis DR, Slavov GT, Yin T, Zhang X, Muchero W, Tuskan GA, DiFazio SP (2012) Identification of Quantitative Trait Loci (QTL) and candidate genes for cadmium tolerance in *Populus*. Tree Physiol 32(5):626–638

Iwano M, Takayama S (2012) Self/non-self discrimination in angiosperm self-incompatibility. Curr Opin Plant Biol 15:78–83

Kereszt A, Kondorosi E (2011) Unlocking the door to invasion. Science 331:865–866

Labbé J, Jorge V, Kohler A, Vion P, Marçais B, Bastien C, Tuskan GA, Martin F, Le Tacon F (2011) Identification of quantitative trait loci affecting ectomycorrhizal symbiosis in an interspecific F1 poplar cross and differential expression of genes in ectomycorrhizas of the two parents: *Populus deltoides* and *Populus trichocarpa*. Tree Genet Genom 7:617–627

Muchero W, Sewell MM, Ranjan P, Gunter LE, Tschaplinski TJ, Yin T, Tuskan GA (2013) Genome anchored QTLs for biomass productivity in Hybrid *Populus* grown under contrasting environments. PLoS ONE 8(1):e54468

Rodgers-Melnick E, Mane SP, Dharmawardhana P, Slavov GT, Crasta OR, Strauss SH, Brunner AM, DiFazio SP (2012) Contrasting patterns of evolution following whole genome versus tandem duplication events in *Populus*. Genome Res 22:95–105

Schwessinger B, Ronald PC (2012) Plant innate immunity: perception of conserved microbial signatures. Annu Rev Plant Biol 63:451–482

Slavov GT, DiFazio SP, Martin J, Schackwitz W, Muchero W et al (2012) Genome resequencing reveals multiscale geographic structure and extensive linkage disequilibrium in the forest tree *Populus trichocarpa*. New Phytologist 196(3):713–725. doi:10.1111/j.1469-8137.2012.04258.x

Tagu D, Rampant PF, Lapeyrie F, Frey-Klett P, Vion P, Villar M (2001) Variation in the ability to form ectomycorrhizas in the F1 of an interspecific poplar (*Populus* spp.) cross. Mycorrhiza 10:237–240

Tuskan GA (1998) Short-rotation forestry: what we know and what we need to know. Biomass Bioenergy 14(4):307–315

Tuskan GA, Walsh M (2001) Short-rotation woody crop systems, atmospheric carbon dioxide and carbon management. For Chron 77:259–264

Tuskan GA et al (2006) The genome of black cottonwood, *Populus trichocarpa* (Torr & Gray). Science 313:1596–1604

Yang X, Kalluri UC, DiFazio SP, Wullschleger SD, Tschaplinski TJ, Cheng Z-M, Tuskan GA (2009) Poplar genomics: state of the science. Crit Rev Plant Sci 28:285–308

Comparative and Evolutionary Genomics of Forest Trees

Andrew Groover and Stefan Jansson

Abstract Comparative and evolutionary genomic approaches can identify genes regulating biological processes, and describe how those genes have been modified through speciation to produce phenotypic variation. These approaches have the potential to address fundamental issues of forest biology, including the regulation of biological traits important to industry and conservation, but have not been widely applied because of technical limitations. Here, we argue that powerful "next generation" DNA sequencing technologies now make comparative and evolutionary genomic approaches not only tractable for basic biological research in trees, but also have the potential to be more informative and cost effective than traditional, one-species-at-a-time approaches. However, designing effective comparative studies for forest trees requires careful consideration of the evolutionary relationships of tree species and biological traits important to forest biology.

This chapter first provides an introduction to comparative and evolutionary genomics, followed by a brief review of some of the general features of the evolution and diversification of forest tree species. Next, two biological processes are discussed that are fundamental to forest trees: wood formation and perennial growth. We examine the varied evolutionary histories of these biological processes, and how these histories relate to the comparative genomic approaches used to research the genes and mechanisms underlying these processes. The chapter is concluded with discussion of practical issues that must be addressed to fully enable this new and powerful direction in forest genomics research, as well as how comparative genomics could support future research and applications for forest management.

A. Groover (✉)
USDA, Davis, CA 95618, USA
e-mail: agroover@fs.fed.us

S. Jansson
Umeå Plant Science Centre, Department of Plant Physiology,
Umeå University, 901 87, Umeå, Sweden
e-mail: stefan.jansson@umu.se

T. Fenning (ed.), *Challenges and Opportunities for the World's Forests*
in the 21st Century, Forestry Sciences 81, DOI 10.1007/978-94-007-7076-8_26,
© US Government 2014

1 Introduction

Forest trees are defined by practical attributes, typically as woody perennial plants with a primary stem of some minimum height. The majority of forest trees share additional attributes. Forest trees typically go through a period of juvenility before undergoing phase change and becoming sexually mature. Forest trees tend to be highly out crossing, genetically heterozygous, and suffer from inbreeding depression. And forest trees are largely undomesticated.

Forest trees do not represent a monophyletic group of plants, however. Tree-like growth has been gained and lost in different seed plant lineages, and extant forest trees can be found among various gymnosperm and angiosperm taxa (Groover 2005; Spicer and Groover 2010). This situation raises important considerations for research aimed at understanding growth and development of the large number of taxonomically diverse tree species that are important to society and to ecosystems. As an example, the genus *Eucalyptus* is more closely related taxonomically to the herbaceous annual, *Arabidopsis*, than to the angiosperm trees ash, ebony, sycamore, and sweet gum (Angiosperm Phylogeny Group 2003). But even though *Eucalyptus* and *Arabidopsis* are more closely related, is *Arabidopsis* a suitable model for helping understand the traits associated with perennial, woody growth of *Eucalyptus*? Even more challenging, gymnosperms such as pines, yews, spruces, and firs are separated from angiosperm trees by more than 300 million years of evolution (Taylor et al. 2009). Does the woody, perennial growth of gymnosperms have a common (homologous) origin with that of angiosperms? If not, the mechanisms underlying growth must be identified independently in gymnosperms and angiosperms. On the other hand if they are homologous, comparative studies between gymnosperms and angiosperms could identify common, ancestral mechanisms regulating tree-like growth, and determine how those mechanisms have been modified during speciation to produce the variation we see today among extant tree species. Such a strategy could be more informative and cost effective than working within individual species, as has largely been the case to date, and would provide insights relevant to all tree species.

Genomic sciences are now providing the means to address previously intractable problems in forest biology. Two decades ago it was a significant feat to clone a single gene from a tree, but the sequencing of the entire genome of *Populus trichocarpa* (Tuskan et al. 2006) supported an explosion of new research and enabled the facile cloning and characterization of genes. Genome sequence also supported the development of powerful functional genomic tools for *Populus* such as microarrays capable of genome-wide assay of gene expression in tissue samples from individual trees (Tsai et al. 2010), and the identification of large numbers of genetic markers useful for association mapping studies that seek correlations between genotypes and phenotypes e.g. (Ingvarsson et al. 2008; Wegrzyn et al. 2010). Full genome sequence is now available for other woody perennial species, including *Eucalyptus* (http://www.phytozome.net/eucalyptus.php), grape (Velasco et al. 2007), and papaya (Ming et al. 2008). Excitingly, more genome sequencing is underway, including several conifers species that represent an important taxonomic group. The expectation is that genome sequence and extensive transcriptome (expressed gene sequence) information will be available in the near future for an increasing number of tree species.

The opportunities afforded by the ability to sequence the genomes and comprehensively genotype large numbers of individuals from various forest tree species should compel us to rethink our strategies for forest biology and forest genomics research. How can we optimize genomic research strategies to better understand the evolutionary histories and relationships among the thousands of forest tree species of interest? How can we use genomics to better predict future responses of forests to climate change, develop better forest-based biofuels feedstocks, or meet the demands for forest-based products? How do we best develop genomic research tools and applications to understand, monitor, and manage the growth, health, and conservation of forests? These and other fundamental questions are being pushed to the front of forest biology by competing pressures on forests by climate change, increasing needs for forest products and ecosystem services, and the desire to use forest trees as a source of bioenergy (FAO 2008).

We propose that comparative and evolutionary genomics will provide the next major technical and methodological advances that will produce fundamental breakthroughs in our understanding of the basic biology underlying traits important to forestry science, conservation, and management. We anticipate that the next chapter in forest biology will leverage approaches of comparative and evolutionary genomics, with experiments designed to explore the diversification of genes and the regulation of biological traits both within and among species. Such approaches are potentially more scientifically powerful than working within individual species. Such strategies can also address some very practical problems, including how a limited research community can make best use of increasingly large sequencing resources, and address complex biological problems in the large number of species of scientific, economic, and ecological interest. In the next section, we provide a brief introduction to the sequencing technologies that enable comparative and evolutionary genomics.

2   Comparative and Evolutionary Genomics in the Age of Next Generation Sequencing

"Next generation sequencing" (NGS) technologies are dramatically changing the strategies for forest genomics research. NGS technologies are developing with an enormous speed and increasingly produce longer and more numerous sequences, while at the same time the price per-base pair of sequencing is decreasing exponentially(Stein 2010). Others have reviewed technologies underlying NGS e.g. (Metzker 2010), which are quickly evolving. Regardless of the specifics of the underlying technologies, in general NGS provides increasingly cost effective methods for new genomics approaches and applications that rely on the ability to sequence the expressed genes or even entire genomes of multiple species and/or individuals within species (Mardis 2008; Schuster 2008). For example, "RNA sequencing" (RNA-seq) using NGS can comprehensively quantify the expression of all the genes in an individual organism for specific developmental time-points, among different tissues, or in response to different environmental conditions or treatments (Wang et al. 2009). NGS is also being used to produce large numbers of genetic markers within species that can be used for population genetic studies of genetic diversity, association mapping of genes regulating traits of interest, or any other marker-based approaches. For example, "genotype by sequencing" approaches using NGS can be used to simultaneously identify hundreds of thousands of genetic markers and genotype hundreds of individuals (Elshire et al. 2011). Such approaches enable research into previously intractable problems in forest biology including the dissection of the genetic regulation of complex biological traits, and understanding population-level genetic factors that underlie adaptation to specific environments (Neale and Kremer 2011).

Comparative genomics and the closely-related evolutionary genomics refer to a variety of approaches that ultimately seek to understand the diversification of genes, genomes, species, and biological traits within an evolutionary and taxonomic framework. There are increasing examples of comparative and evolutionary studies from both herbaceous model plants and forest trees, where analysis of DNA sequence and gene expression data have provided important insights into the evolution of genome properties as well as genes that ultimately underlie phenotypic differences among and within species. For example, gene discovery and annotation is a basic need for making any new forest tree genome sequence usable by the research community, and remains a challenging but increasingly automated task. Comparative genomic approaches played an important part in the annotation and interpretation of the first forest tree genome to be sequenced, *Populus trichocarpa* (Tuskan et al. 2006). Some 45,000 gene models in the *Populus* genome were discovered and annotated in part through comparison to other angiosperm genomes, including *Arabidopsis* (Tuskan et al. 2006). Sequence comparisons both among regions within the *Populus* genome and among other angiosperm genomes enabled the inference of genome duplication events within angiosperm and *Salicaceae* (order containing *Populus*) lineages (Sterck et al. 2005; Tuskan et al. 2006). Duplicated gene pairs (paralogs) are of interest because they provide the opportunity for divergence in expression or function of one or both paralogs, and thus provide fodder for evolutionary change. Use of microarray analysis of expression for duplicated gene paralogs in different tissues of *Populus* provided evidence for significant sub-functionalization of duplicated genes that have acquired distinct expression patterns (Segerman et al. 2007). Similar but more detailed studies of duplicated genes in *Arabidopsis* found that paralogous genes duplicated as part of large-scale events (i.e. duplication of entire or large segments of chromosomes) tend to have more similar expression pattern than genes duplicated on small-scale segments (Casneuf et al. 2006), and that most of the differences in expression between gene pairs occurs shortly after duplication (Ganko et al. 2007). Looking across species, comparison of expression for orthologous genes in *Arabidopsis* versus *Populus* organs found a range of congruity. The percentage of orthologous gene pairs expressed in both species was 60 % for genes expressed in roots, 58 % in young leaves, 69 % in mature leaves and stem nodes, and 76 % in stem nodes (Quesada et al. 2008). These same authors found that genes broadly expressed in multiple plant organs tend to be more conserved in their expression, in contrast to genes that are organ-specific (Quesada et al. 2008).

Changes in protein sequence can also play an important role in the evolution of gene function and phenotypic traits. For example, *Arabidopsis* gene families that contain at least one member that presents a loss-of-function morphological pheno-type were used to examine the relative role of changes in expression pattern versus changes in protein sequence in driving morphological evolution (Hanada et al. 2009). Changes in protein sequence (59–67 % of changes), not changes in gene expression (33–41 % of changes), played the larger role in functional changes to duplicated genes that influenced morphological traits (Hanada et al. 2009). This stands in contrast to evolution of development in animals, where it has been argued that changes in gene expression play the prominent role in evolution of morphologi-cal traits (Carroll 2008). This could be a reflection of the history of prevalent genome duplication in plants versus animals. However, the relative number of examples in which the causative locus underlying morphological evolution in plants is low, and thus the relative contribution of changes in gene expression versus protein function in influencing evolution of traits remains uncertain.

While challenging, progress is being made in moving from comparative studies that primarily examine DNA sequence diversity, to studies that describe causative links between evolution of genes and phenotypic outcomes. For example, compara-tive approaches have been used to study the rapid diversification and speciation in *Aquilegia* (columbines) (Kramer 2009), including identification of loci involved in the diversification flower morphology (Voelckel et al. 2010) and adaptation to diverse habitats, and to identify miRNAs and their cognate target loci (Puzey and Kramer 2009). In *Mimulus* (monkey flowers), comparative genomic approaches were used to identify loci involved in parallel evolution of petal pigmentation in two species, *M. cupresus* and *M.luteus* (Cooley et al. 2011). A fascinating case in tomato illustrates how change in both gene expression and protein function for a single gene influenced morphological diversification (Kimura et al. 2008). In this case, changes in degree of complexity of compound leaves between wild tomato species of the Galapagos Islands has been ascribed to changes in the expression of a single transcription factor, PETROSELINIUM (PTS). PTS is a truncated KNOX transcrip-tion factor that lacks a homeodomain. PTS acts by competing with "normal" KNOX proteins for heterodimerization with a BEL-like Homeodomain protein. But in addition, in species with more complex compound leaves, *PTS* expression is up-regulated by as a result of a single base pair change in the gene's promoter, thus promoting the antagonistic effects of the truncated PTS protein.

NGS sequencing technologies are greatly expanding the previously limited scope of experimental approaches for comparative and evolutionary genomics. But how do we apply NGS-based comparative methods to the study of forest trees? To do so, it is vital to first consider the relationships among tree species. In the next section, we discuss a foundation topic for comparative genomics in forest trees – the evolutionary history and diversification of forest trees.

3 Evolutionary Origins and Diversification of Trees

A defining feature of all extant forest trees is the presence of a meristematic layer of cells inside the stem, the vascular cambium. The cambium is a thin layer of cells that divide over time to produce new cells that can develop into either wood or inner bark, processes collectively known as secondary vascular growth (Larson 1994). What are the advantages to secondary vascular growth and woody stems, and what factors drove the evolution of woody plant forms? Woody stems can confer great advantages in the competition for light, as woody stems allow for dramatic increase in height, and this has undoubtedly been a major factor in the diversification of woody plant forms. But the first appearance of woody growth is now attributed to diminutive plants of the early Devonian (ca 400 MYA), which are likely related to the extinct basal euphyllophyte genus, *Psilophyton* (Gerrienne et al. 2011). While these plants made small amounts of wood from a cambium, they only grew to modest height, supporting the hypothesis that wood may have evolved initially to provide increased water conduction needs rather than mechanical support (Gerrienne et al. 2011). The selective pressures for such innovation could have been driven in part by decreasing CO_2 levels during the early Devonian, which would have increased transpiration rates (McElwain and Chaloner 1995).

Through vascular plant evolution, there have been multiple, independent origins of arborescent plants that contained a vascular cambium. For example, extinct arborescent lycopsids (notably, *Lepidodendron*) were dominant species in swamps and wetlands of the Carboniferous (ca 300–360 MYA) that produced the massive coal deposits mined today (Taylor et al. 2009). In contrast to extant tree species, these plants had a unifacial vascular cambium that produced limited secondary xylem (wood) but no secondary phloem (Cichan 1985a; Eggert 1961). They did have extensive periderm, however, earning the term "bark stem." Similarly, extinct sphenopsid (group that includes extant horsetails) species produced arborescent forms (Cichan 1985b). These stems were tubes characterized by wide pith and a cambium that produced both secondary xylem and phloem (Eggert and Gaunt 1973; Taylor et al. 2009). However, these lineages were evolutionary dead ends, and are not ancestral to any living arborescent species.

The progymnosperms emerged in the Devonian as the first arborescent plants that had a bifacial cambium producing a woody stem similar in structure and anatomy to extant forest trees (Beck 1960). Although there are significant uncertainties about the relationships among progymnosperms, gymnosperms, and angiosperms, it is likely that modern gymnosperm and perhaps angiosperms have their evolutionary origins in progymnosperm lineages (Gifford and Foster 1989). If that is the case, it is possible that the vascular cambium and woody growth found in angiosperms and gymnosperms are homologous (have a shared evolutionary origin).

In extant species, forest trees can be found distributed among both angiosperm and gymnosperm lineages. While there is fossil evidence for extinct gymnosperms with herbaceous habit (Rothwell et al. 2000), all extant and most extinct gymnosperms are characterized by woody growth from a bifacial cambium. For angiosperms, there is a distinct possibility that the ancestral state was woody (but for a more nuanced view, see (Carlquist et al. 1996)). Basal angiosperm lineages include forest trees and woody plants. For example, *Amborella* is a genus of extant basal angiosperms that grow as shrubs or small trees with woody growth supported by a bifacial vascular cambium, and an *Amborella* genome is currently being sequenced (Soltis et al. 2008). Interestingly, the wood of *Amborella* lacks vessel elements (Feild et al. 2000), and given its basal angiosperm phylogenetic position *Amborella* provides insights into early angiosperm wood anatomy. Notably, while there are arborescent monocots (e.g. palms), their stem anatomy is distinctly different from forest trees: monocots lack a true vascular cambium and do not make wood.

Forest trees are found throughout most eudicot orders, and do not represent a monophyletic group (Groover 2005). Interestingly, eudicot tree species can be found that have close relatives with distinctly different growth habits, including herbaceous annuals. Examples of both gain and loss of woody habit can be found within eudicot orders. Gain of woody growth during speciation from herbaceous ancestors (so called secondarily woody species) are relatively common and have occurred independently in various eudicot taxa (Carlquist 2009), suggesting that woody growth form can re-evolve from herbaceous ancestors relatively rapidly.

4   Evolution and Diversification of Wood Formation

The woody growth of stems within angiosperm and gymnosperm taxa is characterized by a bifacial vascular cambium that produces both secondary phloem (inner bark) and secondary xylem (wood) (Esau 1977). Regardless of the evolutionary origins of woody growth in angiosperms and gymnosperms, there are some important differences in woody development and anatomy in these taxa (Esau 1977). Wood of familiar conifers (e.g. pines) is characterized as being comprised of long tracheary elements called tracheids that serve both water conducting and mechanical support functions. In most angiosperm woods, tracheary elements known as vessel elements are the primary water conducting cell type, and can have much wider lumens than tracheids. Mechanical support in angiosperm wood is provided by lignified fibers. Both angiosperm and conifer woods have rays, while only conifer woods contain well developed resin ducts. In response to gravity, angiosperm stems from tension wood to pull leaning stems upright, while conifers form compression wood to push leaning stems upright (Wilson and Archer 1977).

While conifer woods are relatively homogeneous in their anatomies, amazing variation can be found in angiosperm wood anatomy, including presence or absence of vessels, successive cambia, xylem furrowed by xylem, ring porous versus ring diffuse wood, storied versus unstoried cambia, and variation in the presence and structure of rays (Carlquist 2001). Many of these variations have arisen independently in multiple lineages, indicating that the developmental mechanisms regulating the cambium and wood formation in angiosperms are relatively plastic to evolutionary forces.

Important differences can be found between the biochemical makeup of angiosperm and gymnosperm woods that impact applications including pulp and paper production, biofuels, and mechanical properties important for use of wood in construction. For example, lignin is a biopolymer that imparts hydrophobicity and strength to cell walls (Boerjan et al. 2003), and has played a major role in land plant evolution and diversification (Peter and Neale 2004; Weng and Chapple 2010). Lignin is a major component of wood, and is the second most abundant biopolymer on earth. Angiosperm lignin typically consist primarily of guaiacyl and syringyl units, with traces of *p*-hydroxyphenyl units (Boudet et al. 1995). In contrast, gymnosperm lignin is composed primarily of guaiacyl units with significant *p-hydroxyphenyl* units, but lack syringyl units. These differences have major impacts on pulping efficiencies and end uses for softwoods (conifers) versus hardwoods (dicots).

So significant similarities and differences exist between conifer and angiosperm wood, and among angiosperm woods from different species. But how is that reflected by the genetic mechanisms that regulate wood formation among gymnosperm and angiosperm species?

5 Molecular and Genetic Regulation of Wood Formation

Genomic and molecular genetic tools have enabled a rapid increase in our understanding of the genes and mechanisms controlling the cambium and wood formation. Several reviews have summarized the current state of knowledge of mechanisms regulating the developmental aspects of the cambium and wood formation (Du and Groover 2010; Groover et al. 2010; Matte Risopatron et al. 2010), and the biosynthesis of secondary cell walls (Carpita 2011) and lignification (Boerjan et al. 2003; Zhong and Ye 2009). What follows is a brief synopsis of some of the major findings.

One interesting insight into the evolution of the vascular cambium is that some of the important regulatory genes that control the shoot apical meristem have been co-opted during the evolution of the cambium (Groover 2005). Extensive gene expression datasets from wood forming tissues of *Populus* showed that several important transcriptional regulators are expressed in both the shoot apical meristem and the cambium (Schrader et al. 2004). Since the shoot apical meristem predates the cambium in plant evolution (Gifford and Foster 1989), this indicates that these genes acquired a new expression pattern that extended into the meristematic cells of the cambium. Examples of these directly co-opted regulatory genes include the Class I KNOX transcription factors ARBORKNOX1 and 2, which regulate cell differentiation during wood formation (Du et al. 2009; Groover et al. 2006). Interestingly, this small family of transcription factors forms a separate clade in angiosperms and gymnosperm lineages (Guillet-Claude et al. 2004). In other examples, shoot apical meristem regulatory genes have undergone duplication, with a duplicate copy eventually acquiring unique expression in the cambial zone. For example, WUSCHEL (*WUS*) is well characterized for its role in maintaining the stem cells of the shoot apical meristem (Laux et al. 1996). *WUS* is not expressed in the cambial zone, but a related family member, WOX4, is and acts to regulate cambial divisions (Hirakawa et al. 2010). Undoubtedly there are also regulators that are unique to the cambium, but to date there is not a definitive view of the relative roles of co-option versus evolution of unique regulatory modules in the cambium and wood forming tissues.

There is limited insight into the evolution of mechanisms regulating the cambium and wood formation in angiosperms versus gymnosperms. One study of gene expression in wood forming tissues of loblolly pine found evidence for a significant percentage of genes uniquely expressed in pine versus Arabidopsis or *Populus* (Kirst et al. 2003), but such studies are limited in their ability to detect orthologs that have significantly diverged.

6 Evolution and Developmental Regulation of Perennial Growth

The most extensive wood formation is found in perennial plants such as forest trees. The two traits are separable to some degree, however, as there are annual plants that can produce woody tissues from a vascular cambium (e.g. Arabidopsis (Chaffey et al. 2002)) as well as perennial plants that do not produce wood (e.g. red clover). As previously mentioned, most extinct and all extant gymnosperms undergo secondary growth, and are also perennial. With angiosperms, there is a more complex evolutionary history of woody, perennial growth. It has been proposed that early angiosperms were woody perennials, and that herbaceous annuals first arose as angiosperms were experiencing the challenges of harsh winters in higher latitudes during the Tertiary (Sinnott and Bailey 1915). Presumably this trend reflects that herbaceous annuals have the advantage of overwintering underground in the form of seed or roots, in contrast to woody perennials whose persistent above ground vegetation must bear the full brunt of winter weather. This trend is reflected in current plant distributions, where 85–90 % of angiosperm species in alpine regions are herbaceous, while in tropical regions only 25–40 % of angiosperm species are herbaceous (Sinnott and Bailey 1915). Additionally, shifts among growth habits and woodiness have been commonplace in the evolution of many angiosperm taxa (Carlquist 2009).

The growth habit of most forest trees in the temperate parts of the world is rather similar. As temperatures rise after the winter, trees will eventually start to grow laterally (i.e. wood formation will start) and apically (i.e. buds will flush), new leaves will develop and shoots will elongate (see e.g. Hänninen and Tanino 2011 for a review). A major challenge is to time growth initiation in the spring so that the growing season is maximized without risking premature growth and exposure to late winter/early spring cold spells. For example, in boreal forests there are periods when air temperatures during the day could be high while the ground is still frozen, and initiation of vegetative growth and transportation cannot be supported by the root system (Oquist and Huner 2003). To avoid desiccation, dormancy mechanisms must be in place to block growth until conditions are permissive.

The summer season is characterized by photosynthetic carbon fixation driving lateral growth of the stem and root growth. Some trees (e.g. pines and oaks) have a determinate growth in which vegetative buds set the previous season contain all the primordia of the new year's leaves, and are expanded to produce a flush of leaves and elongating stem tissue, before setting new terminal buds. Other species are characterized by an indeterminate apical growth habit in which new lateral buds and leaves continue being produced (Kozlowski and Pallardy 1979). In some species both determinate and indeterminate growth may be present on the same tree, and can often be very easily distinguished from each other. In aspen (*Populus tremula* and *P. tremuloides*), the characteristic rounded, serrated trembling leaves come from determinate growth of mature parts of tree, but if juvenile shoots emerge from the same trunk leaves are typically much larger and not rounded. A microRNA (miR156) seems to have a role in this phase change (Wang et al. 2011).

At the end of the growing season, forest trees react to environmental cues that start several processes; wood formation stops, terminal buds are formed, cold hardiness is acquired and – in deciduous species – leaves will senescence and abscise and dormancy is induced. The obvious cues to these developmental transitions are the shortening of the photoperiod and decreasing temperatures, and trees typically use both to correctly time these events (Garner and Allard 1923). Most trees flower in spring, some very early before vegetative budbreak (e.g. *Populus trichocarpa*), others later. The developmental decision to flower has therefore been taken the year before flowering and the buds that are formed in the autumn are either vegetative or reproductive (for an example in *Populus*, see (Yuceer et al. 2003)). Environmental conditions permissive for flowering may induce the formation of reproductive buds in parts of the tree that have reached maturity, but almost nothing is known about how this is regulated. The physiology of the tree is also a critical factor since many trees – especially in harsher climates – flower periodically, some years most trees in an area may flower, other years almost none. Needless to say, there exist a large amount of within-species variation in all these traits, and exploration of this natural variation has been one fruitful approach to dissect the processes behind different phenological traits e.g. (Frewen et al. 2000; Ma et al. 2010).

7   Regulation of Bud Set and Autumn Senescence

Several molecular details behind the regulation of bud set have been elucidated by studies in *Populus*. Photoreceptors phytochrome A and phytochrome B (Ingvarsson et al. 2008; Olsen et al. 1997) – and orthologs of the downstream components *CONSTANS(CO)* and *FLOWERING LOCUS T(FT)*(Böhlenius et al. 2006), have been shown to regulate bud set and dormancy in *Populus*. Also in *Populus*, orthologs of circadian clock components *LATE ELONGATED HYPOCOTYL1* and *2* and *TIMING OF CAB EXPRESSION1* have been implicated in cold hardiness and bud burst (Ibanez et al. 2010). In annuals, the main function of these components seems to be in the regulation of flowering time, which they also regulate in trees (Bohlenius et al. 2006). It is intriguing that the same system has been recruited for different purposes and in trees – at least in *Populus* – two different outputs are generated that regulates different key processes (see e.g. (Lagercrantz 2009)). Although photoperiodic control of bud set is very common among forest trees in temperate regions, some *Rosaceae* species like *Sorbusaucuparia* may instead rely on temperature cues (Heide and Prestrud 2005), and at extreme latitudes cases diurnal fluctuations in temperature may substitute for photoperiodic cues (Heide 1974). A *PopulusAINTEGUMENTA-LIKE 1* (AIL1) transcription factor appears to act downstream of the *CO*/*FT*regulon, and regulates the expression of key cell cycle genes, e.g. cyclins (Karlberg et al. 2011). Downregulation of the *PopulusAIL1* and/or other homologous genes seem to be a prerequisite for growth cessation. The Arabidopsis *Aintegumenta* has previously been shown to regulate cell cycle genes (Mizukami and Fischer 2000). The fact that in *Populus*, orthologs of *CO*/*FT* and *AIN1* together regulate growth cessation has been suggested to be an example of an "evolutionary mix and match" strategy (Karlberg et al. 2011). Hormones including GA and ABA e.g. (Eriksson and Moritz 2002; Ruttink et al. 2007) have crucial roles during growth arrest and bud set. More recently, changes of auxin response has been implicated in growth cessation, through stabilization of repressor auxin (AUX)/indole-3-acetic acid (IAA) proteins (Baba et al. 2011).

Less is known about the trigger(s) of autumnal senescence in deciduous trees. It is not obvious that photoperiod should trigger leaf senescence in milder climates, precocious senescence in a mild autumn would lead to significant losses in annual photosynthetic yield. Many trees do not shed their leaves until they do not contribute to net photosynthesis. However, there is a potential tradeoff between annual photosynthesis and nutrient status; if the senescence process has not reached completion, valuable nutrients – in particular nitrogen which often is the limiting factor from growth in many forests – are lost when leaves are shed (Keskitalo et al. 2005). Therefore, "safeguarding" by triggering autumnal senescence by photoperiod would be a useful adaptation to nutrient limitation at higher latitudes where photoperiodic cues are good predictors of freezing conditions. Studies in aspen in northern Europe indicates that, provided that the tree has reached a "competence to senescence" which seem to be related to completion of bud set and growth arrest, a second critical photoperiod may be sufficient to trigger onset of senescence (Fracheboud et al. 2009).

8 Dormancy

Meristem dormancy in buds and cambium is an essential trait for trees (see e.g. Cooke et al. 2012 for a review). Dormancy can be of different types, ecodormancy is provoked by certain environmental conditions and endodormancy occurs when the tissue itself has gone into a dormant state that cannot readily be activated by permissive growth conditions (Lang 1987). Dormancy is complex and it has recently been suggested that the terms endo – and ecodormancy need to be revised (Cooke et al. 2012). In any case, during the annual cycle of trees, ecodormancy is typically induced by a shortening of the photoperiod in the autumn, while the transition to endodormancy that occurs later in the season is less well understood. Induction of growth arrest and ecodormancy often happen in parallel suggesting that they are triggered by the same cues, although it is not easy to experimentally separate these processes. When ABA or ethylene signaling is impaired, it is however possible to induce dormancy independent of bud set (Rohde et al. 2002; Ruonala et al. 2006). Release from dormancy is, like vernalization, associated with periods of chilling temperatures. So far, no obvious molecular similarities between the two processes have been identified.

9 Evolutionary Insights from Molecular Mechanisms Regulating Perennial Growth

The intriguing observation that the same or similar molecular mechanisms that regulate flowing in both annuals and perennials, also regulate bud set in *Populus* illustrates well how comparative studies could give insights to the molecular evolution traits relating to the perennial lifestyle. The results of the studies of (Bohlenius et al. 2006) and (Hsu et al. 2011) show how complex the regulatory networks could be and how complicated it may to disentangle them (Ballerini and Kramer 2011). Output signals from the circadian clock can be used to regulate many – potentially all – traits under photoperiodic control, but it is still possible that there are other mechanisms, yet to be discovered, that help trees to accurately get information about the time of the year. Induction of dormancy coincide largely with bud set, it is likely that the initial photoperiodic trigger is the same despite the pathways diverge downstream of the clock. It should also be kept in mind that, while seed dormancy is considered by many plant biologists as a very old trait, it is more recent than the dormancy associated with perennial growth. For dormancy, knowledge transfer from herbs to trees has so far been less successful.

10 Towards the Future: Comparative and Evolutionary Genomic Studies in Trees

The future holds great promise for new insights into forest biology from evolutionary and comparative genomic studies. The advancement of our understanding of fundamental biological processes (e.g. wood formation and perennial growth as discussed here) in a limited number of model species will continue to provide a foundation for extending and testing models of regulatory pathways into other taxa. This work is already underway in *Populus*, eucalyptus, and a handful of conifer species. A promising approach in this regards is the modeling of the gene regulatory networks that control traits of interest. This is the level of complexity at which most traits of interest lie, and network models direct development and testing of hypothesis that can make research more targeted and effective. Once models of gene regulatory networks are established, a next step is to understand how these networks have been rewired and modified over evolutionary history to produce the diversity of form and function seen in extant species. Other approaches will start directly with comparative analyses. For example, comparison of gene expression in cambia across gymnosperm and angiosperm taxa could provide immediate insights into the evolutionary origins of cambia, identify the conserved or ancestral regulatory mechanisms, and characterize how regulatory mechanisms have been modified to produce phenotypic variation. Comparative methods can also be used to identify signatures of selection for genes within and across genomes, and provide insights into adaptive traits that are of fundamental importance to understanding how forest species may respond to climate change. Importantly, next generation sequencing technologies is highly supportive of comparative genomics, and is providing a new foundation of forest genomics-based research.

While genomics will provide new technical advances, comparative approaches also rely on knowledge of anatomy, morphology, physiology, taxonomy and other traditional disciplines. During the design of sequencing-based comparative genomic studies, it is imperative to consider the evolutionary origins, taxonomic distribution, and diversification of the trait under study. Knowledge of the physiological features, anatomical makeup, or developmental stages of a trait can be critical to experimental design. For example, comparative studies of fast-evolving mechanisms underlying disease resistance should take a different approach from studies addressing more evolutionarily conserved mechanisms, like wood formation. Clearly, collaboration among researchers is necessary to address the full range of technical issues surrounding comparative genomic studies. Another unappreciated aspect to comparative genomic studies that can require extensive collaboration is the practical issue of sourcing appropriate plant materials (Groover and Dosmann 2012).

Other practical issues include the fact that the relatively small research communities associated with forest biology must collaborate and communicate effectively. One example is a need to standardize the collection and processing of samples that are used to generate sequence data, and to provide the information about samples and sequencing libraries associated with high throughput sequencing datasets. Following such standards will help ensure that the data collected from different species or labs will be directly comparable, that the growth and other conditions associated with samples is well documented, and also ensure that experiments can be repeated and verified by other researchers. There is also an increasing need for empowering smaller labs with access to computer resources and informatics tools to make use of next generation sequence data, and to perform robust comparative analyses.

11 Conclusions

Sound forest management relies on scientific insights into the biological processes underlying tree growth and survival. An exciting new era of discovery is being ushered in by technological advances in genomics. But to make the most of these advances will require careful planning and coordination by the research community, funding agencies, and stakeholders who may ultimately benefit from the knowledge soon to come.

References

Angiosperm Phylogeny Group (2003) An update of the Angiosperm Phylogeny Group classification for the orders and families of flowering plants: APG II. Bot J Linn Soc 141:399–436

Baba K, Karlberg A, Schmidt J, Schrader J, Hvidsten TR, Bako L, Bhalerao RP (2011) Activity-dormancy transition in the cambial meristem involves stage-specific modulation of auxin response in hybrid aspen. Proc Nat Acad Sci 108(8):3418–23

Ballerini ES, Kramer EM (2011) In the light of evolution: a reevaluation of conservation in the CO-FT regulon and its role in photoperiodic regulation of flowering time. Front Plant Sci 2:81

Beck CB (1960) Connection between Archaeopteris and Callixylon. Science 131:1524–1525

Boerjan W, Ralph J, Baucher M (2003) Lignin biosynthesis. Annu Rev Plant Biol 54:519–546

Bohlenius H, Huang T, Charbonnel-Campaa L, Brunner AM, Jansson S, Strauss SH, Nilsson O (2006) CO/FT regulatory module controls timing of flowering and seasonal growth cessation in trees. Science 312:1040–1043

Boudet AM, Lapierre C, Grima-Pettenati J (1995) Biochemistry and molecular biology of lignification. New Phytol 129:203–236

Carlquist S (2001) Comparative wood anatomy. Systematic, ecological, and evolutionary aspects of dicotyledon wood. Springer, Berlin/Heidelberg

Carlquist S (2009) Xylem heterochrony: an unappreciated key to angiosperm origin and diversifications. Bot J Linn Soc 161:26–65

Carlquist S, Taylor DW, Hickey LJ (1996) Wood anatomy of primitive angiosperms: new perspectives and syntheses flowering plant origin, evolution & phylogeny. Springer, US, pp 68–90

Carpita NC (2011) Update on mechanisms of plant cell wall biosynthesis: how plants make cellulose and other 1/4B-d-Glycans. Plant Physiol 155:171–184

Carroll SB (2008) Evo-Devo and an expanding evolutionary synthesis: a genetic theory of morphological evolution. Cell 134:25–36

Casneuf T, De Bodt S, Raes J, Maere S, Van de Peer Y (2006) Nonrandom divergence of gene expression following gene and genome duplications in the flowering plant *Arabidopsis thaliana*. Genome Biol 7:R13

Chaffey N, Cholewa E, Regan S, Sundberg B (2002) Secondary xylem development in Arabidopsis: a model for wood formation. Physiol Plant 114:594–600

Cichan MA (1985a) Vascular cambium and wood development in carboniferous plants. I. Lepidodendrales. Am J Bot 72:1163–1176

Cichan MA (1985b) Vascular cambium and wood development in carboniferous plants. II. *Sphenophyllum plurifoliatum* Williamson and Scott (Sphenophyllales). Bot Gaz 146:395–403

Cooke JE, Eriksson ME, Junttila O (2012) The dynamic nature of bud dormancy in trees: environmental control and molecular mechanisms. Plant Cell Environ 35:1707–1728

Cooley AM, Modliszewski JL, Rommel ML, Willis JH (2011) Gene duplication in mimulus underlies parallel floral evolution via independent trans-regulatory changes. Curr Biol 21:700–704

Du J, Groover A (2010) Transcriptional regulation of secondary growth and wood formation. J Integr Plant Biol 52:17–27

Du J, Mansfield SD, Groover AT (2009) The *Populus* homeobox gene *ARBORKNOX2* regulates cell differentiation during secondary growth. Plant J 60:1000–1014

Eggert DA (1961) The ontogeny of carboniferous arborescent Lycopsida. Paleontographica 108B:43–92

Eggert DA, Gaunt DD (1973) Phloem of *Sphenophyllum*. Am J Bot 60:755–770

Elshire RJ, Glaubitz JC, Sun Q, Poland JA, Kawamoto K, Buckler ES, Mitchell SE (2011) A robust, simple Genotyping-by-Sequencing (GBS) approach for high diversity species. PLoS One 6:e19379

Eriksson M, Moritz T (2002) Daylength and spatial expression of a gibberellin 20-oxidase isolated from hybrid aspen (Populus tremula x P. tremuloides). Planta 214:920–930

Esau K (1977) Anatomy of seed plants, 2nd edn. Wiley, New York

FAO (2008) Forest and energy. Food and Agriculture Organization of the United Nations FAO Forestry Paper 154

Feild TS, Zweiniecki MA, Brodribb T, Jaffre T, Donoghue MJ, Holbrook NM (2000) Structure and function of tracheary elements in *Amborella trichopoda*. Int J Plant Sci 161:705–712

Fracheboud Y, Luquez V, Björkén L, Sjödin A, Tuominen H, Jansson S (2009) The control of autumn senescence in European aspen. Plant Physiol 49:1982–1991

Frewen BE, Chen THH, Howe GT, Davis J, Rohde A, Boerjan W, Bradshaw HD (2000) Quantitative trait loci and candidate gene mapping of bud set and bud flush in populus. Genetics 154:837–845

Ganko EW, Meyers BC, Vision TJ (2007) Divergence in expression between duplicated genes in Arabidopsis. Mol Biol Evol 24:2298–2309

Garner WW, Allard HA (1923) Further studies in photoperiodism, the response of the plant to relative length of day and night. J Agric Res 23:871–920

Gerrienne P, Gensel PG, Strullu-Derrien C, Lardeux H, Steemans P, Prestianni C (2011) A simple type of wood in two early devonian plants. Science 333:837

Gifford EM, Foster AS (1989) Morphology and evolution of vascular plants. W. H. Freeman and Company, New York

Groover A, Dosmann M (2012) The importance of living botanical collections for plant biology and the next generation? of evo-devo research. Front Plant Sci 3:137

Groover A, Nieminen K, Helariutta Y, Mansfield SD (2010) Wood formation in *Populus*. In: Jansson S, Bhalerao RP, Groover AT (eds), Genetics and genomics of *Populus*. Springer: New York

Groover A, Mansfield S, DiFazio S, Dupper G, Fontana J, Millar R, Wang Y (2006) The *Populus* homeobox gene *ARBORKNOX1* reveals overlapping mechanisms regulating the shoot apical meristem and the vascular cambium. Plant Mol Biol 61:917–932

Groover AT (2005) What genes make a tree a tree? Trends Plant Sci 10:210–214

Guillet-Claude C, Isabel N, Pelgas B, Bousquet J (2004) The evolutionary implications of knox-I gene duplications in conifers: correlated evidence from phylogeny, gene mapping, and analysis of functional divergence. Mol Biol Evol 21:2232–2245

Hanada K, Kuromori T, Myouga F, Toyoda T, Shinozaki K (2009) Increased expression and protein divergence in duplicate genes is associated with morphological diversification. PLoS Genet 5:e1000781

Hänninen H, Tanino K (2011) Tree seasonality in a warming climate. Trends Plant Sci 16:412–416

Heide OM (1974) Growth and Dormancy in Norway Spruce Ecotypes (Picea abies) I. Interaction of photoperiod and temperature. Physiol Plant 30:1–12

Heide OM, Prestrud AK (2005) Low temperature, but not photoperiod, controls growth cessation and dormancy induction and release in apple and pear. Tree Physiol 25:109–114

Hirakawa Y, Kondo Y, Fukuda H (2010) TDIF peptide signaling regulates vascular stem cell proliferation via the WOX4 homeobox gene in Arabidopsis. Plant Cell 22:2618–2629

Hsu C-Y, Adams JP, Kim H, No K, Ma C, Strauss SH, Drnevich J, Vandervelde L, Ellis JD, Rice BM, Wickett N, Gunter LE, Tuskan GA, Brunner AM, Page GP, Barakat A, Carlson JE, dePamphilis CW, Luthe DS, Yuceer C (2011) FLOWERING LOCUS T duplication coordinates reproductive and vegetative growth in perennial poplar. Proc Natl Acad Sci 108:10756–10761

Ibanez C, Kozarewa I, Johansson M, Ögren E, Rohde A, Eriksson ME (2010) Circadian clock components regulate entry and affect exit of seasonal dormancy as well as winter hardiness in Populus trees. Plant Physiol 153(4):1823–1833

Ingvarsson PK, Garcia MV, Luquez V, Hall D, Jansson S (2008) Nucleotide polymorphism and phenotypic associations within and around the phytochrome B2 locus in European Aspen (Populus tremula, salicaceae). Genetics 178:2217–2226

Karlberg A, Bako L, Bhalerao RP (2011) Short day-mediated cessation of growth requires the down regulation of AINTEGUMENTALIKE1 transcription factor in Hybrid Aspen. PLoS Genet 7:e1002361

Keskitalo J, Bergquist G, Gardeström P, Jansson S (2005) A cellular timetable of autumn senescence. Plant Physiol 139:1635–1648

Kimura S, Koenig D, Kang J, Yoong FY, Sinha N (2008) Natural variation in leaf morphology results from mutation of a novel KNOX gene. Curr Biol 18:672–677

Kirst M, Johnson AF, Baucom C, Ulrich E, Hubbard K, Staggs R, Paule C, Retzel E, Whetten R, Sederoff R (2003) Apparent homology of expressed genes from wood-forming tissues of loblolly pine (Pinus taeda L.) with Arabidopsis thaliana. Proc Natl Acad Sci U S A 100:7383–7388

Kozlowski TT, Pallardy SG (1979) Physiology of woody plants. Academic, New York

Kramer EM (2009) Aquilegia: a new model for plant development, ecology, and evolution. Annu Rev Plant Biol 60:261–277

Lagercrantz U (2009) At the end of the day: a common molecular mechanism for photoperiod responses in plants? J Exp Bot 60:2501–2515

Lang G (1987) Dormancy: a new universal terminology. HortSci 22:371–377

Larson PR (1994) The vascular cambium. Springer, Berlin

Laux T, Mayer KF, Berger J, Jurgens G (1996) The WUSCHEL gene is required for shoot and floral meristem integrity in Arabidopsis. Development 122:87–96

Ma X-F, Hall D, St. Onge KR, Jansson S, Ingvarsson PK (2010) Genetic differentiation, clinical variation and phenotypic associations with growth cessation across the Populus tremula photoperiodic pathway. Genetics 186:1033–1044

Mardis ER (2008) The impact of next-generation sequencing technology on genetics. Trends Genet 24:133–141

Matte Risopatron JP, Sun Y, Jones BJ (2010) The vascular cambium: molecular control of cellular structure. Protoplasma 247:145–161

McElwain JC, Chaloner WG (1995) Stomatal density and index of fossil plants track atmospheric carbon dioxide in the Palaeozoic. Ann Bot 76:389–395

Metzker ML (2010) Sequencing technologies [mdash] the next generation. Nat Rev Genet 11:31–46

Ming R, Hou S, Feng Y, Yu Q, Dionne-Laporte A, Saw JH, Senin P, Wang W, Ly BV, Lewis KLT, Salzberg SL, Feng L, Jones MR, Skelton RL, Murray JE, Chen C, Qian W, Shen J, Du P, Eustice M, Tong E, Tang H, Lyons E, Paull RE, Michael TP, Wall K, Rice DW, Albert H, Wang M-L, Zhu YJ, Schatz M, Nagarajan N, Acob RA, Guan P, Blas A, Wai CM, Ackerman CM, Ren Y, Liu C, Wang J, Wang J, Na J-K, Shakirov EV, Haas B, Thimmapuram J, Nelson D, Wang X, Bowers JE, Gschwend AR, Delcher AL, Singh R, Suzuki JY, Tripathi S, Neupane K, Wei H, Irikura B, Paidi M, Jiang N, Zhang W, Presting G, Windsor A, Navajas-Perez R, Torres MJ, Feltus FA, Porter B, Li Y, Burroughs AM, Luo M-C, Liu L, Christopher DA, Mount SM, Moore PH, Sugimura T, Jiang J, Schuler MA, Friedman V, Mitchell-Olds T, Shippen DE, dePamphilis CW, Palmer JD, Freeling M, Paterson AH, Gonsalves D, Wang L, Alam M (2008) The draft genome of the transgenic tropical fruit tree papaya (Carica papaya Linnaeus). Nature 452:991–996

Mizukami Y, Fischer RL (2000) Plant organ size control: AINTEGUMENTA regulates growth and cell numbers during organogenesis. Proc Natl Acad Sci U S A 97:942–947

Neale DB, Kremer A (2011) Forest tree genomics: growing resources and applications. Nat Rev Genet 12:111–122

Olsen JE, Junttila O, Nilsen J, Eriksson ME, Martinussen I, Olsson O, Sandberg G, Moritz T (1997) Ectopic expression of oat phytochrome A in hybrid aspen changes critical day length for growth and prevents cold acclimatization. Plant J 12:1339–1350

Öquist G, Huner NPA (2003) Photosynthesis of overwintering evergreen plants. Annu Rev Plant Biol 54:329–355

Peter G, Neale D (2004) Molecular basis for the evolution of xylem lignification. Curr Opin Plant Biol 7:737–742

Puzey JR, Kramer EM (2009) Identification of conserved Aquilegia coerulea microRNAs and their targets. Gene 448:46–56

Quesada T, Li Z, Dervinis C, Li Y, Bocock PN, Tuskan GA, Casella G, Davis JM, Kirst M (2008) Comparative analysis of the transcriptomes of Populus trichocarpa and Arabidopsis thaliana suggests extensive evolution of gene expression regulation in angiosperms. New Phytol 180:408–420

Rohde A, Prinsen E, Rycke RD, Engler G, Montagu MV, Boerjan W (2002) PtABI3 impinges on the growth and differentiation of embryonic leaves during bud set in poplar. Plant Cell 14:1885–1901

Rothwell GW, Grauvogel-Stamm L, Mapes G (2000) An herbaceous fossil conifer: gymnospermous ruderals in the evolution of Mesozoic vegetation. Palaeogeogr Palaeoclimatol Palaeoecol 156:139–145

Ruonala R, Rinne PLH, Baghour M, Moritz T, Tuominen H, Kangasjärvi J (2006) Transitions in the functioning of the shoot apical meristem in birch (Betula pendula) involve ethylene. Plant J 46:628–640

Ruttink T, Arend M, Morreel K, Storme V, Rombauts S, Fromm J, Bhalerao RP, Boerjan W, Rohde A (2007) A molecular timetable for apical bud formation and dormancy induction in poplar. Plant Cell 19:2370–2390

Schrader J, Nilsson J, Mellerowicz E, Berglund A, Nilsson P, Hertzberg M, Sandberg G (2004) A high-resolution transcript profile across the wood-forming meristem of poplar identifies potential regulators of cambial stem cell identity. Plant Cell 16:2278–2292

Schuster SC (2008) Next-generation sequencing transforms today's biology. Nat Meth 5:16–18

Segerman B, Jansson S, Karlsson J (2007) Characterization of genes with tissue-specific differential expression patterns in *Populus*. Tree Genet Genomes 3:351–362

Sinnott EW, Bailey IW (1915) The evolution of herbaceous plants and its bearing on certain problems of geology and climatology. J Geol 23:289–306

Soltis D, Albert V, Leebens-Mack J, Palmer J, Wing R, dePamphilis C, Ma H, Carlson J, Altman N, Kim S, Wall PK, Zuccolo A, Soltis P (2008) The Amborella genome: an evolutionary reference for plant biology. Genome Biol 9:402

Spicer R, Groover A (2010) The evolution of development of the vascular cambium and secondary growth. New Phytol 186:577–592

Stein L (2010) The case for cloud computing in genome informatics. Genome Biol 11:207

Sterck L, Rombauts S, Jansson S, Sterky F, Rouzé P, Van de Peer Y (2005) EST data suggest that poplar is an ancient polyploid. New Phytol 167:165–170

Taylor TN, Taylor EL, Krings M (2009) Paleobotany. The biology and evolution of fossil plants. Academic/Elsevier, London

Tsai CJ, Ranjan P, DiFazio S, Tuskan G, Johnson V (2010) Poplar genome microarrays. Genetics, genomics, and breeding of crop plants: poplar. Science Publishers, Enfield

Tuskan GA, DiFazio S, Jansson S, Bohlmann J, Grigoriev I, Hellsten U, Putnam N, Ralph S, Rombauts S, Salamov A, Schein J, Sterck L, Aerts A, Bhalerao RR, Bhalerao RP, Blaudez D, Boerjan W, Brun A, Brunner A, Busov V, Campbell M, Carlson J, Chalot M, Chapman J, Chen G-L, Cooper D, Coutinho PM, Couturier J, Covert S, Cronk Q, Cunningham R, Davis J, Degroeve S, Dejardin A, dePamphilis C, Detter J, Dirks B, Dubchak I, Duplessis S, Ehlting J, Ellis B, Gendler K, Goodstein D, Gribskov M, Grimwood J, Groover A, Gunter L, Hamberger B, Heinze B, Helariutta Y, Henrissat B, Holligan D, Holt R, Huang W, Islam-Faridi N, Jones S, Jones-Rhoades M, Jorgensen R, Joshi C, Kangasjarvi J, Karlsson J, Kelleher C, Kirkpatrick R, Kirst M, Kohler A, Kalluri U, Larimer F, Leebens-Mack J, Leple J-C, Locascio P, Lou Y, Lucas S, Martin F, Montanini B, Napoli C, Nelson DR, Nelson C, Nieminen K, Nilsson O, Pereda V, Peter G, Philippe R, Pilate G, Poliakov A, Razumovskaya J, Richardson P, Rinaldi C, Ritland K, Rouze P, Ryaboy D, Schmutz J, Schrader J, Segerman B, Shin H, Siddiqui A, Sterky F, Terry

A, Tsai C-J, Uberbacher E, Unneberg P et al (2006) The genome of black cottonwood, *Populus trichocarpa* (Torr. & Gray). Science 313:1596–1604

Velasco R, Zharkikh A, Troggio M, Cartwright DA, Cestaro A, Pruss D, Pindo M, FitzGerald LM, Vezzulli S, Reid J, Malacarne G, Iliev D, Coppola G, Wardell B, Micheletti D, Macalma T, Facci M, Mitchell JT, Perazzolli M, Eldredge G, Gatto P, Oyzerski R, Moretto M, Gutin N, Stefanini M, Chen Y, Segala C, Davenport C, Demattã L, Mraz A, Battilana J, Stormo K, Costa F, Tao Q, Si-Ammour A, Harkins T, Lackey A, Perbost C, Taillon B, Stella A, Solovyev V, Fawcett JA, Sterck L, Vandepoele K, Grando SM, Toppo S, Moser C, Lanchbury J, Bogden R, Skolnick M, Sgaramella V, Bhatnagar SK, Fontana P, Gutin A, Van de Peer Y, Salamini F, Viola R (2007) A high quality draft consensus sequence of the genome of a heterozygous grapevine variety. PLoS One 2:e1326

Voelckel C, Borevitz JO, Kramer EM, Hodges SA (2010) Within and between whorls: comparative transcriptional profiling of *Aquilegia* and Arabidopsis. PLoS One 5:e9735

Wang J-W, Park MY, Wang L-J, Koo Y, Chen X-Y, Weigel D, Poethig RS (2011) MiRNA control of vegetative phase change in trees. PLoS Genet 7:e1002012

Wang Z, Gerstein M, Snyder M (2009) RNA-Seq: a revolutionary tool for transcriptomics. Nat Rev Genet 10:57–63

Wegrzyn JL, Eckert AJ, Choi M, Lee JM, Stanton BJ, Sykes R, Davis MF, Tsai C-J, Neale DB (2010) Association genetics of traits controlling lignin and cellulose biosynthesis in black cottonwood (*Populus trichocarpa*, Salicaceae) secondary xylem. New Phytol 188:515–532

Weng J-K, Chapple C (2010) The origin and evolution of lignin biosynthesis. New Phytol 187:273–285

Wilson BF, Archer RR (1977) Reaction wood: induction and mechanical action. Annu Rev Plant Physiol 28:23–43

Yuceer C, Land SB Jr, Kubiske ME, Harkess RL (2003) Shoot morphogenesis associated with flowering in Populus deltoides (Salicaceae). Am J Bot 90:196–206

Zhong R, Ye Z-H (2009) Transcriptional regulation of lignin biosynthesis. Plant Signal Behav 4:1028–1034

Part VIII
Bio-Energy, Lignin and Wood

Willows as a Source of Renewable Fuels and Diverse Products

Angela Karp

Abstract Willows (*Salix*) have a long history of diverse uses which reflect the large variation in growth form, physiology and biochemistry that can be found within the genus. Perhaps the three most commonly known uses of willows are basket-making, cricket bats and aspirin, all of which rely on specific attributes of willow wood. Other traditional uses include hurdles, windbreaks, river-bank stabilisers, and riparian filters. Whilst of key importance to the industries they support, these applications of willow have not resulted in extensive and wide-scale plantings. However, since the 1980s, willows have become recognised as a biomass crop, providing renewable feedstock for the energy industry. This includes co-firing with coal in electricity plants as well as using willow feedstock in dedicated biomass boilers of all scales to provide heat and/or power. More recently willow is also being investigated as possible biomass source for biofuels and industrial chemicals. These latter applications have already resulted in increased acreage of willow and could in the near future result in very large scale plantations in many countries world-wide.

This chapter reviews the diversity of willow in growth form, wood structure and biomass composition and then relates this to the potential of this diverse genus for different end-uses, starting with the more traditional but focusing especially on the more recent.

1 Introduction

Willows (*Salix*) have a long history of diverse uses, but most recently have become of interest as a sustainable source of biomass for the bioenergy, biofuel and industrial product industries. The many historical and traditional uses of willow, as well

A. Karp (✉)
Cropping Carbon Institute Strategic Programme, AgroEcology Department,
Rothamsted Research, Harpenden, Herts, AL5 2JQ, UK
e-mail: angela.karp@rothamsted.ac.uk

T. Fenning (ed.), *Challenges and Opportunities for the World's Forests in the 21st Century*, Forestry Sciences 81, DOI 10.1007/978-94-007-7076-8_27, © Springer Science+Business Media Dordrecht 2014

as the current interest, arise from the special growth characteristics of the plant and the properties of willow wood. This chapter introduces willows in the context of the traditional uses and then reviews the anatomical, physical and biochemical properties of willow wood, finally revisiting these aspects in relation to the suitability of willow as a feedstock alternative to fossil fuels.

2 The Genus Salix

Willows (*Salix* spp), together with the genus *Populus* (poplar, aspen, and cottonwood) are classified in the family *Salicaceae*s.str. of the order *Salicales*, class *Magnoliopsida*, subclass *Dilleniidae*. Willows are more diverse than poplars and their taxonomy is notoriously difficult, mainly as a result of their morphological plasticity and ease of natural hybridisation. There are several taxonomic schools which recognise some 330–500 different species, but they can be broadly classified into three main types: the true or tree willows (subgenus *Salix* (*Amerina*)); the sallows and osiers (subgenus *Caprisalix* (*Vetrix*)); and the dwarf and creeping willows (sub genus *Chamaetia*) (Newsholme 1992). Specimens of the genus *Salix* are among the earliest recorded pre-Ice Age plants and fossils dating from 70 to 135 million years ago suggest that dwarf willows were abundant in the flora of the Cretaceous. Their centre of origin is considered to be Eastern Asia, but willows are now found in wide range of environments worldwide. They are most numerous throughout the UK, Europe, Asia, Japan, China, North Americas and Canada, but willows imported into other countries such as New Zealand and Australia have thrived and some have even become weeds.

Willows show enormous diversity in growth form, leaf shape and crown architecture (Fig. 1), from the impressive, large, erect (e.g. *S. nigra*; *S. pentandra*) and weeping trees (e.g. *S babylonica*) through to the shrubby species (e.g. *S. viminalis*, *S. purpurea*) and the slow growing dwarf alpines (*S. yezo-alpina*). Leaves are borne alternatively on the stems and vary in shape and size from the long narrow leaves typified by *S. viminalis* to the large rounder leaves seen in many American species and in extreme form in *S. magnifica* (Karp et al. 2011). All willows bear catkins that are wind and insect pollinated and are dioecious, with ploidy levels varying from diploid to dodecaploid. They have separate flower and vegetative buds, which are characteristically covered with a single scale. There is considerable variation in the length and shape of the catkins and they can appear before leaves emerge (precocious -e.g. *S. daphnoides* and *S. viminalis*), with the leaves (coetaneous – e.g. *S. alba* and S. *babylonica*), or after leaves have fully formed (serotinous (e.g. *S. pentandra* and *S. triandra*).

Willows can be propagated easily as stem cuttings and many species can also be coppiced and grown as pollarded trees or in short rotation coppice (SRC) cycles, all of which have facilitated their cultivation for different uses. In SRC the willow plantations are established as stem cuttings which are cut back after the first year of growth. In the following spring dormant buds in the cut-back stools

Fig. 1 A small selection of the diversity of willows; (**a**) *Salix repens*; (**b**) *S. purpurea*: (**c**) *S. viminalis*; (**d**) *S. magnifica* Hemsl; (**e**) *S. cordata* Muhl.; (**f**) *S. babylonica* L. var. Annularis (Photographs taken by Peter Swatton of plants in the National Willow Collection at Rothamsted Research, Harpenden, Herts AL5 2JQ)

become activated and re-sprout to form multiple shoots (coppiced habit). These new stems are allowed to grow for typically 2–3 years before they are harvested and chipped. Re-sprouting follows next spring and the coppicing cycle can continue in this way for 20+ years. Since the stems are harvested after the leaves have dropped, the majority of nutrients are recycled and stored in the stems and roots in the inter-harvest years. Fertiliser is thus only added after harvest when the stems have been removed and only at rates of circa 60 kg ha^{-1}. Weed control is essential, particularly in the establishment and post- harvest years but is minimised once the planation is matured. SRC thus provides a low-input system which is suitable for all scales of production and results in a shorter time to product than traditional or short rotation forestry.

3 Historical and Traditional Uses of Willow

3.1 Spiritual

There are many legends associated with willows. They have long been recognised as having properties of special spiritual value and are often referred to as the "witches' tree", "tree of enchantment" or "tree of dreaming". Willow rods were used for binding magical and sacred objects and the traditional witches' broom was made with an ash handle and birch twigs bound with willow. In folklore willow is associated with the moon, water and feminism. It was used in rituals by the Celts and Druids for spells of fascination and for gaining insight and inspiration. Willow leaves placed under the pillow at night are claimed to be helpful in recovering energy and sleep. The weeping willow has many legends associating it with grief and a wand of willow is professed to help recovery from personal loss. Willow also has a reputation for being one of the best water-divining woods, along with hazel and birch.

3.2 Basketry

Coracles are flat bottomed boats made of animal hide stretched over broad interwoven and tightly bound willow slats. They are one of the earliest forms of willow basketry, which take advantage of the flexible and durable nature of willow wood. They offer a simple but effective form of water transport and their lightness means that the boats can easily be carried long distances over land. The Ancient Britons used coracles several centuries before the arrival of the Romans. Queen Boadicea's followers navigated the fens using them and centuries later coracles were used by supporters of Hereward the Wake during the Norman invasion of Britain. Coracles are still used today, for example as fishing boats in Wales, although waterproof canvass has replaced the use of hide.

Basketry is an ancient craft. The Roman historian Pliny the Elder specifically mentions the cultivation of willows and wickerwork fragments were unearthed from the UK Glastonbury Lake Village dated 100 BC. In England the earliest record of willow basketry is dated 1381. It was an extremely important craft and by the end of the nineteenth Century there was a basket maker in every large village. In WWI special willow panniers were used to carry individual artillery shells to the front and in WWII willow baskets were used to parachute supplies to the troops. This resulted in high demand for willows and the UK government appointed a Willows Officer in 1922 to ensure the security of future supply for the war effort.

Today basketry remains important, although it is restricted to a cottage industry. It has impacted on the current uses of willow (Sect. 5), since it was responsible for the cultivation and improvement of certain willow species and many basket varieties still exist. Although many indigenous species were used, there was also export and import and much inter-specific hybridisation among the different basket species. The shrubby willows, with their flexible and often coloured stems, and ease of coppicing, are the most suitable. In Europe, *S. triandra*, *S. purpurea* and *S. viminalis* are among the many favourites grown, whilst in North America *S. eriocephala* and *S. cordata* as well as *S. purpurea* are preferred. In Japan the native *S. kinuiyanagi* is used whilst in other countries like Kashmir in India, which has no native species, willows like *S. viminalis* were imported from the UK.

3.3 Cricket Bats

Cricket bats first appeared in a record in 1624 with reference to the death of a fielder who was hit with a bat during a game. At this time the bats were more similar to hockey sticks and although their shape began to be modified by the 1770s, it wasn't until the 1820s that cricket bats started to resemble their modern form. Traditionally, all cricket bats are made from a female clone of the English or White Willow, *Salix alba*, variety *coerula*. This variety was the preferred choice since the early 1800s due to its resilience against the impact of the hard ball, its toughness and also its relative lightness. Other timbers were found to be either too dense, and thus too heavy, or not dense enough, causing them to break too easily on contact. The early bats were constructed from the heartwood of the tree, which is denser and darker in colour and could weigh up to five pounds. In 1890 the English bat manufacturer C.C. Bussey started using the sapwood instead, which he found to be far lighter and more attractive in appearance. Very soon all bats were constructed from 'white' willow wood. Since then, woods of many other species have been tested, and alternatives such as kevlar, carbon fibre and titanium have all been tried, but nothing else competes. White Willow has been imported by countries like India, but there is still a preference for wood sourced from English soils and a small number of UK companies supply the majority of the World market. The willows are carefully grown as single stems for many years before being felled to make the bats.

3.4 Herbal and Medicinal

The analgesic property of White Willow was known as long ago as the fifth century BC, when a Greek physician noted how a bitter powder that came from the bark could be used to ease pains and reduce fever. The active extract of the bark (salicin) was isolated in crystalline form in 1828 by the French pharmacist, Henri Leroux. An Italian chemist, Raffaele Piria, was able to convert it to salicylic acid, which was an effective medicine but also caused digestive problems when consumed in high doses. The French chemist, Charles Frederic Gerhardt, first prepared acetylsalicylic acid in 1853, but it was only later in 1897, that a chemist at Friedrich Bayer and Co. began investigating acetylsalicylic acid as a less-irritating replacement. By 1899 Bayer was marketing it worldwide as "aspirin".

Although most attention has focused on the White Willow, other species have been reputed to have medicinal value. Black Willow (*S. nigra*), is professed to have similar analgesic properties and, has also been used in the past as an aphrodisiac. Similarly Goat Willow (*S. caprea*) is also used to relieve fever and an infusion of bark tea is recommended for indigestion, whooping cough and catarrh. It can also be used as an antiseptic and disinfectant.

3.5 Environmental and Ornamental

Willows are often planted along the banks of rivers that are subject to flooding as their spreading root system helps prevent erosion. The can also act as riparian filters, removing nitrates from farmland before they are leached into rivers, and can be exploited for alleviating environmental problems where nutrients and other elements are in excess. Such environmental applications of willow have recently become increasingly important in providing ways of responding to operational tools such as the EU water framework directive (Mirck et al. 2005). Furthermore, growing willow in this way is more attractive to farmers as the added value that the phytoremediation confers on the energy produced, has the potential to improve the economic sustainability of the crop (Rosenqvist and Dawson 2005).

Certain fast growing varieties of willow are used as natural windbreaks and many basket species are used for static and living sculptures. The diversity of the genus and the attractive growth habit and/or catkins of many species have made them favourites for achieving decorative effects in gardens and parks. Their speed of growth and natural vigour has also resulted in amenity plantings and many roads, especially in urban areas, are lined with planted willow. Willows also harbour a lot of natural biodiversity and add much value to the landscape through attracting insect and bird species.

In the 1980s willows were among a number of non-food crops recognised as having potential as a biomass crop for the energy industry (Sect. 5). The use of willow as an industrial crop has heightened the need for an improved understanding of their characteristics as a feedstock.

4 Properties and Qualities of Willow Wood

4.1 Anatomical

Wood has evolved over millions of years to serve three main functions: the conduction of water from the roots to the leaves; the mechanical support of the plant body and the storage of biochemicals (Wiedenhoeft and Miller 2005).

The trunk of trees is comprised of six layers: outer bark, inner bark (phloem), vascular cambium, sapwood, heartwood and pith. These are all clearly discernible in any willow species (Metcalfe 1939; Sennerby-Forsse 1989; Arihan and Güvenç 2011) (Fig. 2). The inner bark (phloem) is responsible for translocation of photosynthate from the leaves to the growing portions and roots. The sapwood is involved in conducting water and nutrients up the tree and in storage and synthesis of biochemicals. The heartwood originates from the sapwood and is the long-term storage site of most extractives. The pith constitutes the remnants of early growth of the tree before wood was formed.

Trees are classed into softwoods (gymnosperms) and hardwoods (angiosperms), with willow being a hardwood. In both types the outer bark provides mechanical protection and helps to limit water loss through evaporation. Willow bark is very variable in colour and texture. Highly smooth, furrowed and flaky textures are found and colours range from shades of yellow, red and brown to black. Many biochemicals are found in willow bark, which have significance in pest protection and medicinal properties (section "Extractives").

The different cells of wood form integrated systems that are continuous from root to twig and are oriented in axial (parallel with the stem) and radial (pith to bark) planes. The axial systems are concerned with sap and water movement and provide strength. The radial systems are involved in the lateral transport and storage of biochemicals, nutrients and carbohydrates. In hardwoods three main cell types comprise wood (each of which may be present in a diversity of forms): vessel elements that are stacked to form vessels (pores), fibre cells that are shorter and provide support and the ray parenchyma cells (Fig. 2).

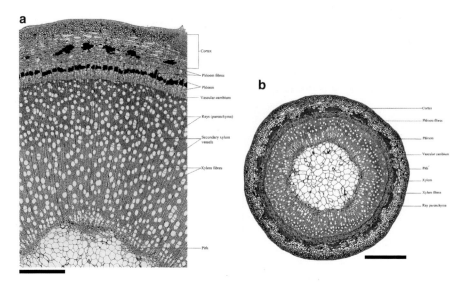

Fig. 2 Anatomy of a willow stem (**a**) Section from cortex to pith; sample taken from the middle height of the stem of a *S. viminalis* × *S. Schwerinii* hybrid, fixed in FAA (formalin, acetic acid and ethanol) and embedded in Spurr's resin. 1 um cross section stained with Toluidine Blue O, pH9; (**b**) young stem of *S. purpurea* collected in the field: fixed in 4 % paraformaldehyde in phosphate buffer, pH 7.2, embedded in LR White resin. 1um cross section stained with Toluidine Blue O, pH9. Scale bars are 500 um (0.5 mm) (Photographs kindly provided by Dr. Cristina Gritsch, Rothamsted Research, Harpenden, Herts AL5 2JQ)

In an early study of the anatomy of the cricket bat willow *S. alba* var. Caerulea (=*coerula*), Metcalfe (1939) described the vessels as "medium size; the end walls are moderately oblique and each contains a large oval simple perforation, the remainder of the end wall being covered by bordered pits". He noted they were numerous, with a limited number of groups of two or three radially contiguous vessels. The fibres were described as typically hexagonal, arranged in regular rows, moderately short and thin walled, with simple pits. The rays were mostly uniseriate (one row). Similar, more detailed descriptions can be found in later studies of *S. caprea* and *S. pentandra* (Sennerby-Forsse 1989), *Salix x rubens* (de Andrada Wagner et al. 2009) and more recently of nine *Salix* species(Arihan and Güvenç 2011, below).

The source of cells in all tree trunks is the vascular cambium which undergoes cell divisions to produce sapwood (or secondary xylem) to the inside and bark (or secondary phloem) to the outside. Divisions of the vascular cambium are also needed as the girth of the tree enlarges. New growth occurs as a sheath covering the main stem, branches and twigs. Trees grow taller from the tip(s) of the stem(s) and from any other point they only grow in diameter. In SRC willow there is the added complexity of a phase of regrowth from the cut stools following harvest, after which "normal" growth occurs in the inter-harvest years.

In most trees large cohorts of cells in the vascular cambium undergo cell divisions over a discrete time interval, resulting in growth rings. The cells formed at the beginning (early wood cells) can normally be distinguished from those produced later (latewood cells). Willows are diffuse porous (with no abrupt distinction between earlywood and latewood), or semi-ring porous (where some differences are recognisable). Thus, in *S. alba* a higher number of vessels per unit area was found to in latewood and vessels were larger in the earlywood (Metcalfe 1939). Similarly, in *S. caprea* and *S. pentandra*, cells were longer and had thicker cell walls in late wood, and the lumen was larger in the earlywood (Sennerby-Forsse 1989).

Differences occur in anatomy with age of the willow tree, which means that sampling strategies have to allow for comparable ages. In *Salix x rubens* young specimens were diffuse porous and mature specimens were semi-porous (de Andrada Wagner et al. 2009). In *S. caprea S. pentandra* (Sennerby-Forsse 1989) and *S. alba* (e.g. Leclercq 1997) vessel and fibre length increased from the pith to the vascular cambium.

Anatomical variation also occurs among species. For example, the fibres and vessels of *S. alba* are longer and wider than those of *S. caprea* and *S. pentandra*. In *S. pentandra* there are a higher proportion of vessels, whilst *S. caprea* had a higher proportion of thick-walled fibres. The most complete comparison of *Salix* species to date can be found in Arihan and Güvenç (2011) who describe nine different species (*S. triandra* L subsp. *triandra*; *S. alba* L., *S. excelsa* J. F. Gmel., *S. fragilis* L., *S. babylonica* L., *S. caprea* L., *S. cinerea* L., *S. pseudomedemii* E. Wolf and *S. amplexicaulis* Bory & Chaub), all sampled as second year stems. Similarities as well as differences were highlighted by these authors. In all species the epidermal cells were one cell thick and rectangular, however, differences occurred in the thickness of the walls and in the presence in the outer layer of cuticle in *S. caprea* and suberin in *S. triandra* and *S. pseudomedemii*. Eglandular trichomes were observed in *S. pseudomedemii* and *S. cinerea* only. In all species the parenchyma under the epidermis comprised rounded, oval, thick walled cells. These often contained starch, and interestingly druses – groups of crystals of calcium oxalate (section "Extractives"). The exceptions were *S. triandra* and *S. amplexicaulis*, where druses were absent. Some form of sclerenchyma within the bark tissue could be seen in all species, but these could be scattered (*S. excelsa, S. babylonica, S. cinerea* and *S. amplexicaulis*), form larger clusters (*S. triandra* L subsp. *triandra, S. fragilis*) or ordered belts (*S. alba, S. caprea*). Similarly phloem sclerenchyma were found in all species except *S. alba* and *S. babylonica*. All ray cells were uniseriate and the cambium was consistently 3–4 cells thick. The pith could be rounded (*S. triandra, S. caprea, S. cinerea, S. amplexicaulis*), almost stellate (*S. alba, S. babylonica, S. pseudomedemii*) or stellate (*S. excelsa, S. fragilis*) and in some cases (*S. excelsa, S. fragilis, S. babylonica, S. caprea* L, *S. cinerea*) the pith also contained druses.

In all willow species studied to date xylem vessels have been described as large and thin-walled, although size differences are evident among species. This variation is most likely of adaptive significance since drought sensitivity in willow is related (at least in part) to the resistance of the xylem to cavitation when water is scarce (Cooper and Cass 2001; Wikberg and Őgren 2007). Phloem cells are smaller, generally irregular and again thin walled. The rapid mobilisation of these conducting systems is essential for new growth. In *S. viminalis*, it has been shown that several years of mature over-wintered vessels become simultaneously functional to allow for rapid mobilisation of resources in initial spring growth (Sennerby-Forsse 1986). Phloem formation starts 2 weeks before flowering and xylem formation (Sennerby-Forsse 1986). Activity in the vascular cambium to make new cells spreads slowly from the apices, taking several weeks to reach the stem base.

4.2   Physical and Mechanical

The thickness of the fibre cell wall in hardwoods is the major factor determining properties associated with density and strength. The fibres of willows are generally arranged in regular radial rows. They are moderately short and thin walled, with the double cell wall of adjacent fibres being less than one quarter the diameter of the fibre lumen. Fibres show only little variation in general properties among the species studied although they do vary with age. (Sennerby-Forsse 1989; Leclercq 1997; Arihan and Güvenç 2011; de Andrada Wagner et al. 2009)

Wood density, (the weight of wood divided by volume at a given moisture content) and the related term specific gravity (the ratio of the density of wood to the density of water) are among the most important physical properties of wood. They are strongly dependent on the volume of cell wall material compared to the volume of the lumina of those cells in a given bulk volume (Wiedenhoeft and Miller 2005). Specific gravity can be determined at any moisture content, but is normally based on oven-dried wood (i.e. Basic or Mean Specific Gravity). In willow values of 0.37–0.45 have been reported for *S. alba* (Leclercq 1997), 0.44–50 for *S. caprea* and 0.43–45 for *S. pentandra*. This is low to mid range in comparison with, for example, 0.20 for balsa wood and 1.00 or more for some tropical hardwoods.

The moisture content of wood is the weight of water in the cell walls and lumina expressed as a percentage of the oven-dry weight. It can exceed 100 %, which occurs when the weight of water in the wood is greater than the weight of the dried cells. At harvest during the winter the moisture content of willow is about 50 % (Ledin 1996) but drops to 25–30 % when the wood is allowed to dry out in a stack.

Moisture content, wood density and specific gravity vary depending on the age of the tree. Sennerby-Forsse (1989) found that in *S. caprea* and *S. pentandra* the mean specific gravity increased from the base to approximately 50 % of the stem height, after which it decreased, and also varied among samples from both within and between stands, although no geographic trend was detectable. Similarly, moisture content increased slightly from the base to 30 % of the stem height and varied within and between stand, with no geographic trend.

4.3 Biochemical

Cell Walls

Plant cell walls consist of the middle lamella, the primary wall and the secondary wall. In each, the three major components are cellulose microfibrils, hemicelluloses and a matrix of cementing material – usually pectin in primary walls and lignin in secondary walls. The secondary cell wall is itself composed of three layers. The first is the thin S_1 layer adjacent to the primary wall, in which the cellulose microfibrils are laid down in a helical fashion. Interior to this is the S_2, which is the thickest layer and has a lower lignin than the S_1 and a low microfibril angle. The final layer, the S_3, is relatively thin. It has the lowest percentage of lignin (a hyrdrophobic molecule) in order that water can adhere and be moved via the transpiration system (Wiedenhoeft and Miller 2005).

Compared with grasses, the composition of the hardwood cell walls differ by having a high amount of hemicelluloses (fucoside xyloglucan (XG), xylans and some mannins) and structural proteins. Hardwood hemicellulose contains primarily glucomannan (5 % per dry wt) and glucoronoxylan, which is the largest component (20–30 % per dry wt). The primary walls have high amounts of pectins (homogalacturonans (HG) rhamnogalacturonan (RG) I and II whilst the secondary walls have high amounts of lignins, which are mainly comprised of trans-coniferyl and trans-sinapyl alcohols forming guaicyl (G) and syringyl (S) units (Sarkar et al. 2009). Studies of biomass willows have revealed that significant variation exists in composition, which could be exploited in improving willow as a feedstock for the biofuel industry (Sect. 5).

Extractives

Extractives are extraneous and not chemically attached to the lignocellulose cell wall – although they can reside within it. Insoluble extractives include crystalline inclusions such as the druses of calcium oxalate mentioned earlier, silica, and starch granules. Soluble extractives are removable with inert solvents such as ether, alcohol-benzene acetone, and cold water. Soluble extractives from primary metabolites include simple sugars, amino acids, simple fats, and various carboxylic acids. They are always present, but amounts vary depending on e.g. the tissue, age, nutritional state and season. Secondary metabolites are highly variable, more complex and show species-specific differences. They may also be limited to specific tissues or conditions. In the Salicaea, secondary metabolites are highly diverse. They are most well studied in poplar, (helped by the availability of the genome sequence), but data for willows are accumulating. Three classes can be distinguished based on their chemical structure and from their origins via the shikimate-phenyl propanoid, terpenoid or lipoxygenase/hydroperoxide pathways(Chen et al. 2009).

Table 1 Phenolic glycosides reported in willow (Data extracted from Boeckler et al. 2011)

Compound	Number of species in which compound has been reported (out of a set of 33 species)
2'-Cinnamoylsalicortin	1
2'-*O*-acetylsalicin	1
2'-*O*-acetylsalicortin	1
Chaenomeloidin	1
Fragilin	18
Lasiandrin	1
Populin	8
Salicin	31
Salicortin	31
Salicyloylsalicin	2
Salicyloyltremuloidin	3
Salireposide	4
Tremulacin	3
Tremuloidin	15
Trichocarposide	1

The shikimate-phenyl propanoid pathway is responsible for some of the most widely studied metabolites as one branch leads to lignin biosynthesis whilst others result in non-structural phenyl-propanoid derivatives (phenolics and polyphenolics) that are subdivided into glycosides, hydroxycinnamates, flavonoids and condensed tannins (Chen et al. 2009).

Phenolic glycosides are among the most widely studied extractives in the Salicaea due to their medicinal value and role in mediating plant-herbivore interactions (Boeckler et al. 2011). They comprise any molecule containing a sugar unit bound to a phenol aglycone but the name phenolic glycoside is commonly applied to compounds with a core structure of salicyl alcohol and β-D-glucopyranose moieties, with an ether linkage between the phenolic hydroxyl group and the anomeric C atom of the glucose (Boeckler et al. 2011). The simplest is salicin, which is also the basic element of many, more complex phenolic glycosides. Others (nigracin, populoside A and salireposide) contain gentisyl alcohol instead of salicylic alcohol. Of 22 phenolic glycosides reported in Salicaea, 15 were reported in at least one (out of 33) species of willow (Table 1), but they vary widely among genotypes, tissues, ages and seasons (Boeckler et al. 2011).

Hydroxycinnamates and their derivatives can make up to 2–8 % of leaf dry weight in poplars and willows and can occur in bud exudates. Their biosynthesis is closely related to the lignin pathway. The principle phenylpropenoic acids are ferulic acid, isoferulic acid, ρ-coumaric acid, caffeic acid, vanillic acid and ρ-hydroxybenzoic acid as well as their esters and aldehyde forms (Chen et al. 2009).

Flavonoids and condensed tannins are a diverse group which act as pigments, pollinator attractants, antioxidants and UV protectors. The flavonoid biosynthetic pathway is well known and initiated by chalcone synthase. Condensed tannins are oligomers or polymers of flavonoid units (flavan-3-ols, also known as catechins). They share the same initial biosynthetic pathway as anthocyanins, but leucoanthocyanidin is then converted to anthocyanidin by anthocyanidin synthase and to catechin by a leucoanthocyanidin synthase. A study of condensed tanning in a range of willow genotypes has shown them to be widely diverse (Falchero et al. 2012).

The terpenoid pathway is the pathway from which the light harvesting pigments for photosynthesis (carotenoids and the phytol chain of chlorophyll) and several plant hormones (including gibberellins, abscisic acid, cytokinins and brassinolides) are derived. It is also responsible for the largest class of secondary metabolites – around 50,000 plant terpenoids have been structurally identified. There are two pathways leading to their formation; the mevalonate pathway (MVA) localised in the cytosol and the methylerythritolphosphate pathway (MEP), which has also been called the 1-deoxy-D-xylulose (DOX) pathway, localised in the plastid. Both pathways lead to the formation of the C5 precursor isopentenyl pyrophosphate (IPP) which can be converted to dimethylallyl pyrophosphate (DMAPP). IPP and DMAPP lead to the formation of geranyl pyrophosphate (GPP) and geranylgeranyl pyrophosphate (GGPP) in the plastid, and to farnesyl pyrophosphate (FPP) in the cytosol. Terpene synthases catalyse the formation of monoterpenes, diterpenes and sesquiterpenes from GPP, GGPP and FPP, respectively.

Isoprene, the most abundant volatile emitted by plants, is converted from DMAPP by isoprene synthase (IPSP). An estimated 5–10 % of the photosynthetically assimilated carbon is utilised in the emission of isoprene. Both poplars and willows are known to be high emitters. Genes encoding ISPS have been isolated from hybrid poplars and subcellular localization studies have shown that ISPS is localized in the plastids, suggesting that the DMAPP is produced by the MEP pathway. In a recent study, use of RNA interference technology (RNAi) to knock out expression of ISPS resulted in isoprene-free poplars which showed no difference in growth or yield, but were more attractive to herbivores and had reduced sensitivity to fungal infection (Behnke et al. 2011). Other derivatives of the terpenoid pathway that have been studied in poplars and willows are the sesquiterpenes, such as (E,E)-α-farnesene and germacrene D.

The final class of secondary metabolites which derive from the lipoxygenase/hydroperoxide pathway are fatty-acid derivatives. These include "green leaf volatiles" such as (Z)-3-hexenylacetate, (E)-2-hexenal, (Z)-3-hexenal and 1-pentenol which result in a distinctive scent when leaves are damaged. There is less knowledge of these in willows, although there is evidence suggesting that changes in fatty acid contents maybe associated with initiation of pre-winter hardening of stems (Hietala et al. 1998).

4.4 Calorific Value and Inorganic Content

The calorific value of willow wood chips has been determined to be 19.7 MJkg^{-1} (Ledin 1996), which places it close to wood from conventional forests (19.4–21.2

MJkg^{-1}). The ash sulphur and phosphorus contents are low (1.0–1.2 %, 0.03 % and 0.09 %, respectively) compared with straw, but similar to other forestry wood. The contents of nitrogen (N), micro nutrients (Zn, Mn, Cu) and heavy metals (Cd, Ni Pb) can also be low but are highly variable, reflecting the land management of the plantation and the fact that willows are efficient in nutrient uptake and are often used for phytoremediation (Sect. 3.5).

4.5   Reaction Wood

Willows show a typical reaction of many angiosperm woody species to displacement of their branches or stems as a result of wind or mechanical stress by forming a specialized type of wood, called tension wood. Tension wood develops on the upper sides of leaning stems/branches where tensile stress is exerted. Cells in this region show altered composition and modified cell wall structure, typically increased cellulose content and higher cellulose crystallinity, reduced lignin content with alterations in monolignol composition and changes in hemicellulose composition (Foston et al. 2011). More specifically, tension wood is often characterised by the formation of a gelatinous layer within the fibre cells (g-fibres) of the secondary xylem. This unique cell wall layer differs from the normal fibre cell wall and is thought to be non-lignified and mainly composed of cellulose with the potential additions of arabinogalactan and xyloglucan. Additionally, as in other woody Angiosperms, here is a polarised antagonistic response on the lower side of the stem, called opposite wood, but less is known about the composition of opposite wood.

5   Willow as a Feedstock for the Energy and Chemical Industries

5.1   Bioenergy

SRC willows first received new notoriety as a biomass crop for renewable energy in the 1980s, when fuel prices spiked. The majority of willow biomass currently used for heat, power, or combined heat and power (CHP) is processed via direct combustion in dedicated biomass boilers or co-fired with coal. Willow woodchips are already among the feedstock being supplied to domestic-scale boilers and large industrial energy plants for this purpose in many countries worldwide. The largest challenge here is meeting the demand for sustainable supplies of biomass in ways that are economically attractive to the industry (to encourage farmers to grow the crops) and do not require the use of the best agricultural land, that is required for food production. Increases in biomass yield are essential to reaching these goals. Modelling shows that a doubling of yield more than doubles the returns to the farmer and a recent report indicated that an annual yield increase would double production volumes by 2050, without any land expansion (NNFCC 2012).

As described earlier, many cultivated willows were originally grown for specialist uses such as cricket bats and basket-making and have not been selected for biomass production. The potential use of SRC willow for large scale supply of biomass feedstock has resulted in increased efforts to breed improved willows for the renewable energy industries. A significant advantage of willow from a breeding perspective is that, unlike most trees (including poplar) many willows will flower within 2 years from a seed, or a cutting, and often within the first year. Willows are also highly diverse and most species will hybridise readily (at least within sub-genera). Seed set can be very high, which is a big advantage for genetic mapping and QTL studies. Progeny can also be easily multiplied and propagated as stem cuttings and most varieties are F_1 hybrids. On the more problematic side, a number of species of interest for breeding are polyploid (Macalpine et al. 2008), and willows cannot be selfed to form inbred lines. Measuring traits is time consuming and trials are expensive to maintain on multiple sites and for successive SRC cycles. There are few traits that can be scored reliably in the nursery and although there are non-destructive methods of estimating yield, based on stem heights and diameters, true (harvested) biomass yield can only be assessed after 4 years and then in 3 year cycles. Application of molecular markers at the nursery stage in marker assisted selection could make a large impact on reducing the time to breed new varieties (Karp et al. 2011).

Conventional willow breeding has been carried out in a few countries, particularly the UK, Sweden, New Zealand and the USA (Kuzovkina et al. 2008). In Europe many breeding pedigrees are based around *S. viminalis* although other species such as *S. schwerinii* have been crossed, in particular to introduce resistance to rust. *S. viminalis* has also been popular in Canadian breeding programmes in addition to *S. eriocephala*. In the US, breeding has mostly concentrated on *S. eriocephala*, *S. miyabeana* and *S. purpurea* (Kuzovkina et al. 2008). As a result of these efforts biomass yields were essentially doubled from 6 to 7 oven dry tonnes per hectare (odt ha^{-1} year^{-1}) to over 12 odt ha^{-1} year^{-1} today. These increases were achieved at a time when there was little understanding of the genetics of many important traits and selections were made solely on the basis of observed stem characteristics (height, diameter, straightness) and coppicing response (number of shoots, shoot vigour), as well as resistance to pests, diseases and environmental stress (Larsson 1998).

From the 1990s genetic studies were initiated for the identification of quantitative trait loci (QTL) affecting key agronomic traits in willow. A number of crosses were established, of which the largest (K8; n=967) was created in 1999 at Long Ashton, UK. This population has since been planted on two further sites, one at Rothamsted and a third on a separate Rothamsted farm (at Woburn) which has nutrient and water limited soils. An additional 11 populations have since been made at Rothamsted for mapping purposes, most of which have in the order of 500 progeny. Several smaller *S. viminalis* crosses have also been generated in Sweden and in the US mapping families have been established for *S. eriocephala* and *S. purpurea*. Using such resources QTLs have been mapped for a large number of traits including rust (Hanley 2003; Tsarouhas et al. 2003) and insect resistance (Rönnberg-Wastljung et al. 2006), shoot height, stem diameter, and stem number (Tsarouhas et al. 2002; Hanley 2003), frost tolerance and phenology (Tsarouhas et al. 2003, 2004), water-use efficiency and drought tolerance (Rönnberg-Wastljung et al. 2005; Weih et al. 2006). Direct alignment of the K8 map to the poplar genome (Hanley et al. 2006) has proven useful for identifying candidates underlying QTLs. Markers closely linked to tightly resolved QTL are now being used at Rothamsted in marker-assisted selection (MAS) and yields of 15–18 odt ha^{-1} year^{-1} are now being attained in experimental trials conducted over wide sites.

In addition to improving yields there are aspects of biomass quality that need consideration in bioenergy generation. Alkali metals (e.g. K and Na) can cause problems with slagging and fouling, whereby ash quality falls and boiler tubes may be coated in harmful deposits and become chemically eroded. The N or S containing compounds in the feedstock form NOx or SOx in the exhaust resulting in GHG emissions or acidification, respectively. Therefore, although a wide range of biomass is used, woody sources, such as willow where such inorganic fractions are lower, are preferable over grass-based feedstocks (Sect. 4.4). Moisture content is also an important consideration beyond transport and storage, as calorific value in combustion is negatively correlated with moisture. Some efforts have also been directed towards optimising the logistics of feedstock supply, including the timing of harvest, methods of drying and storing and also the impact of soils and/or fertilisers on inorganic content.

One of the major challenges now facing future supply of willow biomass is sustaining sufficient yields in challenged environments. This has come to the fore for two reasons: climate change and the increasing need to grow the crops on sub-optimal land to reduce competition with food production. Previous experience suggests that yields can be effectively halved in conditions of severe drought, for example. In recognition of this, current breeding efforts are placing much emphasis on resource-use efficiency (RUE), particularly with regard to water (WUE) and nitrogen (NUE). Willows grow on a wide range of environments, diversity is known to exist and QTLs have been identified for these traits (see above). This provides confidence that breeding programmes should be able to select for genotypes that can sustain yields when resources are limiting (Fig. 3).

5.2   *Biofuels*

The production of biofuels from crop feedstock can be pursued through a variety of routes, which can be grouped into thermal-chemical, biological or chemical conversion. The main chemical conversion process is trans-esterification which results in the production of biodiesel; the fatty acids methyl esters (FAME) produced from different oil-containing crops. This conversion process has less relevance to willow biomass, which is more suited as a feedstock to thermal-chemical and biological conversion.

Fig. 3 Trialling for water-use efficiency. The K8 mapping population grown on water and nutrient limited land at the Woburn Farm of Rothamsted Research. The plants on the *left* have not been irrigated, whilst those on the *right* have received drip irrigation

Thermo-chemical routes included pyrolysis and gasification. Both use woody biomass as a preferred feedstock and willow has much to offer here. Pyrolysis occurs in the complete absence of oxygen at temperatures of 400–800 °C and results in the cellulose and hemicellulose and part of the lignin disintegrating to form gases. As these gases cool, some of the vapours condense to form bio-oil, which can be used as a substitute for fuel oil and as a feedstock for synthetic gasoline or diesel production. The remaining biomass, mostly comprised of lignin, forms charcoal. Gasification requires partial oxidation and temperatures of around 800 °C. The biomass is partially burned to form "producer gas" and charcoal. Producer gas contains 18–20 % H_2, 18–20 % CO, 8–10 % CO_2, 2–3 % methane (CH_4), trace amounts of higher hydrocarbons (e.g. ethane), water, nitrogen (if air is used as the oxidising agent) and various contaminants such as small char particles, ash, tars and oils. Gasification with oxygen or steam produces a medium heating value gas, which is suitable for limited pipeline distribution, as well as a synthesis gas or "syngas" (typically 40 % CO, 40 % H_2, 3 % CH_4 and 17 % CO_2, dry basis). Syngas can be used to make methanol, ammonia and diesel using the process first developed by F. Fischer and H. Tropsch in 1923. Thermo-chemical routes require large volumes of biomass and yield increases (described earlier) are key to the economic feasibility of this process at a large scale, from the feedstock perspective. Although, in principal, aspects of biomass composition are less important, this is not the case when the process is extended to a biorefinery concept where production of high value products is sought in addition to fuels (discussed below in Sect. 5.3). Moreover, the process becomes most effective when all of the bi-products can be utilised and in this regard the exact composition of chars, bio-oils and ash are important to consider. These topics are becoming of increasing importance, but have received little attention in willow up till now.

Fermentation, followed by distillation, is the biological conversion process used for converting sugars to ethanol or, depending on the microbial strain, other low molecular weight alcohols. Most ethanol fermentation is based on Baker's yeast (*Saccharomyces cerevisiae*), which requires simple (monomeric) sugars as raw material. However, willow is a lignocellulosic feedstock in which the sugars comprise carbohydrate polymers that are components of the cell wall (section "Cell Walls"). Additional steps of pre-treatment and hydrolysis are first required to release the sugars for fermentation. There are a large number of pretreatment technologies, but common approaches in willow include dilute acid hydrolysis, steam explosion and organosolvents. Current methodologies for enzymatic saccharification utilise a suite of enzymes (primarily cellulases), usually extracted from fungi to break-down the carbohydrate polymers of the cell wall into their respective monomeric sugars. Both pretreatment and enzymatic conversion are energy intensive and costly steps and thus the target of much research at present. This includes characterisation of variation in biomass composition in relation to processibility.

In an initial study to investigate genetic determinants that affect saccharification potential in willow, 138 genotypes of the K8 mapping population (Sect. 5.1) were used to examine variation in enzymatic glucose release from stem biomass. Significant variation was present and four Enzyme Derived Glucose (EDG) QTL were mapped onto chromosomes V, X, XI and XVI (Brereton et al. 2010) A finding from this work, which has still remained consistent for willow genotypes studied to date, is that saccharification yields were independent of biomass yield. This suggests that cell wall characteristics beneficial to the biofuel process may be novel traits that require independent selection in breeding.

In a subsequent investigation, 35 field grown willows, many of which are commercially grown, were assessed for biomass compositional traits. They showed significant variation in the contents of glucan, xylan, arabinan and lignin. Mannose and galactose were also measured, but variation was not significant. Interestingly, no inverse relationship was found between glucan content and lignin content in the range of 35 genotypes studied, however composition affected sugar release following enzymic saccharification, (Ray et al. 2012). Response to pretreatment was also investigated by studying saccharification potential for biomass with and without pretreatment. This revealed strong correlation between the accessibility of willow cell wall sugars without any pretreatment with cell walls after a dilute acid pretreatment (Ray 2012). If this relationship holds true on further study it will greatly impact on the feasibility of large scale analysis as genotypes can reasonably be screened for beneficial saccharification traits without the need for pretreatment.

Another route to reduce recalcitrance of lignocellulose might be to exploit the production of reaction wood (Sect. 4.5), since the altered composition and cell walls properties of tension wood have been demonstrated to result in reduced recalcitrance and increased enzymic sugar yields compared with opposite or normal wood (Foston et al. 2011). The potential problem here is that the amount of tension wood present in a tree is difficult to quantify, a fact made more difficult by the occurrence of opposite wood (in which some of the properties of tension wood might to some extent be reversed).

In a number of trees, including poplar, alteration of cell wall genes, such as those involved in lignin biosynthesis through transgenic approaches has also demonstrated the potential of reducing recalcitrance through direct manipulation of cell wall genes (e.g. Mansfield et al. 2012). As yet there is no known system for genetic transformation of willow, but once such a system is developed the impact of altering genes in this way can also be tested.

5.3 Routes to Industrial Chemicals

The market for renewable chemicals is growing as volatility in the prices of fossil fuels has impacted on the cost of producing chemicals from oil. Biomass has the potential to form the foundation of a wide range of high value innovative substitutes for the chemicals industry. Examples of markets where interest is already emerging include polymers, solvents, resins and surfactants/detergents. This presents a huge opportunity to increase the value of biomass chains and could alter the economic attractiveness of biomass production for farmers, who currently struggle to achieve favourable returns based on current yields and the feedstock prices that can be offered by the heat and power industries. There are two potential sources of such high value products from willow: (i) the component polymers of biomass and (ii) specific extractives in the bark or heartwood.

Exploiting Biomass Polymers

There are number of challenges that face the development of chemicals from biomass. Naphtha is the main raw material for many petrochemical based products, whereas biomass can be considered a mixed polymer source of principally lignocellulose (lignin, cellulose and hemicelluloses) and also amines, proteins, fats, and oils (Fig. 4). In the case of petrochemicals, their development was not achieved by pre-identification of chemical targets. Instead chemical transformations (from ethylene, propylene and benzene) were first developed leading to synthesis of new diverse chemical species. In contrast, whilst routes for chemical transformations of biomass have been suggested, they have not yet been fully determined for all the different component compounds (Nikolau et al. 2008). The biomass-derived chemicals may need several years of testing and will need to compete with the existing chemicals that have a proven safety track record and are already familiar to industry. They will also need to be competitive in both price and performance with existing products.

There are reasons to be optimistic that these challenges can be met. It is possible that "green" chemicals derived from biomass will be able to hold a price premium due to the marketability of their sustainable production. Moreover, the very diversity and nature of the chemical platforms offered could prove to be an opportunity for real innovation. A way to consider this was presented by Vennestrøm et al. (2011), who compared the H/C ratios of the most common fossil and biomass-based

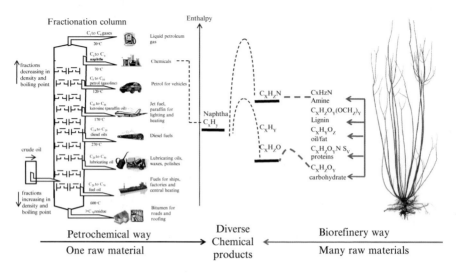

Fig. 4 Schematic depicting the potential of willows as a feedstock for the chemicals industry (Adapted from Sanders et al. 2007)

resources, together with the most common platform, intermediate and target chemicals. Transportation fuels were noted to have an effective H/C ratio similar to crude oil, whilst commodity chemicals ranged over a wider H/C ratio which is more comparable to biomass. Based on their H/C ratios this would argue that sugars, for example, could be a more ideal feedstock than fossil fuels in some cases. These authors further point out that by utilizing biomass for chemicals rather than fuels, the need for deoxygenation (which presents a big challenge in conversion of biomass to liquid fuels) is partially or completely avoided. Ethylene glycol, acetic acid and acrylic acid were given as examples where production from biomass could be more efficient than currently possible from oil. Olefins, conversely, would not be an easy target to replace. Vennestrøm et al. (2011) suggested these as examples where alternatives could be developed for indirect replacement.

In analogy with the existing petrochemical industry, which is based on a relatively low number of hydrocarbon-based platform chemicals, a number of reports have identified building block chemicals that could be produced from biomass via biological or chemical processing of sugars or lignin, fats or oils (Nikolau et al. 2008). For example the Department of Energy (DOE) in the USA identified 12 molecules that can be produced from sugars via biological or chemical conversions. The functional groups of these molecules enable subsequent conversion to high value products. A whole range of viable bulk organic chemicals could then be derived from biomass by industrial biocatalysis and synthetic biology (e.g. DOE 2004). It should be noted that these proposed platform sugars do not have an extensive set of chemical derivatization processes, as is the case for the three petroleum-based platform chemicals (ethylene, propylene and benzene) that are the basis for the current chemicals industry. Therefore, their potential impact would be confined to a limited number of end products unless further derivatization routes can be identified (Nikolau et al. 2008). None-the-less they provide an important starting point from which to encourage further innovative thinking. Lignin also has potential as a platform chemical, although the most effective route is likely to be via the production of syngas (Sect. 5.2) which can then be transformed in to methanol, dimethyl ether, olefins and mixed alcohols. In addition lignin can be converted into aromatic hydrocarbon products that could be used, or are already being used commercially, such as phenol or vanillin. Both the sugar and lignin platforms could have much potential for willow, given the variation in composition found within the genus and its potential for release of sugars following biological and chemical conversions (Sects. 5.1 and 5.2). However it is less clear whether willow will be a competitive feedstock for the variety of chemical and enzymatic reactions that are already well established in the oleochemical industry for the transformation of oils and fats. Nor is it clear whether conversion of amino acids would be a viable option. A better understanding of the diversity of secondary metabolites in willow is needed (section "Extractives" and below) and utilisation of these latter factions may only be realistic in modular biorefineries, where all the different fractions of the biomass are utilised in some way.

Exploiting the Extractives

Willows are harvested after leaf drop. This means that whilst secondary metabolites in leaves have significance in protection against biotic and abiotic stresses, it is the extractives in the bark and stem that are of relevance to the potential exploitation of this crop for high value products. Most knowledge of extractives in willow relates to bark (or whole stems, with bark), encouraged by the early development of aspirin from White Willow (Sect. 3.5). Experience can also be gained from what is known in other trees, including poplar (Chang and Mitchell 1955).

The chemical characteristics of bark are different from those of wood but the calorific value is essentially the same. The bark proportion on a stem is roughly proportional to the area:volume ratio of the stem, thus high bark proportions are found in coppiced systems compared with single stem tree systems. Ash content in bark tends to be higher than wood (4.1–4.8 % cf 0.9–1.0 %), indicating a higher content of nutrient elements. Actively metabolizing bark is especially rich in potassium. Adler et al (2005) estimated harvestable biomass contained 8.4 kg N, 1.1 kg P and 7.3 kg K t^1 in bark and 2.3 kg N, 0.4 kg P and 1.8 kg K t^{-1} in wood. They showed that the bark proportion of young shoots can be between 0.2 and 0.4 whereas older shoots with diameters (at 55 cm height) have a bark proportion of 0.2. These authors proposed that bark proportions could be manipulated by managing shoot size distribution. This could be a useful approach to consider in SRC willow if emphasis is shifted from solely fuels towards the production of high value products.

Willow bark, like willow leaves, contains many metabolites that are associated with anti-herbivory and anti-microbial activity. Some of these are in common but many are specific to tissues, developmental stages and genotypes. Tannins appear to be absent in heartwood but present in bark, for example, and show considerable variation among willow genotypes (section "Extractives"). Acetone and the ethanol extracts of *S. petiolaris* bark were shown to contain the phenolic glycosides salicin, picein, vimalin, salicyloylsalicin, salireposide, grandidentatin, populin, tremulacin and tremuloidin as well as, salicyloylsalicin-2-*O*-benzoate, (+)-catechin, and β-sitosterol (Steele et al. 1972). More recently Kammerer et al. (2005) identified 13 compounds in willow bark (saligenin, salicylic acid, salicin, isosalicin, picein, salidroside, triandrin, salicoylsalicin, salicortin, isosalipurposide, salipurposide, naringenin-7-O-glucoside and tremulacin), whilst Pobłocka-Olech et al. (2007) identified salicylic compounds and other phenolic derivatives in the bark of six willow species: *S. purpurea*, *S. daphnoides*, *S. alba*, *S. triandra*, *S. viminalis*, and *S. herbacea*. Differences in the composition of phenols among the species were identified. Additionally the presence of 2′-acetylsalicortin was revealed for the first time in *S. alba*. Clearly the diversity of chemistry that exists in willow bark provides a potentially rich resource for the extraction of high value products that are at least worthy of further investigation.

   Less is known about willow extractives from the heartwood, but it may be possible to draw from knowledge in poplar. For example saponification and fractionation of the neutrals from the benzene extract of poplar wood gave linoleic acid, oleic acid and C_{12}-C_{28} wax acids. The major alcohols found were glycerol and the wax alcohols. Acetone-base extractions followed by saponification have additionally led to the isolation of α- and β-amyrin, butyrospermol, 24-methylenecycloartanol, lupeol and α-amyrenolol. Steam distillation of the heartwood gave benzyl alcohol, ρ-ethylphenol, β-phenylethanol, n-hexanol and n-heptanol, whilst methanol extractives contained phenylalanine, tyrosine, serines, glycine, sinapaldehyde, sucrose, glucose, fructose, O-α-D-glucopyranosyl –(1→2)-β-D-fructofuranoside and traces of xylose and raffinose. Petroleum ether extraction or acetone extraction of heartwood followed by saponification indicated the presence of C_{16}-C_{28} wax acids and sistosterol glyceride and C_{14}-C_{29} normal paraffins, C_{24}, C_{26} and C_{28} wax alcohols, 3, 5-stigmastadien-7-one (tremulone) and β-sitosterol. As more of these compounds become identified in willow it will become possible to consider their possible exploitation through chemical transformations.

6   Summary and Future Prospects

Willows (*Salix* spp) are part of our cultural history and remain the major woody feedstock for many traditional uses. They are easily propagated and can be cultivated in short rotation coppice (SRC), which has much shorter duration from planting to harvestable product compared with most conventional forestry. SRC also has lower requirements of energy and agrochemical inputs compared with arable farming. These advantages have led SRC willow to be among the sources of sustainable and renewable feedstock for the bioenergy and biofuel industries.

However, willows are hugely diverse and this feature extends to their chemistry. They represent a potentially untapped resource of compounds for substitution of not just solid and liquid fuels but also industrial chemicals. Advances in yield that have already been achieved through conventional and now marker-assisted breeding are testimony to the potential for improvement of this crop given sufficient effort and attention. Over the years our understanding of the growth and physiology of willow has progressed enormously and the underpinning genetic and genomic platforms have been established to enable rapid advancements to be made in the identification of key target genes. The scene is now set for willow to become as important for our future history as it has been in the past.

Acknowledgements The author acknowledges support of the Biotechnological and Biological Sciences Research Council (BBSRC) for funding the Institute Strategic Programme "Cropping Carbon" at Rothamsted Research and the "BBSRC Sustainable Bioenergy Centre (BSBEC): Perennial Bioenergy Crops Programme: BSBEC-BioMASS" (Grant Ref: BB/G016216/1). Rothamsted Research is a national institute of bioscience strategically funded by the BBSRC.

References

Adler A, Verwijst T, Aronsson P (2005) Estimation and relevance of bark proportion in a willow stand. Biomass Bioenergy 29:102–113

Arihan O, Güvenç A (2011) Studies on the anatomical structure of stems of willow (*Salix* L.) species (*Salicacea*) growing in Ankara Province, Turkey. Turk J Bot 35:1–17

Behnke K, Grote R, Brüggermann N, Zimmer I, Zhou G, Elobeid M, Janz D, Polle A, Schnitzler J-P (2011) Isoprene emission-free poplars – a chance to reduce the impact from polar plantations on the atmosphere. New Phytol 194:70–82

Boeckler GA, Gershenzon J, Unsicker SB (2011) Phenolic glycosides of the *Salicacea* and their role as anti-herbivore defences. Phytochemistry 72:1497–1509

Brereton NJ, Pitre FE, Hanley SJ, Ray M, Karp A, Murphy RJ (2010) Mapping of enzymatic saccharification in short rotation coppice willow and its independence from biomass yield. Bioenergy Res 3:251–261

Chang Y-P, Mitchell RL (1955) Chemical composition of common North American pulpwood barks. Tappi 38:315–320

Chen F, Liu C-J, Tschaolinski TJ, Zhao N (2009) Genomics of secondary metabolism in Populus: interactions with biotic and abiotic envrionments. Crit Rev Plant Sci 28:375–392

Cooper RL, Cass DD (2001) Comparative evaluation of vessel elements in Salix spp. (Salicacea) endemic to the Athanasca sand dunes of northern Saskatchewan, Canada. Am J Bot 88 583–587

de Andrada Wagner M, de Chiara Moço MC, Sawczuk AT, Soffiatti P (2009) Wood anatomy of *Salix x rubens* Schrank used for basketry in Brazil. Hoehnea 36:83–87

Falchero L, Brown RH, Mueller-Harvey I, Hanley S, Shield I, Karp A (2012) Condensed tannins in willow (*Salix* spp.): a first step to evaluate novel feeds for nutritionally improved animal products. Grassl Sci Eur 16:434

Foston M, Hubbell CA, Samuel R, Jung S, Fan H, Ding S-Y, Zeng Y, Jawdy S, Davis M, Sykes R, Gjersing E, Tuskan GS, Kalluri U, Ragauskas AJ (2011) Chemical, ultrastructural and supramolecular analysis of tension wood in *Populus tremula* x *alba* as a model substrate for reduced recalcitrance. Energy Environ Sci 4:4962–4971

Hanley SJ (2003) Genetic mapping of important agronomic traits in biomass willow. PhD thesis, University of Bristol

Hanley SJ, Mallott MD, Karp A (2006) Alignment of a *Salix* linkage map to the *Populus* genomic sequence reveals macrosynteny between willow and poplar genomes. Tree Genet Genomes 3:35–48

Hietala T, Hiekkala P, Rosenqvist H, Laakso S, Tahvanainen L, Repo T (1998) Fatty acid and alkane changes in willow during frost hardening. Phytochemistry 47:1501–1507

Kammerer B, Kahlich R, Biegert C, Gleiter CH, Heide L (2005) HPLC-MS/MS analysis of willow bark extracts contained in pharmaceutical preparations. Phytochem Anal 16:470–478

Karp A, Hanley SJ, Trybush SOT, Macalpine W, Pei M, Shield I (2011) Genetic improvement of willow for bioenergy and biofuels. J Integr Plant Biol 53:151–165

Kuzovkina YA, Weih M, Abalos Romero M, Charles J, Hurst S, Mcivor I, Karp A, Trybush S, Labrecque M, Teodorescu TI (2008) *Salix*: botany and global horticulture. Hortic Rev 34:447–489

Larsson S (1998) Genetic improvement of willow for short-rotation coppice. Biomass Bioenergy 15:23–26

Leclercq A (1997) Wood quality of white willow. Biotechnol Agron Soc Environ 1:59–64

Ledin S (1996) Willow wood properties, production and economy. Biomass Bioenergy 11:75–83

Loretta Pobłocka-Olech L, van Nederkassel A-M, Heyden YV, Krauze-Baranowska M, Glód D, Baczek T (2007) Chromatographic analysis of salicylic compounds in different species of the genus *Salix*. J Sep Sci 30(17):2958–2966

Macalpine WJ, Shield IF, Trybush SO, Hayes CM, Karp A (2008) Overcoming barriers to crossing in willow (*Salix* spp.) breeding. Asp Appl Biol 90:173–180

Mansfield SD, Kang K-Y, Chapple C (2012) Designed for deconstruction – poplar trees altered in cell wall lignification improve the efficacy of bioethanol production. New Phytol 194:91–101

Metcalfe G (1939) Observations of the anatomy of the cricket bat willow (*Salix caerulea* SM.). Phytologist 38:150–158

Mirck J, Isebrands JG, Vwewijst T, Ledin S (2005) Development of short-rotation willow coppice systems for environmental purposes in Sweden. Biomass Bioenergy 28:219–228

Newsholme C (1992) Willows. The genus *Salix*. Timber Press, Oregon

Nikolau BJ, Perera MADN, Brachova LL, Brent S (2008) Harnessing plant biomass for biofuels and biomaterials. Platform biochemicals for a biorenewable chemical industry. The Plant J 54:536–545

NNFFCC UK National Non-Food Crops Centre (NNFCC) Briefing document. http://www.nnfcc. co.uk/publications/nnfcc-briefing-document.-the-changing-face-of-the-planet-the-role-of-bioenergy-biofuels-and-bio-based-products-in-global-land-use-change (last accessed 02.07.2012)

Ray MJ, Brereton NJB, Shield I, Karp A, Murphy RJ (2012) Variation in cell wall composition and accessibility in relation to biofuel potential of short rotation coppice willows. Bioenergy Res. doi:10.1007/s12155-011-9177-8

Rönnberg-Wastljung AC, Glynn C, Weih M (2005) QTL analyses of drought tolerance and growth for a *Salix dasyclados* x *Salix viminalis* hybrid in contrasting water regimes. Theor Appl Genet 110:537–549

Rönnberg-Wastljung AC, Ahman I, Glynn C, Widenfalk O (2006) Quantitative trait loci for resistance to herbivores in willow: field experiments with varying soils and climates. Entomol Exp Appl 118:163–174

Rosenqvist H, Dawson M (2005) Economics of using wastewater irrigation of willow in Northern Ireland. Biomass Bioenergy 28:7–14

Sanders J, Scott E, Weusthuis R, Mooibroek H (2007) Bio-refinery as the bio-inspired process to bulk chemicals. Macromol Biosci 7:105–117

Sarkar P, Bosneaga E, Auer M (2009) Plant cell walls throughout evolution: towards a molecular understanding of their design principles. J Exp Bot 60:3615–3635

Sennerby-Forsse L (1986) Seasonal variation in the ultrastructure of the cambium in young stems of willow (*Salix viminalis*) in relation to phenology. Physiol Plant 67:529–537

Sennerby-Forsse L (1989) Wood structure and quality in natural stands of *Salix caprea* and *Salix pentandra* L. Studia Forestalia Suecica 182:1–17

Steele JW, Weitzel PF, Audette RCS (1972) Phytochemistry of the *Salicacea*. IV investigation of the bark of *Salix petiolaris* Sm. J Chromatogr 71:435–441

Tsarouhas V, Gullberg U, Lagercrantz U (2002) An AFLP and RFLP linkage map and quantitative trait locus (QTL) analysis of growth traits in *Salix*. Theor Appl Genet 105:277–288

Tsarouhas V, Gullberg U, Lagercrantz U (2003) Mapping of quantitative trait loci controlling timing of bud flush in *Salix*. Hereditas 138:172–178

Tsarouhas V, Gullberg U, Lagercrantz U (2004) Mapping of quantitative trait loci (QTLs) affecting autumn freezing resistance and phenology in *Salix*. Theor Appl Genet 108:1335–1342

United States Department of Energy (2004) Top value added chemicals from biomassvolume I: results of screening for potential candidates from sugars and synthesis gas produced by staff at the Pacific Northwest National Laboratory (PNNL) and the National Renewable Energy Laboratory (NREL); Werpy Y; Petersen, G. (2004); Contributing authors: Aden A. and Bozell J. (NREL); Holladay J., White J (PNNL); Manheim (DOE-HQ) Publishers: U.S. Department of energy, Office of Scientific and Technical Information, P.O. Box 62, Oak Ridge, TN 37831–0062

Vennestrøm PNR, Osmundsen CH, Christensen CH, Taarning E (2011) Beyond petrochemicals: the renewable chemicals industry. Angew Chem Int Ed 50:10502–10509

Weih M, Rönnberg-Wastljung AC, Glynn C (2006) Genetic basis of phenotypic correlations among growth traits in hybrid willow (*Salix dasyclados* x *S. viminalis*) grown under two water regimes. New Phytol 170:467–477

Wiedenhoeft AC, Miller RB (2005) Structure and function of wood. Handbook of wood chemistry and wood composites. CRC Press-Taylor & Francis Group, 6000 Broken Sound Parkway NW, STE 300, Boca Raton, FL 33487–2742 USA. pp 9–33

Wikberg J, Őgren E (2007) Variation in drought resistance, drought acclimation and water conservation in four willow cultivars used for biomass production. Tree Physiol 27:1339–1346

Woodfuel Production in the UK: Unlocking the Existing Resource and Growing for the Future

Ian Tubby

Abstract Developing energy markets have the potential to transform forestry. Although all UK forests and woodlands have been managed in the past, large areas of woodland have not been thinned or harvested for many years. This lack of management has led to a decline in woodland biodiversity. The emergence of viable renewable heat markets, able to cover the cost of harvesting, extracting and processing wood, have the potential to increase woodland management levels, and reverse the decline in biodiversity. However, the existing forest resource can only supply a fraction of our total energy demand and increased competition for this resource is likely. Increased demand for, and hence value of, wood could lead to more woodlands being established in the UK. Energy markets may lead to an increase in the use of alternative silvicultural systems such as short rotation coppice and short rotation forestry. The way wood is used to produce energy is likely to change in the future. Research continues on using wood to produce liquid fuels whilst combining carbon capture and storage facilities with biomass fired power generation has the potential to deliver carbon negative energy. However, these environmental and economic benefits will only be delivered if bioenergy supply chains are based on established sustainable forest management practices.

1 Introduction

The volume of wood used by energy markets in the UK has increased significantly in recent years and this trend is set to continue for the foreseeable future. The UK Government's commitment to bioenergy is demonstrated in the 2012 Bioenergy Strategy and developing energy markets have the potential to transform forestry in the UK.

I. Tubby (✉)
Forestry Commission England, Alice Holt Lodge, Surrey GU10 4LH, Farnham, UK
e-mail: ian.tubby@forestry.gsi.gov.uk

T. Fenning (ed.), *Challenges and Opportunities for the World's Forests*
in the 21st Century, Forestry Sciences 81, DOI 10.1007/978-94-007-7076-8_28,
© Crown Employees UK 2014

The recent interest in biomass as a fuel source began in the late 1970s following the oil crisis of 1973 and 1979. Early research focused on identifying plant species able to produce high volumes of biomass over a short space of time as governments looked at options to improve energy security. Since then a huge body of research covering everything from the silviculture and management of woody crops through to the development of processing equipment has been carried out. Today biomass, particularly biomass produced by trees, is regarded as a versatile feedstock that can be used to generate heat, power and transport fuels such as ethanol and synthetic diesel.

2 Carbon Savings and Sustainability

Although increased use of domestically sourced biomass could contribute to the UK's energy security, in recent years the main driver behind the deployment of biomass heat, power and fuel generation has been climate change mitigation. Even allowing for emissions from harvesting, processing and transporting woodfuel, the best supply chains can achieve carbon savings of 80 % or more by displacing fossil fuels. However, if supply chains are based on unsustainable forest management or deforestation, then net carbon emissions may be higher than those associated with gas or coal fired energy generation.

Supply chains that include the production and use of wood products as well as woodfuel can offer the very best carbon savings (Matthew et al. 2012). In such scenarios carbon removed from the atmosphere as the tree grows remains 'locked up' for many years in durable products such as roof joists, doors or furniture. As trees are grown to produce sawntimber goods, co-products such as thinnings, branch wood, sawdust and slab wood could be used to produce wood panel products and also provide fibre to displace fossil fuels. At the end of their serviceable lives the wood products could be recycled or burned to generate energy.

Supply chain permutations are endless and the carbon savings they make are dependent on the counterfactuals involved, the woodland management regime in place and the ultimate fate of the wood (i.e. disposed of in landfill with or without energy recovery equipment, disposed of by incineration with or without energy recovery). The time period considered during the life cycle analysis of woodfuel supply chains can also have a significant bearing on results. The geographic boundaries of the analysis i.e. net changes in carbon stocks at the tree, stand, forest, landscape, national and international level must also be considered.

In the meantime, to ensure that the deployment of bioenergy results in carbon savings and does not lead to deforestation and environmental degradation, many countries, including the UK are developing 'sustainability criteria' with which energy generators must comply to remain eligible for government subsidy. The way in which forests are managed to produce fuel is a central element of these criteria. Fortunately, the sustainable management of woodlands and forests is well understood and outlined in regulations and standards such as the UK Forestry Standard and also delivered via market driven voluntary schemes such as the familiar Forest

Stewardship Council (FSC) and Programme for the Endorsement of Forest Certification (PEFC) brands. The forestry sector is making considerable efforts to ensure that the existing government standards and voluntary schemes are considered as bioenergy sustainability criteria are developed, otherwise the additional administrative burden is likely to put potential suppliers off entering the market and could constrain fuel supply.

These opening paragraphs suggest that although a simple concept, growing and managing trees to produce fuel at a scale able to make meaningful contributions to our energy needs and emissions reduction commitments is not straight forward. Wood is a finite resource, even when supply chains are based on sustainable forest management practices. Its production is ultimately limited by the availability of suitable land and its use as fuel depends on the establishment of economically viable markets able to pay for growing the feedstock, harvesting and transporting the wood from the forest or field, processing the raw material into fuel and finally generating energy. Both the grower and generator must receive sufficient economic reward if the supply chain is to be robust and reliable and this is ultimately decided by the price energy consumers are willing or able to pay.

In turn these constraints are influenced by other factors; land available for forestry and energy crops will be limited by the need to grow food and our desire to protect habitats such as moorland, heaths and wetlands. Existing markets for wood require a steady flow of raw material to ensure they are able to produce the myriad sawn timber, panel, and paper products that society relies on. This demand will influence the price, and hence the quantity, of wood available to energy markets. Globally, the interactions between food production, diet, woodland area, fuel and non-fuel timber markets are dynamic and are often hard to predict. The availability and price of both food and wood are dependent on weather conditions, pest and disease outbreaks, currency exchange rates and energy prices. Market speculation can also have profound effects on commodity prices. On top of this, the presence or absence of government incentives designed to encourage the uptake of low carbon energy technology is likely to influence market confidence in the woodfuel supply chain and the economic return available to grower, generator and user.

3 Woodfuel Markets

Against this backdrop, woodfuel markets in the UK have begun to develop into significant elements of both the forestry and energy sectors. In recent years three distinct woodfuel markets have emerged and all are showing signs of growth.

Perhaps the simplest of these markets, and one familiar to most people, is the use of wood in log burning stoves (Fig. 1). These items have become very popular in recent years with around 160,000 home owners installing stoves between July 2008 and June 2009 alone and sales remain strong. The carbon savings they provide is perhaps debatable and dependant on how they are used in conjunction with other heat sources in the building. However, their popularity has seen an increase in

Fig. 1 Log burning stoves have stimulated demand for fire wood

firewood production and some firewood business have evolved into quite substantial processing facilities, kiln drying several thousand tonnes of firewood each year (Fig. 2). Some companies are even importing firewood from overseas.

The second market also focuses on heat but at a slightly larger scale (Fig. 3). This market uses wood chip and wood pellets as fuel (Figs. 4 and 5). Various types of boilers are available to suit applications as small as individual houses through to very large heat users such as hospitals, nursing homes and community heat projects. This market is small but has enjoyed some growth with installed capacity increasing in England by 70 % between 2009 and 2013. Most of the fuel used in this sector is UK grown although some pellets may be imported.

The number of power stations using biomass as fuel has also increased. Around 20 large scale power plants are in operation in the UK at present and more co-fire biomass with coal. Supply chains for these plants are generally international and fuels such as olive cake, straw and even sewage sludge are burned are well as wood. In 2010/2011 around 5.5 million tonnes of biomass was burned in these power plants, much of it imported.

4 Woodfuel and the Woodland Environment

The benefits and risks that this presents to the environment are wide ranging. From an ecological perspective, if developing markets for wood products lead to an increase in the area of actively managed established woodland it might help protect

Fig. 2 Firewood can be big business – drying split logs in wood fired kiln

Fig. 3 Estate scale fuel production

Fig. 4 Mobile chipper processing small round wood

Fig. 5 Wood pellets have become an internationally traded commodity

or even enhance biodiversity. In the UK all woodland has been managed by man at some point to produce fuel and fibre. This management has shaped the plant and animal communities present today. However, in the second half of the twentieth century, active, productive woodland management levels declined sharply. Of those woods under management, many are capable of supplying larger volumes of wood products to the market than they currently do. This low level of management is clearly associated with a decline in the populations of woodland birds, insects and butterflies and changes in the structure of woodland plant communities. Simplistically, these changes in biodiversity reflect changes in temperature and light levels at the forest floor and at different levels through the woodland canopy. Benign neglect leads to woodlands becoming darker and colder places.

To restore and conserve biodiversity, suitable habitats need to be made available throughout the landscape and over long periods of time. Ideally these habitats should be linked or in close proximity to each other. This pattern can be achieved by increasing production of fuel and other goods. Increased production will result in a mosaic of discreet areas of woodland with different canopy conditions, different light levels and different temperatures. The size of these areas will vary with the prevailing management regime but at a landscape scale the effect remains similar whether the management is clearfell or coppicing. Areas of woodland floor receiving high levels of sunlight will support a different range of species compared to areas receiving lower levels of sunlight. If thinning, felling and coppicing activities stop or are reduced, then the canopy will become dense and uniform across all areas. This colder, darker environment alters the number of ecological niches available for plants, animals and fungi to exploit and threatens many of the species we value today.

The role that increased demand for woodfuel, and hence increased woodland management, can play in conserving woodland species has been recognized by a number of wildlife and conservation organisations (Position Statement by Wildlife and Countryside link on the Forestry Commission's Woodfuel Strategy for England 2009). But what caused the decline in woodland management in the first place? Perhaps it comes as no surprise to find that there is no single cause but instead a number of contributing factors. These include increased availability of imported wood, the creation of large conifer plantations and changes in the rural workforce. Together these factors made it progressively more difficult for owners of estate and farm woodlands to make woodland management profitable. The long term decline in the value of timber is well documented (Timber Price Indices Data to September 2011) (Charts 1 and 2). In real terms the prices paid for standing timber in 2011 were roughly half those paid in 1985.

Over the same period the number of sawmills has reduced significantly with numbers falling from 498 in 1994 to 189 in 2010 (Forestry Commission Sawmill Survey 2011). As the industry has moved to fewer larger mills, often sited close to extensive conifer plantations, so local markets for wood, especially hardwood, declined. Changes in other markets also occurred in this period, it is perhaps ironic that a consequence of pulpmills switching from virgin to recycled fibre was a decline in the management and vitality of the woodlands that once supplied them.

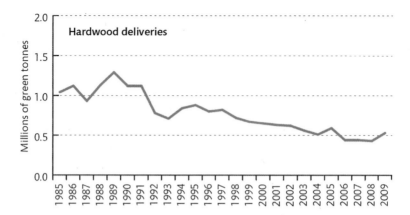

Chart 1 Hardwood deliveries to market over time. Note: All woodland in England has been managed for fuel or fibre production at some stage. Timber prices have been low for the last few decades, as a result less wood is being harvested, more woods have fallen out of management. FC estimates suggest that woodfuel accounts for around 75 % of hard wood deliveries in the UK – the woodfuel sector is key in keeping woodland management ticking over in the private sector (Newer version here http://www.forestry.gov.uk/pdf/FCFS213.pdf/$FILE/FCFS213.pdf)

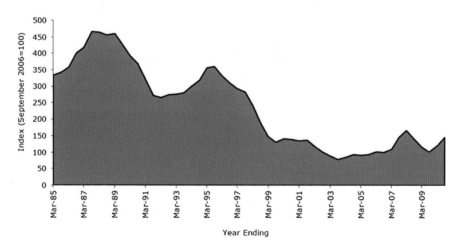

Chart 2 Coniferous Standing Sales Price Index for Great Britain (Fisher Index year ending September 2006 = 100, real terms) (From FC 'Timber Price Indicies' published 11 November 2010 Newer version here http://www.forestry.gov.uk/pdf/tpi201303.pdf/$FILE/tpi201303.pdf)

In theory, if these trends could be reversed and new local markets, created then levels of woodland management would increase. Could energy markets fulfill this role? To get an understanding of this, the scale of the 'unmanaged woodland resource' as well as the timber market and the energy market must be considered.

5 Resource Availability from Existing Woodlands

In England the Forestry Commission estimates that each year unmanaged woodlands produce around 4 million tonnes of wood that is not harvested (A Woodfuel Strategy for England 2007). Of this, the increment in broadleaved woodland accounts for 2.7 million tonnes, the rest coming from conifer woodlands. However, it is unlikely that all of this material could be harvested, some woodland owners may be unwilling to remove any material from their woodlands regardless of a paying market. Harvesting costs may be prohibitive in small blocks of woodland, on steep slopes or on very wet ground. Environmental considerations could further reduce the volume that could be harvested annually on a sustainable basis.

Taking these limitations into account, in England the Forestry Commission estimates that 2 million green tonnes of wood could be delivered from private woodlands to energy market each year by 2020. In addition it is estimated that Scotland's forests could provide a further 2 million tonnes and Welsh woodlands 0.2 million tonnes. Viewed in isolation this sounds like a lot of wood but how does this compare to total UK forestry production?

Forestry Commission data (UK Wood Production and Trade (provisional figures) 2010) show that in 2012 just over 10 million tonnes of round wood were harvested, of which the vast majority was softwood, while deliveries of wood from broadleaved woodlands amounted to only 0.5 million tonnes. Delving further into the data reveals that fuel markets accounted for 80 % of hardwood (broadleaved) roundwood deliveries and around 10 % of softwood (conifer) round wood deliveries.

Between 2006 and 2012 deliveries of softwood to energy markets increased by a factor of 10 from around 0.1 million tonnes to 1 million tonnes. Over the same period hardwood deliveries to energy markets increased from 250,000 to 400,000 tonnes. Despite these changes deliveries to other markets like pulp, panel and sawmills remained fairly constant.

From this it is clear that energy markets are already making themselves felt in forestry circles. In the case of hardwood deliveries, woodfuel is now the only significant market in most parts of the country. Without this market the forestry sector would support fewer jobs.

To estimate how much energy an additional 4.2 million tonnes of woodfuel could provide, some assumptions must be made on energy content and how the woodfuel is used. When freshly felled, moisture content of wood varies between species. Conifers generally have a higher moisture content that hardwoods. A very rough 'average' moisture content of 55 % is often assumed for freshly felled wood regardless of species when estimating potential energy yields. Green wood is generally unsuitable for use in the type of stoves, boilers and power stations encountered in the UK and wood is usually seasoned or dried to a moisture content of between 10% and 30 % before use. At 30% moisture content, wood has an energy content of around 3.5 MWh per tonne with very small variations between species.

6 The Contribution of Existing Woodlands to Energy Markets

Applying these assumptions to the estimated 4.2 million 'green tonnes' of UK wood that may be available to energy markets, gives 2.7 million tonnes of seasoned wood with a total energy content of 9.5 TWh. If used to generate heat in modern, efficient boilers this could deliver around 7.6 TWh of heat. This would be sufficient to heat around 420,000 houses. Although this is a significant amount of energy it pales against the 711 TWh of heat used in the UK in 2008, to satisfy this demand around 200 million tonnes of seasoned wood is needed!

If used to produce electricity, 4.2 million green tonnes of wood could generate around 4 TWh of electrical power, or roughly 1 % of present demand. From this it is clear that woodfuel sourced from existing UK forests will never be able to meet our energy needs and perhaps suggests that UK forestry needs energy markets more than energy markets need UK forestry!

However, this view is over simplistic and overlooks the benefits that woodfuel supply chains can bring to woodland owners and energy users. The renewable heat sector in particular can provide the conditions required to increase fuel production in currently neglected woods. A future scenario in which a few tens of thousands of boilers are in use across the country, each requiring a few tens or hundreds of tonnes of wood fits well with the scattered distribution of UK woodlands. Furthermore this market is more likely to offer a price able to cover the cost of producing fuel from small blocks of currently unmanaged broadleaved woodland.

Fuel grade wood chip production costs (including seasoning and local delivery) are generally in the region of £80–90 per tonne when working small woods. Site conditions, standing volume, tree size, price of standing timber, extraction and delivery distances will all influence this and production costs in large scale clear fell conifer forests will be much lower. However, a cost of around £90 per tonne is roughly equivalent to a fuel price of 2.6p per kWh. Power generators buy coal and gas at around 1.5p per kWh. Perhaps this difference in fuel cost is one reason why biomass co-firing projects and bioenergy power plants have not made more of an impact on the UK forestry sector over the last 10 years, despite the availability of support via the Renewables Obligation.

On the other hand in the heat market users are paying around 5p per kWh for gas, and 6p per kWh for oil. This is equivalent to paying between £170 and £210 per tonne of seasoned woodfuel. Where electricity is used to heat buildings, at current prices, switching to woodfuel and paying anything less than £500 per tonne of seasoned wood chips could provide the user with fuel cost savings.

Once government support such as the Renewable Heat Incentive (RHI) are added to the potential fuel savings, then the economics of woodfired heating look very attractive indeed. Where appropriately specified boilers are used to heat offices, blocks of flats and similar buildings returns on capital investment of at least 12 % are achievable. Even better returns may be possible where woodfuel is used in commercial heating situations such as horticulture, poultry or fishing farming. At the time of writing the RHI supports around 500 MW of biomass boiler capacity in the UK and uptake of the scheme is growing.

If the end user also owns woodlands and can produce their own fuel then even more impressive savings may be achievable. In some parts of the UK some support is available to woodland owners establishing the access tracks and hard standing necessary to enable the harvesting and extraction of woodfuel and other products. This type of infrastructure is often missing from woodlands that have not been managed for several decades. Once in place new tracks could last for many years and have benefits that extend beyond the woodfuel sector. Some woodland owners are already supplying timber to non fuel markets following positive experiences with woodfuel. This is an encouraging sign and may signal the beginning of a return to silvicultural thinnings designed to improve the quality of trees destined to provide sawn timber products.

7 Spreading the Word and Building Confidence

Communicating these potential economic opportunities to heat users, woodland owners and fuel producers remains perhaps the most important task that needs to be addressing before the full potential of the woodfuel market can be realised. The issue of communicating and engaging with woodland owners has become more complex during recent decades as woodlands have fallen out of management. Without local markets to supply to, many owners have no reason to be members of forestry trade associations or communicate with the Forestry Commission. Over time the culture of woodland management has become lost or diluted on many farms and estates around the country. This is something that the forestry sector is trying to address, but with an estimated 100,000 woodland owners in the UK it is not straightforward. Effective engagement now is required to help ensure current investment in the sector delivers returns over the next few decades.

Effective communication is also required to develop new ways of working within the forestry sector. Due to the generally small areas of woodland involved, the traditional model of farm and estate woodland owners working in isolation can often lead to high harvesting costs and limit profitability. If neighbouring woodland owners work together to an agreed management plan that covers several blocks of woodland, then the potential economies of scale would probably lead to reduced management costs and improved margins. This approach is currently being tested in South West England where several clusters of owners have begun to work together and are beginning to see the benefits of this approach.

8 Energy Markets and Woodland Creation

As well as increasing management levels within existing woodlands, energy markets have the potential to drive the creation of new woodlands and plantations. Land owners including local authorities, utilities companies and farmers are beginning to look at their land holdings and reappraise whether the current land use is providing them with the greatest long term benefit. As energy prices increase, the ability to 'self supply' fuel used to heat, power or cool buildings becomes more attractive and provides a degree of energy security. The prospect of reform of the Common Agricultural Policy is being mooted by some commentators and if this comes to pass then the competitiveness of woodland creation may improve.

There are several different types of woodland creation options open to land owners and some are likely to provide first harvests or thinnings much more quickly than conventional, low density woodland planting. At one extreme, and perhaps stretching the term 'woodland' a little too far, Short Rotation Coppice (SRC) offers the potential to take a harvest 3–5 years after establishment. This system is close to agricultural production and generally involves fast growing willow clones planted at densities in excess of 10,000 plants per hectare. Up take of this crop has been slow and it's image was rather unfairly tarnished by association with a failed biomass gasifier based power station in the late 1990s. A few years later the price of grain and other agricultural crops strengthened and this discouraged farmers from planting perennial energy crops.

More recently some potential end users have considered Short Rotation Forestry (SRF) as a feedstock and a number of field trials at various locations around the UK have been established. The term SRF is generally used to describe plantations of single stem trees planted at high stocking densities and managed on a clearfell rotation of between 8 and 20 years. Compared with SRC it enables the land manager and end user to use more conventional forestry equipment and infrastructure.

A range of native and exotic species can be used in this system, depending on site conditions, and much of the recent interest in SRF has focused on various species of eucalyptus. Eucalypts are widely used in plantations supplying the pulp and papers markets in the southern hemisphere and are also encountered commercially in Southern Europe. In these climates very high yields are possible and rotation lengths may be as short as 5 years. However, the hard winter of 2010/2011 in the UK highlighted the frost susceptibility of many of these species in the UK and so sycamore or chestnut may be more reliable performers at many sites. Species choice is becoming increasingly important as the climate changes and pest and disease pressures evolve. The widespread decline of Corsican pine following the inadvertent introduction of Dothistroma Needle Band (also known as Red Band Needle Blight) is a recent example of how invasive diseases can have a large impact on commercially important species. This disease, along with Phytophthora spp. on Larch, will have a much larger impact on UK wood processors than Chalara die back of ash, despite the latter receiving more media attention.

Climate change and disease issues notwithstanding, turning the current interest in woodland creation into action on the ground remains difficult. For example, cash flow and other economic constraints mean that many land owners require a steady annual return from their land. Conventional forestry does not always deliver this. Woodland creation bonds and other financial instruments may be able to address this and future carbon markets could also play a role in attracting investments in woodland creation.

If these financial barriers could be removed then the potential for increased woodland cover and deployment of perennial energy crops is very large, with studies suggesting that in the region of 0.9– 3.5 million hectares of land could be made available in England and Wales whilst avoiding the best agricultural land and land with high conservation value (Domestic Energy Crops and Potential and Constraints Review 2012). This level of potentially available land fits well with a recent review of the role that UK forests could play in climate change mitigation (Combating Climate Change – a role for UK forests 2009) which suggested that planting 23,000 ha or woodland each year for 40 years would deliver cost effective climate change abatement measures including increased availability of woodfuel and wood products. The total land resource required to deliver this aspiration would be just over 0.9 million hectares. Although ambitious, this would increase the area of woodland in the UK from 12 % to 16 %, still well below the European average of 37 %.

Even with significantly increased woodland cover, the extent to which UK grown biomass can meet our energy needs is limited and unlikely to contribute more than a few percent towards total energy demand. Constraints such as high population density (third highest in the world after Bangladesh and the Netherlands) limit land availability and the temperate climate does not support the high levels of growth seen in the tropics. It is inevitable that the UK will remain a net importer of biomass in the future. Provided international supply chains are based on sustainable land management, there is no reason why this cannot deliver environmental and economic benefits. However, any residual doubts about this should not be used an as excuse to not increase woodland cover as, done sensitively, increased areas of woodland could deliver many other benefits from water management, including flood alleviation and improved water yield and water quality, to biodiversity.

9   Future Technologies

The technologies that use woodfuel from yet to be planted woodlands will also change and potentially offer enhanced carbon savings. The use of Carbon Capture and Storage (CCS) technology in conjunction with biomass fired power generation is perhaps the pinnacle of bioenergy technology. Carbon Capture and Storage removes CO_2 from exhaust gases produced by power stations. The CO_2 is then piped to depleted gas or oil fields or saline aquifers where it can be stored indefinitely. When used in conjunction with fossil fuel power stations this could result in power generation with very low carbon emissions.

However, if this technology is linked to a biomass fired power station, because the biomass fuel has removed carbon from the atmosphere during photosynthesis and because no CO_2 is emitted back to the atmosphere following combustion, the overall carbon emissions from the supply chain will be negative. Very few options exist that are able to produce carbon negative electricity. As a result, a very strong case can be made to invest in the development and deployment of CCS equipped biomass fired power stations. In the UK the commercialization of CCS linked to conventional power generation is being encouraged and the use of CCS linked to bioenergy plant being researched. It is likely that the commercial realization of this combination is still several years away.

Other technologies that are in the process of advanced research or commercial demonstration include the production of ethanol from lignocellulosic feedstock. This process typically involves enzyme or acid hydrolysis of hemicelllulose and cellulose into saccharides and then a fermentation process to convert these to ethanol. The requirements of this process could influence species choice or even tree breeding programmes as the emphasis in this pathway is on sugar yield rather than standing volume and the two are not always linked.

The Fischer-Tropsch process that can be used to produce synthetic diesel from wood also has some promise and was used in Germany during the Second World War to produce diesel substitutes from coal. This process requires a steady supply of 'synthesis gas' (a mixture of carbon monoxide and hydrogen) produced by the gasification of solid biomass. This is proving difficult to achieve reliably at industrial scales at present and is the subject of much research and development.

Pyrolysis, the precursor to gasification in the combustion process, also offers some promise and enables processors to increase the energy density of biomass derived fuels prior to transportation. Pyrolysis oil is a versatile energy vector and can be burned to produce heat and power or upgraded via gasification and Fischer-Tropsch to a transport fuel.

Although these technologies offer considerable promise, it is the deployment of current, proven technologies is most likely to support the development of woodfuel supply chains over the next 30 years or so. As energy prices rise and fuel production businesses become more efficient, it is likely that the sector will become less reliant on incentives and other support measures. This will be a sure sign that we are moving away from fossil fuels towards green economy and will pave the way towards the adoption of more advanced energy production technologies.

In conclusion, woodfuel markets are beginning to unlock the latent timber resource in UK woodlands are all also increasing interest in woodland creation. Woodfuel markets are also helping to improve woodland habitats through increased levels of woodland management. Although UK grown wood will never meet all of our energy needs it can play an important contribution alongside other forms of biomass grown both domestically and sourced from international markets. The exposure of more woodland owners, end users and members of the public to woodfuel supply chains based on sustainable woodland management could help bring about a step change in woodland management and how we think about energy procurement. It is likely that in the longer term energy markets will lead to a more robust, higher profile forestry sector.

References

A Woodfuel Strategy for England (2007) http://www.forestry.gov.uk/pdf/fce-woodfuel-strategy.pdf/$file/fce-woodfuel-strategy.pdf. Accessed 30 Sept 2013

Combating Climate Change – A Role for UK Forests (2009) http://www.forestry.gov.uk/readreport. Accessed 30 Sept 2013

Domestic Energy Crops; Potential and Constraints Review (2012) https://www.gov.uk/government/uploads/system/uploads/attachment_data/file/48342/5138-domestic-energy-crops-potential-and-constraints-r.PDF. Accessed 30 Sept 2013

Forestry Commission Sawmill Survey (2011) http://www.forestry.gov.uk/website/forstats2011.nsf/0/6322930083F37DA88025731E0047F672. Accessed 30 Sept 2013

Matthew R, Mortimer N, Mackie E, Hatto C, Evans A, Mwabonje O, Randle T, Rolls W, Tubby I (2012) Carbon impacts of using biomass in bioenergy and other sectors. https://www.gov.uk/government/uploads/system/uploads/attachment_data/file/48346/5133-carbon-impacts-of-using-biomanss-and-other-sectors.pdf. Accessed 30 Sept 2013

Position Statement by Wildlife and Countryside Link on the Forestry Commission's Woodfuel Strategy for England (2009) http://www.forestry.gov.uk/pdf/eng-woodfuel-linkstatement-030709.pdf/$file/eng-woodfuel-linkstatement-030709.pdf. Accessed 30 Sept 2013

Timber Price Indices Data to September 2011 (2012) http://www.forestry.gov.uk/pdf/tpi201109.pdf/$FILE/tpi201109.pdf. Accessed 06 June 2012

UK Wood Production, in Forestry Commission Statistics 2013 (2013) http://www.forestry.gov.uk/website/forstats2013.nsf/LUContents/88BDD8FEA0D881448025734E004F27BB. Accessed 30 Sept 2013

Bioenergy Opportunities from Forests in New Zealand

Peter Hall and Michael Jack

Abstract An analysis on New Zealand's bioenergy resources found that plantation forest derived and other wood residues are the largest bioenergy resource now and into the future (2040).

New Zealand's forest and wood processing industries are based almost entirely on intensively managed plantations (~1.75 million hectares) of introduced species (*Pinus radiata* (89 %)), Douglas Fir (6 %) and a mixture of other species including eucalypts (1.5 %). The location, productivity and age class distribution of the resources is described in a national dataset, allowing prediction of future harvest volumes.

From the data available it is possible to determine supply trends at a national and regional level. There is a clear increase in wood supply between 2010 and 2030. Given the mix of logs that will occur, a large increase in the volume of chip grade logs is expected (up to 3.0 million tonnes per annum). Further, there is not expected to be a commensurate increase in uptake of these logs by the pulp and paper or reconstituted panel industries.

The highest priority for New Zealand in terms of fossil fuel substitution is liquid fuels for transport, and specifically a drop-in diesel. We have other options for electricity and heat production. Given that 97.7 % of our existing plantation resource is softwood this presents challenges for some biomass to liquid fuel conversion routes, such as enzymatic hydrolysis.

The volumes of residues and pulp logs potentially available could theoretically produce sufficient volumes of liquid fuels to meet 5–6 % of total liquid fuel demand or 15–16 % of diesel demand.

P. Hall (✉) • M. Jack
Scion, Rotorua, New Zealand
e-mail: Peter.Hall@scionresearch.com

T. Fenning (ed.), *Challenges and Opportunities for the World's Forests in the 21st Century*, Forestry Sciences 81, DOI 10.1007/978-94-007-7076-8_29,
© Springer Science+Business Media Dordrecht 2014

   Studies of the potential for forestry derived bioenergy found that there would be significant environmental benefits from establishing a larger forest resource on low productivity hill country grazing lands, and that this resource could provide a very substantial part of New Zealand's future liquid fuel demand.

1   Introduction

The combination of energy challenges, population density and land resources in New Zealand suggests that one of the most important future renewable energy opportunities for New Zealand is the large-scale production of transportation fuels from woody biomass.

   New Zealand is a small (26.7 million hectare, population 4.4 million) island nation, with a developed market economy in the South Pacific. The economy is largely based on production and export of primary produce (dairy, meat, forestry, fishing, wool, fruit and vegetables). Primary produce makes up three quarters of our exports and forestry is the third largest earner, behind dairy and meat (Statistics New Zealand 2010). Per capita energy consumption is high (similar to France and South Korea) especially in transport fuels (ranked 11th globally).

   Energy supply and environmental issues facing New Zealand are similar to the rest of the world including;

– Tight oil supply and rising prices
– Concerns over our greenhouse gas emissions and climate change

   New Zealand has a large (~70 %) part of its electricity supply produced from renewable resources (hydro, geothermal, wind) with a government strategy to increase this to 90 % over the next 20 years (MED 2011a).

   Heat energy is largely (56 %) supplied from gas and coal resources, with 28 % from biomass (mostly in the wood processing industry) and ~10 % from geothermal (MED 2010). Coal resources are large and coal is a relatively cheap fuel for industrial heat. Gas supply has been relatively cheap but prices are rising as the giant Maui field declines (de Vos and Heubeck 2009). New gas discoveries are being sought.

   New Zealand has a number of small oilfields. Production is mostly exported as the one domestic oil refinery is not suited to the type of crude they produce. The currently producing wells are expected to peak in the near future and domestic oil production will decline rapidly unless new discoveries are made (de Vos and Heubeck 2009). The government is encouraging new exploration (MED 2011a).

   Beyond domestic resources of energy, New Zealand is dependent on imported oil (~85 % of liquid fuels is derived from imports) (MED 2010), and we are a small market in a remote part of the South Pacific at the end of a long supply chain, which implies potential vulnerability in a supply constrained world.

   Of New Zealand's greenhouse gas (GHG) emissions around half come from agriculture (ruminant animals and fertiliser). Of our energy related GHG emissions 54 % are from liquid fuels, 29 % from natural gas and 12 % from coal, (MED 2011b).

There is clearly scope to reduce energy related GHG emissions from greater use of renewable energy, including biofuels, especially in transport.

This energy and GHG emissions structure leads to a focus on finding alternatives to oil and gas for liquid fuel and heat supply.

New Zealand's low population density also means that there are significant opportunities for bioenergy to play a significant role in meeting renewable transport fuel challenges.

Residual and waste biomass resources would only be able to meet a small percentage of transport fuel demand (Scion 2008a). However, use of effluents and municipal wastes for energy has significant environmental upsides in terms of reduced GHG emissions and reduced waste discharge (Scion 2008b), and so should be a short term development priority.

New Zealand's largest potential source of biomass for energy is New Zealand's sustainably managed plantation forestry estate (New Zealand's remaining native forests are protected from logging). In addition, New Zealand has a significant quantity of low to moderate productivity land that could be afforested to provide significant biomass resource without significantly impacting New Zealand's most valuable export industries which utilize high-value grazing and cropping land. It is estimated that utilizing 1.0 million hectares or 12 % of the low productivity land could produce a biomass resource capable of supplying 30 % of New Zealand's transport fuel by 2040, using reasonably conservative conversion factors (BANZ 2010).

We present a summary of a recent analysis of New Zealand's potential future energy challenges and the potential of bioenergy to meet some of these challenges. As suggested above, this analysis points strongly towards a strategy of large-scale production of transportation fuels from woody biomass and wastes.

Forestry for biofuels is seen as a low risk option as the wood can be used for fuel, lumber, reconstituted wood products, carbon sequestration or a mix of these. There are environmental benefits from afforestation of hill country and there is potential to have food production (cattle grazing) of the same land for at least part of the rotation.

In Sect. 2 we briefly outline New Zealand's current and future energy supply and demand options. Section 3 discusses New Zealand's residual biomass. Sections 4 and 5 covers New Zealand's largest biomass resource, wood from plantation forests, and the potential for greater future afforestation, including environmental benefits. Section 6 gives a brief summary of wood to liquid fuel conversion technologies trialled in New Zealand. Sections 7 and 8 present the challenges we face and a summary.

2   Energy Demand / Supply

Electricity demand in New Zealand (NZ) is ~149 PJ per annum or 41,392 GWh (MED 2010), and on average around 60 % of this demand is met by hydroelectric generation. Whilst demand is increasing, NZ also has significant renewable

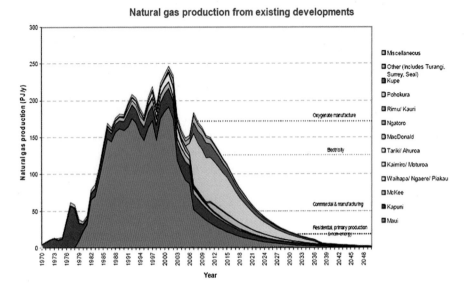

Fig. 1 New Zealand's proven gas resources (de Vos and Heubeck 2009)

resources that can be harnessed to generate more electricity including; further hydro development as well as extensive geothermal, wind and marine (wave and tide) resources which are yet to be fully exploited (NIWA 2009). Recent major developments have been from geothermal and wind, with investigation of marine projects underway. Burning biomass to generate electricity is a low priority at a national level, and will probably only occur where there are co-generation opportunities or site specific drivers such as remote communities or off-grid demands.

Heat demand is circa 195 PJ per annum. Most of this is met from coal and gas, with 55 PJ coming from wood and wood processing, including pulp mill black liquor (MED 2010) Black liquor is the lignin stripped from the wood during pulping; this is distilled and then burned in a recovery boiler to extract chemicals and to provide process heat. Firewood for domestic heating is a small contributor, at around 2–3 PJ per annum. New Zealand has extensive coal resources and this is not expected to a constraint in the medium term. Gas is used extensively for heat and for electricity generation. Gas supply is an issue in the near term and may be a driver of bioenergy development unless current exploration for new resources is successful. New Zealand's gas reserves are shown in Fig. 1. A key energy resource for New Zealand has been the massive Maui gas field which has provided the bulk of the gas supply for 30 years; this field is now in decline.

New Zealand has a small amount of domestic oil production, and most of it is exported, as the single oil refinery in New Zealand is tuned to run on imported crudes that are heavier than those we produce domestically. Consumption of liquid fuels is ~215 PJ per annum. We import over 90 % of our oil needs and around two thirds of our transport fuels are refined in New Zealand with the rest imported as refined product.

In an analysis of liquid biofuel use in NZ the Parliamentary Commissioner for the Environment (PCE 2010) concluded that in New Zealand we should;

- Focus on biofuels that benefit the environment
- Invest in biofuels that can be produced in large quantities, and in New Zealand's case this means wood and other ligno-cellulosics (straw residues and possibly miscanthus) as the resource
- Focus on drop-in biodiesel because;
  - Diesel for heavy machinery and freight drives our primary production industries and is thus more important to the economy than petrol for private transport
  - Drop in substitutes will be possible and desirable as blending and infrastructural incompatibility issues are avoided

We support these findings and would add jet fuel as product to focus on as it is critical to the air transport industry.

3 Residual Biomass Resources

In 2007, major studies of New Zealand's current and potential energy resources were commissioned by the government – Bioenergy Options for New Zealand and New Zealand's EnergyScape.

A summary of the biomass/ bioenergy resources derived from these studies is presented in Table 1. These figures represent those residues and wastes that are not currently being used.

Table 1 New Zealand's potential bioenergy resources, PJ per annum (Scion 2008)

Residual waste type/ source	2005 PJ p.a.	2030 PJ p.a.
Forest residues	14.6	34.4
Wood process residues	7.0	9.1
Municipal wood waste	3.5	2.2
Horticultural wood residues	0.3	0.3
Straw	7.3	7.3
Stover	3.0	3.0
Fruit and vegetable culls	1.2	1.2
Municipal biosolids	0.6	0.7
Municipal solid waste, landfill gas	1.9	2.3
Farm dairy	1.2	1.2
Farm piggery	0.1	0.1
Farm poultry	0.0	0.0
Dairy industry	0.4	0.4
Meat industry (effluent only)	0.5	0.5
Waste oil	0.2	0.2
Tallow	3.6	3.6
Total	**45.9**	**66.5**
Available biomass as % of consumer energy	8.5	9.2
Available biomass as % of primary energy	6.6	7.3

Residues Are Not Enough for A Bioenergy Future

As energy demand (especially for oil) rises in the future, biomass wastes and residuals are insufficient to meet more than a small amount of the energy demand.

Woody biomass is the largest biomass resource, and forest and wood processing residues are the largest contributors.

These insights led to a concept strategy being developed that outlined the potential of wood from new purpose-grown energy forests. It became apparent that energy from forests could be a huge contributor to low carbon energy in a New Zealand context.

This concept envisioned 3.2 million hectares of forests, providing 100 % of New Zealand's liquid fuels and some heat fuel. NZ has 9.6 million hectares of hill country grazing, of which 0.8–1.0 million hectares is highly vulnerable to erosion.

Whilst there are a wide range of biomass resources available to use, and many of these such as municipal wastes and industrial effluents, have significant environmental benefits if consumed for energy, clearly the largest opportunity comes from woody biomass, which is over half the current bioenergy resource and with a rising volume and percentage contribution out to 2030.

There is some current use of forest harvest residues for bioenergy, but it is a small proportion (estimated at 12–14 %) of the total resource.

If the existing residual wood resources and future volumes from existing forests are to be used then a market needs to be developed. Currently wood energy (including black liquor from pulp and paper mills) provides ~55 PJ per annum of primary energy, mostly as heat in the wood processing industry. Historically increased use of wood as fuel outside of the wood processing industry has been constrained by several key issues including;

– Cheap fossil energy from coal and gas
– Variable quality of biomass fuels provided for heat production (moisture and ash content)
– Uncertainty of supply (as forest and wood processing residue production is dependent on the health of domestic and international lumber and log markets)
– Cost of infrastructure change

4 New Zealand's Forests

New Zealand's forest and wood processing industries are based almost entirely (over 99.9 %) on intensively managed plantation forests of introduced species. The plantation forest estate is approximately 1.75 million hectares, with 89.3 % of the area in *Pinus radiata*, 6.3 % Douglas Fir, 2.1 % other softwoods, 1.5 % *Eucalyptus* species and 0.7 % other hardwoods. Thus, 97.7 % of the wood resource (by area) is

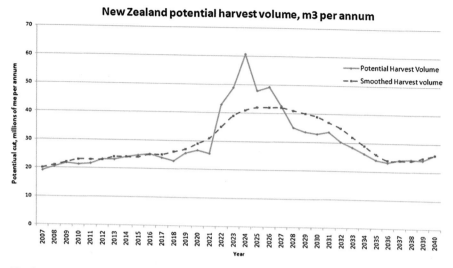

Fig. 2 New Zealand's potential forest harvest volume (MAF 2010)

softwoods. This resource is well described in terms of its area, location, yield and age class distribution, in the National Exotic Forest Description (NEFD) a national dataset maintained by the Ministry of Forests (MAF 2010) and associated maps and data sets (Ministry for the Environment 2007). This data can be used to predict the volume of material that is potentially available for harvest; at national (Fig. 2), regional and territorial authority levels. The most common species, *Pinus radiata*, is typically grown on a ~28 year rotation (25–32).

For the year to March 2010 the forest harvest volume was 20.5 million cubic metres and to March 2011 the figure will be around 25.5 million cubic metres.

New Zealand's forest industry faces some significant challenges in the next 30 years as represented by Fig. 2. The first challenge is 2020–2025, when the volume of wood potentially available to harvest trebles in a space of 3–4 years.

The second challenge is that post 2025; the potential cut goes into steep decline, and is back to current (2011) levels by around 2035.

These changes reflect the age class distribution of the established forests and are largely driven by a planting boom in the mid 1990s. It is unlikely that the forest will (or even could) be harvested at the peak of the potential rate. Whilst some smoothing of the harvest volume (dotted line = approximation) will occur and may ease the supply glut/infrastructure bottleneck situation, we still face a harvest volume that could double and halve over a 15 year period.

However, there will still be a substantial increase in forest harvest (a total cut of up to 40 million cubic metres per annum is realistic), with a consequent increase in harvest residues (as indicated in Table 1) and there will also be an increase in the volume of low grade/low value logs (pulp/ chip grade). This increased volume of low grade logs could be in the order of 4.0 million tonnes per annum (~20 % of the harvest), with some doubt over the ability of the existing pulp, paper and reconstituted panel industries to consume this material. The existing New Zealand pulp mills are ageing, at capacity, and there are no plans to expand these or build new ones. This is a reflection of the dropping demand worldwide for all paper products except tissue.

The conclusion from this is that there is a potential resource of 3–4 million cubic metres per annum of softwood biomass available for non-traditional uses. This material has an energy equivalence of up to 27.7 PJ (3.6 % of New Zealand's primary energy demand of 776 PJ).

5   Potential for Future Afforestation

An analysis of the potential for further afforestation for production of biomass for energy was undertaken as part of the Bioenergy Options for New Zealand project (Scion 2009, various authors).

This study looked at the energy supply volume, cost, land use change and associated environmental and macro-economic impacts of four large-scale afforestation scenarios, for liquid biofuels production. These afforestation scenarios ranged from 0.8 to 4.9 million hectares.

These figures need to be seen in the context of other national level data; NZ has 9.6 million hectares of hill country grazing land, of which 0.8–1.0 million hectares has been identified as being at high risk of erosion, and 2.5 million hectares of which is earning less than $100 per ha per annum in its current use (typically sheep and beef grazing).

The scenario seen as realistic but still with significant impact is scenario 2 (Table 2, shaded), with ~1.8 million hectares of new afforestation (Map – Fig. 3).

In Table 2;

- LPe = litres of petrol equivalent
- TEB = total extractable biomass = total recoverable stem volume + bark + (branch biomass × 0.8) + (0.8 of the estimated 15 % of the above ground biomass in currently unmerchantable stem breakage)

The large-scale afforestation scenarios were based on the assumption that the crop would be radiata pine. This does not mean that all the afforestation would or should be this species. It is however the species that has the most information available at a national level on its productivity, across a wide range of sites, thus allowing more detailed and accurate predictions than is possible for other species.

The forest management regime (high stocking, minimum tending) assumed in the new afforestation scenarios gives market options for the logs produced other than 100 % to energy, for example, 56 % sawlog and 44 % chip grade logs. It also gives high volumes of carbon sequestered/ stored (Fig. 4).

There are substantial variations in afforestation areas between regions, for a wide range of reasons. New afforestation areas are limited in Waikato, Bay of Plenty (Central North Island) and the West Coast of the South Island. A large percentage of Waikato land area is high value cropping or dairying suitable. The Bay of Plenty already has a high proportion of land under forest and the West Coast has significant areas in the conservation estate and their small plantation forests have performed poorly for climatic, soil drainage and fertility reasons.

Table 2 Summary of potential biomass and liquid fuel production (assumes sustained yield harvest on 25-year rotation with earliest harvest in 2037)

Region	Scenario 1 / 0.8		Scenario 2 / 1.8		Scenario 3 / 3.3		Scenario 4 / 4.9	
	TEB p.a. m³ millions	LPe, p.a. millions	TEB p.a. m³ millions	LPe, p.a. millions	TEB p.a. m³ millions	LPe, p.a. millions	TEB p.a. m³ millions	LPe, p.a. millions
Northland	0.29	25.2	1.08	94.2	3.07	267.1	8.38	728.8
Auckland	0.01	0.9	0.51	44.3	1.15	100.6	2.47	214.8
Waikato	0.23	20.4	4.39	382.0	11.35	987.4	16.88	1,468.3
Bay of Plenty	0.02	2.3	0.44	39.4	1.24	107.8	2.29	199.2
Gisborne	0.26	22.9	6.26	544.8	10.93	950.7	13.26	1153.6
Hawke's Bay	0.51	44.9	8.47	736.8	16.86	1,466.3	20.12	1,750.1
New Plymouth	0.52	45.4	2.60	226.5	3.83	333.6	4.84	421.5
Manawatu-Wanganui	1.35	117.7	16.08	1,389.2	25.93	2,252.2	29.80	2,591.4
Wellington	0.36	31.4	5.73	499.0	7.97	693.2	9.76	849.4
Tasman	0.10	8.8	0.81	710.4	1.24	108.3	1.70	148.4
Nelson	0.00	0.1	0.11	9.3	0.13	11.7	0.14	12.9
Marlborough	0.88	77.2	3.24	288.1	4.16	362.0	5.58	485.7
West Coast	0.14	12.5	0.34	30.1	0.94	81.9	1.29	112.5
Canterbury	9.90	861.2	12.14	1055.7	18.86	1,640.2	27.16	2,361.7
Otago	6.47	563.4	8.27	714.3	13.12	1,141.5	17.54	1,525.4
Southland	1.49	129.9	3.00	261.0	5.79	503.7	7.39	642.9
Total*	22.59	1964.2	73.55	7,039.1	126.63	11,011.2	168.67	14,666.1

In Table 2;
- LPe = litres of petrol equivalent
- TEB = total extractable biomass = total recoverable stem volume + bark + (branch biomass x 0.8) + (0.8 of the estimated 15% of the above ground biomass in currently unmerchantable stem breakage)

However, other regions (East Coast, Hawkes Bay, Manawatu-Wanganui – East and lower North Island) have large areas that could go into forests as these regions have large areas of hill country grazing on land that is prone to erosion.

Table 3 summarises the environmental impacts of the afforestation scenarios on some key variables (erosion, water yield, N leaching and Carbon) (Scion 2009a).

For Scenario 2 the net gain in carbon stock, in 2050 versus 2005, was 651.1 million tonnes of CO_2 equivalent (Fig. 4). Potential displaced emissions were 3.4 million tonnes of CO_2 equivalent per annum by 2035.

All the afforestation scenarios provide increased carbon stocks. The two mid-range scenarios (2 and 3) increase carbon stocks by 650 million tonnes and 1,188 million tonnes respectively. At the lower scale of planting scenario 1 increases CO_2 equivalent stocks by 207 million tonnes and at the higher end scenario 4 increases CO_2 equivalent stocks by 2039 million tonnes. The two mid-range scenarios might also provide reductions in agricultural GHG emissions of around 3.3 and 7.3 million tonnes of CO_2 equivalent per annum, and transport GHG emissions reductions of 12.1 and 21.9 million tonnes of CO_2 equivalent per annum.

New Zealand's net GHG emissions in 1990 were 41.299 million tonnes of CO_2 equivalent and in 2006 were 54.951 million tonnes (MED 2008), an increase of 13.655

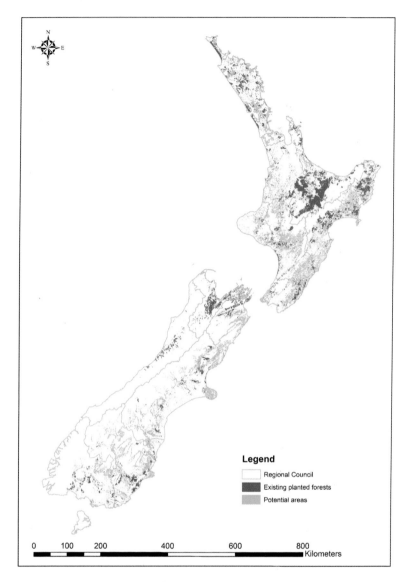

Fig. 3 Example afforestation scenario map - showing existing forest estate and potential additional ~1.7–1.8 million hectares of plantation forest

million tonnes of CO_2 equivalent. The increase in carbon stocks from the afforestation scenarios is substantial in comparison to these figures. For example, the increase in carbon stocks for Scenario 2 of 651 million tonnes of CO_2 equivalent is equal to; about 11 years of net emissions, or 47 years of the 1990–2006 difference in net emissions.

There are also beneficial impacts on indigenous biodiversity. The land being targeted for afforestation is typically in exotic grasses, with low biodiversity and the

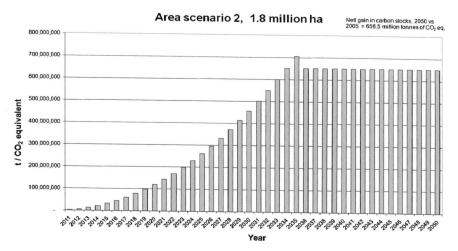

Fig. 4 CO$_2$ equivalent of carbon stock increase for forest area scenario 2, 1.8 million hectares

Table 3 Environmental impacts (national level) of afforestation scenarios

Scenario No. & millions of ha of new forest	GHG impacts, Reduced emissions, millions of tonnes, CO$_2$e	Stored carbon, millions of tonnes, CO$_2$e	% reduction in erosion	% reduction in N leaching	% reduction in water yield
1. 0.8	5.02	207.8	1.1	0.3	0.9
2. 1.8	15.49	651.1	8.0	3.4	2.6
3. 3.2	29.21	1188.5	16.6	8.4	5.1
4. 4.9	37.29	2039.7	20.2	12	7.2

flora and fauna is typically introduced species. The establishment of forests, even those with introduced tree species as the crop, has been shown to lead to greater biodiversity, with many of the species in the sub-canopy being native scrubs with associated indigenous birds and insects (Pawson 2009).

Further, for a given estate area, some of the land could be retained as permanent carbon forests, some logged, and there are a range of options for marketing the material produced. These market options reduce the financial risk involved in growing the crop.

5.1 Softwood Conversion

The nature of our forest resource is that it is largely, and will be for the foreseeable future, dominated by softwoods (principally *Pinus radiata*). Softwoods have higher lignin contents and are more recalcitrant than hardwoods when subject to a biochemical conversion process (enzymatic hydrolysis to ethanol). Thus we need (and have) a research program focussed on developing processes and treatments that overcome this recalcitrance.

Table 4 Comparison of possible *Pinus radiata* regimes and crops (Radiata Pine Calculator Version 3.0 Pro)

Regime type	Initial and final stockings	Total merchantable. Volume, m³/ha (at age 25)	Piece size, m³	% sawlog	% pulp
Conventional[a] sawlog focussed	833 / 364	771	2.11	90	10
Compromise	700 / 518	876	1.69	86	14
Biomass, volume focussed	833 / 667	910	1.36	56	44

[a] Thin to 400 sp/ha at mean crop height of 14 m, approximately age 7

5.2 Yields

There are many items that contribute to the cost of a log product, or a product derived from wood. A major influence is the cost of capital, other influences are crop and conversion yields. Identification and development of conversion processes that give the maximum energy yield from a given amount of wood is critical to the success of a biofuel development.

Currently many New Zealand plantation forest management regimes target the growth of large diameter, high quality sawlogs, which gives greater market value in the current paradigm, but in doing so these regimes sacrifice total volume production. Some compromise on the traditional approach needs to be considered in order to fully enable a biofuels future. Achieving this will not be easy given the time frames of forest rotations and the risk involved in taking a new approach to forest growing. However, modelling suggests that yield gains are possible with minor changes to the approach to forest management (Table 4).

The biomass focussed regime gives a volume increase of 18 % over the conventional regime, but only ~4 % over the compromise regime. The compromise regime would give ~14 % more volume than the conventional regime. Modelling runs for the figures in Table 4 assumed a productive Central North Island site with re-establishment on harvested forest cutover.

5.3 Land Use Change

New Zealand has seen significant land use change over the last 800 years due to two waves of immigration and settlement. Prior to human occupation most of the land mass was covered in forest (with the exception of high altitude grasslands, scrublands and mountain ranges). Much of this forest has been removed and converted to grazing land (Fig. 5).

The remaining indigenous (native) forest cover (Fig. 5) is now largely protected and is managed by the Department of Conservation as national parks and reserves. It will not be available for timber or energy use. The expansion of the plantation

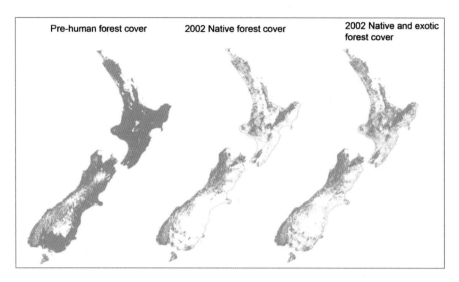

| Pre-human forest cover | 2002 Native forest cover | 2002 Native and exotic forest cover |

Fig. 5 Historical and current forest cover

forest estate would get much of the previously forested land which was cleared for grazing back into a ground cover much more like its natural state. Further, stock grazing (typically beef cattle) within plantations forests is possible and has been used throughout New Zealand on a variety of plantation forest sites. Thus hill country forests can produce food as well as logs, for at least part of the rotation.

5.4   Economic Welfare

One of the most important questions around bioenergy would be: what are the macro economic effects of large scale forestry for energy on economic welfare measures, such as standard of living?

In most of the scenarios considered biofuels lead to a decrease in productive efficiency and this implies a reduction in economic welfare. However, the production and use of biofuels reduces CO_2 emissions, so if there is an international price on carbon, New Zealand's liability to purchase offshore emission units is ameliorated. This generates a gain in Real Gross National Disposable Income. In addition, the increase in allocative efficiency reflected in increases in the terms of trade for high oil prices also leads to increases in economic welfare.

Further, economically it is better to use only the lower value logs for energy and higher grade logs for sawn lumber as opposed to all of them for energy. Forests should be regarded as having multiple values, including environmental and recreational, and many potential end-uses for the wood, including Carbon. These values offer risk mitigation for the afforestation for energy strategy.

Volatility of oil prices also has an impact on the viability of a biofuels industry, and its potential for producing economic gains. Specifically, it was shown (Stroombergen and McKissack 2009) that when oil prices exceed the costs of biofuels the contribution of biofuels to economic welfare could potentially quickly outweigh the losses to economic welfare that would occur if regulation force the blending of biofuels and fossil fuels when oil prices are low enough to undercut the cost of biofuels.

The costs of not developing a biofuels option may be high if a sudden increase in oil price occurs, as in 2008 and 2011. New Zealand's exchange rate versus the US dollar is also an important factor in fuel price at the pump, for much of the last 2 years (2009–2011) the New Zealand dollar has been at 0.7–0.8 well above the long run average of 0.65, insulating consumers from some of the impact of the recent increases in oil price.

The Bioenergy Association of New Zealand recently published a Bioenergy Strategy for New Zealand (BANZ 2010), that aims for biofuels from forests providing 30 % of New Zealand's liquid fuel demand in 2040, equivalent to 3.243 billion litres. An analysis of this strategy by BERL (2011) estimates that this strategy would increase GDP by 1.2 %, or $6.1 billion dollars.

Additional Wood Processing Potential

New Zealand exports significant volumes of saw logs in unprocessed form. In the year to June 2011 this figure peaked at ~12.2 million tonnes p.a. or 50 % of the total annual harvest.

If New Zealand's solid wood processing industry expanded to process even half this log export volume then there would be a supply of sawmill residues of ~3 million cubic metres (~20 PJ) per annum available for use as bioenergy feedstock.

Summary

Taking all these findings together we get to the point where we could look to forests to provide the feedstock for liquid fuels (biodiesel/jet fuel) production as a major focus of bioenergy development.

6 Processes Suitable for Converting Woody Biomass to Liquid Fuels Investigated in New Zealand

6.1 Acid Hydrolysis (AH)

Acid hydrolysis of wood to produce ethanol is an established technology and New Zealand had a significant research programme including a pilot plant in the late 1970s. However, after the 1970s oil shock, research and development on this topic ended in New Zealand, although development has continued elsewhere (Bluefire Ethanol).

6.2 Enzymatic Hydrolysis (EH)

Production of ethanol from wood and other ligno-cellulosic biomass via enzymatic hydrolysis is a topic of global R&D. Scion in New Zealand is working on EH of softwoods, principally *Pinus radiata*, which poses significant challenges due to the recalcitrance and higher lignin contents inherent in softwoods. Innovative pre-treatments are used to improve cellulose availability. The lignin left over after EH could be used for biochemical/ biopolymer production or for production of heat and electricity for the process.

6.3 Gasification and FT (GFT)

Production of advanced biofuels (biodiesel) via gasification of woody biomass and Fischer Tropsch catalysis has potential in a New Zealand context, especially if the economic scale of the plant can be reduced. R&D on this topic is underway in NZ (University of Canterbury). Feasibility studies of biomass GFT plants or co-firing of biomass with lignite or coal have been undertaken, but no developments are currently planned. NZ has a very large lignite resource (6.3 billion tonnes recoverable) in Southland and Otago and using this resource for liquid fuel production via GFT has been considered. Co-firing of biomass and lignite would help with the poor GHG profile of a pure lignite fired GFT plant.

6.4 Pyrolysis and Hydro-treating / Refining

Several groups have, or are, looking at pyrolysis of woody biomass to produce bio-crudes, liquid boiler fuel or advanced liquid fuels (typically jet fuel or diesel replacements).

6.5 Competing Conversion Technologies

Globally there is significant R&D on creating liquid biofuels from woody and ligno-cellulosic biomass. There is no clear technology winner and there is a need for unbiased analysis of the fundamentals (thermodynamics and efficiencies) of a range of technologies to determine which will have the greatest potential in a New Zealand context (feedstock, costs, fuel type, economic scale) – which may assume a wood to liquid fuel (biodiesel) paradigm.

7 Challenges

New Zealand thus has a number of challenges and opportunities that could be brought together to provide an energy solution based on forests.

Energy Challenges;

– High per capita liquid fuel consumption
– Limited domestic oil resources
– Rising oil/ liquid fuel prices
– Volatile oil prices
– Potential constraints on oil supply

Environmental Challenges

– High GHG emissions
– Erosion on steep grazing lands
– Water quality from grazing land run-off

Forest Challenges

– Expanding resource with significant increase in residues and low grade logs (2011–2030)
– Inconsistent wood supply volumes (declining harvest post 2030)

Technology development or identification

– Determining which wood to liquid fuel conversion technology is the most cost effective for New Zealand.

Making it happen

– Establishment of a large forest resource / biofuels supply will not happen without significant long-term buy-in and commitment from industry, landowners and government.

Opportunities

– To meet some of the national liquid fuel demand / increasing fuel cost issues, with low carbon liquid biofuels made from existing and potential domestic forest resources, with environmental and economic benefits.

8 Summary

New Zealand has a large and well established forest industry, based on exotic softwood plantations. This resource will provide an expanding harvest over the next 15–20 years.

Use of woody biomass as a heat fuel is well established in the wood processing industry and is slowly expanding, but there is still a significant but geographically distributed wood residue resource that is unutilised.

There is significant potential to utilise marginal lands to expand the forest resource and stabilise the harvest volume at a level of 40 million cubic metres per annum. This material could provide a base for a domestic biofuels industry.

Changes to the wood processing industry (expansion) could increase the wood processing residue volume, creating a substantial bioenergy feedstock.

A key future energy need is likely to be liquid fuel and the priority is diesel for primary industries.

Technology development in New Zealand needs to focus on identifying which conversion route will best take the very large potential of the woody biomass resource through to the priority demand of an advanced biodiesel.

Forestry for bioenergy represents a low risk option, in that the crop, whilst slow to establish and mature can be used for many other products; lumber, reconstituted wood products, solid fuels and Carbon storage.

Establishment of forests on hill country grazing land currently providing low economic returns could improve returns to land owners as well as providing improved environmental outcomes (reduced erosion etc.).

Multiple-use management of forests is common and there are many instances of stock grazing within plantation forests and thus the food versus fuels issue which occurs when farm land is converted to energy crops is mitigated by further developing a food and fuels paradigm in a forestry context.

References

BANZ (2010) Bioenergy Association of New Zealand – New Zealand bioenergy strategy. Sep 2010

BERL (2011) Business and Economic Research Limited. Preliminary economic impact assessment of the New Zealand bioenergy strategy

Bioenergy Options for New Zealand. http://www.scionresearch.com/general/science-publications/science-publications/technical-reports/bioenergy/bioenergy-options

de Vos R, Heubeck S (2009) New Zealand's energy situation – medium term supply–demand tension review. In Bioenergy options for New Zealand – transition analysis, Scion 2009, pp 151–210

MAF (2010) A national exotic forest description as at 1 April 2009. Ministry of Forestry. http://www.maf.govt.nz/news-resources/statistics-forecasting/statistical-publications/national-exotic-forest-description-provisional-rel.aspx

MED (2008) Ministry of Economic Development. New Zealand's energy green house gas emissions 1990–2007

MED (2010) New Zealand's energy data file. Ministry of Economic development, 2009

MED (2011a) Ministry of Economic Development. New Zealand energy strategy 2011–2021. Developing our energy potential

MED (2011b) Ministry of Economic Development. New Zealand's energy green house gas emissions 2011

Ministry for the Environment (2007) The New Zealand land cover database. LCDB2, 2007

New Zealand's EnergyScape. http://www.niwa.co.nz/our-science/energy/research-projects/all/energyscape/deliverables

NIWA (2009) New Zealand's EnergyScape basis review, Section 2. Renewable resources. National Institute of Water and Atmospheric Research

Parliamentary Commissioner for the Environment (2010) Some biofuels are better than others – thinking strategically about biofuels. http://www.pce.parliament.nz/publications/all-publications/some-biofuels-are-better-than-others-thinking-strategically-about-biofuels

Pawson S (2009) In Bioenergy options for New Zealand – analysis of large scale bioenergy from forestry; Productivity, land use and environmental and economic implications, Chapter 2 environmental impacts of large-scale forestry for bioenergy, biodiversity, Scion 2008, pp 97–107

Scion (2008a) Bioenergy options for New Zealand – situation analysis, biomass resources and conversion technologies, Peter Hall and John Gifford

Scion (2008b) Bioenergy options for New Zealand – pathways analysis; energy demand, pathways evaluation, life cycle analysis of biomass resources to consumer energy. Scion 2008 – various authors

Scion (2009a) Bioenergy options for New Zealand – analysis of large-scale bioenergy from forestry; land use, environmental and economic impacts, Scion 2009, various authors

Scion (2009b) Bioenergy options for New Zealand – transition analysis: the role of woody biomass from existing plantation forests, species options and drivers for change in energy supply, Scion 2009

Statistics New Zealand (2010) http://www.stats.govt.nz/infoshare/

Stroombergen A, McKissack D (2009) General equilibrium analysis of bioenergy supply from New Zealand's forest estate sand the impacts of volatile fuel prices. In: Scion, 2009b. Bioenergy options for New Zealand; transition analysis – the role of woody biomass from existing plantation forests, species options & drivers for change in energy supply, Chapter 4, pp 132–150

Creating the Wood Supply of the Future

Barry Gardiner and John Moore

Abstract Global demand for wood as a raw material is growing with a projected annual increase in industrial roundwood consumption of between 1.3 % and 1.8 % up to 2030. This rise is driven by the projected growth in the world's population and economic activity. Much of the increased consumption will be in the rapidly expanding economies of China, India and south-east Asia and the escalating use of wood for biomass, particularly in Europe. The potential to expand forestry will be limited in the region of highest growth (Asia and the Pacific rim, with the exception of China) because of competing land-uses and high population densities. In addition there is an ever increasing requirement for forests to provide a range of environmental services such as helping to provide clean air and water, protecting existing biodiversity and this has led to an ever expanding area of protected forests across the globe.

The result of these pressures on forestry is that there will be ever more reliance on managed forests, particularly planted forests, in order to satisfy this increasing demand for wood. While much of the product demand will remain as at present there will be an additional need for more wood in rapidly expanding sectors such as biomass and engineered wood products. This means that forests need to produce more wood per unit area and these wood products need to be more carefully designed to meet the increasing expectations of end users around product performance. This is partly driven by the performance levels of competing materials. Although this sounds daunting the methods and tools are available to make this a reality, but it will require forestry and the forest/wood chain to respond by adopting technologies and techniques that modernise the production and allocation of wood products along the whole production chain from forest to final end use.

B. Gardiner (✉)
Forest Research, Northern Research Station, Roslin, Midlothian, Scotland, UK

INRA-Unité EPHYSE, 33140 VillenaveD'Ornon, France
e-mail: barry.gardiner@bordeaux.inra.fr

J. Moore
Scion, Private Bag 3020, 3046 Rotorua, New Zealand
e-mail: john.moore@scionresearch.com

T. Fenning (ed.), *Challenges and Opportunities for the World's Forests in the 21st Century*, Forestry Sciences 81, DOI 10.1007/978-94-007-7076-8_30, © Springer Science+Business Media Dordrecht 2014

Expanding wood production in line with predicted global demand is entirely possible with the use of genetically improved material and management focussed on production and reducing losses from biotic and abiotic agents. The challenge is to do this in a manner that allows production to remain "sustainable" for the foreseeable future. At the same time the technology exists to make much more focussed use of the material from the forest with allocation decisions taking place as early as possible in the wood chain. In addition information on physical properties will be "tagged" to the material so that informed decisions can be made at every stage along the processing chain. The technologies available include aerial and satellite remote sensing (in particular with LiDAR), ground based scanning, acoustic technology, x-ray scanning, NMR scanning and Fourier Transform Infrared spectroscopy (FTIR). In this chapter we discuss how by combining improved productivity and improved allocation within the wood chain through the use of modern information systems it will be possible to meet the wood supply demands of the twenty-first century. The forest will become an integrated part of the wood chain with the volume and properties of the material well characterised and available at every stage of the journey from the forest to the final product.

1 Introduction/Background

In 2008 global wood usage was around 4.6 billion cubic meters. Of this around 40 % was used for wood fuel and 33 % as roundwood. The market in wood-based products has increased from $60 billion to $257 billion in the 20 years up to 2008 (FAO 2009a), mainly in wood panels and secondary processed wood products (SPWP). All the evidence suggests that global wood consumption will continue to increase (Fig. 1) despite the recent short-term reduction in demand due to the global financial crisis. Predictions suggest a trade of $450 billion by 2020 of which 40 % will be in SPWP. At the same time FAO (2010a) report that between 2000 and 2010 there was a net loss of forested area of 5.2 million hectares per annum from a global forest area of 4 billion hectares. This was mainly due to a large loss in primary forest (13 million ha/year), which has been partly compensated for by large scale commercial planting. The current area of planted forests is approximately 264 million hectares and is increasing by approximately 5 million hectares per annum.

Currently, 30 % of the world's forests are primarily used for wood and non-wood products (FAO 2010a). In the future the majority of wood will come from managed planted forests (plantations)[1] and dependence on natural forests will decline. This switch to planted forests is due to the fact that this appears to be the only way

---

[1] New forests are usually established through the planting of seedlings or the sowing of seeds. Subsequent regeneration of the forest may either be in the same way or through natural regeneration if the conditions are suitable. In the highly productive forests that have been recently established in many parts of the world, planting of seedlings is the primary method of establishment and regeneration, in part because this method can be used to introduce genetically improved material. However, natural regeneration is the traditional method of regeneration in the older managed forests that historically have provided a large proportion of the world's wood supply, such as Central Europe and North America, Fennoscandia, the Baltic States and Russia.

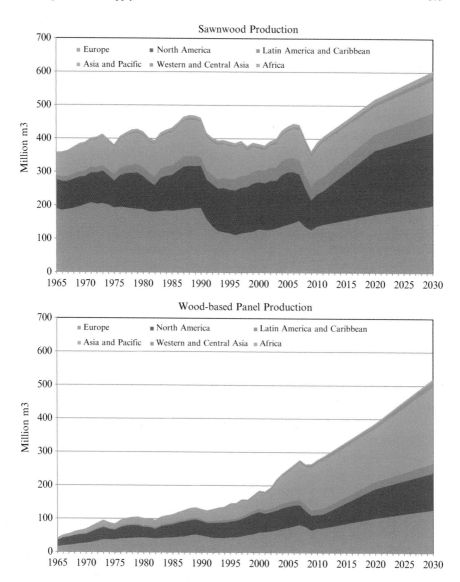

Fig. 1 Past and projected consumption of sawn wood and wood-based panels (reproduced from data in Jonsson and Whiteman 2008; FAO 2009a)

enough wood can be provided to meet the world's future requirements (Fenning and Gershenzonand 2002; Fenning et al. 2008). It is also partly due to mounting pressure to protect natural forests, which are increasingly valued for the ecosystem services they provide (e.g. biodiversity, and soil and water conservation) rather than for their provisioning services (FAO 2009a), even though some people argue that it is possible to maintain productivity from natural forests without deleterious effects (e.g. Putz et al. 2008). Already 50 % of the entire world's wood fibre comes from planted eucalyptus forests and it is forecast that the total area of planted forests will reach almost 450 million hectares by 2020 with 255 million hectares in the tropics (FAO 2007). In addition the protective and social role of forestry is increasing but it is difficult to quantify exactly what impact this will have on forest productivity. Ownership is also changing with a shift from public ownership, which is currently at 80 %, to communities, individuals and private companies (FAO 2009a).

In principle, the additional area of planted forests could meet global wood demand assuming no change in bioenergy consumption (Fig. 2). Currently, woody biomass provides 50 EJ (~10 % of world energy needs), which it makes it the fourth most important source of energy after coal, oil and natural gas. It is estimated that biomass could provide 270 EJ (Ladanai and Vinterbäck 2009). Annual global primary wood production is equivalent to 4,500 EJ for comparison.

Altogether this suggests that forestry is going to have major challenges to continue to meet the world's demands for wood for both those areas that have traditionally dominated (e.g. solid timber, pulp and paper, and panel products) and emerging end-uses (e.g. biomass, biomaterials and chemicals). An increasing percentage of this supply will come from planted forests, the majority of which will be in the tropics. The major issues are how to provide this wood material at competitive prices in a sustainable manner. Since an increasing level of material must come from planted forests the management of these forests and the utilisation of the material from them will have increasing focus.

2 Increasing Wood Production

There is a clear need for increased wood production, increased use of recycled wood and more efficient use of the wood that is produced. Sedjo (1999) sets out the possibilities for increasing the worldwide production of wood. He argued that planted forests offer the possibility of providing much of the world's wood needs while at the same time protecting natural forests. He identified 11 major regions that had actual or potential promise for industrial planted forests: Pacific Northwest, US south, Brazil (Amazonia), Southern Brazil, Chile, Australia, New Zealand, South Africa, Gambia-Senegal, Nordic-Sweden, Finland, Borneo and Indonesia. All of these except Gambia-Senegal have emerged as important areas containing planted forests and in addition there have been significant developments in the Iberian Peninsula and Uruguay. Planted forests are now actively being developed in China, Japan, Korea and Indonesia (Table 1). Most of these new forests are on former agricultural land.

Sedjo (1999) believed that planted forests will increase in importance and the major impediment is the need for long term investment. The trend at the moment seems to be for internal local investment rather than external overseas investment in those countries most rapidly developing their planted forests. A second impediment is environmental concerns and planted forests needs to show that they can offset the need to harvest natural forests and increase the provision of ecosystem services, such as enhanced biodiversity, improved water quality and erosion control, to degraded ex-agricultural land.

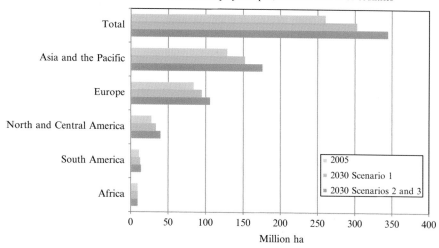

Current and projected planted forest area in 61 countries

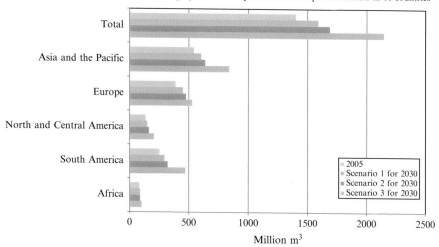

Current and projected wood production from planted forests in 61 countries

Fig. 2 Current and projected areas of planted forest and wood production from planted forests (Carle and Holmgren 2008)

Table 1 Planted forest area by country (From Sedjo 2010)

Country	Forest plantation (1,000 ha)
China	45,083
India	32,578
Russia	17,340
United States	16,238
Japan	10.682
Indonesia	9,871
Brazil	4,982
Thailand	4,920
Ukraine	4,425
Iran	2,284

If bioenergy consumption continues to rise as predicted, then the increasing area of planted forests may not provide on their own enough wood raw-material to meet world demand (see Ince et al. 2011) for an assessment of the impact on the US forest sector of an expansion in wood energy consumption). In addition the increasing need to protect natural forests puts additional pressure on planted forests. Therefore, in the same manner as agricultural production (FAO 2012) there is likely to be a squeeze between raw material demand and the land resource available to provide this raw material and the pressure is likely to be most intense in the poorer countries of the world (Tilman et al. 2011).

When there is an increasing pressure on resources there appear to be a number of philosophies that people take regarding the issue (Pretty 1997):

1. Business as Usual Optimism: The market will provide based on biotechnology, increasing productive area and a slowing down of demand as the world's population begins to stabilise.
2. Environmental Pessimism: The ecological limits to growth have already been reached and we will be unable to sustainably supply the increasing and changing demand.
3. Industrialised World to the Rescue: The developing world cannot manage and the modernised processes of the industrialised world will provide what is needed and help protect the natural environment in the developing world. This suggests that production will move to the industrialised world leaving the developing world with forest reserves.
4. New Modernism: There will be intensification in the use of existing land with much more focus on a "science-based" approach. The system will be more sustainable because it will be focussed on a smaller area.
5. Sustainable Intensification: Integrated use of a range of methods and technologies to manage pests, nutrients, soil and water. Degraded or marginal lands will be reintroduced into production. There will be increasing emphasis on using natural processes to substitute for external inputs (e.g. fertilisers).

The first philosophy appeared to be correct until recent developments when it became clear that this is not a sustainable solution in the long term and can only work when there is an essentially unlimited supply of raw material. The second philosophy may be correct on this occasion but it has consistently underestimated the ability of society to deal with resource demands over many decades. The third philosophy is no longer relevant with the huge industrial developments taking place in countries like Brazil and Indonesia. Therefore, it appears reasonable to concentrate on the final two philosophies: New Modernism and Sustainable Intensification.

2.1 New Modernism/Industrial Forestry

The argument of the new modernist is that by using the very best techniques available it will be possible to provide our wood raw material requirements from a reduced land area. This approach focuses primarily on the supply end of the forest-wood chain and maximising what can be produced from the land using a very technological approach.

As discussed above much of the increasing demand for wood is being provided by planted forests. These planted forests are intensively managed, have a high input/ high output approach making use of increased fertiliser input, improved genotypes of fast-growing trees (often deployed as clones), and potentially transgenic material. In agriculture the approach had led to transgenic crops being routinely grown (typically in soy, cotton, maize and rapeseed) and although transgenic trees are less common in forestry, they are now being commercially planted in, for example, China (poplar) and Brazil (eucalyptus). The New Modernism approach has been adopted by most of the intensively managed planted forests that have been developed in the past few decades and which are currently helping to provide the world's increased demand for wood. Its proponents argue that it is more sustainable overall because it is focussed on a smaller productive area and helps protect the world's natural forests (Gladstone and Ledig 1990).

Some of the most impressive developments in a very scientific approach to increasing production have taken place in countries with a relatively recent history of planted forest management such as Brazil, Chile and New Zealand. Campinhos (1999) discusses the successful development of planted eucalyptus forests around the world and, in particular, the use of clones of good hybrids of *E. grandis* and *E. urophylla* by the Aracruz Celulose S. A. Company. In Brazil, the productivity of four million ha of eucalyptus forests has tripled in the 35 years since the 1970s as a result of intensive research, improved operations (i.e. site preparation, fertilization, weed and pest control), improved seed selection and deployment of clonal material (Goncalves et al. 2008). Fast growing species that can be managed on short rotations offer significant advantages and in many tropical and subtropical regions eucalyptus is the preferred genus because:

1. It adapts well to different ecosystems
2. Naturally occurring native populations already exist
3. It matures quickly and is well shaped
4. Species can be crossed to produce hybrids with added vigour and different wood properties
5. It can be cloned at species and hybrid level
6. It produces wood with good properties for fibre, sawn-timber, chipboard, charcoal, posts and civil construction

Demand for eucalyptus wood has grown by over 10 % per year since 1982 and has been responsible for a shift in pulp production from temperate and sub-temperate to tropical and sub-tropical countries. Countries planting eucalyptus are Brazil, India, Spain, Portugal, South Africa, Angola, China, Ethiopia, Chile, Uruguay and Argentina with rotation lengths of 5–25 years.

The other genus that has been responsible for the large increase in clonal forestry, particularly in the Southern Hemisphere (New Zealand, Australia, Chile and South Africa) is pine (Sutton 1999). Pine and eucalyptus are particularly popular genera because of the choice of a wide genetic base, rapid growth under a range of environmental conditions and the availability of technology to grow them quickly and cost-effectively. Despite the huge increase in eucalyptus and pine planting the supply is still not keeping pace with demand (e.g. Barreiro and Tomé 2012) leading to increasing prices. Haynes (2007) suggests that these increases in stumpage prices were the encouragement for systematic forest management, shortened rotations and practices designed to speed development of well managed forests in the USA. Slowing of the price trend in the 1990s reflected the growing importance of managed forests with higher volumes per ha. The recent increase again in global prices should continue to encourage owners to manage their forests more intensively. Therefore, we should anticipate a large increase in both harvest and inventory volumes although these managed forests will remain a small part of the total global forest base. Haynes (2007) concludes that the majority of forests will be lightly managed, if at all, while a small minority will be heavily (or actively) managed and provide the bulk of world timber needs.

There are concerns in the sector whether management practices designed to improve sustainability may be discouraging for owners because of cost and reduced financial return. The key question is whether increasing productivity is compatible with long-term sustainability. The problems with the New Modernist approach is that it requires high inputs (energy, fertiliser, etc.) so it may struggle to have a long-term future, particularly if the cost of inputs increases at a faster rate than the value of outputs. Social and environmental pressures may also limit the type and intensity of practices that can be employed (e.g. the use of herbicides and pesticides, and clearcut harvesting).

2.2   Sustainable Intensification

This is an idea that first developed in agriculture (Pretty 1997) and focusses much more on the whole chain. The argument is that if all the knowledge we have is implemented in the management of forests then it is possible to boost yields and to bring marginal areas into production without compromising the long term sustainability of the forest. It, therefore, addresses some of the concerns over the New Modernist approach by attempting to integrate a wide range of methods and techniques to manage pests, nutrients, soil and water (Table 2). It is a system that relies heavily on the engagement of all participants and the careful integration of a range of techniques and methods.

It is not as straightforward to implement as the New Modernist approach and has yet to be implemented on any scale. The concept of High Yield Forestry introduced by the Weyerhaeuser Company (Heninger et al. 1997) approaches the concept through use of growth and yield studies and computer simulations, tree improvement, tree propagation, forest regeneration and silvicultural methods while at the same time maintaining soil productivity and biodiversity. This is the system of management that has been commonly used in many parts of the industrialized world with clearly defined productivity and environmental targets. However, Sustainable Intensification relies much more on managing the forest as part of the ecosystem

Table 2 Comparison of different forest types

	Industrial forestry (New modernism)	Low input forestry	Sustainable intensification forestry
Forest size	Extremely large (thousands of hectares)	Generally small scale forestry	Mixture of scales from small to industrial
Ownership	Corporate	Individual	Mix of ownership
Fertilizer	High input of industrial fertilizer	None unless certified organic	Avoid fertilizer and rely on silviculture, agroforestry, etc.
Pest control	Chemical pesticides	No chemical pesticides	Use integrated pest management
Biodiversity impact	Reduced	Enhanced	Stable
Yield	High	Low	High but probably lower than industrial forestry
Cost	Low	High	Potentially lower than industrial forestry

and depends heavily on understanding the biological processes at work and utilising these to increase productivity and to reduce reliance on pesticides and fertilisers. The closest expression of this philosophy is contained through various third party certification organisations such as FSC (FSC 2002) and (PEFC 2010), although the focus is on maintaining social and environmental benefits and there is no focus on encouraging intensification of production. In addition the FAO outline how plantation forests when sustainably managed can provide a range of social, environmental and economic benefits particularly in the developing world (FAO 2010b). Sustainable Intensification, therefore, appears to fall somewhere between the philosophies of High Yield forestry and Forest Stewardship in that it seeks to apply many of the principles of Forest Stewardship to increase productivity.

A further difficulty for the Sustainable Intensification approach is the lack of research on the whole system. Individual components of the methodology have been studied in detail but there have been few attempts to date to pull everything together and to determine what is possible both practically and economically (for an attempt to identify knowledge gaps for real sustainable forest management see Hickey and Innes 2005). Therefore, much of our knowledge of such systems has to come from agriculture where considerably more work has been undertaken and where systems are being implemented (e.g. Tilman et al. 2011; Godfray et al. 2010).

2.3 Forest Management

The key to increasing productivity will be through forest management. There have been a huge numbers of papers and books on all aspects of the management of planted forests (e.g. Evans 2001, 2009). The difficulty is that planted forest management is relatively recent and there is little experience of managing these forests over multiple rotations or of fully implementing the many techniques and approaches required to increase productivity in a sustainable manner. However, the evidence that does exist suggests that forests can be managed over multiple rotations and no fertilizer input with no loss of productivity (Woollons 2000; Evans 2009

In the past there has been a tendency to adopt four main management approaches to forestry but this is now being supplemented by forestry for biomass (Duncker et al. 2012). These are as follows with increasing levels of management intensity and inputs:

1. Forest nature reserve: Unmanaged forest to allow development of natural processes without human intervention.
2. Close to nature: Produce timber by mimicking or emulating natural processes
3. Extensive management with combined objectives: Combine production and ecological objectives at the stand level.
4. Intensive even-aged planted forests: Optimise wood production.
5. Short rotation forestry for biomass production: Produce the highest amount of wood biomass

Intensive even-aged planted forests and short rotation forestry for biomass fit the Industrial Forestry (New Modernist) approach, whereas the close to nature and extensive management with combined objectives forestry are closer to the Sustainable Intensification approach. However, for Sustainable Intensification to take place and be effective there will need to be a synthesis between the approaches.

However, there are other difficulties. In many countries the difficulty will be competing with imported timber from other parts of the world where costs may be lower and/or productivity higher. This may result in a divergence in forest management between public and private forestry such as has occurred in the Pacific Northwest of the US and New Zealand. Public forests will tend to be managed on longer rotations mainly (or in some cases exclusively) for environmental and social benefits with low inputs and potentially lower costs (e.g. much more use of natural regeneration) while private forestry will use shorter rotations and more intensive management. This tendency to zoning could have important impacts on the long-term sustainability of forestry and the ability of forestry to supply the increased wood production required.

There will clearly be a requirement for a number of management operations to be applied for the Sustainable Intensification approach to be successful. It will rely heavily on methods for reducing inputs, costs and for optimising management and the value of products produced:

1. Increased use of natural regeneration where possible.
2. Increased use of mixtures or agroforestry in order to maintain productivity without increased use of fertiliser (e.g. inter-cropping with legumes: Sankaran et al. 2005)
3. Adoption of techniques for integrated pest management in order to reduce the use of pesticides
4. Increased use of modern technologies for optimizing management inputs (e.g. targeting fertilizer use where most required) and the value of the products recovered from the forest. This is discussed in more detail in the following sections.

2.4   Climate Change and Threats to Forests

A major concern for future forestry and managing the increasing demand for wood is the threats from abiotic and biotic hazards and, in particular, how these will change with a changing climate (Lindner et al. 2010). Sedjo (2010) raises the question as to whether trees can migrate fast enough with climate change. The evidence is that many species will struggle (Pearson 2006; Zhu 2012) and some tree species are likely to "die back". The interior of continents will become drier during the summer increasing the probability of drought and winters are likely to be milder in many places increasing the likelihood of problems from pests because their numbers are less affected by mild winters. For example, there has been a huge recent outbreak of mountain pine beetles in Western Canada that is probably due to milder winters (Kurz et al. 2008). Overall climate change is expected to increase the frequency and severity of disturbance (e.g. Dale et al. 2001; Gardiner et al. 2010).

Historically, much of the world's timber supply came from the natural forests of North America, Russia and northern Europe but now the move is towards planted forests in south-east Asia, South America and Equatorial Africa. Global climate change that threatens and puts pressure on northern latitude forests is likely to hasten this transition. Potentially timber will be more abundant with an increased area of planted forests and faster growth rates (longer growing seasons and warmer temperatures) but forestry may no longer be profitable in some areas due to increasing levels of disturbance.

Another concern is the increasing use of clonal material in many planted forests. For example, Aracruz Celulose S. A. relies on clones of good hybrids of *E. grandis* and *E. urephylla* and there is a potentially high risk of major disturbance if the particular clones chosen are not adapted to a particular pest and are highly susceptible. We already know that some eucalyptus clones are more susceptible to abiotic risks such as wind damage (Garcia 2011). The focus on a small number of clones also runs the risk of reducing the genetic base for future improvement. A mosaic distribution of different clones at sites is recommended to reduce the risk associated with low genetic variability and of poor matching of a particular clone to a particular site (DeBell and Harrington 1993).

2.5   Substitution

An alternative to increasing the production of primary wood is to substitute other materials for wood (Wolcott 2003), to use forest and mill residues to make wood composites (Maloney 1996) or to make better use of what material is available. For example, corewood (i.e. wood formed in the first 10–15 annual rings from the pith) can be used for the central portion of laminates, defect cutting and green-gluing of knotty timber can produce strong engineered timber, waste fibres from annual plants substituted for wood fibre (e.g. Yang et al. 2003) and the innovative combination of wood and other materials, such as cement, lime and plastics, can produce material with combined benefits (Karade 2010). Haynes and Weigand (1997) suggested that there will be a reduced reliance in solid wood and an increasing reliance on engineered wood and recycled materials over the next 50 years. For example, advances in light timber frame construction, including the use of engineered wood products such as I-joists, means that the average house built today uses less wood per unit area than it did previously. At the same time the decline in available volume has meant that there has been a switch from just considering quantity and increased focus on value and increasing value (Whittenbury 1997; Murphy 1998). This means making use of low or variable quality wood, which is discussed in more detail later in this chapter.

A key question is what are the impacts of substitution on various measures of sustainability (Petersen and Solberg 2005)? Also can wood-based products incorporating other materials still offer the advantages of bio-materials but have a low enough impact on the environment. For example, particle boards which are increasing in global volume are usually produced using resins that rely on formaldehyde as a solvent and there has been a huge effort to reduce the in-service emissions of formaldehyde. Finally, if there is an increasing reliance on recycled or waste biomaterials and less critical requirements placed on virgin wood and fibre, what are the implications for forest management practice? Will forest management no longer be profitable and will such a change direct managers to a very low input type of forestry. These topics will be discussed at the end of the chapter.

3   Integration of the Wood Supply Chain

Although increasing the volume of wood produced is one option for meeting increasing demand another approach is to make more optimal use of the material that is produced. Murphy et al. (2010) state that: "Optimal allocation of the wood fibre resource is vital if wood and bio-energy are to obtain the material most suited to their needs and wood suppliers are to obtain the best return for their investment in forest land". They found 50 % gains in net value recovery using optimised solutions. More modest gains of 3–7 % were obtained by Mitchell (2004),but higher values have been obtained for New Zealand and Chilean forests (Murphy et al. 2010). Optimization is currently a means for adding value to forest processing but it could also help to satisfy the increasing demand for wood material by reducing wastage and allocating material to the most appropriate end-use. The techniques for rapidly and non-destructively measuring wood properties at all stages of the wood chain and for calculating the best allocation strategies for the complex forest-wood chain are now available due to advances in operations research and artificial intelligence.

Focussing on optimization will also help to re-establish and improve the link between growers, processors and end users. Currently there appears to be little understanding by growers of the detailed requirements of the final end-users and end-users have little comprehension of what is both physically and economically possible by growers (Moore 2012). Processors, who are in the middle of the supply chain, could be engaged in feeding information in both directions along the chain but are often primarily engaged in maximising volume throughput and recovery and pay less attention to ensuring the material properties of their products are specific to the intended end-use. Researchers and breeders have also focussed on increasing volume production and in the future it is going to be vital that improvements in genetics and silviculture need to be focussed not only on productivity but on the suitability of the wood for a particular end-use. The question is whether the current wood production model can survive with improvements in the connectivity of the supply chain or whether we need fully vertically integrated companies or partnerships. This is essentially the choice that needs to be made between the New Modernist and the Sustainable Intensification approaches and it is most likely that different approaches will be required in different parts of the world and in different parts of the wood market.

3.1 Optimisation

In order to improve the efficiency of the forest-to-customer supply chain it is possible to make use of management science techniques. Such techniques have been used to support different aspects of forest management since at least the 1960s (Bare et al. 1984; Martell et al. 1998). These include strategic forest management at the scale of thousands of hectares with long time scales through to operational management at small spatial scales and with relative short term time scales. At the operational level, techniques have been applied to the selection of stands to be harvested (e.g. Papps and Manley 1992), cross-cutting of tree stems into logs (e.g. Deadman and Goulding 1979; Eng et al. 1986; Lembersky and Chi 1986), the allocation of these logs to various mills in order to satisfy demand (e.g. Mendoza and Bare 1986), the transport of these logs to their destination (e.g. Weintraub et al. 1995; Murphy 2003) and the processing of these logs in order to maximise value and/or to produce products with certain characteristics (e.g. Todoroki and Rönnqvist 2002).

In many of these studies linear or integer programming techniques have been used and the focus has been on optimizing a particular component of the forest-to-mill supply chain. However, this is unlikely to yield the same benefits as global optimization of the whole supply chain (Faaland and Briggs 1984; Mendoza and Bare 1986). In addition, optimization studies have tended to focus on external tree attributes such as size (diameter and height), shape (taper, sweep and circularity) and the characteristics of branches (number, size, and angle). However, end-users requirements and product grading specifications are shifting from being based on visual characteristics (e.g. ring width, stem shape, branch size) to being based on material properties such as strength, stiffness, fibre dimensions, and density. Fortunately within the last decade a large number of techniques have become available, which enable some of these properties to be measured non-destructively and rapidly at different points in the supply chain (see section below).

While classical optimization techniques such as linear or integer programming are good at producing exact and accurate solutions to problems that can be formulated in exact mathematical terms, they are not designed to work with inaccurate, noisy or complex data. Neither are they suitable for problems which change over time. Furthermore, the sheer size of the search-space that must be explored for large problems often renders the techniques intractable on many practical problems. In many cases, the use of linear programming techniques in forest management problems has required a considerable simplification of management objectives and constraints (Garcia 1984). The field of meta-heuristics is a relatively new field which although not guaranteeing optimality, reliably achieve "good enough, fast enough" solutions to problems. In general they are simple to implement, computationally cheap, robust to changes in data and have been proven to be highly effective. They are particularly suited to problem areas such as transportation and inventory management where the data being used to attempt to solve a problem is in itself an estimation and the problem has many variables and constraints which rule out any brute-force type approach to finding a solution (Lourenco 2005).

Heuristic techniques are just now being applied to the forestry sector, such as the use of applied tabu-search algorithms for locating machinery for optimizing forest harvesting (Diaz-Legues 2007) and simulated annealing used to develop solutions for balancing harvesting targets against the risk of wind damage (Zeng et al. 2007). Further developments are inevitable in helping to optimize the forest-to-customer supply chain by ensuring that the right material is supplied to the right customer in a cost effective manner and that efficient use is made of capital and distribution networks (Frayret et al. 2007). Ultimately, this could lead to build-to-order supply chains which can deliver considerable cost savings and improved customer satisfaction (Mansouri et al. 2012).

3.2   Wood Property Measurement

Forest Inventory

In order to improve the allocation of wood material in the future it is essential that there is detailed information on the properties of the wood available and this information is linked to material as it passes along the chain. Fundamental information required is tree size, numbers and species, but increasingly information on quality features is also required. New remote sensing techniques now allow information on the forest resource to be obtained in extraordinary detail. In particular, recent developments in airborne laser scanning technology (LiDAR – Light Detection and Ranging) have enabled information on tree heights, crown width, and stand density to be obtained (Hyyppä et al. 2012; Pirotti et al. 2010; Suárez et al. 2008 from virtually (>95 %) of all trees in the forest. Airborne LiDAR in combination with field sampling is now seen as an effective method for operational forest inventories (e.g. Andersen et al. 2006; Wynne 2006). In addition to providing this kind of detailed

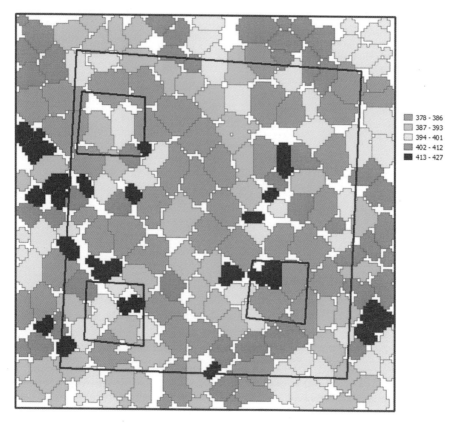

Fig. 3 Map of individual Sitka spruce trees in Aberfoyle Forest, Scotland with predicted wood densities (kg m⁻³). Based on airborne LiDAR measurements and the wood density model of Gardiner et al. (2011), Suárez (2010) (Courtesy of Juan Suarez, Forest Research)

mensurational information at the individual tree or stand scale, LiDAR allows the mapping of structural differences in the forest at a very fine scale (metres) and over very large (>100 km) areas. Such surveys have already been conducted or are in progress for a number of countries. Much of the current research focus is now directed towards the prediction of internal wood properties from metrics obtained using LiDAR (e.g. van Leeuwen et al. 2011; Fig. 3).

In addition to airborne LiDAR, Terrestrial Laser Scanning (TLS) technology has shown the potential to provide information on tree and stand characteristics such as stand density, tree diameter, height and crown shape (Dassot et al. 2011); stem profiles (Aschoff and Spiecker 2004), and some branch attributes (Schutt et al. 2004; Keane 2007). Stem data collected by terrestrial scanners can also be stored in standard harvester data formats (e.g., StanForD: Skogforsk 2011), and used to assist forest managers with optimizing harvest scheduling and cross-cutting into logs (Keane 2007).

Fig. 4 Measuring wood
stiffness in a standing tree
using a hand-held acoustic
tool (photograph B. Gardiner)

Acoustic Testing

The use of stress (acoustic) waves to determine the properties of solids is a well-established technique in the field of materials science (Kolsky 1963). Of particular value is the fact that the stress wave velocity in a material is proportional to the material stiffness (modulus of elasticity) and inversely proportional to its density. This technique has been applied to wood since the 1950s (Jayne 1959), initially in the laboratory, but more recently as commercially-available machines for strength grading sawn structural timber in a mill.

Stress wave techniques have also been applied to identify those logs and standing trees with the higher stiffness wood necessary to produce high quality structural timber (e.g. Ross et al. 1997; Tseheye et al. 2000; Wang et al. 2007a; Auty and Achim 2008). In recent years a large number of studies involving the use of stress waves to measure the wood stiffness in standing trees and logs have been undertaken, in part because the forest products industries are under increasing economic pressure to maximise the value obtained from the resource (Wang et al. 2007b). A number of portable tools have been developed to make these measurements (e.g. Fig. 4) and their operating principles are described in Huang et al. (2003) and Huang (2005). Results from numerous studies show that there is a strong correspondence between stress wave velocity measurements made on standing trees and those made on logs (e.g. Wang et al. 2005; Moore et al. 2013), and with these measurements and the yield of structural grade timber (Carter et al. 2005; Moore et al. 2013). Stress

Fig. 5 X-ray scan of Sitka spruce log (3 mm slice) clearly showing heartwood, sapwood, growth rings and knots. Image obtained with a Somaton Esprit CT scanner (Siemens, Forchheim, Germany)

wave velocity measurements have also been use to segregate logs for veneer (Carter et al. 2005) and pulp production (Albert et al. 2002).

Measurement of the stress wave velocity in standing trees within the forest is also of value to forest owners as it not only allows the resource to be characterised with respect to wood properties (e.g. Moore et al. 2009), but it also allows the effect of different silvicultural treatments, such as thinning, to be evaluated for their impact of wood stiffness without having to fell the trees (e.g. Lasserre et al. 2005. The relationships between timber stiffness and stress wave velocity measurements can also be used to inform tree breeding efforts focussed on improving this timber property (e.g. Kumar 2004; Cherry et al. 2008; Vikram et al. 2011). Recent developments have seen acoustic measurements implemented in the cutting head of harvesting machines (Amishev and Murphy 2008), which provides the possibility of thinning selectively on the basis of the wood stiffness of trees and optimise cross-cutting (bucking) of trees during harvesting. Such cross-cutting can be linked with forest inventory data (see section above) and customer requirements to cut logs for different end products in the forest rather than later in the mill.

X-Ray CT Log Scanning

Interest in within-tree distributions of knot/branch and wood properties has increased in recent decades and is driven by the desire to achieve greater value recovery from logs. However, historical methods for analysing the internal tree stem structure were destructive and time consuming. Since the pioneer work performed in Sweden (Lindgren 1991; Grundberg and Grönlund 1997), the use of X-ray CT scanners enables more detailed investigation of the internal structure of the tree stem (see Fig. 5) in order to assess the wood density variations, and the shape and size of the included part of the branches (knots) (Longuetaud et al. 2005; Sepulveda et al. 2002).

The knowledge of the distribution of heartwood and sapwood in trees stems is also important as their physical and technological properties often differ in terms of colour (Münster-Swendsen 1987; Espinoza et al. 2005), natural durability (Cruz et al. 1998; Björklund 1999; Kärenlampi and Riekkinen 2003), impregnation properties and durability (Wang and De Groot 1996), and moisture content (Fromm et al. 2001). The use of X-ray log scanners allows the measurement and the modelling of the within tree stems distribution of sapwood (Longuetaud et al 2007) even when there is no difference in colour between heartwood and sapwood (e.g. Norway spruce).

X-ray scanners have been available for boards for a number of years and in recent years there has been considerable research into the development of commercial X-ray scanners for logs, with the first machines now available (e.g. www.microtec.eu).

Growth and Wood Properties Modelling

The use of robust models is a key component of forest management where they are used to support activities such as the development of silvicultural regimes and forecasting future timber yields. For a long time, forest models focused on the prediction of stand and tree-level attributes, such as dominant height, basal area and volume or tree diameter at breast height as a function of age, stand density and site quality (Clutter et al. 1983). Data collected from empirical studies were used to construct yield models and tree and stand-level growth models (e.g. Houllier et al. 1991; Goulding 1994). The connections of these models with wood quality were often very limited: the main information that they provided about the yields and attributes of wood products was related to the size of the trees (for the mean and dominant trees with stand models or for every tree in the stand with tree models: Lemoine 1991), tree taper through stem profile equations (e.g. Newnham 1992) and branch models (e.g. Grace et al. 1999). Decision support systems such as StandPak (Whiteside 1990) integrated several of these models together in order to understand the impacts of forest management on log quality and stand value. In some cases results from sawing studies on logs of known quality were incorporated into these decision support systems, which enabled product quality to be linked back to forest management (e.g. Whiteside and McGregor 1987).

More recently, the need for management tools that integrate both growth and wood quality information led to a new generation of more detailed models which use stand– and tree-level information to predict wood properties and quality attributes such as branch characteristics (i.e. number, position, size, insertion angle and status), fibre dimensions, wood density, microfibril angle, and wood stiffness (e.g. Mitchell 1988; Maguire et al. 1991; Briggs 1992; Leban et al. 1997; Houllier et al. 1995; Longuetaud et al. 2007; Gardiner et al. 2011; Auty et al. 2013; Meredieu et al. 1998). Together these stem and wood properties are important determinants of the quality of end-products in terms of characteristics such as strength (bending and crushing), stiffness, colour, shrinkage and distortion. End product quality can either be modelled empirically (e.g. Johansson et al. 2001) or numerically from information on key wood properties (e.g. Ormarsson et al. 1999). These models have mostly been developed for coniferous trees but are now being developed for fast-growing hardwoods such as eucalyptus (Downes et al. 2009) and poplar (Jiang et al. 2007) and a number of tropical species being grown in planted forests (e.g. Kokutse et al. 2010; Perera et al. 2012). These models operate at the tree level but may be either average tree models, distance-independent or distance-dependent individual tree models and are often used in conjunction with sawmill conversion simulation systems (Vaisanen et al. 1989; Leban and Duchanois 1990; Barbour and Kellogg 1990; Briggs 1992; Leban et al. 1997) in order to provide a link between product recovery and different silvicultural strategies, raw material sources, or bucking and sawing patterns. This enables predictions of suitability for different end-products and product performance to be made prior to being put in-service rather than being discovered in-service. It also makes it possible to study the utilisation of timber in terms of alternative decisions in the whole forest-wood chain.

The Forest Resource

The traditional description of the forest resource emphasises factors such as available wood volume, and trees size and status. For the future, information is required on internal stem properties including the property differences between various types of stands, trees and parts of trees. Such information can be provided with measurements on the standing trees using non-destructive methods as discussed above or by model estimation using traditional or extended monitoring data as inputs. Recently the concept of Regional Resource Databases has been developed, to support optimal allocation and processing of available wood for different products (Grahn and Lundqvist 2008; Lundqvist et al. 2008). Forest resources large enough to be representative of a region or the resource catchment of a mill are simulated, providing information about the properties and volumes of trees, parts of trees and potential products from the trees. The simulation is performed in a stepwise manner with a set of integrated models, starting with measured breast height diameters and stand data, estimating (i) the age, height and taper of the tree, (ii) the interior growth pattern, (ii) the wood and fibre properties at different heights (wood density, heartwood, knot properties, fibre dimensions, etc.), and (iii) the properties and volumes of parts of the tree related to potential products (pulpwood logs, sawlogs, parts which will become sawn products and sawmill chips, etc.).

The forest, therefore, becomes the warehouse in an integrated processing system. Modern measurement methods and IT systems allow continual updating of the wood available or moving through the system and the properties of this material. Artificial intelligence can then optimize this flow of material against a set of clearly defined targets and to adjust processes as the material properties or availability changes (e.g. Frayret et al. 2007).

4 Implications for Forestry

Wood is in a prime position to be one of the materials of choice in the twenty-first century with its combination of availability, sustainability and low cost. However, to ensure that this vision is realised a big shift in the way the whole forest wood chain operates has to take place. All parts of the process have to be seen as integral and not just as individual components with only a limited local importance. It probably means a tendency to more companies operating across the entire chain from forest to end-product or much closer strategic partnerships in order for knowledge to be freely shared between all parts of the chain.

In order for increasing demand for wood products to be met there has to be a twofold approach. Firstly, there needs to be increased production through increased use of planted forests and increased productivity of existing planted forests as we discussed at the beginning of the chapter. This requires the use of the very latest techniques and an integrated approach to management in order to boost productivity without sacrificing long-term sustainability. Forests can no longer be managed with a single focus, such as maximizing volume, without taking into consideration the impact on soil status, water quality, biodiversity, etc. The result is almost certainly a separation of forest types with increased emphasis on identifying those forests (or areas to be reforested or afforested) that have the potential to supply the increasing demand and those forests that are unable to do so for environmental, social or economic reasons and which will be managed with a focus on something other than wood production.

Secondly, there needs to be much better information available on the forest resource and the properties of this resource and the properties of material within the chain comprehensively and continuously tracked (see an idealised concept for an optimized wood chain in Fig. 6). The forestry sector is significantly behind other

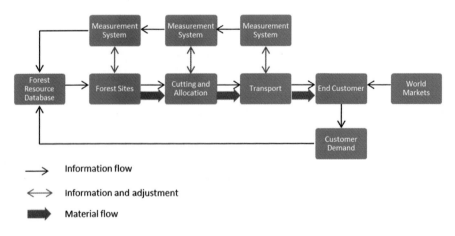

Fig. 6 Stylised flow diagram of an integrated and optimised wood chain

sectors in its knowledge of the material it is processing and many decisions on material suitability are made much later than is ideal. In addition there is limited feedback at each processing stage which is required to inform decision making early and later in the chain to adjust and match end-product requirements and material availability. However, the tools for characterizing the material properties, to track these properties, to transfer and manage this information are all already available. The difficulties of implementing such a plan are partly the tight margins that operate in the forest products sector and especially for growers but also the innate conservatism in an industry that has focussed too often on volume and has learnt to adapt and make-do with whatever material is available. However, unless the sector changes to one primarily focussed on matching end-product requirements against raw material properties it will continue to struggle to compete with other sectors producing material to much tighter specifications.

Alongside all this are unknowns such as changing levels of risk, changes in economics and market requirements. In order to deal with these challenges forestry and the silvicultural systems will need to be flexible and resilient (Moore 2012). This is enormously difficult because competition is fierce and global and the temptation is to find the solutions that provide the highest profit margins. These have a tendency to be focussed on single species and a limited set of silvicultural systems but the dangers of such an approach are evident in the high levels of damage that have been recently experienced in Europe and North America from both biotic and abiotic agents. Silviculture needs to be dynamic and flexible enough to incorporate these changes and to anticipate the problems of the future and to adapt the management to these threats before rather than after events.

5   Summary and Conclusions

For the forests of the world to continue to be able to provide the increasing demand for wood and wood-based products requires an increasing reliance on existing and new planted forests. These forests will need to be managed using the very latest understanding of the biological processes at work in forest ecosystems in order to increase their productivity while maintaining their sustainability similar to the requirement for crop production (FAO 2009b). In addition the silvicultural systems will need to be designed to be resilient and to incorporate risk mitigation strategies under a changing climate.

Wood fibre produced from the worlds' forests needs to be used in a more optimized manner with material matched to end-use requirements and wood-based products being designed to utilise the mix of virgin and recycled material that will be available in the future. To do this requires a much greater level of integration of the forest-wood chain and proper exchange of information on material dimensions and properties between all stages of the process.

The knowledge and tools for implementing systems to increase the productivity of our forests and to make the best use of the material they produce are already available. The question is whether the sector can respond to the challenge and modernise the system in a way that allows wood-based products to become the materials of choice in the future. It will require investment, co-operation of all stakeholders from growers to processors and researchers, and hard decisions on the management focus for every forest. With increasing pressure on land-use from urbanisation and agriculture the forest-based sector has to make much more targeted and efficient use of the land it has available in order to continue to provide the world's wood needs.

References

Albert DJ, Clark TA, Dickson RL, Walker JCF (2002) Using acoustics to sort radiata pine pulp logs according to fibre characteristics and paper properties. Int Forest Rev 4(1):12–19

Amishev D, Murphy GE (2008) Implementing resonance-based acoustic technology on mechanical harvesters/processors for real-time wood stiffness measurement: opportunities and considerations. Int J Forest Eng 19(2):48–56

Andersen H-E, Reutebuch SE, McGaughey RJ (2006) A rigorous assessment of the height measurements obtained using airborne lidar and conventional field methods. Can J Remote Sens 32(5):355–366

Aschoff T, Spiecker H (2004) Algorithms for the automatic detection of trees in laser scanner data. ISPRS 36:66–70

Auty D, Achim A (2008) The relationship between standing tree acoustic assessment and timber quality in Scots pine and practical implications for assessing timber quality from naturally regenerated stands. Forestry 81(4):475–487

Auty D, Gardiner B, Achim A, Moore JR, Cameron AD (2013) Models for predicting microfibril angle variation in Scots pine. Ann Forest Sci. doi:1007/s13595-012-0248-6

Barbour RJ, Kellogg RM (1990) Forest management and end-product quality-a Canadian perspective. Can J Forest Res 20(4):405–414

Bare BB, Briggs DG, Roise JP, Schreuder GF (1984) A survey of systems analysis models in forestry and the forest products industry. Eur J Oper Res 18:1–18

Barreiro S, Tomé M (2012) Analysis of the impact of the use of eucalyptus biomass for energy on wood availability for eucalyptus forest in Portugal: a simulation study. Ecol Soc 17(2):14, http://dx.doi.org/10.5751/ES-04642-170214

Björklund L (1999) Identifying heartwood-rich stands or stems of *Pinus sylvestris* by using inventory data. Silva Fenn 33:119–129

Briggs DG (1992) Models linking silviculture, wood quality and product value: a review and example for US coastal Douglas-fir. In: IUFRO all division 5 conference forest products, Nancy, 23–28 Aug 1992. ARBOLOR Bd, pp 285–294

Campinhos E (1999) Sustainable plantations of high-yield shape eucalyptus trees for production of fiber: the Aracruz case. New Forest 17:129–143

Carle J, Holmgren P (2008) Wood from planted forests: a global outlook 2005–2030. Forest Prod J 58(12):6–18

Carter P, Briggs D, Ross RJ, Wang X (2005) Acoustic testing to enhance western forest values and meet customer wood quality needs. PNW-GTR-642, productivity of western forests: a forest products focus. USDA Forest Service, Pacific Northwest Research Station, Portland, pp 121–129

Cherry ML, Vikram V, Briggs D, Cress DW, Howe GT (2008) Genetic variation in direct and indirect measures of wood stiffness. Can J Forest Res 38:2476–2486

Clutter JL, Forston JC, Pienaar LV, Brister GH, Bailey RL (1983) Timber management, a quantitative approach. Wiley, New York, p 333

Cruz H, Nunes L, Machado JS (1998) Update assessment of Portuguese maritime pinetimber. Forest Prod J 48:60–64

Dale VH, Joyce LA, McNulty S, Neilson RP, Ayres MP, Flannigan MD, Hanson PJ, Irland LC, Lugo AE, Peterson CJ, Simberloff D, Swanson FJ, Stocks BJ, Wotton BM (2001) Climate change and forest disturbances. BioScience 51:723–734

Dassot M, Constant T, Fournier M (2011) The use of terrestrial LiDAR technology in forest science: application fields, benefits and challenges. Ann Forest Sci 68:959–974. doi:10.1007/s13595-011-0102-2

Deadman MW, Goulding CJ (1979) A method for the assessment of recoverable volume by log types. N Z J Forest Sci 9:225–239

DeBell DS, Harrington CA (1993) Deploying genotypes in short-rotation plantations: mixtures and pure cultures of clones and species. Forest Chron 69:705–713

Diaz-Legues A, Ferland JA, Ribeiro CC, Vera JR, Weintaub A (2007) A tabu search approach for solving a difficult forest harvesting machine location problem. Eur J Oper Res 179:788–805

Downes GM, Drew D, Battaglia M, Schulz D (2009) Measuring and modelling stem growth and wood formation: an overview. Dendrochronologia 27:147–157, 2009

Duncker PH, Barreiro S, Hengeveld G, Lind T, Mason W, Ambrozy S, Spiecker H (2012) Classification of forest management approaches: a new methodological framework and its applicability to European forestry. Ecol Soc 17(4):50, http://dx.doi.org/10.5751/ES-05066-170450

Eng G, Daellenback HG, Whyte AGD (1986) Bucking tree-length stems optimally. Can J Forest Res 16:1030–1035

Espinoza GR, Hernandez R, Condal A, Verret D, Beauregard R (2005) Exploration of the physical properties of internal characteristics of sugar maple logs and relationships with CT images. Wood Fiber Sci 37:591–604

Evans J (ed) (2001) The forest handbook, volume 2: applying forest science for sustainable forest management. Blackwell Science, Oxford, p 382, ISBN 0-632-04823-9

Evans J (ed) (2009) Planted forests: uses, impacts and sustainability. Published jointly by FAO and CAB International. Wallingford, UK. ISBN 978-1-84593-564-1

Faaland B, Briggs D (1984) Log bucking and lumber manufacturing using dynamic programming. Manag Sci 30:245–257

FAO (2007) Global wood and wood products flow: trends and perspectives. Advisory Committee on Paper and Wood Products, Shanghai, 6 June 2007

FAO (2009a) State of the world's forests 2009, Rome, Italy. Also available at www.fao.org/docrep/011/i0350e/i0350e00.HTM

FAO (2009b) Increasing crop production sustainably: the perspective of biological processes, Rome, Italy. Also available at http://www.fao.org/docrep/012/i1235e/i1235e00.htm

FAO (2010c) Global forest resources assessment 2010: main report. Forestry Paper 163, Rome, Italy, 378 pp. Also available at http://www.fao.org/forestry/fra/fra2010/en/

FAO (2010b) Planted forests in sustainable forest management: a statement of principles, Rome, Italy, 16pp. Also available at http://www.fao.org/docrep/012/al248e/al248e00.pdf

FAO (2012) The state of food insecurity in the world, Rome, Italy, 65pp. Also available at http://www.fao.org/docrep/016/i3027e/i3027e00.htm

Fenning TM, Gershenzonand J (2002) Where will the wood come from? Plantation forests and the role of biotechnology. Trends Biotechnol 20:291–296

Fenning TM, Walter C, Gartland KMA (2008) Forest biotech and climate change. Nat Biotechnol 26:615–617

Frayret J-M, D'Amours S, Rousseau A, Harvey S, Gaudreault J (2007) Agent-based supply-chain planning in the forest products industry. Int J Flex Manuf Syst 19:358–391. doi:10.1007/s10696-008-9034-z

Fromm JH, Sautter I, Matthies D, Kremer J, Schumacher P, Ganter C (2001) Xylem water content and wood density in spruce and oak trees detected by high-resolution computed tomography. Plant Physiol 127:416–425

FSC (2002) FSC principles and criteria for forest stewardship. FSC-STD-01-001 (version 4-0) EN, Bonn, Germany, p 13. http://ic.fsc.org/download.fsc-std-01-001-v4-0-en-fsc-principles-and-criteria-for-forest-stewardship.315.pdf

Garcia O (1984) FOLPI, a forestry-orientated linear programming interpreter. In: Nagumo H, et al (eds) Proceedings IUFRO symposium on forest management planning and managerial economics. University of Tokyo, Japan, pp 293–305

Garcia J (2011) Biomechanics of young grown eucalypts trees and pulling test relationship. IUFRO 8.03.06 – Impact of wind on forests 6th international conference on wind and trees, Athens, 31 July to 4 Aug 2011

Gardiner B, Blennow K, Carnus J-M et al (2010) Destructive storms in European forests: past and forthcoming impacts. Final report to EC DG environment http://ec.europa.eu/environment/forests/fprotection.htm

Gardiner B, Leban J-M, Auty D, Simpson H (2011) Models for predicting the wood density of British grown Sitka spruce. Forestry 84(2):119–132. doi:10.1093/forestry/cpq050

Gladstone WT, Ledig FT (1990) Reducing pressure on natural forests through high-yield forestry. Forest Ecol Manag 35:69–78

Godfray HCJ, Beddington JR, Crute IR, Haddad L, Lawrence D, Muir JF, Pretty J, Robinson S, Thomas SM, Toulmin C (2010) Food security: the challenge of feeding 9 billion people. Science 327:812–818

Goncalves JLM, Stape JL, Laclau J-P, Bouillet J-P, Ranger J (2008) Assessing the effects of early silvicultural management on long-term site productivity of fast-growing eucalypt plantations: the Brazilian experience. South Forest 70:105–118

Goulding CJ (1994) Development of growth models for *Pinus radiata* in New Zealand – experience with management and process models. Forest Eco Manag 69:331–343

Grace JC, Pont D, Goulding CJ, Rawley B (1999) Modeling branch development for forest management. N Z J Forest Sci 29:391–408

Grahn T, Lundqvist S-O (2008) Use of inventory data, simulation and regional resource databases for selection of wood to different end-products. In: Proceedings of sixth IUFRO workshop on modelling of wood quality, Koli, Finland, 9–13 June 2008

Grundberg S, Grönlund A (1997) Simulated grading of logs with an X-ray log scanner–grading accuracy compared with manual grading. Scand J Forest Res 12:70–76

Haynes RW (2007) Integrating concerns about wood production and sustainable forest management in the United States. J Sustain Forest 24(1):1–18

Haynes RW, Weigand JF (1997) The context for forestry economics in the 21st century. In: Kohn K, Franklin J (eds) Creating a forestry for the 21st century. Island Press, Washington D.C., U.S.A., pp 285–302

Heninger RL, Terry TA, Dobowski A, Scott W (1997) Managing for sustainable site productivity: Weyerhaeuser's forestry perspective. Biomass Bioenergy 13:255–267

Hickey GM, Innes JL (2005) Scientific review and gap analysis of sustainable forest management. Criteria and indicators initiatives: Forrex forest research extension partnership, Kamloops, B.C. Forrex Series 17 www.forrex.org/publications/FORREXSeries/FS17.pdf

Houllier F, Bouchon J, Birot Y (1991) Modelisation de la dynamique des peuplements forestiers: etat et perspectives. Rev Forest Fr 43:87–108

Houllier F, Leban J-M, Colin F (1995) Linking growth modelling to timber quality assessment for Norway spruce. For Ecol Manag 74:91–102

Huang C-L (2005) System and method for measuring stiffness in standing trees. Acoust Soc Am J 118(5):2763–2764

Huang C-L, Lindström H, Nakada R, Ralston J (2003) Cell wall structure and wood properties determined by acoustics: a selective review. HolzalsRoh– und Werkstoff 61:321–335

Hyyppä J, Yu X, Hyyppä H, Vastaranta M, Holopainen M, Kukko A, Kaartinen H, Jaakkola A, Vaaja M, Koskinen J, Alho P (2012) Advances in forest inventory using airborne laser scanning. Remote Sens 4:1190–1207. doi:10.3390/rs4051190

Ince PJ, Kramp AD, Skog KE, Yoo D, Sample VA (2011) Modeling U.S. forest sector trade impacts and expansion in wood energy consumption. J For Econ 17:142–156

Jayne BA (1959) Vibrational properties of wood as indices of quality. Forest Prod J 9(11):413–416

Jiang ZH, Wang XQ, Fei BH, Ren HQ, Liu XE (2007) Effect of stand and tree attributes on growth and wood quality characteristics from a spacing trial with *Populus xiaohei*. Ann Forest Sci 64:807–814. doi:10.1051/forest:2007063

Johansson M, Perstorper M, Kliger R, Johansson G (2001) Distortion of Norway spruce. Part 2. Modelling twist. HolzalsRoh– und Werkstoff 59:155–162

Jonsson R, Whiteman A (2008) Global forest product projections. Food and Agricultural Organization (FAO) of the United Nations, Rome

Karade SR (2010) Cement-bonded composites from lignocellulosic wastes. Constr Build Mater 24:1323–1330

Kärenlampi PP, Riekkinen M (2003) Prediction of the heartwood content of pine logs. Wood Fiber Sci 35:83–89

Keane E. (2007) The potential of terrestrial laser scanning technology in pre-harvest timber measurement operations. COFORD CONNECTS: Harvesting/Transportation No. 7

Kokutse AD, Stokes A, Kokutse NK, Kokou K (2010) Which factors most influence heartwood distribution and radial growth in plantation teak? Ann Forest Sci 67:407. doi:10.1051/forest/2009127

Kolsky H (1963) Stress waves in solids, 2nd edn. Dover, U.K., pp 213

Kumar S (2004) Genetic parameter estimates for wood stiffness, strength, internal checking, and resin bleeding for radiata pine. Can J Forest Res 34:2601–2610

Kurz WA, Dymond CC, Stinson G, Rampley GJ, Neilson ET, Carroll AL, Ebata T, Safranyik L (2008) Mountain pine beetle and forest carbon feedback to climate change. Nature 452:987–990

Ladanai S, Vinterbäck (2009) Global potential of sustainable biomass for energy. Report 013, ISSN 1654-9406. Swedish University of Agricultural Sciences Department of Energy and Technology, Uppsala, http://pub.epsilon.slu.se/4523/1/ladanai_et_al_100211.pdf

Lasserre J-P, Mason EG, Watt MS (2005) The effect of genotype and spacing on *Pinus radiata* [D. Don] corewood stiffness in an 11-year old experiment. Forest Ecol Manag 205:375–383

Leban JM, Duchanois G (1990) SIMQUA : modelling wood quality – New software –SIMQUA. Ann Sci Forest 47(5):483–493

Leban JM, Daquitaine R, Houllier F, Saint André L (1997) Linking models for tree growth and wood quality in Norway spruce. Part I: validation. IUFRO Working party S5.01-04, biological improvement of wood properties. Second workshop connection between silviculture and wood quality through modelling approaches and simulation softwares, Kruger National Park, South Africa, 26–31 Aug, pp 220–228

Lembersky M, Chi U (1986) Weyerhaeuser decision simulator. Interfaces 16:6–15

Lemoine B (1991) Growth and yield of maritime pine (*Pinus pinaster* Ait.): the average dominant tree of the stand. Ann Sci Forest 48:593–611

Lindgren LO (1991) Medical CAT-scanning: x-ray absorption coefficients CT-numbers and their relation to wood density. Wood Sci Technol 25:341–349

Lindner M, Maroschek M, Netherer S, Kremer A, Barbati A, Garcia-Gonzalo J, Seidl R, Delzon S, Corona P, Kolström M, Lexer MJ, Marchetti M (2010) Climate change impacts, adaptive capacity, and vulnerability of European forest ecosystems. Forest Ecol Manag 259:698–709

Longuetaud F, Saint André L, Leban JM (2005) Automatic detection of whorls on *Picea abies* logs using an optical and an X-ray scanner. J Non-Destruct Eval 24(1):29–43

Longuetaud F, Mothe F, Leban JM (2007) Automatic detection of the heartwood / sapwood limit from stacks of CT images of Norway spruce logs. Comput Electron Agric 58(2):100–111

Lourenco H (2005) Logistics management: an opportunity for meta-heuristics. In: Rego C, Alidaee B (eds) Metaheuristic optimization via memory and evolution: Tabu search and scatter search. Kluwer Academic Publishers, Boston/Dordrecht/London, pp 329–35630

Lundqvist S-O, Grahn T, Olsson L et al (2008) Modelling and simulation of properties of forest resources and along the paper value chain. In: Proceedings of sixth IUFRO workshop on modelling of wood quality, Koli, Finland, 9–13 June 2008

Maguire DA, Kershaw JA, Hann DW (1991) Predicting the effects of silvicultural regime on branch size and crown wood core in Douglas-fir. Forest Sci 37:1409–1428

Maloney TM (1996) The family of wood composite materials. Forest Prod J 46(2):19–26

Mansouri-Afshin S, Gallear D, Askariazad MH (2012) Decision support for build-to-order supply chain management through multi objective optimization. Int J Prod Econ 135:24–36

Martell DL, Gunn EA, Weintraub A (1998) Forest management challenges for operational researchers. Eur J Oper Res 104:1–17

Mendoza GA, Bare BB (1986) A two-stage decision model for bucking and allocation. Forest Prod J 36(10):70–74

Meredieu C, Colin F, Hervé J-C (1998) Modelling branchiness of Corsican pine with mixed-effect models (*Pinusnigra*Arnold spp. Laricio (Poiret) Maire). Ann Sci Forest 55:359–374

Mitchell KJ (1988) Sylver: modelling the impact of silviculture on yield, lumber value, and economic return. Forest Chron 64:127–131

Mitchell SA (2004) Operational forest harvest scheduling optimisation—a mathematical model and solution strategy. PhD thesis, University of Auckland, New Zealand. pp 278

Moore J (2012) Growing fit-for-purpose structural timber. What is the target and how do we get there? N Z J Forest 57(3):17–24

Moore JR, Lyon AJ, Searles GJ, Vihermaa LE (2009) The effects of site and stand factors on the tree and wood quality of Sitka spruce growing in the United Kingdom. Silva Fenn 43(3):383–396

Moore JR, Lyon AJ, Searles GJ et al (2013) Within– and between-stand variation in selected properties of Sitka spruce sawn timber in the United Kingdom: implications for segregation and grade recovery. Ann Forest Sci (in press)

Münster-Swendsen M (1987) Index of vigour in Norway spruce (*Picea abies* Karst.). J Appl Ecol 24:551–561

Murphy GE (1998) Allocation of stands and cutting patterns to logging crews using a Tabu search heuristic. Int J Forest Eng 9(1):31–38

Murphy G (2003) Reducing trucks on the road through optimal route scheduling and shared log transport services. South J Appl Forest 27:198–205

Murphy G, Lyons J, O'Shea M, Mullooly G, Keane E, Devlin G (2010) Management tools for optimal allocation of wood fibre to conventional log and bio-energy markets in Ireland: a case study. Eur J Forest Res 129:1057–1067. doi:10.1007/s10342-010-0390-3

Newnham RM (1992) Variable-form taper functions for four Alberta species. Can J Forest Res 22:210–223

Ormarsson S, Dahlblom O, Petersson H (1999) A numerical study of the shape stability of sawn timber subjected to moisture variation part 2: simulation of drying board. Wood Sci Technol 33:407–423

Papps SR, Manley BR (1992) Integrating short-term planning with long-term forest estate modelling using FOLPI. In: Proceedings of the conference on integrating forest information over space and time, Canberra, Australia, 13–17 Jan, pp 188–198

Pearson RG (2006) Climate change and the migration capacity of species. Trends Ecol Evol 21(3):111–113

PEFC (2010) Sustainable forest management – requirements: PEFC ST 1003:2010. PEFC, Geneva, p 16, http://www.pefc.org/standards/technical-documentation/pefc-international-standards-2010/676-sustainable-forest-management-pefc-st-10032010

Perera PKP, Amarasekera HS, Weerawardena NDR (2012) Effect of growth rate on wood specific gravity of three alternative timber species in Sri Lanka; *Swietenia macrophylla*, *Khaya senegalensis* and *Paulownia fortune*. J Trop Forest Environ 2(01):26–35

Petersen AK, Solberg B (2005) Environmental and economic impacts of substitution between wood products and alternative materials: a review of micro-level analyses from Norway and Sweden. Forest Policy Econ 7:249–259

Pirotti F, Guarnieri A, Vettore A (2010) Analisi di dati lidar waveform Sulla struttura Della vegetazione e Sulla morfologia costiera. Ital J Remote Sens 42((2):117–127, http://dx.doi.org/10.5721/ItJRS20104229

Pretty JN (1997) The sustainable intensification of agriculture. Nat Resour Forum 21(4):247–256

Putz FE, Sist P, Fredericksen T, Dykstra D (2008) Reduced-impact logging: challenges and opportunities. Forest Ecol Manag 256:1427–1433. doi:10.1016/j.foreco.2008.03.036

Ross RJ, McDonald KA, Green DW, Schad KC (1997) Relationship between log and lumber modulus of elasticity. Forest Prod J 47(2):89–92

Sankaran KV, Chacko KC, Pandalai RC, Mendham DS, Grove TS (2005) Sustaining productivity of eucalypt plantations in Kerala, India. Int For Rev 7(5):14

Schutt C, Aschoff TT, Winterhalder D et al (2004) Approaches for recognition of wood quality of standing trees based on terrestrial laser scanner data. In: ISPRS proceedings of laser-scanners for forest and landscape assessment, Freiburg, Germany 3–6 Oct, pp 179–182

Sedjo RA (1999) The potential of high-yield plantation forestry for meeting timber needs. New Forest 17:339–359

Sedjo RA (2010) The future of trees: climate change and the timber industry. Resources for the future: Winter/Spring 2010, Number 174:29–33. http://www.rff.org/RFF/Documents/Resources_174_2009_Future_of_Trees.pdf

Sepulveda P, Oja J, Gronlund A (2002) Predicting spiral grain by computed tomography of Norway spruce. J Wood Sci 6(48):479–483

Skogforsk (2011) StanForD. Listing of variables by category. http://www.skogforsk.se/PageFiles/60712/AllVarGrp_ENG_110504.pdf. Accessed Oct 13 2011

Suárez J (2010) An analysis of the consequences of stand variability in Sitka spruce plantations in Britain using a combination of airborne LiDAR analysis and models. PhD Thesis, University of Sheffield, p 285

Suárez J, Rosette J, Nicoll B, et al (2008) A practical application of airborne LiDAR for forestry management in Scotland. In: Hill R, Rosette J, Suárez J (eds) 8th international conference on LiDAR applications in forest assessment and inventory, Edinburgh. ISBN 978-0-85538-774-7

Sutton WRJ (1999) The need for planted forests and the example of radiata pine. New Forest 17:95–109

Tilman D, Balzer C, Hill J, Befort BL (2011) Global food demand and the sustainable intensification of agriculture. PNAS 108(50):20260–20264, www.pnas.org/cgi/doi/10.1073/pnas.1116437108

Todoroki C, Rönnqvist M (2002) Dynamic control of timber production at a sawmill with log sawing optimization. Scand J Forest Res 17:79–89

Tseheye A, Buchanan AH, Walker JCF (2000) Sorting of logs using acoustics. Wood Sci Technol 34:337–344

Vaisanen H, Kellomaki S, Oker-Blom P, Valtonen E (1989) Structural development of *Pinus silvestris* stands with varying initial density: a preliminary model for quality of sawn timber as affected by silvicultural measures. Scand J Forest Res 4:223–238

van Leeuwen M, Hilkera T, Coops NC, Frazer G, Wulder MA, Newnham GJ, Culvenor DS (2011) Assessment of standing wood and fiber quality using ground and airborne laser scanning: a review. Forest Ecol Manag 261:1467–1478

Vikram V, Cherry ML, Briggs D, Cress DW, Evans R, Howe GT (2011) Stiffness of Douglas-fir lumber: effects of wood properties and genetics. Can J Forest Res 41:1160–1173

Wang JZ, De Groot R (1996) Treatability and durability of heartwood. In: Ritter MA, Duwadi SR, Lee PDH (eds) National conference on wood transportation structures. Department of Agriculture, Forest Service, Forest Products Laboratory, Madison, pp 252–260

Wang X, Ross RJ, Carter P (2005) Acoustic evaluation of standing trees – recent research and development. In: Proceedings of the 14th international symposium on nondestructive testing of wood, Eberswalde, Germany, pp 455–465

Wang X, Ross RJ, Carter P (2007a) Acoustic evaluation of wood quality in standing trees part I. Acoustic wave behaviour. Wood Fiber Sci 39(1):28–38

Wang X, Carter P, Ross RJ, Brashaw BK (2007b) Acoustic assessment of wood quality of raw materials – a path to increased profitability. Forest Prod J 57:6–15

Weintraub A, Magendzo A, Magendzo A, Malchuck D, Jones G, Meacham M (1995) Heuristic procedures for solving mixed-integer harvest scheduling-transportation planning models. Can J Forest Res 25:1618–1626

Whiteside ID (1990) STANDPAK modelling system for radiata pine. In: James RN, Tarlton GL (eds) Proceedings of the IUFRO symposium on new approaches to spacing and thinning in plantation forestry. FRI Bull 151. New Zealand Ministry of Forestry, Forest Research Institute, Rotorua, pp 106–111

Whiteside ID, McGregor MJ (1987) Radiata pine sawlog evaluation using the sawing log yield model. In: Kininmonth, JA (ed) Proceedings of the conversion planning conference. FRI Bulletin 128. New Zealand Ministry of Forestry, Forest Research Institute, Rotorua, pp 124–146

Whittenbury CG (1997) Changes in wood products manufacturing. In: Kohn K, Franklin J (eds) Creating a forestry for the 21st century. Island Press, Washington D.C., U.S.A., pp 303–314

Wolcott M (2003) Production methods and platforms for wood plastic composites. In: Proceedings of the non-wood substitutes for solid wood products: new technologies and opportunities for growth conference, Melbourne, Australia

Woollons RC (2000) Comparison of growth of *Pinus radiata* over two rotations in the central north island. Int Forest Rev 2:84–89

Wynne RH (2006) Lidar remote sensing of forest resources at the scale of management. Photogramm Eng Remote Sens 72(12):1310–1314

Yang H-S, Kim DJ, Kim HJ (2003) Rice straw–wood particle composite for sound absorbing wooden construction materials. Bioresour Technol 86:117–121

Zeng H, Pukkala T, Peltola H (2007) The use of heuristic optimization in risk management of wind damage in forest planning. Forest Ecol Manag 241:189–199

Zhu K, Woodall CW, Clark JS (2012) Failure to migrate: lack of tree range expansion in response to climate change. Glob Chang Biol 18(3):1042–1105. doi:10.1111/j.1365-2486.2011.02571.x

Part IX
Forest Science, Including Ecological Studies

Community Genetics Applications for Forest Biodiversity and Policy: Planning for the Future

Adam S. Wymore, Helen M. Bothwell, Zacchaeus G. Compson,
Louis J. Lamit, Faith M. Walker, Scott A. Woolbright,
and Thomas G. Whitham

Abstract In many ecosystems, the genetic variation within foundation tree species drive key ecological processes. Here we present four key findings from community genetics research that can be applied to the preservation of forest biodiversity and improvement of management policy. (1) Different tree genotypes support different communities and different ecosystem processes; (2) primary productivity is, in part, genetically-based and is linked to biodiversity; (3) with changing climate, gene by environment interactions will affect forests and their dependent communities; and (4) minimum viable interacting population theory and analyses of species interaction networks provide a framework for integrating genetics into each of the above topics. Inclusion of community genetics in forest management is important because species evolve, exist, and interact within the context of a community. This approach allows for the creation of policies that are less susceptible to pitfalls inherit to single species management. Specific policy suggestions include the incorporation of a community genetics perspective in federally-funded reforestation and restoration

A.S. Wymore • H.M. Bothwell • Z.G. Compson • L.J. Lamit
F.M. Walker
Department of Biological Sciences, Northern Arizona University,
Flagstaff, AZ 86011, USA
e-mail: asw54@nau.edu

S.A. Woolbright
Department of Biological Sciences, Northern Arizona University,
Flagstaff, AZ 86011, USA

Institute of Genomic Biology, University of Illinois,
Urbana-Champaign, IL 61801, USA

T.G. Whitham (✉)
Department of Biological Sciences, Northern Arizona University,
Flagstaff, AZ 86011, USA

Merriam Powell Center for Environmental Research, Northern Arizona University,
Flagstaff, AZ 86011, USA
e-mail: Thomas.Whitham@nau.edu

T. Fenning (ed.), *Challenges and Opportunities for the World's Forests
in the 21st Century*, Forestry Sciences 81, DOI 10.1007/978-94-007-7076-8_31,
© Springer Science+Business Media Dordrecht 2014

projects, the use of non-local genotypes where climate change is predicted, and the development of provenance trials in a community context to identify superior genotypes and mixes of genotypes that will perform best in a changing environment. Such amendments to policy would unite the efforts of research, production, and conservation to maximize tax-payer investment.

1   Introduction

Genetic variation is the cornerstone of a species' ability to persist and to potentially adapt to environmental change (Fisher 1930). Numerous ecologically-based studies have shown the positive effects of increased genetic variation and diversity. These studies include increased pathogen control in rice agriculture (Zhu et al. 2000), increases in annual net primary productivity (Fig. 1; Crutsinger et al. 2006), resistance to disturbance (Hughes and Stachowicz 2004), and improved estimates of species response to global climate change (O'Neill et al. 2008). Although much

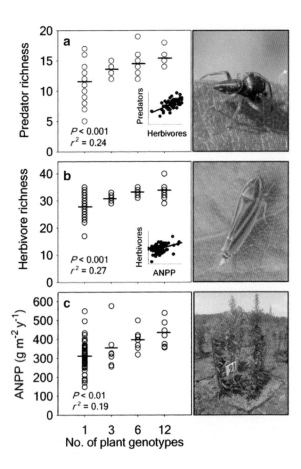

Fig. 1 Genetic diversity is important to consider when assessing the influence of genetics on community and ecosystems processes. As the number of goldenrod (*Solidago*) genotypes increases rates of aboveground net primary productivity (*ANPP*), herbivore richness and predator richness all increase significantly (Crutsinger et al. 2006)

progress has been made in understanding the fundamental principles of how whole communities respond to genetic variation in foundation species, their applications to management, conservation, and climate change are just beginning to emerge. For example, many current conservation efforts focus on federally listed and rare species that might affect relatively few other species; however, there is a growing appreciation and understanding of the importance of focusing on foundation species that create habitat and largely structure whole communities and ecosystems (Dayton 1972; Ellison et al. 2005). Genetic variation in these species is especially important to conserve as they interact with numerous other species and these interactions play a major role in defining their respective ecosystems (Whitham et al. 2010). Failure to account for genetically-based interactions can lead to degraded or simplified communities and ecosystems (Soulé et al. 2003). Without recognition of the evolutionary and ecological role of genetic variation, future conservation efforts risk failure (Noss 2001; Laikre et al. 2010; Laikre 2010).

In this chapter we explore several ways in which community genetics can be incorporated into policies regarding forest management. First we provide a brief review of community genetics and how genetic variation within foundation tree species can support and stabilize higher levels of biological diversity. Second, we provide examples of how the maintenance of genetic variation can be merged with agriculture and commercial production while simultaneously supporting conservation goals. Third, we outline how genetic variation should be managed in the context of climate change. Fourth, we describe Minimal Viable Interacting Populations (MVIP) and network theory as a conceptual framework to integrate genetics into the above topics. We conclude with a series of policy recommendations that combine the efforts of research, conservation and production.

2 Fundamental Principles of Forest Community Genetics

Genetic variation within a population can have profound effects on community structure and ecosystem processes. However, such a statement begs the question: genetic variation of *which* species deserves attention and priority? Focusing efforts on all species within a given ecosystem is currently impractical. The field of community genetics, defined as the study of genetic interactions between species and the abiotic environment within a community or ecosystem context (Whitham et al. 2003, 2006; Wymore et al. 2011), offers an approach to manage this complexity. Community genetics often focuses on foundation species, organisms that "structure a community by creating locally stable conditions for other species, and by modulating and stabilizing fundamental ecosystem processes" (Dayton 1972; Ellison et al. 2005). In many forest ecosystems, dominant tree species are likely to be foundation species.[1] By focusing on the genetic variation within foundation tree species,

---

[1] The term foundation species overlaps with other concepts such as keystone species, ecosystem engineer and dominant species. All of these terms refer to species that have a major influence on the local ecosystem. For consistency within this chapter, and with the field of community genetics,

researchers can better account for the ecological and evolutionary consequences resulting from these interactions. Examples of foundation tree species include eastern hemlock (*Tsuga canadensis*), whitebark pine (*Pinus albicaulis*), American chestnut (*Castanea dentata*), narrowleaf cottonwood (*Populus angustifolia*), one-seed juniper (*Juniperus monosperma*), and Tasmanian blue gum (*Eucalyptus globulus*) (Ellison et al. 2005; Schweitzer et al. 2008; Barbour et al. 2009; Kane et al. 2011).

Different genotypes within foundation tree species can support different plant, microbial and animal communities. For instance, within individual cottonwood species, genetic factors have been shown to impact plant and arthropod communities (Shuster et al. 2006; Keith et al. 2010), soil microbes (Schweitzer et al. 2008), bark lichens (Lamit et al. 2011), twig endophytes (J. Lamit, unpublished data), trophic interactions (Bailey et al. 2006), and ecosystem processes such as decomposition and nutrient cycling (Schweitzer et al. 2008), source-sink dynamics (Compson et al. 2011), and productivity (Lojewski et al. 2009; Grady et al. 2011). These communities and their associated ecosystem processes can be thought of as "community and ecosystem phenotypes" (Whitham et al. 2006) because they are genotype dependent. Several of the above studies have shown that these phenotypes exhibit broad-sense heritability, where related individuals tend to support similar communities or ecosystem processes (see Shuster et al. 2006; Keith et al. 2010; Compson et al. 2011; Lamit et al. 2011). Such community and ecosystem phenotypes of foundation trees have been observed across a wide range of forest ecosystems including temperate deciduous forests (Madritch et al. 2006; Donaldson and Lindroth 2007), pine-oak forests (Tovar-Sanchez and Oyama 2006), pinyon-juniper woodlands (Sthultz et al. 2009), tropical forests (Zytynska et al. 2011), broadleaf eucalypt forests (Barbour et al. 2009), mallee forests (Dungey et al. 2000) and riparian zones of the northeastern (Hochwender and Fritz 2004) and southwestern United States (Bangert et al. 2008).

The effect of increasing genotypic diversity can also be observed at larger scales, including the patch and stand level. Genotypic diversity has been shown to increase biodiversity of associated communities in herbaceous plants such as goldenrod (*Solidago altissima*) (Genung et al. 2010), evening primrose (*Oenothera biennis*) (Cook-Patton et al. 2011) and sea grass (reviewed in Duffy 2006). A similar influence of tree genetic diversity has also been shown for stands of trees in hybrid systems including cottonwoods (Wimp et al. 2004) and oaks (Tovar-Sanchez and Oyama 2006) (but see Kanaga et al. 2009; Tack and Roslin 2011). The process of genotypic diversity supporting higher levels of community diversity may be due to a diverse array of niche space created by the presence of different and multiple host genotypes (e.g., Booth and Grime 2003; Crutsinger et al. 2006; Hughes et al. 2008). A second mechanism is facilitation, where one plant species or genotype positively influences a neighboring species or even a whole community. For example, Michalet et al. (2011) argued that genetically-based differences in the architecture of the cushion plant, *Geum rossii*, had a dramatic impact on the rest of the alpine plant community. Plants with open cushions hosted more species than cushions with a

---

we use the term, foundation species (Dayton 1972; Ellison et al. 2005). Soule et al. (2003) emphasize that foundation species are often extremely abundant and/or ecologically dominant.

tight architecture. Furthermore, even though this plant occupied only 15 % of the available habitat, all species were found growing within the stable environment of the cushion and six species were only found in this association.

In conclusion, three major findings of community genetics have emerged with fundamental implications for the forests of the world. First, different tree genotypes within a species can support different communities of organisms. Second, common garden studies show these community and ecosystem phenotypes represent heritable plant traits. Third, these effects are likely most important in foundation species, such as trees, exhibiting significant genetic variation. Non-foundation species can also exhibit significant community consequences (reviewed by Whitham et al. 2012), but their overall impacts on the community and ecosystem are likely much less than those of foundation species. Based on these findings, the choice of individual genotypes and source populations used in forest plantations, biofuels production, reforestation, and restoration will likely have major community, ecosystem and biodiversity consequences. We argue that management policy needs to be cognizant of these effects, especially in a world where increasing demands on land use and climate change will affect future forest productivity and distribution. The remainder of this chapter emphasizes these fundamental consequences.

3 Tree Productivity, Tree Genetics and Biodiversity

The previous section highlighted how communities are sensitive to variation in tree genetics. Here we narrow the focus to specifically discuss how a genetically based ecological and economically important trait, tree productivity, influences associated communities. Variation in plant growth is often connected to variation in associated organisms (e.g., Price 1991; Gange et al. 2007; Lamit et al. 2011). Work with pinyon pine (*Pinus edulis*) in the southwestern U.S.A. reveals that although many organisms have associations with tree growth, the nature of the relationship varies. For instance, richness and abundance of communities of foliar arthropods increases linearly with radial trunk growth (Fig. 2; Stone et al. 2010), whereas the abundance of ectomycorrhizal fungi colonizing pinyon pine roots exhibits a curvilinear relationship with radial trunk growth (Swaty et al. 2004). Plant productivity is under partial genetic control, and several studies indicate that the relationship between variation in tree productivity and associated organisms has an underlying genetic basis (e.g., Korkama et al. 2006; Crutsinger et al. 2008; Lamit et al. 2011). Given the importance of tree productivity to natural forests and plantations, and the demonstrated link between productivity and biodiversity, it would seem that production and biodiversity goals are not necessarily mutually exclusive in commercial forests. Furthermore, commercial forests should be studied to help understand the links between tree genetics, productivity and associated organisms to better conserve biodiversity.

Plant genetic influences on associated organisms can have strong ecological consequences at the stand or plot level due to positive effects of genetic diversity on productivity (Hughes et al. 2008). Mixtures of plant genotypes can be more productive

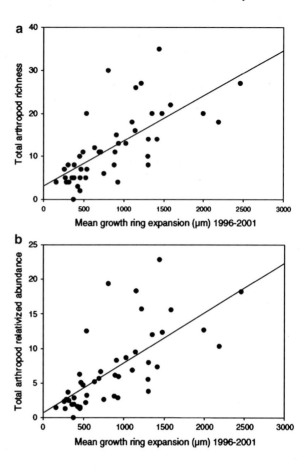

Fig. 2 Arthropod species richness (**a**) and relativized abundance (**b**) show positive relationships with average growth of individual trees (Stone et al. 2010)

relative to monocultures of fast growing genotypes, because of greater resource use efficiency and facilitative interactions among genotypes (see Hughes et al. 2008). For example, Crutsinger et al. (2006) showed that arthropod diversity in plots of tall goldenrod (*Solidago altisima*) increased as goldenrod genotypic diversity increased. The positive relationship between goldenrod genotypic and arthropod species diversity is a product of greater goldenrod productivity in plots with higher genotype richness and increasing diversity of plant resources in genotype mixtures relative to genotype monocultures (Crutsinger et al. 2006). The increase in productivity is not simply a result of highly productive genotypes dominating multi-genotype mixtures, but is partially due to positive interactions among genotypes (Crutsinger et al. 2006). Similar relationships are seen in a handful of other systems (Hughes et al. 2008), although more work needs to be done with stands of adult trees.

Due to the strong emphasis on productivity in agriculture, biofuels programs, and tree plantations, the broader consequences of growth and related traits (e.g., herbivore resistance) are also important in commercial forests. For example, in a tree productivity trial with Norway spruce (*Picea abies*), fast growing genotypes had higher diversity of primarily mutualistic fungi than slow growing genotypes (Korkama et al. 2006). Importantly, some fungal species preferentially associated with either slow or fast growing genotypes, suggesting that a mixed stand of fast and slow growing genotypes would likely support the highest diversity of these fungi. Ectomycorrhizal fungi can have strong influences on plant growth and nutrient uptake, which can vary among fungal species and levels of fungal diversity (Baxter and Dighton 2001; Smith and Read 2008). Alteration of mycorrhizal fungal community composition and diversity through selection for high productivity tree genotypes may feed back to affect tree performance in unpredictable ways, and may also decrease overall fungal diversity in a stand.

Breeding or genetic modification often involves "genes of large effect" that alter fundamental plant traits such as productivity and defenses, it seems likely that unintended community and ecosystem phenotypes will often emerge that affect biodiversity. In a study with Bt (*Bacillus thuringiensis*) corn (*Zea mays*), which is genetically modified to increase herbivore resistance, Rosi-Marshall et al. (2007) demonstrated that Bt corn detritus can have strong negative impacts on aquatic invertebrates. Corn bi-products are frequently carried into waterways adjacent to fields, indicating a potentially large negative effect of Bt genes on biodiversity of organisms beyond targeted pests. In contrast, positive effects on non-target pest organisms have been found with Bt hybrid aspen trees (*Populus tremula x tremuloides*) (Axelsson et al. 2011). Our aim here is not to take a specific stance on tree breeding and genetic modification. However, due to the prevalence of breeding and genetic modification for these traits, and the possibility of plant material, propagules and genes having influences beyond the intended location (Rosi-Marshall et al. 2007; Langdon et al. 2010; Axelsson et al. 2011; Steinitz et al. 2011), it is important for managers and conservation biologists to have a clear understanding of the ecological consequences for the world's forests. This is necessary in order to adapt policy where needed to ensure the protection of biodiversity via the management and reduction of unintended consequences.

The extended effects of stand-level genetic diversity have important implications for intensively managed tree plantations and natural forests managed for production. Mixtures of tree genotypes may be as productive, or more so, relative to monocultures of fast growing genotypes, due to more efficient resource use and facilitative interactions as discussed previously (Hughes et al. 2008). Mixtures of genotypes may also decrease the possibility of a single pest species or genotype laying waste to an entire stand (Zhu et al. 2000; Henery 2011). Although the yield improvements due to genotype mixtures in many studies with annual crops are typically modest (Smithson and Lenne 1996), we hypothesize that the benefits may be amplified by the longer time scales of forest production where pathogens can build up over many years (e.g., Lung-Escarmant and Guyon 2004). Using genotype mixtures to control for pests that negatively affect yield and productivity may also reduce the need for expensive pesticides that have extensive effects on non-target organisms (Zhu et al. 2000). Furthermore, employing genotype mixtures in both tree plantations and post-logging re-vegetation efforts in natural forests will maximize support of native biodiversity through their positive influence on diversity (Hughes et al. 2008) and negative influence on invasion by exotic species (Crutsinger et al. 2008). Because many forest managers are asked to manage for production *and* biodiversity, including tree genetic diversity in management plans and planting designs can help satisfy these seemingly opposing goals.

4 Genetic Diversity and Climate Change

A rapidly changing climate poses significant challenges to the management of our world's forests. Climate change has already altered many species' distributions. Using repeat photography, Hastings and Turner (1965) showed that plant communities shifted upward in elevation by approximately 1,000 ft during the previous 100 years. Recent studies further illustrate the effects of climate change. For example, Ault et al. (2011) reported dates of spring onset arriving at a rate of 1.5 days earlier per decade for three western North American woody shrub species. And climate models of British Columbia predict that the climate optima of many tree species (e.g., Douglas-fir, western hemlock, spruce, and ponderosa pine) will move approximately 100 km north per decade (Fig. 3; Hamann and Wang 2006). Most of these shifts in distribution are hypothesized to be the result of climate and environmental optima moving northward and up in elevation. Yet, only a small body of scientific literature has begun to examine the interactions between the effects of genetic variation and climate change (but see O'Neill et al. 2008; Sthultz et al. 2009).

Pinyon pine (*Pinus edulis*) provides an example of how a genetically-determined trait (i.e., insect resistance) can interact with climate to produce changes at the community level. Pinyon pine genotypes resistant to a stem-boring moth support significantly greater abundance of fungal-decomposers, ectomycorrhizal fungi, birds and mammals compared to their susceptible counterparts (Whitham et al. 2003). However, under a severe climatic event, these interactions changed dramatically. During a record drought year, moth-resistant genotypes suffered 68 % mortality compared to only 21 % mortality for moth-susceptible genotypes (Sthultz et al. 2009). Thus, over the span of a few years, drought selected against moth-resistant trees and their associated communities. Even ecosystem-level processes were affected in which resistant and susceptible trees had different microclimates, litter quality, and nutrient cycling (Classen et al. 2007). Such findings suggest that the impact of climate change on foundation tree species will cascade far beyond the tree itself to affect the whole community and ecosystem.

Provenance trials (e.g., transplanting trees across a wide range of climates) provide a powerful tool for studying local adaptation and the response of species to climate change by allowing researchers to measure performance of specific tree genotypes and source populations under different environmental conditions. Provenance trials can mimic the effects of climate change by moving genotypes into warmer or drier climates and are advantageous in providing long-term data. One trial consisting of 140 populations of lodgepole pine (*Pinus contorta*) was planted across 60 test sites in western Canada (Illingworth 1978). Nearly four decades later, researchers are still learning from these provenance trials. For example, O'Neill et al. (2008) developed an integrative method known as the Universal Transfer Function (UTF), which combines environmental and genetic effects in order to measure how key traits, such as productivity, change for particular genotypes when planted in new locations. As a result, this technique identifies particular genotypes that perform better under different climates.

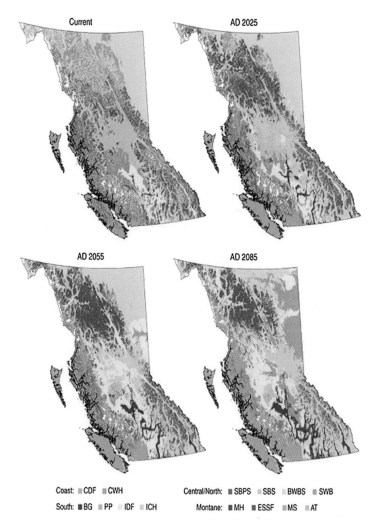

Coast: ▪CDF ▪CWH   Central/North: ▪SBPS ▪SBS ▪BWBS ▪SWB
South: ▪BG ▪PP IDF ▪ICH   Montane: ▪MH ▪ESSF ▪MS ▪AT

Fig. 3 Climate envelopes for several key ecological zones of British Columbia are predicted to shift northward at approximately 100 km per decade during this century (Hamann and Wang 2006). Each *color* represents a unique ecological zone (see key)

Forest planting policies typically mandate the use of nearby seed stock to ensure locally-adapted native genotypes. However, we now know that many species are already experiencing decreases in their optimum performance in their current locations. This may be due to the degree of local adaptation within certain populations. The more locally adapted populations are to current or past conditions, the more maladapted they may become to future climatic conditions (Grady et al. 2011). Results from provenance trials combined with climate model predictions can assist planting efforts by identifying appropriate seed stock and genotypes that will perform best in specific areas over the next century, especially in those locations where climate models predict local extinction due to unsuitable habitat.

Predictive climate models are important for identifying suitable future habitats and areas of conservation concern. Combined with data gathered from provenance trials, these techniques can greatly improve the success of restoration and forestry plantings. However, transfer of non-local genotypes into new regions could potentially have unintended consequences, including invasive behavior, greater susceptibility to local pests and disease, and loss of important symbioses such as mycorrhizal fungi associations that significantly increase tree productivity. Pollen from plantings will also be released, spreading novel genes through adjacent natural stands. Caution should be exercised and experiments undertaken to test performance in a community context in more complex wild systems. One such effort is underway in the American Southwest in which the National Science Foundation has funded the establishment of ten common gardens along an elevation gradient in two different soils types (Southwest Experimental Garden Array, SEGA). The goal of this array is to reciprocally transplant the same genotypes of individual species along with important associated community members to identify the genotypes and associated interactions that might best be adapted to ongoing climate change in one of the most strongly impacted regions in North America (Seager et al. 2007). In short, this effort is focused on identifying the genetic by abiotic environment by biotic environment interactions that will be crucial for successful restoration, as well as for understanding and mitigating community consequences of climate change.

Adapting policy to permit the use of non-local genotypes gives foresters and land managers greater flexibility to maintain production levels while concurrently managing for genetic diversity and the dependent biotic community. Facilitated migration offers a cost-effective and commercially successful means for coping with the effects of climate change on the timber industry (Province of British Columbia, Ministry of Forest and Range 2006). Recent British Columbia planting policy increased the distance seed stock can be moved 'uphill' to 500 m. This coincides with seed stock adapted to climates approximately 2 °C warmer than a given planting site currently experiences, an increase that corresponds to moderate climate change that forecasts trees are predicted to experience over their lifetimes (Marris 2009).

Without a shift in paradigm to allow increased flexibility in planting policies, we can expect some species to go locally extinct in many of their current locations. The Nature Conservancy, National Park Service, and the U.S. Forest Service have been proactive in revising their planting policy to allow non-local stock with regard to the above-mentioned SEGA project. The capacity to identify appropriate genotypes for facilitated migration is likely vital for the perpetuity of certain tree species and their associated communities. This issue is of immediate concern for parks, many of which were established to protect populations of specific focal species. Scott et al. (2002) found little predicted overlap between present and future forest tree species distributions in Canada's National Parks. And without intervention, Joshua Tree National Park in the American Southwest is predicted to see wholesale extirpation of its namesake (Cole et al. 2011).

Resource managers can achieve optimal survival and productivity by using models to match tree genotypes to locations they are best adapted (Hamann and Wang 2006). The response of a species to climate change varies throughout its range and is influenced not only by its genetic history, but also by the community of organisms it interacts with, and so management protocols that incorporate population-level variation within a community context are needed.

5   Minimum Viable Populations and Species Interactions

Estimates of the rate of overall current species extinctions are as much as 100 times higher than rates observed in the fossil record and are expected to increase in the future by as at least another order of magnitude (Mace et al. 2005). Habitat fragmentation and isolation are increasing at equally alarming rates, which will only exacerbate the problem. In an effort to combat species losses, much focus has been directed toward estimating the number of individuals necessary for species persistence within a specific geographic area (often socio-political boundaries such as national parks and preserves). The 'Minimum Viable Population' (MVP) concept has been proposed as a guideline for determining the minimum number of individuals necessary for ensuring species survival. It is defined as the number of individuals necessary to maintain an effective population size at equilibrium. This includes adequate genetic diversity needed to avoid inbreeding depression and to contend with and rebound from stochastic events (Shafer 1981). The inclusion of genetic data has important implications not only for the species in question, but also for the enhanced survival of associated organisms. Given the focus on individual species, policies that rely on the MVP, such as the U.S. Endangered Species Act, take a very limited approach to conservation, overlooking the value of a holistic process that includes evaluating the community- and ecosystem-level interactions for enhancing species survival.

Despite the apparent popularity of the MVP in the literature, it suffers from two major drawbacks. First, species differ dramatically in life history traits relating to reproductive strategy and success (e.g., time to maturity, fecundity, dispersal ability, mortality rate, etc.) and there is no "magic number" applicable to all species (Thomas 1990). Thus, the MVP must be estimated individually for each species in question. Second, and more significantly, no species "evolves in a vacuum" (Whitham et al. 2010; see also Thompson 2005; Wade 2007), and interactions with other organisms may be as important or more important as the MVP of a target species for its long-term survival. Until recently, the importance of interactions and community and ecosystem based efforts at conservation have been largely overshadowed by single species approaches.

Species interactions can have cascading effects on other associated organisms. In a review on the necessity of including species interactions for conservation goals, Soulé et al. (2003) showed how the extirpation of gray wolves from the northern Rocky Mountains, USA resulted in the degradation of forest and riparian habitats by elk. Prior to wolf reintroduction, preferential browsing by elk on willows led to a 60 % reduction in beavers (Soulé et al. 2003), a foundation species that provides habitat for numerous other species including fish, aquatic plants, waterfowl, wading birds and others. In addition, 25 % of breeding birds in the Yellowstone region specialize on these forest and riparian communities (Hansen et al. 1999). The effect on aspen has been the complete elimination of recruitment in the canopy over the last 80 years (Ripple and Larson 2000). Thus, removal of wolves resulted in effects that cascaded through multiple trophic levels to include the removal of a foundation species and a dramatic decline in habitat quality for birds and other organisms. This example illustrates the shortcomings of the MVP and shows the need for a new strategy that includes species interactions in a community context. Soulé and collaborators have proposed that the MVP be replaced by 'Ecologically Effective Population Densities' (EEPDs), defined as "densities that maintain critical interactions and help ensure against ecosystem degradation" (Soulé et al. 2003).

Since wolf reintroduction, restoration of riparian ecosystems, beaver pond communities (Ripple and Beschta 2003), aspen forests (Ripple and Beschta 2007), and songbird assemblages (Berger et al. 2001) in Yellowstone and other parks have been observed. However, efforts to de-list the status of wolves under the Endangered Species Act have set target numbers much lower than the suggested EEPD (Morell 2008). Thus, the resuming of wolf hunting and active depredation by land managers to keep populations at target levels may have dire consequences for aspen forests and riparian habitats, with cascading effects on beavers, birds, and numerous other species that depend on these ecosystems for survival.

The major differences between the MVP and the EEPD are the absence of species interaction perspectives in the former, and the lack of a genetic component in the latter. As we have seen in previous sections, genetic diversity, particularly in foundation species, can affect the distribution, abundance, and even the evolution of dependent species. To remedy this situation, Whitham et al. (2003, 2010) proposed the Minimum Viable Interacting Population (MVIP) defined as "the size of a population needed to maintain genetic diversity at levels required by other interacting species in the community". Thus, the MVIP combines the genetic approach of the MVP while recognizing the importance of species interactions and EEPDs to community and ecosystem conservation.

Interactions among individual species may not be readily apparent from simple observations of community composition. Recent developments in network theory provide tools for analyzing host genetic diversity and the resulting distribution and abundance of dependent organisms in a community context. Such analyses provide a means to tease apart the various ways species interact (directly or indirectly), allowing for a more in-depth understanding of the relationships between hosts, dependent species and other members of the community (i.e., the number of organisms interacting, the strength of interactions among groups, and how important specific organisms/species/genotypes are to the cohesion of the community). A hypothetical network analysis is illustrated in Fig. 4 (adapted from Lau et al. 2010). It demonstrates how both genetic and community-level data can be combined in order to effectively apply the MVIP concept. As illustrated, the inclusion of all

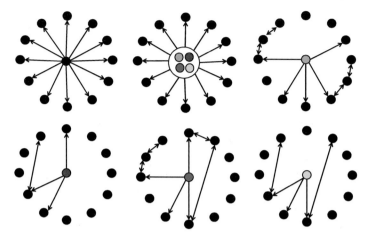

Fig. 4 Genetic Data, Species Interaction Networks and the MVIP (Adapted from Lau et al. 2010). The figure at *top left* represents a hypothetical population of a host species (*center dot*), with dependent organisms from the community represented by *outer dots*. The *arrows* suggest that each species is directly affected by the presence of the host species, and at this scale genotypic diversity is absent. The figure *top middle* includes hypothetical genotypes that differ in genetic make up (*colored dots*), each of which may support a different community. By partitioning the genotypic classes, we see that different genotypes not only directly affect certain members of the community (i.e. resistance vs. susceptibility), but that those members may in turn affect the distribution and abundance of other species indirectly affected by plant genes

genotypic classes of the host tree maximizes genetic diversity, and, as genotypes are removed, survival of dependent organisms decreases.

The above example illustrates the importance of plant genetic variation in conservation efforts and the MVIP concept. When environmental factors account for most of the observed differences among communities of different genotypes, the MVIP will approach the MVP. However, as community traits become more tightly linked to tree genetics in which different tree genotypes support different communities, the differences in the MVIP and MVP will also increase. In other words, when community patterns are largely the product of tree genotype, the MVIP is likely to be larger than the MVP needed to ensure survival of the foundation tree population (Whitham et al. 2003, 2010). Thus, conservation efforts should focus on maximizing the genetic diversity of foundation plant species such as trees in order to capture the maximum amount of biodiversity in the dependent community. However, it should also be noted that including differing tree genotypes in the MVP may still be inadequate for the minimum viable interacting population unless enough trees of similar genotypes are included to provide sufficient ecologically effective population densities for maintaining the MVPs of dependent species.

When conservation or restoration efforts are constrained by the amount of available habitat (e.g., within park or preserve boundaries), the MVP will necessarily represent the lower bound of the MVIP, and maximization of plant genetic variation should be a priority (which is by definition part of the MVP in any case). Community-wide surveys and network analyses may not be economically feasible for many communities. In such cases, DNA technologies (Sunnucks 2000) provide tools for maximizing the genetic diversity for the MVP, creating a useful proxy for the MVIP and for conserving the greatest amount of biodiversity in a given area.

6 Community Genetics and Land Management Policy

Studies highlighted in this chapter demonstrate that genetic variation within populations is a critical consideration for the management of forests. Individual tree genotypes can differ in three ecologically and evolutionarily important ways: (1) different tree genotypes can support unique biotic communities, from microbes to insects and vertebrates, (2) they can respond uniquely to environmental factors, and (3) they can vary with respect to growth rates and metrics of productivity. Ultimately, these interactions can affect biodiversity and the stability of ecosystems. In the concluding paragraphs we illustrate briefly these three major themes finishing with a series of recommendations for future policy and forest management protocols.

Genetic variation within foundation tree species can facilitate the differentiation of dependent species. Evidence for this comes from herbivore populations adapted to individual cottonwood genotypes (Evans et al. 2008) and such local adaptation has been shown in other species (see review by Mopper 1996; Mopper et al. 2000), Furthermore, diverse communities are now known to differ among genotypes or populations of the same plant species (reviewed by Whitham et al. 2012). This argues that conserving tree genetic diversity can simultaneously conserve genetic diversity within individual dependent species as well as greater biodiversity.

Genotype by environment interactions can fundamentally alter key traits such as resistance and susceptibility. For instance, some silky willow genotypes (*Salix sericea*) showed increases in herbivore abundance with increases in soil fertilizer, while other genotypes showed reduced levels of herbivores under the same fertilizer treatments (Orians and Fritz 1996). These final examples highlight the importance of considering genotype by environment interactions when predicting how tree species may respond to certain management protocols or environmental change.

These dynamic interactions resulting from genetically-based processes can be better accounted for by using an approach such as Minimal Viable Interaction Population (MVIP). A concept such as MVIP inherently includes the ecological reality that no species, foundation or otherwise, evolves or exists in a vacuum. Rather, management protocols built around the concept of MVIP can incorporate the influence of genetically-based processes while focusing on the conservation of communities and ecosystem processes.

Based on these findings, we suggest a series of revisions to land and forest management policies. First, we recommend mandating the incorporation of genetic diversity into future reforestation and restoration projects. This is especially important for long-lived species that will encounter changing climatic conditions this century (e.g., O'Neill et al. 2008; Grady et al. 2011). By increasing genetic diversity into production stands and planting provenances best adapted to predicted future climatic conditions, foresters could still meet production requirements while increasing local biodiversity, stabilizing ecosystem processes, and protecting stands against climate change. This requirement of genetic diversity into forestry protocols should be required for federally funded programs where plantings are currently focused on restoring for the current local climatic conditions rather than long-term climatic conditions. Failure to utilize genotypes and source populations that can survive in the likely future climatic conditions could result in the failure of expensive restoration and reforestation projects.

Second, we advocate a new policy to allow federal agencies (e.g., National Park Service) to use genotypes from non-local source populations for replanting and restoration projects. Genotypes that are better adapted for a changing climate can be identified via provenance trials and can be used to maintain and promote stable populations, communities and ecosystems into the future. Such provenance trials have been widely used and successfully applied in forestry. We recommend more complex experimental designs that incorporate multiple species and their interactions (e.g., Grady et al. 2011), which will provide increased resolution and allow researchers to better identify the best provenances to use at specific sites undergoing climate change.

Third, we suggest that Minimal Viable Interacting Population (MVIP) be used as a framework when designing protocols as it considers the influence of a population's genetic diversity on the local community and not just the species of interest. This community genetics approach explicitly includes the thousands of species associated with foundation tree species and their associated ecosystem processes. Because of the expense of reforestation and restoration projects, we propose that future forest management policies adopt a dual purpose model requiring research to be designed in tandem with forest projects. Because restoration and research are both expensive, we cannot afford to conduct each separately. By merging the two, we will leverage limited funds and both will benefit from the shared knowledge.

References

Ault TR, Macalady AK, Pederson GT, Betancourt JL, Schwartz MD (2011) Northern hemisphere modes of variability and the timing of spring in western North America. J Clim 24:4003–4014

Axelsson PE, Hjältén J, Whitham TG, Julkunen-Tiitto R, Pilate G, Wennström A (2011) Leaf ontogeny interacts with Bt modification to affect innate resistance in GM aspens. Chemoecology 21:161–169

Bailey JK, Wooley SC, Lindroth RL, Whitham TG (2006) Importance of species interactions to community heritability: a genetic basis to trophic-level interactions. Ecol Lett 9:78–85

Bangert RK, Lonsdorf EV, Wimp GM, Shuster SM, Fischer D, Schweitzer JA, Allan GJ, Bailey JK, Whitham TG (2008) Genetic structure of a foundation species: scaling community phenotypes from the individual to the region. Heredity 100:121–131

Barbour RC, O'Reilly-Wapstra JM, De Little DW, Jordan GJ, Steane DA, Humphreys JR, Bailey JK, Whitham TG, Potts BM (2009) A geographic mosaic of genetic variation within a foundation tree species and its community-level consequences. Ecology 90:1762–1772

Baxter JW, Dighton J (2001) Ectomycorrhizal diversity alters growth and nutrient acquisition of grey birch (Betula populifolia) seedlings in host–symbiont culture. New Phytol 152:139–149

Berger J, Stacey PB, Bellis L, Johnson MP (2001) A mammalian predator–prey imbalance: grizzley bear and wolf extinction affect avian neotropical migrants. Ecol Appl 11:947–960

Booth RE, Grime JP (2003) Effects of genetic impoverishment on plant community diversity. J Ecol 91:721–730

Classen AT, Chapman SK, Whitham TG, Hart SC, Koch GW (2007) Genetic-based plant resistance and susceptibility traits to herbivory influence needle and root litter nutrient dynamics. J Ecol 95:1181–1194

Cole KK, Ironside K, Eischeid J, Garfin G, Duffy PB, Toney C (2011) Past and ongoing shifts in Joshua tree distribution support future modeled range contraction. Ecol Appl 21:137–149

Compson ZG, Larson KC, Zinkgraf MS, Whitham TG (2011) A genetic basis for the manipulation of sink-source relationships by the galling aphid, Pemphigus betae. Oecologia 167:711–721

Cook-Patton SC, McArt SH, Parachnowitsch AL, Thaler JS, Agrawal AA (2011) A direct comparison of the consequences of plant genotypic and species diversity on communities and ecosystem function. Ecology 92:915–923

Crutsinger GM, Collins MD, Fordyce JA, Gompert Z, Nice CC, Sanders NJ (2006) Plant genotypic diversity predicts community structure and governs an ecosystem process. Science 313:966–968

Crutsinger GM, Souza L, Sanders NJ (2008) Intraspecific diversity and dominant genotypes resist plant invasion. Ecol Lett 11:16–23

Dayton PK (1972) Toward an understanding of community resilience and the potential effects of enrichments to the benthos at McMurdo Sound, Antarctica. In: Parker BC (ed) Proceedings of the colloquium on conservation problems. Allen, Lawrence, pp 81–96

Donaldson JR, Lindroth RL (2007) Genetics, environment, and their interaction determine efficacy of chemical defense in trembling aspen. Ecology 88:729–739

Duffy JE (2006) Biodiversity and the functioning of seagrass ecosystems. Mar Ecol Prog Ser 311:233–250

Dungey HS, Potts BM, Whitham TG, Li H-F (2000) Plant genetics affects arthropod community richness and composition: evidence from a synthetic eucalypt hybrid population. Evolution 54:1938–1946

Ellison AM, Bank MS, Clinton BD, Colburn EA, Elliott K, Ford CR, Foster DR, Kloeppel BD, Knoepp JD, Lovett GM, Mohan J, Orwig DA, Rodenhouse NL, Sobczak WV, Stinson KA, Stone JK, Swan CM, Thompson J, Von Holle B, Webster JR (2005) Loss of foundation species: consequences for the structure and dynamics of forested ecosystems. Front Ecol Environ 3:479–486

Evans LM, Allan GJ, Shuster SM, Woolbright SA, Whitham TG (2008) Tree hybridization and genotypic variation drive cryptic speciation of a specialist mite herbivore. Evolution 62:3027–3040

Fisher RA (1930) The genetical theory of natural selection. Clarendon, Oxford

Gange AC, Dey S, Currie AF, Sutton BC (2007) Site- and species-specific differences in endophyte occurrence in two herbaceous plants. J Ecol 95:614–622

Genung MA, Lessard JP, Brown CB, Bunn WA, Cregger MA, Reynolds WN, Felker-Quinn E, Stevenson ML, Hartley AS, Crutsinger GM, Schweitzer JA, Bailey JK (2010) Non-additive effects of genotypic diversity increase floral abundance and abundance of floral visitors. PLoS One 5:e8711

Grady KC, Ferrier SM, Kolb TE, Hart SC, Allan GJ, Whitham TG (2011) Genetic variation in productivity of foundation riparian species at the edge of their distribution: implications for restoration and assisted migration in a warming climate. Global Climate Change 17:3724–3735

Hamann A, Wang T (2006) Potential effects of climate change on ecosystem and tree species distribution in British Columbia. Ecology 87:2773–2786

Hansen AJ, Rotella JJ, Kraska MPV, Brown D (1999) Dynamic habitat and population analysis: an approach to resolve the biodiversity manager's dilemma. Ecol Appl 9:1459–1476

Hastings JR, Turner RM (1965) The changing mile: an ecological study of vegetation change with time in the lower mile of an arid and semiarid region. University of Arizona Press, Tucson

Henery ML (2011) The constraints of selecting for insect resistance in plantation trees. Agric For Entomol 13:111–120

Hochwender CG, Fritz RS (2004) Plant genetic differences influence herbivore community structure: evidence from a hybrid willow system. Oecologia 138:547–557

Hughes AR, Stachowicz RR (2004) Genetic diversity enhances the resistance of a seagrass ecosystem to disturbance. Proc Natl Acad Sci, USA 101:8998–9002

Hughes AR, Inouye BD, Johnson MTJ, Underwood N, Vellend M (2008) Ecological consequences of genetic diversity. Ecol Lett 11:609–623

Illingworth K (1978) Study of lodgepole pine genotype–environment interaction in B.C. In: Proceedings of International Union of Forestry Research Organizations (IUFRO) joint meeting of working parties: Douglas-fir provenances, Lodgepole pine provenances, Sitka spruce provenances, and Abies provenances, Vancouver, BC, Canada, pp 151–158

Kanaga MK, Latta LC, Mock KE, Ryel RJ, Lindroth RL, Pfrender ME (2009) Plant genotypic diversity and environmental stress interact to negatively affect arthropod community diversity. Arthropod-Plant Interactions 3:249–258

Kane JM, Meinhardt KA, Chang T, Cardall BL, Michalet R, Whitham TG (2011) Drought-induced mortality of a foundation species (Juniperus monosperma) promotes positive afterlife effects in understory vegetation. Plant Ecol 212:733–741

Keith AR, Bailey JK, Whitham TG (2010) A genetic basis to community repeatability and stability. Ecology 11:3398–3406

Korkama T, Pakkanen A, Pennanen T (2006) Ectomycorrhizal community structure varies among Norway spruce (Picea abies) clones. New Phytol 171:815–824

Laikre L (2010) Genetic diversity is overlooked in international conservation policy implementation. Conserv Genet 11:349–354

Laikre L, Allendorf FW, Aroner LC, Baker CS, Gregovich DP, Hansen MM, Jackson JA, Kendall KC, McKelvery K, Neel MC, Olivieri I, Ryman N, Schwartz MK, Short Bull R, Stetz JB, Tallmon DA, Taylor BL, Vojta CD, Waller DM, Waples RS (2010) Neglect of genetic diversity in implementation of the conservation of biological diversity. Conserv Biol 24:86–88

Lamit LJ, Bowker MA, Holeski LM, Næsborg RR, Wooley SC, Zinkgraf M, Lindroth RL, Whitham TG, Gehring CA (2011) Genetically-based trait variation within a foundation tree species influences a dominant bark lichen. Fungal Ecol 4:103–106

Langdon B, Pauchard A, Aguayo M (2010) Pinus contorta invasion in the Chilean Patagonia: local patterns in a global context. Biol Invasion 12:3961–3971

Lau MK, Whitham TG, Lamit LJ, Johnson NC (2010) Ecological and evolutionary interaction network exploration: addressing the complexity of biological interactions in natural systems with community genetics and statistics. JIFS 7:17–25

Lojewski NR, Fischer DG, Bailey JK, Schweitzer JA, Whitham TG, Hart SC (2009) Genetic basis of aboveground productivity in two native Populus species and their hybrids. Tree Physiol 29:1133–1142

Lung-Escarmant B, Guyon D (2004) Temporal and spatial dynamics of primary and secondary infection by Armillaria ostoyae in a Pinus pinaster plantation. Phytopathology 94:125–131

Mace G, Massundire H, Baillie J, Ricketts T, Brooks T, Hoffmann M, Stuart S, Balmford A, Purvis A, Reyers B et al (2005) Biodiversity. In: Hassan H, Scholes R, Ash N (eds) Ecosystems and human well being: current state and trends, vol 1. Island Press, Washington, DC, pp 77–122

Madritch MD, Donaldson JR, Lindroth RL (2006) Genetic identity of Populus tremuloides litter influences decomposition and nutrient release in a mixed forest stand. Ecosystems 9:528–537

Marris E (2009) Planting the forest of the future. Nature 459:906–908

Michalet R, Xiao S, Touzard B, Smith DS, Cavieres LA, Callaway RM, Whitham TG (2011) Phenotypic variation in nurse traits and community feedbacks define an alpine community. Ecol Lett 14:433–443

Mopper S (1996) Adaptive genetic structure in phytophagous insect populations. Trends Ecol Evol 11:235–238

Mopper S, Stiling P, Landau K, Simberloff D, Van Zant P (2000) Spatiotemporal variation in leaf miner population structure and adaptation to individual oak trees. Ecology 81:1577–1587

Morell V (2008) Ruling protects wolves. Science 25:475

Noss RF (2001) Beyond Kyoto: forest management in a time of climate change. Conserv Biol 15:578–590

O'Neill GA, Hamann A, Wang T (2008) Accounting for population variation improves estimates of the impact of climate change on species' growth and distribution. J Appl Ecol 45:1040–1049

Orians CM, Fritz RS (1996) Genetic and soil-nutrient effects on the abundance of herbivores on willow. Oecologia 105:388–296

Price PW (1991) The plant vigor hypothesis and herbivore attack. Oikos 62:244–251

Province of British Columbia, Ministry of Forests and Range (2006) Preparing for climate change: adapting to impacts on British Columbia's forests and range resources. www.for.gov.bc.ca/mof/Climate_Change/Preparing_for_Climate_Change.pdf. 12 Jan 2012

Ripple WJ, Beschta RL (2003) Wolf reintroduction, predation risk, and cottonwood recovery in Yellowstone National Park. For Ecol Manag 184:299–313

Ripple WJ, Beschta RL (2007) Restoring Yellowstone's aspen with wolves. Biol Conserv 138:514–519

Ripple WJ, Larson EJ (2000) Historic aspen recruitment, elk, and wolves in northern Yellowstone National Park, USA. Biol Conserv 95:361–370

Rosi-Marshall EJ, Tank JL, Royer TV, Whiles MR, Evans-White M, Chambers C, Griffiths NA, Pokelsek J, Stephen ML (2007) Toxins in transgenic crop byproducts may affect headwater stream ecosystems. Proc Natl Acad Sci USA 104:16204–16208

Schweitzer JA, Bailey JK, Fischer DG, LeRoy CJ, Lonsdorf EV, Whitham TG, Hart SC (2008) Plant-soil-microorganism interactions: heritable relationship between plant genotype and associated soil microorganisms. Ecology 89:773–781

Scott D, Malcolm JR, Lemieux C (2002) Climate change and modeled biome representation in Canada's national park system: implications for system planning and park mandates. Glob Ecol Biogeogr 11:475–484

Seager R, Ting M, Held I, Kushnir Y, Lu J, Vecchi G, Huang H, Harnik N, Leetmaa A, Lau N, Li C, Velez J, Naik N (2007) Model projections of an imminent transition to a more arid climate in southwestern North America. Science 316:1181–1184

Shafer ML (1981) Minimum population sizes for species conservation. Bioscience 31:131–134

Shuster SM, Lonsdorf EV, Wimp GM, Bailey JK, Whitham TG (2006) Community heritability measures the evolutionary consequences of indirect genetic effects on community structure. Evolution 60:991–1003

Smith SE, Read DJ (2008) Mycorrhizal symbiosis, 3rd edn. Academic, London

Smithson JB, Lenne JM (1996) Varietal mixtures: a viable strategy for sustainable productivity in subsistence agriculture. Ann Appl Biol 128:127–158

Soule ME, Estes JA, Berger J, del Rio CM (2003) Ecological effectiveness: conservation goals for interactive species. Conserv Biol 17:1238–1250

Steinitz O, Robledo-Arnuncio JJ, Nathan R (2011) Effects of forest plantations on the genetic composition of conspecific native Aleppo pine populations. Mol Ecol 21:300–313

Sthultz CM, Gehring CA, Whitham TG (2009) Deadly combination of genes and drought: increased mortality of herbivore-resistant trees in a foundation species. Glob Chang Biol 15:1949–1961

Stone AC, Gehring CA, Whitham TG (2010) Drought negatively affects communities on a foundation tree: growth rings predict diversity. Oecologia 164:751–761

Sunnucks P (2000) Efficient genetic markers for population biology. Trends Ecol Evol 15:199–203

Swaty RL, Deckert RJ, Whitham TG, Gehring CA (2004) Ectomycorrhizal abundance and community composition shifts with drought: predictions from tree rings. Ecology 85:1072–1084

Tack AJM, Roslin T (2011) The relative importance of host-plant genetic diversity in structuring the associated herbivore community. Ecology 92:1594–1604

Thomas CD (1990) What do real population dynamics tell us about minimum viable population size? Conserv Biol 4:324–327

Thompson JN (2005) The geographic mosaic of coevolution. University of Chicago Press, Chicago

Tovar-Sánchez E, Oyama K (2006) Effect of hybridization of the Quercus crassifolia x Quercus crassipes complex on the community structure of endophagous insects. Oecologia 147:702–713

Wade MJ (2007) The coevolutionary genetics of ecological communities. Nat Rev Genet 8:185–195

Whitham TG, Young WP, Martinsen GD, Gehring CA, Schweitzer JA, Shuster SM, Wimp GM, Fischer DG, Bailey JK, Lindroth RL, Woolbright S, Kuske CR (2003) Community and ecosystem genetics: a consequence of the extended phenotype. Ecology 84:559–573

Whitham TG, Bailey JK, Schweitzer JA, Shuster SM, Bangert RA, LeRoy CJ, Lonsdorf EV, Allan GJ, DiFazio SP, Potts BM, Fischer DC, Gehring CA, Lindroth RL, Marks JC, Hart SC, Wimp GM, Wooley SC (2006) A framework for community and ecosystem genetics: from genes to ecosystems. Nat Rev Genet 7:510–523

Whitham TG, Gehring CA, Evans LM, LeRoy CJ, Bangert RK, Schweitzer JA, Allan GJ, Barbour RC, Fischer DG, Potts BM, Bailey JK (2010) A community and ecosystem genetics approach to conservation biology and management. In: DeWoody A, Bickham J, Michler C, Nichols K, Rhodes G, Woeste K (eds) Molecular approaches in natural resource conservation and management. Cambridge University Press, New York, pp 50–73

Whitham TG, Gehring CA, Lamit LJ, Wojtowicz T, Evans LM, Keith AR, Smith DS (2012) Community specificity: life and afterlife effects of genes. Trends in Plant Science. doi:10.1016/j.tplants.2012.01.005

Wimp GM, Young WP, Woolbright SA, Martinsen GD, Keim P, Whitham TG (2004) Conserving plant genetic diversity for dependent animal communities. Ecol Lett 7:776–780

Wymore AS, Keeley ATK, Yturralde KM, Schroer ML, Propper CR, Whitham TG (2011) Genes to ecosystems: exploring the frontiers of ecology with one of the smallest biological units. New Phytol 191:19–36

Zhu Y, Chen H, Fan J, Wang Y, Li Y, Chen J, Fan J, Yang S, Hu L, Leung H, Mew TW, Teng PS, Wang Z, Mundt CC (2000) Genetic diversity and disease control in rice. Nature 406:718–722

Zytynska SE, Fay MF, Penney D, Preziosi RF (2011) Genetic variation in a tropical tree species influences the associated epiphytic plant and invertebrate communities in a complex forest ecosystem. Philos Trans Roy Soc B 366:1329–1336

The Long Reach of Biogenic Emissions in the Atmosphere

Sanford Sillman

Abstract This chapter provides a brief summary of the impacts of biogenic emissions in the atmosphere, focusing on reactive organics. These species have a significant impact on the chemical balance of the troposphere and also affect the formation of ozone and other secondary species in polluted regions. Methane, though slowly reacting, also plays an important role in the chemistry of the global troposphere. Methane and aerosol production from biogenic organics both affect climate. Two distinctions are emphasized: (i) between naturally occurring emissions from largely unperturbed ecosystems (usually forests) as opposed to emissions associated with agriculture and other human activities; and (ii) between reactive emissions (including slowly reacting species) and completely nonreactive species such as carbon dioxide. The reactive species are eventually removed from the atmosphere, but carbon dioxide and other nonreactive species accumulate in the atmosphere over time and are only removed on time scales of 100 years or longer.

It is widely known that emissions from the biosphere have a large effect on the evolution of the atmosphere. Most immediately, respiration of CO_2 and exhale of O_2 by plants causes a noticeable seasonal atmospheric cycle, with CO_2 decreasing during the northern summer at the time of maximum net photosynthesis on a global scale. In addition to the global cycle, the biosphere represents a reservoir of carbon that can be exchanged with atmospheric CO_2, with time scales ranging from years (for most plant species) to centuries (for stored carbon in soils). On geologic time scales the evolution of the biosphere is directly associated with release of free oxygen (O_2) to the atmosphere, and even today the reservoir of atmospheric oxygen turns over on a time scale of approximately 25 million years. The increase in atmospheric

S. Sillman (✉)
Department of Atmospheric, Oceanic and Space Sciences,
University of Michigan, Ann Arbor, MI 48109-2143, USA
e-mail: sillman@umich.edu

T. Fenning (ed.), *Challenges and Opportunities for the World's Forests
in the 21st Century*, Forestry Sciences 81, DOI 10.1007/978-94-007-7076-8_32,
© Springer Science+Business Media Dordrecht 2014

oxygen over geologic time (from <0.1 % in the early Earth to approximately 13 % at the start of the Cambrian era and 21 % today) was associated initially with the breakdown of H_2O to form free oxygen and subsequently (including the current long-term cycling of O_2) with the conversion of sulfate and carbonate rocks to organic plant matter.

Over the past 30 years there has been renewed interest in the release of shorter-lived chemically reactive species from the biosphere and its impact on the photochemical balance of the atmosphere. The reactive species include reactive nitrogen (NO and NO_2, collectively known as NO_x) and a range of biogenic hydrocarbons, each with reaction time scales of a few hours in the atmosphere. These species (NO_x and reactive organics) are generally associated with the chemistry of air pollution. The initial studies of the chemistry of polluted air in Los Angeles in the 1950s identified reactions of organics and NO_x that lead to the formation of ozone (O_3), and the reaction sequences identified at that time remain central to the chemistry of polluted environments today. Beginning in the late 1970s it was found that analogous organics are released into the atmosphere by trees, a discovery that received a brief burst of publicity during the 1980 U.S. presidential campaign. These biogenic emissions have a major role in the photochemistry of the atmosphere in remote locations and on global scales. In the late 1980s it was also discovered that biogenic emissions also affected the formation of ozone in polluted regions and that accounting for these emissions had a major impact on the selection of control strategies for ozone. There is still considerable uncertainty about the photochemical breakdown of biogenic organics in the atmosphere, its impact on ozone and reactive nitrogen, and its role in producing organic aerosols. Biogenic emissions also are a major source of atmospheric methane (CH_4), which, although it reacts relatively slowly (on a time scale 9–14 years), is one of the main factors in the photochemistry of the global troposphere. Biogenically produced methane and organic aerosols both have a large impact on the transfer of radiation in the atmosphere. These have a significant impact on climate and are of special concern for their possible impact on climate change today.

In analyzing biogenic emissions it is useful to make a distinction between ***naturally occurring*** emissions that are part of the natural ecosystems and the more general category of biogenic emissions. The term 'biogenic emissions' is often regarded as naturally occurring and unrelated to human activities, as opposed to anthropogenic, or human-related emissions. However, the term is also used to refer to emission of species (such as NO_x and methane) that are biological in origin but associated largely with human agriculture and animal husbandry and are strongly influenced by specific practices such as fertilization. A more important distinction is that between naturally occurring emissions, which I will define as emissions from biological organisms in naturally occurring and relatively unperturbed ecosystems, and anthropogenic emissions, which include emissions associated with agricultural practices and other artificial or semi-artificial ecosystems as well as activities related to modern industry and technology. Even this distinction is not absolute: emissions from naturally occurring ecosystems (forests, natural wetlands) are also impacted by the global extent of those ecosystems (which is heavily influenced by human activities) and changes associated with runoff of fertilizer and waste generated by human activities, atmospheric pollution, acid deposition and so on.)

Another important distinction for biogenic emissions involves the time scale. Species are regarded as ***fast-reacting*** (or ***reactive***) if they react on time scales from approximately 10 min to 10 days. Aerosols and semivolatile species (which partition between gas-phase and aerosol components) are also removed from the atmosphere (via surface deposition and usually also by wet deposition during rain events) on a time scale of 5–10 days even if the species are completely nonreactive. The fast-reacting species may be contrasted with ***slow-reacting*** gaseous species such as carbon monoxide (CO) and CH_4, which have atmospheric lifetimes ranging from 2 months (CO) to 14 years (CH_4) with respect to photochemical removal. Finally, special attention must be given to emitted species that are effectively ***nonreactive***, including CO_2 and the original chloro-fluorocarbons (CFC's). There are no immediate removal processes for these species in the troposphere. They participate in global cycles with effective time scales ranging from 75 years (CFCs) to several thousand years or more (CO_2). These nonreactive species have almost no impact in the immediate vicinity of the emission source, but they are associated with some of the most difficult and intractable environmental problems. These are the major drivers of climate change.

The time scale for reactive species also translates directly into a spatial scale of influence. Fast-reacting species often have a large influence on local atmospheric chemistry and are important in the development of ***urban-scale*** pollution evens (with spatial extent approximately 100 km) and ***regional-scale*** pollution events (with spatial scale 1,000 km). Biogenic emissions are especially important for regional events because the latter include a combination of emissions from rural environments (possibly including agriculture, naturally occurring forest ecosystems and wooded suburbs) in combination with on-site and transported emissions from anthropogenic sources. These urban and region-wide pollution events can persist for several days and have detrimental impacts on human health, agricultural productivity and forest ecosystems.

The other major spatial scale for atmospheric pollution involves the ***global-scale*** chemistry for the troposphere as a whole. The global tropospheric balance is dominated by chemistry in remote locations, over oceans and in the 'free troposphere' – which refers to the troposphere above 3,000 m, where the air is generally not in direct contact with surface emissions. The circulation of the atmosphere results in transport and distribution of surface emission sources throughout the troposphere, but the time scales for transport is usually 1 month or more. Chemical balances in the remote troposphere are strongly influenced by two slow-reacting species: carbon monoxide (CO, with atmospheric lifetime 2 months) and methane (CH_4, with lifetime 9–14 years). The fast-reacting species are present in the remote troposphere at much lower concentrations relative to near-source locations, but some fast-reacting species (notably, reactive nitrogen) have a major influence on remote chemistry despite the lower concentrations. Whereas the urban and regional-scale chemistry in polluted regions has an immediate effect in terms of human health and agriculture/ecology, the global-scale processes can affect longer-term atmospheric balances relating to climate. This difference is related to the time scale of months or years (global scale) as opposed to a time scale of days (urban and regional-scale).

The most critical distinction to make with regard to time scale for atmospheric processes is the distinction between these relatively tractable processes (urban, regional or global scale, with temporal response scales from days up to 10 years) and semi-permanent processes that involve long-term changes on a temporal scale that equals or exceeds the human lifespan. The most important long-term species is CO_2, which is also the principal driver of climate change. CO_2 cycles between four major reservoirs – the atmosphere, the biosphere, soils and the surface ocean (to a depth of 100 m) on time scales of a few years, and the biosphere, soil and surface ocean have magnitudes comparable to the atmosphere as reservoirs of atmospheric carbon (which includes both CO_2 and organic carbon in the biosphere which is readily converted to CO_2). However, the only way to remove carbon from these four reservoirs taken as a whole is through transport to the deep ocean, which is a much larger carbon reservoir than the atmosphere or biosphere. Transfer of carbon to the deep ocean (which is associated with the decay of organic matter that has fallen from the near-surface ocean ecosystem) occurs on a time scale of 1,000 years or longer and provides the ultimate sink for CO_2 released from fossil fuels. On shorter time scales little can be done to reduce the levels of atmospheric CO_2 or eliminate the excess carbon that results from fossil fuel usage.

For this reason, the problem of atmospheric CO_2 must be considered in a separate category from other anthropogenic impacts, even though other species also contribute to climate change. The other species that impact climate (most notably methane, black carbon and other aerosols) have atmospheric lifetimes ranging from a few months up to 14 years. These are all short-term in comparison to CO_2. The other species are also less central to modern human activity and lifestyles than CO_2, which unavoidably emitted by the burning of any fossil fuel.

The more reactive species, in contrast to CO_2, are more likely to have immediate rather than long-term impacts. Also in contrast to CO_2, the reactive emissions and their products (primarily ozone and aerosols) have a direct impact on human health and agricultural productivity. These effects are reversible: atmospheric conditions will change in response to changes in emissions. Again, in contrast to CO_2: the impact of reactive emissions reflects the current state of emissions rather than the cumulative total since the beginning of the industrial era. Because the immediate impacts are felt locally, they are more likely to elicit an effective response that addresses the problem.

The major reactive biogenic emissions are as follows:

- **Reactive organics**: The largest organic emissions are isoprene (C_5H_8) and terpenes such as α-pinene ($C_{10}H_{10}$). Both of these are naturally emitted from forests. Isoprene is emitted primarily by oak and other deciduous trees, while the terpenes are emitted primarily by conifers. These are short-lived species (with lifetimes 30 min-2 h), but they lead to complex sequences of reactions that may include long-lived products.
- **Methane (CH_4)**: Methane is associated with a variety of natural emissions (associated primarily with bacterial processes in wetlands) and from a variety of human agricultural activities (most notably, digestion in cattle and other

ruminants and anaerobic decomposition associated with rice cultivation). There are also significant anthropogenic sources, including emissions associated with natural gas production and emission from landfills. A common feature of many biogenic sources of methane is production under anaerobic conditions. In contrast to the reactive organics, methane is long-lived (9–14 years). Due to its long lifetime methane is fairly uniformly distributed throughout the troposphere. Methane is a major factor in the chemistry of the remote troposphere, especially in the mid – and upper troposphere (where there is little direct contact with surface emissions).

- **Nitrogen oxides (NO and NO$_2$, or NO$_x$):** These emissions are mainly associated with agricultural fertilizers. NO$_x$ is short-lived (2–4 h) but plays central role in the chemistry of the troposphere, so that even a small amount in remote locations and in the mid – and upper troposphere can have a large impact.
- **Nitrous oxide (N$_2$O):** In contrast to the nitrogen oxides, nitrous oxide is a non-reactive species. Its long lifetime in the atmosphere (120 years) makes it more comparable to CO$_2$ than to the reactive (even slow-reacting) emissions. N$_2$O is emitted from aerobic bacterial processes in soil, a process that is enhanced by agricultural fertilizers. Once in the atmosphere N$_2$O accumulates until it reaches the stratosphere, where it is photolyzed by ultraviolet radiation to form NO and NO$_2$. The resulting nitrogen oxides are an important source for the upper troposphere. Nitrous oxide is a greenhouse gas, and its total contribution to the greenhouse effect is comparable to methane and approximately one-third that of carbon dioxide. Nitrous oxide in the atmosphere has increased since pre-industrial times, but only by 16 %, which is far less than the increase in CO$_2$ (50 %). The contribution to climate change associated with the increase in N$_2$O since preindustrial times is therefore an order of magnitude lower than the impact of the increase in CO$_2$.

The nitrogen oxides have a major impact on the chemistry of the troposphere. Sources include both natural (biogenic N$_2$O and NO$_x$ created by lightning) and anthropogenic emissions (primarily from automobiles and coal-fired industry and power plants).

The biogenic hydrocarbons are noteworthy because global emissions (primarily isoprene) account for approximately 70 % of total emission of reactive organics. As such, the biogenic species have a large impact on the chemistry of the remote troposphere. Chemistry at remote forest sites is often dominated by just two species: isoprene and NOx. At unperturbed sites NO$_x$ is very low (100 ppt or less) and the resulting chemistry does not create dangerous reaction products and may even cause a decrease in harmful atmospheric ozone. However, a moderate input of anthropogenic NO$_x$ can result in significant photochemical production of ozone. These conditions occur widely in the eastern and southeastern US, where the combination of biogenic isoprene and NO$_x$ (the latter being emitted in large quantities from coal-fired power plants) contributes significantly to ozone production in region-wide pollution events. Similar problems could occur in tropical forested regions as modernizing economic conditions lead to increased NO$_x$ emissions (Fig. 1).

Fig. 1 The Sammis power plant, Steubenville, Ohio. Nitrogen oxides from coal-fired power plants and other anthropogenic sources, in combination with natural emissions of isoprene from deciduous forests, result in ozone formation and are a major contributor to ozone air pollution in the eastern U.S. Government regulation lead to a reduction in nitrogen oxide emissions in the early 2000s, leading to a significant reduction in ozone (http://www.eenews.net/assets/2012/02/08/photo_cw_02.jpg, http://www.eenews.net/public/climatewire/2012/02/08/1)

Ironically, the most important result pertaining to isoprene is its impact on pollution episodes in urban areas. The initial studies of urban smog in the 1950s focused on anthropogenic organic emissions and their role in the formation of ozone during pollution events. Isoprene emissions typically represent just 10 % of total organic emissions in urban locations, although their impact on the chemistry of smog events is enhanced by the rapid rate of reaction of both isoprene and many of its reaction products. In 1988 it was found that when isoprene was included in models for urban pollution formation, results suggested that reductions in emissions of anthropogenic hydrocarbons would be much less effective as a strategy for lowering ozone. These findings were amplified in the 1990s in a series of studies suggesting that in most of the eastern US meaningful reductions in ozone could only be achieved if there were reductions of NOx emissions. At the time major efforts had been launched to reduce emission of anthropogenics, but there was little change in atmospheric ozone. As the role of biogenic hydrocarbons was recognized, policy initiatives shifted to place greater emphasis on reductions of NO_x emissions. The shift to focus on NOx was largely credited for leading to reduced ozone in the eastern U.S. in the 2000s.

Isoprene contributes to issues of atmospheric composition and air quality in four distinct ways. It affects urban air quality directly, largely by determining whether rates of ozone formation in urban locations are sensitive to anthropogenic hydrocarbon emissions, NO_x, or both. It enhances ozone formation in rural areas, which is a function of available NO_x. It is also a dominant factor in the chemistry of air in remote forested regions where NO_x is very low. Lastly, isoprene is a significant atmospheric

Fig. 2 Haze over the Smoky Mountains, Tennessee. A characteristic "blue haze" can be formed by aerosols produced from terpenes, which are emitted from a variety of fragrant plants and especially from coniferous forests. However, today the main constituents of haze, even in the Smoky Mountains, are anthropogenically produced sulfate and nitrate aerosols http://wncmountainliving.blogspot.com/2010/07/blue-haze-on-mountains.html

source of carbon monoxide (CO), which has a long enough lifetime (2 months) to be widely distributed in the troposphere and participate in global-scale chemistry.

Other biogenic organic emissions (primarily terpenes, emitted largely by conifers) can lead directly to aerosol formation. The resulting "blue haze" can be entirely biogenic, although its formation is influenced by the presence of nitrogen oxides. The naturally occurring aerosols may be overshadowed locally by aerosols from anthropogenic sources (Fig. 2).

There is currently a large uncertainty about the impact of isoprene at the global scale, especially with regard to aerosols. The immediate photochemical reaction products of isoprene are all short-lived, like isoprene itself. However some of the subsequent products have lifetimes of a few days or more, long enough to influence chemistry through much of the troposphere. At the same time there is uncertainty about the sources of organic aerosols in the atmosphere. The aerosols are important because they have a detrimental impact on human health and because they have a complex influence on climate. It is known that some aerosols (sulfates) have a significant cooling effect on climate, while others (soot, or black carbon) have a major warming effect by blocking outgoing radiation. Aerosols also affect cloud reflectivity and cloud formation rates, each of which has climate implications. The organic aerosols add an additional level of complexity because it is not clear to what extent

Fig. 3 Rice paddies in Indonesia. Human agricultural activities (primarily rice cultivation and the raising of cattle and other ruminants) represent approximately 25 % of global methane emissions, and the expansion of agriculture (along with other anthropogenic activities) has contributed to a more-than-doubling of the concentration of methane in the atmosphere over the past 200 years (http://media.web.britannica.com/eb-media/97/20297-004-73911B76.jpg)

these species are sensitive to anthropogenic or biogenic emissions. The various isoprene reaction products (the peroxides in particular) are currently being investigated as possible sources of organic aerosols.

Methane also has impacts on local pollution effects and on the global scale, but in a different way from isoprene. Because it reacts slowly, methane has no direct impact on formation of pollutants during urban or regional-scale episodes. Instead, methane is distributed fairly uniformly throughout the troposphere, where it has a major impact on chemistry. This does not mean that methane has no impact on pollution events, however. It has been argued that controlling methane is a cost-effective strategy for reducing ozone. By reducing the global 'background' ozone, a reduction in methane would simultaneously lower ozone in cities and polluted regions around the world. The practical difficulty is that controls on methane can only be put into effect through international agreements. Methane is an important greenhouse gas, and emission reductions are being sought primarily for their impact on climate (Fig. 3).

The major uncertainties for methane are related less to its behavior in the atmosphere than to the possibility of large releases of stored methane as a result of climate change Concern has focused on methane releases from permafrost in response to warming, and also on releases from the continental shelves of oceans. This could have the effect of hastening climate change and of increasing ozone worldwide (Fig. 4).

A final topic worth considering is the stability of the atmospheric cycling of biogenics with respect to chemistry. Some of the most difficult environmental problems involve unstable systems that might suddenly transform to a different mode of

Fig. 4 The Lena River delta in northeastern Siberia. Natural emissions from wetlands associated with anaerobic bacteria is responsible for approximately 40 % of methane emissions to the atmosphere. It is uncertain how climate change will affect emissions from high-latitude wetlands and permafrost (http://www.awi.de/fileadmin/user_upload/News/Press_Releases/2010/2._Quartal_ 2010/2007_Eispolygone_KPiel_out_p.jpg, http://www.awi.de/en/news/press_releases/detail/item/ higher_wetland_methane_emissions_caused_by_climate_warming_40000_years_ago/?cHash=5 4bc4b14397e0ae4778c9dd7b0d20c80

behavior, often associated with runaway feedback cycles. Climate in general has unstable features: soil moisture and rainfall, snow cover and temperature are each coupled and reinforce each other. The chemistry of both isoprene and methane also involves positive feedback that amplifies the impact of changes in emissions, but there are also factors that are likely to insure stability.

Tropospheric chemistry is controlled largely by the OH radical, which initiates reaction sequences that convert organics (including isoprene and methane) to oxidized forms (ultimately CO and CO_2). At the same time, isoprene (in forest environments) and methane (in the larger free troposphere) act as major sinks of OH, so that increased emissions leads to reduced OH and longer atmospheric lifetimes for themselves. In the case of isoprene, it sometimes happens that local emissions exceed the source of OH, causing a large increase in chemical lifetimes and increased atmospheric concentrations by a factor of five or more. These large increases usually happen in forests dominated by oak, which are high emitters of isoprene. These localized events have little impact on the troposphere as a whole, since the excess isoprene is rapidly removed once it is dispersed beyond the immediate forest environment. Positive feedback on the global scale is important for methane, which (along with CO) represents a large sink for OH in the troposphere as a whole. The direct removal rate for methane through reaction with OH implies a lifetime of approximately 9 years based on ambient OH concentrations. When the removal rate for methane is calculated using complete representation of tropospheric chemistry, including the decrease in concentrations of OH resulting from an increase in methane, the lifetime of methane increased to 14 years.

There is little danger of a major shift in the atmospheric response to increases in isoprene or methane emissions because of many stabilizing features associated with OH chemistry. OH is created partly through photolysis of ozone followed by reaction with water vapor, a process that is not likely to change significantly. The reactions that remove isoprene and CH_4 often regenerate OH and subsequent reaction products are often an additional source of OH. This provides a level of stability to tropospheric chemistry that, sadly, is lacking in other issues relating to climate and atmospheric composition.

Changes in anthropogenic emissions and in biogenics associated with human activities, in combination with naturally occurring biogenic emissions, may affect air quality and impact human health, agricultural productivity and ecosystems. But these are gradual rather than catastrophic changes, and respond relatively rapidly to changes in human practices. The same cannot be said of many other challenges associated with the atmosphere and climate.

Acknowledgments This article is based on research activities supported by the National Science Foundation (grant #ATM-0454838), Environmental Protection Agency (STAR grant #R8333770) and by the University of Michigan. This article has not been subjected to peer and administrative review by either NSF or EPA, and therefore may not necessarily reflect the views of the agencies, and no official endorsement should be inferred.

References and Further Reading

Bloomer BJ, Stehr JW, Piety CA, Salawitch RJ, Dickerson RR (2009) Observed relationships of ozone air pollution with temperature and emissions. Geophys Res Lett 36:L09803. doi:10.102 9/2009GL037308

Chameides WL, Lindsay RW, Richardson J, Kiang CS (1988) The role of biogenic hydrocarbons in urban photochemical smog: Atlanta as a case study. Science 241:1473–1474

Dentener FJ, Krol MC (2002) Stability of tropospheric hydroxyl chemistry. J Geophys Res 107(D23):4715. doi:10.1029/2002JD002272

Fiore AM, Jacob DJ, Logan JA, Yin JH (1998) Long-term trends in ground level ozone over the contiguous United States, 1980–1995. J Geophys Res 103:1471–1480

Fiore AM, Jacob DJ, Field BD, Streets DG, Fernandes SD, Jang C (2002) Linking ozone pollution and climate change: the case for controlling methane. Geophys Res Lett 29(19):1919. doi:101.1029/2002GL015601

Forster P, Ramaswamy V, Artaxo P, Berntsen T, Betts R, Fahey DW, Haywood J, Lean J, Lowe DC, Myhre G, Nganga J, Prinn R, Raga G, Schulz M, Van Dorland R (2007) Changes in atmospheric constituents and in radiative forcing. In: Solomon S, Qin D, Manning M, Chen Z, Marquis M, Averyt KB, Tignor M, Miller HL (eds) Climate change 2007: the physical science basis. Contribution of working group I to the fourth assessment report of the intergovernmental panel on climate change. Cambridge University Press, Cambridge, UK/New York

Frost GJ et al (2006) Effects of changing power plant NOx emissions on ozone in the eastern United States: proof of concept. J Geophys Res 111:D12306. doi:10.1029/2005JD006354

Harvey LDD, Huang Z (1995) Evaluation of the potential impact of methane clathrate destabilization on future global warming. J Geophys Res 100(D2):2905–2926. doi:10.1029/94JD02829

Ito A, Sillman S, Penner JE (2009) Global chemical transport model study of ozone response to changes in chemical kinetics and biogenic volatile organic compounds emissions due to increasing temperatures: Sensitivities to isoprene nitrate chemistry and grid resolution. J Geophys Res 114:D09301. doi:10.1029/2008JD011254

Lin G, Penner JE, Sillman S, Taraborrelli D, Lelieveld J (2011) Global mechanistic model of SOA formation: effects of different chemical mechanisms. Atmos Chem Phys Discuss 11:26347–26413

O'Connor FM et al (2010) Possible role of wetlands, permafrost, and methane hydrates in the methane cycle under future climate change: a review. Rev Geophys 48:RG4005. doi:10.1029/2010RG000326

Pierce T, Geron C, Bender L, Dennis R, Tonnesen G, Guenther A (1998) Influence of increased isoprene emissions on regional ozone modeling. J Geophys Res 103:25611–25630

Prather MJ (2007) Lifetimes and time scales in atmospheric chemistry. Philos Trans R Soc Math Phys Eng Sci 365:1705–1726

Tarnocai C, Canadell JG, Schuur EAG, Kuhry P, Mazhitova G, Zimov S (2009) Soil organic carbon pools in the northern circumpolar permafrost region. Global Biogeochem Cycles 23:GB2023. doi:10.1029/2008GB003327

Wang JS, Logan JA, McElroy MB, Duncan BN, Megretskaia IA, Yantosca RM (2004) A 3-D model analysis of the slowdown and interannual variability in the methane growth rate from 1988 to 1997. Global Biogeochem Cycles 18:GB3011. doi:10.1029/2003GB002180

West JJ, Fiore AF, Horowitz LW, Mauzerall DL (2006) Mitigating ozone pollution with methane emission controls: global health benefits. Proc Natl Acad Sci 103(11)

Ecological Interactions of the Host-Insect System *Quercus robur* and *Tortrix viridana*

Hilke Schroeder and Riziero Tiberi

Abstract The interaction between herbivorous insects and their host plants is a never-ending race related to evolutionary adaptation. Plants have developed an armament against herbivore attacks including indirect defences which besides others are comprised of volatile substances, as well as toxic secondary metabolites act directly against feeding herbivores. Insects, however, can rapidly evolve mechanisms to adapt to these compounds to prevent being harmed by them. Thus, herbivorous insects represent a fascinating feeding guild that are comprised of several arthropod groups, and especially the lepidopteran genera that feed on tree species. One of these lepidopteran species is the green oak leaf roller, *Tortrix viridana* L., a major pest on oaks throughout Europe. Its' defoliating larvae use different species of the genus *Quercus* and cause severe damage to the oaks. Defoliation leads to decrease of wood formation and fructification, and to an increase of vulnerability to secondary pathogens (fungi, viruses and other insects). This tree-insect-system serve as a model system for a specialised herbivorous insect and its' host plant from as well an ecological as a molecular point of view combined with modelling aspects.

1 The Nature of Herbivory

The presence and activity of phytophagous (herbivorous) insects in forest habitats can very often be traced to the conditions which the tree populations are living in; these arthropods are in fact influenced both by climatic and seasonal conditions and

H. Schroeder (✉)
Thünen Institute of Forest Genetics, Grosshansdorf, Germany
e-mail: hilke.schroeder@vti.bund.de

R. Tiberi
Agro-Biotechnology and Plant Protection Department, University of Florence, Florence, Italy

T. Fenning (ed.), *Challenges and Opportunities for the World's Forests*
in the 21st Century, Forestry Sciences 81, DOI 10.1007/978-94-007-7076-8_33,
© Springer Science+Business Media Dordrecht 2014

by the host plants and other animals (Grison 1973; Dahlsten and Dreistadt 1984). Normally the continuous changes which occur in forest communities are the basis of the very dynamism of such populations. Such changes relate to the development of the trees, the composition and structure of the phytocenosis and its function, and may occur in limited areas or over vast territories, often without man's intervention (Waters and Stark 1980).

Plants ability to react to insect infestations depends to a large extent on abiotic and biotic environmental factors, with their specific vulnerability depending on the type of population they belong to (Tiberi et al. 1993). The extensive and changing face of forest landscapes in various European regions clearly offers suitable habitats for many species of phytophagous insects to develop. Insect fauna characteristics differ in relation to the specific characteristics of each ecosystem (physical aspects of the position and age of the trees, species diversity and population structure, growing methods used etc.). Such insect fauna also includes species able to periodically outbreak and capable of causing serious damage (Bovey 1970; Covassi 1989). The occurrence and repetition of attacks may vary greatly and in part relates to the complexity of the ecosystem.

It is well established that the biotic communities present in natural forest ecosystems are systematically more diversified than in simplified systems such as those which are purely agricultural or single species plantations grown for the industrial production of timber. The high species diversity typical of natural forests ensures efficient biological stability, which is the result of the dense network of relations which the organisms (animal and plant) habituate to each other as well as to abiotic factors. These relationships regulate the size of the various populations which in conditions of equilibrium have an essentially constant mean density.

In practice, a dynamic and preordained sequence of biocenotic equilibriums deriving from the interactions of the plant and animal component may be observed, in which the arthropods are the most highly represented group. As heterotrophic organisms, insects depend on plants but in turn play a decisive role in starting, and subsequently in sustaining, the mechanisms behind the dynamics of the plant component of forest systems. Through their action they help to regulate the productivity of the plants, and by transforming plant matter to animal matter perform a crucial step in the energy flow through the subsequent links in the trophic chain. Again, through the continual albeit uneven supply of organic residues of various types to organisms living in the soil, they influence the cycle of nutritional elements. Insects have an impact on the specific composition of plant communities in different ways; by selecting the host plants they influence or modify competitive relations among plants, thereby permitting the development of species that would otherwise be hindered in their growth and the spread of others that are better able to exploit the resources offered by the environment (Covassi 1989; Tiberi et al. 1993).

The specific composition of insect communities and population density of the insect species varying over time and space is subject to more or less marked variations during the developmental phases of the tree stand. The group of species prevailing on young trees is thereby gradually modified in the intermediate ages and again, in the long term, in the adult phase and then from maturity as far as senescence. In a natural system all the phytophagous insects of the various feeding groups are present at the same time as they are sure to find a place for feeding, and thereby developing, in the variegated plant component. Species with great ability to extract cellular liquids, phloem and xylem, depending on their specialisation, prevail at the expense of young trees. These sap-sucking insects are out and out "parasites" of the plants in that they demand high quality food and have no interest in interfering with the vegetative cycle of the host and thereby depressing the vegetative state of the host plant. The leaves, vegetating tips or lignifying parts are excellent sites for their development so that as the tree continues to grow, the vegetative growing parts continue to act as their habitat. Other species of different feeding groups, such as endophytic defoliators (leaf miners) or some coleopterans which need leaves or vegetating tips for maturation of their gonads may feed on even these same parts of the tree. Phytophagous insects which consume moderate amounts of plant matter are extremely demanding in terms of food quality, in the form of leaves, buds, shoots and green bark. Xylophagous insects, mainly phloem boring insects are usually only found on trees already subject to stress, of various origins or types.

As the youthful development of the trees progresses, the presence of true defoliating insect species increases. The vegetative vigour of the tree and increased leaf mass in this developmental phase assures prevalence of defoliators and their efficiency at transforming plant matter into animal matter. These insects belong mainly to the Lepidoptera order, but Coleoptera and Hymenoptera are also frequently present, the latter especially in hilly and mountain environments.

The ability of the trees to survive removal of their photosynthesising parts, not only leaves, varies according to tree species, the time of year, the succession of the various phytophagous species over the course of the season and, above all, on the age of the tree. It has been ascertained that defoliation is less well-tolerated by conifers than by broad leaved trees and that of these, evergreens are less tolerant than deciduous trees. Together with the defoliators however, the sap-sucking insects and miners can also be found on the crown even though confined to the vegetative tip or young buds, while on the lignified parts a progressive increase of xylophagous insects, mainly phloem boring insects may be observed. These are insects which generally require trees which are already stressed for their reproduction, often by the very same defoliators or sap-suckers and whose newly emerged adults in some cases feed on the green or lignifying organs of more vigorous specimens. It should however be emphasised that the removal of organic matter performed by phytophagous insects, and by phyllophagous insects in particular, should be seen in a broader context and only deemed damaging when the negative effects prevail over the positive. Carlisle et al. (1966) clearly showed the importance of the role played by defoliators in regulating the level of nutritive elements in the litter and in encouraging a faster turnover which according to Ghilarov (1971), lies at the basis of the productivity of forest stands.

Weighing up such factors is however, always a complex matter since the influence of the defoliating insects in the transformation and turnover process of the litter needs to be considered from different points of view and always in relation to the intensity of defoliation (Battisti et al. 1986). However, the activity of phytophagous

Fig. 1 (**a**) Sap sucking insects: neanid and nymph of Aphidoidea (Homoptera); (**b**) Defoliating insects: larva of *Erannis defoliaria* (Lepidoptera: geometridae); (**c**) Xylophagous insects: larva of Cerambycidae (Coleoptera); (**d**) Seed eating insects: nymph and adult of Coreidae (Hemiptera)

insects may have negative effects on the equilibriums of the plant community in that it may reduce fruit or seed set to the point of eliminating it entirely and thereby reducing the supply of food to many other animals that habitually or occasionally depend on this food source. In addition, the lack of or decrease in the production of seeds has inevitable consequences on species renewal and consequently may eventually produce significant modifications to the structure of forest populations.

Extraordinary catastrophic defoliation events which leave profound and lasting traces on the organisation of the plant community and greatly influence the productivity of the forest for human needs are not unheard of.

As trees approach maturity and later in the senescent phase, they are colonised mainly by xylophagous insects with the progressive contribution of wood boring insects. These phytophagous insects which feed on the woody parts influence the longevity of the trees: the phloem borers cause ecological damage which usually results in the death of the affected individuals, while the wood borers progressively reduce the structural stability of the trees and facilitate the settlement of other demolition organisms, including insects, fungi and bacteria.

By way of example, a multiyear study in central Italy on mixed woods of conifers and broadleaves has been conducted (Tiberi et al. 2009). The phytophagous insects observed were divided into three feeding groups (sap-sucking insects, defoliating insects and xylophagous insects, Fig. 1) and according to the age of the plants attacked (up to 20 years, 50 years and over 50 years) as well as according to the population frequency and density.

The result of surveys carried out in Italy to verify the ability of plant-eating insects to spread fungal pathogens of cypresses, oaks, pines, plane trees and chestnut trees, showed the reliability of such theories for species closely connected to the permanent structures of the trees. However other cases have been observed for insects and fungi which inhabit the reproductive organs of cypresses, pines or the vegetative parts of the plane tree. In all these cases it is important to remember that such relations are to be deemed optional in that the pathogen is most likely conveyed by (1) simple adhesion of the propagules to the body and/or to the appendages of the insect; (2) through the excrement of mites and bark beetles, after transit in their digestive systems.

There are innumerable insects which develop at the expense of the oak and which on account of their biology, behaviour and relations with the host plants may play a decisive role in the transport of pathogenic fungi (Tiberi and Ragazzi 1998; Ragazzi and Tiberi 2011).

2   The Host-Insect System *"Tortrix-Quercus"*

2.1   *The Genus* Quercus

The genus *Quercus* is native to the northern hemisphere. The number of species varies greatly (from 280 to 400 species) as a result of taxonomic uncertainties resulting from the significant genetic diversity, polymorphism and the high degree of hybridization which characterises this species group (Bernetti 1995). *Quercus* includes deciduous and evergreen species, which are trees, often of considerable size, or shrubs, and they always have ovoidal buds and lobed or indented leaves.

In the European Palaearctic region the species occupying the largest area is the Pedunculate oak (*Q. robur*), extending from the Urals to northern Spain, and from the British Isles and the southern part of Scandinavia to mainland Italy; according to some botanists even appearing in Sicily. This extensive distribution of the Pedunculate oak in the Palaearctic has enabled the development of various sub-species or varieties variously located in the vast area of *Q. robur*. Usually Pedunculate oak buds burst in spring and the leaves appear at the same time as the flowers. There are some varieties however, such as the *Q. r. tardissima* which sprout new leaves as much as a month later than the other Pedunculate oaks, thereby avoiding late frosts and the attack of early defoliators. The Sessile oak (*Q. petraea*) grows in portions of the area occupied by the Pedunculate oak, with clear geographic limitations related to the different climatic conditions and is in fact absent from the area of the Ural Mountains. Budburst occurs on average 10–15 days before the Pedunculate oak, but such difference is less marked in the northernmost latitudes, with the leaves also appearing at the same time as the flowers. Most other prominent species of the genus are distributed in different parts of Southern and South-Eastern Europe. Only rare artificial stands of oak species other than *Q. robur* and *Q. petraea* can be found

Fig. 2 Second instar
of *T. viridana* on a just
opened bud

in Northern Europe. These two species are also the most economically important species of the genus due to their high quality timber.

The genus *Quercus* is one of the tree taxa with the highest number of herbivorous insects feeding on it (Yela and Lawton 1997). In Europe *Tortrix viridana*, the green oak leaf roller, is one of the most widespread defoliating insects among those associated with deciduous and evergreen oaks and can cause significant ecological, economic and environmental damage. It is considered the most harmful spring defoliator and is common throughout Europe, North Africa and the Middle East, wherever there are trees of the genus *Quercus*, which is it's natural host.

2.2   *Life Cycle of* Tortrix viridana

The green oak leaf roller completes one generation in a year with a life cycle characterised by rapid juvenile development (from larva to pupa) in spring and by the long duration of the egg stage (summer, autumn and winter). The adults appear from the end of May to early June and outbreaks at higher altitude and latitudes may last until July (Tiberi and Roversi 1990). After mating, the eggs are laid on peripheral twigs of the crown where they remain until the following spring. A significant preference for 2-year old twigs located in the upper part of the crown was observed in studies in Italy (Tiberi et al. 2005a, b). The eggs open at the moment of budburst and hatching of the larvae is completed over a period varying from several days to about a month (Du Merle 1983). The younger larvae attack the buds, getting in between the bud scales and their activity can be noted by the presence of holes, excrement and silky threads (Du Merle and Mazet 1983). The larvae feed first on the young leaves which they join with silken threads (Fig. 2); subsequently, as the host tree vegetation develops, they erode the leaf which is folded back to form a case where they shelter (Fig. 3). IV and V instar larvae are extremely mobile and agile so that if disturbed they move rapidly or drop from silken threads (Fig. 4).

Fig. 3 Fifth instar
of *T. viridana* building
a typical leaf roll on oak

Fig. 4 Fifth instar
of *T. viridana*

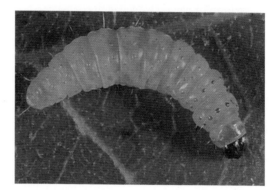

The results of a study conducted by Bogenschütz (1978), and confirmed by those of Tiberi et al. (2005a), indicates that the first instar lasts from 4 to 8 days, the II from 2 to 5, the III from 2 to 6, the IV from 4 to 11 and the V from 4 to 8. The pupal stage lasts rather longer (12–20 days) especially for the females; while the males emerge from their cocoons earlier and are therefore actively flying by the time the females appear. Adults (Fig. 5) fly mainly during the early part of the afternoon and a minority percentage of the population is capable of migrating over large distances; thereby enabling the species to spread to other environments including those located at different latitudes and altitudes from the area of provenance (Du Merle 1985a, b; Tiberi and Roversi 1989). For *T. viridana*, ecophysiological and genetic adaptations to phenological different hosts within the genus *Quercus* have been shown (Du Merle 1999).

The abundance of *T. viridana* is subject to fluctuations in population size, as typically observed for herbivorous insects: periods of small population sizes (latent periods), alternate with periods of high population sizes (outbreaks) (Horstmann 1984; Hunter et al. 1997; Johnson et al. 2006). In addition, its appearance is influenced by the phenology of oaks. The first instar larvae need the physiological

Fig. 5 Adult female
of *T. viridana*

conditions of newly opened buds for successful development (Hunter 1990) and the later immature stages depend on the growth rate of the leaves of the host plant for successful development (Ivashov et al. 2002). The duration and recurrence of massive infestations of the green oak leaf roller varies considerably over the different areas of its range. Generally speaking, the culminating phase lasts 2–4 years and is repeated every few years. Various authors note that in cork forests especially, the expansion phases of the infestations last from 3 to 4 years and are repeated every 8–10 years while others speak of 2 years of teeming infestations at intervals of 5–6 years. These two different patterns have been found, the first in the cork forests of Sardinia (Luciano and Roversi 2001) and the second in turkey oak and downy oak forests on the Italian mainland (Tiberi 1991; Tiberi et al. 2005a).

2.3   Damage Caused by the Green Oak Leaf Roller

Outbreaks of *T. viridana* often leads to the defoliation of oaks in spring (Figs. 6 and 7). Such defoliation causes a decrease of wood productivity, due to re-directing nutrients to the production of new foliage (e.g. Jüttner 1959). Furthermore, this re-allocation of resources also leads to a lack of fructification (e.g. Crawley 1985) and an increase of vulnerability of the affected trees to other pests and diseases, because the defoliation establishes the basis for further pest insects (for example buprestids) and fungi. Attacks on the buds affect the tree phenology, causing a reduction in the trees leaf area and canopy, irregular blossom and poor fructification. This leads to serious consequences for the natural renewal of oak forests and on the food source for many warm-blooded vertebrates, especially for those dependant on acorns for food. The attack on the young leaves reduces the production of elaborated sap, influences the growth of shoots, the transpiration and heat regulation processes and, in general, the growth of the crown; it also alters the normal translocation of the nutrient substances and phyto-hormones (Tiberi 1991; Tiberi et al. 1993). In the year of attack as well as the following year, the defoliated trees show an irregular and

Fig. 6 Leaf rolls of fourth and fifth instars of *T. viridana* in an oak stand in Southern Finland

Fig. 7 Oaks in North Rhine-Westphalia (Germany) defoliated by the green oak leaf roller

modest longitudinal and, especially, radial growth due to the reduced production of summer wood.

In the 1920s, at least 7,250 ha of oak forests in North Rhine-Westphalia (Western Germany) were severely damaged as consequence of a *T. viridana* outbreak, as described above (Gasow 1925). In Central Europe, frequent outbreaks of the green oak leaf roller is thought to be one reason for oak decline events reported during the last century (Erler 1939; Röhrig 1950; Altenkirch 1966; Thomas 2008; Denman and Webber 2009). In Germany, the first important oak decline was observed from 1911 to 1920, when oaks were observed to die after repetitive defoliation by *T. viridana*. Later, in the 1980s, oaks of all age groups died at a rate of 2–5 trees per hectare, per year (Hartmann and Blank 1992). The results of surveys of cork oaks in various Mediterranean countries and in Sardinia (Italy) showed that total defoliation caused a reduction of height increment of over 63 % and of radial growth of 45 % (Magnoler and Cambini 1968a, b). In addition, total defoliation led to losses of cork production of 60 %, while with 50 % defoliation of the crown the reduction in cork produced was around 42 % (Cambini 1971; Tiberi et al. 1993).

3   Host-Insect Interaction

3.1   General Characteristics of the Order Lepidoptera

Among the various insect orders that live on tree crowns at the expense of the green parts (leaves, buds and green bark), the Lepidoptera stand out both in terms of their number and for their ecological and economical importance. In forest environments, this group of species are defoliators *par excellence*, with the exception of some species which attack wood, seeds or buds and a few carnivorous species. The predominance of the Lepidoptera in our broad-leaved and temperate conifer forests is indisputable, while in conifer forests in mountainous or colder areas, such phyllophagous insects may be outnumbered in terms of importance by the Hymenoptera.

The fierceness which distinguishes the Lepidoptera from other defoliators depends to a large extent on ecological characteristics commonly found in this group. Only a few species appear to be restricted to using a limited number of trees as their food source, the majority being polyphagous. However, sometimes clear host preferences are shown, as highlighted for example in *Malacosoma neustrium* L. in central Europe (Schwenke 1978) and in *Lymantria dispar* L., two polyphagous species which have a preference for oaks.

The winter moth, *Operophthera brumata* L., has a broad spectrum of host plants, but may show preferences depending on the type of plant community present (Kudler, in Schwenke 1978). Depending on the composition of the plant communities present, the Lepidoptera defoliating broadleaved trees have widely recognised "refuge" hosts, from which they spread during the expansive phase of infestation. In thermophilic vegetations such as the Mediterranean maquis, their "refuge" hosts are the evergreen oaks, while deciduous oaks or beech are chosen in the mesophilic vegetation. The absence of constraints of dependency with the host tree and consequent wide range of feeding choices, enable the Lepidoptera to make the most of the available environmental resources, unlike the phytophagous Hymenoptera (or Coleopterans and Dipterans in other ways), which need to establish close relations with the host plant from oviposition onwards. For instance, the eggs of the Symphyta (Hymenoptera), depend on the host plant physiology, since the females insert them in incisions made in the leaf tissue with their terebra, and the development of the embryo is therefore strictly related to the vegetative cycle of the plant. Conversely, most Lepidoptera lay their eggs, in the manner particular to the species, on the surfaces of various plant organs and therefore, in this phase at least, their survival does not depend either on the presence and abundance or vital functions of the chosen trees.

In addition, oviposition on different plant species enables the eggs, and later the young larvae, to more easily save themselves from a deficiency of any one host, as well as the ravages of specialised and generic predators and parasitoids. These find it more difficult to locate their victims as a result of having to decipher the different chemical messages emitted by the various trees. Along with the method of oviposition (eggs laid individually or in groups), the protective strategies adopted by many Lepidoptera are an important factor in defining the relations between the defoliator and its natural enemies (Roversi et al. 1991). Again, in some species the egg cluster is essential for inducing larval aggregation, as larvae often live in groups for all or part of their development so as to more efficiently counter environmental adversities.

In Europe there are over 200 species of defoliating insects belonging to various orders reported as harmful to oak crowns, but the following Lepidoptera are considered the most important in terms of the number of species as well as the ecological and economic damage they inflict: Tortricidae, Lymantriidae, Geometridae, Thaumetopoeidae and Lasiocampidae; damage by some Noctuidae and Notodontidae may be frequent but is usually limited. However, only some species are capable of large-scale and repeated infestations, showing conspicuous and sudden demographic increases which may last several years and repeat themselves at regular intervals (Tiberi et al. 1993).

The tendency of the Lepidoptera to colonise forest ecosystems may be rooted in some aspect of the biological characteristics peculiar to this group. Their ability to overwinter at different stages of development (as eggs, larvae or pupae) enables various species to use the crown of the tree in succession for instance. Apart from some groups, the absence of retardation or postponement in the sequence of generations caused by diapause phenomena, enables the Lepidoptera to actively re-appear on the tree year after year. The sexual reproduction typical of the entire order, on the one hand may limit the size of the populations, while on the other it guarantees their capacity to adapt to the often rapid changes in environmental conditions, and so is crucial to their survival. The ability of the adults of many species to undertake long-distance flights enables them to spread to new areas, as observed for various Tortricids, including *Tortrix viridana* (Du Merle 1985; Schroeder et al. 2010).

The relations which adult Lepidoptera establish with the rest of the animal community are more diversified than those of the other herbivorous forest pests. The adults of species with functioning mouthparts occur on trees infested by those Rhynchota producing honeydew, which is a significant and sought-after food supply. The adults in turn, like the larvae of many species, are a widely available and ideal food source at certain times of year, not just for predatory insects but above all for insectivorous vertebrates; the observations of Betts (1955) on the food supplies for the brood by *Parus major* L. and *P. caeruleus* L. are of interest.

From an ecological point of view, the Lepidoptera's low feeding efficiency is to be noted, deriving both from their inability to cut into the cell walls to reach the contents and by their poor digestion of the same (Evans 1939). As a result, the organisms living on the ground receive greenfall material rich in nutrients mixed with frass droppings which can be rapidly utilised, transformed and then returned to the trees within a short space of time as fertiliser. In a study of the impact of defoliators in forests, Pitman et al. (2010) found a significant amount of nitrogen, phosphorous and potassium in the frass falling to the ground, while the leaf fragments add high concentrations of potassium to the soil, otherwise only received during winter leaf abscission. The same authors report reductions of 40–55 % of annual net carbon exchange in oak forests in New Jersey, after defoliation by *Lymantria dispar*.

3.2   Insect-Plant Synchrony

The larvae of many defoliating Lepidoptera depend on newly grown leaves in spring. If the first instar larvae hatch before budburst, the larvae of most species can only survive starvation for a couple of days, after which the mortality rate rises rapidly; there are however some exceptions as in the case of *Tortrix viridana* which can survive without food for up to 10 days (van Asch and Visser 2007). Nor is the late emergence of the larvae beneficial to the species in that the concentrations of defence compounds, such as tannins, are higher in the maturing leaves, whereas nitrogen and water content are lower. As these leaves are less nutritious, this causes a reduction in larval survival, in the growth rate, as well as in the fecundity of the adults (van Asch and Visser 2007).

Three different models of development can be identified for the defoliating Lepidoptera, depending on the vegetative phases of the oak and the presence of larvae in action on the crown of the trees: early season species, early-mid season species and late season species.

The early season species include those which attack the host plant immediately after budburst and are capable of causing very serious damage. Such species show many biological and behavioural similarities: overwintering at the egg stage or as first instar larvae, marked synchrony between hatching of larvae or their resumption of feeding activity and budburst and, for those species overwintering as eggs, rapid larval development (maturity being reached within a period of 4–6 weeks). The brief period in the larval stage is a result of the high concentration of nitrogenous substances (proteins and amino acids) in the food ingested (Feeny 1970).

Tortrix viridana is undoubtedly the most damaging of the early season species on oaks. When present in the same forest, deciduous oaks are preferred to evergreens (Du Merle 1983); even among deciduous trees, defoliation of varying intensity is reported depending on the synchrony with the insects' development cycle (Schwerdtfeger 1961). Normally, in mixed forests, the late season oaks host larger larval populations (Bogenschütz 1978); in mixed forests of *Quercus cerris* and *Q. pubescens* in central Italy, more intensive activity of the green oak leaf roller has been constantly reported on *Q. pubescens* (Tiberi and Roversi 1989; Tiberi et al. 2005a). The minor presence of *T. viridana* on the turkey oak may in part be attributed to the higher larval mortality immediately after hatching on account of sudden halts in development of the buds and the abnormal worsening of climatic conditions (Schwerdtfeger 1971; Du Merle and Mazet 1983). In mixed stands of *Q. robur* and *Q. petraea* in Germany, a strong preference of the green oak leaf roller for the pedunculate oak has been observed (Schröder and Ziegler 2006). Moreover for the newly hatched larvae, the importance of the quality of the food has been seen to be important (van Asch and Visser 2007). Depending on the altitude, latitude and species of oak present, there are *T. viridana* populations which have adapted perfectly to the vegetative cycle of their host plants (Dajoz 1980; Du Merle 1983; Tiberi et al. 2005).

It is known that marked seasonal variations in temperate region climates result in a brief period of the year which is optimal for plant and animal development. For many herbivores, including insects, the phenology of the host plant determines such period and it is therefore crucial that their feeding and reproduction is synchronised with the host plants' phenology. Asynchronous development of the herbivore may have extremely serious physical consequences for it, with regard to its fecundity and survival. The most favourable period for defoliating Lepidoptera is as soon as the leaves of the host plant appear, since the young leaves usually provide the best food for them. Considering that the phenology of the trees varies from 1 year to another, it follows that the time of year in which such optimal period occurs also varies.

Generally speaking, the degree of synchrony between herbivores and plants is a result of the influence of many factors among which the most important are temperature and photoperiod. Many herbivores use adaptive responses depending on the climatic conditions in the various areas of their range to ensure synchrony with their host plants. The results of surveys conducted in mixed turkey and downy oak forests in central Italy show that the development cycle of *T. viridana* is strictly correlated to the phenology of the oak species on which the eggs have been laid (Tiberi et al. 2005). The hatching of the larvae on the turkey oak actually occurred 2 weeks earlier than on the downy oak and this phenological shift persisted until the appearance of the adults, confirming the findings of Du Merle and Mazet (1983) in France for the downy and holm oak, and of Altenkirch (1966) in Portugal for the holm and cork oaks. The authors of both works hypothesise the existence of two biological races well adapted to the phenology of each host plant. Also in Italy, the results of a survey by Tiberi and Roversi (1989) of a Tuscan forest of turkey and downy oaks, located at 100–540 m above sea level, demonstrated that altitude also affects budburst of the two oaks and therefore the development of the green oak leaf roller: adults located at higher altitudes appeared from 7 to 12 days later than those from lower elevations (Fig. 8).

3.3 Plant Vulnerability to Defoliating Insects

The consequences of defoliation will vary greatly both as regards the individual tree in relation to its age and species and as regards the population, in relation to the type, composition and structure. Juvenile trees undergoing rapid growth and mature trees are the most sensitive for different reasons. In the first case, leaf damage, in

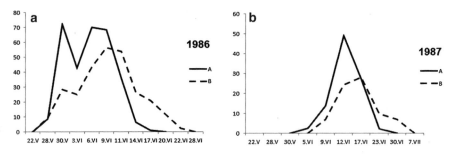

Fig. 8 Flight curve of *Tortrix viridana* in Tuscany (Italy) in 1986 and 1987. (**a**): altitude from 140 to 540 m above the sea level. (**b**) Altitude from 340 to 730 m above the sea level (Tiberi and Roversi 1989)

spring especially, reduces longitudinal and radial growth, causes physiological imbalances in the tree and has a negative influence on the development of the plant organs which will become active the following year. The damage will reduce the competitive position of worst affected juvenile trees versus those that are less affected, and thus eventually affect which tree survives to reproduce most success-fully. Mature or senescent trees suffer the consequences of defoliation with chronic states of vegetative stress and therefore decline, given that the reduced rate of growth does not permit a prompt and effective reaction.

The repetitive effects of leaf damage produced by Lepidopteran larvae and primarily by the green leaf roller moth, causes different effects depending on whether the trees attacked are evergreen or deciduous. Generally speaking the former suffer greater damage in that the leaf mass is composed of elements of varying age, the replacement of which requires a number of vegetative cycles and, therefore, evergreens have to invest more resources in their leaves than deciduous trees. However, repeated defoliation by *T. viridana* may also have very serious consequences on deciduous oaks in the form of a progressive reduction of the crown dimension and sprouting of new shoots limited to the trunk and main branches only. The more visible consequences of defoliation are accompanied by longer term effects not immediately obvious such as changes in the phenology, chemical composition and structure of the leaves. These can induce a follow-on effect of increased mortality of the larval population and therefore have a marked influence on fluctuations in number of the phyllophagous insects (Benz 1974; Baltenswalter et al. 1977; Haujioja 1980).

The onset and development of infestations by defoliating Lepidoptera in forest environments including oaks, is influenced by their specific composition and structure. It is known that mixed populations of trees, especially if of mixed age, tolerate defoliation better than pure, even-aged formations. Species diversity in the forest represents an obstacle to the propagation of attacks by monophagous and oligophagous Lepidoptera (such as *T. viridana*), since egg-laying adults encounter difficulties in locating the preferred trees and the most suitable portions of the crown. This may prove even harder as a result of the different heights of the trees, as observed for the egg-laying adults of the green oak leaf roller (Tiberi 1991). These difficulties also apply to the newly hatched larvae when their dispersal depends on the wind (such as for some Geometridae and Lymantriidae).

3.4   Molecular Genetics of Oaks and the Green Oak Leaf Roller

During the last ice age, many forest tree species survived in refuges in South and South-East Europe. For the genus *Quercus*, three glacial refuges are recognized on the basis of palynological and molecular phylogeographic analyses: The Iberian Peninsula, The Apennine Peninsula of Italy and the Balkans (Petit et al. 2002). Postglacial recolonisation routes and seed transfers made by humans gave rise to the current distribution of the genetic variation within the genus *Quercus* (König et al. 2002). The chloroplast genome is known to be inherited maternally in oaks (Dumolin et al. 1995) and, because of limited dispersal of the seeds, reveals higher levels of variation among populations than does the analysis of nuclear markers (Petit et al. 1993, Schroeder and Degen 2008a). Variation in chloroplast DNA is geographically structured in oaks, because long-distance dispersal events occurred rarely after the last ice age and so related haplotypes often have a similar distribution (Dumolin-Lapegue et al. 1997). In general, tree species are remarkable for their high intra-population diversity and low interpopulation differences for nuclear genes. Two possible explanations for these phenomena are a reasonably high level of pollen flow and the characteristics of the life cycles of trees (Austerlitz et al. 2000). Pollen-mediated gene flow is possible up to a distance of about 500 m, but oak pollen can be physically transported up to a distance of 100 km (Schueler et al. 2005).

In contrast to the case of the oaks, examination of maternally inherited mitochondrial markers of *T. viridana* revealed that the genetic variation was high within populations and low between populations (Schroeder and Scholz 2005; Schroeder and Degen 2008a). Compared with other insect species, the intrapopulation genetic variation in *T. viridana* is relatively high showing 14 haplotypes in 10 populations (Fig. 9, Schroeder and Degen 2008a). Within populations of the milkweed beetle, genus *Chrysochus*, only 1–5 haplotypes are found in large areas in the USA (Dobler and Farrell 1999) and only two haplotypes of *Aedes aegypti* can be found in several regions of Argentina (Rondan Duenas et al. 2002). Analysis of biparentally inherited nuclear markers, for the insect as well as for the oaks, found only limited genetic diversity among populations which has been interpreted as an indicator of a high gene flow and therefore high flight activity (Schroeder and Degen 2008a, Schroeder et al. 2010).

Analyses of the within population genetics of the green oak leaf roller revealed spatially limited structures for small areas within a radius of up to 40 m indicating a close relationship of individuals from trees next to each other (Schroeder and Degen 2008b). Thus, there should be a mechanism stabilising the genetic similarity at this small spatial scale. Such a mechanism could be the mating behaviour of the males of *T. viridana*. Indeed, Simchuk et al. (1999) observed

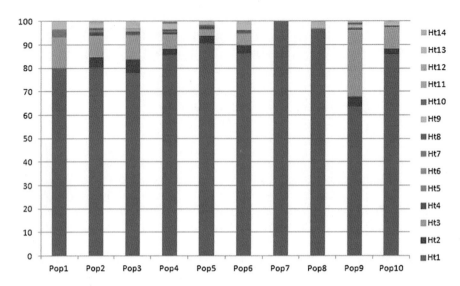

Fig. 9 Percentage of *T. viridana* mitochondrial haplotypes (cytochromoxidase I) in ten popula-tions in North Rhine-Westphalia (Germany). *Ht* haplotype, *Pop* population (Schroeder and Degen 2008a)

that males of *T. viridana* mate first with females that developed at the same tree and then start flying over some greater distance (up to a few kilometres). In addition, the males of *T. viridana* have been the subject of intensive search activity. As for most lepidopteran species, females of the green oak leaf roller also produce pher-omones to attract males (Knipple et al. 2002). Although it is not known how far *T. viridana* males are attracted by sex pheromones, for other lepidopteran species the distances males will fly to find mates varies from several metres up to several kilometres (Schneider 1992). Such behaviour explains the observed family struc-ture within populations because of the first mating of closely related individuals within one or only few trees.

A host and its herbivorous insect are often species with extremely different generation times and mobility (Hamilton et al. 1990). Trees have generation times of up to 60 years, whereas insects generally have one or more generations per year. Trees are sessile and insects are mobile, with mobility varying among species. The effect of these factors is manifested in the organism-specific geo-graphic patterns of genetic variation and differences of the involved species. Therefore, the differences in the current genetic patterns of *Q. robur* and *T. viri-dana* may be due to different metapopulation dynamics. For the oaks the pattern of chloroplast DNA diversity has been proven to be clearly the result of post glacial recolonisation (Petit et al. 2002). But for *T. viridana* as an insect pest species, it is likely that there have been many extinction and recolonisation events after the post-glacial recolonisation. The fluctuations within the popula-tions of *T. viridana* are temporally asynchronous along its distribution area

(Schwerdtfeger 1971). Thus the genetic pattern of the insect maybe dominated by these more recent recolonisation events from remaining populations after latency periods (Schroeder and Degen 2008a).

3.5 Population Dynamics

Forecasting Models

The fluctuations in numbers reported in the populations of various phytophagous species and the many factors involved are fundamental for studying the population dynamics of animals. Demo-ecological information on such organisms is very difficult to acquire, but is essential for defining a valid programme of integrated pest management for limiting their economic and ecological impact. In long term situations such as forest environments, a thorough knowledge of the factors causing numerical variations of the pests population is required (Waters 1969). It is now widely accepted that the increase, stability or decrease of a population over time and space is the result of a continuous struggle between the biotic potential of the species and the resistance opposed by the environment. To quantify the probability of a demographic explosion occurring it is therefore essential to consider such factors as climate, soil, orography, structure and position of the tree population at risk, growth rates and the vegetative vigour of the trees, and how they affect the population dynamics of the most serious pests. Such findings are all used to define models forecasting the likely behaviour of pest populations (Waters and Stark 1980). At the same time however, to define an effective forecasting model, the factors most directly affecting the density variations of the species and relative parameters must be known. A thorough knowledge of the effects deriving from intra and inter species competition, of the effect of natural antagonists and of the availability of plentiful, suitable quality food is therefore needed.

Such models, even if theoretically perfect, are not always reliable in practice especially in the long term, since they are difficult to apply and also on account of the incomplete knowledge of some biological and ecological aspects of the individual species, which may be fundamental to their fluctuations in number. One of the critical elements in demo-ecological studies that is frequently underestimated in these models, is the method of dispersal of the species concerned: their ability to move and method of movement, and the specific composition and structure of phytocenosis which also affects dispersal (Campbell and Sloan 1978). The build-up of damage produced by plurivoltine phytophagous pests or by the succession of species using the host tree at different times affects the tolerance of the trees and this is not always adequately taken into account by forecasting models. Another factor not to be underestimated is the ability of insects to use the three dimensions of space.

This factor considerably complicates the surveyor's task in choosing the right sampling techniques, given the different distribution of the individuals on the tree, which is influenced over the course of the year not only by the quantity but above all by the quality of the food available at the various levels of the crown. Consequently, considering all the factors influencing the demographic variations of insects and the different acquisition techniques needed to formulate efficient and reliable forecasting models, the difficulties encountered when operating over an extensive environment such as that of forests become apparent. For these reasons, identification of the areas most favourable to the pest species, followed by an in depth study of the factors and mechanisms regulating the quantitative variations of the herbivorous insects in these "risk areas" has been proposed (Covassi and Tiberi 1994). According to Waters and Stark (1980), the elements which still seem insufficient for developing appropriate forecasting models are the quantification of the global impact of pests and the reliability of the information gathered from forecasting models on infestations in various environments.

The population densities of many defoliating insects, among which several early season pests including *Operophtera brumata* and *Tortrix viridana*, show clear fluctuations in spring, as a result of a range of ecological factors. Other species show less marked population fluctuations while others still enjoy phases of expansion at regular intervals even if only in specific regions (van Asch and Visser 2007).

The population fluctuations of the green oak leaf roller rarely show a regular trend over the course of the years in single and mixed oak forests, inasmuch as they are essentially dependent on the fertility of the females, the ability of the larvae to survive and the conditions existing during their development. One important factor influencing the juvenile development of the leaf roller is the availability of suitable food in terms of quality and the more or less favourable climate, especially in terms of temperature; the limiting effect posed by natural enemies (insect pathogens, arthropods and birds) also affects population density.

The population density is believed to be influenced by synchronism with the phenology of the host trees (Feeny 1976; Hunter 1990). If the phytophagous insects are well synchronised with their host plants, large numbers of larvae survive the first instars and the population density therefore remains at very high levels; however not many studies have been conducted to support this theory and so far the results are not conclusive.

Despite not being the only determining factor, synchrony seems to play an important role in the population fluctuations of herbivorous insects. This can be confirmed by the fact that the effect of temperature on the survival of the larvae also directly influences the budburst and subsequent vegetative development of the trees.

For infestation forecasts of *T. viridana* it would be useful to develop a specific time series in relation to: (1) capture trends of adults using pheromone traps; (2) the percentage of females with eggs and (3) the percentage of buds containing newly hatched larvae. As regards the first point, while it is true that only males are caught in the pheromone traps, if appropriately interpreted, capture trends and their intensity may be used to draw a flight curve and through logical analysis, to assess the risk of infestation in the short term. As far as the other two points are concerned, a number of studies conducted in various parts of Europe have shown that the risk threshold corresponds to a ratio of 1 egg per bud or 0.6 larvae per bud. Clearly, the sampling of peripheral twigs from the canopy and their thorough analysis in the laboratory are crucial operations and, though far from easy to carry out, they offer reliable information.

Population Cycles Models

A major question of population dynamics is "Why do animal populations fluctuate as they do?" (Royama 1992). The answer to this question has fascinated ecologists for a long time (Reynolds and Freckleton 2005). A well known is the textbook example of lynx-hare interactions in Northern Canada (Krebs et al. 1995), where predation by the lynx is supposed to lead to stable 9–10-year cycles among snow-shoe hares. Nevertheless, until now surprisingly little consensus has been reached in the ecological community as to what exactly are the factors that shape these, some-times startlingly regular cycles. Even in the very well documented example above, field experiments failed to confirm the proposed predator-prey interactions to be the main cause of the observed cycles (Krebs et al. 1995; Inchausti and Ginzburg 2009 and references therein).

Meanwhile a multitude of factors has been suggested, to affect animal population cycles. These factors are generalist and specialist predators (Dwyer et al. 2004, Elkinton et al. 2004; Bjørnstad et al. 2010), parasites and parasitoids (Berryman 1996; McCullough 2000; Turchin 2003), pathogens (Myers 2000), food quality and plant induced defence mechanisms on herbivore populations (Selås 2003; Högstedt et al. 2005; Haynes et al. 2009), food limitation (Abbott and Dwyer 2007), maternal (Hunter 2002; Beckerman et al. 2006; Inchausti and Ginzburg 2009) and genetic effects (Chitty 1960; Sinervo et al. 2000; Sinclair et al. 2003), as well as environ-mental conditions (Elton 1924; Hunter and Price 1998; Selås et al. 2004).

Theoretical biologists have developed a multitude of models to describe and reconstruct population cycles in order to understand the underlying mechanisms and predict future population behaviour. Mostly, they concentrated on single factors known or supposed to influence population dynamics (Ginzburg and Taneyhill 1994; Dwyer 2000; Kapeller et al. 2011). Some authors combined several factors to study possible interactions (Dwyer et al. 2004; Abbott and Dwyer 2007; Bjørnstad et al. 2010), or applied statistical testing to identify major regulating factors in experimental data sets (Frago et al. 2011). Others attempted to understand the math-ematical processes leading to the observed patterns without any a priori assump-tions about underlying biological mechanisms (Nedorezov and Sadykova 2008).

As also discussed in the above studies, many investigations into the cyclical changes of populations focus on trophic interactions and environmental conditions. Only in recent years have researchers begun to address maternal or genetic effects, even though both of these were proposed quite early in the debate on population cycles (Chitty 1960; Ginzburg and Taneyhill 1994). Several of these studies are concerned with maternal effects (Myers et al. 1998; Erelli and Elkinton 2000; Benton et al. 2005; Beckerman et al. 2006; Plaistow and Benton 2009), whereas only few focussed on genetic effects (Simchuk et al. 1999; Sinervo et al. 2000;

Sinclair et al. 2003). One major maternal effect, changing fecundity in the course of the population's cycle have been reported from several forest defoliating Lepidoptera: *Choristoneura pinus pinus* (McCullough 2000), *Euproctis chrysorrhoea* (Frago et al. 2011), *Lymantria dispar* (Rossiter 1991) *Malacosoma californicum pluviale* (Myers 2000), *Syntypistis punctatella* (Kamata 2000), and also for *Tortrix viridana* (Simchuk et al. 1999). The most interesting observation in all these species was that fecundity decreases, when population density increases.

An example of genetic effects was reported by Simchuk et al. (1999), who noted a change in heterozygosity during different cycle phases of *T. viridana*. High genetic heterogeneity was observed during the outbreak phase favouring the possibility of using different food micro-niches and therefore also favouring increasing of the population density. Their explanation is that during phases of low population density the adult moths mate randomly leading to a higher heterozygosity, and when population density is high they switch to assortative mating resulting in reduced heterozygosity, with the resultant inbreeding depression leading to lower capability to react to the environmental changes (Simchuk et al. 1999). Kapeller et al. (2011) formulated a model to show the population dynamics trend of this pest species and its interaction with the different oak species (Fig. 10). Specifically, the dispersal ability of adult moths, the mortality rate of the larvae in relation to the host tree and different oak-types (eco-/genotypes) preferred for oviposition were considered. While dispersal ability did not prove significant in influencing the population dynamics of the green oak leaf roller, larval mortality proved to be very important, while the preference for specific oak-types has yet to be considered.

4 Defence Response Against Herbivores

What Ehrlich and Raven (1964) wrote nearly 50 years ago about plant-insect interactions is still true: Secondary plant substances play the leading role in determining patterns of utilization. And, when some of these secondary compounds serve to reduce or destroy the palatability of the plant in which they are produced, then such a plant would in a sense have entered a new adaptive zone.

Trees suffer constantly from the pressures of herbivory, and so it should be no surprise that their defence responses are both highly evolved and excellently regulated. As a first barrier constitutive chemical and physical mechanisms may reduce access or availability of the plant resources to the herbivorous insects. As a second barrier, inducible defences may involve a broad range of molecules whose synthesis is under tight temporal and spatial regulation (Walling 2000). We know of at least two types of inducible defence responses: direct defences which inhibit the growth or development of herbivorous insects and indirect defences including plant volatiles for example, which attract the herbivore's parasitoids and predators (Boland et al. 1995; Thaler 1999).

Many herbivores are specialised to feed on a single (monophagous) or only few (oligophagous) plant species, whereas a minority are polyphagous (Bernays 1998).

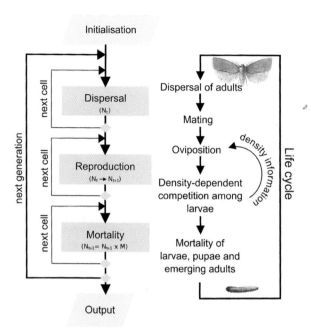

Fig. 10 *Left side*: Flowchart illustrating the course of three procedures dispersal, reproduction and mortality within a model run. After dispersal, adults (N_t) reproduce with reproductive rates determined by the source tree. Selection acts among competing larvae (N_{t+1}). And finally, the population is reduced by mortality factor (M). Each of these procedures runs in an inner loop for each cell and then in an outer loop for each simulated generation. *Right side*: Respective life cycle stages. Density dependent competition is acting among foraging larvae. It is assumed that population density affects fertility of emerging adults at this stage (From Kapeller et al. 2011)

Induced plant defence responses to herbivory are characterised by a level of specificity that is difficult to study in depth. Some induced defence responses in wild radish (*Raphanus sativus*) protect against generalist, but not specialist, herbivorous insects (Agrawal 1999). Recent developments in genomic transcript profiling methods have significantly improved the possibility of studying defence response in plants (e.g. Reymond 2001; Reymond et al. 2004; Poelman et al. 2008), while differently expressed genes allow for a detailed characterisation of the relevant system during plant-insect interaction (e.g. Hui et al. 2003; Fischer et al. 2008; Cha et al. 2011). Thus, a lot of work has been done in this rapidly moving area in the last decades, for example from an evolutionary point of view (e.g. Kessler and Heil 2011; Agrawal 2011) or concerning the matter of herbivore-induced plant volatiles (e.g. Staudt and Lhoutellier 2007; Dicke and Baldwin 2010; Bruce and Pickett 2011) and much more research will follow. Overall, defence responses in plants against herbivores are highly complex and currently not much is known about these complex interactions in oaks (Forkner et al. 2004; Schroeder et al. 2011; Ghirardo et al. 2012).

References

Abbott KC, Dwyer G (2007) Food limitation and insect outbreaks: complex dynamics in plant-herbivore models. J Anim Ecol 76:1004–1014

Agrawal AA (1999) Induced responses to herbivory and increased plant performance. Science 279:1201–1202

Agrawal AA (2011) Current trends in the evolutionary ecology of plant defence. Funct Ecol 25:420–432

Altenkirch W (1966) Zum Vorkommen von *Tortrix viridana* L. in Portugal. Zeits Zool 53:403–415

Austerlitz F, Mariette S, Machon N, Gouyon PH, Godelle B (2000) Effects of colonization processes on genetic diversity: differences between annual plants and tree species. Genetics 154:1309–1321

Baltensweiter W, Benz G, Bovey P, Delucchi V (1977) Dynamics of larch bud moth populations. Annu Rev Entomol 22:79–100

Battisti A, Dell'Agnola G, Masutti L (1986) L'attività di *thaumetopoea pityocampa* (Denis et Schiffermüller) nel ciclo della sostanza organica in popolamenti artificiali di *Pinus nigra* Arnold. Frustula Entomologica 7–8:507–520

Beckerman AP, Benton TG, Lapsley CT, Koesters N (2006) How effective are maternal effects at having effects? Proc Roy Soc Lond B 273:485–493

Benton TG, Plaistow SJ, Beckerman AP, Lapsley CT, Littlejohns S (2005) Changes in maternal investment in eggs can affect population dynamics. Proc Roy Soc Lond B 272:1351–1356

Benz G (1974) Negative Rückkoppelung durch Raum - und Nahrungskonkurrenz sowie zyklische Veränderung der Nahrungsgrundlage als Regelprinzip in der Populationsdynamik des Grauen Lärchenwicklers, *Zeiraphera diniana* (Guénée). Z Angew Entomol 76:196–228

Bernays EA (1998) Evolution of feeding behaviour in insect herbivores. Bioscience 48:35–44

Bernetti G (1995) Selvicoltura speciale. Utet, Turin, 415 pp

Berryman AA (1996) What causes population cycles of Lepidoptera? Trends Ecol Evol 11:28–32

Betts MM (1955) The food of titmice in oak woodland. J Anim Ecol 24:282–323

Bjørnstad ON, Robinet C, Liebhold AM (2010) Geographic variation in North American gypsy moth cycles: subharmonics, generalist predators, and spatial coupling. Ecology 91:106–118

Bogenschütz H (1978) Tortricinae. In: Schwenke W (ed) Die Forstschädlinge Europas. Band III. Paul Parey, Hamburg und Berlin, pp 55–89

Boland W, Hopke J, Donath J et al (1995) Jasmonsäure- und Coronatin-induzierte Duftproduktion von Pflanzen. Angew Chem 107:1715–1717

Bovey P (1970) Impact de l'insecte déprédateur sur la forêt. Rev. for. fran., 12, n.spec., "La lutte biologique en forêt": 199–204

Bruce TJA, Pickett JA (2011) Perception of plant volatile blends by herbivorous insects – finding the right mix. Phytochemistry 72:1605–1611

Cambini A (1971) Valutazione dei danni causati dagli insetti defogliatori alla quercia da sughero. In: Atti del 1° Convegno Regionale del Sughero, Tempio Pausania, 14–16 Ottobre 1971: 327–339

Campbell RW, Sloan RJ (1978) Natural maintenance and decline of gypsy moth outbreaks. Environ Entomol 7:389–395

Carlisle A, Brown AHF, White EJ (1966) Litter fall, leaf production and the effects of defoliation by *Tortrix viridana* in a sessile oak woodland. J Ecol 54(1):65–85

Cha DH, Linn CE Jr, Teal PEA, Zhang A, Roelofs WL, Loeb GM (2011) Eavesdropping on plant volatiles by a specialist moth: significance of ratio and concentration. PLoS One 6:e17033. doi:10.1371/journal.pone.0017033

Chitty D (1960) Population processes in the vole and their relevance to general theory. Can J Zool 38:99–113

Covassi M (1989) Gli insetti e l'alterata dinamica degli ecosistemi di foresta. Criteri per il riassetto delle entomocenosi. In: Atti del Convegno sulle Avversità del bosco e delle specie arboree da legno, Firenze 15–16 ott. 1987: 405–447

Covassi M, Tiberi R (1994) Interventi integrati di controllo dei fitofagi forestali. – In: Atti XVII Congresso Nazionale Italiano di Entomologia, Udine 13–18 Giugno 1994: 723–738

Crawley MJ (1985) Reduction of oak fecundity by low-density herbivore populations. Nature 314(14):163–164

Dahlsten DL, Dreistadt SH (1984) Forest insect pest management. Bull Entomol Soc Am 30(4):19–21

Dajoz R (1980) Écologie des insectes forestiers. Bordas, Paris, 1980, ed. Gauthier-Villars: XI-489

Denman S, Webber J (2009) Oak declines: new definitions and new episodes in Britain. Q J Forest 103:285–290

Dicke M, Baldwin IT (2010) The evolutionary context for herbivore-induced plant volatiles: beyond the 'cry for help'. Trends Plant Sci 15:167–175

Dobler S, Farrell BD (1999) Host use evolution in *Chrysochus* milkweed beetles: evidence from behaviour, population genetics and phylogeny. Mol Ecol 8:1297–1307

Du Merle P (1983) Phénologies comparées du chêne pubescent, du chêne vert et de Tortrix viridana L. (Lep., Tortricidae). Mise in évidence chez l'insecte de deux popolations sympatriques adaptées chacune a' l'un des chênes. Acta aecol Oecol Applic 4(1):55–74

Du Merle P, Mazet R (1983) Stades phénologiques et infestation par *Tortrix viridana* L. (Lep., Tortricidae) des bourgerons du chêne pubescent et du chêne vert. –. Acta aecol Oecol Applic 4(1):47–53

Du Merle P (1985a) Piégeage sexuel de *Tortrix viridana* L. (Lep., Tortricidae) en montagne méditerranéenne. I. Epoque de vol et dispersion des adultes. Z Angew Entomol 100:146–163

Du Merle P (1985b) Piégeage sexuel de *Tortrix viridana* L. (Lep., Tortricidae) en montagne méditerranéenne. II. Relation entre le nombre des captures et le niveau de popolation. Rendement des piéges. Z Angew Entomol 100:272–289

Du Merle P (1999) Egg development and diapause: ecophysiological and genetic basis of phenological polymorphism and adaptation to varied hosts in the green oak tortrix, Tortrix viridana L. Lepidoptera: Tortricidae. J Insect Physiol 45:599–611

Dumolin S, Demesure B, Petit RJ (1995) Inheritance of chloroplast and mitochondrial genomes in pedunculate oak investigated with an efficient PCR method. Theor Appl Genet 91:1253–1256

Dumolin-Lapegue S, Demesure B, Fineschi S, Le Corre V, Petit RJ (1997) Phylogeographic structure of white oaks throughout the European continent. Genetics 146:1475–1487

Dwyer G, Dushoff J, Elkinton JS, Levin SA (2000) Pathogen-driven outbreaks in forest defoliators revisited: building models from experimental data. Am Nat 156:105–120

Dwyer G, Dushoff J, Harrell Yee S (2004) The combined effects of pathogens and predators on insect outbreaks. Nature 430:341–345

Ehrlich PR, Raven PH (1964) Butterflies and plants: a study in co-evolution. Evolution 18:586–608

Elkinton JS, Liebhold AM, Muzika R-M (2004) Effects of alternative prey on predation by small mammals on gypsy moth pupae. Popul Ecol 46:171–178

Elton CS (1924) Periodic fluctuations in the numbers of animals: their causes and effects. J Exp Biol 2:119–163

Erelli MC, Elkinton JS (2000) Maternal effects on gypsy moth (Lepidoptera: Lymantriidae) population dynamics: a field experiment. Environ Entomol 29:476–488

Erler E (1939) Beobachtungen zur Ökologie und Bekämpfung des Eichenwicklers (*Tortrix viridana* L.) in Westfalen. Anz Schädlingskunde 8:85–93

Evans AC (1939) The utilization of food by the larvae of the buff-tip, *Phalera bucephala* L. Proc R Entomol Soc Lond (A) 14:25–30

Feeny P (1970) Seasonal changes in oak leaf tannins and nutrients as a cause of spring feeding by winter moth caterpillars. Ecology 51:565–581

Feeny P (1976) Plant apparency and chemical defense. Rec Adv Phytochem 10:1–40

Fischer HM, Wheat CW, Heckel DG, Vogel H (2008) Evolutionary origin of a novel host plant detoxification gene in butterflies. Mol Biol Evol 25:809–820

Forkner RE, Marquis RJ, Lill JT (2004) Feeny revisited : condensed tannins as anti-herbivore defences in leaf-chewing herbivore communities of *Quercus*. Ecol Entomol 29:174–187

Frago E, Pujade-Villar J, Guara M, Selfa J (2011) Providing insides into browntail moth local outbreaks by combining life table data and semi-parametric statistics. Ecol Entomol 36: 188–199

Gasow H (1925) Der grüne Eichenwickler (*Tortrix viridana* Linné) als Forstschädling. Arb Biol Reichsanstalt 12:355–508

Ghilarov MS (1971) Invertebrates which destroy the forest litter and ways to increase their activity. In: Productivity of forest ecosystems, UNESCO, Proceedings, Brussels Symposium, 1969, pp 433–442

Ghirardo A, Heller W , Fladung M, Schnitzler JP, Schroeder H (2012) Function of defensive volatiles in pedunculate oak (*Quercus robur*) is tricked by the moth *Tortrix viridana*. Plant Cell Environ 35(12):2192–2207

Ginzburg LR, Taneyhill DE (1994) Population cycles of forest Lepidoptera: a maternal effect hypothesis. J Anim Ecol 63:79–92

Grison P (1973) Lutte intégrée en forêt. Phyt Phytopharm 22:229–248

Hamilton WD, Axelrod R, Tanese R (1990) Sexual reproduction as an adaptation to resist parasites. Proc Natl Acad Sci USA 87:3566–3573

Hartmann G, Blank R (1992) Winterfrost, Kahlfraß und Prachtkäferbefall als Faktoren im Ursachenkomplex des Eichensterbens in Norddeutschland. Forst und Holz 15:443–452

Haujioja E (1980) On the role of plant defenses in the fluctuation of herbivore populations. Oikos 35:202–213

Haynes KJ, Liebhold AM, Johnson DM (2009) Spatial analyses of harmonic oscillation of gypsy moth outbreak intensity. Oecologia 159:249–256

Högstedt G, Seldal T, Breistøl A (2005) Period length in cyclic animal populations. Ecology 86:373–378

Horstmann K (1984) Untersuchungen zum Massenwechsel des Eichenwicklers, *Tortrix viridana* L. Lepidoptera: Tortricidae), in Unterfranken. Z Angew Entomol 98:73–95

Hui D, Iqbal J, Lehmann K et al (2003) Molecular interactions between the specialist herbivore *Manduca sexta* (Lepidoptera, Sphingidae) and its natural host *Nicotiana attenuata*. V. Microarray analysis and further characterization of large-scale changes in herbivore-induced mRNAs. Plant Physiol 131:1877–1893

Hunter MD (1990) Differential susceptibility to variable plant fenology and its role in competition between two insect herbivores on oak. Ecol Entomol 15:401–408

Hunter MD (2002) Maternal effects and the population dynamics of insects on plants. Agric For Entomol 4:1–9

Hunter MD, Price PW (1998) Cycles in insect populations: delayed density dependence or exogenous driving variables? Ecol Entomol 23:216–222

Hunter MD, Varley GC, Gradwell GR (1997) Estimating the relative roles of top-down and bottom-up forces on insect herbivore populations: a classic study revisited. Proc Natl Acad Sci USA 94:9176–9181

Inchausti P, Ginzburg LR (2009) Maternal effects mechanism of population cycling: a formidable competitor to the traditional predator-prey view. Philos Trans R Soc B 364:1117–1124

Ivashov AV, Boyko GE, Simchuk AP (2002) The role of host plant phenology in the development of the oak leaf roller moth, *Tortrix viridana* L. Lepidoptera: Tortricidae. Forest Ecol Manag 157:7–14

Johnson DM, Liebhold AM, Bjørnstad ON (2006) Geographical variation in the periodicity of gypsy moth outbreaks. Ecography 29:367–374

Jüttner O (1959) Ertragskundliche Untersuchungen in wicklergeschädigten Eichenbeständen. Forstarchiv 30:78–83

Kamata N (2000) Population dynamics of the beech caterpillar, *Syntypistis punctatella*, and biotic and abiotic factors. Popul Ecol 42:267–278

Kapeller S, Schroeder H, Schueler S (2011) Modelling the spatial population dynamics of the green oak leaf roller (*Tortrix viridana*) using density dependent competitive interactions: effects of herbivore mortality and varying host-plant quality. Ecol Model 222:1293–1302

Kessler A, Heil M (2011) The multiple faces of indirect defences and their agents of natural selection. Funct Ecol 25:348–357

Knipple DC, Rosenfiels C-L, Nielsen R, You KM, Jeong SE (2002) Evolution of the integral membrane desaturase gene family in moths and flies. Genetics 162:1737–1752

König AO, Ziegenhagen B, van Dam BC et al (2002) Chloroplast DNA variation of oaks in western Central Europe and genetic consequences of human influences. Forest Ecol Manag 156:147–166

Krebs CJ, Boutin S, Boonstra R, Sinclair ARE, Smith JNM, Dale MRT, Martin K, Turkington R (1995) Impact of food and predation on the snowshoe hare cycle. Science 269:1112–1115

Kudler J (1978) Geometroidea. In: Schwenke W (ed) Die Forstschädlinge Europas. Band III. Paul Parey, Hamburg und Berlin, pp 218–263

Luciano P, Roversi PF (2001) Oak defoliators in Italy. Edizioni Poddinghe, Sassari, p 161

Magnoler A, Cambini A (1968a) Accrescimento radiale della quercia da sughero ed effetti delle defogliazioni causate da larve di *Lymantria dispar* L. e di *Malacosoma neustria* L. I. Indagini su piante non decorticate. Mem. Staz. Sper. Sughero Tempio Pausania 27:1–6

Magnoler A, Cambini A (1968b) Accrescimento radiale della quercia da sughero ed effetti delle defogliazioni causate da larve di *Lymantria dispar* L. e di *Malacosoma neustria* L. II. Indagini su piante in produzione. Mem. Staz. Sper. Sughero Tempio Pausania 28:1–16

McCullough DG (2000) A review of factors affecting the population dynamics of jack pine budworm (*Choristoneura pinus pinus* Freeman). Popul Ecol 42:243–256

Myers JH (2000) Population fluctuations of the western tent caterpillar in southwestern British Columbia. Popul Ecol 42:231–241

Myers JH, Boettner G, Elkinton J (1998) Maternal effects in gypsy moth: only sex ratio varies with population density. Ecology 79:305–314

Nedorezov LV, Sadykova DL (2008) Green oak leaf roller moth dynamics: an application of discrete time mathematical models. Ecol Model 212:162–170

Petit RJ, Csaikl UM, Bordács S et al (2002) Chloroplast DNA variation in European white oaks phylogeography and patterns of diversity based on data from over 2600 populations. Forest Ecol Manag 156:5–26

Petit RJ, Kremer A, Wagner DB (1993) Finite island model for organelle and nuclear genes in plants. Heredity 71:630–641

Pitman RM, Vanguelova EI, Benham SE (2010) The effects of phytophagous insects on water and soil nutrient concentrations and fluxes through forest stands of the Level II monitoring network in the UK. Sci Total Environ 409:169–181

Plaistow SJ, Benton TG (2009) The influence of context-dependent maternal effects on population dynamics: an experimental test. Philos Trans R Soc B 364:1049–1058

Poelman EH, Broekgaarden C, van Loon JJA, Dicke M (2008) Early season herbivore differentially affects plant defence responses to subsequently colonizing herbivores and their abundance in the field. Mol Ecol 17:3352–3365

Ragazzi A, Tiberi R (2011) Interazioni insetti fitofagi-funghi patogeni delle specie arboree ornamentali. In: Convegno "Gestione delle emergenze parassitarie nel verde urbano e periurbano". Grugliasco (TO), 24 febbraio 2011. Arbor 30: 7–11

Reymond P (2001) DNA microarrays and plant defence. Plant Physiol Biochem 39:313–321

Reymond P, Bodenhausen N, Van Poecke RMP et al (2004) A conserved transcript pattern in response to a specialist and a generalist herbivore. Plant Cell 16:3132–3147

Reynolds JD, Freckleton RP (2005) Population dynamics: growing to extreme. Science 309:567–568

Röhrig E (1950) Geographische Verbreitung und Schadgebiete des Eichenwicklers. Allg Forstz 51:554–555

Rondan Duenas JC, Panzetti-Dutari GM, Blanco A, Gardenal CN (2002) Restriction fragment length polymorphism of the mtDNA A+T rich region as a genetic marker in *Aedes aegypti*. Diptera: culicidae. Ann Entomol Soc Am 95:352–358

Rossiter MC (1991) Environmentally-based maternal effects: a hidden force in insect population dynamics? Oecologia 87:288–294

Roversi PF, Tiberi R, Bin F (1991) I parassitoidi oofagi dei principali lepidotteri defogliatori del gen. Quercus in Italia. In: Aspetti fitopatologici delle Querce, Atti del Convegno. Problematiche fitopatologiche del gen. *Quercus* in Italia., Firenze, 19–20 nov. 1990: 316–330

Royama T (1992) Analytical population dynamics. Chapman & Hall, London

Schneider D (1992) 100 years of pheromone research. Naturwissenschaften 79:241–250

Schroeder H, Degen B (2008a) Genetic structure of the green oak leaf roller (*Tortrix viridana* L.) and one of its hosts, *Quercus robur* L. Forest Ecol Manag 256:1270–1279

Schroeder H, Degen B (2008b) Spatial genetic structure in populations of the green oak leaf roller *Tortrix viridana* L. (Lepidoptera, Tortricidae). Eur J Forest Res 127:447–453

Schroeder H, Ghirardo A, Schnitzler JP, Fladung M (2011) Tree-insect interaction – defence response against herbivorous insects. BMC Proc 5(Suppl 7):P101

Schroeder H, Scholz F (2005) Identification of PCR-RFLP haplotypes for assessing genetic variation in the green oak leaf roller *Tortrix viridana* L. (Lepidoptera, Tortricidae). Silvae Genet 54(1):17–24

Schroeder H, Yanbaev Y, Degen B (2010) A very small and isolated population of the green oak leaf roller, *Tortrix viridana* L., with high genetic diversity – How does this work? J Hered 101(6):780–783

Schröder H, Ziegler C (2006) Die Situation der Eiche in NRW im Frühjahr 2005. AFZ-Der Wald 6:320–321

Schueler S, Schlünzen KH, Scholz F (2005) Viability of oak pollen and it's implications for long distance gene flow. Trees 19:154–161

Schwenke W (1978) Die Forstschädlinge Europas. Band III. Parey, Hamburg und Berlin, p 467

Schwerdtfeger F (1961). Das Eichenwickler-Problem. Auftreten, Schaden, Massenwechsel und Möglichkeiten der Bekämpfung von *Tortrix viridana* L. in Nordwestdeutschland. Forsch. Berat. Landesaussch. landw. Forsch. Minist. Ernähr. Nordrh. Westf.(c) pt.: 1-174.

Schwerdtfeger F (1971) Vergleichende Untersuchungen an der Kronenfauna der Eichen in Latenz- und Gradationsgebieten des Eichenwicklers (*Tortrix viridana* L.). 3. Die Bedeutung der Parasiten für den lokalen Fluktuationstyp des Eichenwicklers. Z Angew Entomol 67:296–304

Selås V (2003) Moth outbreaks in relation to oak masting and population levels of small mammals: an alternative explanation to the mammal-predation hypothesis. Popul Ecol 45:157–159

Selås V, Hogstad O, Kobro S, Rafoss T (2004) Can sunspot-activity and ultraviolet-B radiation explain cyclic outbreaks of forest moth pest species? Proc R Soc Lond B 271:1897–1901

Simchuk AP, Ivashov AV, Companiytsev VA (1999) Genetic patterns as possible factors causing population cycles in oak leaf roller moth, *Tortrix viridana* L. Forest Ecol Manag 113:35–49

Sinclair ARE, Chitty D, Stefan CI, Krebs CJ (2003) Mammal population cycles: evidence for intrinsic differences during snowshoe hare cycles. Can J Zool 81:216–220

Sinervo B, Svensson E, Comendant T (2000) Density cycles and an offspring quantity and quality game driven by natural selection. Nature 406:985–988

Staudt M, Lhoutellier L (2007) Volatile organic compound emission from holm oak infested by gypsy moth larvae: evidence for distinct responses in damaged and undamaged leaves. Tree Physiol 27:1433–1440

Thaler JS (1999) Jasmonate-inducible plant defences cause increased parasitism of herbivores. Nature 399:686–688

Thomas FM (2008) Recent advances in cause-effect research on oak decline in Europe. CAB Rev Perspect Agric Vet Sci Nutr Nat Res 37:1–12

Tiberi R, Roversi PF (1989) Osservazioni sull'impiego di trappole a feromone sessuale di *Tortrix viridana* L. in querceti della Toscana (Italia Centrale) (Lepidoptera, Tortricidae). Redia, LXXII(1): 277–290

Tiberi R, Roversi PF (1990) Leaf roller moths on oak in Italy (Preliminary note). In: Oak decline in Europe. Proceedings of the international symposium, Kornik, Poland, 1518 May 1990, pp 343–347

Tiberi R (1991). I lepidotteri defogliatori delle querce decidue: bioecologia e danni. In: Atti del Convegno, Aspetti fitopatologici delle Querce, Firenze 19-20 Novembre 1990: 239–250.

Tiberi R, Prota R, Masutti L (1993) Esigenze, prospettive e proposte di nuovi criteri di intervento per il controllo dei lepidotteri defogliatori delle foreste. In: M.A.F. - Convegno "Piante forestali", Firenze, 1992, (ed.) Ist. Sper. Pat. Veg., Roma: 19–34

Tiberi R, Ragazzi A (1998) Association between fungi and xilophagous insects of declining oak in Italy.Redia LXXXI:83–91

Tiberi R, Benassai D, Niccoli A (2005a) Influence of different host plants on the biology and behaviour of the green oak leaf roller, *Tortrix viridana* L.: first results. Integrated Protection in Oak Forest. IOBC/WPRS Bull 28(8):211–217

Tiberi R, Benassai D, Niccoli A (2005b) Influenza della pianta ospite sulla biologia e comportamento della tortrice della quercia, *Tortrix viridana* L. - Linea Ecologica, XXXVII (6): 42–46

Tiberi R, Bracalini M, Panzavolta T (2009) I dinamismi entmologici in boschi gestiti, non gestiti e percorsi dal fuoco. In proceedings of congress Getione sostenibile dei boschi in area mediterranea. Monte S. Angelo, 9th–10th Oct 2008, pp 41–49

Turchin P (2003) Complex population dynamics. Princeton University Press, Princeton

Van Asch M, Visser ME (2007) Phenology of forest caterpillars and their host trees: the importance of synchrony. Annu Rev Entomol 52:37–55

Walling LL (2000) The myriad plant responses to herbivores. J Plant Growth Regul 19:195–216

Waters WE (1969) The life table approach to analysis of insect impact. J Forest 67(5):300–304

Waters WE, Stark RW (1980) Forest pest management: concept and reality. Annu Rev Entomol 25:479–509

Yela JL, Lawton JH (1997) Insect herbivore loads on native and introduced plants: a preliminary study. Entomol Exp Appl 85:275–279

Forests Have Survived Climate Changes and Epidemics in the Past. Will They Continue to Adapt and Survive? At What Cost?

Stephanos Diamandis

Abstract The global climate change that we are now experiencing is not a new event in the long course of our planet. Evidence shows that past changes in climate have caused the extinction of a wide diversity of life forms. What is new is that the current climate change is due to human activity, while in the past climate change was caused by natural events.

Climate change is predicted to have profound effects on the health status, distribution, function and productivity of forests. The impact of climate change on forests is a complicated and multilateral subject and cannot be exhaustively discussed in a short book chapter. A consideration of the principle effects of what may be expected, however, will help the reader to understand the extent of the problems likely to be encountered. Predicting the impact of climate change on forests is a challenge, taking into account the complex ecological and biological interactions as well as the potential loss to society and the economy.

Forests, in past periods, were definitely different from what they are today in coverage, structure and diversity. It is presumed that they were always able to adjust to the climate and human interference and other biotic agents, although not necessarily without negative effects. The important point is that whatever changes they went through, they survived. The current climate change is bound to affect and have more severe impacts on forests of "sensitive" climatic zones first. The temperate and the boreal zones are surely in this category. Warmer and drier conditions are expected to cause stress on trees and render them more sensitive to secondary infections and infestations by pathogens and insects, as well as to forest fires. These pests have complex interactions with hosts, vectors and natural enemies. Moreover, the ecology of all of these organisms is likely to be affected by a changing climate.

S. Diamandis (✉)
Forest Research Institute, 570 06 Vassilika Thessaloniki, Greece
e-mail: diamandi@fri.gr; sdiamand@otenet.gr

T. Fenning (ed.), *Challenges and Opportunities for the World's Forests in the 21st Century*, Forestry Sciences 81, DOI 10.1007/978-94-007-7076-8_34,
© Springer Science+Business Media Dordrecht 2014

Movement of alien invasive pathogens and pests is mainly due to human activity. Alien invasive pathogens and pests may be particularly aggressive when they meet different host trees and different climatic conditions. Examples of catastrophes to forests and forest trees by alien invasive pests in the twentieth century are the Chestnut blight in the USA and Europe, Dutch elm disease in Europe, cypress canker, sudden oak death in the USA, the red palm weevil in the Mediterranean region, kudzu in the USA and numerous others.

Today's forests, therefore, are threatened by two major factors (a) climate change caused by scientifically unsound and materialistic policies and (b) alien invasive pathogens and pests spread yet again, by human activity. International organizations, governments and NGOs have become active over the last 20 years with policies and actions aimed at reducing carbon dioxide emissions and mitigating climate change, as well as to prevent the introduction of alien invasive pathogens and pests. The questions that arise from all this are: Is there sufficient political will at the national and international levels and are there strong enough mechanisms in place to enforce any of the decisions that may be reached? Are the decisions taken and the protocols signed timely, prompt and scientifically sound? Is the extent of the measures enough to stop the deviation from "normal"? Only history will be able to say if humanity had the will.

1   Forests and Climate Change

In the long existence of Earth, life has never remained the same, and has constantly needed to evolve to cope with changing climate as well as other factors. Large-scale climate changes have occurred many times in the past due to natural factors such as changes in the tilt of the Earth's axis and tectonic plate movements and have greatly disturbed our planet by affecting life and biodiversity.

In the mid Cretaceous, about 100 million years ago, evidence suggests that subtropical conditions extended to Alaska and Antarctica and there were no polar ice caps. The planet was warmer than today by 6–8 °C and carbon dioxide (CO_2) levels in the atmosphere were about five times higher than today. These warm conditions lasted for tens of millions of years before the climate started cooling.

One of the fastest changes in Earth's temperature occurred during the Palaeocene-Eocene epoch 55 million years ago, when global temperatures rose 6 °C over a period of 20,000 years or less, probably due to the rapid release of greenhouse gasses due to volcanism in that case, but not unlike the situation that is occurring today with anthropomorphic emissions. This ancient period of rapid climate change caused major ecosystem changes and the extinction of many organisms. The current period of change is likely to do the same.

Over the past two million years, evidence shows that the climate has fluctuated dramatically between ice ages and warm interglacial periods, similar to today's climate. An inescapable conclusion of all the above is that the Earth's climate is unstable and many large changes in climate have already occurred.

Forests have always played an important role in life and its evolution as an enormous and variable source of food, as hot-spots for biodiversity, as regulators of water and carbon balance and as producers of wood fuel, which may prove to be an important alternative to fossil fuels. Because the global human population in the past was very small and widely dispersed, most of the disturbances were caused by acute weather events and general climate change. Forests were burnt mainly by natural fires while trees were killed by local pathogen infections and insect infestations. Species extinction and fluctuation in species domination were primarily driven according to the prevailing climate and natural conditions. When climate changes were gradual, life adjusted and the loss of biodiversity was not dramatic, but when the changes were rapid, many more major disturbances occurred.

It is now widely accepted that (a) climate change is a global problem, (b) carbon dioxide emissions are the main cause of climate change and (c) that deforestation is currently responsible for almost 20 % of annual global emissions of carbon dioxide.

Over the past 150 years the atmospheric concentration of carbon dioxide has increased significantly as a direct result of human activities, mainly through the use of fossil fuels and changing land use. There is mounting evidence that human-induced climate change could create impacts on our environment that may be substantial, abrupt, unpredictable and irreversible, which will almost certainly not be to our advantage.

Deposits of air pollutants on forests is now considered a major stress factor which damages leaves of broadleaved trees and the needles of conifers alike, changes soil and water conditions and, as a result, affects forest tree health, ground vegetation and ecosystem stability. Forests could contribute in reducing emissions through carbon sequestration in their biomass and forest soil. In the last few centuries, however, we have removed more than half of the world's forest cover. We need to reduce this tide of deforestation, but we can make a much greater impact if we also put back some of our lost forests. The industrialized countries which signed the Kyoto Protocol may reach their greenhouse gas reduction commitments either by decreasing emissions or by promoting sinks in carbon (Schulze et al. 2003). Planting more trees can lock up more carbon, reduce the rate of global warming, improve the environment and also people's lives. Many regions and countries have already made efforts to restore much of their forest, but are such fragmentary initiatives enough to hold back climate change?

Throughout this changing environment, forests have continually adapted and survived to the present day in the structure, extent and health status we now experience. Not, however, without losses. Loss of biodiversity which is irreversible has been immense. In our day, there is increasing concern about the scale of the losses of forest area and biodiversity because of rapid climate change. Evidence shows that species are now becoming extinct at an alarming rate almost entirely as a direct consequence of human activities. Can humanity afford to continue this behavior ignoring the consequences? Can international organizations, governments, societies and we as individuals do anything to reduce the rate of climate change with all its adverse effects? Do we envisage leaving a better earth to future generations?

2 Impact of Climate Change on the Health Status of Global Forests. What Is to Be Expected

Knowledge on climate change is driven by long term research into weather data and patterns. The conclusion reached demonstrates that after a long period of gradual warming, probably due to carbon emissions resulting from the rise of agriculture and the associated loss of forest, Earth's climate has warmed rapidly in the twentieth century (Canadell et al. 2007). Land has warmed faster than the oceans (Solomon et al. 2007). Therefore, the impact of global warming is expected to be more severe on forests and terrestrial ecosystems and biodiversity than on the marine environment.

The effects of climate change on forests will depend not only on climatic factors, but also on stresses such as pollution, future trends in forest management practices, including fire control and demand for timber, and land-use changes. It is difficult to separate the influence of climate change from these other pressures.

Climate change effects on forests are very likely to include changes in forest health and productivity and changes in the geographic range of certain tree species. These effects will in turn alter timber production, water quality, wildlife, outdoor recreational activities and rates of carbon storage (Boko et al. 2007).

In general, forests are sensitive to climatic variability and change. Climatic factors that influence forest health are air temperature, precipitation and its seasonal distribution, atmospheric levels of carbon dioxide and other greenhouse gases, extreme weather and the frequency and severity of fire events. All these factors are changing and are expected to continue to change due to human activities. Environmental variables, however, are interwoven in relatively complex ways. Warmer temperatures, for example, can generally be expected to increase plant productivity, but such changes in temperature may also produce other effects. Plants require both water and carbon dioxide for photosynthesis and growth, and plants also use water to maintain heat balance. Warming can cause greater evaporation of water from soils which would limit its availability to plants, potentially causing drought stress and reducing plant productivity. However, this effect can be ameliorated by increased atmospheric carbon dioxide, which allows plants to use water more efficiently. In addition, plants at different stages in their life cycles often have different levels of resistance to extreme environmental conditions. Therefore, any attempt to understand in depth and conclude possible impacts will be uncertain.

Understanding the long term changes in local and global climate that will occur in future will help to model the way that forests may respond. Apart from global warming, which is strongly supported by the existing evidence and the factors which are causing it, there is a remarkable divergence in the speculations on what will happen in the case of global precipitation, which areas will be most affected and to what degree. For instance, temperature increases are widespread over the globe with the greatest increases occurring at the higher northern latitudes. Total global precipitation has increased since 1906 (Nicholls et al. 1996) but not all areas are experiencing increases. Lower elevations are receiving more precipitation in

Fig. 1 Climate change characterized by long dry seasons results in water stress of pines which makes them more susceptible to bark beetle attacks

the form of rain and less in the form of snow (Barnett et al. 2008), whereas areas affected by drought appear to have been on the increase since the 1970s (Solomon et al. 2007). Increased global warming is linked to changes in precipitation patterns, decrease in snowfall and changes in the frequency and intensity of acute weather events.

Changes in temperature and precipitation are expected to change forest location, composition, and productivity. Climate change is likely to drive the migration of tree species, resulting in changes in the geographic distribution of forest types and new combinations of species within forests. In the Northern Hemisphere for example, many tree species may shift northward or to higher elevations.

Climatic changes may increase the spread and distribution of pests and diseases by removing abiotic constraints that previously limited their geographic distribution. Increased temperatures are also likely to increase fire risk in areas that suffer from hotter, longer and drier summer seasons (IPCC 2007). Examples of higher fire frequency and severity in recent years come from Russia, the Western and Southwestern United States, the Mediterranean region and Australia. Such extensive forest disturbance regimes could have a negative impact on forest coverage and the market for forest products and especially timber. In addition, climate change could promote the rapid increase in diseases and pests that attack tree species (Fig. 1). Although such disturbances may be detrimental to forests, in the short term they may have little impact on the market due to salvage operations that harvest timber from dying forests (Shugart et al. 2003). Already, alien invasive organisms are causing terrible disturbances in terrestrial ecosystems in many parts of the world.

Climate change effects that influence tree growth will also alter rates of carbon storage (or sequestration) in trees and soils. Increased carbon sequestration will remove more carbon dioxide from the atmosphere (a negative feedback that lessens climate change), whereas carbon losses due to forest disturbances will result in more carbon dioxide entering the atmosphere (a positive feedback that strengthens climate change). The IPCC (2007) predicts that "net carbon uptake by terrestrial ecosystems is likely to peak before mid-century and then weaken or even reverse, thus amplifying climate change". It is strongly and generally supported, however, that in times of acute industrialization and carbon emissions, forests have the potential to absorb about one-third of global carbon emissions into their biomass, soils and products and to store them – in principle in perpetuity.

In regard to tree growth and productivity, understanding just how they will change is quite complicated. Plant-level experimental evidence suggests that carbon fertilization is likely to increase individual tree growth for some period of time, at least for those species or areas where carbon dioxide is the limiting factor for growth. Some evidence also suggests that the carbon dioxide effect makes trees use water more efficiently, thereby making them less vulnerable to drought. Other evidence, however, suggests that the effects of carbon fertilization decline as trees age and at wider spatial scales where forest losses from competition, disturbance, and nutrient limitations become important.

Overall and taking into account numerous serious studies, it is widely accepted that the impacts of climate change on the forestry sector will not be distributed evenly throughout the world. In the USA for example, the Southeast, which is currently a dominant region for forestry, is likely to experience losses as tree species migrate northward and tree productivity declines. Meanwhile, the north is likely to benefit from tree migration and longer growing seasons. The disturbance, however, in conservation of forests and biodiversity under the current circumstances will be immense.

3   Impact of Alien Invasive Pathogens and Pests on Forests

It is also widely accepted that under the global warming scenarios, insect and pathogen outbreaks will likely increase in frequency and severity. As forest ecosystems change and move in response to climate changes, they will become more vulnerable to disturbances by biotic agents. In addition to the local pests and pathogens which are already expected to gradually become more destructive, alien invasive pests which are being increasingly spread around the globe may rapidly cause major losses.

One commonly used definition of alien invasive species is those species which occur as the result of an introduction, intentionally or accidentally by humans, and then reproduce in considerable numbers when away from their parental range

(Stenlid 2011). Invasive species are organisms that have been introduced into an environment where they have no co-evolution or co-adaptation history and whose introduction causes, or is likely to cause, economic or environmental harm. Typically, these species have few or no natural enemies to limit their reproduction and spread. The ease and speed with which commodities and people can travel between countries and continents, especially in recent decades, has created new and faster mechanisms for the spread of alien invasive pests and diseases than ever existed before.

To illustrate the magnitude of this problem, the current economic impact caused by all invasive species in the USA has been estimated at over \$137 billion per year. The Federal/State cost-share program for gypsy moth suppression in the eastern USA alone during the years 1990–2001 ranged from between \$4.8 and \$22.5 million per year and totalled \$123.3 million.

The ecological impact has also been tremendous. Invasive insects and pathogens can rapidly establish themselves in a forest, weaken or kill native plants, degrade wildlife habitat, alter nutrient cycles, and threaten basic ecosystem structure and functions. In recent years, pathogens of unknown origin such as sudden oak death in the USA and insects of known origin such as the red palm weevil in the Mediterranean pose new imminent threats to many urban and rural forests, plantations and parks. Invasive plants can cause great harm to forest ecosystems by altering basic ecological processes such as the role of fire. Needless to say, forests that are damaged by invasive species are expected to yield far fewer goods and services such as timber and recreation, and also have a lower wildlife value.

Last but not least, societal impacts caused by invasive species include the annoyances associated with the presence of large numbers of invasive species, especially insects in urban and recreation areas. They may also include skin irritations and allergies. Defoliation and other types of damage caused to plants may threaten the extinction of native plants which are used for medicinal purposes, both current and/or future, and traditional uses.

Climate change in the form of milder winters will enable increased winter survival of plant pathogens and pests as well as accelerated vector and pathogen life cycles. Climate change is expected to enable plants and pathogens to survive outside their historic ranges. A few examples of relatively recent movements of such pathogens can illustrate the magnitude of the potential impacts.

Chestnut blight caused by the fungus *Cryphonectria parasitica* is probably the most striking case of an alien pathogen responsible for the virtual extinction of the gorgeous American chestnut, *Castanea dentata*, in the twentieth century. The fungus entered North America from China, and from 1904 until 1945 infected and killed the extensive chestnut forests of the eastern United States. The fungus was also transferred to Europe in 1938 and spread over all the chestnut growing regions of southern Europe, causing terrible disturbance to the chestnut orchards which continues to this day. Luckily in Europe, viruses with ds-RNA seemed to spread and convert the virulent strains of the fungus to hypovirulent forms, increasing hopes for reversing the effects of this decline (Fig. 2).

Similar stories apply to the diseases of cypress canker, caused by the alien pathogenic fungi *Seridium cardinale* threatening several *Cupressus* species, *Ophiostoma*

Fig. 2 (**a**) Lethal canker (*left*), (**b**) Converted, superficial cankers (shown by *arrows*) on chestnut coppice (*right*)

novo-ulmi threatening European elms and canker stain disease of *Platanus* caused by *Ceratocystis platani*. This last pathogen, after having caused great loss to urban *P. accerifolia* in Italy and Western Europe, is spreading to the Eastern Mediterranean threatening the more susceptible, autochthonous *Platanus orientalis*.

The introduction of *Phytophthora cinnamomi* to Australia had a devastating effect on *Eucalyptus marginata* and many other herbaceous species (Weste and Marks 1987).

Phytophthora ramorum, the cause of sudden oak death in western North America which has now started spreading in Europe, provides a current example, dramatically illustrating the potential of these pathogens for rapid ecological and economic damage to world forests.

Destructive alien insects also have the potential to make sudden impact on forests and forest trees. In recent years, the horse chestnut leaf miner, *Cameraria ohridella*, spread from the Balkans to almost all of Europe, causing severe defoliation to horse chestnut and upsetting urban areas while the chestnut wasp, *Dryocosmus kuriphilus*, is spreading rapidly to the European chestnut region from Italy where it entered in 2001 from S. Korea. The red palm weevil, *Rhynchophorus ferrugineus*, has put on alert all Southern European countries along the Mediterranean coast (Fig. 3). The gypsy moth, *Lymantria dispar*, is undoubtedly the most serious insect pest of broadleaf trees in the eastern USA. This insect was introduced into Massachusetts from Europe in 1869 and is now distributed throughout most of the northeast and is rapidly spreading south into Virginia and west into Ohio, Michigan and Wisconsin. The gypsy moth can feed on at least 500 species of trees, shrubs and vines while older larvae are capable of feeding on conifers (McManus 1980).

The pinewood nematode, *Bursaphlenchus xylophilus*, is another recent example of human interference with extreme destructive effect to forests. The nematode,

Fig. 3 Palm trees recently infested by the red palm weevil

which is native to the USA and does not threaten native forests there, was reported only in 1979 to induce pine wilt disease of non-native pines in Missouri. The nematode was introduced to Japan and it was not identified as the causal agent of pine wilt disease until 1971. Since the pinewood nematode was introduced into Japan, it has extensively damaged Japanese red pines (*Pinus densiflora*) and black pines (*P. thunbergii*). The pinewood nematode has also been reported in Canada and Mexico.

Asian countries other than Japan began to report presence of the pinewood nematode in the mid- to late-1980s. Taiwan reported the pinewood nematode in Japanese black pine and luchu pine (*P. luchuensis*) in 1985. By 1989, China and S. Korea had also reported pinewood nematode in Japanese red and black pine. In 1999, Portugal reported the pinewood nematode present in declining maritime pine (*P. pinaster*) in the Iberian Peninsula, putting the phytosanitary system of the European Union on alert. Possible spread of the nematode into Europe will threaten some of the most important pine species such as Scots pine (*P. sylvestris*), Austrian pine (*P. nigra*) and maritime pine.

Not only insects and pathogens, but any invasive alien organism may out-compete native species, repressing or excluding them and, therefore, fundamentally change the ecosystem. They may indirectly transform the structure and species composition of the ecosystem by changing the way in which nutrients are cycled through the ecosystem (McNeeley et al. 2001). Entire ecosystems may be placed at risk through knock-on effects. Given the critical role biodiversity plays in the maintenance of essential ecosystem functions, invasive alien species may cause changes in environmental services, such as flood control and water supply, water assimilation, nutrient recycling, conservation and regeneration of soils.

Invasive plants can become major problems to forest ecosystems. A well known example is kudzu, *Pueraria montana* var. *lobata*, which was introduced to the USA in 1876 from Japan as an ornamental plant. Nowadays, it has spread over seven

Fig. 4 Kudzu covering entire young pine trees (State of Tenn.)

million acres in the South-eastern U.S. The problem is that it just grows too well. The climate of the South-eastern U.S. is perfect for kudzu. The vines grow as much as a foot per day during summer months, climbing trees and destroying valuable forests by preventing trees from receiving sunlight (Fig. 4). Under ideal conditions kudzu vines can grow 60 ft each year.

Africa is now home to hundreds of alien invasive species, but the magnitude of the problem varies from country to country and from ecosystem to ecosystem. Invasive alien species are a problem in diverse ecosystems from savannahs to tropical forests (UNEP 2004). Disturbed ecosystems are particularly vulnerable to invasion by alien species. In Tanzania, for example, the large African forest tree *Maesopsis eminii* has become dominant in logged forests. It is also capable of regenerating in natural forests, particularly where there are large gaps caused by tree-falls (Bingelli et al. 1998). South Africa seems to be the leading country in numbers of alien invasive pests in the African continent.

The European Union, as well as many other countries, has created lists of quarantine organisms and has introduced strict legislation to keep the borders closed to such threatening agents for its forests. Although inappropriate importation of timber and dunnage has been of major importance in the introduction of some pathogens, it is strongly believed that commercial movement of living plants, together with unlicensed specialists or amateur plant collectors is the pathway of highest risk (Brasier 2008). Inspections are enforced on all relevant products at national entry points and surveys are carried out by phytosanitary inspectors on a national level. Despite all the measures taken, however, the introduction of alien invasive species has not been halted. According to Brasier 2008 and based on new incidents in the plant trade and on regular breaches worldwide, some tenets underlying the protocols must now be viewed as outdated and seriously flawed. The scenario of climate

Fig. 5 Galls on chestnut tree caused by the chestnut wasp

change has increased fears that there will be even more acute damage to world forests by such organisms.

The question that arises at this point in relation to movement of organisms between countries and continents is to what extent such movements occurred in the past. Travel of humans, animals and agricultural products which seem to be the most common entry route for alien invasive organisms started to increase sharply in the twentieth century. The number of introductions which we have been experiencing is obviously linked to the ease of travel and globalization of commerce.

The management and control of alien invasive species present some important challenges for decision-makers. Globally, preventing their introduction is seen as the cornerstone of effective measures for dealing with them. This approach is believed to be the most cost-effective and environmentally-sound approach, as once invasive species become established, eradication may be impossible and the ecological damage irreversible (Shrine et al. 2000).

Therefore, only coordinated action on a global scale can reduce the risk of new introductions. The matter, however, is so delicate that even one single event of negligence is enough to put whole tree species at risk as well as extensive tracts of forest, not only in a single country but across whole continents. A recent example of such negligence is the introduction of the chestnut wasp, *Dryocosmus kuriphilus*, from S. Korea to N. Italy in 2001 (Fig. 5). Not only has its eradication proved impossible, but the insect has now spread all over Italy, France, Slovenia and Croatia, further threatening the chestnut production along the Mediterranean European coast.

The IUFRO Unit 7.03.12 has been established to examine global forestry issues related to the unwanted international movement of alien invasive species, including fungi, insects, nematodes, and plants. The collaborative work of such a group can provide valuable information to decision makers on how to deal with this increasing threat due to increasing movement pathways. Strict national legislation, coordination on international scale and public awareness are critical points towards reducing the threat from alien invasive organisms.

4   Is There Political Will to Work Towards Halting Climate Change and the Movement of Alien Invasive Organisms in Order to Conserve the World Forests?

Certain products and services produced by forests naturally belong to the country where they are located, however, certain others such as the effect on world climate and carbon sequestration are global. The protection and conservation of world forests therefore, is of international concern.

Almost two decades ago, in 1992, countries joined an international treaty, the United Nations Framework Convention on Climate Change (UNFCCC), to cooperatively consider what they could do to reduce the average global temperature increases and the resulting climate change, and to cope with whatever impacts were then predicted. By 1995, countries realized that emission reductions provisions in the Convention were inadequate. They launched negotiations to strengthen the global response to climate change and, 2 years later, adopted the Kyoto Protocol.

The Kyoto Protocol legally binds developed countries to emission reduction targets. The Protocol's first commitment period started in 2008 and ends in 2012. The UNFCCC secretariat supports all institutions involved in the international climate change negotiations, particularly the Conference of the Parties (COP), the subsidiary bodies (which advise the COP), and the COP Bureau which deals mainly with procedural and organizational issues arising from the COP and also has technical functions.

What happens beyond 2012 is one of the key issues governments of the 195 Parties to the Convention are currently negotiating. At the time of writing this article, there is no concrete plan for what will happen after 2012 even though the Kyoto Protocol is very near its expiration date. A few industrialized countries are against any further action while the majority of developing countries declared that they will not sit and negotiate in case there is no agreement.

Climate change is a complex problem which, although environmental in nature, has consequences for all aspects of our existence on our planet. It either impacts upon, or is impacted by global issues, including poverty, economic development, population growth, sustainable development and resource management. It is not surprising, then, that the solutions must come from all disciplines and fields of research and development.

The issue of alien invasive pests and pathogens also has a global dimension in threatening the world's forests. As potentially invasive organisms move from the location of their origin to a new location, they may find suitable environments without their natural enemies and susceptible hosts which have no co-evolved defences against them. Travelers should always comply with the relevant legislation and exported goods should comply with all necessary phytosanitary requirements (Brasier 2008). Countries of destination should have effective inspection mechanisms to halt alien invasive species before they arrive because preventing the entrance of such organisms is the key issue in avoiding minor or major disturbance of natural ecosystems.

The European Union has issued directives to all member states in an effort to keep the European natural environment and forests safe from alien invasive pests and pathogens. It appears that although there is political will and action is being taken, the arrival and spread of alien species at this time is still not a rare phenomenon.

5 Discussion-Conclusions

Global climate change and alien invasive pests and pathogens should be considered to be an emerging threat, with potentially severe implications for forests and biodiversity the world over. The evidence shows that the survival of forests at this time is threatened by two major factors (a) the climate change caused by unscientific and materialistic policies and (b) the alien invasive pathogens and pests caused again by human activity. The question that arises is whether there is political will to work towards the conservation of the world forests which are the basis of life on the Earth.

The three Rio Conventions-on Biodiversity, Climate Change and Desertification-which are intrinsically linked, operating in the same ecosystems and independently addressing those issues which derive directly from the 1992 UN Conference on Environment and Development Earth Summit.

The United Nations has been taking drastic action on a global scale over the last 20 years. At the very heart of the response to climate change is the need to reduce emissions. In 2010, governments agreed that emissions need to be reduced so that global temperature increases are limited to below 2 °C. The world has been eager to see measures agreed upon by main industrial and developing countries targeting to reduce the impact of global climate change, as such measures would relieve the impact on the current forest distribution, stabilize the climate and conserve biological diversity. However, although the concepts of *biodiversity*, *sustainable management of resources*, *conservation of forests* and *climate change* are increasingly familiar to policy-makers, opinion-formers and the concerned public, forests, especially those in tropical regions, are still being grossly overexploited and the rate of global warming has not been reduced.

Individual countries need to respond to these requirements by drawing up comprehensive domestic strategies for climate change, which compliment global efforts. Implementing such strategies will ensure that forestry policies and practices allow woodlands to withstand the rigours of climate change and even help to mitigate it.

We need to understand what the changes might be and how we can plan management for robust forests that will be sustainable in the long term. We also have a role in helping forestry play a part in alleviating the impacts of climate change in the wider landscape by, for example, providing wildlife refuges and reducing the impact of flooding.

Research carried out worldwide is an important first step towards understanding the potential effects of global warming on the world's forest ecosystems. Still, we have a very limited understanding of the interactions of the multiple stresses on forest ecosystems and much less about what will happen in a warmer world. In addition to research targeted to understand the mechanisms associated with climate change impacts on forests, there must be an improved and enhanced emphasis on long-term monitoring of forest composition and growth. Biotechnological research focused on identifying and distributing fast growing, short rotation trees that were adapted to areas targeted for reforestation may represent an excellent solution for tackling climate change while also contributing to wider conservation, energy security and development objectives (Fenning et al. 2008).

The role of the international scientific community is to investigate all aspects of climate change and improve the understanding of how changes to the natural environment may influence forests and woodlands in the future and to provide authoritative advice to decision makers as to what policies to use and how woodland management can adapt to the changes that have been predicted.

Some future climate changes are inevitable but anything we can do to reduce the scale of this change will be worthwhile. Are the commitments taken on international and national levels implemented into concrete action? Is there strong political will on the national level and are there international mechanisms to enforce decided actions? Are the decisions taken and the protocols signed timely and prompt? Are the agreed measures sufficient to stop the deviation from what we have come to consider "normal" climatic conditions? Much of the biology is unknown, but only history will be able to say whether humanity had the will. The answer is in our hands.

References

Barnett TP, Pierce DW (2008) When will Lake Mead go dry? Water Resource Res 44:W03201. DOI:10.1029/2007WR06704

Binggeli P, Hall JB, Healey JR (1998) An overview of invasive woody plants in the tropics. School of agricultural and forest sciences. University of Wales, Bangor, http://www.safs.bangor.ac.uk/IWPT

Boko M, Niang I, Nyong A, Vogel C, Githeko A, Medany M, Osman-Elasha B, Tabo R, Yanda P (2007) Africa. Climate change 2007: impacts, adaptation and vulnerability. In: Parry ML, Canziani OF, Palutikof JP, van der Linden PJ, Hanson CE (eds) Contribution of working group II to the fourth assessment report of the intergovernmental panel on climate change. Cambridge University Press, Cambridge, pp 433–467

Brasier CM (2008) The biosecurity threat to the UK and global environment from international trade of plants. Plant Path 57:792–808

Canadell JG, Le Quéré C, Raupach MR, Field CB, Buitenhuis ET, Ciais P, Conway TJ, Gillett NP, Houghton RA, Marland G (2007) Contributions to accelerating atmospheric CO_2 growth from economic activity, carbon intensity, and efficiency of natural sinks. PNAS 104(47):18866–18870

Fenning T, Walter C, Gartland KMA (2008) Forest biotech and climate change. Nat Biotechnol 26(6):615–616

IPCC (2007) Climate change 2007: impacts, adaptation, and vulnerability. In: Parry, Martin L, Canziani, Osvaldo F, Palutikof JP, van der Linden PJ, Hanson CE (eds) Contribution of working group II to the fourth assessment report of the intergovernmental panel on climate change. Cambridge University Press, Cambridge, p 1000

McManus ML (1980) The gypsy moth. USDA forest service, forest insect and disease leaflet 162, p 10

McNeeley JA, Mooney HA, Neville LE, Schei P, Waage JK (2001) Global strategy on invasive alien species. IUCN – the World Conservation Union, Gland

Nicholls NN, Gniza GV, Jouzel J, Karl TR, Ogallo LA, Parker DA (1996) Observed climate variability and change. In: Houghton JT, Meira Filho LG, Callander BA, Harris N, Kattenberg N, Maskell K (eds) Climate change in 1995: the science of climate change. Cambridge University Press, Cambridge, pp 135–192

Schulze E-D, Mollicone D, Archard F, Matteucci G, Federici S, Evans HD, Valentini R (2003) Making deforestation pay under the Kyoto protocol. Science 299:1669

Shrine C, Williams N, Gundling L (2000) A guide to designing legal and institutional frameworks on alien invasive species. IUCN – the World Conservation Union, Gland

Shugart H, Sedjo R, Sohngen B (2003) Forests & global climate change: potential impact on U.S. forest resources. Pew Center on Global Climate Change, Arlington VA, p 50

Solomon S, Qin D, Manning M, Chen M, Marquis M, Averyt KB, Tignor M, Miller HL (eds) (2007) Climate change 2007-the physical science basis. Contribution of working group I to the fourth assessment report of the inter-governmental panel on climate change. Cambridge University Press, Cambridge, UK/New York, p 996, http://www.ipcc.ch/ipccreport/ar4-wgl.htm. (Dec. 2009)

Stenlid J (2011) Alien and invasive fungi-What can be expected from a changing climate? Book of Abstracts, XVI CEM, Thessaloniki, pp 132–134

UNEP (2004) World conservation monitoring centre. Annual report, p 17

Weste G, Marks GC (1987) The biology of Phytophthora cinnamomi in Australasian forests. Ann Rev Phytopathol 25:207–229

Horse Chestnut Bleeding Canker: A Twenty-First Century Tree Pathogen

S. Green, B.E. Laue, R. Nowell, and H. Steele

Abstract European horse chestnut is an important amenity tree species which has recently been devastated by an emerging epidemic of bleeding canker disease. Symptoms include bleeding cankers on the stem and branches which can lead to tree mortality. The causal agent of this disease is the pathogenic bacterium, *Pseudomonas syringae* pv. *aesculi*, which is believed to have originated in India on Indian horse chestnut. This bacterium probably spread to Europe in the early 2000s via an unknown pathway, with the epidemic in Britain resulting from the introduction of a single bacterial strain. The development of a real-time PCR diagnostic test for *P. syringae* pv. *aesculi* has facilitated its rapid detection in symptomatic trees and provides a useful tool to study host infection and survival outside the host. Lesions caused by *P. syringae* pv. *aesculi* developed on woody branches of horse chestnut during the host dormant season and were centred mainly on lenticels, leaf scars and nodes. The lesions developed in the cortex and phloem and extended into the cambium to cause longitudinally spreading cankers. To better understand the evolutionary history of *P. syringae* pv. *aesculi* and its genetic adaptation on to a novel host species, draft genome sequences were generated for a strain from Britain and a type strain from India. Genomic comparisons with other *P. syringae* pathovars showed that *P. syringae* pv. *aesculi* has readily gained and lost genes during its recent past. Potentially important genetic gains include a pathway for the enzymatic degradation of plant-derived aromatic products and two genes involved in the metabolism of nitric oxide which may enable the bacterium to disable an important host defence response. Future genomic comparisons combined with functional

S. Green (✉) • B.E. Laue • H. Steele
Forest Research, Northern Research Station, Roslin, Midlothian,
Scotland EH25 9SY, UK
e-mail: sarah.green@forestry.gsi.gov.uk

R. Nowell
Institute for Evolutionary Biology, University of Edinburgh, Kings Buildings,
Ashworth Laboratories, West Mains Road, Edinburgh EH9 3JT, UK

T. Fenning (ed.), *Challenges and Opportunities for the World's Forests in the 21st Century*, Forestry Sciences 81, DOI 10.1007/978-94-007-7076-8_35,
© Crown Employees UK 2014

analyses of genetic pathways will help unravel the key host-pathogen interactions which underlie bacterial diseases of trees.

1   Introduction

European horse chestnut (*Aesculus hippocastanum*), which is native to the Balkan region, is an important amenity tree species throughout much of Europe and is planted widely in parks and gardens in both urban and rural areas, often in avenues bordering roads. Horse chestnut is highly regarded for its qualities as a shade tree, for its showy white flowers in spring and the production of its fruits or 'conkers'. In 2002/2003 a highly damaging new disease of horse chestnut was first reported in The Netherlands, Belgium and Britain. The disease, called horse chestnut bleeding canker because it affects woody stems and branches, became rapidly widespread in several countries of northwest Europe, including Britain, The Netherlands, Belgium and northern Germany. The causal agent of this horse chestnut bleeding canker epidemic is the bacterium, *Pseudomonas syringae* pv. *aesculi*, originally isolated in 1969 from leaf lesions on Indian horse chestnut (*Aesculus indica*) in India (Durgapal 1971; Durgapal and Singh 1980; Webber et al. 2008; Green et al. 2009). Recent isolations of *P. syringae* pv. *aesculi* from diseased horse chestnut in Ireland (B. Laue), Norway (V. Talgø, personal communication), Czech Republic (I. Pánková, personal communication) and Saxony in eastern Germany (S. Hilgert, personal communication) suggest that the pathogen is extending its range in Europe. Horse chestnut bleeding canker is just one of an increasing number of highly damaging new tree diseases to have emerged around the world in the last two decades, many of which can be directly linked to the burgeoning international trade in live plants and inadequate international plant health protocols (Brasier 2008). Until its arrival in Europe, almost nothing was known of *P. syringae* pv. *aesculi*. This is typical for translocated plant pathogens which have escaped notice in their geographical centres of origin because they tend to cause little damage to their native hosts which have co-evolved resistance. When a pathogen which has not previously been the subject of scientific study arises in a new geographical location on a susceptible new host not previously exposed to that pathogen, there is a need to generate information very quickly. In this case, high-throughput sequencing technology can be used to perform genome-wide surveys of the genetic make-up of the emerging pathogen, providing important taxonomic information, identifying appropriate loci for the development of diagnostic markers, and giving insights into the genetic basis enabling fitness in a new environment and the potential for further adaptive change. Given that *P. syringae* pv. *aesculi* has only recently emerged in Europe, has spread rapidly, is highly virulent and evidently capable of attacking the woody parts of the tree directly, it is an ideal model organism for the study of an emerging tree disease. Furthermore, since very little is currently known about bacterial tree diseases in general, biological information gained from studying *P. syringae* pv. *aesculi* can be applied to tackle future emerging bacterial

Fig. 1 Symptoms of horse chestnut bleeding canker

pathogens of trees. This chapter provides an overview of disease symptoms caused by *P. syringae* pv. *aesculi*, the development of a diagnostic tool to detect the pathogen in a variety of substrates, and summarises our current knowledge of disease epidemiology on European horse chestnut. Also discussed is how the rapidly expanding field of comparative genomics and its associated repertoire of tools was used to gain insights into the evolutionary background and genetic make-up of this important, newly emerging bacterial tree pathogen.

2 Symptoms and Disease Management

Symptoms of horse chestnut bleeding canker include rust-coloured or blackened liquid oozing from cracks in the bark located on the main stem and branches (Fig. 1), necrotic phloem underlying the outer bark, blackened and cracked cankers on small diameter branches and dieback often leading to tree death. Thousands of horse chestnut trees in Britain have exhibited these symptoms since 2003. A survey of horse chestnuts conducted in Britain in 2007 found that over 70 % of trees surveyed in parts of England had bleeding canker symptoms, with 36 % and 42 % of surveyed trees showing these symptoms in Wales and Scotland, respectively (Forestry Commission 2008). Observations of the disease suggest that there may be variability in the horse chestnut population in terms of resistance to infection, with the most susceptible individuals killed outright during the initial wave of the epidemic and the remaining population exhibiting a range of disease symptoms from mild to severe which may appear worse in some years than others. A useful aspect for further study would be to investigate whether such observations of perceived resistance have a genetic basis and also whether apparent annual fluctuations in symptom severity are driven by environmental factors. Typically, infected horse chestnut trees which remain live have bleeds and cracks on one or more branches or on part of the stem, partial crown dieback, and premature leaf senescence in the autumn as a result of damage to the phloem in the stem and branches. In Britain, *P. syringae* pv. *aesculi* is now considered to be endemic throughout England, Wales and south-central Scotland, and therefore little can be done to prevent a newly planted tree from becoming infected, and there is no known cure for the disease.

3 Diagnosing *Pseudomonas syringae pv. aesculi* on Horse Chestnut

Initially, the recent epidemic of horse chestnut bleeding canker was thought to be due to infection by *Phytophthora* spp., which were first reported causing similar symptoms on horse chestnut in southern parts of Britain during the 1970s (Brasier and Strouts 1976). However, the incidence of *Phytophthora* on horse chestnut is very low compared with *P. syringae* pv. *aesculi*. Nonetheless, one of the early obstacles to diagnosing *P. syringae* pv. *aesculi* on horse chestnut was the lengthy procedure associated with confirming its presence on the host given that *Phytophthora* pathogens cause similar symptoms and because damaged tree bark can be colonised by a large number of non-pathogenic bacterial species. Therefore, a quantitative real-time PCR assay was developed to detect *P. syringae* pv. *aesculi* on horse chestnut (Green et al. 2009). This is a rapid molecular test which uses DNA primers to recognise and amplify a short section of target DNA with a sequence of base pairs specific to *P. syringae* pv. *aesculi*. To develop this assay, bacteria were isolated from necrotic bark on diseased horse chestnut from around Britain. Bacterial isolates were also collected from the bark of healthy horse chestnut. Real-time PCR primers were developed based on gyrase B gene sequence data for the specific detection of *P. syringae* pv. *aesculi*. Primer specificity was tested on isolates of the target pathogen as well as on a broad range of non-target *P. syringae* pathovars and numerous other bacterial spp. which were found to inhabit the bark of horse chestnut, including six bacteria of known species (*Pseudomonas marginalis*, *P. mediterranea*, *P. fluorescens*, *Erwinia rhapontici*, *Erwinia billingiae* and *Yersinia frederiksenii*) and several previously undescribed bacteria not represented in the publically accessible NCBI GenBank DNA database (Green et al. 2009). The real-time primers reliably amplified *P. syringae* pv. *aesculi* down to 1 pg of extracted DNA, and also amplified unextracted DNA in whole cells of the bacterium down to at least 160 colony forming units (Green et al. 2009). Thus, rapid diagnosis of the pathogen in infected trees can now be achieved by sampling necrotic inner bark, incubating small bark fragments overnight on nutrient agar and testing bacterial growth the next day for the presence of *P. syringae* pv. *aesculi* using the real-time PCR assay. This assay has also provided the international plant health community with a useful tool for screening horse chestnuts for the bacterium. For example, plant health staff in Ohio, USA, are using the real-time PCR assay to screen native buckeye (*Aesculus* spp.) for *P. syringae* pv. *aesculi* in case the bacterium may have entered the USA on *Aesculus* imports from Europe (J. Fisher, pers. comm.). In addition to enabling rapid diagnosis of the target pathogen for phytosanitary purposes, this quantitative real-time PCR assay provides the facility to study aspects of the biology of *P. syringae* pv. *aesculi* on horse chestnut, for example detecting routes of host infection and the potential for survival in the environment outside the host.

4 Infection of Horse Chestnut by *P. syringae* pv. *aesculi*

Branches and whole trees of horse chestnut naturally infected with *P. syringae* pv. *aesculi* at three locations in Scotland were subjected to detailed morphological and histological examination to identify the primary infection sites on the woody parts of the tree, the year and season that infection had occurred, and the patterns of subsequent lesion expansion within the host (Steele et al. 2010). The real time PCR diagnostic test was used to confirm the presence of *P. syringae* pv. *aesculi* in host tissues. The study found that *P. syringae* pv. *aesculi* caused lesions on woody branches of various ages towards the end of the growing season, with the lesions centred mainly on lenticels, leaf scars and nodes (Steele et al. 2010). Horse chestnut saplings with young, extending shoots were also inoculated in late-April by spraying them with a suspension of *P. syringae* pv. *aesculi* cells in water. Disease symptoms appeared within a few weeks as black, necrotic flecks which were generally associated with lenticels on current season leading shoots and petioles (Green et al. 2009). Primary infection of woody tree parts via lenticels has not been reported for other *Pseudomonas* pathogens. Thus *P. syringae* pv. *aesculi* may be uniquely adapted amongst *Pseudomonas* tree pathogens to penetrating the lenticels of its woody host, and given that lenticels represent a very high number of potential infection sites, the success of *P. syringae* pv. *aesculi* as a tree pathogen and the causal agent of a large-scale epidemic may be down to an ability to infect the aerial woody parts of its host directly (Steele et al. 2010).

Infection of woody branches via leaf scars is well documented for other bacterial canker pathogens, including *P. syringae* pv. *morsprunorum* on stone fruit trees (Crosse 1951). The susceptibility of leaf scars is highest in the autumn immediately after leaf fall as cortical and phloem tissues are directly exposed to inoculum before suberisation of the leaf scar occurs (Crosse 1955). Strong winds, which cause premature defoliation, combined with rainfall are the ideal conditions for leaf scar infection by bacterial pathogens of trees (Crosse 1966). In the case of another bacterial pathogen, *Erwinia amylovora*, the causal agent of fire blight disease of fruit trees, thunderstorms with heavy rain and hail or strong winds may predispose trees to infection as leaves are torn off, exposing unprotected leaf traces and allowing direct entry of the bacterium into the shoots (Billing 2010). Frost cracks and other freeze injuries may also provide infection routes for other *P. syringae* pathogens of trees (Scortichini 2002; Kennelly et al. 2007). However, injuries of this nature have not been reported on horse chestnut infected with *P. syringae* pv. *aesculi*.

On branches of horse chestnut, *P. syringae* pv. *aesculi* caused lesions to develop in the cortex and phloem which extended into the cambium to form a canker in the period between the cessation of active host growth in late summer and the onset of host growth the following spring (Steele et al. 2010). Other studies involving *P. syringae* pv. *morsprunorum* on plum, *P. syringae* pv. *actinidiae* on kiwifruit and *P. avellanae* on hazelnut showed that woody host parts are most susceptible to primary infection during host dormancy (Wormald 1931; Crosse 1966; Serizawa et al. 1994; Scortichini 2002). This may be due in part to the fact that dormant trees are unable to initiate an active structural defence response, which requires meristematic activity (Crosse 1966). This allows bacterial pathogens to colonise and multiply in the infection court, whereas actively growing trees are able to seal off infections rapidly by the production of a necrophylactic periderm impregnated with lignin and suberin.

Pseudomonas syringae pv. *aesculi* survived and remained active for several years within localised lesions on branches, extending tangentially into the phloem and cortex after subsequent secondary growth in the host. Indeed, the bacterium was isolated in 2009 from discrete lesions which had killed the cambium as far back as the 04/05 dormant season (Steele et al. 2010). A closely related pathogen, *P. syringae* pv. *morsprunorum*, which causes bacterial canker of plum and cherry, can also remain localised within infection sites in branches if bacterial numbers are not high enough to overwhelm the host's natural defence responses (Crosse 1966). In this case, *P. syringae* pv. *morsprunorum* survives in infection sites during the summer before becoming active again in the following autumn. The fact that *P. syringae* pv. *aesculi* survived within discrete branch lesions over several years is significant since it indicates that the pathogen may retain the potential to expand within the host for several years after initial colonisation (Steele et al. 2010).

Examination of two whole 30 year old horse chestnut trees from Glasgow revealed extensive cankers in the phloem and cambium which had formed within a single growing season in 2004 (Steele et al. 2010). The original infection sites in these trees were not determined since the cankers were continuous and extended from the small-diameter branches to the lower main stem. However, it is possible that the cankers may have originated from numerous infection sites on the woody parts of the tree where there were bark discontinuities, including lenticels, leaf scars, nodes and side shoots in the crown, and branch insertion points or wounds on the main stem (Steele et al. 2010). The morphology of the cankers, which spiralled up the stems of the trees following the line of the vascular tissues, suggests that *P. syringae* pv. *aesculi* had spread within the phloem. There was no evidence from any of these histological studies that *P. syringae* pv. *aesculi* colonised and spread in the xylem (Steele et al. 2010). The observation that the cankers on the Glasgow trees extended during the host's growing season contrasts with the observations of discrete lesions on branches in which the cambium was always killed during host dormancy. This suggests that *P. syringae* pv. *aesculi* may have a disease cycle similar to other *Pseudomonas* tree diseases which initially invade woody branches during host dormancy and may remain localised in the bark during the winter before spreading systemically within the host when active growth resumes in the spring (Crosse 1966; Scortichini 2002; Steele et al. 2010).

The processes which allow *P. syringae* pv. *aesculi* to extend within the woody parts of horse chestnut from the original infection foci remain uncertain. By the mechanism of quorum sensing, pathogenic bacteria are able to sense the density of their population and will only activate genes involved in motility and virulence once a threshold population has been reached (Cha et al. 1998; Elasri et al. 2001; Quiñones et al. 2005; Dulla and Lindow 2008). It is possible that such a mechanism operates during infection of horse chestnut by *P. syringae* pv. *aesculi*, with canker extension occurring once the bacterium has multiplied to a sufficient cell density to allow it to overcome the host defences (Steele et al. 2010).

5 Dispersal of *P. syringae* pv. *aesculi*

In Scotland it was found that *P. syringae* pv. *aesculi* first started causing lesions on horse chestnut during the 2004/2005 dormant season. Thereafter the number of new lesions increased in each subsequent year, reflecting the arrival and establishment of the pathogen in the area during the mid 2000s and a subsequent increase in local inoculum levels (Steele et al. 2010). *Pseudomonas* diseases of trees are most prevalent in regions with cool, wet climates and the causal pathogens are thought to be spread mainly in wind-blown rain (Crosse 1966; Kennelly et al. 2007). The *P. syringae* pv. *aesculi* epidemic on horse chestnut appears to be geographically centred on northwestern Europe where cool, wet climatic conditions prevail, and the bacterium may well be disseminated in wind or rain since lesions on branches were almost certainly initiated by aerial inoculum. It has been suggested that plant pathogenic strains of *Pseudomonas syringae* can be disseminated across very long distances at high altitudes, with deposition in precipitation (Morris et al. 2006), and a similar mode of dispersal by *P. syringae* pv. *aesculi* might explain its rapid spread across northern Europe (Steele et al. 2010) and its recent establishment in southern Norway.

To understand more about how *P. syringae* pv. *aesculi* is dispersed, it would be useful to know how the bacterium can survive outside the host, for example in soil or water. Experiments are currently underway to determine for how long, and at what concentrations *P. syringae* pv. *aesculi* can survive in the soil under a range of controlled conditions, and to also determine whether it retains pathogenicity after this time period (Laue, B. and Steele, H., unpublished). Initial results indicate that the bacterium is still present and pathogenic after 11 months of incubation in the soil (unpublished data). If it survives for significant periods of time, there is also the possibility that *P. syringae* pv. *aesculi* could be spread in soil, either in potted plants, or boots, tools and on vehicle tyres. This knowledge has important implications for disease management.

6   Exploring the Evolutionary History of *P. syringae* pv. *aesculi* Through Comparative Genomics

The horse chestnut bleeding canker epidemic has highlighted several gaps in the general understanding of the biology of bacterial diseases of trees. In particular, very little is known about the genetic basis underlying the association between bacterial pathogens and their tree hosts, or the genes required for infection and pathogenesis in the woody parts of trees. The disease was officially recorded circa 2002 in the UK, thus it is a modern, '21st century disease'. Correspondingly, some of the methods, tools and ideological approaches used to study its epidemiology and evolution are correspondingly up to date, and incorporate very recent advances in both data production and analysis. In this section, we discuss the role of 'next-generation' DNA sequencing technologies and the subsequent application of comparative genomics methods in the study of bacterial pathogens, and examine in particular how such methods can provide valuable insights into the evolution of pathogenicity and adaptation in *P. syringae* pv. *aesculi*.

The *Pseudomonas syringae* species complex consists of at least 50 different pathovars, generally delineated into approximately four major 'phylogroups', or taxonomic subunits based on DNA sequence similarity (Sarkar and Guttman 2004). Currently, there exists publicly available whole-genome sequence data for at least 27 different pathovars pathogenic on a wide range of host species, and includes the sequences for two strains of the *aesculi* pathovar as well as some other pathovars known to be pathogenic on other woody hosts. In line with the general exponential upwards trend in the production of whole-genome data, this number is likely to increase substantially over the next few years. Thus, for *P. syringae* just as for other bacterial pathogens, comparative genomics methods that are able to comprehend such data and make informed insights regarding the biology or evolution of these organisms are becoming increasingly useful as tools for understanding the processes of both pathogenesis and adaptation in bacterial species.

In 2010 the genome sequences for four strains of *P. syringae* pv. *aesculi* were obtained; three strains isolated from the diseased woody parts of infected European horse chestnut (*A. hippocastanum*) from around Britain and another that had been isolated from a leaf-spot lesion on an Indian horse chestnut (*A. indica*) in India in 1969 (Green et al. 2010). Initial analysis of these data revealed that the three British strains were virtually identical to each other and had most likely descended from a single, recent introduction into Britain, highlighting the risks posed by the accidental introduction of exotic bacteria to new geographical locations. In addition, the British strains were also very closely related to the Indian strain, suggesting that the *aesculi* pathovar may have originated in India and that the current epidemic of bleeding canker is the result of a recent introduction of this pathovar into Europe, where it has found a new host (Green et al. 2010). The Green et al. (2010) analysis also revealed a number of candidate loci putatively involved in the adaptation of pv. *aesculi* on to European horse chestnut. Of particular interest were a number of *aesculi*-specific genome regions containing genes with functions putatively involved in pathogenicity and/or adaptation, including genes involved in the catabolism of plant derived aromatic compounds, iron sequestration and nitric oxide metabolism (Green et al. 2010). Some of these genetic pathways can be found in soil-dwelling bacterial species such as *Pseudomonas putida* and *Acinetobacter* spp., known for their ability to break down a wide range of aromatic compounds including those derived from plants. It is possible that these pathways enable *P. syringae* pv. *aesculi* to utilize aromatic substrates specifically derived from the tissues of woody plants as carbon sources (Kozlowski and Pallardy 1997). Two genes present in *P. syringae* pv. *aesculi* that have a predicted function in breaking down nitric oxide are of interest because nitric oxide is an antimicrobial signalling molecule produced by the plant's immune response, which is thought to play a key role in plant disease resistance by acting as a signal which induces plant genes to synthesise defence-related products (Delledonne et al. 1998). It is thought that these two genes might be important to survival of *P. syringae* pv. *aesculi* during host infection by compromising the plant's disease-resistance response and promoting bacterial growth *in planta*. Work is currently underway to construct a strain of *P. syringae* pv. *aesculi* with a deletion mutation in these two genes, which will be characterised to understand their role in pathogenicity.

Since 2010 the number of available whole genome-sequences for *P. syringae* pathovars has increased substantially, enabling a more comprehensive assessment of gene acquisition in pv. *aesculi* and broadening the scope and applicability of comparative genomic analyses within *P. syringae* generally. This allows for an assessment of the differences and similarities in gene content among pathovars with different phenotypes. Here, the underlying hypothesis is that differences in pheno-types among pathovars are represented (at least in part) by genomic differences in terms of the presence and absence of specific genes. In this way, loci that are func-tionally responsible for certain traits can be identified if they correlate with that trait. An obvious example is if a pathovar expresses a unique trait that is not shared by close relatives – candidate genes that could be functionally responsible for that trait are likely to also be unique to that pathovar and are thus referred to as 'lineage-specific' genes. Differences in genome content among closely related lineages (such as pathovars) are most likely due to a high rate of gene gain and loss; processes that lead to significant 'genome fluctuation' through time (Treangen and Rocha 2011). In prokaryotes, the gain of genes is mediated primarily through the mechanism of horizontal gene transfer (HGT) (Ochman et al. 2000; Koonin et al. 2001). HGT involves the transfer of genetic material between two lineages that may be only distantly related (i.e., transcending the species boundary), and is a well-studied process known to play a pivotal role in the evolution of pathogenicity in a number of bacterial species (Touchon et al. 2009).

Investigation of recently acquired genes in *P. syringae* pv. *aesculi* compared with 27 other *P. syringae* genomes has revealed that many of the pathogenicity-related genes originally identified by Green et al. (2010) were also found to be present in other *P. syringae* pathovars, except for the suite of ~10 linked genes encoding proteins involved in the degradation of protocatechuate via the protocatechuate 4,5-dioxygenase pathway and the synthesis of the enterobactin siderophore, which remained specific to the *aesculi* pathovar (British and Indian strains) (Nowell et al. unpublished). Although not much is currently known regarding this particular catabolic pathway, it may have a role in the breakdown of lignin-derived compounds, based on homology to genes found in other bacterial species known to possess this ability (Masai et al. 2007; Green et al. 2010). However, it remains to be seen if these particular *aesculi*-specific loci confer the singular ability to inhabit and infect the horse chestnut host, or whether, as is more likely, these genes act in combination with other enzymatic pathways encoded within the pv. *aesculi* genome to colonise the woody parts of its host.

At least eight of the pathovars included in the whole-genome comparative analysis are reported to have been isolated from woody hosts. However, there were no genes found to be uniquely shared only by these pathovars, suggesting that the ability to inhabit the woody parts of plant host-species is not underpinned by a single common set of genes (Nowell et al. unpublished). Thus, considerable diversity may exist in the genetic and genomic bases underlying pathogenicity and adaptation onto woody hosts. Future work will hopefully elucidate further on this diversity of genomic mechanisms, and will aim to highlight potential gene pathways that may functionally underlie this ability in pv. *aesculi* as well as other *P. syringae* pathovars of other tree species.

In summary, comparative analyses of *P. syringae* pathovars using whole-genome data has revealed a complex evolutionary history for this species. Rates of gene gain and loss appear to be high, a process that undoubtedly leads to the ability of this species to quickly adapt on to new host species, and is probably a reflection of the ubiquity of *P. syringae* across a multitude of different habitats and lifestyles. Through a quantitative analysis of gene gain and loss along the phylogenetic lineage leading to the horse chestnut pathogen *P. syringae* pv. *aesculi*, a number of candidate loci have been identified that have been gained in the recent evolutionary history of this pathovar and could underlie pathogenicity and/or adaptive functions. In the future, collaborative efforts will aim to elucidate which loci in particular appear to be key in these processes, and an expansion of the comparative dataset will allow for an investigation into the phylogenetic distribution of candidate gene pathways, allowing further insights into the evolution of *P. syringae* on to tree hosts to be made.

References

Billing E (2010) Fire blight. Why do views on host invasion by *Erwinia amylovora* differ? Plant Pathol 60:178–189

Brasier CM (2008) The biosecurity threat to the UK and global environment from international trade in plants. Plant Pathol 57:792–808

Brasier CM, Strouts RG (1976) New records of Phytophthora on trees in Britain. I. Phytophthora root rot and bleeding canker of horse chestnut (*Aesculus hippocastanum* L.). Eur J Forest Pathol 6:129–136

Cha C, Gao P, Chen YC, Shaw PD, Farrand SK (1998) Production of acyl-homoserine lactone quorum-sensing signals by gram-negative plant-associated bacteria. Mol Plant Microbe Interact 11:1119–1129

Crosse JE (1951) The leaf scar as an avenue for infection for the cherry bacterial canker organism, *Pseudomonas mors-prunorum* Wormald. Nature 168:560

Crosse JE (1955) Bacterial canker of stone fruits. I. Field observations on the avenues of autumnal infection of cherry. J Hortic Sci 30:131–142

Crosse JE (1966) Epidemiological relations of the pseudomonad pathogens of deciduous fruit trees. Annu Rev Phytopathol 14:291–310

Delledonne M, Xia Y, Dixon RA, Lamb C (1998) Nitric oxide functions as a signal in plant disease resistance. Nature 394:585–588

Dulla G, Lindow SE (2008) Quorum size of *Pseudomonas syringae* is small and dictated by water availability on the leaf surface. Proc Natl Acad Sci 105:3082–3087

Durgapal JC (1971) A preliminary note on some bacterial diseases of temperate plants in India. Indian Phytopathol 24:392–395

Durgapal JC, Singh B (1980) Taxonomy of pseudomonads pathogenic to horse-chestnut, wild fig and wild cherry in India. Indian Phytopathol 33:533–535

Elasri M, Delorme S, Lemanceau P, Stewart G, Laue B, Glickmann E, Oger PM, Dessaux Y (2001) Acyl-homoserine lactone production is more common among plant-associated *Pseudomonas* spp. than among soilborne *Pseudomonas* spp. Appl Environ Microbiol 67:1198–1209

Forestry Commission (2008) Report on the national survey to assess the presence of bleeding canker of horse chestnut trees in Great Britain. Forestry Commission, Edinburgh

Green S, Laue B, Fossdal CG, A'Hara SW, Cottrell JE (2009) Infection of horse chestnut (*Aesculus hippocastanum*) by *Pseudomonas syringae* pv. *aesculi* and its detection by quantitative real-time PCR. Plant Pathol 58:731–744

Green S, Studholme DJ, Laue B, Dorati F, Lovell H, Arnold D, Cottrell JE, Bridgett S, Blaxter M, Huitema E, Thwaites R, Sharp PM, Jackson RW, Kamoun S (2010) Comparative genome analysis provides insights into the evolution and adaptation of *Pseudomonas syringae* pv. *aesculi* on European horse chestnut. PLoS ONE 5(4):e10224. doi:10.1371/journal.pone.0010224

Kennelly MM, Cazorla FM, de Vicente A, Ramos C, Sundin GW (2007) *Pseudomonas syringae* diseases of fruit trees. Progress towards understanding and control. Plant Dis 91:4–16

Koonin EV, Makarova KS, Aravind L (2001) Horizontal gene transfer in prokaryotes: quantification and classification. Annu Rev Microbiol 55:709–742. doi:10.1146/annurev.micro.55.1.709

Kozlowski TT, Pallardy SG (1997) Physiology of woody plants, 2nd edn. Academic Press, San Diego, California, USA, pp 160–167

Masai EE, Katayama YY, Fukuda MM (2007) Genetic and biochemical investigations on bacterial catabolic pathways for lignin-derived aromatic compounds. Biosci Biotechnol Biochem 71:1–15

Morris CE, Kinkel LL, Xiao K, Prior P, Sands D (2006) Surprising niche for the plant pathogen *Pseudomonas syringae*. Infect Genet Evol 7:84–92

Ochman H, Lawrence JG, Groisman EA (2000) Lateral gene transfer and the nature of bacterial innovation. Nature 405:299–304. doi:10.1038/35012500

Quiñones B, Dulla G, Lindow S (2005) Quorum sensing regulates exopolysaccharide production, motility, and virulence in *Pseudomonas syringae*. Mol Plant Microbe Interact 18:682–693

Sarkar SF, Guttman DS (2004) Evolution of the core genome of *Pseudomonas syringae*, a highly clonal, endemic plant pathogen. Appl Environ Microbiol 70:1999–2012

Scortichini M (2002) Bacterial canker and decline of European hazelnut. Plant Dis 86:704–709

Serizawa S, Ichikawa T, Suzuki H (1994) Epidemiology of bacterial canker of kiwifruit. 5. Effect of infection in fall to early winter on the disease development in branches and trunk after winter. Ann Phytopathol Soc Jpn 60:237–244

Steele H, Laue BE, MacAskill GA, Hendry SJ, Green S (2010) Analysis of the natural infection of European horse chestnut (*Aesculus hippocastanum*) by *Pseudomonas syringae* pv. *aesculi*. Plant Pathol 59:1005–1013

Touchon M, Hoede C, Tenaillon O, Barbe V, Baeriswyl S et al (2009) Organised genome dynamics in the *Escherichia coli* species results in highly diverse adaptive paths. PLoS Genet 5:e1000344. doi:10.1371/journal.pgen.1000344.t006

Treangen TJ, Rocha EPC (2011) Horizontal transfer, not duplication, drives the expansion of protein families in prokaryotes, Moran, NA, editor. PLoS Genet 7:e1001284. doi:10.1371/journal. pgen.1001284.t002

Webber JF, Parkinson NM, Rose J, Stanford H, Cook RTA, Elphinstone JG (2008) Isolation and identification of *Pseudomonas syringae* pv. *aesculi* causing bleeding canker of horse chestnut in the UK. Plant Pathol 57:368

Wormald H (1931) Bacterial diseases of stone-fruit trees in Britain. III. The symptoms of bacterial canker in plum trees. J Hortic Sci 9:239–256

The Use of Small Footprint Airborne LiDAR for the Estimation of Individual Tree Parameters in Sitka Spruce Stands

Juan C. Suárez

Abstract A new canopy delineation using Definiens Developer has been developed to detect individual tree crowns in two areas of the UK: Aberfoyle (Scotland) and Kielder Forest (North of England). This method is building from the lessons learned in the use of other methods used in the past over the same areas. The new method aims at the location of tree tops, using a canopy height model (CHM) derived from airborne LiDAR, by looking at the degree of exposure of each cell compared to its neighbours. Then, the model uses a pouring method constrained by ground classes, a maximum canopy expansion threshold and the location of sheltered spots in the CHM. The results are evaluated at a stand level using field plots. The method shows a good detection of individual trees (c 90 % accuracy) and top height (c. 100 %). Basal area and volume were overpredicted by 10–20 % due to crown clumping by small canopies growing in the proximity of the larger ones.

1 Introduction

The development of airborne laser scanners with increasing ranging frequencies has enabled the detection and measurement of small objects on the ground (Heurich et al. 2004). Since the mid 1990s, some authors have demonstrated a large potential for airborne scanning in the detection and measurement of stand parameters (Vauhkonen et al. 2012). Currently, there are two main approaches to estimate forest resources using airborne laser scanning (ALS): the area-based approach and the individual tree detection. The area-based methods were firstly pioneered by Naesset et al. (1997a, b) and, in time, they have become fully operational for forest inventory. These authors

J.C. Suárez (✉)
Centre for Forest Resources and Management, Forest Research,
Northern Research Station, Roslin, Midlothian EH25 9SY, Scotland, UK
e-mail: juan.suarez@forestry.gsi.gov.uk

T. Fenning (ed.), *Challenges and Opportunities for the World's Forests in the 21st Century*, Forestry Sciences 81, DOI 10.1007/978-94-007-7076-8_36,
© Crown Employees UK 2014

fitted allometric models to predict stand volume as a function of mean height and estimates of crown closure. Results suggested that a large proportion of stand volume variation could be explained for Picea abies (L.) Karst., (R2 = 0.84) but not for Pinus sylvestris L., (R2 = 0.47). Standard errors were 20.9 % and 42.7 %, respectively. Crown closure was computed as the mean of the values resulting from dividing the number of laser pulses, classified as canopy hits, by the total number of pulses transmitted within each cell.

Single tree-based inventories have aimed at the estimation of important biophysical parameters that ultimately can provide a comprehensive and more intensive picture of the forest stand community. This approach aims at estimating structural parameters such as tree height and diameter distributions and potentially can offer better estimations of timber volume and biomass than standard field-based methods or area-based analysis. Likewise, the capability of being able to spatially locate each individual tree within the stand community provides an important step forward in our understanding about the ecological processes intervening on tree growth (Suárez 2010).

Different methods have been developed by many authors in the area of single tree detection and they have been summarised by Hyyppä et al. (2008) and more recently by Vauhkonen et al. (2012). The first studies conducted to detect individual trees used high-resolution aerial image analysis (Gougeon 1999; Erikson 2001; Dralle and Rudemo 1996; Brandtberg and Walter 1998; Brandtberg 1999; Pollock 1996). However, Hyyppä and Inkinen (1999) and later Persson et al. (2002) showed that tree crowns could be delineated with LiDAR using similar principles. Also, they found that crown position and height could be estimated using LiDAR data only. In order to estimate timber volume using individual tree volume equations that related volume (as a dependent variable) to height and diameter (as explanatory variables), diameter was calculated by linear regression (considering crown diameter and height as independent variables). Hyyppä et al. (2001) followed the same approach to obtain very accurate results for the estimation of forest parameters in large areas, particularly for volume, which was estimated from the detection of individual trees using LiDAR. The standard error obtained for volume estimates in this study was 18.5 $m^3\,ha^{-1}$ (c. 10.5 %).

Since then, several canopy delineation algorithms have been developed by many authors:

- Tree delineation algorithm using eCognition by Bunting and Lucas (2006),
- FOI, by Persson et al. (2002) and Elmquist et al. (2001),
- ITC, by Gougeon (2005),
- Tree finder (TF), by Straub (2003),
- Joanneum Research, by Hirschmugl et al. (2005),
- Metla, by Pitkänen et al. (2004) and Pitkänen (2005),
- TreeVaW using IDL, by Popescu et al. (2003) and Popescu (2008),
- Morsdorf et al. (2004),
- TreesVIS, by Weinacker et al. (2004).

Normally, the detection of individual trees is undertaken by locating maximum height differences in a canopy height model (CHM) derived from the difference between ground level and a surface model interpolated from the ALS point clouds. Watershed or pouring algorithms (as well as derivations of these algorithms) are used for the delineation of single tree crowns (Gougeon 2005; Brandtberg et al. 2003; Popescu et al. 2003; Weinacker et al. 2004; Koch et al. 2006; Maltamo et al. 2006). However, latterly, approaches are becoming increasingly based on raw data using clustering or blob detection methods (Morsdorf 2006; Wang et al. 2008).

Individual tree parameters, such as the position of each individual height and sometimes its crown area, provide a powerful synoptic view of stand structure that is difficult to model otherwise (Hyyppä and Inkinen 1999; Holmgren et al. 2003; Hyyppä et al. 2001; Lim et al. 2001; Persson et al. 2002; Popescu et al. 2003; Andersen et al. 2001). Hidden parameters to direct LiDAR observation, such as dbh (diameter at breast height), have to be modelled using allometries between crown dimensions and tree height (Suárez et al. 2008a, b). The accurate estimation of individual dbh values are then being use to create better assessments of basal area, which also improves the estimation of standing volume and biomass (Suárez 2010).

Most authors have focused on coniferous trees (spruce and fir) in boreal forests as these species are morphologically simpler than broadleaves and mature pines (Heurich et al. 2004). The results evidenced that LiDAR could detect most of the crowns of dominant and co-dominant trees in mature stands dominated by coniferous species, but found difficult to differentiate individual crowns in young stands and deciduous trees (Brandtberg et al. 2003). They found that individual tree detection becomes more difficult in those stands because of their more spherical crown shape compared to conifers and the higher probability for the occurrence of more than one apex in each tree (Heurich et al. 2004) and frequent crown clumping. Published studies showed good results with 65–90 % of correct tree counts within conifer stands (Hyyppä et al. 2001; Persson et al. 2002; Koch et al. 2006), while for broadleaves the results were less accurate (Hyyppä et al. 2001; Koch et al. 2006). An underestimation of volume is normally reported (van Aardt et al. 2006; Korpela et al. 2007).

The aim of this work is the development of a new canopy delineation method based on the experience drawn from the application of three mainstream methods in a study area in Scotland (Suárez et al. 2008b). This study evidenced important limitations of these models for the location of tree tops and accurate delineation of individual canopies. The new method aims at solving those problems in order to provide better estimates at the lower end of the height and diameter distributions and use this information to refine the estimation of stand parameters. This is regarded as a necessary step for future predictions of yield and volume.

2   Study Areas

Two study areas were measured in the field and with airborne LiDAR. The Aberfoyle study area is located at 56°10′ N, 4°22′ W, within the Trossachs-Ben Lomond National Park in SW Scotland. Kielder Forest is situated in Northern England inside

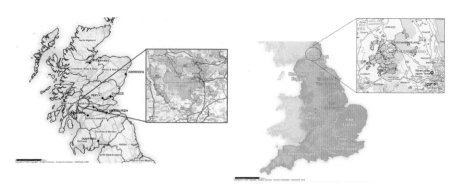

Fig. 1 Location of the Aberfoyle (*left*) and Kielder Forest (*right*) study areas

the Northumberland National Park at 55°14′ N, 2°35′ W. The combination of high wind, high rainfall and soil types with restricting rooting depth limited management and the choice of species for plantation to mainly Sitka spruce (Picea sitchensis Bong. Carr.) (c. 70 % of total planting area) (Fig. 1).

Most of these areas have been managed in planted and clearfelled rotations of 40–60 years. In Aberfoyle, around 10 % of the stands initiated the transition to Continuous Cover Forestry (CCF) about 10 years ago. CCF aims at a permanent forest structure, avoiding clearfellings, expanded diameter distributions and the use of natural regeneration. The decisive factor for the transformation of some stands into CCF was the risk of windthrow (Mason and Kerr 2004), in particular the location of each plot according to the Windthrow Hazard Classification (WHC), as defined by Miller (1985). Other criteria included site conditions (Pyatt et al. 2001) and seed bed availability (Nixon and Worrell 1999).

Kielder Forest is the largest man-made forest in Europe. Planting started in 1926 with the primary objective of producing high volume of timber as fast as possible. The planted area increased steadily until the end of the World War II. Thereafter, about 50 % of all the woodland areas were planted between 1945 and 1960. So, by the early 1970s, there were 70,000 ha of plantations, 86 % of them with spruce, 13 % with pine and only 1 % broadleaves. Apart from the important transformations to the landscape that the large introduction of spruce has created, the district has been struggling in the last few years to maintain a consistent forest structure. The massive afforestation during the 1960–1970s means that nowadays most of Kielder forest is reaching the end of the rotation. This implies that most of the timber has to be extracted now by clear felling, which is likely to transform substantially the landscape of this area. In a context of changing forest policies, where timber production shares interests with conservation and an incipient recreation industry, a scenario of large deforestation is no longer an option. Therefore, foresters in this area are looking for new inventory methods that can provide them with a better description of stand structure in order to plan long term retentions or to initiate the process of transformation to CCF.

3 Materials and Methods

3.1 Mensuration Plots

Mensuration data were taken for 11 plots in Aberfoyle and 7 plots Kielder Forest. The field survey methods aimed to estimate both stand and individual tree parameters in mature Sitka spruce stands (see Fig. 2).

The selection of plots in Aberfoyle aimed to choose stands with a similar age (around 30 years old with 3 years difference), in areas of relatively flat terrain (no slopes above 10 %), containing a single component (one species component planted as a Sitka spruce monoculture) and similar soil conditions (brown earth with some iron pan in certain areas). The plots distributed into three Yield Class (YC) categories:

- Low productivity: YC16,
- Medium productivity: YC20,
- High productivity: YC24.

All the plots in this area had been thinned recently, which in theory facilitated the visibility of individual tree crowns.

In Kielder Forest the criteria were that plots should be as variable as possible in order to test all possible effects likely to affect tree detection. So, age classes ranged

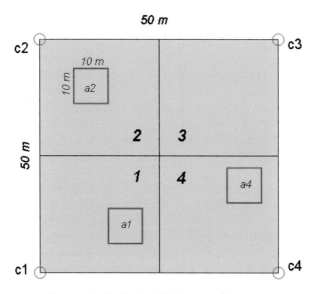

Fig. 2 Configuration of the sample plots in the field. Plots are oriented to the magnetic north. Each corner is labelled with a 'c' and a number starting from the Southwest corner in sequence and finishing with the one to the Southeast. Each quadrant is labelled in the same sequence as the corresponding corners. Small plots of 10 m × 10 m are labelled with an 'a' and the number of the quadrant they lie in

between 29 and 61 years old, YC8-20 and different thinning regimes. Thinned stands were not always possible to find due to the fact that most of Kielder forest had thinning restrictions due to high risk of wind damage. Plots 10–12 were chosen because they were permanent plots established about 30 years ago for testing different Sitka spruce provenances. Therefore, there were extensive records for these plots about height and diameter distributions. At the time of the survey, these three plots had been chemically thinned, although trees had not started to die back yet. Also, plots 6–7 had been line thinned with c. 30 % removal.

The field survey comprised plots of 50 m × 50 m in each study area. The objective was to gather enough information for the validation of analysis at both stand and individual tree level. At stand level, all plots were measured according to an adaptation of the Mensuration Handbook guidelines (Matthews and Mackie 2006). The parameters being captured were: dbh, location of all trees in X and Y, top height (TH), basal area (G) and volume (V). At the tree level, the 50 m × 50 m plots were divided into four quadrants. Each quadrant of about 25 × 25 m² was labelled 1–4, starting from the SW corner. Then, three of them were randomly selected and a 10 × 10 m² small plot was located inside each one in an equally random manner (see Fig. 2). Inside those small plots, the height of all trees was measured with a Sonic Vertex (Haglöf, Sweden) and canopy dimensions were taken (canopy length as the height to first living whorl, and canopy width in two axes: N-S and E-W).

The location of the corners was done in sequence, starting at c1 (Fig. 2). Navigation inside the stand in between corners was carried out with compass and measuring tape. Navigation from the second to the third corner was established by a prism, looking at a 90° angle. The big and the small plot corners were marked with permanent wooden posts in the ground and painted in reflective colours. All the corners were linked with plastic tape in order to determine which standing trees belonged to the plot. Trees on the boundary were included inside the plot if at least the centre of the stem was inside.

The locations of the big plots were at least 10 m away from the stand boundary to avoid edge effects. All plots were surveyed shortly after each LiDAR survey to minimise temporal decorrelation due to growth. So, field campaigns were undertaken in Aberfoyle during April–June 2006 to coincide with the LiDAR surveys on the 31st of May 2006.

In Kielder Forest, the field data collection took place between May and June 2003. The LiDAR survey took place in this area on 26th March 2003. However, the contractors had to repeat the survey on the 25th of April 2004 because the delivered product did not fulfil the desired density of returns (data were delivered at only 1 return per square metres). The new LiDAR survey was undertaken before the start of the next growing season. This might cause decorrelation problems between field data (taken at the beginning of the previous growing season) and the new laser scanning period. However, it was not possible to repeat field data collection.

Data gathered in the field for each tree were stored on a PocketGIS (Pocket Systems Ltd., 1996-, Bedfordshire, UK) operated on a Husky portable computer. The data were downloaded at the end of each working day to an Excel spreadsheet.

Table 1 Stand parameters in Aberfoyle in 2006. *TH* top height, *G* basal area, *V* volume

Plot	No trees	TH (m)	G (m²)	Mean DBH (cm)	V (m³)	Plot size (ha)	Spacing (m)	Trees/ha	G/ha (m²)	V/ha (m³)
2	79	25.54	7.72	33.41	85.25	0.256	5.69	309	30.15	333.02
3	106	24.96	8.04	29.67	86.73	0.250	4.86	424	32.16	346.90
4	60	28.42	6.18	35.04	76.21	0.244	6.38	246	25.34	312.34
5	154	24.25	9.07	26.16	94.95	0.245	3.99	629	37.01	387.53
6	129	25.61	9.83	30.16	108.92	0.268	4.56	481	36.69	406.42
7	95	27.99	9.73	35.45	118.05	0.242	5.05	393	40.19	487.82
8	104	31.22	12.12	37.46	164.46	0.252	4.92	413	48.08	652.63
9	144	28.32	11.12	30.15	136.61	0.249	4.16	578	44.67	548.62
10	137	27.06	10.60	30.71	124.22	0.245	4.23	559	43.25	507.03
11	139	26.69	8.35	26.67	96.45	0.250	4.24	556	33.38	385.79
12	124	27.38	8.93	29.49	106.00	0.246	4.45	504	36.31	430.88

All trees were located in Eastings and Northings using a combination of Laser relascope (Criterion laser, Laser Technology Inc. 1992-, Englewood, Colorado, US) and Robotic Total Station – TS (Trimble 5600, Trimble Navigation Limited, 935 Stewart Drive, Sunnyvale, California 94085, US). The main limitation in taking the XY position of individual trees was the difficulty of detecting a reliable GPS signal on the ground. Therefore, laser-guided systems were used to traverse inside the forest canopy, taking ground-based stations as starting points, and to locate individual trees. A prism was placed at the centre of each tree and used by both systems to enhance their precision.

Ground based stations were taken by Leica 500 DGPS (Leica Geosystems AG, Heinrich-Wild-Strasse, CH-9435 Heerbrugg, Switzerland). The DGPS was operated for 20 min to collect GPS raw data in the vicinity of each plot, where clear sky and good reception allowed it to be taken. Then, the data were post-processed to achieve sub-centimetre precision using the nearest OS Net RINEX GNSS reference stations (KILN in Aberfoyle; 32 km distance; and KELS in Kielder Forest; 37 km distance). The data were processed in Real-time RTK software from Leica (Tables 1 and 2).

3.2   LiDAR Data

LiDAR data was capture for both sites by the same contractor: the English Environment Agency (Lower Bristol Road, Twerton, Bath, BA2 9ES). The contractor used two Optech scanners: ALTM2033 and ALTM 3100 (Optech Incorporated, 100 Wildcat Road Toronto, Ontario, Canada). Details of these surveys are provided in Table 3.

Position and elevation accuracy was calculated by the contractor by setting up ground measurements with DGPS on targets located in flat areas. Additional checks,

Table 2 Stand parameters in Kielder Forest in 2003. *TH* top height, *G* basal area, *V* volume

Plot	No trees	TH (m)	G (m²)	Mean DBH (cm)	V (m³)	Plot size (ha)	Spacing (m)	Trees/ha	G/ha (m²)	V/ha (m³)	Age (yr)
6	238	32.53	21.43	32.50	303.36	0.261	3.31	913	82.23	1164.07	61
7	292	27.50	15.68	24.97	186.95	0.255	2.95	1,146	61.58	734.02	61
8	507	19.84	15.16	18.36	128.98	0.239	2.17	2,120	63.39	539.46	35
9	493	19.94	15.15	18.48	129.66	0.246	2.23	2,006	61.65	527.52	35
10	493	17.32	16.06	19.96	118.70	0.271	2.34	1,819	59.27	438	30
11	682	20.52	16.94	17.30	149.32	0.239	1.87	2,856	70.95	625.3	33
12	550	21.00	18.07	19.94	163.10	0.250	2.13	2,204	72.39	653.45	32

Table 3 LiDAR configurations for the study areas

Parameter	Aberfoyle 2006	Kielder 2004
Sensor	Optech ALTM2100	Optech ALTM2033
Date	31 May 2006	25 April 2004
Laser pulse frequency	100,000 Hz	33,000 Hz
Flying altitude	950 m	1,000 m
Beam divergence	0.3 mrad	0.3 mrad
Scanning angle	10°	10°
Sampling intensity	8–11 returns per m²	8–10 returns per m²
Position accuracy	X,Y < 40 cm	X,Y < 40 cm
Elevation accuracy	Z < 9–15 cm	Z < 9–15 cm
Data format	First and last returns, two intermediate returns plus intensity (8-bit)	First, last returns and intensity (8-bit)

also with DGPS, were undertaken independently in the Forest District car park in Aberfoyle for each survey and Caperburn peninsula in Kielder, with similar results.

In both areas, data were distributed by the contractor in ASCII XYZ format, where first and last returns, with their corresponding intensities, were located to the OS national grid. The vertical accuracies were about 15 cm RMSE while the horizontal accuracy was approximately twice the footprint size (20 cm). Data pre-processing consisted of data filtering for noise, differential correction and assembly of data into flight lines. This work was undertaken by the contractor.

The two LiDAR surveys were identically processed for the two study areas. First of all, a routine was written in Delphi 5.0 (Borland, 8310 North Capital of Texas Highway Building 2, Suite 100 Austin, TX 78731, USA) to select first (FR) and last (LR) return signals inside the boundaries of each field surveyed plot. The second laser echo was exported in ASCII XYZ format and processed in Terrascan 007.008 Software (Terrasolid Ltd, Sohlberginkatu 10, Fin 40530 Jyväskylä, Finland) in order to filter ground hits (Fig. 3). First returns were directly classified as high vegetation after the elimination of outliers (most of them were already eliminated by the contractor). Then, both the ground and the vegetation classes were exported to Surfer 8.0 (Golden Software, Inc. 809 14th Street Golden,

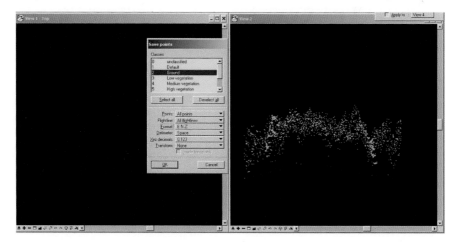

Fig. 3 The filtering process of ground hits in Terrascan. Most of the last return points are intercepted by the forest canopy (in *green*), whereas a minor proportion are deemed to represent the underlying ground (in *blue*). Those points, once classified as ground, will be exported to Surfer in order to create a DTM using a surface interpolation routine

Colorado 80401-1866, U.S.A.), where they were gridded into a Digital Terrain Model (DTM) and a Digital Surface Model (DSM) respectively. This procedure was undertaken by a kriging interpolation method using a linear model without anisotropy, at a grid resolution of 0.5 m. After this, the DTM was subtracted from the DSM to create a normalised model of the tree canopy. The resulting Canopy Height Model (CHM) created the raw material for the next level of processing: the delineation of individual tree canopies.

3.3　Canopy Delineation Algorithm in Definiens

In order to address some of the limitations of other algorithms, a new method for canopy delineation was developed using Definiens Developer 7.0 (Definiens Imaging GmbH, 2001; Trappenstreustrasse 1, 80339 Muenchen, Germany). This new algorithm was applied to delineate individual tree canopies in the two monitoring areas solely from the LiDAR-derived 0.5 m resolution CHM.

The Integrated Development Environment (IDE) in Definiens Developer makes use of the functionality available in the Definiens Cognition Network Language (Definiens, User Guide, 2007). This environment implements a series of routines to be followed in sequence. The main concept behind this language is the progressive transformation of object primitives (CHM cells) into meaningful entities which are semantically consistent, i.e. tree canopies.

All the logical procedures along the workflow of the canopy delineation algorithm were grouped into a series of routines: data preparation or pre-processing,

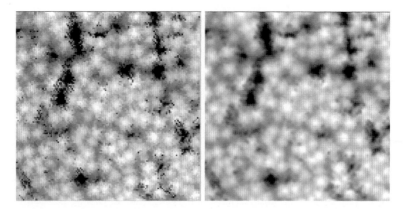

Fig. 4 Original CHM created for plot 8 in Aberfoyle (*left*) and the smoothed version created with a Gaussian kernel (*right*). This operation eliminates noise by reducing extreme values

segmentation, canopy delineation and refinement. Data export created two types of shapefiles: a point file with the location of tree tops and a polygon file showing the area extent of delineated canopies. A complete description of every step is explained below:

1. Pre-processing

    This group of procedures aimed at the creation of working images to be used by the rest of the methods. The key ideas were derived from the lesson's learned in the use of the other algorithms: noise reduction and enhancement of edges (Suárez et al. 2008a).

    1.1. *Smoothing*. The original CHM was smoothed with a Gaussian kernel of 3 × 3 to eliminate noise in the interpolated surface. This noise was produced by laser penetration through the canopy, which created a succession of peaks that could be confused with tree tops. This effect is observed in Fig. 4 when comparing the original CHM (on the left) to the smoothed one (right). Areas of low values, depicted in black, appeared inside the canopy, adding a salt-and-pepper effect. The smoothed image eliminated most of them, allowing an easier interpretation of boundaries and peaks. Also, though not so clear in the image, multiple peaks in each canopy became unified in the smoothed version of the CHM, which allowed an easier detection of the apices. These were part of the lessons learned in the use of other segmentation techniques, in particular TreesVIS (Weinacker et al. 2004).

    1.2. *Edge detection*. Additional routines were used to highlight exposed (bright) and secluded (dark) areas in the canopy. This operation was performed over the smoothed CHM to highlight tree tops and areas in between canopies. The routine is called in Definiens 'Edge Extraction Lee Sigma', after the Biased Sigma Filter originally developed by Lee (1983). The Sigma filter is an edge preserving and edge enhancing filter. This method searches those pixels in a square neighbourhood at a certain range above or below the

Fig. 5 Chessboard
segmentation

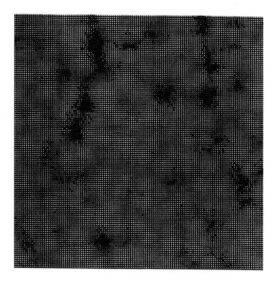

mean, measured in standard deviations. Basically, this routine highlights the outliers above or below the distribution of pixel values. The Sigma value describes how far away a data point is from the mean (x), in standard deviations (σ) to be considered an edge. So, this value is used as a threshold to locate the brighter and the darker pixels (i.e. the most exposed and the most sheltered parts of the canopy). A higher Sigma value results in stronger edge detections. For a given window, the Sigma value is computed as:

$$\Sigma = \sqrt{\frac{\sigma^2}{\left(\bar{X}\right)^2}} \tag{1}$$

In this case, a Sigma value of '5' was used to detect bright (above the mean) and dark (below) edges. The size of the searching box in Definiens is unknown.

2. *Segmentation.* The *Chessboard segmentation* procedure transformed the pixel images into objects of equal dimensions for the classification process (Fig. 5). This step is a prerequisite for the rest of the object-oriented classification procedures. Unlike other segmentation methods, no knowledge was applied in this operation apart from the definition of the dimension of the objects. In this case, an object size equal to '1' created objects with identical dimensions to each pixel in the original image (0.5 × 0.5 m²).

Unlike the original pixels, these objects can contain more than one value associated to it. In this case, each object stored information from the four data layers used in the classification such as CHM, the smoothed CHM and the two Sigma Lee layers. In addition, each object contained additional information like the topology with neighbour objects (e.g. close to, at a certain distance from, etc.), class relations, etc.

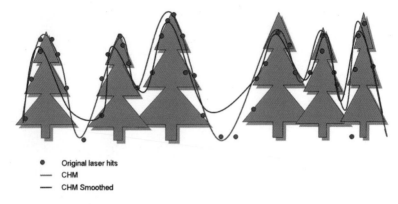

●   Original laser hits
—   CHM
—   CHM Smoothed

Fig. 6 The interpolation process can mask the location of gaps in the canopy

3. *Canopy delineation*. The objectives of this group of procedures were the discrimination of canopies and gaps, identification of individual tree tops and the delineation of individual canopies.

    3.1. *Gap delineation*. This step classified objects as 'ground' class if the elevation in the original CHM was less than or equal to 7 m. This threshold was the product of several trial and error attempts. This value appeared to be adequate for eliminating unwanted errors in the interpolation of the laser data and the smoothing process. Figure 6 shows how the interpolation and the smoothing processes can eliminate gap areas that become confused with the lower parts of the canopy. This procedure tends to eliminate extreme values above and below the kernel.

    In the next step, areas that were not classified as 'ground' were temporarily assigned to the 'canopy' class (Fig. 7). The 'ground' class is shown in white, the 'unclassified' in brown and the temporary 'canopy' class appears in green. After the 'canopy' class assignment, the 'unclassified' class disappears completely.

    3.2. *Find local extrema*. This procedure identified the top and lower parts within the canopy. It used the values of the smoothed CHM, the bright and the dark edge images to classify image objects fulfilling a local extrema condition (maximum or minimum). Image objects with either the smallest or the largest feature value within a specific neighbourhood were classified. In this case, the search range was defined in pixels. All image objects belonging to the temporary 'canopy' class included within the search range were part of the *search domain*. After several trial and error attempts, the search range was set to '1'. This procedure compared each pixel with the surrounding 8 pixels and reclassifies it as 'top tree' class if the pixel was completely exposed (Fig. 8).

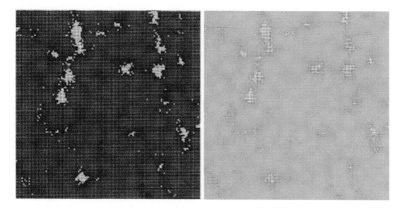

Fig. 7 The 'ground' class appears in *white* colour, whereas the 'canopy' class is shown in *green*. The *brown* colour represents the 'unclassified' class (*left* image) prior to its conversion to 'canopy' (*right* image)

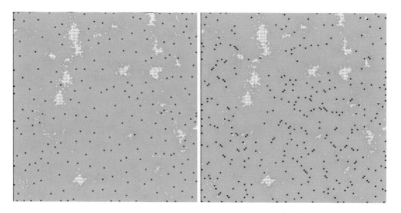

Fig. 8 The 'find extrema' procedure identifies tree tops (*left*, in *blue*) and canopy intersections (*right*, in *black*)

The same procedure was applied in order to find the local minima. This method located the point of intersection between neighbouring canopies. These points were classified as 'temporary border'. The local minima acted as a barrier for the expansion procedure from 'tree top' in the following step.

The information generated in the pre-processing part for locating bright and dark edges was used to narrow down the number of candidate objects classified as 'canopy'. A threshold value of ≥0.1 was selected for each one. The most exposed objects in the 'canopy' class that also showed a bright edge value of ≥0.1 were classified as 'top tree'. The most sheltered

Fig. 9 The excluded areas at
a distance of ≥4 m from the
detected tree tops are
displayed in *red*

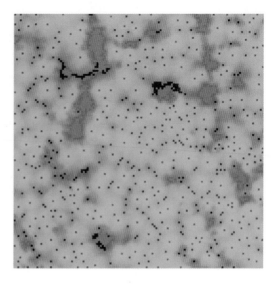

objects in the 'canopy' class with a similar value in the dark edge layer
were classified as 'temporary border' class. The classified tree tops were
exported as a shape file showing their location in X and Y and its elevation
value in metres.

3.3. *Elimination of extreme values.* This procedure excluded parts of the canopy
at a certain distance from the newly located tree tops. The aim of this step
was to restrict canopy expansion in the next process beyond a reasonable
distance. Without this restriction, the pouring algorithm in the next step
would expand continuously, creating impossible crowns. This was a problem detected with the ITC algorithm (Gougeon 2005).

Therefore, a heuristic approach was followed next. This method looked for
maximum Canopy Width values measured in the field inside the small 10 m
× 10 m plots. The results were averaged. This value constituted the average
limit for the canopy expansion in the pouring algorithm. This threshold was
set to 4 m for all plots. This procedure, once implemented in the ruleware,
was also found to eliminate unwanted effects derived by the interpolation
method or by loose branches hanging outside the core of the canopy (Fig. 9).

The threshold value was implemented in the code as a variable. So, it
could be easily modified by other users interactively by moving a sliding
bar. This approach was valuable in the processing of some plots in Kielder
forest, which presented larger canopies (Plots 6 and 7).

3.4. *Canopy delineation.* This procedure was achieved by means of the 'grow
region' algorithm. This method enlarged image objects defined in the image
object domain by merging them with neighbouring image objects ("candidates") that matched the criteria specified in the parameters. In this case, the
criterion was the proximity of the objects classified as 'temporary canopy'
class to a classified 'tree top'.

Fig. 10 The result of
applying the grow region
algorithm. Individual tree
canopies are delineated

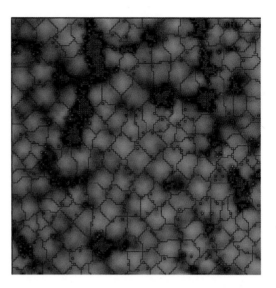

The grow region algorithm worked in sweeps. This means that each execution of the algorithm merged all direct neighbouring image objects from the 'canopy' class while they shared a common border. To grow image objects into a larger space, the procedure was operated in a loop 'while something changes' (i.e. while canopies were classified into tree tops) until there were no more objects to be classified that met the criterion.

This procedure worked like a pouring algorithm that progressively incorporated new objects into the 'tree top' class until there was nothing else to be incorporated or barriers were encountered (i.e. ground, borders or excluded areas) defined as a different class in the previous procedures (Fig. 10).

Refinements. The remainder of the classes were re-assigned to class 'ground' and aggregated to reduce their number. The 'tree top' class that represented each individual tree was pruned from dangling segments by means of the 'morphology' procedure. This operation is a pixel-based binary morphology operation that makes use of a circular 4 × 4 mask that excludes those parts outside the boundaries of an object. The result was a group of canopies with smoother borders (Fig. 11).

The results of the classification were exported as a polygon shapefile containing a list of generated attributes for each canopy such as:

- Maximum height (m), that corresponds to the tree Height, previously exported as a point shape file,
- Shape indexes like: compactness, ratio width/length, main direction of the crown ellipses (degrees), radius of the longest and smallest axis of each crown ellipse (m), roundness and area (in m),
- Position of the centroid of each polygon in X and Y,
- Object ID

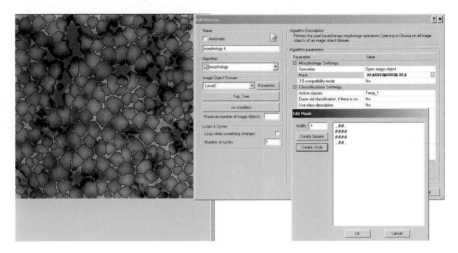

Fig. 11 The result of applying the morphology procedure and the shape mask. This mask works by assigning more weight to those parts of the object that could fit into a hypothetical circle. The *dots* in the corners mean that those parts have more probability of being excluded

The ellipse of the crown was the result of fitting the polygon that represented each canopy into an ellipse. Then, the longest and smallest axis could be calculated.

4   Results

4.1   Aberfoyle

The delineation algorithm was run to locate tree tops and individual tree crowns inside each 50 m × 50 m plot. Three plots were randomly selected out of the 11 plots for training allometric functions to estimate tree height and dbh (plots 5, 10 and 12). The selection of trees was done manually in ArcGIS by linking detected individuals with measured trees in the field (80 trees in total between the three plots). This operation was undertaken manually to minimise the chances of confusing target trees with their neighbours.

The retrieved tree canopies contained information about maximum and minimum pixel height (i.e. the height of the tree and the length of the canopy, in m) and canopy area (in m²). Tree height and canopy area were used to create a recovery model to estimate the true height and the dbh for each tree individual. The height and dbh models were run in SAS (SAS Institute Inc. 100 SAS Campus Drive Cary, NC 27513-2414 USA) to look at all possible combinations and to select those models which were statistically more significant (Table 4).

Table 4 Height and DBH recovery models from LiDAR (n = 80), where H is tree height and A is canopy area

Model	R^2	Equation	P-value intercept	P-value first param.	P-value second param.
Height	0.93	$-15.4879 + 2.2927*H - 0.0237*H^2$	0.016	<0.0001	0.014
DBH	0.67	$-9.7156 + 1.5525*H + 0.49152*A$	0.032	<0.0001	<0.0001

The models were applied to the rest of the plots in Aberfoyle and the results were aggregated at stand level in order to facilitate the comparison to field measurements.

The estimation of volume in a stand followed the standard Forest Enterprise method described in Matthews and Mackie (2006). As taper functions are not yet available in the UK for individual trees, it was not possible to calculate volume for each detected individual and aggregate the results at stand level. Instead, 'crop form' as defined in the FE method, was used as a proxy for aggregated taper functions within a forest stand. Therefore, the total volume at stand level was the product of multiplying the crop form (F) by basal area (G):

$$V = G \cdot F \tag{2}$$

$$F = a_1 + \left(a_2 \cdot TH\right) + \left(a_3 \cdot TH^2\right) \tag{3}$$

where,

F is crop form,
V is volume,
TH is top height,
a_1, a_2 and a_3 are parameters in the crop form function that, for Sitka spruce, are:

$a_1 = -0.314044,$
$a_2 = 0.444794,$
$a_3 = 0.0$

Volume estimations were calculated for field-measured and LiDAR-detected trees, using top height and basal area. Results were presented as percentage of accuracy between detected and observed stand values. The overall performance across stands was assessed as average accuracy, relative RMSE (rRMSE) and relative bias (rBias) (Table 5).

The results showed a good detection of top height in all stands with values closed to the ones measured in field. The number of individual trees was generally underestimated by 8 % with a variability of c. 15 % across individual stands. Volume and basal area produced an overestimation of 10 % with higher variability across stands (c. 20 %) and more substantial bias (c. 6.5 %).

It became evident that stand structure had a large influence on the capability of this method for detecting individual tree canopies. The algorithm was capable

Table 5 Derived stand parameters for 11 plots in Aberfoyle

Plot	Trees	TH	G	V
2	92.41	97.10	97.81	94.90
3	103.77	102.74	109.43	112.51
4	111.67	100.49	119.61	120.20
5	68.83	101.98	89.65	91.47
6	89.15	98.41	103.28	101.59
7	99.08	101.55	121.45	123.38
8	100.91	100.75	137.82	138.87
9	79.17	99.60	108.20	107.75
10	94.89	102.38	113.97	116.75
11	80.58	98.87	95.42	95.57
12	96.77	98.20	116.59	114.44
Average accuracy	**92.47**	**100.19**	**110.29**	**110.68**
rRMSE	**15.80**	**1.74**	**20.22**	**21.61**
rBias	**4.76**	**0.52**	**6.10**	**6.51**

of detecting the majority of trees with large diameters (i.e. large crown areas), whereas small trees were mostly undetected and their canopies clumped with neighbouring ones in the delineation process. This was mostly evident in the case of Plot 5, where the detection process underestimated the number of trees by more than 30 % (Table 5). This area was a second rotation plot with a very heterogeneous diameter distribution and growing trees close to each other in an irregular planting scheme.

Another evident limitation of this method was that it tended to produce normal diameter distributions. Large trees were effectively detected by the algorithm, but the last process of pruning of the canopies and the expansion threshold that limited canopies within 4 m radius also reduced the probability of measuring accurate values for the largest canopies. The example depicted in Plot 2 (Fig. 12, bottom) illustrates the combined effect of small trees being missed and larger trees having their values reduced by these methods. As a result, although nearly all trees were detected in this stand, both extremes of the distribution were under represented.

However, in stands where canopies were more regularly spaced the canopy detection worked considerably better. In the case of Plot 6, the 'grow region' procedure in the algorithm seemed to benefit from clear separations between canopies (Fig. 12, top). Although some small trees were undetected, the final estimations of volume and basal area were very close to the field measured values.

4.2 Kielder Forest

The seven plots tested in Kielder forest were more variable in terms of age, ranging from 30 to 61 years old, and management regime. Only two plots were line thinned, whereas the majority of them remained unthinned. As a result, tree densities were

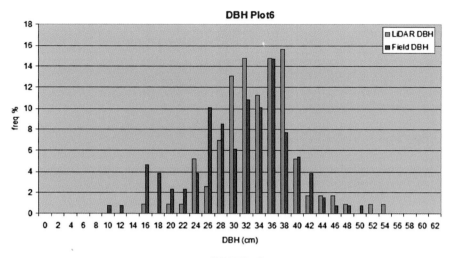

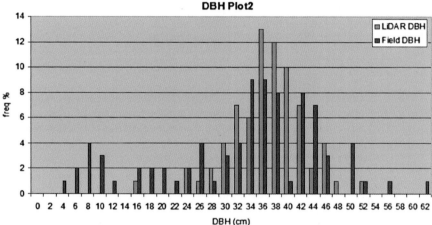

Fig. 12 Stand diameter distributions derived from LiDAR (*blue*) compared to field observations (*purple*). Examples taken from plots 6 (*above*) and 2 (*below*) in Aberfoyle

higher with spacings ranging from 1.87 to 2.34 m in the unthinned plots and 2.95–3.31 in the two thinned plots (plots 6 and 7). In order to improve canopy discrimination a grid resolution of 0.25 m was used. The rest of the parameters in the canopy delineation procedures remained similar to the ones used in Aberfoyle for consistency.

The construction of dbh and height recovery models involved a similar technique than in Aberfoyle. So, 39 trees were manually selected within the 7 plots for training models (Table 6).

Contrary to Aberfoyle, the recovery model for individual tree height did not require the use of square LiDAR height (P = 0.69). The relationship was perfectly linear with a level of underestimation of 3 % compared to field measurements. R^2 was 0.99.

Table 6 Recovery models for height and DBH in Kielder with the 2004 dataset

Year	Model_25	R^2	Equation	P-value intercept	P-value first param.	P-value second param.
2004	Height	0.99	0.4624 + 1.0095*H	0.194	<0.0001	N/A
	DBH	0.79	−3.3845 + 1.3058*H +0.2085*A	0.315	<0.0001	0.0034

Table 7 Derived stand parameters in Kielder forest

Plot	Trees	TH	G	V
6	147.17	101.56	185.64	188.60
7	59.15	102.20	119.54	122.23
8	85.45	100.94	116.33	117.47
9	86.85	102.27	114.58	117.28
10	102.20	105.83	104.03	110.35
11	82.02	102.84	138.32	142.39
12	73.94	105.56	120.08	126.99
Average accuracy	**90.97**	**103.03**	**128.36**	**132.19**
rRMSE	**21.86**	**3.16**	**44.36**	**62.19**
rBias	**8.26**	**1.19**	**16.77**	**23.50**

Different dbh models were tested in SAS for different independent variables. The best results were achieved with a model similar to Aberfoyle, using canopy area and height as detected by the sensor. However, the intercept was less significant (P = 0.315). R^2 was 0.79, which marked a significant improvement in comparison with Aberfoyle (Table 7).

The results showed a good estimate of top height and number of detected trees in the majority of the plots with the exception of plot 6 and 7. General detection of trees missed an average of 10 % of the canopies, whereas top height showed a slight overprediction of c. 3 %. These results were consistent with the ones obtained in Aberfoyle. However, basal area and volume were significantly worse in comparison, with values overpredicted by 20 % (if plot 6 was not taken into consideration).

The largest discrepancies were observed in Plot 6. This stand showed another important limitation of the tree delineation algorithm. The use of the 3 × 3 kernel smoothed the transition between canopies in a clearly defined removal of trees following a line thinning. As a result, bogus tree tops were located in those areas of transition that were later on assigned a canopy area. The consequence of this was a considerable level of overprediction in the number of trees detected in this plot, which increased the basal area and volume estimates (Fig. 13).

It looked like the larger number of points in this survey did not contribute to improve stand estimates. However, it was necessary to create a finer resolution CHM to discriminate trees in these dense plots. A previous attempt to discriminate trees using a 0.5 m grid created substantial amounts of trees missing (40–50 %) in all plots. So, the 0.25 m grid was the best compromise.

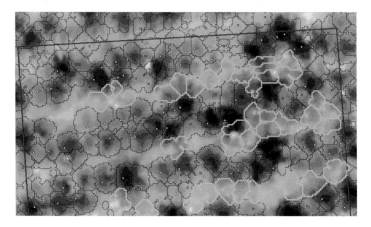

Fig. 13 Plot 6, bogus tree canopies (*light blue*) between thinning rows. The *yellow dots* indicate the location of each tree in the field

5   Conclusion

A new tree canopy delineation algorithm was developed for estimating the inventory of stands and individual tree parameters. This method was implemented using the Integrated Development Environment (IDE) in Definiens Developer. It followed a simple set of rules that progressively transformed the original CHM cells into tree tops and individual tree canopies. The results were exported as shapefiles to ArcGIS, showing the spatial location of each crown and its horizontal dimensions. This information was then combined with allometric functions derived from the analysis of field-measured plots to produce estimates of height and dbh for each detected tree.

The results were validated at stand level for 11 plots in Aberfoyle and 7 plots in Kielder Forest. Although the results were variable across the stands, the average results demonstrated that it was possible to detect 80–90 % of all trees in a stand, close to 100 % of the top height values, but with a 10–20 % overestimation of both basal area and volume. The results in Kielder forest were slightly worst due to high tree densities.

This procedure has been shown to have a good potential to be used as a tool for forest inventory assessments as it compared very well with field methods. However, it must be considered that field methods are normally undertaken as a 1 % ha sample (Matthews and Mackie 2006). In this study the field sample in fact represented a 25 % ha. Instead, LiDAR analysis did not work with a representative sample but it used the whole population. Therefore, these results showed a good potential to creating more accurate forest inventories than the ones normally obtained by field plots. For instance, top height is usually time consuming and difficult to measure in the field with accurate precision, especially in dense forest cover. Nevertheless,

LiDAR was capable of producing an almost perfect match with field measurements in all the test plots used here.

The estimation of basal area and volume suggested that this algorithm needs to improve the problem of the aggregation of small canopies to the largest ones. As the method for calculating dbh was highly dependent upon canopy area and volume was calculated as a function of dbh, the final estimates for volume and basal area were overpredicted in almost all stands. Especially in those stands with larger number of individuals, this problem was likely to get worse.

This method, despite its limitations, creates the possibility of spatially locating individual trees of different dimensions as well as estimating their height and diameter distributions. This new knowledge provides a good understanding of stand structure beyond the traditional flat polygon representation in the Forest Enterprise Sub-Compartment Database. In theory, forest managers now have the possibility of focusing attention at the sub-stand level for a much better understanding on the effects of stand structure over forest production. Potentially, this method will allow them to formulate more solid assumptions about the characteristics of their crops and to choose the best management option. Especially in those areas, like CCF, with a forest structure that is not fully covered by current stand models.

Acknowledgments This research was funded by the Corporate Forest Services from the Forestry Commission and the Forest Research CE discretionary fund. Special gratitude to our colleagues: Steve Osborne and Carlos Ontiveros for their abnegate dedication during data collection in the field and to Dr. Jackie Rosette (University of Swansea) for her good advice and suggestions during all the stages of this work and her inputs in the timber quality modelling.

References

Andersen HE, Reutebuch SE, Schreuder GF (2001) Automated individual tree measurement through morphological analysis of a LiDAR-based canopy surface model. Proceedings of the first international precision forestry cooperative symposium, Seattle, USA, pp 11–22

Brandtberg T (1999) Automatic individual tree-based analysis of high spatial resolution remotely sensed data. Ph.D. thesis. Acta Universitatis Agriculturae Sueciae, Silvestria. Swedish University of Agricultural Sciences, Uppsala, Sweden

Brandtberg T, Walter F (1998) Automated delineation of individual tree crowns in high spatial resolution aerial images by multiple-scale analysis. Mach Vis Appl 11:64–73

Brandtberg T, Warner T, Landenberger R, McGraw J (2003) Detection and analysis of individual leaf-off tree crowns in small footprint, high sampling density lidar data from the eastern deciduous forest in North America. Remote Sens Environ 85:290–303

Bunting P, Lucas RM (2006) The delineation of tree crowns in Australian mixed species forests using hyperspectral Compact Airborne Spectrographic Imager (CASI) data. Remote Sens Environ 101(2):230–248

Dralle K, Rudemo M (1996) Stem number estimation by kernel smoothing of aerial photos. Can J Forest Res 26:1228–1236

Elmquist M, Jungert E, Lantz F, Persson A, Söderman U (2001) Terrain modelling and analysis using laser scanner data estimations of laser data using shape models. Int Arch Photogramm Remote Sens 34–3(W4):219–227

Erikson M (2001) Structure-keeping colour segmentation of tree crowns in aerial images. In: Proceedings from the Scandinavian conference on image analysis (SCIA), Bergen, Norway, pp 185–191, 11–14

Gougeon FA (1999) Automatic individual tree crown delineation using a valley-following algorithm and a rule-based system. In: Hill DA, Leckie DG (eds) Proceedings of automated interpretation of high spatial resolution digital imagery for forestry, Victoria, British Columbia, Canada, February 10-12, 1998. Natural Resources Canada, Canadian Forest Service, Pacific Forestry Centre, Victoria, pp 11–23

Gougeon F (2005) The Individual Tree Crown (ITC) suite manual. Pacific Forestry Centre, Canadian Forest Service, Natural Resources Canada, Victoria

Heurich M, Persson Å, Holmgren J, Kennel E (2004) Detecting and measuring individual trees with laser scanning in mixed mountain forest of central Europe using an algorithm developed for Swedish boreal forest conditions. Int Arch Photogramm Remote Sens Spat Inf Sci XXXVI:307–312, Commission 8

Hirschmugl M, Franke M, Ofner M, Schardt M, Raggam H (2005) Single tree detection in very high resolution remote sensing data. Proceedings of ForestSat, 31 May–3 June, Borås, Sweden. ISSN:1100-0295

Holmgren J, Nilsson I, Olsson H (2003) Estimation of tree height and stem volume on plots using airborne laser scanning. Forest Sci 49:419–428

Hyyppä J, Inkinen M (1999) Detecting and estimating attributes for single trees using laser scanner. Photogramm J Finl 16:27–42

Hyyppä J, Kelle O, Lehikoinen M, Inkinen M (2001) A segmentation-based method to retrieve stem volume estimates from 3-D tree heights models produced by laser scanners. IEEE Trans Geosci Remote Sens 39(5):969–975

Hyyppä J, Hyyppä H, Leckie D, Gougeon F, Yu X, Maltamo M (2008) Review of methods of small-footprint airborne laser scanning for extracting forest inventory data in boreal forests. Int J Remote Sens 29:1339–1366

Koch B, Heyder U, Weinacker H (2006) Detection of individual tree crowns in airborne lidar data. Photogramm Eng Remote Sens 72:357–363

Korpela I, Dahlin B, Schäfer H, Bruun E, Haapaniemi F, Honkasalo J, Ilvesniemi S, Kuutti V, Linkosalmi M, Mustonen J, Salo M, Suomi O, Virtanen H (2007) Single-tree forest inventory using Lidar and aerial images for 3D treetop positioning, species recognition, height and crown width estimation. In: Proceedings of ISPRS III/3, III/4, V/3 and VIII/11. "Laserscanning 2007 and Silvilaser 2007", vol XXXVI, Part 3/W52, 12–14 Sept 2007, Espoo, Finland, pp 227–234

Lee J (1983) Digital image smoothing and the sigma filter. Comput Vis Graph Image Process 24:255–269

Lim K, Treitz P, Groot A, St-Onge B (2001) Estimation of individual tree heights using LiDAR remote sensing. In: Proceedings of the 23rd annual Canadian symposium on remote sensing, Quebec, Canada

Maltamo M, Eerikainen K, Packalén P, Hyyppa J (2006) Estimation of stem volume using laser scanning-based canopy height metrics. Forestry 79:217–229

Mason WL, Kerr G (2004) Transforming even-aged conifer stands to continuous cover forestry. Information Note 40. FC Publications, Edinburgh

Matthews RW, Mackie ED (2006) Forest mensuration. A handbook for practitioners. Forestry Commission Publications, Edinburgh

Miller KF (1985) Windthrow hazard classification. Forestry Commision Publications, Edinburgh 85, 14 pp

Morsdorf F (2006) LIDAR remote sensing for estimation of biophysical vegetation parameters. PhD thesis, University of Zurich

Morsdorf F, Meier E, Kötz B, Itten KI, Dobbertin M, Allgöwer B (2004) Lidar-based geometric reconstruction of boreal type forest stands at single tree level for forest and wildland fire management. Remote Sens Environ 3(92):353–362

Naesset E (1997a) Determination of mean tree height of forest stands using airborne laser scanner data. ISPRS J Photogramm Remote Sens 52:49–59

Naesset E (1997b) Estimating timber volume of forest stands using airborne laser scanner data. Remote Sens Environ 61:246–253

Nixon CJ, Worrell R (1999) The potential for the natural regeneration of conifers in Britain. Forestry Commission Bulletin 120. Forestry Commission, Edinburgh

Persson Å, Holmgren J, Söderman U (2002) Detecting and measuring individual trees using an airborne laser scanner. Photogramm Eng Remote Sens 68(0):1–8

Pitkänen J (2005) A multi-scale method for segmentation of trees in aerial images. Forest inventory and planning in Nordic countries. In: Proceedings of the SNS-meeting at Sjusjøen, Norway, 6–8 Sept 2004. Norwegian Institute of Land Inventory, NIJOS-report 09/05, s. 207–216

Pitkänen J, Maltamo M, Hyyppä J, ja Yu X (2004) Adaptive methods for individual tree detection on airborne laser based canopy height model. Int Arch Photogramm Remote Sens Spat Inf Sci XXXVI(part 8/W2):187–191

Pollock R (1996) The automatic recognition of individual trees in aerial images of forests based on a synthetic tree crown image model. Ph.D. thesis, University of British Columbia, Vancouver, Canada

Popescu SC (2008) TREEVAW (Tree Variable Window). http://www-ssl.tamu.edu/personnel/s_popescu/TreeVaW/. Latest Accessed Jan 2008

Popescu S, Wynne R, Nelson R (2003) Measuring individual tree crown diameter with lidar and assessing its influence on estimating forest volume and biomass. Can J Remote Sens 29(5):564–577

Pyatt G, Ray D, Fletcher J (2001) An ecological site classification for forestry in Great Britain, vol 124, Forestry Commission Bulletin. Forestry Commission, Edinburgh

Straub B (2003) A top-down operator for the automatic extraction of trees - concept and performance evaluation. In: Proceedings of the ISPRS working group III/3 workshop '3-D reconstruction from airborne laser scanner and InSAR data', 8–10 Oct 2003, Dresden, Germany, pp 34–39

Suárez JC (2010) An analysis of the consequences of stand variability in Sitka spruce plantations in Britain using a combination of airborne LiDAR analysis and models. PhD thesis at the Department of Applied Mathematics, University of Sheffield

Suárez JC, García R, Gardiner BA, Patenaude G (2008a) The estimation of wind risk in forest stands using ALS. J Forest Plann 13:165–186

Suárez J, Rosette J, Nicoll B, Gardiner BA (2008b) A practical application of airborne LiDAR for forestry management in Scotland. In: Hill R, Rosette J, Suárez J (eds) 8th international conference on LiDAR applications in forest assessment and inventory. Forestry Commission, Edinburgh. ISBN:978-0-85538-774-7

Van Ardt JAN, Wynne RH, Oderwald RG (2006) Forest volume and biomass estimation using small footprint Lidar-distributional parameters on a per-segment basis. Forest Sci 52(6):636–649(14)

Vauhkonen J, Liviu E, Sandeep G, Heinzel J, Holmgren J, Pitkänen J, Solberg S, Wang Y, Weinacker H, Hauglin KM, Lien V, Packalén P, Gobakken T, Koch B, Naesset E, Tokola T, Maltamo M (2012) Comparative testing of single-tree detection algorithms under different types of forest. Forestry 85(1):28–40

Wang Y, Weinacker H, Koch B, Sterenczak K (2008) Lidar point cloud based fully automatic 3D single tree modelling in forest and evaluations of the procedure. Int Arch Photogramm Remote Sens Spat Inf Sci 38:45–51

Weinacker H, Koch B, Weinacker R (2004) TREESVIS – a software system for simultaneous 3D-Real time visualisation of DTM, DSM, laser raw data, multispectral data, simple tree and building models. In: Thies M, Koch B, Weinacker H (eds) Laser scan workshop: ISPRS: 'Laser scanners for forest and landscape assessment'. University of Freiburg, Freiburg, 3–6 Oct 2004, pp 90–95. ISSN 1962-1750

Index

T. Fenning (ed.), *Challenges and Opportunities for the World's Forests
in the 21st Century*, Forestry Sciences 81, DOI 10.1007/978-94-007-7076-8,
© Springer Science+Business Media Dordrecht 2014